Aluminum—Lithium Alloys

Aluminum—Lithium Alloys

Processing, Properties, and Applications

N. Eswara Prasad

Amol A. Gokhale

R.J.H. Wanhill

AMSTERDAM • BOSTON • HEIDELBERG • LONDON
NEW YORK • OXFORD • PARIS • SAN DIEGO
SAN FRANCISCO • SINGAPORE • SYDNEY • TOKYO
Butterworth-Heinemann is an imprint of Elsevier

Acquiring Editor: *Stephen Merken*
Development Editor: *Jeffrey Freeland*
Project Manager: *Jason Mitchell*
Designer: *Mark Rogers*

Butterworth-Heinemann is an imprint of Elsevier
The Boulevard, Langford Lane, Kidlington, Oxford, OX5 1GB, UK
225 Wyman Street, Waltham, MA 02451, USA

Library of Congress Cataloging-in-Publication Data
A catalog record for this book is available from the Library of Congress

British Library Cataloguing-in-Publication Data
A catalogue record for this book is available from the British Library

ISBN: 978-0-12-401698-9

Contents

Foreword   xv
Preface   xvii
About the Editors   xix
A Note of Gratitude from the Editors   xxi
List of Contributors   xxiii

Part I
Introduction to Al-Li Alloys

1. **Historical Development and Present Status of Aluminum−Lithium Alloys**   3

Edgar A. Starke, Jr.

1.1   **Introduction**   4
1.2   **Lithium Additions to Aluminum Alloys: Early Days**   4
     1.2.1   History of the Development of the First Modern Al−Li Alloys   5
     1.2.2   Development of Alcoa's 2020   6
     1.2.3   The Ductility Problem of 2020   7
     1.2.4   Development of Al−Li Alloys in the Soviet Union   7
1.3   **Development of Modern Aluminum−Lithium Alloys**   8
     1.3.1   The Second Generation of Aluminum−Lithium Alloys   10
     1.3.2   Manufacturing Issues   11
     1.3.3   Understanding the Precipitate Structure in Al−Li−X Alloys   11
     1.3.4   The Effect of Prior Deformation on Precipitation During Ageing   12
     1.3.5   Deformation Behavior in Aged Al−Li−X Alloys   15
     1.3.6   Predicting Strain Localization in Al−Li−X Alloys   17
     1.3.7   Applications and Problems of the Second-Generation Al−Li Alloys   19
     1.3.8   The Third Generation of Aluminum−Lithium Alloys   19
     1.3.9   Background Information That Led to Improvements in Al−Li−X Alloys   20
1.4   **Closure**   22
Acknowledgments   23
References   23

2. Aerostructural Design and Its Application to
Aluminum–Lithium Alloys   27
R.J.H. Wanhill and G.H. Bray

2.1 Introduction   28
2.2 Aircraft Structural Property Requirements   28
    2.2.1 Fuselage/Pressure Cabins   29
    2.2.2 Wings   29
    2.2.3 Empennage (Tail)   30
2.3 Engineering Property Requirements for Al–Li Alloys
    in Aircraft Structures   30
    2.3.1 The Key Role of Lithium Content: Density and Stiffness 31
    2.3.2 Strength and Ductility   34
    2.3.3 Damage Tolerance   35
2.4 Families of Al–Li Alloys   39
    2.4.1 Second-Generation Alloys   39
    2.4.2 Third-Generation Alloys   40
2.5 Examples of Third-Generation Alloy Property Developments
    and Trade-Offs   44
    2.5.1 Fuselage/Pressure Cabins   44
    2.5.2 Upper Wings   47
    2.5.3 Lower Wings   49
    2.5.4 Spars, Ribs, and Other Internal Structures   52
2.6 Service Qualification Programs   53
2.7 Summary and Conclusions   55
References   56

Part II
Physical Metallurgy

3. Phase Diagrams and Phase Reactions in Al–Li Alloys   61
N. Eswara Prasad and T.R. Ramachandran

3.1 Introduction   62
    3.1.1 Alloy Development History   62
    3.1.2 The Influence of Alloying Elements and Composition   63
3.2 Nature of Phases   65
3.3 Al–Li Binary System   66
3.4 Ternary Systems   71
    3.4.1 Al–Li–Mg System   71
    3.4.2 Al–Li–Cu System   73
    3.4.3 Al–Cu–Mg System   76
    3.4.4 Al–Li–Zr System   78
3.5 Quaternary Al–Li–Cu–Mg System   79
    3.5.1 Phase Equilibria   79
    3.5.2 Overview of Precipitate Phases in Commercial
          Al–Li Alloys   80

3.6   Minor Alloying Additions to Al–Li Alloys   85
      3.6.1   Minor Alloying Additions for Grain Refinement   85
      3.6.2   Minor Alloying Additions for Strengthening   86
3.7   Impurities and Grain Boundary Precipitation   87
      3.7.1   Impurities   88
      3.7.2   Precipitates and PFZs   88
3.8   Chapter Summary   89
Acknowledgments   90
References   91

4. Microstructure and Precipitate Characteristics of Aluminum–Lithium Alloys   99

K. Satya Prasad, N. Eswara Prasad, and Amol A. Gokhale

4.1   Introduction   99
4.2   Microstructures in the Solution-Treated Condition   101
      4.2.1   Alloys with Zr   102
4.3   Age Hardening Behavior   102
      4.3.1   Age Hardening Curve   103
      4.3.2   Microstructures at Different Stages of Aging   104
      4.3.3   Microstructures After High-Temperature Exposure   105
4.4   Characteristics of Precipitates   106
      4.4.1 Al_3Li (δ') Phase   107
      4.4.2 Al_2Cu (θ') Phase   111
      4.4.3 Al_2CuLi (T_1) Phase in Ternary Al–Li–Cu Alloys   112
      4.4.4 Al_2CuLi (T_1) Phase in Quaternary Al–Li–Cu–Mg Alloys   115
      4.4.5 Al_2CuMg (S') Phase and the Equilibrium S Phase   116
      4.4.6 Al_3Zr (β') Phase   121
      4.4.7   AlLi (δ) Phase   125
      4.4.8 Al_6CuLi_3 (T_2) Phase   127
4.5   Summary   130
Acknowledgments   131
References   131

5. Texture and Its Effects on Properties in Aluminum–Lithium Alloys   139

K.V. Jata and A.K. Singh

5.1   Introduction   140
5.2   Texture in Al–Li Alloys   142
5.3   Texture Evolution During Primary Processing   145
      5.3.1   Beta Fiber Components (Bs, S, and Cu) in Binary
              and Ternary Al–Li Alloys   146
      5.3.2   Alpha Fiber Components (Goss and Cube/Rotated Cube)   149
      5.3.3   Role of Precipitates and Slip Character During Hot
              and Cold Forming   151
      5.3.4   Influence of Non-octahedral Slip   152

5.4 Macroscopic Anisotropy of Yield Strength 152
5.5 Practical Methods to Reduce Texture in Al–Li Alloys 154
    5.5.1 Processing Methods 154
    5.5.2 Thick Plate Anisotropy 156
    5.5.3 Sheet Product Fatigue Crack Deviations 157
    5.5.4 Example Microstructures for Third-Generation
           Al–Li Alloys 159
5.6 Summary 160
Acknowledgments 160
References 160

Part III
Processing Technologies

6. Melting and Casting of Aluminum–Lithium Alloys 167

Vijaya Singh and Amol A. Gokhale

6.1 Introduction 167
6.2 Melt Protection from the Atmosphere 168
    6.2.1 Lithium Reactivity 168
    6.2.2 Addition of Lithium via an Al–Li Master Alloy 170
    6.2.3 Li Addition and Further Melt Processing Under
           Flux Cover 170
    6.2.4 Melting Under Inert Atmosphere 171
6.3 Crucible Materials 172
6.4 Hydrogen Pickup and Melt Degassing 174
6.5 Grain Refinement 174
6.6 Casting Practices 178
    6.6.1 Metal/Mold Reactions, Castability, and Shape Casting 178
    6.6.2 DC Casting 180
6.7 Summary 181
References 182

7. Mechanical Working of Aluminum–Lithium Alloys 187

G. Jagan Reddy, R.J.H. Wanhill, and Amol A. Gokhale

7.1 Introduction 187
    7.1.1 Overview 189
Part 1: Workability 191
7.2 Introduction 191
7.3 General Review of Hot Deformation Characteristics
    of Al–Li Alloys 192
    7.3.1 Hot Workability of a Binary Al–Li Alloy 192
    7.3.2 Hot Workability of Multicomponent Al–Li Alloys 193
    7.3.3 Hot Deformation Regions Favoring Superplasticity 195
7.4 Hot Deformation and Process Mapping of Al–Li Alloy UL40 196
    7.4.1 Process Maps and EBSD–OIM 196

7.5 Summary of the Hot Deformation Investigations 203
Part 2: Processing of Al–Li Alloys 203
7.6 General Considerations 203
   7.6.1 Rolled Products 205
   7.6.2 Extrusions and Forgings 209
7.7 Industrial-Scale Processing 211
   7.7.1 Introduction 211
   7.7.2 Review and Discussion of Available Information
       for Third-Generation Alloys 212
7.8 Summary 216
References 216

8. Superplasticity in and Superplastic Forming of Aluminum–Lithium Alloys 221

S. Balasivanandha Prabu and K.A. Padmanabhan

8.1 Introduction 221
8.2 Superplasticity 225
   8.2.1 Phenomenon of Superplasticity 225
   8.2.2 Experimental Investigations 230
   8.2.3 Low-Temperature Superplasticity 234
   8.2.4 Effects of Strain Rate and Strain Rate Sensitivity
       (m value) on Superplasticity 237
8.3 Superplastic Forming 239
   8.3.1 Cavity Nucleation and Growth 240
8.4 Role of Friction Stir Processing on Superplastic Forming 245
   8.4.1 Superplasticity in FSP Materials 245
   8.4.2 Superplastic Behavior and Deformation
       Mechanism 246
   8.4.3 Cavity Density and Size Distribution 250
8.5 Applications 251
8.6 Concluding Remarks 253
References 253
Further Reading 258

9. Welding Aspects of Aluminum–Lithium Alloys 259

G. Madhusudhan Reddy and Amol A. Gokhale

9.1 Introduction 260
9.2 Weld Metal Porosity 260
9.3 Solidification Cracking 264
   9.3.1 General Considerations 264
   9.3.2 Al–Li Alloy Solidification Cracking Guidelines 265
   9.3.3 Development of Weldable Al–Li Alloys 267
9.4 Liquation Cracking 272
9.5 EQZ Formation and Associated Fusion Boundary Cracking 273
   9.5.1 Experimental Observations 273

9.5.2   Hypotheses of EQZ Formation and Their Evaluation   276
9.5.3   Fusion Boundary Cracking   278
9.6   **Modification of Fusion Zone Microstructures**   278
   9.6.1   Inoculation   279
   9.6.2   Pulsed Current   280
   9.6.3   Magnetic Arc Oscillation   282
9.7   **Mechanical Properties**   284
   9.7.1   Al−Li 1420 (First-Generation Al−Li Alloy)   286
   9.7.2   Al−Li 1441, AA 8090, and AA 2090
            (Second-Generation Al−Li Alloys)   286
   9.7.3   Al−Li AA 2195 (Third-Generation Al−Li Alloy)   289
9.8   **Corrosion**   289
9.9   **Solid-State Welding Processes**   292
   9.9.1   Friction Welding   292
   9.9.2   Friction Stir Welding   295
9.10   **Summary**   296
Acknowledgments   296
References   297
Further Reading   302

Part IV
Mechanical Behavior

10. Quasi-Static Strength, Deformation, and Fracture Behavior of Aluminum−Lithium Alloys   305

T.S. Srivatsan, Enrique J. Lavernia, N. Eswara Prasad,
and V.V. Kutumbarao

10.1   **Introduction**   306
10.2   **Mechanisms of Strengthening**   307
   10.2.1   Strengthening by δ' Precipitates in Al−Li Alloys   307
   10.2.2   Strengthening by Other Phases   310
10.3   **Ductility and Fracture Toughness**   312
   10.3.1   The Nature and Occurrence of Planar
             Slip Deformation   313
   10.3.2   Methods for Reducing Planar Slip and
             Strain Localization   314
   10.3.3   Thermal Treatments for Improving Strength
             and Fracture Toughness   315
10.4   **Anisotropy of Mechanical Properties**   318
10.5   **Tensile Properties of Selected Aluminum−Lithium Alloys**   318
   10.5.1   AA 2020: A First-Generation Al−Li Alloy   319
   10.5.2   AA 8090: A Second-Generation Al−Li Alloy   322
   10.5.3   AA 2198: A Third-Generation Al−Li Alloy   330
10.6   **Summary and Conclusions**   331
Acknowledgments   334
References   334

11. Fatigue Behavior of Aluminum–Lithium Alloys 341

N. Eswara Prasad, T.S. Srivatsan, R.J.H. Wanhill, G. Malakondaiah,
and V.V. Kutumbarao

11.1 Introduction 342
11.2 The Phenomenon of Fatigue 342
Part A: Low Cycle Fatigue (LCF) 344
11.3 LCF Behavior 344
11.4 Test Methods and Analyses 345
    11.4.1 Characterization of Cyclic Stress
        Response Behavior 347
    11.4.2 Characterization of CSS Behavior 347
11.5 LCF Behavior of Aluminum–Lithium Alloys 348
    11.5.1 General Survey/Microstructural
        and Environmental Effects 348
    11.5.2 Fatigue Life Power Law Relationships 349
    11.5.3 CSR Behavior 355
    11.5.4 CSS Behavior 357
    11.5.5 Fatigue Toughness 359
    11.5.6 The LCF Resistance of Aluminum–Lithium Alloys 362
Part B: High Cycle Fatigue (HCF) 363
11.6 Introduction to the HCF Behavior of Aluminum Alloys 363
11.7 Background on Test Methods 366
11.8 HCF Behavior of Aluminum–Lithium Alloys 366
    11.8.1 General Survey 366
    11.8.2 Effects of Lithium Content, Aging, and Cold Work 366
    11.8.3 HCF Behavior of AA 2020-T651—A First-Generation
        Al–Li Alloy 367
    11.8.4 HCF Behavior of AA 8090-T651—A Second-Generation
        Al–Li Alloy 369
    11.8.5 HCF Behavior of AA 2098—A Third-Generation
        Al–Li Alloy 370
    11.8.6 The HCF Resistance of Al–Li Alloys: Smooth and
        Notched Properties, and a Note of Caution 371
11.9 Summary and Conclusions 371
11.10 Final Remarks 374
List of Symbols 375
References 375

12. Fatigue Crack Growth Behavior of Aluminum–Lithium Alloys 381

R.J.H. Wanhill and G.H. Bray

12.1 Introduction 382
12.2 Background on Test Methods and Analysis 383
    12.2.1 Testing 383
    12.2.2 Analysis 384

12.3 Survey of FCG of Al—Li Alloys | 384
12.3.1 Long/Large Cracks: CA/CR Loading | 385
12.3.2 Long/Large Cracks: Flight Simulation Loading | 386
12.3.3 Short/Small Cracks | 387
12.4 FCG Comparisons of Al—Li and Conventional Alloys I: Long/Large Cracks, CA/CR Loading | 387
12.4.1 First-Generation Al—Li Alloys | 387
12.4.2 Second-Generation Al—Li Alloys | 388
12.4.3 Third-Generation Al—Li Alloys | 392
12.5 FCG Comparisons of Al—Li and Conventional Alloys II: Long/Large Cracks, Flight Simulation Loading | 395
12.5.1 Second-Generation Al—Li Alloys: Gust and Maneuver Spectrum Loading | 395
12.5.2 Third-Generation Al—Li Alloys: Gust Spectrum Loading | 401
12.6 FCG Comparisons of Al—Li and Conventional Alloys III: Short/Small Cracks | 402
12.6.1 CA Loading | 402
12.6.2 Flight Simulation Loading | 402
12.7 Differing FCG Behaviors and Advantages for Second- and Third-Generation Al—Li Alloys | 405
12.8 Summary and Conclusions | 406
12.8.1 Al—Li Alloy FCG | 406
12.8.2 Practice-Related Crack Growth Regimes | 408
References | 409

13. Fracture Toughness and Fracture Modes of Aerospace Aluminum—Lithium Alloys — 415

S.P. Lynch, R.J.H. Wanhill, R.T. Byrnes, and G.H. Bray

13.1 Introduction | 416
13.2 Test Methods for Determining Fracture Toughness (and Terminology) | 418
13.2.1 Plane-Stress/Plane-Strain Considerations | 419
13.2.2 Testing Thick Products | 420
13.2.3 Testing Sheet and Thin Plate | 422
13.3 Effects of Microstructural Features on Fracture Toughness and Fracture Modes | 423
13.3.1 Extrinsic Inclusions and Porosity | 424
13.3.2 Constituent Particles | 424
13.3.3 Alkali-Metal Impurity Phases | 426
13.3.4 Dispersoids | 427
13.3.5 Matrix Precipitates | 427
13.3.6 Grain-Boundary Precipitates and Precipitate-Free Zones | 430
13.3.7 Grain-Boundary Segregation | 430
13.3.8 Overview and Microstructural Design of Third-Generation Al—Li Alloys | 433

13.4 Fracture Toughness of Second-Generation Al–Li Alloys
    Versus Conventional Al Alloys 435
    13.4.1 Room-Temperature Data 435
    13.4.2 Effects of Testing Temperature and Strain Rate 439
13.5 Fracture Toughness of Third-Generation Al–Li Alloys 442
    13.5.1 Thick Plate and Other Thick Products 442
    13.5.2 Sheet and Thin Plate 444
13.6 Uses and Potential Uses of Third-Generation Al–Li Alloys 448
13.7 Conclusions 449
References 451
Specific references 451
Other references 455

14. Corrosion and Stress Corrosion of Aluminum–Lithium Alloys 457

N.J.H. Holroyd, G.M. Scamans, R.C. Newman, and A.K. Vasudevan

14.1 Introduction and Historical Background 457
14.2 Localized Corrosion of Al–Li Based Alloys 461
    14.2.1 Al–Li Binary Alloys 461
    14.2.2 Al–Li–Cu and Al–Li–Cu–Mg Alloys 462
    14.2.3 Al–Mg–Li Alloys 478
14.3 Stress Corrosion Cracking 479
    14.3.1 Al–Li Binary Alloys 482
    14.3.2 Al–Li–Cu and Al–Li–Cu–Mg Alloys 483
    14.3.3 Mechanistic Implications 490
14.4 Summary and Conclusions 492
Acknowledgment 493
References 493

Part V
Applications

15. Aerospace Applications of Aluminum–Lithium Alloys 503

R.J.H. Wanhill

15.1 Introduction 503
15.2 Weight Savings 504
    15.2.1 Density 506
    15.2.2 Density and Elastic Moduli: Specific Stiffnesses 506
    15.2.3 Modern/Innovative Structural Concepts
        and a Case Study 509
15.3 Materials Selection 514
    15.3.1 Aluminum Alloys, CFRPs, and FMLs: Advantages
        and Disadvantages 514
    15.3.2 Materials for Aircraft Structures 517

15.3.3   Aluminum Alloys for Spacecraft Structures   519
15.3.4   Materials Qualification: an Example   521
15.4   Applications of Al–Li Alloys (Third Generation)   521
15.4.1   Aircraft   521
15.4.2   Spacecraft   525
15.5   Summary and Conclusions   530
Acknowledgments   531
References   531

16. Airworthiness Certification of Metallic Materials 537

B. Saha, R.J.H. Wanhill, N. Eswara Prasad, G. Gouda,
and K. Tamilmani

16.1   Introduction   537
16.2   Aviation and Airworthiness Regulatory Bodies   538
16.2.1   Civil Aviation   538
16.2.2   Military Aviation   539
16.3   Airworthiness of Metallic Materials   540
16.3.1   Fatigue Design Philosophies   541
16.3.2   Materials and Structures Certification Methodology   544
16.4   Example of Certification of an Al–Li Alloy   549
16.4.1   Certification Methodology   549
16.4.2   Certification of Al–Li Alloy 1441M Sheet   550
16.5   Summary   553
References   553
Specific References   554

Appendix 1: Interconversion of Weight and Atomic Percentages
    of Lithium and Aluminum in Aluminum–Lithium Alloys   555

Selected Conversion Factors For SI Units   559

Index   561

The demand for lightweight structural elements is perennial, particularly for aerospace
applications. Lithium is the lightest metallic
element, and thus offers stellar advantages as
an alloying element in aluminum, the primary
structural material for light-alloy based
aerospace structures. Each unit addition of Li
reduces the density of Al alloys by about 3%
and enhances the elastic modulus by about
6%, besides contributing to enhanced tensile
strength accompanied by improved high cycle
fatigue as well as fatigue crack growth
resistance. The strength−fracture toughness

combination at cryogenic temperatures is also superior in Al−Li alloys.
However, there are considerable challenges in practically realizing the
above-stated advantages. As an alkali metal, Li is highly reactive and harms
any surface with which it comes into contact. Therefore, innovative melting
and casting technologies must be used to make Al−Li alloy products.
Moreover, Al−Li alloys require stringent composition control and have
narrow process windows for obtaining optimum properties.

Intense international R&D efforts over the past two decades have
addressed a whole range of issues related to making, shaping, and treating
Al−Li alloys. As a result, the following have been notably established:
(i) production technologies for large scale melting and casting Al−Li alloys
with optimized chemistry, (ii) advanced processing based on process modeling, (iii) thermal and thermomechanical treatments to achieve the desired
microstructure−mechanical property (strength and fracture toughness as well
as fatigue and crack growth resistance) combinations, and (iv) fabrication
technologies, including superplastic forming and friction stir welding. As a
consequence, numerous patented Al−Li alloys have been commercially produced in different product forms. These alloys possess, in comparison to the
traditional 2XXX and 7XXX series Al alloys, higher modulus (15−25%) and
higher specific strength (8−15%), and therefore provide aeronautical designers
an opportunity to significantly reduce the weight of aeronautical and space
structures for ensuring enhanced fuel efficiencies and higher payloads.

The present volume systematically documents how a vast array of problems have been overcome to enable significant progress in diverse technical

areas to realize production and deployment of Al−Li alloys. All the authors have had long-standing experience with this relatively more difficult class of Al alloys, with noteworthy original findings to their credit. Beginning with a superb outline of the historical development, the introductory part (first 2 chapters) summarizes the present status of Al−Li alloys and describes aspects of aerostructural design in the context of their application. In the subsequent four parts, various aspects of physical metallurgy (3 chapters), processing (4 chapters), mechanical behavior (5 chapters), and applications (2 chapters) of Al−Li alloys are dealt with. It is particularly pleasing for me to see that several contributing authors and two of the three editors are from the Indian Defence Metallurgical Research Laboratory with which I was associated three decades ago, just about when Al−Li alloys were beginning to attract international attention.

This thoughtfully edited book of informative articles on Al−Li clearly promises to be one of the most important reference volumes on these scientifically intriguing and technologically challenging alloys.

P. Rama Rao
FNAE, FNA, FASc., FNASc, FREng., Foreign Associate NAE (US),
FTWAS, FUAS
Presently Chairman, Governing Council, ARCI, PO Balapur,
Hyderabad, India
Former Director, Defence Metallurgical Research Laboratory, DRDO,
Hyderabad, India
Former Secretary, Department of Science and Technology (DST),
Government of India, New Delhi, India
Former Chairman, Atomic Energy Regulatory Body, Government of India,
Mumbai, India, Former VC, University of Hyderabad (UoH),
Hyderabad, India
Former President of IIM, ISCA, IAS, INAE, MRSI, ICF, IUMS and
Present President, Indian Nuclear Society, India

The history of aluminum—lithium (Al—Li) alloys goes back to the 1920s, but it is only since the 1990s that the fundamental understanding of these alloys has matured and enabled the development of a family of alloys with excellent combinations of engineering properties. These are the so-called third generation Al—Li alloys, and excellent property combinations are required for them to compete in the "high-tech" field of aerospace structures with established conventional aluminum alloys and carbon-fiber composites.

This book surveys the knowledge about Al—Li alloys and their development, concentrating on the second- and third-generation alloys. The second-generation alloys were largely unsuccessful from an engineering property viewpoint, but the lessons learned from their development in the 1970s and 1980s have been essential in defining and establishing the third-generation Al—Li alloys.

The book is divided into the following main parts:

- Part I: Introduction to Al—Li Alloys (Chapters 1 and 2)
- Part II: Physical Metallurgy (Chapters 3—5)
- Part III: Processing Technologies (Chapters 6—9)
- Part IV: Mechanical Behavior (Chapters 10—14)
- Part V: Applications (Chapters 15 and 16)

Each chapter discusses a specific topic, which is summarized in the last section of the chapter. References to the literature are added to each chapter rather than collectively at the end of the book. The scope is such that many authors were needed, generally more than one author for each chapter, and their contributions are acknowledged by the editors.

The current pace of Al—Li alloy commercial development is such that this book will inevitably become outdated within the next few years. Nevertheless, we hope it will provide a valuable reference work for the foreseeable future.

N Eswara Prasad

Amol A Gokhale

RJH Wanhill

Republic of India and the Netherlands
September, 2013

N. Eswara Prasad obtained B.Tech. (in 1985) and Ph.D. (in 1993) degrees in Metallurgical Engineering from Indian Institute of Technology (BHU), Varanasi, India, and joined the Indian Defence Research and Development Organisation (DRDO) in 1985. Since then, Dr. Prasad worked in the fields of Design, Development, Life Prediction, and Airworthiness Certification leading to the production of aero-materials, namely Al and Al−Li alloys; Mo and Ti intermetallics; aero Steels; Ti- and Ni-based high temperature alloys; monolithic ceramics such as structural alumina, graphite, and SiC; carbon, silica, and SiC-based *C*ontinuous *F*iber-reinforced, *C*eramic-matrix *C*omposites (CFCCs). His research contributions are principally on the deformation behavior, fatigue power-law relationships, creep−fatigue interactions (CFI), and micromechanisms of fatigue and fracture. Dr. Prasad has authored 140 peer-reviewed publications and nearly 200 confidential reports and certification documents. Dr. Prasad is the recipient of several awards and recognitions, including "Young Scientist" from Indian Science Congress Association in 1991, "Young Metallurgist" of Indian Institute of Metals for 1994, "Humboldt Research Fellowship" from Alexander von Humboldt-Stiftung, Bonn, Germany (1998−1999), "Visiting Scientist" at Max-Planck-Institut fuer Metallforschung, Stuttgart, Germany (1998−1999), *Binani Gold Medal* of Indian Institute of Metals (2006), "Metallurgist of the Year for 2010" by the Ministry of Steel, Government of India. Dr. Prasad is a Fellow of Institution of Engineers (FIE − 2009), Fellow of Indian Institute of Metals (FIIM − 2011), and Fellow of Andhra Pradesh Akademi of Sciences (FAPASc − 2011). Dr. Prasad is presently the Regional Director of the Regional Centre for Military Airworthiness (Materials), CEMILAC at Hyderabad, India.

Amol A. Gokhale obtained a B.Tech. in Metallurgical Engineering, from the Indian Institute of Technology, Bombay, in 1978, and M.S. and a Ph.D. in Metallurgical Engineering from the University of Pittsburgh, Pittsburgh, USA, in 1980 and 1985, respectively. Since joining the Defence Metallurgical Research Laboratory (DMRL) in 1985, Dr. Gokhale led the development of second-generation Al−Li alloys for Indian defence for about 20 years, which included collaborative efforts with VIAM and KUMZ in Russia, and ADA and HAL in India to manufacture deep formed clad sheet products. He also led the development of aluminum alloy castings for torpedoes and aluminum foams for shock absorption applications. His most recent contributions are in the development of materials and modules for hot structures of hypersonic vehicles. He is working in association with Belarus Powder Metallurgy Laboratory to develop Ni-based foams. Important awards received by Dr. Gokhale include the National Research and Development Corporation "Technology Innovation" Award (1994), the "Metallurgist of the Year Award" by the Ministry of Steel (2000), the "DRDO Scientist of the Year" award (2008), and the "GD Birla Gold Medal" from the Indian Institute of Metals (2010). He has been a Fellow of the Indian National Academy of Engineering since 2011. He has published about 80 technical papers and filed 1 patent. Presently, Dr. Gokhale is the Director of the Defence Metallurgical Research Laboratory, Hyderabad, India.

Russell Wanhill has a Ph.D. (1968) in Metallurgy from the University of Manchester Institute of Science and Technology, Manchester, UK, and a Doctor of Technical Science degree (1995) from Delft University of Technology, Delft, The Netherlands. He joined the National Aerospace Laboratory (NLR) of the Netherlands in 1970 and since then has investi- gated fatigue and fracture of all classes of aerospace alloys, including many service failures. From 1978 to 1996 Russell was Head of the Materials Department  of the NLR, and in 1979–1980 adjunct Professor of materials at Delft University of Technology. He is the co-author of the book "Fracture Mechanics" (1984), which is now into a second edition, and has written more than 320 reports and publications. From November 2009 to May 2010, Russell was a "Visiting Academic" at the Defence Science and Technology Organisation, DSTO, Melbourne, Australia. His collaboration with DSTO col- leagues has resulted in journal publications, four book chapters (including one in the present book), a monograph, and a 1-day failure analysis course. Since 1994 Russell has investigated the embrittlement of ancient silver and published a number of papers on this seemingly esoteric but serious problem. "In 2002 the Board of the Foundation NLR awarded Russell the first Dr. ir. B.M. Spee Prize for outstanding contributions in the field of aerospace materials". Russell is presently Emeritus Principal Research Scientist in the Aerospace Vehicles Division of the NLR.

A Note of Gratitude from the Editors

The present monograph deals with the past, present, and future of one of the most exiting alloys to emerge in recent decades that have a vast potential for aeronautical applications—namely the aluminum−lithium (Al−Li) alloys—and highlights the aspects of alloy processing, mechanical behavior, and properties and aerospace applications. Such an arduous task is not easy to accomplish without the total support and whole-hearted participation of all the authors and members of book editorial offices at Hyderabad, India, and Elsevier, New York. We wish to profoundly thank each one of them—Professor Edgar A. Starke, Jr., Professor T.R. Ramachandran, Dr. K. Satya Prasad, Dr. Kumar V. Jata, Dr. Vijaya Singh, Dr. G. Jagan Reddy, Dr. S.B. Prabu, Professor K.A. Padmanabhan, Dr. G. Madhusudhan Reddy, Professor T.S. Srivatsan, Professor Enrique J. Lavernia, Professor V.V. Kutumbarao, Dr. G. Malakondaiah, Professor S.P. Lynch, Dr. Gary H. Bray, Dr. N.J. Henry Holroyd, Professor Roger C. Newman, Dr. A.K. Vasudevan, Mr. B. Saha, Mr. G. Gouda, Dr. K. Tamilmani, and Members, Hyderabad Book Secretariat Mr. Y. Balaji, Mr. Ch.V.A. Narasayya, Mrs. M. Swarna Bai, Mrs. C. Poornia, and Ms. P. Varsha. We are deeply beholden to Professor Palle Rama Rao, Distinguished Scientist, who led the Al−Li alloy development program in India, for the exquisite Foreword.

Most of the authors of this book had in their past authored or guided doctoral theses based on works on Al−Li alloys, which include several years of research, development, production, and certification. Some of them have also painstakingly reviewed the book chapters. Notable in this effort are: Professor Ed Starke reviewing the book chapter on Aero-Structural Design; Dr. Kota Harinarayana and Mr. P.S. Subramanyam on Aero-Structural Design and Aero Certification; Professor David Laughlin on Phase Diagrams and Phase Reactions; Professor T.R. Ramachandran on Microstructure Evolution and Melting and Casting; Dr. Vikas Kumar on Fatigue Crack Growth, and Dr. S.V. Kamat on Fracture Toughness. We owe our profound gratitude to all these contributing reviewers.

The scientific, technological, and design data of several aluminum−lithium alloys have been shared magnanimously by several authors, especially those concerned with third generation Al−Li alloys. In fact, there was a stage in the book project that there could be no further progress without the inputs on third generation Al−Li alloys. The support of Professor Ed Starke and

Dr. Gary H. Bray at this crucial juncture proved critical as this very support alone had advanced the book project further to its completion. In this context, we also thank all the authors of cited publications, who have generously allowed us to include their data and schematics in the present book.

Finally, it is a great pleasure to acknowledge the professional support by the Elsevier Team—particularly, Mr. Stephen P. Merken, Acquisitions Editor (person having involved at all stages of book publication), Ms. Amorette Padersen, Vice President, Marketing, and all the members of Book Design Team, headed by Mr. Jeffrey M. Freeland, Editorial Project Manager.

N Eswara Prasad

Amol A Gokhale

RJH Wanhill

Republic of India and the Netherlands
September, 2013

Dr. S. Balasivanandha Prabu Professor, Department of Mechanical Engineering College of Engineering, Guindy Anna University Chennai − 600 025, India

Dr. G.H. Bray Head, Aerospace and Defence Alloys Alcoa Technical Center 100 Technical Dr. Alcoa Center, PA 15069, USA

Dr. R.T Byrnes Defence Science and Technology Organisation (DSTO) 506 Lorimer Street, Fishermans Bend Melbourne, Victoria-3207, Australia

Dr. N. Eswara Prasad Scientist G and Regional Director, Regional Centre for Military Airworthiness (Materials), CEMILAC, [DRDO, Min. Defence, Govt. India] PO Kanchanbagh, Hyderabad − 500 058, India

Dr. Amol A. Gokhale Outstanding Scientist and Director, Defence Metallurgical Research Laboratory, [DRDO, Min Defence, Govt. India] PO Kanchanbagh, Hyderabad − 500 058, India

G. Gouda Group Director (Propulsion), Centre for Military Airworthiness and Certification, CEMILAC [DRDO, Min. Defence, Govt. India] PO Marathahalli Colony, Bangalore − 560 037, India

Dr. N.J.H. Holroyd Consultant Riverside California, CA 92506, USA

Dr. G. Jagan Reddy Scientist F, Defence Metallurgical Research Laboratory, [DRDO, Min Defence, Govt. India] PO Kanchanbagh, Hyderabad − 500 058, India

Dr. K.V. Jata Materials and Manufacturing Directorate Air Force Research Laboratory Wright-Patterson Air Force Base Ohio-45433-7750, USA

Professor V.V. Kutumbarao Visiting Scientist, Defence Metallurgical Research Laboratory, [DRDO, Min Defence, Govt. India] PO Kanchanbagh, Hyderabad − 500 058, India

Professor Enrique J. Lavernia Executive Vice-Chancellor, University of California Davis Davis, CA-95616-5294, USA

Dr. S.P. Lynch Defence Science and Technology Organisation (DSTO) 506 Lorimer Street, Fishermans Bend Melbourne, Victoria-3207, Australia

Dr. G. Madhusudhana Reddy Scientist G and Head, MJG, Defence Metallurgical Research Laboratory, [DRDO, Min Defence, Govt. India] PO Kanchanbagh, Hyderabad − 500 058, India

Dr. G. Malakondaiah Distinguished Scientist & CCR&D (HR&M), DRDO, DRDO Bhavan, Rajaji Marg, New Delhi − 110 105, India

Professor R.C. Newman Department of Chemical Engineering & Applied Chemistry University of Toronto Ontario, M5S 3E5, Canada

Professor K.A. Padmanabhan University Chair Professor, School of Engineering Sciences and Technology (SEST) University of Hyderabad (UoH) Gachibowli Hyderabad – 500 046, India

Professor T.R. Ramachandran Consultant, Non-Ferrous Technology Development Centre (NFTDC), PO Kanchanbagh Hyderabad – 500 058, India

B. Saha Scientist F and Additional Regional Director, Regional Centre for Military Airworthiness (Materials), CEMILAC, [DRDO, Min Defence, Govt. India] PO Kanchanbagh, Hyderabad – 500 058, India

Dr. K. Satya Prasad Consultant, International Advanced Research Centre for PM and New Materials (ARCI), PO Balapur, Hyderabad—-500 005, India

Dr. G.M. Scamans Innoval Technology Limited Banbury Oxon, OX16 1TQ, United Kingdom

Dr. Vijaya Singh Scientist G and Head, LACG, Defence Metallurgical Research Laboratory, [DRDO, Min Defence, Govt. India] PO Kanchanbagh, Hyderabad – 500 058, India

Professor T.S. Srivatsan Division of Materials Science and Engineering Department of Mechanical Engineering The University of Akron Akron, Ohio 44325-3903, USA

Professor Edgar A. Starke, Jr. University Professor Emeritus, Department of Material Science and Engineering University of Virginia, PO Box 400745, 395 McCormick Road, Charlottesville, VA 22904-4745, USA

Dr. K. Tamilmani Distinguished Scientist and CC R&D (Aero), DRDO, DRDO Bhavan, Rajaji Marg, New Delhi – 110 105, India

Dr. A.K. Vasudevan Office of Naval Research Arlington, VA-22203, USA

Dr. R.J.H. Wanhill Emeritus Principal Research Scientist, Aerospace Vehicles Division National Aerospace Laboratory NLR, PO Box 153, 8300 AD Emmeloord 1006 BM, The Netherlands

Part I

Introduction to Al–Li Alloys

Historical Development and Present Status of Aluminum–Lithium Alloys

Edgar A. Starke, Jr.

Department of Materials Science and Engineering, University of Virginia, Charottesville, USA

Contents

1.1 Introduction   4
1.2 Lithium Additions to Aluminum Alloys: Early Days   4
   1.2.1 History of the Development of the First Modern Al–Li Alloys   5
   1.2.2 Development of Alcoa's 2020   6
   1.2.3 The Ductility Problem of 2020   7
   1.2.4 Development of Al–Li Alloys in the Soviet Union   7
1.3 Development of Modern Aluminum–Lithium Alloys   8
   1.3.1 The Second Generation of Aluminum–Lithium Alloys   10
   1.3.2 Manufacturing Issues   11
   1.3.3 Understanding the Precipitate Structure in Al–Li–X Alloys   11
1.3.4 The Effect of Prior Deformation on Precipitation During Ageing   12
1.3.5 Deformation Behavior in Aged Al–Li–X Alloys   15
1.3.6 Predicting Strain Localization in Al–Li–X Alloys   17
1.3.7 Applications and Problems of the Second-Generation Al–Li Alloys   19
1.3.8 The Third Generation of Aluminum–Lithium Alloys   19
1.3.9 Background Information That Led to Improvements in Al–Li–X Alloys   20
1.4 Closure   22
Acknowledgments   23
References   23

Aluminum–Lithium Alloys.

1.1 INTRODUCTION

Although "age hardening" was discovered in 1902 by Alfred Wilm at the Center for Scientific Research in Germany (Wilm, 1911), it was not until 1919 that the phenomenon of "age hardening" was explained (Starke and Hornbogen, 2008). Merica et al. (1919) were conducting research on Wilm's alloy duralumin at the National Bureau of Standards in the United States and concluded that the cause of age hardening was due to precipitation from a supersaturated solid solution. They also stated that the key step in the process was a decrease in solid solubility of alloying elements with decreasing temperature and speculated that quenching from the high temperature and ageing at a lower temperature would produce a very fine dispersion of small particles within the matrix. They then introduced the idea of a critical particle size associated with maximum strengthening but made no attempt to describe how these particles resulted in the strength increase. Later, Jeffries and Archer (1921) proposed that the very fine particles formed during ageing interfered with the slip process. They recognized the crystallographic nature of slip and therefore supported the ideas proposed by Merica and his colleagues. These ideas led to searches for other alloys which could age harden, notably in the United States, Germany, and Japan in the 1920s and 1930s.

1.2 LITHIUM ADDITIONS TO ALUMINUM ALLOYS: EARLY DAYS

Around the same time that Merica and his colleagues were explaining the phenomenon of age hardening, researchers in Germany were exploring aluminum alloys containing lithium. Considering the high solubility of lithium in aluminum at high temperatures and its decreasing solubility as the temperature is lowered, it is not surprising that lithium additions were included in these investigations. Balmuth and Schmidt (1980) discussed this work in their overview of the early development of aluminum–lithium alloys that they presented at the First International Aluminum–Lithium Conference held at Stone Mountain, GA. The first commercial aluminum alloy containing lithium was the German alloy "Scleron" that had a nominal composition of Al−12Zn−3Cu−0.6Mn−0.1Li (Reuleaux, 1924). The claims of the alloy were that it had great resistance to wear, relative cheapness, high tensile strength and resistance to corrosion and oxidation, and was superior to other aluminum alloys because it could be worked into a wide variety of forms. However, research by Assmann (1926) on the strengthening effect of lithium additions to aluminum was inconclusive, and other aluminum alloys that did not contain lithium showed better properties, so Scleron production was discontinued.

The development of modern aluminum–lithium alloys can be traced to the discovery by I.M. LeBaron in 1942 that lithium could be a major strengthening element in aluminum−copper alloys. LeBaron was granted a patent on Al−Cu−Li−Mn−Cd alloys in 1945 (LeBaron, 1945). However, the introduction of 7075 by Alcoa in 1943 established the dominance of the Al−Zn−Mg−Cu system for high-strength applications, and further work on LeBaron's alloys were discontinued at that time. Subsequent work by Hardy and Silcock in England identified the lithium-containing strengthening phases in Al−Cu−Li alloys, and their research led to an increased interest in the alloy system (Hardy, 1955−56; Hardy and Silcock, 1955−56; Silcock, 1959−60). Hardy and Silcock contributed significantly to the scientific understanding of these complex materials.

1.2.1 History of the Development of the First Modern Al−Li Alloys

An abbreviated history of the use of lithium in aluminum alloys up to 1982 is shown in Figure 1.1 (Quist et al., 1983). A number of important events not noted in Figure 1.1 include the substantive contributions of several universities, funding agencies, and research laboratories.

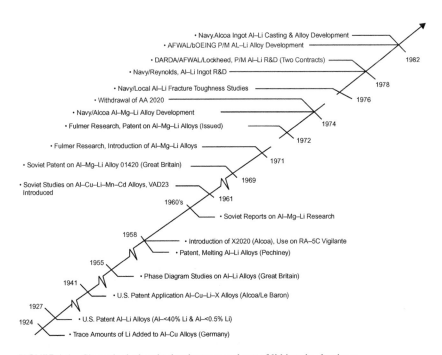

FIGURE 1.1   Chronological early development and use of lithium in aluminum.

1.2.2 Development of Alcoa's 2020

In the 1950s, metallurgists at Alcoa recognized that lithium also increased the elastic modulus of aluminum and developed the high strength Al−Cu−Li alloy 2020 in 1957 (Criner, 1957, 1959). The nominal composition of 2020 was Al−4.5Cu−1.1Li−0.5Mn−0.2Cd, and in addition to high strength at 300−400°F (∼150−200°C), it was claimed that the alloy was resistant to creep at these temperatures. Subsequently, 2020 was commercially produced and used on the United States Navy RA 5C Vigilante aircraft. It performed well for over 20 years without any documented failures (Balmuth and Schmidt, 1980). Aluminum−lithium alloys are attractive for aerospace applications because they have lower density and higher modulus than conventional aluminum aerospace alloys. Each weight percent of lithium lowers the density of aluminum by approximately 3% and increases the modulus by approximately 6% (Starke and Staley, 1996). However, the perceived brittleness of 2020 and the production problems thwarted further use, and it was withdrawn from commercial production in the 1960s. A plot of contemporary data reveals that, although the 2020-T6 may have exhibited low toughness, it was being used in a very strong condition and was not dissimilar in its relative performance to other high-strength aerospace alloys of that period (Figure 1.2) (Peel, 1990).

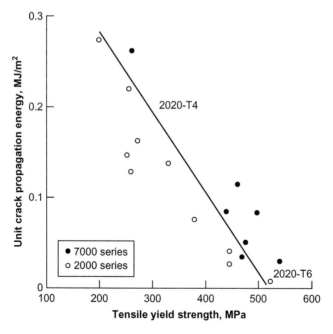

FIGURE 1.2 Toughness of 2020 alloy in the T4 and T6 tempers in comparison to contemporary 1960s aerospace alloys (Peel, 1990).

1.2.3 The Ductility Problem of 2020

The iron and silicon contents of a typical 2020 alloy were relatively high when compared to modern aerospace alloys. During ingot solidification and subsequent processing these impurities precipitate as the insoluble constituent phases $Al_{12}(FeMn)_3Si$ and Al_7Cu_2Fe, which vary in size from 1 to 10 μm (Starke and Staley, 1996). The detrimental effect of large constituent phases on the ductility and fracture toughness of aluminum alloys was first identified in 1961 (Anderson and Quist, 1966; Quist, 1963) and documented by many investigators (Smith, 1991). Not only do these large particles initiate cracks, but they also produce an inhomogeneous strain distribution during working operations, which enhances the possibility of recrystallization during subsequent heat treatments (Cotterill and Mold, 1976). Alloy 2020 had a partially recrystallized structure with some very large recrystallized grains. That feature in combination with planar slip associated with shearable strengthening precipitates and precipitate-free zones along grain boundaries can lead to low-ductility problems (Starke and Lin, 1982). The planar slip and the tendency toward strain localization lead to slip bands impinging upon grain boundaries and can cause stress concentrations, the magnitude of which depends on the slip length and therefore the grain size (Starke et al., 1981). Thus the planar slip and large recrystallized grains had a negative effect on ductility and fracture toughness. Although manganese was added to form fine $Al_{20}Cu_2Mn_3$ dispersoids to control grain structure and to improve the microscopic homogeneity of strain, some of the Mn is lost to the coarse $Al_{12}(FeMn)_3Si$, and the resulting reduced volume fraction was insufficient in 2020 to prevent recrystallization and grain growth during the solutionizing treatment that followed working operations. Keeping the Fe and Si content low was an important step in subsequent aluminum–lithium alloy development.

Following the early work of LeBaron and the research results of Silcock et al. (1955–56) that showed that Cd can enhance the nucleation and precipitation of strengthening precipitates in aluminum alloys, Cd was added for this purpose in AA 2020 (Criner 1957, 1959). Cadmium segregates to the θ'/matrix interface and decreases the interfacial energy: however, it may segregate to all interfaces, including grain boundaries. The solid solubility of Cd in aluminum at the solutionizing temperature of 2020 is less than 0.14 wt% (2020 contained 0.2 wt%), and the excess Cd can segregate to grain boundaries and to the interfaces of particles stable at this temperature. Consequently, the excess Cd in 2020 may also have a negative effect on ductility. So the problems associated with the brittleness of 2020 were most likely due to planar slip, some large recrystallized grains, high Fe and Si content forming coarse constituents, and excessive Cd.

1.2.4 Development of Al–Li Alloys in the Soviet Union

In the early 1960s and following the work in the United States on Al–Cu–Li–Cd alloys, there was significant research and development on aluminum—lithium alloys in the erstwhile Soviet Union, most notably by academician I.N. Fridlyander and his colleagues (Fridlyander, 2006). The Russian alloy VAD23 that was extensively studied at that time had a composition very similar to 2020, i.e., Al−4.8−5.8%Cu, 0.9−1.4%Li, 0.4−0.8% Mn, 0.1−0.25%Cd and not more than 0.15%Ti, 0.3%Fe, 0.3%Si, 0.1%Zn, and 0.05%Mg (Altman, 1974). The alloy saw limited use in the Soviet Union but was employed in the manufacture of wing stabilizers for antitank missiles (Fridlyander, 2006).

In 1952 F.I. Samray studied a number of alloys of the Al−Mg−Li system and established that alloying Al−Mg alloys with lithium had little effect on their mechanical properties and did not result in improvements during thermal treatment (Altman, 1974). He concluded that developing new industrial compositions on this system was not promising (Altman, 1974; Archakova et al., 1960). However, in 1965 I.N. Fridlyander, M.V. Shiryaeva, B.V. Tyurin, and V.S. Sandler discovered the effect of hardening in a vast group of alloys in the Al−Li−Mg system (Fridlyander, 2006). This work led to the development of alloy 01420 containing 2% Li, 5.5% Mg, and 0.1% Zr, which is 10−12% lighter than 2024 and was patented in a number of countries (England, 1967; Japan, 1967; Italy, 1967; France, 1967). It was claimed that the alloy had high corrosion resistance, good weldability, high elastic modulus, and static strength. Fridlyander et al. (1992) also developed a modification of 01420 containing scandium, i.e. 01421. Scandium has a more potent effect in preventing recrystallization than Zr, Cr, and Mn. It aids in the formation of a fine subgrain structure in wrought products and provides additional strengthening due to the formation of stable Al_3Sc dispersoids. Its maximum atomic solubility in aluminum exceeds Zr by approximately 1.7 times (0.5 vs 0.089 at.%) and has about equal solubility by weight (0.25 vs 0.28 wt%) (Fridlyander et al., 1992).

Alloy 01420 exhibits one of the lowest densities available for a commercial alloy. In 1970−71, the alloy was used in the vertical-takeoff and landing aircraft Âk36 and Âk38. This was the first time welded-aluminum alloys were used in aircraft. The alloy was also used for liquid oxygen tanks having a diameter of 4.5 m. In the 1980s the Soviet Union had plans to build hundreds of welded aluminum—lithium MiG29s, but after the confrontation with the United States ended the production was stopped.

1.3 DEVELOPMENT OF MODERN ALUMINUM−LITHIUM ALLOYS

The modern Al−Li alloy development program, which began in the late 1970s and 1980s, was one of the largest single alloy-development programs in the history of aluminum metallurgy and involved hundreds of scientists and engineers. Much of this research and development work will be covered in the later chapters of this book. During the 1970s, fuel costs, the market value associated with increased range, and landing weight fees resulted in a technical focus on weight reduction. Trade-off studies were made to determine which property improvements had the greatest impact on weight savings. These studies showed that a reduction in density was the most advantageous (Figure 1.3), and lithium, being the lightest metal, would have the greatest influence on reducing the density of aluminum (Ekvall et al., 1982). The aluminum companies were also interested in developing low-density alloys due to their concern about the competition from nonmetallic composites for aerospace materials and the belief that there must be major advancements in aluminum metallurgy for aluminum alloys to stay competitive. Although carbon-fiber and boron-fiber nonmetallic composites offer a considerable density advantage over all other structural materials used in aircraft, improvements in the properties of aluminum alloys seemed desirable due to their relatively low acquisition cost and the aircraft community's extensive design and manufacturing experience with these materials.

Aluminum–lithium alloy development programs were initiated in Great Britain, the United States, and France and continued in the Soviet Union. The objective of most of these programs was to develop gauge-for-gauge substitutes for the standard alloys with similar properties while maintaining manufacturability. Detailed design studies predicted that aluminum–lithium

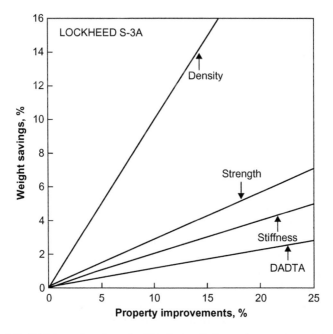

FIGURE 1.3 Weight savings in a Lockheed aircraft for various property improvements. *Courtesy of George Weld, Lockheed Corporation.*

alloys, meeting predefined alloy development targets, would be able to produce weight savings of the order of 8–15% by a combination of density reduction and stiffness enhancement (Peel, 1990). The approach of most of these Al–Li alloy development programs was to use lessons learned from previous studies of aluminum metallurgy. These included decreasing the iron and silicon content to the minimum economically feasible for high toughness and ductility; replacing manganese with zirconium to form Al_3Zr dispersoids for grain refinement since large Mn-rich dispersoids may be detrimental to ductility by nucleating voids; and not using cadmium for nucleating strengthening precipitates since that element seemed to enhance intergranular fracture in alloy 2020. These research programs resulted in the "second generation" of aluminum–lithium alloys.

1.3.1 The Second Generation of Aluminum–Lithium Alloys

There were two different approaches used for the development of the second generation of Al–Li alloys: one involving powder metallurgy (P/M) and the other involving ingot metallurgy (I/M) (Quist et al., 1983). Each initially appeared to offer both advantages and disadvantages. For P/M the advantages were greater composition and microstructural flexibility and several possible production approaches, for example rapid solidification and mechanical alloying. However, the P/M process would cost more and have small production capabilities that would result in small ingot sizes. The I/M process would cost less, could produce large ingot sizes, and for the most part use existing production equipment. The casting facilities for Al–Li would be limited and costly, and the chemical composition would be more restricted than with P/M.

The largest P/M activity involved a team led by R.E. Lewis and Lockheed Missile and Space Company and had the goals of producing an Al–Li–X alloy with an increase of 30% in modulus-to-density and an increase of 20% in strength-to-density ratio compared to AA 7075-T76 (Lewis, 1978). The goals of the program were not achieved, primarily due to manufacturing difficulties, and it became clear during the program that the better route for success would be that using the I/M approach.

The major aluminum producers, Alcoa (US), Pechiney (France), and British Alcan (with help from the Royal Aircraft Establishment), were all involved in the development of aluminum–lithium alloys using the I/M approach. Alcoa focused on a 7075-T6 replacement, Pechiney focused on a substitute for 2024-T3 sheet and light gauge products, as did British Alcan (Rioja and Liu, 2012). Each of the initial alloys from these producers contained approximately 2% or greater Li, around 2% or more Cu, some Mg, and Zr to control grain structure. The compositions, along with the specific gravity, of some of the initial alloys produced are given in Table 1.1.

TABLE 1.1 Compositions of Selected Second-Generation Aluminum−Lithium Alloys (wt%)

Alloy	Li	Cu	Mg	Si	Fe	Zr	Specific Gravity
2090	1.9−2.6	2.4−3.0	0.25	0.10	0.12	0.08−0.15	2.60
2091	1.7−2.3	1.8−2.5	1.1−1.9	0.20	0.30	0.04−0.10	2.58
8090	2.1−2.7	1.0−1.6	0.6−1.3	0.20	0.30	0.04−0.16	2.53
8091	2.4−2.8	1.8−2.2	0.5−1.2	0.30	0.5	0.08−0.16	2.54
8092	2.1−2.7	0.5−0.8	0.9−1.4	0.10	0.15	0.08−0.15	2.53
8192	2.3−2.9	0.4−0.7	0.9−1.4	0.10	0.15	0.08−0.15	2.51

1.3.2 Manufacturing Issues

One of the major issues that had to be initially addressed was the cost of manufacture of aluminum−lithium alloys. This included the cost of special casting facilities associated with the reactivity of lithium-containing aluminum alloys, and manufacturers had to develop methods to control the reaction of lithium with the atmosphere in the casting process. The reaction of molten Al−Li alloys with atmospheric oxygen and water can cause considerable production difficulties (Peel, 1990). These include the very rapid reaction at melt surfaces with the potential for inclusion of oxides, nitrides, and carbonates in the melt; the reaction of the melt with foundry materials, crucibles, linings, and molds; and an increased explosion risk (Page et al., 1987). This necessitated the use of modified refractories and degassing procedures, protective inert cover gases and/or surface fluxes during melting and casting, and special ingot cooling techniques (some employing organic coolants in place of water). In addition, there was the concern of scrap from Al−Li alloys getting mixed with scrap from other aluminum alloys and the effect such contamination would have on properties. Consequently, there was a major effort to develop a scrap recovery system and for reclaiming the valuable ingredients of the alloys for reuse (Wilson et al., 1987). The primary aluminum producers developed proprietary methods to address these issues during the early stage of alloy development and, therefore, those methods will not be covered in this chapter.

1.3.3 Understanding the Precipitate Structure in Al−Li−X Alloys

Work on 2090-T8 verified the negative effect that iron content has on the fracture toughness of aluminum−lithium alloys (Figure 1.4) (Ashton et al.,

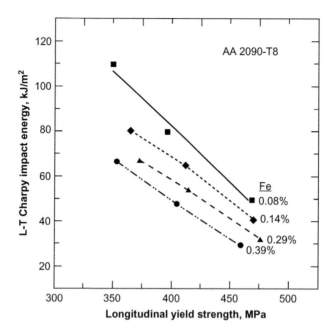

FIGURE 1.4 The effect of Fe content on strength–toughness relationship of 2090.

1986). Depending on other alloying additions, Li can form the coherent Al_3Li (δ'), which can be a major strengthening phase during ageing. Depending on the Cu:Li ratio, Cu additions can form either Al_2Cu (θ') or Al_2CuLi (T_1) with T_1 being the major strengthening phase (for details, refer to Chapter 3). Zirconium additions form the coherent Al_3Zr phase, which is effective in preventing recrystallization but results in a strong deformation texture. In Al–Mg–Li alloys, the Al_2MgLi and δ' phases may form but in complex Al–Li alloys containing Cu, Al_2CuMg (S') forms and can suppress the formation of Al_2Cu. Precipitation during ageing can be very complex with many phases forming in competition with one another. Figure 1.5 is a schematic illustration of precipitate phases that form in Al–Li–X alloys (Narayanan et al., 1982), and Figure 1.6 illustrates the complexity of precipitation in Al–Cu–Li–X alloys (Csontos and Starke, 2000).

1.3.4 The Effect of Prior Deformation on Precipitation During Ageing

In Al–Li alloys containing Cu, a combination of deformation prior to ageing and low-temperature ageing can be used to promote copious matrix precipitation, inhibit, to some extent, grain boundary precipitation and improve ductility and fracture toughness without sacrificing strength (Ashton et al., 1985). Pre-age plastic deformation enhances the ageing kinetics, strength,

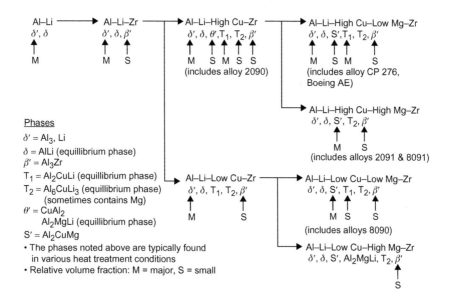

FIGURE 1.5   Schematic of the precipitate phases that form in Al–Li–X alloys.

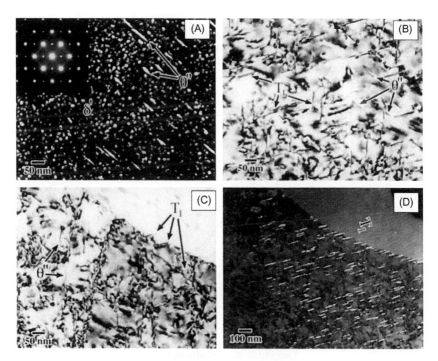

FIGURE 1.6   Micrographs showing the complexity of precipitation in Al–Cu–Li–X alloys: (A) δ', (B) θ'', and (C and D) T_1 precipitates.

and number density of fine strengthening precipitates through the introduction of heterogeneous matrix nucleation sites. The effect of stretch and temperature on the strength—toughness relationship for alloy 2090 is shown in
Figure 1.7, and the effect of stretch on the T_1 number density as a function
of stretch and ageing time is shown in Figure 1.8 (Cassada et al., 1987).
Since strength is related to the precipitate structure (size, volume fraction,
number density), there is a corresponding effect on strength.

As observed in Figure 1.6, the precipitate structure of Al—Cu—Li—X
alloys can be quite complex and can contain two different Cu-containing
precipitates, T_1 (Al$_2$CuLi) with a {111} habit plane and θ''/θ' (Al$_2$Cu) with a
{100} habit plane. Modeling by Zhu et al. (1999) suggests that the strengthening effect of matrix precipitation is optimized through the presence of two
plate precipitates with {111} and {100} habit planes. In quantitative
experiments on the effect of stretch on the volume fractions of T_1 and θ''/θ'
(precipitates in an Al—2.7Cu—1.8Li—0.6Zr—0.3Mg—0.3Mn—0.08Zn alloy),
Gable et al. (2001) showed that the magnitude of pre-age deformation significantly affects the competitive matrix precipitation between T_1 and θ''/θ'. The
difference in ΔG_v for the two precipitates and the accommodation of the
{111}-shear strain for T_1 versus that for θ' aid in T_1's preferential nucleation

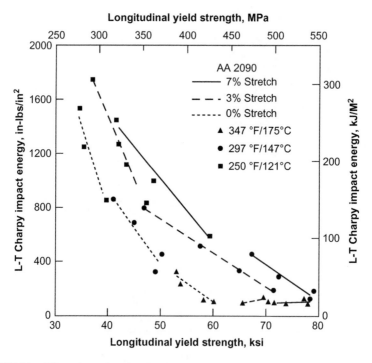

FIGURE 1.7 Effect of stretch and ageing time on the strength—toughness of 2090.

on dislocations. The relative number density, precipitate diameter, and volume fraction of T_1 to θ'', for various degrees of stretch prior to ageing at 150°C for 24 h, is given in Table 1.2, and the corresponding strength data are given in Table 1.3. Therefore, for a given ageing treatment on a particular Al–Cu–Li–X alloy, the relative amounts of the two Cu-containing strengthening precipitates may be varied by the amount of stretch prior to ageing.

1.3.5 Deformation Behavior in Aged Al–Li–X Alloys

The strengthening response of age-hardenable Al–Li–Cu–X alloys depends on the ability of the various precipitates to resist dislocation motion during deformation. All of the aforementioned age-hardenable particles exert some type of force on gliding dislocations and thus contribute to strength. The effectiveness of these precipitates on strengthening depends upon the type, size, morphology, spacing, distribution, coherency, order, volume fraction and number density, and whether these particles are sheared or looped (bypassed) by dislocations. Gliding dislocations either cut or bypass age-hardenable precipitates and play a critical role on the overall slip behavior of the alloy. Precipitate shearing promotes coarse planar slip by locally work softening the operative slip plane due to the reduced particle cross-section,

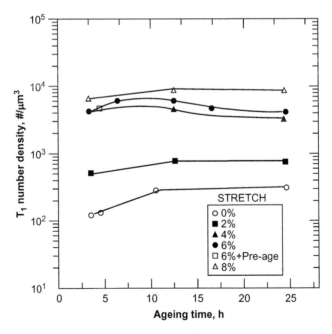

FIGURE 1.8   Effect of cold work (stretch) prior to ageing on the number density and strength of an Al–2.4Li–2.4Cu–0.18Zr alloy.

TABLE 1.2 Quantitative Precipitate Data of Al−Cu−Li−X Alloy After 150°C for 24 h

Quantitative Analysis	Stretch (%)	T_1 (Al$_2$CuLi)	Standard Deviation	θ''/θ' (Al$_2$Cu)	Standard Deviation	δ' (Al$_3$Li)	Standard Deviation
Number density (#/μm^3)	0	286	50	6277	1023	15030	5677
	4	916	280	494	383	15691	866
	8	3035	184	176	106	12585	638
Precipitate diameter (nm)	0	65	2.2	34	0.9	12	2.0
	4	80	5	63	8.4	12	1.7
	8	50	4.6	38	3.9	12	1.9
Volume fraction (%)	0	0.52	0.17	1.14	0.14	11	2.25
	4	2.23	0.23	0.54	0.27	11	1.63
	8	2.84	1.0	0.07	0.04	10	1.5

TABLE 1.3 Tensile Data of Al−Cu−Li−X Alloy After 150°C for 24 h

Quench Media	Pre-Age Stretch (%)	Yield Strength (MPa)	Standard Deviation	Tensile Strength (MPa)	Standard Deviation	Elongation (%)	Standard Deviation
Water Quench (250°C/s)	0	324	12.6	434	12.8	12.9	1.2
	2	415	9.8	473	15.3	10.3	0.9
	4	471	7.0	513	1.5	10.4	0.6
	6	488	15.5	519	12.8	9.9	0.6
	8	506	19.6	531	21.1	10.6	1.5

thus decreasing the strengthening effectiveness of the deformed precipitate. The coarseness of this planar slip from the sheared particles can then be described and quantified as the number of dislocations in the planar pileup. Conversely, bypassing or looping of the precipitates leads to fine homogeneous slip due to the work hardening that occurs on the operating slip plane. Furthermore, the maximum strength is achieved when the precipitates reach a critical size, d_c, where the transition from shearing to looping begins. Figure 1.9A shows an example of the strain localization due to shearing of coherent Al_3Li (δ'), and Figure 1.9B is a dark field micrograph showing the shearing of the individual particles (Cassada et al., 1986). Figure 1.10 shows the shearing of both δ' and T_1 in an Al−Li−Cu−X alloy aged at 150°C for 24 h to the T86 condition (Csontos and Starke, 2005). Although strain localization has an adverse effect on both ductility and fracture toughness, it appears to have a beneficial effect on fatigue crack propagation resistance (Jata and Starke, 1986).

1.3.6 Predicting Strain Localization in Al−Li−X Alloys

Duva et al. (1988) proposed a method to quantify strain localization in Al−Li−X alloys having shearable precipitates and when a grain boundary is the major barrier to slip. They suggested that the best way to quantify the slip intensity in deformed microstructures is to calculate the number, N, of dislocations expected in a slip band that had impinged upon a grain boundary. Calculating the slip intensity for a particular microstructure is straightforward if the size and distribution features are known. A slip intensity relationship has been derived based on the assumption that if slip initially weakens a plane (by reducing the cross-sections of strengthening precipitates), slip will continue on that plane until it regains its initial strength (by work hardening). For slip plane softening (i.e. for shearable precipitates), the

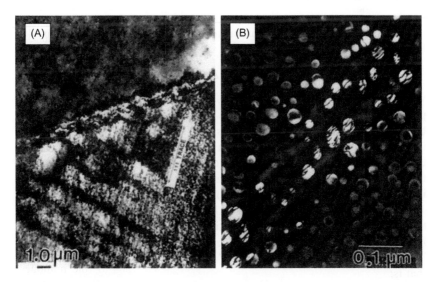

FIGURE 1.9   Slip behavior in an Al−2.0Li alloy. (A) Strain localization due to shearing of δ', (B) dark field micrograph, shearing of individual δ' precipitates.

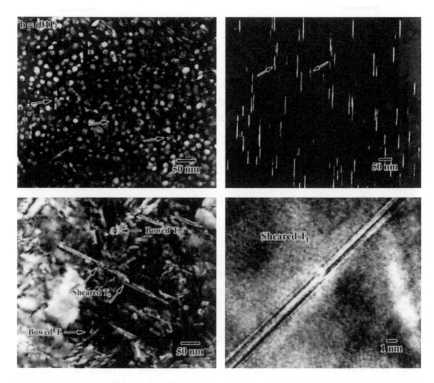

FIGURE 1.10 Shearing of both δ' and T_1 after 2% ε in an Al–Li–Cu–X alloy aged at 150°C for 24 h to the T86 condition.

number of dislocations on a particular slip plane, N, for a particular slip length is given by the equation:

$$N = f^{1/2}r^{1/2}L(C_p/C_B b) \qquad (1.1)$$

where N is the number of dislocations, f is the volume fraction of shearable precipitates, r is the shearable particle radius, L is the slip length (and may be associated with the grain size), b is the Burgers vector, C_p is a constant that depends on the intrinsic properties of the particles, and C_B is a constant that depends on the elastic properties of the matrix. The salient features of this model include increasing slip intensity for increasing volume fractions, increasing particle diameters (below d_c), and increasing slip length.

Knowing the slip intensity caused by precipitates, one can then quantitatively describe the stress concentration at the tip of a dislocation pileup impinging on a grain boundary as the shear stress concentration, τ_C, acting at the grain boundary, which is N times the resolved shear stress, τ_{rss}, that is:

$$\tau_C = N \tau_{rss} \qquad (1.2)$$

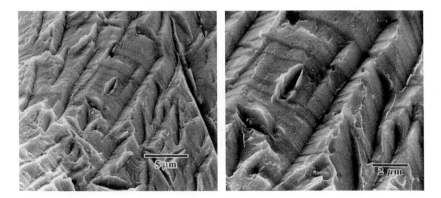

FIGURE 1.11   SEM fractograph of the fracture surface of an Al—Li—Cu—X alloy with the microstructure shown in Figure 1.10.

Crack nucleation at the grain boundary can then proceed when the stress concentration exceeds the fracture strength of the boundary. When fracture occurs the dislocations that were in the pileup will then egress from the surface (owing to image forces) and leave a step on the surface that will be equal to the magnitude of the Burgers vector times the number of dislocations that have egressed. Evidence of intergranular fracture associated with dislocation pileups cracking the grain boundary is shown in Figure 1.11, from an Al—Li—Cu—X alloy aged for 24 h at 150°C and containing both shearable δ' and T_1 precipitates (Csontos and Starke, 2005).

1.3.7 Applications and Problems of the Second-Generation Al—Li Alloys

The second generation of aluminum—lithium alloys saw limited applications, mostly in secondary structure on aircraft, for example, C-17 cargo transport that used 2090, the A340 that used 8090 and 2090, the EH101 helicopter that used 8090, the Titan payload adapter that used 8090, and the Atlas Centaur adaptor that used 2090. The major negative performance attributes for the second generation of Al—Li—X alloys that limited their usage were high anisotropy of mechanical properties that was associated with sharp crystallographic textures, low short transverse properties, crack deviations, and delamination problems during manufacture of parts. Strain localization can have a negative effect on both ductility and fracture toughness and so was one of the problems that needed to be addressed in the development of the third generation of aluminum—lithium alloys (Sugamata et al., 1993).

1.3.8 The Third Generation of Aluminum—Lithium Alloys

The development of the third generation of aluminum—lithium alloys can be traced to the late 1980s when Pickens et al. (1989), working for the Martin

Marietta Corporation (later Lockheed Martin Corporation), set out to design a weldable aluminum base alloy having low density for use in aerospace launch vehicles and cryogenic tankage. They used alloy 2219 as a base, since it had good cryogenic properties, and the phase diagram of Hardy and Silcock (1955–56) to select a composition in the T_1 phase field, hoping to minimize the presence of δ'. Lithium was added until the strength peaked at 1.3 wt% Li. Silver and Mg were added as nucleating agents for the T_1 phase based on the work of Polmear (1986). Zirconium was added for grain structure control and to refine the grain structure in the weld zone. The nominal composition of this alloy, Weldalite™, was Al–(4.6–6.3)Cu–1.3Li–0.4Ag–0.4Mg–0.14Zr–0.06Fe–0.03Si. The alloy could reach a yield strength of 700 MPa through a uniform distribution of T_1. Reynolds aluminum purchased the production rights from Lockheed Martin, but the two companies worked together for future alloy development and production capabilities.

The third generation of Al–Li alloys contained less than 2% Li. Subsequent versions of the Weldalite™ family of alloys contained Zn for improved corrosion resistance (Rioja et al., 1990). Zinc goes into solid solution within the grains and shifts the pitting potential of the matrix to less noble and decreases the electrochemical potential difference between the grain boundary and the matrix, thus improving static and dynamic corrosion properties (Kobayashi et al., 1991). Research led to the development of 2195 that was used on the Super Lightweight Tank of the Space Shuttle that was first flown in 1998. The new alloy and some design changes reduced the tank weight by 3175 kg/7000 lb over the lightweight tank, previously used, and provided a significant increase in the performance required for the shuttle to reach the International Space Station. The compositions of some other third-generation Al–Li alloys are given in Table 1.4.

1.3.9 Background Information That Led to Improvements in Al–Li–X Alloys

Anisotropy of mechanical properties can play an important role in the general use of aluminum alloys for aerospace applications. All of the early Al–Li alloys that contained Zr as a grain refiner showed elongated pancake-shaped grains with a pronounced crystallographic texture (Bull and Lloyd, 1986). The Zr forms coherent Al_3Zr dispersoids during ingot preheat, which are very efficient in inhibiting recrystallization during normal ingot breakdown and subsequent processing. This type of structure can lead to delamination at both ambient and cryogenic temperatures when shearable precipitates, precipitate-free zones, and strain localization are present, leading to significant in-plane anisotropy in tensile, fatigue, and fracture properties (Eswara Prasad, 1993; Jata and Starke, 1988). The delamination has been associated with periodical events of high to low relative misorientations of grains (Roven et al., 1990). Polmear et al. (1986) showed that a fine recrystallized grain structure can be

TABLE 1.4 Chemical Composition of Some Third-Generation Al−Li−X Alloys (wt%)

Alloy	Li	Cu	Mg	Ag	Zr	Mn	Zn
2195	1.0	4.0	0.4	0.4	0.11		
2196	1.75	2.9	0.5	0.4	0.11	0.35 max	0.35 max
2297	1.4	2.8	0.25 max		0.11	0.3	0.5 max
2397	1,4	2.8	0.25 max		0.11	0.3	0.10
2198	1.0	3.2	0.5	0.4	0.11	0.5 max	0.35 max
2099	1.8	2.7	0.3		0.09	0.3	0.7
2199	1.6	2.6	0.2		0.09	0.3	0.6
2050	1.0	3.6	0.4	0.4	0.11	0.35	0.25 max
2060	0.75	3.95	0.85	0.25	0.11	0.3	0.4
2055	1.15	3.7	0.4	0.4	0.11	0.3	0.5

obtained in Al−Li−X alloys by thermomechanical processing, and the recrystallized structure in their studies exhibited almost isotropic tensile properties in an unstretched, peak-aged condition.

Rioja (1998) has discussed the relationship between processing, composition, texture, and properties of aerospace aluminum alloys. He noted that the "Brass" texture component in unrecrystallized plate results in anisotropy in mechanical properties both in-plane and through thickness. Reducing the intensity of crystallographic texture can usually reduce the observed anisotropy. This may be accomplished by recrystallization during solution heat treatment or during an intermediate step when processing wrought alloys, reducing the amount of deformation, replacing coherent dispersoids with semi-coherent/incoherent dispersoids, i.e. by composition control and thermomechanical processing (Rioja et al., 2008). Rioja et al. (2008) also noted that the dependence of properties on texture is different for T3 and T8 tempers in 2XXX alloys. The "Goss" texture seems to improve the strength−toughness relationship in recrystallized fuselage sheet in the T3 temper, and the "Brass" texture improves the fatigue crack growth resistance of plate products in the T3 temper for lower wing applications.

In 1992 the US Air Force Wright Laboratory Materials Directorate initiated a program with the University of Dayton Research Institute, with Alcoa as a subcontractor, to significantly reduce the in-plane anisotropy in aluminum alloys containing more than 2 wt% Li. The initial approach was to introduce an intermediate recrystallization anneal between rolling stages in order to reduce the sharp deformation texture and thereby decrease the alloy's anisotropy (Jata et al., 1996). The designed two-step process to both inhibit and promote recrystallization proved highly successful, and the anisotropy of both the modulus and the yield strength was reduced significantly from 20−25% for earlier Al−Li alloys to less than 10% for the Air Force alloy. In addition, the short transverse fracture toughness was improved nearly fourfold over alloys with greater than 2.0 wt% Li. The Air Force alloy was designated AF/C-489 with a nominal composition of Al−2.7Cu−2.05Li−0.6Zn−0.3Mn−0.3Mg−0.04Zr.

The Air Force and Alcoa developed a derivative of AF/C-489 designated AF/C-458 with a nominal composition of Al−2.58Cu−1.73Li−0.6Zn−0.25Mn−0.26Mg−0.09Zr. The Cu and Li contents were selected to form a combination of Al_3Li, Al_2Cu, and Al_2CuLi precipitates in order to achieve high strength. The 0.6 wt% of Zn was added primarily for its effect on corrosion resistance, but coupled with the 0.26 wt% Mg, these two additions act to drive the δ' solvus boundary lower, thus promoting the precipitation of Al_3Li and possibly the Al_2CuLi strengthening phase (Baumann and Williams, 1984). Magnesium is also added to possibly give some solid solution strengthening (Dinsdale et al., 1981). Zirconium and Mn additions were designed to form Al_3Zr, $Al_{20}Cu_2Mn$, and Al_6Mn dispersoids in an effort to control the recrystallization of the alloy. Furthermore, the 0.25 wt% of Mn was added to minimize the strain localization effects and reduce the sharp texture associated with higher Zr concentrations in typical Al−Li−Cu−X alloys. Greater amounts of Mn result in large Mn-rich primary particles that form during casting and homogenization heat treatments and decrease the ductility and fracture toughness (Walsh et al., 1989). The selected composition is essentially the same as Alcoa's 2X99 alloys.

1.4 CLOSURE

There have been a large number of research and development activities focused on Al−Li−X alloys, in many countries, universities, and industrial and government laboratories, that are not covered in this chapter. Most of those activities will be covered in the following chapters. Although the early Al−Li−X alloys and the second generation of Al−Li−X alloys had undesirable performance and manufacturing characteristics, fundamental studies identified the root causes of those problems and led to the improved third generation of Al−Li−X alloys, having high strength/fracture toughness, fatigue, and corrosion resistance. Rioja and Liu (2012) have reviewed more details of the technical achievements involved in these improved properties. Finally, a number of the new alloys are currently being used, for example:

- The F16 Fighter Aircraft (Weldalite™, 2297)
- The A380 Airbus (2196)
- The Boeing 787 Dreamliner (2099/2199)
- Being considered for the A350 Twin-Engine Aircraft (Alcan Alloys 2098/2198).

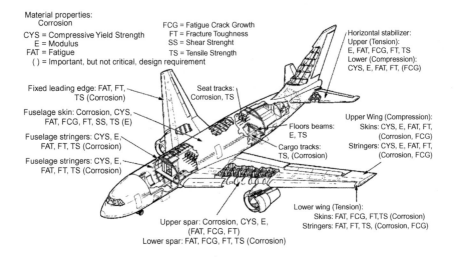

Material properties:
Corrosion
CYS = Compressive Yield Strength
E = Modulus
FAT = Fatigue
( ) = Important, but not critical, design requirement

FCG = Fatigue Crack Growth
FT = Fracture Toughness
SS = Shear Strenght
TS = Tensile Strength

Horizontal stabilizer:
Upper (Tension):
E, FAT, FCG, FT, TS
Lower (Compression):
CYS, E, FAT, FT, (FCG)

Fixed leading edge: FAT, FT,
TS (Corrosion)

Fuselage skin: Corrosion, CYS,
FAT, FCG, FT, SS, TS (E)

Fuselage stringers: CYS, E,
FAT, FT, TS (Corrosion)

Fuselage stringers: CYS, E,
FAT, FT, TS (Corrosion)

Seat tracks:
Corrosion, TS

Floors beams:
E, TS

Cargo tracks:
TS, (Corrosion)

Upper Wing (Compression):
Skins: CYS, E, FAT, FT,
(Corrosion, FCG)
Stringers: CYS, E, FAT, FT,
(Corrosion, FCG)

Upper spar: Corrosion, CYS, E,
(FAT, FCG, FT)
Lower spar: FAT, FCG, FT, TS (Corrosion)

Lower wing (Tension):
Skins: FAT, FCG, FT,TS (Corrosion)
Stringers: FAT, FT, TS, (Corrosion, FCG)

FIGURE 1.12   Property requirements for Jetliner and military transport applications.

The properties of the third generation of Al−Li−X alloys can be tailored to meet a variety of needs of future aircraft and spacecraft for weight savings, performance enhancement, and reduced inspection and maintenance (Rioja and Liu, 2012). Because aluminum alloys are used in a variety of product forms including sheet, plate, extrusions, forgings, and tubes, different processing procedures may be necessary. In addition, different components of an aircraft require different sets of properties (Figure 1.12) (Staley and Lege, 1993), and this may require a variety of aluminum−lithium alloys to be used on a particular aircraft, along with other materials, for example, composites and titanium alloys.

ACKNOWLEDGMENTS

I would like to acknowledge my many former colleagues and students that I worked with on Al−Li−X projects for over 30 years. In particular, I should acknowledge Charles Blankenship, William Cassada, Aladar Csontos, Mark Duva, Brian Gable, Earhard Hornbogen, Kumar Jata, Richard Lewis, Fu-Shiong Lin, William Quist, Hans Roven, Thomas Sanders, Gary Shiflet, and Aiwu Zhu.

REFERENCES

Altman, M.B., 1974. High-Strength, Heat-Resistant and Structural Alloys of Aluminum with Lithium. AD/A-005 977, Army Foreign Science and Technology Center, Charlottesville, VA, Translation from ASlyuminiyevyyesplavy, 1972, Moscow, Chapter 7, pp. 204−230.

Anderson, W.E., Quist, W.E., 1966. Aluminum Alloy, U.S. Patent No. 3,284,193.

Archakova, Z.N., Romanova, O.A., Archakova, Z.I., 1960. Bulletins of AN SSSR OTN (Akademiyanauk SSSR, Academy of Sciences of the USSR, Otdaleniyetechnicheskikhnauk, Department of Technical Sciences), No. 4, p. 106.

Ashton, R.F., Thompson, D.S., Starke Jr., E.A., Lin, F.S., 1985. Processing Al−Li−>Cu−(Mg) alloys. In: Baker, C., Gregson, P.J., Harris, S.J., Peel, C.J. (Eds.), Aluminum−Lithium Alloys III. The Institute of Metals, London, pp. 66−86.

Ashton, R.F., Thompson, D.S., Gayle, F.W., 1986. The effect of processing on the properties of Al−Li alloys. In: Starke Jr., E.A., Sanders Jr., T.H. (Eds.), Aluminum Alloys—Their Physical and Mechanical Properties. EMAS, Warley, England, pp. 403−417.

Assmann, P., 1926. Age hardened aluminium−lithium alloys. Z. Metallkunde 18, 51.

Balmuth, E.S., Schmidt, R., 1980. A perspective on the development of aluminum−lithium alloys. In: Sanders Jr., T.H., Starke Jr., E.A. (Eds.), Aluminum−Lithium Alloys. The Metallurgical Society of AIME, pp. 69−88.

Baumann, S.F., Williams, D.B., 1984. The effect of ternary additions on the δ'/α misfit and the δ' solvus line Al−Li alloys. In: Sanders, T.H., Starke, E.A. (Eds.), Proceedings of the Second International Conference on Aluminum−Lithium Alloys II. The Metallurgical Society of AIME, Warrendale, PA, pp. 17−29.

Bull, M.J., Lloyd, D.J., 1986. Textures developed in Al−Li−Cu−Mg alloy. In: Baker, C., Gregson, P.J., Harris, S.J., Peel, C.J. (Eds.), Proceedings of the Third International Conference on Aluminium−Lithium Alloys III. The Institute of Metals, London, pp. 402−410.

Cassada, W.A., Shiflet, G.J., Starke, E.A., 1986. The effect of germanium on the precipitation and deformation behavior of Al−2Li alloys. Acta Metall. 34, 367−378.

Cassada, W.A., Shiflet, G.J., Starke Jr., E.A., 1987. The effect of plastic deformation on T_1 precipitation. In: Champier, G., Dubost, B., Miannay, D., Sabetay, L. (Eds.), Proceedings of the Fourth International Conference on Aluminium−Lithium Alloys. J. Phys., 48, C3. 397−C3.406.

Cotterill, P., Mold, P.R., 1976. Recrystallization and Grain Growth in Metals. University of Surrey Press, p. 180.

Criner, C.B., Aluminum Base Alloy, U.S. Patent No. 2,784,126, Issued March 5, 1957.

Criner, C.B., Aluminum Base Alloy, U.S. Patent No. 2,915,391, Issued December 11959.

Csontos, A.A., Starke, E.A., 2000. The effect of processing and microstructure development on the slip and fracture behavior of the 2.1 wt pct Li AF/C-489 and 1.8 wt pct Li AF/C-458 Al−Li−Cu−X alloys. Metall. Mater. Trans. A 31A, 1965−1977.

Csontos, A.A., Starke Jr., E.A., 2005. The role of inhomogeneous plastic deformation on the fracture behavior of age-hardenable Al alloys. Int. J. Plast. 21 (6), 1097−1118.

Dinsdale, K., Harris, S.J., Noble, B., 1981. Relationship between microstructure and mechanical properties of aluminium−lithium−magnesium alloys. In: Sanders, T.H., Starke, E.A. (Eds.), Proceedings of the First International Conference on Aluminium−Lithium Alloys. The Metallurgical Society of AIME, Warrandale, PA, pp. 101−118.

Duva, J.M., Daubler, M.A., Starke Jr., E.A., Lütjering, G., 1988. Large shearable particles lead to coarse slip in particle reinforced alloys. Acta Metall. 36, 585−589.

Ekvall, J.C., Rhodes, J.E., Wald, G.G., 1982. Methodology for evaluating weight savings from basic material properties. ASTM STP 761, Design of Fatigue and Fracture Resistant Structures, Abelkis, P.R., and Hudson, C.M. (Eds.), American Society for Testing and Materials, pp. 328−343.

Eswara Prasad, N., 1993. In-Plane Anisotropy in the Fatigue and Fracture Properties of Quaternary Al−Li−Cu−Mg Alloys, Doctoral thesis. Banaras Hindu University, Varanasi, India.

Fridlyander, I.N., 2006. Memories on the Establishment of a Aerospace and Nuclear Technology Aluminium Alloy, second ed. Russian Academy of Sciences, Department of Chemistry and Materials Sciences (in Russian).

Fridlyander, I.N., Kolobnev, N.I., Berezina, A.L., Chuistov, K.V., 1992. The effect of scandium on decomposition kinetics in aluminum—lithium alloys. In: Peters, M., Winkler, P.-J. (Eds.), Aluminium—Lithium, vol. 1. Informationsgesellschaft, Verlag, pp. 107—112.

Gable, B.M., Zhu, A.W., Csontos, A.A., Starke Jr., E.A., 2001. The role of plastic deformation on the competitive microstructural evolution and mechanical properties of a novel Al—Li—Cu—X alloy. J. Light Met. 1, 1—14.

Hardy, H.K., 1955. Trace-element effects in some precipitation-hardening aluminum alloys. J. Inst. Metall. 84, 429—439.

Hardy, H.K., Silcock, J.M., 1955. The phase sections at 500°C and 350°C of aluminium-rich aluminium—copper—lithium alloys. J. Inst. Metall. 84, 423—428.

Jata, K.V., Starke, E.A., 1986. Fatigue crack growth and fracture toughness behavior of an Al—Li—Cu alloy. Metall. Trans. 17A, 1011—1026.

Jata, K.V., Starke, E.A., 1988. Fracture toughness of Al—Li—X alloys at ambient and cryogenic temperatures. Scr. Metall. 22, 1553—1556.

Jata, K.V., Hopkins, A.K., Rioja, R.J., 1996. The anisotropy and texture of Al—Li alloys. Mater. Sci. Forum 217—222, 647—652.

Jeffies, Z., Archer, R.S., 1921. Slip interference theory of hardening of metals. Chem. Met. Eng. 24, 1057—1067.

Kobayashi, K., Ohsaki, S., Kamio, A., Tsuji, Y., 1991. Effect of Zn addition on corrosion resistance of 2090 and 2091 alloys. In: Peters, M., Winkler, P.-J. (Eds.), Aluminum—Lithium, vol. 2. Informationsgesellschaft, Verlag, pp. 673—678.

LeBaron, I.M. U.S. Patent No. 2,381,219, Application date 1942, granted 1945.

Lewis, R.E., 1978. Lockheed Missile and Space Company, Palo Alto, CA, AFML Contract F33615-78-C-5203, sponsored by DARPA and managed by AFML, 1978.

Merica, P.D., Waltenbert, R.G., Freeman, J.R., 1919. Scientific Paper, U.S. Bureau of Standards, p. 347; Trans. AIME, 1921, 64, p. 3, Merica, P.D., Waldenberg, R.G. and Scott, H., Trans. AIME, 1921, 64, p. 41.

Narayanan, G.H., Quist, W.E., Wilson, B.L., Wingert, A.L., 1982. Low density aluminum alloy development. First Interim Technical Report, AFWAL Contract No. F33615-81-C-5053, Air Force Wright Aeronautical Laboratories, Dayton, Ohio, USA.

Page, F.M., Chamberlain, A.T., Grimes, R., 1987. The safety of molten aluminium—lithium alloys in the presence of coolants. In: Champier, G., Dubost, B., Miannay, D., Sabetay, L. (Eds.), Proceedings of the Fourth international Conference on Aluminum Lithium Alloys. J. Phys., Colloque 48, C3.63—73.

Peel, C.J., 1990. The development of aluminium—lithium alloys: an overview. In: AGARD Lecture Series No. 174, New Light Alloys, North Atlantic Treaty Organization, pp. 1—55.

Pickens, J.R., Heubaum, F.H., Langan, T.J., Kramer, L.S., 1989. Al—(4.5—6.3) Cu—1.3Li—0.4Ag—0.4Mg—0.14Zr alloy weldalite 049. In: Sanders, T.H., Starke, E.A. (Eds.), Aluminium—Lithium Alloys, Proceedings of the Fifth International Conference on Aluminum—Lithium Alloys. Vol. 3, Materials and Component Engineering Publications Ltd., Birmingham, UK pp. 1397—1411.

Polmear, I.G., Miller, W.S., Lloyd, D.J., Bull, M.J., 1986. Effect of grain structure and texture on mechanical properties of Al—Li base alloys. In: Baker, C., Gregson, P.J., Harris, S.J., Peel, C.J. (Eds.), Aluminium—Lithium Alloys III. The Institute of Metals, London, pp. 565—575.

Polmear, I.J., 1986. Develop of an experimental wrought aluminum alloy for use at elevated temperatures. In: Starke Jr., E.A., Sanders Jr., T.H. (Eds.), Aluminum Alloys, Their Physical and Mechanical Properties. EMAS, West Midlands, UK, pp. 661—674.

Quist, W.E., 1963. Effect of Composition on the Fracture Properties of Aluminum Alloy 7178, MSc Thesis. University of Washington.

Quist, W.E., Narayanan, G.H., Wingert, A.L., 1983. Aluminum–lithium alloys for aircraft structure—an overview. In: Sanders Jr., T.H., Starke Jr., E.A. (Eds.), Aluminum–Lithium Alloys II. The Metallurgical Society of AIME, pp. 313–334.

Reuleaux, O., 1924. Scleron alloys. J. Inst. Metall. 33, 346.

Rioja, R.J., 1998. Fabrication methods to manufacture isotropic Al–Li alloys and products for space and aerospace applications. Mater. Sci. Eng. A 257, 100–107.

Rioja, R.J., Liu, J., 2012. The evolution of Al–Li base products for aerospace and spaceapplications. Metall. Mater. Trans. A 43, 3325–3337.

Rioja, R.J., Cho, A., Pretz, P.E., 1990. Al–Li Alloys Having Improved Corrosion Resistance Containing Mg and Zn. U.S. Patent No. 4,961,792.

Rioja, R.J., Giummarra, C., Cheong, S., 2008. The roll of crystallographic texture on the performance of flat rolled aluminum products for aerospace applications. In: DeYoung, D.H. (Ed.), Light Metals. TMS, pp. 1065–1069.

Roven, H.J., Starke Jr., E.A., Sφdahl, Φ., Hjelen, J., 1990. Effects of texture on delamination behavior of a 8090-type Al–Li alloy at cryogenic and room temperature. Scr. Metall. 24, 421–426.

Silcock, J.M., 1959. The structural ageing characteristics of Al–Cu–Li alloys. J. Inst. Metall. 88, 357–364.

Silcock, J.M., Heal, T.J., Hardy, H.K., 1955. J. Inst. Metall. 84, 23.

Smith, R.K., 1991. The quest for excellence. In: Greenwood, J.T. (Ed.), Milestones of Aviation. Crescent Books, New York, NY, pp. 222–296.

Staley, J.T., Lege, D.J., 1993. Advances in aluminum alloy products for structural applications in transportation. J. Phys. IV, Colloque C7 3, 179–190.

Starke, E.A., Lin, F.S., 1982. The influence of grain structure on the ductility of the Al–Cu–Li–Mn–Cd alloy 2020. Metall. Trans. 13A, 2259–2269.

Starke Jr., E.A., Staley, J.T., 1996. Application of modern aluminum alloys to aircraft. Prog. Aerosp. Sci. 32, 131–172.

Starke, E.A., Sanders, T.H., Palmer, I.G., 1981. New approaches to alloy development in the Al–Li system. J. Met. 33, 24–36.

Sugamata, M., Blankenship Jr., C.P., Starke Jr., E.A., 1993. Predicting plane strain fracture toughness of Al–Li–Cu–Mg alloys. Mater. Sci. Eng. A 163, 1–10.

Walsh, J.A., Jata, K.V., Starke Jr., E.A., 1989. The influence of Mn dispersoid content and stress state on ductile fracture of 2134 type Al alloys. Acta Metall. 37, 2861–2871.

Wilm, A., 1911. Metallurgie 8, 223.

Wilson, W.R., Worth, J., Short, E.P., Pygall, C.F., 1987. Recycling of aluminium–lithium process scrap. In: Champier, G., Dubost, B., Miannay, D., Sabetay, L. (Eds.), Proceedings of the Fourth International Conference on Aluminium–Lithium Alloys. J. Phys. 48, C3. 75–C3.83.

Zhu, A.W., Csontos, A.S., Starke Jr., E.A., 1999. Computer experiments of superposition of strengthening effects of different particles. Acta Mater. 47, 1713–1731.

Chapter 2

Aerostructural Design and Its Application to Aluminum−Lithium Alloys

R.J.H. Wanhill* and G.H. Bray†
*NLR, Emmeloord, the Netherlands, †Alcoa Technical Center, Alcoa Center, USA

Contents

2.1 Introduction	**28**
2.2 Aircraft Structural Property Requirements	**28**
2.2.1 Fuselage/Pressure Cabins	29
2.2.2 Wings	29
2.2.3 Empennage (Tail)	30
2.3 Engineering Property Requirements for Al−Li Alloys in Aircraft Structures	**30**
2.3.1 The Key Role of Lithium Content: Density and Stiffness	31
2.3.2 Strength and Ductility	34
2.3.3 Damage Tolerance	35
2.4 Families of Al−Li Alloys	**39**
2.4.1 Second-Generation Alloys	39
2.4.2 Third-Generation Alloys	40
2.5 Examples of Third-Generation Alloy Property Developments and Trade-Offs	**44**
2.5.1 Fuselage/Pressure Cabins	44
2.5.2 Upper Wings	47
2.5.3 Lower Wings	49
2.5.4 Spars, Ribs, and Other Internal Structures	52
2.6 Service Qualification Programs	**53**
2.7 Summary and Conclusions	**55**
References	**56**

2.1 INTRODUCTION

Aircraft structural alloys must have outstanding combinations of engineering properties and also enable lightweight and durable structures to be manufactured. Aluminum alloys have been the predominant materials of choice since the introduction of the Boeing 247 (1933) and Douglas DC-2 (1934), although composites (mainly carbon fiber) and titanium alloys nowadays provide competition for certain applications. There is a notable trend toward using composites extensively in the airframes of transport (passenger and cargo) aircraft (e.g. the Boeing 787 *Dreamliner*), but for most transport aircraft, including the Airbus A380, aluminum alloys still account for about 60% of the structural weight.

Modern aluminum–lithium (Al–Li) alloys must compete against established and recently improved conventional aluminum alloys and also the upsurge of interest in composite structures. Composites are rightly acclaimed for their intrinsic lightness, strength, and stiffness, but they have disadvantages. Large and complex composite structures are difficult to design and manufacture to give reliable properties, and some of the weight advantage is reduced by the need to use high safety factors. Composites are also susceptible to property degradations owing to impact damage and adverse environmental conditions ("hot and wet"); and any damage that is incurred may be very difficult or even impossible to repair such that the structural integrity is guaranteed.

While Al–Li alloys cannot achieve the *intrinsic* weight savings *apparently* offered by composites, the final structural weight of composite structures and metallic structures using Al–Li alloys may be similar. Also, Al–Li alloys offer some significant property improvements over conventional aluminum alloys for new and derivative aircraft metallic structures. The main driver for development and use of Al–Li alloys has been the improvements in specific strength (strength/density) and stiffness (elastic modulus/density) offered by the addition of lithium. However, as will be discussed briefly in this chapter, and in more detail in subsequent chapters, many other properties must be taken into account. We note here that Al–Li alloys cost more to produce than conventional aluminum alloys, and this is also important when considering their use. This aspect is discussed briefly in Chapter 15.

2.2 AIRCRAFT STRUCTURAL PROPERTY REQUIREMENTS

Figure 2.1 illustrates the engineering property requirements for several of the main structural areas in a transport aircraft. This is based on a similar illustration by Staley and Lege (1993). The rankings of the requirements differ for different areas, but there is obviously much commonality.

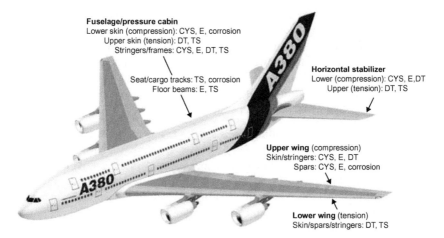

FIGURE 2.1   Engineering property requirements for main structural areas in a transport aircraft. CYS, compressive yield strength; E, elastic modulus; TS, tensile strength; DT, damage tolerance properties (fatigue, fatigue crack growth, and fracture toughness).

2.2.1 Fuselage/Pressure Cabins

The fuselage of a transport aircraft is a cylindrical shell consisting of the skin, longitudinal stringers and longerons, and transverse frames and bulkheads. The skin carries the cabin pressurization (tension) and shear loads; the stringers or longerons carry longitudinal tension and compression loads; the circumferential frames maintain the fuselage shape and redistribute loads into the skin; and bulkheads carry concentrated loads (Mouritz, 2012; Starke and Staley, 1996).

The fuselage can be divided into three areas: crown, sides, and bottom. During flight the predominant loads are tension in the crown, shear in the sides, and compression in the bottom. The main engineering property requirements are strength (TS, CYS), stiffness (E), damage tolerance (DT: fatigue, fatigue crack growth, fracture toughness), and corrosion (general and stress corrosion) (Figure 2.1).

2.2.2 Wings

Aircraft wings experience the most complex and highest loads and stresses. They are essentially beams that transmit the air loads to the central wing/fuselage attachments. Additional loads come from the fuel pressure in the internal fuel tanks, leading and trailing edge loads from the flaps and ailerons, and landing gear loads (Mouritz, 2012; Starke and Staley, 1996).

The *wing box*, which is the part of the wing that carries the loads, consists of the upper and lower covers (skins and lengthwise stringers that stiffen the skin) that take the bending loads; lengthwise spars that form the sides, and sometimes the middle, of the wing box and take torsion loads to counteract wing twisting during flight; and transverse ribs that maintain the shapes and minimize buckling of the covers (Starke and Staley, 1996).

During flight, the upper wing structure is predominantly subjected to compression loads, whereas the lower wing structure undergoes tension loads. This difference is indicated in Figure 2.1, and it is convenient to consider the upper and lower wings separately:

1. Upper wing: in-flight bending loads stress the upper wing surface in compression. The main property requirements are high strength (CYS) and stiffness (E), and this includes the upper parts of the spars. The damage tolerance (DT) properties and corrosion resistance are also important.
2. Lower wing: the same in-flight bending loads stress the lower wing surface in tension. The main property requirements are damage tolerance (DT) and strength (TS), though stiffness (E) and corrosion resistance are also important. These requirements also hold for the lower parts of the spars.

2.2.3 Empennage (Tail)

The empennage is the entire tail assembly, consisting of the horizontal and vertical stabilizers, the rear section of the fuselage to which they are attached, and the elevators and rudders (Mouritz, 2012). Structural design of the horizontal and vertical stabilizers is essentially the same as for the wing. The horizontal stabilizers are like upside-down wings, so the property requirements are reversed (Figure 2.1).

2.3 ENGINEERING PROPERTY REQUIREMENTS FOR Al–Li ALLOYS IN AIRCRAFT STRUCTURES

As discussed in Section 2.2 and shown in Figure 2.1, the engineering properties required for aircraft structures are strength (TS, CYS), stiffness (E), damage tolerance (DT: fatigue, fatigue crack growth, fracture toughness), and corrosion (general and stress corrosion). Also very important is the material density (ρ). This has been shown by quantification of the weight savings potentially achievable from improvements of various properties (Ekvall et al., 1982; Peel et al., 1984). Figure 2.2 summarizes the results obtained by Ekvall et al. (1982). Reducing density is the most effective way of saving weight. Next are strength and stiffness increases, which combine with reduced density to give improvements in specific strength and stiffness. Finally, improvements in DT properties have the least potential for saving weight, though even small amounts of weight savings can be important.

Additions of lithium to aluminum alloys decrease the density and increase the stiffness, thereby having a synergistic effect on the specific

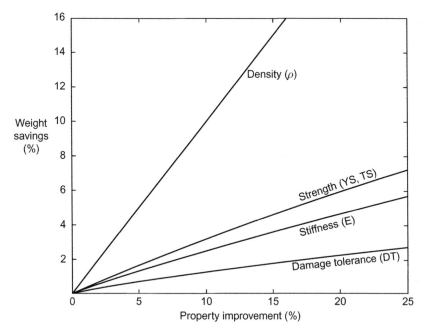

FIGURE 2.2   Effects of property improvements on potential weight savings for aircraft structures (Ekvall et al., 1982). Note: this diagram is also given in slightly different form as Figure 1.3.

stiffness (E/ρ). Thus Al–Li alloy development may already be successful from an engineering property viewpoint—certainly with respect to equivalent conventional alloy products—if other properties are simply maintained. This is attractive to commercial alloy producers, since there is the possibility of obtaining families of Al–Li alloys to replace conventional alloys for a variety of applications. This possibility has been a main driver for developing the second- and third-generation Al–Li alloys, as will be discussed in Section 2.4.

2.3.1   The Key Role of Lithium Content: Density and Stiffness

Table 2.1 gives a survey of the chemical compositions and densities of many of the commercially developed Al–Li alloys. The first use goes back to the late 1950s, when AA 2020 was used for the wing skins and tails of the Northrop RA-5C Vigilante aircraft (Starke and Staley, 1996). In the mid-1960s the Soviet Union developed the much lighter but relatively low-strength 142X-type alloys, which contained no copper and were weldable. These first-generation Al–Li alloys were followed by the second and third generations.

TABLE 2.1 Compositions and Densities of Al–Li Alloys

Alloys	Li	Cu	Mg	Ag	Zr	Sc	Mn	Zn	Density ρ (g/cm^3)	Introduction
First Generation										
2020	1.2	4.5					0.5		2.71	Alcoa 1958
1420	2.1		5.2		0.11				2.47	Soviet 1965
1421	2.1		5.2		0.11	0.17			2.47	Soviet 1965
Second Generation (Li ≥ 2wt%)										
2090	2.1	2.7			0.11				2.59	Alcoa 1984
2091	2.0	2.0	1.3		0.11				2.58	Pechiney 1985
8090	2.4	1.2	0.8		0.11				2.54	EAA 1984
1430	1.7	1.6	2.7		0.11	0.17			2.57	Soviet 1980s
1440	2.4	1.5	0.8		0.11				2.55	Soviet 1980s
1441	1.95	1.65	0.9		0.11				2.59	Soviet 1980s
1450	2.1	2.9			0.11				2.60	Soviet 1980s
1460	2.25	2.9			0.11	0.09			2.60	Soviet 1980s
Third Generation (Li < 2wt%)										
2195	1.0	4.0	0.4	0.4	0.11				2.71	LM/Reynolds 1992
2196	1.75	2.9	0.5	0.4	0.11		0.35 max	0.35 max	2.63	LM/Reynolds/ McCook Metals 2000
2297	1.4	2.8	0.25 max		0.11		0.3	0.5 max	2.65	LM/Reynolds 1997
2397	1.4	2.8	0.25 max		0.11		0.3	0.10	2.65	Alcoa 2002
2098	1.05	3.5	0.53	0.43	0.11		0.35 max	0.35	2.70	McCook Metals 2000
2198	1.0	3.2	0.5	0.4	0.11		0.5 max	0.35 max	2.69	Reynolds/ McCook Metals/Alcan 2005
2099	1.8	2.7	0.3		0.09		0.3	0.7	2.63	Alcoa 2003
2199	1.6	2.6	0.2		0.09		0.3	0.6	2.64	Alcoa 2005
2050	1.0	3.6	0.4	0.4	0.11		0.35	0.25 max	2.70	Pechiney/ Alcan 2004[a]

(*Continued*)

TABLE 2.1 (Continued)

Alloys	Li	Cu	Mg	Ag	Zr	Sc	Mn	Zn	Density ρ (g/cm^3)	Introduction
2296	1.6	2.45	0.6	0.43	0.11		0.28	0.25 max	2.63	Constellium Alcan 2010[a]
2060	0.75	3.95	0.85	0.25	0.11		0.3	0.4	2.72	Alcoa 2011
2055	1.15	3.7	0.4	0.4	0.11		0.3	0.5	2.70	Alcoa 2011
2065	1.2	4.2	0.50	0.30	0.11		0.4	0.2	2.70	Constellium 2012
2076	1.5	2.35	0.5	0.28	0.11		0.33	0.30 max	2.64	Constellium 2012

[a]*Pechiney acquired by Alcan 2003; Constellium formerly Alcan Aerospace.*
(Source: After Rioja and Liu (2012) and other sources.)

The second-generation alloys were developed primarily in the United States and Europe, beginning in the 1970s and continuing through the 1980s, and also in the Soviet Union in the 1980s and 1990s. The aim was to obtain alloys 8−10% lighter (and stiffer) than equivalent conventional alloys by the addition of about 2 wt% lithium. Table 2.1 shows that the densities of the second-generation alloys range from 2.54 to 2.61 g/cm^3. These densities are indeed 8−10% less than those of conventional AA 2XXX (2.78 g/cm^3) and AA 7XXX (2.81−2.86 g/cm^3) alloys.

Unfortunately, it was subsequently found that lithium contents of 2 wt% or more are linked to several disadvantages. These disadvantages—which are characteristic of second-generation alloys (Rioja and Liu, 2012)—include a tendency for strongly anisotropic mechanical properties, low short-transverse ductility and fracture toughness, and loss of toughness owing to thermal instability (Eswara Prasad et al., 2003; Lynch et al., 2003; Rioja and Liu, 2012). These adverse links between lithium content and properties are why the third-generation alloys have been developed with reduced lithium contents (and higher densities), as given in Table 2.1.

The third-generation alloys have been developed mainly in the United States. Developments started in the late 1980s and early 1990s and have continued through the first decade of this century. Table 2.1 shows that the densities of the third-generation alloys range from 2.63 to 2.72 g/cm^3. These densities are 2−8% less than those of conventional alloys, more favorable comparisons being obtained with the denser AA 7XXX alloys. However, even a 2% density advantage, which translates directly into a 2% weight savings (Figure 2.2), is worthwhile for aerospace structures, especially launch vehicles, satellites, and helicopters.

2.3.2 Strength and Ductility

Simple additions of lithium to aluminum alloys cannot provide satisfactory combinations of strength and ductility: a judicious combination of alloying elements is essential. This was well recognized before the second-generation alloys were developed. Figure 2.3 shows the strength—ductility relationships established for several generic types of Al—Li alloys at the beginning of the 1980s (Starke et al., 1981). It is evident that the best mechanical properties were (are) obtainable from Al—Li—Cu—Mg—Zr alloys, and Table 2.1 shows that the majority of the second-generation alloys were of this type.

Table 2.1 also shows that the third-generation alloys are basically of the Al—Li—Cu—Mg—Zr type, though with generally lower lithium contents for the reasons mentioned in Section 2.3.1. However, there are some significant additions to optimize the properties. Silver and zinc have direct influences on strength, and zinc improves the corrosion resistance; and manganese is added, besides zirconium, to control recrystallization and texture (Rioja and Liu, 2012).

A notable example of the importance of strength and ductility is the high-strength Al—Li—Cu—Mg—Ag—Zr alloy AA 2195. This alloy is weldable and

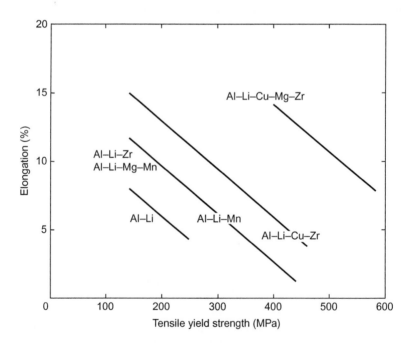

FIGURE 2.3 Schematic of strength—ductility relationships for various types of Al—Li alloys (Starke et al., 1981). The trends were obtained from longitudinal tensile tests on material samples.

has excellent combinations of strength and fracture toughness at ambient and cryogenic temperatures (Pickens and Tack, 1995; Pickens et al., 1994). It was developed as a plate alloy to replace the conventional alloy AA 2219 used for the Space Shuttle external tank. Combined with a structural redesign, the higher specific strength of 2195 enabled a weight reduction of 12.5% (Williams and Starke, 2003). For launch vehicles this is a major achievement.

Finally, it is worth remarking on an engineering rule of thumb concerning the tensile ductility. In general, an aerospace structural alloy should have a tensile elongation of at least 5%. Experience has shown that this level of ductility, and higher, can compensate for both anticipated and unanticipated stress concentrations by local yielding without cracking. The importance of this is illustrated by the occurrence of cracking during the insertion of high-interference fit fasteners into fastener holes in second-generation Al−Li products (Rioja and Liu, 2012). Cracking occurred because the alloys could not sustain the plastic deformation (hole expansion) resulting from the interference fit. This problem has been solved for third-generation alloys by restricting the level of interference fit and improving the ability to undergo plastic deformation without cracking (Rioja and Liu, 2012).

2.3.3 Damage Tolerance

In the present context it is convenient to consider the DT properties to include fatigue (low-cycle fatigue, LCF, and high-cycle fatigue, HCF) with fatigue crack growth and fracture toughness. Table 2.2 gives an overall view of the property and analysis requirements for damage tolerance design of aerospace structures. It is evident that damage tolerance design is a major undertaking requiring an extensive database.

The individual topics of fatigue, fatigue crack growth, and fracture toughness are treated in detail in Chapters 11−13, and so only summaries are given in this Section.

Fatigue

Table 2.2 shows that there are two types of fatigue to consider: LCF and HCF. These two fatigue regimes are discussed in Parts A and B of Chapter 11. LCF is characterized by significant cyclic plastic strains (typically 0.05−2%), while HCF occurs under nominally elastic conditions. The transition from LCF to HCF occurs at fatigue lives greater than about 10^4 cycles, depending on the material.

LCF data for some experimental, first- and second-generation Al−Li alloys suggest that their fatigue properties are inferior to those of conventional aerospace aluminum alloys (see Section 11.5.6). This could be a problem for using Al−Li alloys in LCF-limited applications. However, most

TABLE 2.2 Aluminum Alloy Fatigue and Fracture Data Requirements for Aerospace Structures

Properties	Types of Data	Special Considerations
Fatigue life and strength	• LCF[a] of smooth and notched specimens • HCF[a] of notched coupons, joints, and components • CA[a] and VA[a] load histories • Stress levels for adequate crack-free lives • Crack-free lives for specific components	• Notch stress concentration factors, K_t[a] • Environmental effects • Corrosion protection (coatings and primers) • Scatter factors on life • Modeling and prediction of lives
Fatigue crack growth	• Long and short cracks • CA and VA load histories • Component and full-scale tests	• Texture and anisotropy effects • Environmental effects • Modeling and prediction of crack growth
Fracture	• Plane strain fracture toughness, K_{Ic} • Nominally plane stress fracture toughnesses, K_{app} • K_R and R-curves • Residual strengths of components and structures	• Texture and anisotropy effects • Temperature: 225–295 K; also cryogenic (space) • Thermal stability during service • Possible dynamic effects on toughness

[a]$K_t = 1$ for smooth specimens: higher values, ranging from 1.5 to 4, are relevant to aerospace components and structures.
LCF, low-cycle fatigue; HCF, high-cycle fatigue; CA, constant amplitude; VA, variable amplitude.
(Source: After Wanhill (1994a,b).)

aerospace components and structures experience predominantly HCF load histories, and so the HCF long-life properties are much more important.

It has long been known that the HCF long-life *fatigue strengths* of aerospace aluminum alloys are difficult to improve (Kaufman, 1976). This is especially true for notched fatigue, which is more relevant than the data obtained from smooth specimen tests (notch stress concentration factor, $K_t = 1$). Notched fatigue data are appropriate because fatigue cracks in aircraft components and structures generally start at stress concentrations, especially rivet and bolt holes, but also at cutouts, corners, and other necessary geometry changes. HCF data for K_t values ranging from about 1.5 to 4 are used for most design purposes (see the footnote of Table 2.2); and fastener holes, of which there are very many in aircraft structures, have K_t values typically of the order of 3.

In light of the foregoing remarks, it should be no surprise that the notched HCF fatigue strengths of Al–Li alloys are generally equivalent to, but not

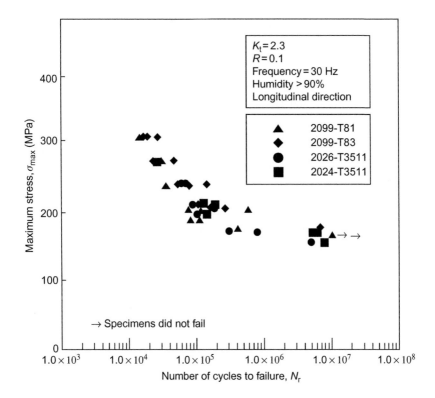

FIGURE 2.4   HCF notched specimen data comparisons for the third-generation Al–Li extrusion alloys AA 2099-T81 and AA 2099-T83 and the conventional alloys AA 2024-T3511 and AA 2026-T351 (Alcoa Aerospace Technical Fact Sheet, 2008). Note: this diagram is also given as Figure 11.20.

significantly better than those of conventional alloys. This is true of both the second- and third-generation Al–Li alloys. Figure 2.4 shows an example comparison between the third-generation Al–Li extrusion alloys AA 2099-T81 and AA 2099-T83 and the conventional AA 2024-T3511 and AA 2026-T351 alloys. At long lives $\sim 10^7$ cycles the fatigue strengths of all four alloys tend to similar values of σ_{max}.

Fatigue Crack Growth

Many factors influence the fatigue crack growth properties of aluminum alloys (see Section 12.7 and Figure 12.14). In practice this has generally meant that alloys have been optimized for other properties, for example strength, ductility, fracture toughness, and the fatigue crack growth properties are subsequently determined.

Most Al–Li alloy fatigue crack growth data have been obtained for second-generation alloys, specifically under constant amplitude (CA) and constant stress ratio (CR) loading, and for long/large cracks. These data show advantages over conventional alloys owing to the development of rough fracture surfaces that cause high levels of crack closure in the wakes of the fatigue cracks and greatly reduce the crack tip driving force. These advantages essentially disappear under variable amplitude (VA) load histories and are also absent for short/small cracks. Furthermore, these fatigue crack growth advantages are due to inhomogeneous plastic deformation owing to slip planarity, the development of intense slip bands, and strong crystallographic textures. These features are linked to the relatively high lithium contents of second-generation alloys and have undesirable consequences for other important mechanical properties (Section 2.3.1).

Fatigue crack growth data for third-generation Al–Li alloys are becoming more available. As discussed in Chapter 12, the long/large crack CA fatigue crack growth properties of third-generation Al–Li alloys are at least equivalent and mostly superior to those of the best conventional alloys they are intended to replace. This trend appears to be maintained under VA loading: an example is provided in Section 2.5.3. It should be noted that many of the issues associated with second-generation alloys have been eliminated or greatly alleviated for third-generation alloys. This means that (i) the fatigue crack growth behavior of third-generation alloys is more similar to that of well-established conventional alloys, and (ii) the long/large crack growth advantages so far demonstrated by third-generation alloys are likely to be maintained.

It is presently unknown whether third-generation Al–Li alloys will show improvements in short/small fatigue crack growth properties compared to conventional alloys. This topic is important for analyses and predictions of fatigue crack growth behavior (Molent et al., 2011).

Fracture Toughness

Fracture toughness and crack resistance (K_R and R-curve) data are useful in two general ways. Firstly, the data may be used to screen and choose between candidate materials for classes of applications. Secondly, they may be used as design parameters to determine the residual strengths of components and structures, as indicated in Table 2.2.

As discussed in Chapter 13, developments and improvements in fracture toughness have a long history, beginning in the late 1960s with the conventional AA 2XXX and AA 7XXX alloys. These developments are continuing, but in the meantime there have been developments for the second-generation Al–Li alloys from the mid-1970s to early 1990s, and now the recent and ongoing fracture toughness developments for the third-generation Al–Li alloys.

Much information on fracture toughness is available for second-generation Al–Li alloys. This information falls into two categories, plane strain (thick section) fracture toughness, and plane stress/mixed plane strain−plane stress (thinner sections) fracture toughness and crack resistance. The results may be summarized as follows:

1. Plane strain fracture toughness, K_{Ic}: The optimized longitudinal (L−T crack plane) plane strain fracture toughnesses of second-generation Al−Li alloys compare well with the toughnesses of their contemporary conventional alloys (Wanhill, 1994a,b). However, a contributing factor to the relatively good performance of some of the Al−Li alloys was through-thickness delamination during testing (Venkateswara Rao and Ritchie, 1989). This susceptibility to delamination corresponded to low-energy intergranular fracture and the disadvantage of low short-transverse fracture toughness already mentioned in Section 2.3.1.

    Explanation of the low-energy intergranular fracture has been a long-running issue. It is most probably associated with lithium segregation to grain boundaries (rather than a planar slip mode), as discussed in a recent review (Pasang et al., 2012) and in Chapter 13.

2. Nominally plane stress fracture toughness, K_{app}: This definition covers full and predominantly plane stress fracture, appropriate to sheet and thinner plate materials. Damage-tolerant (lower strength) second-generation Al−Li alloy products had fracture toughness and R-curve characteristics equivalent to those of conventional damage-tolerant alloys but at slightly lower strength levels. However, medium- and high-strength Al−Li alloys possessed combinations of fracture toughness, R-curve characteristics, and yield strength significantly inferior to those of conventional alloys (Wanhill, 1994a,b).

Improved fracture toughness has been a major incentive for the development of third-generation Al−Li alloys (Rioja and Liu, 2012). The results are demonstrably successful, and several examples are given in Section 2.5.

2.4 FAMILIES OF Al−Li ALLOYS

2.4.1 Second-Generation Alloys

The second-generation alloys were developed with the intention of obtaining families of Al−Li alloys 8−10% lighter (and stiffer) than equivalent conventional alloys and to replace them for a variety of applications. This is illustrated for the Western Al−Li alloys in Table 2.3 (Kandachar, 1986). This early classification includes the X8092 and X8192 alloys, which were never developed to commercial status, and AA 8091, which had even higher lithium content (2.6 wt% vs 2.4 wt%) and also more copper (1.9 wt% vs 1.2 wt%) than 8090 (Tite et al., 1988).

The two most promising second-generation alloys, 2090 and 8090, have been used in "niche" applications (Rioja and Graham, 1992; Smith, 1987;

TABLE 2.3 Families of Second-Generation Al–Li Alloys *Intended* to Replace Conventional Aluminum Alloys

Conventional Alloys to Be Replaced	Alcoa		Alcan		Cegedur-Pechiney	
	Internal Code	AA Code	Internal Code	AA Code	Internal Code	AA Code
7075-T6: high strength	Goal B	2090	Lital B	8091	CP 276	
2014-T6: medium/high strength		8090	Lital A	8090	CP 271	8090
2024-T3: damage tolerant	Goal A	2091 8090A	Lital C	8090-T81	CP 274	2091
7075-T73: corrosion resistant	Goal D	X8092				
Minimum density: general purpose	Goal C	X8192				

Starke and Staley, 1996) but have not found widespread application owing to the disadvantages mentioned in Section 2.3.1. These problems will be discussed in more detail in subsequent chapters, notably Chapters 5, 10, and 13. However, it is illustrative to show the on of material choices for a notable— and arguably unadvisable (Pasang et al., 2012)—"niche" use of Al–Li alloys in the AgustaWestland EH101 helicopter. The decision to use Al–Li alloys was strongly motivated by the necessity to reduce weight, which is always a premium for helicopters since they must be able to take off and land vertically.

Table 2.4 lists the early (1984) and final selections of second-generation Al–Li alloys for use in the EH101 helicopter, and Figure 2.5 is a schematic of the product forms and usage locations on the helicopter. The AA 2091 alloy was initially envisaged as a contender, but ultimately only a "family" of 8090 variants was selected. Virtually all the aluminum alloy structures in the airframe use 8090, thereby saving 8–10% weight. Unfortunately, early service experience has shown that the fracture toughness and fatigue properties are giving cause for concern (Merati, 2011). This fracture toughness problem, mentioned in Sections 2.3.1 and 2.3.3, is discussed in Chapter 13.

2.4.2 Third-Generation Alloys

A number of third-generation alloys have already found use. Early examples are AA 2195 plate (Section 2.3.2), which was used for the Space Shuttle

TABLE 2.4 Early (1984) and Final Selections of Second-Generation Al–Li Alloys for the EH101 Helicopter

Product	Alloy/Temper	Substitute for	Applications
Early Selections (Smith, 1984)			
Sheet	8090: unrecrystallized medium strength	2014-T6	—
	2091/8090: recrystallized medium strength	2014-T6	
	2091/8090: recrystallized damage tolerant	2024-T3	
Forgings	8090: unrecrystallized medium strength	7010-T7452 (forgings) 7010-T736 (forgings) 7010-T7451 (plate)	—
Extrusions	8090: unrecrystallized medium strength	7075-T7411	—
Final Selections (Merati, 2011)			
Sheet	8090-T3: as received	—	Floor installations, brackets, stiffeners, frames, spars, stringers, longerons, ribs, bulkheads, tail cone skin
	8090C-T8: medium strength	2014-T6	Flying control systems and avionics bay structures, nose cap, cabin roof frames
	8090C-T81: damage tolerant	2024-T3, 2024-T42	Flying control structures, skin panels (lower fuselage), cabin roof frames, flat roof panels
	8090-T84: medium strength	—	Z-stiffeners
	8090C-T621: damage tolerant	2024-T42	Sponsons, repairs
Forgings	8090-T852: medium/high strength	7010-T7451	Cabin roof and side frames (frames, stringers, joints, intercostals)
Extrusions	8090-T8511: medium strength	7075-T73511	Frames, brackets, stringers, bulkheads, door rails, seat tracks

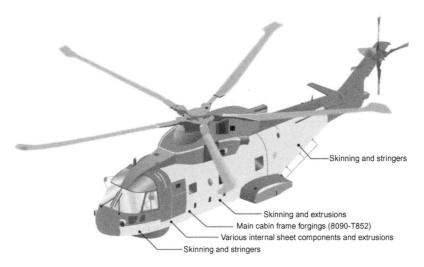

Skinning and stringers

Skinning and extrusions

Main cabin frame forgings (8090-T852)

Various internal sheet components and extrusions

Skinning and stringers

FIGURE 2.5 Use of second-generation Al–Li alloys (8090 variants) on the AgustaWestland EH101 (Pasang et al., 2012).

external tank (Williams and Starke, 2003), and AA 2297 and AA 2397, which have been used for Lockheed Martin F-16 bulkheads and other parts in military aircraft (Acosta et al., 2002; Balmuth, 2001). More recently, several alloys have been qualified for applications in commercial transport aircraft, notably the Airbus A380, and there are ongoing developments. This may be seen from Table 2.5 and Figure 2.6, which have been compiled from several sources (Daniélou et al., 2012; Denzer et al., 2012; Giummarra et al., 2007; Karabin et al., 2012; Lequeu, 2008; Lequeu et al., 2007, 2010; Magnusen et al., 2012; Pacchione and Telgkamp, 2006; Rioja and Liu, 2012; Tchitembo Goma et al., 2012) and Alcoa and Constellium in-house knowledge.

There is a clear trend to develop basically similar third-generation alloys for different product forms and applications, i.e. the development of a family of third-generation alloys. Table 2.5 shows that the potential exists for replacing virtually all the current conventional alloys by third-generation Al–Li alloys, and this is emphasized by Figure 2.6. A major contribution to this potential is the development of a range of alloy tempers that allow optimizations and trade-offs of properties, and hence considerable flexibility in matching the alloys to particular applications. Another important aspect is the generally good-to-excellent corrosion and stress corrosion resistance (Denzer et al., 2012; Karabin et al., 2012; Rioja and Liu, 2012). This is especially of interest for sheet and plate applications, since the low-strength cladding layers needed for imparting corrosion resistance to conventional AA 2XXX alloys and some AA 7XXX alloys can (sometimes) be omitted, resulting in an extra weight saving. Cladding also decreases the fatigue durability of mechanically fastened joints (e.g. riveted splices) (Wanhill, 1986).

TABLE 2.5 Proposed and Actual Uses of Third-Generation Al–Li Alloys to Replace Conventional Alloys

Product	Alloy/Temper	Substitute for	Applications
Sheet	2098-T851,2198-T8, 2199-T8E74, 2060-T8E30: damage tolerant/ medium strength	2024-T3, 2524-T3, 2524-T351	Fuselage/ pressure cabin skins
Plate	2199-T86, 2050-T84,2060-T8E86: damage tolerant	2024-T351, 2324-T39, 2624-T351, 2624-T39	Lower wing covers
	2098-T82P (sheet/plate): medium strength	2024-T62	F-16 fuselage panels
	2297-T87, 2397-T87: medium strength	2124-T851	F-16 fuselage bulkheads
	2099-T86: medium strength	7050-T7451, 7X75-T7XXX	Internal fuselage structures
	2050-T84,2055-T8X, 2195-T82: high strength	7150-T7751, 7055-T7751, 7055-T7951, 7255-T7951	Upper wing covers
	2050-T84: medium strength	7050-T7451	Spars, ribs, other internal structures
	2195-T82/T84: high strength	2219-T87	Launch vehicle cryogenic tanks
Forgings	2050-T852, 2060-T8E50: high strength	7175-T7351, 7050-T7452	Wing/fuselage attachments, window and crown frames
Extrusions	2099-T81, 2076-T8511: damage tolerant	2024-T3511, 2026-T3511, 2024-T4312, 6110-T6511	Lower wing stringers Fuselage/ pressure cabin stringers
	2099-T83, 2099-T81, 2196-T8511, 2055-T8E83, 2065-T8511: medium/high strength	7075-T73511, 7075-T79511, 7150-T6511, 7175-T79511, 7055-T77511, 7055-T79511	Fuselage/ pressure cabin stringers and frames, upper wing stringers, Airbus A380 floor beams and seat rails

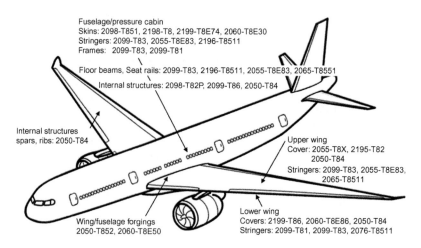

Fuselage/pressure cabin
Skins: 2098-T851, 2198-T8, 2199-T8E74, 2060-T8E30
Stringers: 2099-T83, 2055-T8E83, 2196-T8511
Frames: 2099-T83, 2099-T81

Floor beams, Seat rails: 2099-T83, 2196-T8511, 2055-T8E83, 2065-T8551
Internal structures: 2098-T82P, 2099-T86, 2050-T84

Internal structures
spars, ribs: 2050-T84

Upper wing
Cover: 2055-T8X, 2195-T82
       2050-T84
Stringers: 2099-T83, 2055-T8E83,
           2065-T8511

Wing/fuselage forgings
2050-T852, 2060-T8E50

Lower wing
Covers: 2199-T86, 2060-T8E86, 2050-T84
Stringers: 2099-T81, 2099-T83, 2076-T8511

FIGURE 2.6 Proposed use of third-generation Al—Li alloys for main structural areas in a transport aircraft.

2.5 EXAMPLES OF THIRD-GENERATION ALLOY PROPERTY DEVELOPMENTS AND TRADE-OFFS

Examples of third-generation Al—Li developments for specific structural areas have been published by Daniélou et al. (2012), Denzer et al. (2012), Karabin et al. (2012), Lequeu (2008), Lequeu et al. (2007), Magnusen et al. (2012), Rioja and Liu (2012), and Tchitembo Goma et al. (2012).

2.5.1 Fuselage/Pressure Cabins

Fuselage Skins

Transport aircraft fuselage skins are made from sheet materials with optimum combinations of strength and fracture toughness. Figure 2.7 shows the evolution of specific strength—toughness combinations for conventional and Al—Li sheet alloys used in several aircraft types (Magnusen et al., 2012). The latest Al—Li alloy, AA 2060 in a T8 temper, provides significant gains in both strength and toughness.

Table 2.5 lists the three Al—Li sheet alloys, AA 2198, AA 2199, and 2060, evaluated for fuselage skin applications (Giummarra et al., 2007; Magnusen et al., 2012; Pacchione and Telgkamp, 2006; Rioja and Liu, 2012), and Figure 2.8 is a "spider chart" comparing two of these alloys with the baseline conventional alloy Alclad AA 2524-T3. (Note that this alloy is Alclad, i.e. clad with thin layers of pure aluminum to impart corrosion resistance. The cladding contributes to the overall weight and density of the sheet material but not to the other properties, i.e. it is disadvantageous except for corrosion resistance.)

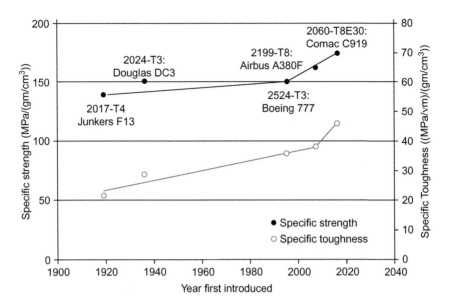

FIGURE 2.7 Improvements in specific tensile strength and fracture toughness in AA 2XXX sheet products used for fuselages (Magnusen et al., 2012).

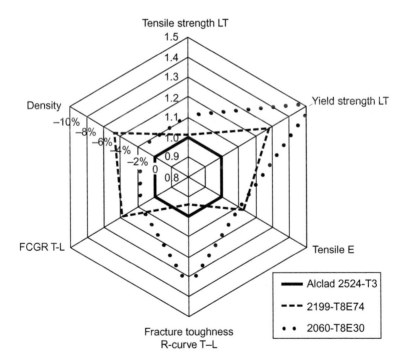

FIGURE 2.8 Property comparisons for the Al—Li sheet alloys 2199-T8E74 and 2060-T8E30 and the conventional alloy Alclad 2524-T3 (Magnusen et al., 2012). FCGR, fatigue crack growth resistance; LT, long transverse loading direction; T—L, transverse loading direction and longitudinal crack growth.

Figure 2.8 expresses the Al−Li alloy density reductions, tensile yield and ultimate strengths, elastic modulus, fracture toughness (R-curve), and fatigue crack growth resistance (FCGR) as ratios of the equivalent properties of Alclad 2524-T3. There are several points to note:

1. The key attributes of 2199-T8E74 are a 5% density reduction and significantly higher fatigue crack growth resistance.
2. 2060-T8E30 offers a 2% density reduction with much higher strength and fracture toughness.
3. Both Al-Li alloys give overall property improvements compared to Alclad 2524-T3.
4. A spider chart presentation enables visualizing trade-offs between density, strength, and fracture toughness when the two Al−Li alloys are considered for similar applications.

Fuselage Stringers and Frames

Figure 2.9 presents spider charts comparing candidate Al−Li extrusion alloys for stringers and frames with conventional extrusion alloys.

1. Stringers: Figure 2.9A shows the most relevant properties of AA 2099 and AA 2055 extrusions compared to the baseline AA 7055-T6511. The 2099-T83 alloy has significantly lower strength than 7055-T6511, while 2055-T8E83 has similar strength. Both Al-Li alloys offer significant density reductions (7% and 5%, respectively) and higher stiffness (compressive E). Not shown on the chart is the important fact that the Al−Li alloys have better corrosion resistance than 7055-T6511. Note also that 2055-T8E83 is a candidate for upper wing stringers (Section 2.5.2).
2. Frames: Figure 2.9B is similar to Figure 2.9A, this time comparing the properties of 2099 extrusions in two temper options with those of the

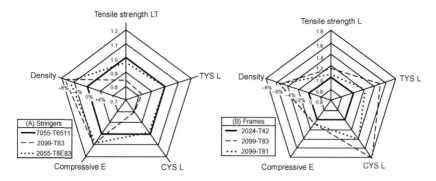

FIGURE 2.9 Property comparisons for the Al−Li extrusion alloys 2099-T83/T81 and 2055-T8E83 and the conventional alloys 7055-T6511 and 2024-T42 (Magnusen et al., 2012). TYS, tensile yield strength; CYS, compressive yield strength; L, longitudinal loading direction.

baseline AA 2024-T42 alloy. 2099-T83 is the high-strength temper, with much higher yield strengths than 2024-T42. The 2099-T81 temper trades strength for improved damage tolerance (Magnusen et al., 2012), while still having significant strength advantages over 2024-T42. There is also a density reduction of 5.4%.

2.5.2 Upper Wings

Upper Wing Covers

Transport aircraft upper wing covers are typically sized by compressive strength and stiffness to resist buckling. They must also have adequate fracture toughness (Denzer et al., 2012). Figure 2.10 presents specific strength–stiffness combinations for conventional AA 7XXX plate alloys used in several aircraft types and also the latest high-strength Al–Li plate alloy, 2055-T8X (Denzer et al., 2012).

The specific yield strengths of high-strength 7XXX alloys have steadily increased over the last half-century, but Figure 2.10 shows that the specific stiffness is essentially unchanged. This leads to the most important point about the candidature of 2055: this alloy offers significant increases in both strength and stiffness.

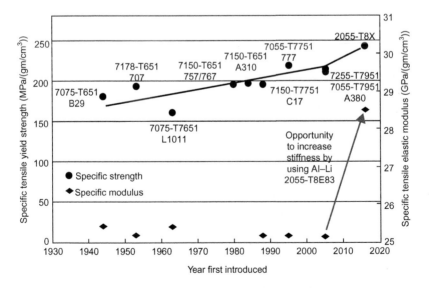

FIGURE 2.10 Specific yield strengths and elastic moduli for high-strength AA 7XXX plate products and the new Al–Li alloy AA 2055-T8X (Denzer et al., 2012). Note: in this diagram the tensile yield stress is used as a convenient substitute for the compressive yield strength.

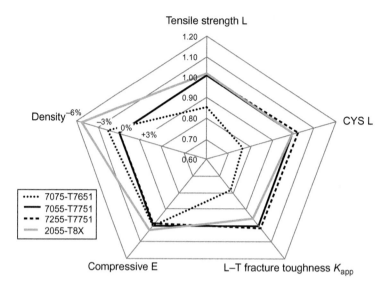

FIGURE 2.11 Property comparisons for the Al—Li high-strength plate alloy 2055-T8X and the conventional alloys 7075-T7651, 7055-T7751, and 7255-T7751 (Denzer et al., 2012). CYS, compressive yield strength; L, longitudinal loading direction; L—T, longitudinal loading direction and transverse crack growth; K_{app}, plane stress fracture toughness.

(Here it is appropriate to mention that increases in specific stiffness, arising from higher elastic moduli, must be carefully considered when proposing to substitute Al—Li alloys for conventional alloys. Simple substitution may not be an option, since higher elastic moduli can alter load paths and distributions within an airframe (Merati, 2011). In particular, higher stiffness attracts more load and hence higher stresses, which can cause problems in areas subjected to fatigue loading.)

Figure 2.11 is a spider chart comparing the Al—Li 2055-T8X plate alloy with conventional 7XXX high-strength plate alloys. The most important points to be made from these comparisons are:

1. The more modern conventional alloys 7055-T7751 and 7255-T7751 have similar static and fracture properties (7255-T7751 offers improvements in fatigue performance). Both alloys provide large (25—30%) increases in compressive yield strength and fracture toughness compared to 7075-T7651, with a density increase of about 2%.

2. The tensile strength and compressive yield strength of 2055-T8X are similar to those of 7055-T7751 and 7255-T7751, with a 5% decrease in fracture toughness. However, the compressive elastic modulus of 2055 is 4% higher, and a 5.6% density reduction results in the significant gains in specific strength and stiffness shown in Figure 2.10.

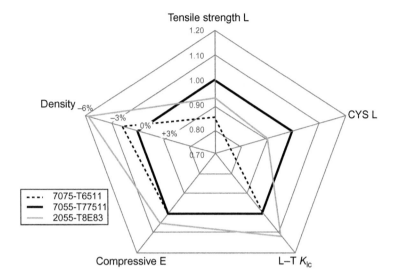

FIGURE 2.12   Property comparisons for the Al–Li high-strength extrusion alloy 2055-T8E83 and the conventional alloys 7075-T6511 and 7055-T77511 (Denzer et al., 2012). CYS, compressive yield strength; L, longitudinal loading direction; L–T, longitudinal loading direction and transverse crack growth; K_{Ic}, plane strain fracture toughness.

Upper Wing Stringers

Figure 2.12 is a spider chart comparing the Al–Li 2055-T8E83 extrusion alloy with two conventional high-strength 7XXX extrusion alloys. 2055-T8E83 is generally superior to the older 7075-T6511 alloy. In comparison to 7055-T7751, the higher compressive elastic modulus, higher fracture toughness, and lower density of the 2055 alloy are combined with lower tensile strength and compressive yield strength. However, since the most important property for upper wing stringers is their stiffness, the 2055-T8E83 alloy—with its lower density as an additional benefit—is a very good candidate for this application. As discussed in Section 2.5.1 and illustrated in Figure 2.6, 2055-T8E83 is also a candidate for fuselage stringers.

2.5.3 Lower Wings

Lower Wing Covers

Transport aircraft lower wing covers are plate materials with optimum combinations of strength and damage tolerance. The traditional materials are AA 2X24 alloys in naturally aged (T3XX) tempers. However, there are currently three candidate Al–Li damage-tolerant plate alloys, AA 2199-T86 and AA 2060-T8E86 from Alcoa and AA 2050-T84 from Alcan (Table 2.5 and

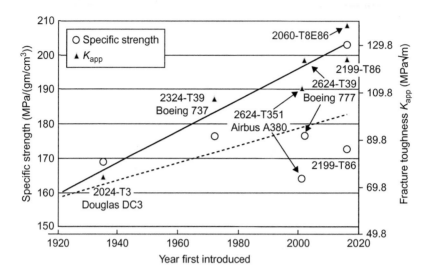

FIGURE 2.13 Improvements in specific tensile strength and plane stress fracture toughness in AA 2XXX plate products for lower wing covers (Karabin et al., 2012).

Figure 2.6). The 2199 and 2060 plate alloys have been developed specifically for lower wing covers (Karabin et al., 2012), while 2050 was initially developed to replace AA 7050-T7451 medium-to-thick plate (Lequeu, 2008; Lequeu et al., 2010).

Figure 2.13 shows the evolution of specific strength and toughness combinations for conventional 2X24 alloys and also the new 2199 and 2060 plate alloys (Karabin et al., 2012). Compared to 2024-T3, major increases in toughness are obtainable from the more modern alloys. It is also possible to achieve large increases in strength, as the 2060-T8E86 data demonstrate.

The Al–Li plate alloys have different combinations of strength and toughness and are instructive for discussing property trade-offs. The properties and property trade-offs will be discussed with the aid of Figure 2.13, the spider charts in Figure 2.14, and the fatigue crack growth data in Figure 2.15.

1. Figure 2.14 shows that 2199-T86 and 2060-T8E86 have superior strength, stiffness, toughness, and spectrum fatigue crack growth resistance compared with the baseline conventional 2X24-T3XX alloys. Furthermore, the Al–Li alloys offer density reductions of 5% and 2%, respectively.
2. Figure 2.15 shows that under spectrum loading (gust spectrum flight simulation), the lower-strength 2199-T86 alloy gives a longer fatigue crack growth life than 2060-T8E86.

From the strength and toughness results in Figures 2.13 and 2.14, it might appear that 2060 should be consistently preferred to 2199. However, if the

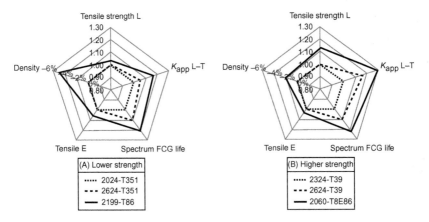

FIGURE 2.14 Property comparisons for the Al–Li plate alloys 2199-T86 and 2060-T8E86 and the baseline conventional 2X24-T3XX alloys (Karabin et al., 2012). L, longitudinal loading direction; L–T, longitudinal loading direction and transverse crack growth; K_{app}, plane stress fracture toughness; FCG, fatigue crack growth.

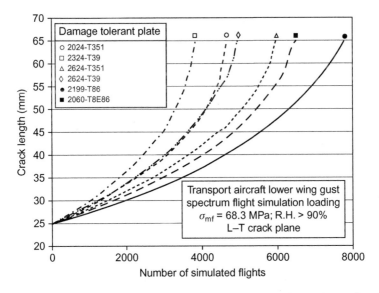

FIGURE 2.15 Gust spectrum FCG curves in the in-service inspectable crack growth regime for the Al–Li plate alloys 2199-T86 and 2060-T8E86 and the baseline conventional 2X24-T3XX alloys, specimen thickness 12 mm (Karabin et al., 2012). Notes: σ_{mf}, mean stress in the simulated flight load history; this diagram is also presented as Figure 12.11, where more details are given.

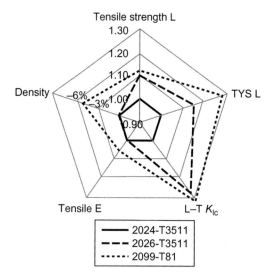

FIGURE 2.16 Property comparisons for the Al–Li damage-tolerant extrusion alloy 2099-T81 and the conventional alloys 2024-T3511 and 2026-T3511. TYS, tensile yield strength; L, longitudinal loading direction; L–T, longitudinal loading direction and transverse crack growth; K_{Ic}, plane strain fracture toughness.

strength and toughness of 2199 are deemed sufficient for particular applications, then this alloy's better fatigue crack growth resistance (Figure 2.15), lower density (2.64 g/cm^3 compared to 2.72 g/cm^3 for 2060; Table 2.1), and higher elastic modulus may make it more attractive.

Lower Wing Stringers

Figure 2.16 is a spider chart comparing the damage-tolerant Al–Li 2099-T81 extrusion alloy with two conventional damage-tolerant 2XXX extrusion alloys. The 2099-T81 and 2026-T3511 alloys have much better tensile yield and ultimate strengths and fracture toughness than the older 2024-T3511 alloy; and 2099-T81 has the best combination of properties, including a density reduction of 5.4%.

2.5.4 Spars, Ribs, and Other Internal Structures

Table 2.5 and Figure 2.6 show that the Al–Li plate alloy 2050-T84 is a candidate for wing spars and ribs and other internal structures in wings and fuselages. As medium-to-heavy gauge plate this alloy is a low-density alternative to conventional 7XXX alloys, in particular 7050-T7451 (Lequeu, 2008; Lequeu et al., 2010). The direct benefits of lithium are a 4.6% density reduction and an 11% increase in elastic modulus (Lequeu et al., 2010).

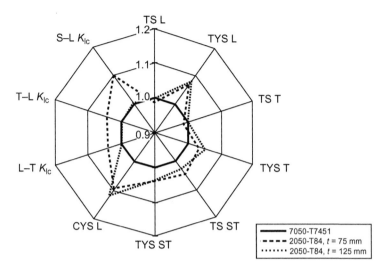

FIGURE 2.17   Property comparisons for the Al−Li medium-strength plate alloy 2050-T84 and the conventional alloy 7050-T7451 (Lequeu 2008). TS, tensile strength; TYS, CYS, tensile and compressive yield strengths; L, T, ST, longitudinal, transverse, and short-transverse loading directions; L−T, longitudinal loading direction and transverse crack growth; T−L, transverse loading direction and longitudinal crack growth; S−L, short-transverse loading direction and longitudinal crack growth; K_{Ic}, plane strain fracture toughness; t, thickness.

Figure 2.17 is a spider chart comparing the strength and fracture toughness properties of 2050-T84 and 7050-T7451 plate. The chart shows that:

1. The longitudinal and transverse tensile strengths of all three products are similar, but 2050-T84 has better short-transverse strength.
2. The 2050-T84 yield strengths are generally higher than those of 7050-T7451.
3. The fracture toughnesses of the 125-mm thick 2050-T84 plate and 7050-T7451 plate are similar, while the 75-mm thick 2050-T84 plate has higher fracture toughnesses.

An additional and important benefit of 2050-T84 for medium-to-heavy gauge plate applications is a significantly better stress corrosion resistance than 7050-T7451 (Lequeu et al., 2010).

2.6 SERVICE QUALIFICATION PROGRAMS

The previous sections of this chapter have concentrated on straightforward descriptions of the engineering property requirements for major structural areas in transport aircraft, which are seen as the most important contenders for the widespread use of third-generation Al−Li alloys. However,

TABLE 2.6 Example of a Service Qualification Program for a Damage-Tolerant Sheet/Plate Alloy

Properties	Property Details	Special Considerations
Mechanical	• TYS, CYS, TS, E • shear and bearing strengths	• multiangle (texture and anisotropy effects) • temperature: 225–424 K; also cryogenic (space)
Fatigue strength	• HCF of notched coupons and structural joints • CA and VA load histories	• texture and anisotropy effects • environmental effects • corrosion protection (coatings and primers)
Fatigue crack growth	• long and short cracks • CA and VA load histories	• texture and anisotropy effects • environmental effects • modeling and prediction of crack growth
Manufacturing and process technologies	• plane strain/plane stress fracture toughness • K_R and R-curves • residual strength of stiffened panels	• texture and anisotropy effects • temperature: 225–295 K; also cryogenic (space) • thermal stability during service • possible dynamic effects on toughness
Corrosion	• pitting and exfoliation • stress corrosion • accelerated and natural environments	• intergranular attack • microbiological attack in fuel tanks • structural joints
Manufacturing and process technologies	• forming: stretch and spin forming • machining, cutting, and hole drilling • mechanical fastening: rivets and bolts • welding: FSW and LBW • surface treatments: chemical milling, etching, anodizing, priming, and adhesive bonding	• texture and anisotropy • superplasticity • combinations of forming and heat treatment • automated 3D welding; post-weld heat treatments and stress relief

HCF, high-cycle fatigue; CA, constant amplitude; VA, variable amplitude; FSW, friction stir welding; LBW, laser beam welding.

qualification of an alloy, or alloys, for service usage involves numerous considerations relating to these properties as well as other requirements.

As an example, Table 2.6 outlines a service qualification program for a damage-tolerant sheet alloy (Wanhill, 1994b) with some updates (Lequeu et al., 2010; Pacchione and Telgkamp, 2006; Rioja and Liu, 2012). This table is a more comprehensive version of Table 2.2. Many of the topics listed in these tables will be discussed in subsequent chapters.

2.7 SUMMARY AND CONCLUSIONS

This chapter has discussed the structural and engineering property requirements for widespread deployment of aluminum−lithium (Al−Li) alloys in aircraft. In particular, attention has been paid to the requirements for commercial transport aircraft, which are the most important and likely contenders for the use of Al−Li alloys.

The development of Al−Li alloys has been driven mainly by the fact that additions of lithium to aluminum alloys lower the density and increase the elastic modulus, thereby offering the potential of significant weight savings with respect to conventional (non-lithium containing) alloys.

The first use of Al−Li alloys in aircraft goes back to the late 1950s (alloy AA 2020) and mid-1960s (alloys 1420 and 1421). These materials are referred to as the first-generation Al−Li alloys. Since then there have been two major development programs resulting in the second- and third-generation alloys.

Development of the second-generation Al−Li alloys began in the 1970s and continued through the 1980s. Attempts were made to develop families of Al−Li alloys for widespread replacement of conventional alloys. Ultimately this program was unsuccessful, except for so-called niche applications, one of which—the AgustaWestland EH101 helicopter—is discussed in this chapter. The cause of failure to find widespread application was associated largely with the too-high lithium contents of the alloys. This resulted in serious disadvantages, including mechanical property anisotropy, low short-transverse ductility and fracture toughness, and thermal instability.

Development of the third-generation Al−Li alloys began in the late 1980s and is ongoing. These alloys have significantly reduced lithium contents (0.75−1.8 wt%) compared to the second generation (typically more than 2 wt%), and there are other important compositional changes. Silver and zinc have been added for strength, and zinc improves the corrosion resistance. Manganese is added besides zirconium, which was already present in the second-generation alloys, to control recrystallization and texture. These differences and improved knowledge about thermomechanical processing and heat treatment have resulted in a family of alloys with significant property advantages covering all major structural areas and applications for transport aircraft.

More information on actual and potential aerospace applications of Al–Li alloys is presented in Chapter 15.

REFERENCES

Acosta, E., Garcia, O., Dakessian, A., Aung Ra, K., Torroledo, J., Tsang, A., et al., 2002. On the effect of thermomechanical processing on the mechanical properties of 2297 plates. Mater. Sci. Forum 396–402, 1157–1162.

Alcoa Aerospace Technical Fact Sheet, 2008. Alloy 2099-T83 and 2099-T8E67 extrusions. In: AEAP-Alcoa Engineered Aerospace Products. Lafayette, IN.

Balmuth, E.S., 2001. Application of aluminum alloy 2297 in fighter aircraft structures. In: Kim, N., Lee, C.S., Eylon, D. (Eds.), Proceedings of the International Conference on Light Materials for Transportation Systems (LiMAT—2001), vol. 2. Pohang Institute of Science and Technology, Pohang, Korea, pp. 589–596.

Daniélou, A., Ronxin, J.P., Nardin, C., Ehrström, J.C., 2012. Fatigue resistance of Al–Cu–Li and comparison with 7XXX aerospace alloys. In: Weiland, H., Rollett, A.D., Cassada, W.A. (Eds.), Proceedings of the 13th International Conference on Aluminum Alloys (ICAA13). The Minerals, Metals and Materials Society (TMS) and John Wiley & Sons, Hoboken, NJ, pp. 511–516.

Denzer, D.K., Rioja, R.J., Bray, G.H., Venema, G.B., Colvin, E.L., 2012. The evolution of plate and extruded products with high strength and toughness. In: Weiland, H., Rollett, A.D., Cassada, W.A. (Eds.), Proceedings of the 13th International Conference on Aluminum Alloys (ICAA13). The Minerals, Metals and Materials Society (TMS) and John Wiley & Sons, Hoboken, NJ, pp. 587–592.

Ekvall, J.C., Rhodes, J.E., Wald, G.G., 1982. Methodology for evaluating weight savings from basic material properties. Design of Fatigue and Fracture Resistant Structures. American Society for Testing and Materials, Philadelphia, PA, ASTM STP 761, pp. 328–341.

Eswara Prasad, N., Gokhale, A.A., Rama Rao, P., 2003. Mechanical behaviour of aluminium–lithium alloys. Sādhanā 28 (1 & 2), 209–246.

Giummarra, C., Thomas, B., Rioja, R.J., 2007. New aluminum lithium alloys for aerospace applications. In: Sadayappan, K., Sahoo, M. (Eds.), Proceedings of the Third International Conference on Light Metals Technology. CANMET, Ottawa, ON, pp. 41–46.

Kandachar, P.V., 1986. Applications for advanced aluminium alloys in aircraft. In: Wanhill, R.J. H., Bunk, W.J.G., Wurm, J.G. (Eds.), Advanced Materials Research and Developments for Transport: Light Metals. Les Éditions de Physique Editions is with an accent only, not a circonflex underneath it as well, Les Ulis, France, pp. 23–32.

Karabin, L.M., Bray, G.H., Rioja, R.J., Venema, G.B., 2012. Al–Li–Cu–Mg–(Ag) products for lower wing skin applications. In: Weiland, H., Rollett, A.D., Cassada, W.A. (Eds.), Proceedings of the 13th International Conference on Aluminum Alloys (ICAA13). The Minerals, Metals and Materials Society (TMS) and John Wiley & Sons, Hoboken, NJ, pp. 529–534.

Kaufman, J.G., 1976. Design of aluminum alloys for high toughness and high fatigue strength. Specialists Meeting on Alloy Design for Fatigue and Fracture Toughness, AGARD Conference Proceedings No. 185. Advisory Group for Aerospace Research and Development, Neuilly-sur-Seine, France, pp. 2-1 – 2-26.

Lequeu, P., 2008. Advances in aerospace aluminum. Adv. Mater. Processes 166 (2), 47–49.

Lequeu, P., Lassince, P., Warner, T., 2007. Aluminum alloy development for the Airbus A380—Part 2. Adv. Mater. Processes 165 (7), 41–44.

Lequeu, P., Smith, K.P., Daniélou, A., 2010. Aluminum—copper—lithium alloy 2050 developed for medium to thick plate. J. Mater. Eng. Perform. 19 (6), 841–847.

Lynch, S.P., Shekhter, A., Moutsos, S., Winkelman, G.B., 2003. Challenges in developing high performance Al—Li alloys. LiMAT 2003, Third International Conference on Light Materials for Transportation Systems. Center for Advanced Aerospace Materials, Pohang University of Science and Technology, Pohang, Korea.

Magnusen, P.E., Mooy, D.C., Yocum, L.A., Rioja, R.J., 2012. Development of high toughness sheet and extruded products for airplane fuselage structures. In: Weiland, H., Rollett, A.D., Cassada, W.A. (Eds.), Proceedings of the 13th International Conference on Aluminum Alloys (ICAA13). The Minerals, Metals and Materials Society (TMS) and John Wiley & Sons, Hoboken, NJ, pp. 535–540.

Merati, A., 2011. Materials replacement for aging aircraft. "Corrosion Fatigue and Environmentally Assisted Cracking in Aging Military Vehicles," RTO AGARDograph AG-AVT-140. Research and Technology Organisation (NATO), Neuilly-sur-Seine, France, pp. 24-1–24-22.

Molent, L., Barter, S.A., Wanhill, R.J.H., 2011. The lead crack fatigue lifting framework. Int. J. Fatigue 33, 323–331.

Mouritz, A.P., 2012. Introduction to Aerospace Materials. Woodhead Publishing Ltd, Cambridge, UK.

Pacchione, M., Telgkamp, J., 2006. Challenges of the metallic fuselage. Paper ICAS 2006-4.5.1, 25th International Congress of the Aeronautical Sciences (ICAS 2006). German Society for Aeronautics and Astronautics (DGLR), Bonn, Germany.

Pasang, T., Symonds, N., Moutsos, S., Wanhill, R.J.H., Lynch, S.P., 2012. Low-energy inter-granular fracture in Al—Li alloys. Eng. Fail. Anal. 22, 166–178.

Peel, C.J., Evans, B., Baker, C.A., Bennet, D.A., Gregson, P.J., Flower, H.M., 1984. The development and application of improved aluminium—lithium alloys. In: Sanders, T.H., Starke Jr., E.A. (Eds.), Aluminium-Lithium Alloys II: Proceedings of the Second International Conference. The Metallurgical Society of AIME, Warrendale, PA, pp. 363–392.

Pickens, J.R., Tack, W.T., 1995. Al—Cu—Li Alloys with Improved Cryogenic Fracture Toughness, U.S. Patent No. 5,455,003, October 3, 1995.

Pickens, J.R., Tack, W.T., Gayle, F.W., Maisano, J.R., 1994. High strength Al—Cu—Li alloys for launch systems. In: Bhat, B.B., Bales, T.T., Vesley Jr., E.J. (Eds.), Aluminum—Lithium Alloys for Aerospace Applications Workshop. NASA Conference Publication 3287, Government Printing Office, Washington D.C., USA, pp. 57–72.

Rioja, R.J., Graham, R.H., 1992. Al—Li alloys find their niche. Adv. Mater. Processes 6, 23–26.

Rioja, R.J., Liu, J., 2012. The evolution of Al—Li base products for aerospace and space applications. Metall. Trans. A 43A, 3325–3337.

Smith, A.F., 1984. Uses and properties of Al—Li on the new EH101 helicopter. "New Light Alloys," AGARD Conference Proceedings No. 444. Advisory Group for Aerospace Research and Development, Neuilly-sur-Seine, France, pp. 19-1–19-19.

Smith, A.F., 1987. Aluminium—lithium alloys for helicopter structures. Met. Mater. 3 (8), 438–444.

Staley, J.T., Lege, D.J., 1993. Advances in aluminium alloy products for structural applications in transportation. J. Phys. IV 3, 179–190.

Starke Jr., E.A., Staley, J.T., 1996. Application of modern aluminum alloys to aircraft. Prog. Aerosp. Sci. 32, 131–172.

Starke Jr., E.A., Sanders Jr., T.H., Palmer, I.G., 1981. New approaches to alloy development in the Al−Li system. J. Met. 33, 24−33.

Tchitembo Goma, F.A., Larouche, D., Bois-Brochu, A., Blais, C., Boselli, J., Brochu, M., 2012. Fatigue crack growth behavior of 2099-T83 extrusions in two different environments. In: Weiland, H., Rollett, A.D., Cassada, W.A. (Eds.), Proceedings of the 13th International Conference on Aluminum Alloys (ICAA13). The Minerals, Metals and Materials Society (TMS) and John Wiley & Sons, Hoboken, NJ, pp. 517−522.

Tite, C.M.J., Gregson, P.J., Pitcher, P.D., 1988. Further precipitation reactions associated with β' (Al$_3$Zr) particles in Al−Li−Cu−Mg−Zr alloys. Royal Aircraft Establishment Technical Memorandum MAT/STR 1119, Farnborough, UK.

Venkateswara Rao, K.T., Ritchie, R.O., 1989. Mechanical properties of Al−Li alloys: Part 1. Fracture toughness and microstructure. Mater. Sci. Technol. 5, 882−895.

Wanhill, R.J.H., 1986. Effects of cladding and anodising on flight simulation fatigue of 2024-T3 and 7475-T761 aluminium alloys. In: Barnby, J.T. (Ed.), Fatigue Prevention and Design. Engineering Materials Advisory Services Ltd, Warley, UK, pp. 323−332.

Wanhill, R.J.H., 1994a. Fatigue and fracture properties of aerospace aluminium alloys. In: Carpinteri, A. (Ed.), Handbook of Fatigue Crack Propagation in Metallic Structures. Elsevier Science Publishers, Amsterdam, The Netherlands, pp. 247−279.

Wanhill, R.J.H., 1994b. Status and prospects for aluminium−lithium alloys in aircraft structures. Int. J. Fatigue 16, 3−20.

Williams, J.C., Starke Jr., E.A., 2003. Progress in structural materials for aerospace systems. Acta Mater. 51, 5775−5799.

Part II

Physical Metallurgy

Chapter 3

Phase Diagrams and Phase Reactions in Al–Li Alloys

N. Eswara Prasad* and T.R. Ramachandran†

*Regional Centre for Military Airworthiness (Materials), CEMILAC, Hyderabad, India
†Non-Ferrous Materials Technology Development Centere, Hyderabad, India

Contents

3.1 Introduction	62	**3.6 Minor Alloying Additions to**	
3.1.1 Alloy Development		**Al–Li Alloys**	85
History	62	3.6.1 Minor Alloying	
3.1.2 The Influence of		Additions for Grain	
Alloying Elements and		Refinement	85
Composition	63	3.6.2 Minor Alloying	
3.2 Nature of Phases	65	Additions for	
3.3 Al–Li Binary System	66	Strengthening	86
3.4 Ternary Systems	71	**3.7 Impurities and Grain**	
3.4.1 Al–Li–Mg System	71	**Boundary Precipitation**	87
3.4.2 Al–Li–Cu System	73	3.7.1 Impurities	88
3.4.3 Al–Cu–Mg System	76	3.7.2 Precipitates and	
3.4.4 Al–Li–Zr System	78	PFZs	88
3.5 Quaternary Al–Li–Cu–Mg		**3.8 Chapter Summary**	89
System	79	**Acknowledgments**	90
3.5.1 Phase Equilibria	79	**References**	91
3.5.2 Overview of Precipitate			
Phases in Commercial			
Al–Li Alloys	80		

Aluminum–Lithium Alloys.

3.1 INTRODUCTION

3.1.1 Alloy Development History

Interest in aluminum—lithium (Al—Li) alloys arises from three important considerations: as the lightest metal, lithium additions reduce density ($\sim$3% decrease for every weight percent), increase elastic modulus ($\sim$6% increase for every weight percent) (Pickens, 1985, 1990), and introduce precipitation hardening by the formation of the metastable δ' (Al$_3$Li) phase. The increase in *specific* strength (strength/density) and *specific* modulus (E/density) combined with attractive fatigue and cryogenic properties offers possibilities for the use of Al—Li alloys in aerospace and cryogenic applications (e.g., fuel tanks in launch vehicles, like the external tank of the US space shuttle). Developmental activities in the field started from the 1920s, but the first commercial alloy AA 2020 (Al—1.1Li—4.5Cu—0.5Mn—0.2Cd) (all compositions are given in weight percent in this chapter, unless otherwise stated) was introduced only in 1958. This alloy was successfully used for the wing skins and tails of the Northrop RA-5C Vigilante aircraft, but concerns about its fracture toughness led to its abandonment. Significant research work in the former Soviet Union led to the development of VAD 23 with nominal composition Al—1.1Li—5.3Cu—0.6Mn—0.17Cd and 1420 (Al—2.0Li—5.3Mg—0.5Mn) in the 1960s. All three alloys are customarily referred to as belonging to the first generation of Al—Li alloys.

The potential (and actual) threat of replacement of aluminum alloys by resin-based composites resulted in extensive research work on new generations of Al—Li alloys, beginning in the 1970s. This research has seen the (largely unsuccessful) development of a second generation of Al—Li alloys, and the emergence of a third generation. The overall development of Al—Li alloys can be grouped into three categories:

- First-generation alloys developed in the 1950s to 1970s, including 2020, VAD 23, and 1420. 2020 was characterized by high strength but low ductility and toughness, while the Soviet alloys lacked strength. In the 1970s it was found that additions of copper, magnesium, and small amounts of zirconium improved the strength and ductility of Al—Li alloys, and all subsequent alloys are basically of the Al—Li—Cu—Mg—Zr type.
- Second-generation alloys produced in the 1980s, including AA 2090, AA 2091, and AA 8090. These alloys had relatively high additions of lithium, about 2—2.4%, which led to several disadvantages, including a tendency for strongly anisotropic mechanical properties, low short-transverse ductility and fracture toughness, and loss of toughness owing to thermal instability. These adverse links between lithium content and properties are the reasons why the second-generation alloys have been largely unsuccessful and why the third-generation alloys have been developed with lower lithium contents.

- Third-generation alloys that started to be introduced in the early 1990s (Weldalite, 2094, 2095, etc.), but mainly reached commercial status in the present century. These alloys have lithium contents mostly between 1% and 1.8%. This reduction, combined with better understanding about the effects of alloying elements and improved thermomechanical processing and heat treatment, has resulted in a family of Al−Li alloys potentially suitable for widespread application in the aerospace industry (see Chapters 2 and 15).

Classification of the alloys can also be based on chemical composition, as detailed below:

- Al−Cu−Li, with small amounts of Mn, Cd (2020, VAD 23).
- Al−Li−Mg (Russian alloys 1420, 1421, and 1423).
- Al−Li−Cu−Mg−Zr (2091, 8090, Weldalite 049, CP276, 1441Rus, 1464Rus, 2094, 2095, 2195, 2196, 2197, 2198, 2099, 2199, 2055); some of these alloys contain small amounts ($\sim 0.5\%$) of Ag.

Table 3.1 lists typical compositions of some Al−Li alloys (Belov et al., 2005; Eswara Prasad et al., 2003; Giummarra et al., 2007; Kostrivas and Lippold, 1999; Röyset, 2007; The Aluminium Association, 2009; Wang and Starink, 2005). Several phases (metastable and equilibrium) that are present in these alloys are similar to those found in the binary and ternary systems, Al−Li, Al−Cu, Al−Mg, Al−Cu−Mg. The inevitable impurities in aluminum, Fe and Si, result in the formation of phases deleterious to the fracture toughness. Intentional additions of Mn, Zr, Ag, and Sc lead to the formation of both desirable and undesirable phases. This duality signifies that the characteristics of these phases have to be carefully considered.

3.1.2 The Influence of Alloying Elements and Composition

As mentioned above, alloys developed prior to and in the early 1980s (e.g., 2020, 1420, 8090, 2090, and 2091) tended to suffer from anisotropic mechanical properties, low short-transverse ductility and fracture toughness, and loss of toughness owing to thermal instability. All of these disadvantages are linked to the Li content (Eswara Prasad et al., 2003; Lynch et al., 2003; Rioja and Liu, 2012), but other elements also play important roles.

Low ductility and fracture toughness have been associated with tramp elements such as Na and K segregating at grain boundaries, formation of coarse intermetallic compounds containing Fe and Si during solidification from the melt, strain localization resulting from the easy shear of the metastable δ' phase, and the presence of δ' precipitate-free zones (PFZs) near grain boundaries and coarse precipitating phases (Blankenship and Starke, 1992). Alloying additions of Cu, Mg, Zr, Ag, etc., minimize these problems by the introduction of new phases and changes in the nucleation

TABLE 3.1 Compositions of Some Al–Li Alloys: Fe and Si Are Not Shown, but Are Normally Below 0.1% Each

Alloy	Alloying Elements (%)				
	Li	Cu	Mg	Zr	Others
2020	1.2	4.4			0.5 Mn, 0.2 Cd
VAD23Rus	1.15	5.15			0.6 Mn, 0.18 Cd
1420	2.0		5.0	0.10	0.5 Mn
1421	2.1		5.2		0.15 Sc
1423	1.9		3.5		
1430	1.5–1.9	1.5–1.8	2.5–3.0	0.11	
1440	2.3	1.5	0.9	0.15	
1441	1.8–2.1	1.5–1.8	0.7–1.1	0.04–0.16	0.06 Be
1450	1.9	2.0	0.9	0.09	
1460	2.0	3.0			0.1 Sc
1464	1.7	3.0	0.5		Sc&Zr
2090	1.9–2.6	2.4–3.0	0.25	0.08–0.15	
CP276	1.9–2.6	2.5–3.3	0.2–0.8	0.04–0.16	
2091(CP274)	1.70–2.30	1.8–2.5	1.1–1.9	0.04–0.16	
8090	2.2–2.7	1.0–1.6	0.6–1.3	0.04–0.16	
8091	2.4–2.8	1.8–2.2	0.5–1.2	0.08–0.16	
Weldalite 049	1.3	5.4	0.4		0.4 Ag
2094	0.7–1.4	4.4–5.2	0.25–0.8	0.04–0.18	0.25–0.6 Ag
2095	0.7–1.5	3.9–4.6	0.2–0.8	0.04–0.18	0.25–0.6 Ag
2195	0.8–1.2	3.7–4.3	0.25–0.8	0.08–0.16	Mn 0.25, Ag 0.25–0.6
2196	1.4–2.1	2.5–3.3	0.25–0.8	0.04–0.18	Mn 0.35, 0.25–0.6 Ag
2197	1.3–1.7	2.5–3.1	0.25	0.08–0.15	0.1–0.5 Mn
2198	0.8–1.1	2.9–3.5	0.25–0.8	0.04–0.18	0.5 Mn, 0.1–0.5 Ag
2099	1.6–2.0	2.4–3.0	0.1–0.5	0.05–0.18	0.4–1.0 Zn,0.1–0.5 Mn
2199	1.4–1.8	2.3–2.9	0.05–0.4	0.05–0.12	0.2–0.9 Zn, 0.1–0.5 Mn
2050	0.7–1.3	3.2–3.9	0.20–0.6	0.06–0.14	0.2–0.5 Mn, 0.2–0.7 Ag

characteristics of δ'. More information and discussion on this topic are given in Chapter 13.

The nature, structure, size and distribution of the phases, and the properties of the matrix-precipitate interfaces, are influenced by chemical composition and heat-treatment parameters. In turn, all these microstructural features have a profound effect on the mechanical properties. This emphasizes the need for extensive and detailed studies of phase equilibria and nucleation and growth characteristics.

In the present chapter we shall examine the nature and precipitation characteristics of various phases in Al–Li alloys of commercial interest. The binary Al–Li system will also be covered, since the δ' phase is encountered in many of these alloys. Various aspects have been extensively reviewed in the literature: the recent ones include phase studies in Al–Cu–Mg–Li alloys by Wang and Starink (2005) and some general aspects by Rioja and Liu (2012).

3.2 NATURE OF PHASES

It is well known that the phases present in aluminum alloys can be classified into three groups: constituent particles, dispersoids, and precipitates.

Intermetallic constituent particles form during solidification of the molten alloy at relatively high temperatures, and their sizes lie in the range of one to several tens of microns. Some of them, such as Al_2Cu, Mg_2Si, and Al_2CuMg, may dissolve during subsequent thermal treatments (homogenization), while others, such as Al_7Cu_2Fe (in 2090, 2091, and 2095), $Al_{12}Fe_3Si$ (in 2091), and Al_3Fe (in Al–Li binary alloys and 8090) (Soni et al., 1992; Tiryakioglu and Staley, 2003), are relatively unaffected by thermal treatments. The coarse insoluble constituent particles have minor effects on strength but adversely affect ductility and fracture toughness. Some of these phases sequester solute elements such as Cu and therefore decrease the volume fraction of the precipitate particles that contribute to improved strength. Consequently Fe and Si, the inevitable impurities in Al, should be kept low in order to offset this effect in aerospace alloys. The use of ALCOA P 0404 grade aluminum (0.04% each of Si and Fe) or equivalent metal is recommended for this purpose.

Dispersoid particles form predominantly during ingot homogenization, with sizes in the range of about $0.05-0.5$ μm. They are usually Mn-, Zr-, or Cr-containing particles such as $Al_{20}Cu_2Mn_3$, Al_3Zr, and $Al_{12}Mg_2Cr$. They have a strong influence on retarding recrystallization and grain growth and consequently on grain size control. The fine grain size contributes to improved strength. Alloying elements such as Li and Mg partition between the matrix and some of the dispersoid particles, for example, $Al_3(Zr,Sc)$, resulting in improved ductility.

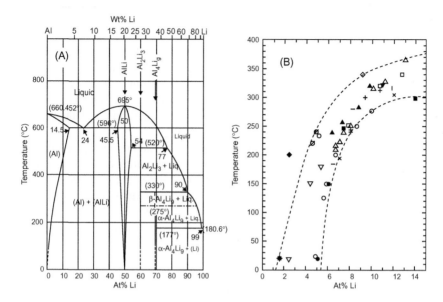

FIGURE 3.1 Al–Li binary system: (A) complete phase diagram and (B) α-δ' solvus line in the phase diagram with data from several investigators summarized by Noble and Bray (1998).

Precipitate particles and their nature depend on the alloying elements present, their concentration, and the thermal treatment conditions. It is therefore necessary to consider the phase diagrams of various systems and to examine the possibilities of formation of metastable phases. This aspect will be considered in detail in the following sections of this chapter.

3.3 Al–Li BINARY SYSTEM

Various aspects of the Al–Li phase diagram have been extensively reviewed in the literature, and several versions of the diagram are published. The recently reported ones are due to Hallstedt and Kim (2007) and Butts and Gale (2004). A typical phase diagram of the system is shown in Figure 3.1A (Butts and Gale, 2004). There is a eutectic reaction at ∼600°C at the Al-rich end, leading to the formation of the Al-rich α solid solution and δ phase (AlLi, cubic, space group Fd3m, with a lattice parameter, $a = 0.637$ nm). The solid solubility of Li varies from ∼4.0% at this temperature to <1% at 100°C. The equilibrium δ phase is preceded by the metastable δ' (Al₃Li, L1₂ structure with lattice parameter, $a = 0.401$ nm). The solvus line for the metastable phase lies in the temperature range 150–250°C for alloys of practical interest.

The similarities in the structure and lattice parameters of aluminum ($a_{Al} = 0.405$ nm) and the δ' phase result in a small lattice mismatch and hence reduced strain effects. Values for misfit strain obtained by TEM studies lie in the range -0.30% to -0.08%, while X-ray studies indicate even smaller ones (Flower and Gregson, 1987). Consequently, the metastable phase forms as coherent spherical particles which retain this shape up to large sizes (typically 0.3 μm) (Williams, 1981). The surface energy of the α/δ' interface is very small, $< 30\ \text{mJ/m}^2$ and the antiphase boundary energy of the δ' phase (180 mJ/m²) is low in comparison with that of metastable Al₃Zr (447 mJ/m²) (Sainfort and Guyot, 1986). These comparatively low energies are the reasons why the δ' particles are easily sheared by dislocations, resulting in strain localization during deformation.

Over the last 30 years there have been many attempts at measuring the solid solubility of lithium in aluminum, and most of these have been concerned with defining the α/δ' solvus boundary; the results are summarized in Figure 3.1(B) (Noble and Bray, 1998). There is some agreement in the data reported by different investigators for lithium concentrations of 7−8 at.%; but even in this range the temperature for a given alloy composition exhibits a scatter of $\sim 50°C$. The wide variation in the lower and higher Li concentrations is mainly due to slow precipitation kinetics in low-lithium alloys and competing formation of δ phase at high lithium concentrations. The temperature dependence of the concentration, C_e, at the α/δ' boundary is closely defined over the range 2−13 at.% Li by the equation (Noble and Bray, 1998),

$$\ln C_e(\text{at.\%}) = 4.176 - 9180/RT \qquad (3.1)$$

The decomposition of the as-quenched disordered solid solution to the ordered δ' phase has been extensively investigated, both theoretically and using thermal analysis, electrical resistivity measurements, X-ray and electron diffraction, and high-resolution transmission electron microscopy (HREM). The free energy change involved in the decomposition process is a function of both Li concentration and order parameter (degree of ordering) of the quenched solid solution.

A phenomenological model developed by Sigli and Sanchez (1986), based on the cluster variation method for description of the S−L and S−S phase equilibria in the phase diagram, indicates a miscibility gap in the $\alpha + \delta'$ field. However, this aspect has not been experimentally confirmed. The occurrence of a miscibility gap for the disordered phase has also been postulated (Gayle et al., 1984a), with a monotectoid configuration connecting the gap with the ordered phase region (Figure 3.2A).

To understand the interaction between ordering and clustering processes, Soffa and Laughlin (1989) used free energy-composition diagrams to theoretically deduce the loci of thermodynamic instability defining regions of continuous transformation. Khachaturyan et al. (1988) adapted a mean field model to calculate the subphase fields in the $\alpha + \delta'$ region. Their results

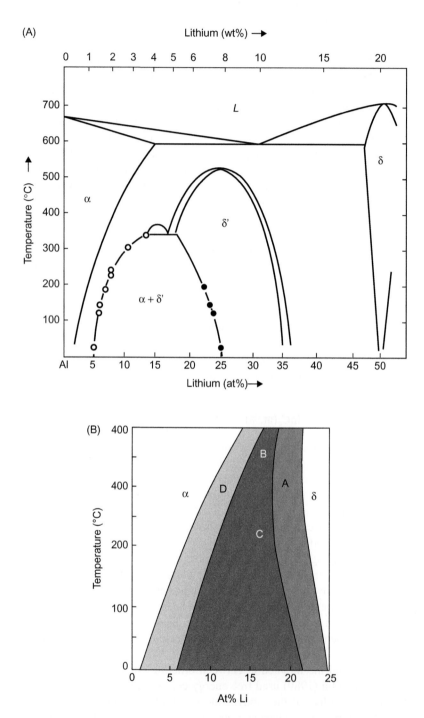

FIGURE 3.2 Al-rich end of the Al–Li phase diagram showing (A) miscibility gap (Gayle and Van der Sande , 1984a), and (B) subregions in the $\alpha + \delta'$ phase for various modes of decomposition (Khachaturyan et al., 1988).

indicate that the decomposition process can follow several mechanisms depending on composition and temperature of aging (Figure 3.2B):

1. In region A, the quenched disordered solid solution will order first, congruently at higher temperatures and continuously at lower temperatures, before decomposition by nucleation and growth.
2. The quenched disordered solid solution undergoes congruent ordering in region B and continuous ordering in region C. In both cases the Li atoms move into the appropriate lattice sites of the ordered L12 structure without any change in local composition.
   - In region B, the disordered solid solution is metastable with respect to an ordered phase of the same composition, and hence ordering occurs through nucleation of ordered domains.
   - In region C, the solid solution is completely unstable and follows continuous or spinodal ordering.
   - In both regions B and C, spinodal decomposition follows congruent ordering. When the concentration of the Li-depleted region reaches a certain value, complete disordering occurs, resulting in the formation of the disordered solid solution and ordered δ' phase.

The conclusions drawn from theoretical studies are supported by several experimental observations based on small-angle X-ray scattering, thermal analysis, electrical resistivity measurements, and HREM. Osamura and Okuda (1993) have shown that the formation of δ' in Al−11.8 at.% Li alloy is by nucleation and growth at 533 K and through spinodal decomposition of the ordered phase at 413 K. Radmilovic et al. (1989) found positive evidence for spinodal decomposition during aging of a series of quenched Al−Li alloys. Their HREM studies on underaged and overaged samples indicated that the ordered $\delta'(Al_3Li)$ precipitate results from lithium enrichment of the ordered regions and lithium depletion of the disordered regions in the as-quenched spinodally decomposed alloys. Schmitz et al. (1994) studied the early stages of decomposition of Al−7.8 at.% Li, Al−10.4 at.% Li, and Al−11.8 at.% Li alloys by HREM. Domains of L1$_2$ structure were present in the as-quenched alloys, but no clear evidence for congruent ordering could be obtained. Two-stage quenching (from the solution treatment temperature to a temperature slightly above the δ' solvus and then down to room temperature) gave rise to faint order spots in the diffraction pattern which cannot be unambiguously attributed to congruent ordering. These observations suggest that ordering and spinodal decomposition progress concomitantly.

Floriano et al. (1996) modeled the small-angle X-ray scattering curve obtained from an Al−8.49 at.% Li alloy and showed that the composition of the precipitating phase is the same, ~ 21 at.%, right from its formation into the growth stage. They concluded that precipitation was by nucleation and growth rather than by spinodal decomposition. Spowage and Bray (2011) used thermal analysis and electrical resistivity measurements and summa-rized the decomposition sequence in Al−8.7 at.% alloy as follows:

Spinodal Ordering → Congruent ordering + spinodal decomposition

+ dissolution of small spinodally ordered regions

→ Growth of δ'

→ Dissolution of δ'

→ Nucleation and growth followed by dissolution

of the δ phase

There is some evidence that excess quenched-in vacancies influence the pre-cursor structures (Noble and Trowsdale, 1995). This would imply that quench rate, holding time at room temperature, and T6 temper conditions are important parameters in controlling the decomposition process. Extensive cold working (95%) does not completely destroy the ordered structure of the as-quenched alloy. However, the high-diffusivity paths resulting from the presence of dislo-cations influence both the precursor events and δ' precipitation.

The strengthening due to δ' precipitates decreases drastically when Al−Li alloys are aged at temperatures above 200°C. The strength loss is due to a decrease in δ' volume fraction as well as disordering that accompanies the coarsening of δ' precipitate particles. The coarsening kinetics follow the well-known Lifshitz−Wagner kinetic model (Lifshitz and Slyozov, 1961; Noble and Thompson, 1971; Wagner, 1961; Williams and Edington, 1975) and can be expressed by the relationship:

$$[\text{average } \delta' \text{ precipitates size } (r)] \propto [\text{aging time } (t)]^{1/3} \qquad (3.2)$$

The rate of coarsening of δ' depends on temperature and lithium content: an increase in either or both of these parameters increases the coarsening kinetics (Jha et al., 1987). Further, preferential coarsening of δ' is found to occur at dislocations owing to rapid pipe diffusion (Williams and Edington, 1975) and at grain boundaries owing to a discontinuous coarsening reaction that results in lamellar $\alpha + \delta'$ growing at the expense of fine homogeneous δ' (Dhers et al., 1986; Williams and Edington, 1976).

The final aspect concerning the binary Al−Li alloy system is the precipita-tion of equilibrium δ (AlLi) phase. The proposed mechanisms include: (i) for-mation of δ from the metastable δ' (Nozato and Nakai, 1977; Venables et al., 1983) and (ii) δ formation by a martensite-type shear process from δ'. However, both of these mechanisms may be discounted because of the large difference in the lattice parameters ($\sim 50\%$) of δ and δ' precipitates, and also because of the small α/δ' interfacial energy even at high temperatures. On the other hand, Sanders and Balmuth (1978), and Jha et al. (1987) have provided evidence for the nucleation of δ whether or not δ' is present and for its growth at the expense of the surrounding metastable δ'.

The nucleation of δ at the grain boundaries and its growth at the expense of lithium in the aluminum solid solution, as well as from fine δ', result in the formation of δ'-PFZs. The widths of these PFZs (w) follow a simple relation:

$$w = k_p \, t^{1/2} \tag{3.3}$$

where the kinematic constant k_p increases with increase in Li content (Jha et al., 1987). PFZs and grain boundary δ have deleterious effects on the mechanical properties of Al−Li alloys.

3.4 TERNARY SYSTEMS

3.4.1 Al−Li−Mg System

Extensive work was carried out on the ternary Al−Li−Mg system by Fridlyander and his coworkers at the Institute of Aviation Materials (VIAM), Moscow. Their work has resulted in a patented commercial alloy 1420 (Al−5Mg−2Li−0.5Mn), its variant 1421, and the simple ternary alloy 1423, which contains 3.5 Mg and 1.9 Li (see Table 3.1).

A detailed study was conducted by Levinson and McPherson (1956) on the equilibrium phase reactions in Al−Li−Mg alloys. Ghosh (1993) has reviewed the solidus and solvus surfaces and isothermal sections at 400°C, 300°C, and 200°C, of this ternary system. The Al-rich end of the Al−Li−Mg ternary system is shown in Figure 3.3 for two temperatures, 500°C and 200°C, which are typical solution treatment and aging temperatures, respectively (Drits et al., 1977). α (Al-rich solid solution), δ, and Al_8Mg_5 are the first solid phases to form from the melt. These phases participate in peritectic reactions to form Al_2LiMg (T phase) and $Al_{12}Mg_{17}$. It should be noted that the high solubility of Mg in Al is relatively unaffected by the presence of Li in solid solution. However, Mg reduces the solubility of Li in Al, and this results in the formation of a larger volume fraction of δ' at the aging temperature.

The phases of interest in aged alloys are equilibrium δ (and metastable δ') and T. The T phase is cubic, with lattice parameter 2.031 nm. The precipitation sequence depends on the alloy composition and is as follows (Noble and Thompson, 1973):

α(Supersaturated) $\rightarrow \delta'$ (Al_3Li) $\rightarrow \delta$ (AlLi)   [for high Li/Mg ratio]

or

α(Supersaturated) $\rightarrow \delta'$ (Al_3Li) $\rightarrow$ T(Al_2MgLi)   [for low Li/Mg ratio]

No experimental evidence is available for existence of the precursor phases of either the equilibrium Al_2MgLi [the existence of GP zones of this phase has been suggested by Fridlyander (1965)] or the metastable T phases [phases designated as S'' and S' by Shchegoleva and Rybalko (1980) could possibly be the precursor phases].

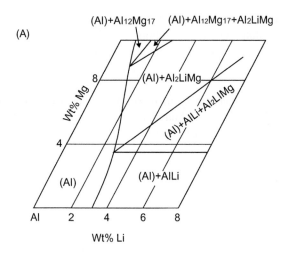

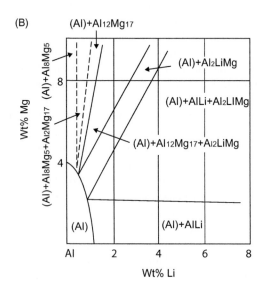

FIGURE 3.3 Isothermal sections of the Al–Li–Mg system at (A) 500°C and (B) 200°C (Drits et al., 1977).

According to Noble and Thompson (1973), strengthening in Al–Li–Mg alloys is mainly due to the metastable δ' phase. However, increasing the Mg content (>2 wt%) increases the likelihood of formation of the ternary T phase. Thus Mg plays a less direct role in improving the strength of Al–Li–Mg alloys, namely by (i) enhanced solid solution strengthening, (ii) partially substituting for Li in the δ' precipitates, and (iii) reducing the Li solid solubility in aluminum and thereby enhancing the volume fraction of δ'

precipitates for a given Li content (Dinsdale et al., 1981; Sanders and Niskanen, 1981; Vasudevan et al., 1986a).

Deschamps et al. (2012) and Gault et al. (2012) used atom probe tomography, transmission electron microscopy (TEM), differential scanning calorimetry (DSC), small-angle X-ray scattering, and density functional calculations to study the effect of Mg on the solubility of Li in Al, and also the possible distribution of Mg in the δ' phase in an Al−5%Mg−1.8%Li alloy aged at 150°C, 8 h, and 120°C, 24 h. Their results suggest that the main effect of Mg is to reduce the solubility of Li in Al, thereby resulting in an increased volume fraction of δ': the interfacial energy and diffusivity of Li in Al are unaffected. Also, the δ' phase forms with an Mg content similar to that of the alloy, and a Li content of 18.5 at.%. There is thus a strong indication of Mg substituting for Li in the δ' phase.

3.4.2 Al–Li–Cu System

The alloys 2020, VAD 23, and 1460 belong to this system (see Table 3.1). Apart from the three major elements, they have one or more additions of Mn, Cd, Zr, or Sc. The addition of Cu to Al−Li alloys has a similar effect as Mg, whereby these additions decrease the maximum solubility of Li in Al solid solution at all temperatures (Silcock, 1959, 1960).

Some aspects of phase equilibria in the Al−Li−Cu system were reported by Raghavan (2010). Two sections of the Al-rich end of this system at 500°C and 350°C are shown in Figure 3.4 (Drits et al., 1977).

The phases formed under equilibrium conditions are $CuAl_2$ (θ, tetragonal, $a = 0.6063$ nm, $c = 0.4872$ nm); AlLi (δ); Al_2CuLi (T_1, hexagonal, $a = 0.497$ nm, $c = 0.935$ nm); Al_6CuLi_3 (T_2, icosahedral); and $Al_{7.5}Cu_4Li$ (T_B, cubic $a = 0.583$ nm) which is the metastable θ' phase stabilized by replacement of Al atoms by those of Li (Drits et al., 1977; Harmelin and Legendre, 1991). All the metastable phases found in Al−Cu and Al−Li systems, GP zones, θ'', θ', and δ', are also found in these alloys. The strengthening phases are θ, δ', and T_1.

Jo and Hirano (1987) have summarized the precipitation reactions in the Al−Li−Cu ternary system as a function of Cu/Li ratio based on the results of several investigations (Ludwiczak and Rioja, 1991; Noble and Thompson, 1972; Noble et al., 1970; Sankaran and O'Neal, 1984). These reactions are as follows:

$$\text{For Cu/Li} > 4, \quad \alpha(\text{SS}) \rightarrow \text{GP zones} \rightarrow \theta'' \rightarrow \theta'$$

$$\text{For Cu/Li} = 2.5 - 4, \quad \alpha(\text{SS}) \rightarrow \text{GP zones} \rightarrow \text{GP zones} + \delta'$$
$$\rightarrow \theta'' + \theta' + \delta' \rightarrow \delta' + T_1 \rightarrow T_1$$

$$\text{For Cu/Li} = 1 - 2.5, \quad \alpha(\text{SS}) \rightarrow \text{GP zones} + \delta' \rightarrow \theta' + \delta' \rightarrow \delta' + T_1 \rightarrow T_1$$

$$\text{and for Cu/Li} < 1, \quad \alpha(\text{SS}) \rightarrow \delta' + T_1 \rightarrow T_1$$

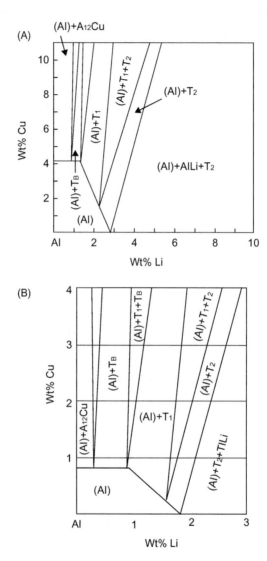

FIGURE 3.4 Isothermal sections of the Al–Li–Cu system at (A) 500°C and (B) 350°C (Drits et al., 1977).

Cu additions have no effect on the position of the α/δ' solvus boundary. However, there is some disagreement on the effect of Cu in solid solution on the precipitation of δ': Baumann and Williams (1984) conclude that δ' precipitation behavior is not affected, while Katsikis (2001) indicates that the kinetics are accelerated through the formation of GP zones acting as nucleation centers.

The influence of the other phases of the Al–Li–Cu system on the mechanism of δ' phase precipitation has been studied in detail by several researchers (Huang and Ardell, 1986; Tosten et al., 1986, 1988). These studies show that δ' can grow on the facets of θ'-plates. At high Cu/Li ratios the T_1 phase competes with the δ' phase for the available lithium. The presence of grain boundary T_1 leads to δ'-PFZs. No such PFZs are observed in alloys with low Cu/Li ratios. Thus the formation of δ' and δ'-PFZs is a function of the Cu/Li ratio in Al−Li−Cu ternary alloys.

Development of precipitates at various stages of annealing in alloys with two different Li contents (Al−1.6%Li−3.2%Cu and Al−2.4%Li−3.2%Cu) was studied by Yoshimura et al. (2003a,b). The δ' phase formed in the early stages of aging in the higher-Li alloy. GP zones formed in both the alloys after aging at 100°C for 3 h. The GP zones offered sites for heterogeneous nucleation of δ' in the lower-Li alloy. The resulting composite structure consisted of GP-I zones flanked by pairs of lenticular particles. Yoshimura et al. (2003a,b) also found evidence that the θ' phase formed in medium- to high-Cu Al−Li alloys has slightly different lattice parameters. The lattice parameters of θ' in the Al−1.6%Li−3.2%Cu aged at 220°C are the same as those in Al−Cu alloys, but the c value of θ' in the concentrated alloy is ~ 0.61 nm (as compared to 0.58 nm in Al−Cu binary alloys).

HREM images by Yoshimura et al. (2003b) reveal typical θ' plates with even larger c values (0.64 nm) and having a habit plane parallel to $\{100\}_\alpha$ matrix planes, although there is some evidence of a few particles growing away from $\{100\}_\alpha$. HREM studies by Howe et al. (1988) on the atomic structure of the composite θ'-δ' precipitate forming in an Al−1%Li−2%Cu alloy indicate that the interface between the two phases is atomically flat and coherent and that the δ' particles on opposite faces of the θ' plate maintain an antiphase relationship. These observations were explained in terms of conservative mechanisms of precipitate growth involving dislocations with Burgers vectors a/2[100] and a/2[010].

The T_1 phase forms as high aspect ratio plates on the $\{111\}_\alpha$ matrix planes during aging in the temperature range 135−260°C. The orientation relationship with the aluminum matrix is expressed as follows (Wang and Starink, 2005):

$$\{111\}_{Al}//\{0001\}_{T_1};\ <\bar{1}10>_{Al}//<10\bar{1}0>_{T_1}$$

T_1 is the major strengthening phase. Since the c parameter for the T_1 phase is a multiple of the $\{111\}_\alpha$ spacing of Al (0.935 nm $\approx 4 \times 0.233$ nm) and the orientation relationship shows the $\{111\}_\alpha$ matrix planes to be parallel to the basal planes of T_1, the structure of this phase is regarded as being made up of 4 $\{111\}_\alpha$ planes with specific occupation of the sites by Al, Li, and Cu to give the overall Al_2CuLi composition. Recent high-angle annular dark field scanning transmission electron microscopy (HAADF-STEM) studies by Donnadieu et al. (2011) indicate that the currently accepted straightforward stacking of $\{111\}$ planes for the T_1 phase may be less likely than the corrugated layer structure derived from X-ray studies by Van Smaalen et al. (1990). There is also some evidence that at very small sizes the particles may exhibit off-stoichiometric composition.

The nucleation sites for the T_1 phase have been investigated by Noble and Thompson (1972) for two Al–Li–Cu alloys with different compositions. In the alloy with higher Cu (3.5%) and lower Li (1.5%), T_1 nucleates on GP zones. On the other hand, in the alloy containing lower Cu (2.5%) and higher Li (2.0%), the T_1 phase precipitates heterogeneously on stacking faults formed by the dissociation of unit dislocations ($b = a/2 <110>$) into Shockley partials ($b = a/6 < 112>$). These precipitates grow by the glide of ledges composed of Shockley partial dislocations on the $\{111\}_\alpha$ matrix planes. Deformation of the quenched alloy after solution treatment and before artificial aging (T8 temper) leads to finer precipitates, with a 100-fold increase in number density and with considerably accelerated aging kinetics. The net result is an increase in yield stress of ~ 100 MPa (Cassada et al., 1991a,b).

3.4.3 Al–Cu–Mg System

Some of the phases found in commercial Al–Li–Cu–Mg quaternary alloys are the same as those found in the Al–Cu–Mg system. Therefore, we shall briefly review the Al–Cu–Mg ternary system in this section. Cross-sections of the Al-rich end of the Al–Mg–Cu system are shown in Figure 3.5 for temperatures of 460°C and 190°C (Brook, 1963; Gregson and Flower, 1986). The equilibrium phases are the α and S (Al$_2$CuMg) phases at the higher temperature, and α, θ (CuAl$_2$), S (Al$_2$CuMg), and T (Al$_6$CuMg$_4$) at the lower temperature.

The precursor phases in the Al–Cu (GP zones, θ'', θ') are also formed along with GPB zones and S' in the early stages, depending on the

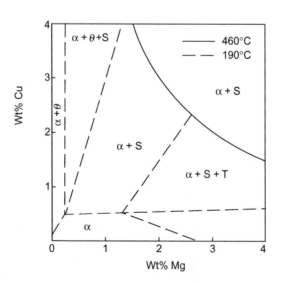

FIGURE 3.5 Sections of the Al–Cu–Mg system at 460°C and 190°C at the Al-rich end.

composition of the alloys. The S' phase has an orthorhombic crystal structure with $a = 0.400$ nm, $b = 0.923$ nm, and $c = 0.714$ nm. It has a {210} habit plane and a lath-like morphology with the longitudinal axis parallel to the $<100>$ directions of the matrix.

The orientation relationship of the S' phase with the aluminum matrix is given by (Bagaryatsky, 1948; Silcock, 1960):

$$\{100\}_{S'}//\{100\}_{\alpha}, \; <010>_{S'}//<021>_{\alpha}; \; <001>_{S'}//<0\bar{1}2>_{\alpha}$$

The S' phase nucleates heterogeneously on quenched-in defects, namely. dislocation loops and helices. S' has a strong hardening potency and is reported to be more effective in homogenizing slip than the T_1 phase. Since the density of quenched-in defects in Li-containing alloys is low, S' precipitation is mainly on low- and high-angle grain boundaries.

The decomposition processes in these alloys depend on the relative amounts of Cu and Mg and can be represented by one of the following reactions (Wang and Starink, 2005):

a. *For high Mg and low Cu alloys*:

Supersaturated solid solution $\alpha \rightarrow$ Cu $-$ Mg co-clusters (GPB or GPB1 zones)

$$\rightarrow \text{GPB2}/S''$$
$$\rightarrow S'/S$$

b. *For low Mg and high Cu Alloys*:

Supersaturated solid solution $\alpha \rightarrow$ GPI (Al$_9$Cu, Al$_7$Cu, Al$_5$Cu, Al$_3$Cu)

$$\rightarrow \text{GPII}/\theta'' \; (\text{Al}_3\text{Cu})$$
$$\rightarrow \theta' \; (\text{Al}_4\text{Cu})$$
$$\rightarrow \theta \; (\text{Al}_2\text{Cu})$$

The GPB2/S'' phase is orthorhombic and coherent with the matrix (probable composition Al$_{10}$Cu$_3$Mg$_3$). S phase is the equilibrium Al$_2$CuMg phase. GPB2/S'' can coexist with the S phase before its formation is complete. Three-dimensional atom probe (3DAP) studies indicate that the co-clusters are formed within a short time during natural or artificial aging and are responsible for the first hardness peak in the aging curve of Al–Cu–Mg alloys (Polmear, 2004). The second stage of hardening is associated with formation of the S'/S phase.

Additions of Ag to Al–Cu–Mg alloys with high Cu/Mg ratio cause nucleation of the Ω phase in preference to S' and θ'. The Ω phase is coherent with the matrix and forms as uniformly dispersed thin hexagonal plates on the $\{111\}_{\alpha}$ matrix planes. It is a modified form of the θ phase in Al–Cu alloys and has the orthorhombic structure with unit cell parameters of

$a = 0.496$ nm, $b = 0.859$ nm, and $c = 0.848$ nm. Ω has the following orientation relationship with the aluminum matrix (Knowles and Stobbs, 1988):

$$\{001\}_\Omega // \{111\}_\alpha; \ [010]_\Omega // [10\bar{1}]_\alpha; \ [100]_\Omega // [1\bar{2}1]_\alpha$$

An alternative structure, hexagonal, has been proposed for this phase, with $a = 0.496$ nm and $c = 0.701$ nm and with the orientation relationship similar to that of the T_1 phase in Al–Li–Cu alloys, but this structure is not accepted.

The Ω phase is more resistant to coarsening than θ' and imparts good thermal stability to commercial Al–Cu–Mg and Al–Li–Cu–Mg alloys containing it. However, with increase in Mg content Ω is replaced by the metastable T phase, $Al_6(Cu,Ag)Mg_4$, which is subsequently replaced by S.

It should be noted here that Ag additions to binary Al–Mg and Al–Cu alloys do not cause Ω phase formation; i.e. it appears that both Mg and Cu are necessary. More specifically, it is now well established that small amounts of Mg are essential for the precipitation of Ω, and the role of Ag is to enhance the precipitation kinetics at the expense of the θ phase. Homogeneous nucleation of Ω phase is observed in Al–4.0Cu–0.5Mg and Al–4.0Cu–0.5Mg–0.5Ag alloys with no apparent preference for formation on defects (Garg and Howe, 1991).

Three dimensional atom probe (3DAP) studies on the evolution of Ω phase reveal that Mg and Ag atoms form co-clusters after several seconds of artificial aging, and these clusters evolve into the Ω phase with further aging. Also, segregation of Ag and Mg atoms is found in the Ω/α interface, contributing to relief of the coherency strain at this interface. Strain energy considerations dictate that growth of Ω occurs along the preferred $\{111\}_\alpha$ matrix planes. This contrasts with the cube planes on which θ' forms in the absence of Ag and Mg (Murayama and Hono, 1998; Reich et al., 1998).

3.4.4 Al–Li–Zr System

Additions of small amounts of Zr, $\sim 0.1\%$, have important effects on the properties of many Al–Li alloys. Nearly all the commercial alloys listed in Table 3.1 contain Zr owing to its influence on controlling recrystallization and crystallographic texture (Rioja and Liu, 2012).

The equilibrium diagram at the aluminum-rich end of the Al–Zr system (Hatch, 1984) shows a peritectic reaction. The maximum solid solubility of Zr in aluminum is very limited (0.28%).

At the Al-rich end of the Al–Li–Zr system, two binary phases, AlLi (δ) and $Al_3Zr(\beta)$, exist in equilibrium with the Al solid solution. The β phase has the DO_{23} structure with $a = 0.401$ nm and $c = 1.732$ nm (Villars and Calvert, 1985; Vecchio and Williams, 1987). Although the solubility of Zr in AlLi is negligible, the Al_3Zr phase can take in up to 1.3 at.% Li (Saunders, 1989). The precursor to the β phase is the metastable β_0 (Al_3Zr with $L1_2$ structure, a50.408

nm) (Nes, 1972). The β_0 phase is resistant to shear by dislocations during plastic deformation and homogenizes.

The precursor to the β phase is the metastable β' (Al$_3$Zr with L1$_2$ structure, $a = 0.408$ nm) (Nes, 1972). The β' phase is resistant to shear by dislocations during plastic deformation and homogenizes slip, thereby contributing to improved ductility. β' also plays an important role in retarding recrystallization and grain growth by pinning down the migration of grain boundaries. During homogenization or solution treatment, a significant amount of Li gets absorbed into the β' phase with the formation of Al$_3$(Li$_x$Zr$_{1-x}$). This phase has a cubic structure of LI$_2$ type (similar to cubic Al$_3$Zr and Al$_3$Li) with a lattice parameter of 0.401 nm very close to that of the metastable cubic Al$_3$Zr and Al$_3$Li phases. Heterogeneous nucleation of δ' on Al$_3$(Li$_x$Zr$_{1-x}$), during aging leads to the "bull's eye" or "donut" structure of composite particles with cores of Al$_3$Zr and shells of Al$_3$Li. The same mechanism is valid for Al-Li-Sc and Al-Li-Sc-Zr alloys (Miura et al., 1994). Zr reduces the diffusivity of Li in Al and hence retards coarsening of the δ' phase (Miura et al., 1994).

3.5 QUATERNARY Al–Li–Cu–Mg SYSTEM

Most of the alloys developed so far and all the third-generation alloys belong to this system (see Table 3.1). These include the later alloys of the Russian 14XX series and those developed in the United Kingdom, France, and the United States since the early 1970s. Many alloys in this family have small additions of Ag ($\leq 0.5\%$) which contribute to strengthening and thermal stability. The presence of two light elements (Li and Mg) and the introduction of additional strengthening phases contribute to improved specific properties and potential weight savings in aerospace structures.

3.5.1 Phase Equilibria

Phase equilibria studies on the Al–Cu–Li–Mg system are somewhat limited. Fridlyander et al. (1993) and Rokhlin et al. (1994) established isothermal sections at 400°C for well-defined Cu concentrations and with variations in Mg and Li content, simulating the compositions of some of the second-generation alloys (8090, 1441, 2090, CP276 and 1464). Their results indicate that the equilibrium phases are the same as those in the ternary systems; i.e., Al$_2$Cu (θ), AlLi (δ), Al$_2$CuMg (S), Al$_2$LiMg, Al$_{7.5}$Cu$_4$Li (T$_B$), Al$_2$CuLi (T$_1$), and Al$_6$CuLi$_3$ (T$_2$). Satya Prasad et al. (1994) studied the formation of δ phase in an 8090-type alloy. Their results showed that precipitation of this phase occurred on grain boundaries over a wide temperature range (170–375°C) and at the T$_2$/Al matrix interface within the grains.

Dorward (1988) established the solidus and solvus isotherms for Al—Li (2.0—2.7%)—Cu(0.5—2.8%) alloys with 0—1.5% Mg. These composition ranges cover several second-generation alloys, including 2090, 2091, 8090, and 8091. The solvus and solidus temperatures can be expressed as a function of concentration of alloying elements by the following equations:

$$T_{\text{solvus}}(^{\circ}\text{C}) = 284 + 67.0(\%\text{Li}) + 34.6(\%\text{Cu}) + 36.5(\%\text{Mg}) \qquad (3.4)$$

and

$$T_{\text{solidus}}(^{\circ}\text{C}) = 730 - 24.5(\%\text{Li}) - 33.2(\%\text{Cu}) - 49.0(\%\text{Mg}) + 10.0(\%\text{Mg})^2$$
$$(3.5)$$

Montoya et al. (1991) studied the effect of alloying elements on the solidus temperatures of Weldalite-type alloys containing 2.5—6.5%Cu, 0—2.0%-Li, 0—0.5%Ag, and 0—0.8%Mg. (The 0%Li alloy had high Cu and low Mg content and small amounts of V, corresponding to the conventional alloy AA 2219.) The results led to four important conclusions:

1. For a fixed Li content of 1.3% the solidus temperature (~510°C) is relatively unaffected by variation in Cu content in the range 4.0—6.3%. For lower values of Cu there is a steady rise to about 570°C at ~3.0%Cu.
2. In the range 4.0—6.3%Cu, variation in Li content gives a minimum in solidus temperature (~510°C) at 1.3%Li.
3. There is a slight decrease (~10°C) in solidus temperature with increase in Mg content.
4. The solidus temperature is hardly altered by variations in Ag content from 0% to 0.5%.

Another important point is that the existence of low-melting eutectics in Weldalite-type alloys should be taken into account when choosing the correct regime of solution treatment.

Based on all the above information, it may be concluded that the solution treatment temperature for Al—Li—Cu—Mg—alloys is in the range 500—540°C. Typical aging temperatures are in the range 160—200°C.

3.5.2 Overview of Precipitate Phases in Commercial Al—Li Alloys

The information available on the phases present in various alloys is summarized in Table 3.2 (Eswara Prasad, 1993; Kulkarni et al., 1989) and Figure 3.6 (Kostrivas and Lippold, 1999). Table 3.2 includes crystal structure, lattice parameters, orientation relationships, and characteristic features of the phases.

Some details on precipitation in commercial alloys of the Al—Li—Cu—Mg system: Figure 3.7 summarizes the information relating to precipitation of

TABLE 3.2 Precipitate Phases in Commercial Al–Li Alloys

Phase	Crystal Structure/ Lattice Parameters (nm)	Orientation Relationship	Characteristic Features
AlLi (δ)	Cubic (NaCl), $a = 0.638$	$(100)_p//(110)$Al $(011)_p//(111)$Al $(011)_p//(112)$Al	Equilibrium phase; plate morphology; addition of Cu or Mg does not alter the lattice parameter
Al_3Li (δ')	Cubic (L1$_2$), $a = 0.401$	Cube–Cube	Metastable, coherent, and ordered phase; spherical in shape
Al_2CuLi (T$_1$)	Hexagonal, $a = 0.4965$, $c = 0.9345$	$(0001)_{T_1}//(111)_{Al}$ $[11\bar{2}0]_{T_1}//[2\bar{1}\bar{1}]_{Al}$ $[10\bar{1}0]_{T_1}//[\bar{1}\bar{1}0]_{Al}$	Partially coherent, equilibrium phase
Al_6CuLi_3 (T$_2$)	Cubic, $a = 1.3914$	–	Displays icosahedral symmetry
$Al_{15}Cu_8Li_2$ (T$_B$)	Cubic (CaF2), $a = 0.583$	$(100)_p//(110)$ Al $(001)_p//(001)$ Al	–
Al_2CuMg (S'); S phase has nearly the same lattice parameters and the same orientation relationship	Orthorhombic, $a = 0.401$, $b = 0.925$, and $c = 0.715$	$\{100\}_{S'}//\{100\}_\alpha$, $<010>_{S'}//<021>_\alpha$; $<001>_{S'}//<0\bar{1}2>_\alpha$	Semicoherent; Rods grow along $<110>$Al which widen forming laths in {210}Al
Al_3Zr (β) Al_3(Zr,Li)	Cubic (L1$_2$), $a = 0.405$	Cube–Cube	Spherical, coherent, and ordered dispersoids
θ' (Al_2Cu)	Tetragonal, $a = 0.404$ nm, $c = 0.58$ nm	$(100)_{Al}//(100)_{\theta'}$ $[001]_{Al}//[001]_{\theta'}$	Rectangular or octagonal plates on {100} planes
Ω	Orthorhombic, $a = 0.496$ nm, $b = 0.859$ nm, and $c = 0.848$ nm	$\{001\}_\Omega//\{111\}_\alpha; [010]_\Omega//$ $[10\bar{1}]_\alpha; [100]_\Omega//[1\bar{2}1]_a$	Hexagonal thin plates in high Cu: Mg alloys

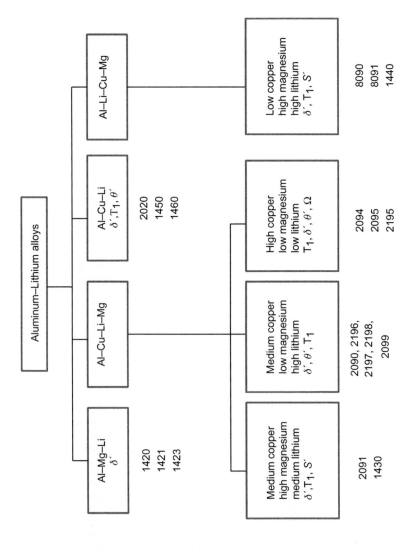

FIGURE 3.6 Strengthening phases in ternary and quaternary Al–Li alloys.

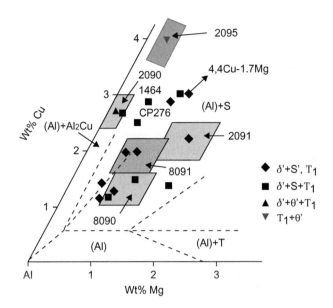

FIGURE 3.7 Precipitating phases in Al−Li−Cu−Mg alloys on aging at 190°C. Important alloys are marked in the figure. Ω phase (present in some third-generation alloys) is not shown in the figure.

various phases in the Al−Li−Cu−Mg system (Flower and Gregson, 1987), with the addition of information on precipitation in AA 2095 alloy. The main points to note are:

- Despite Li contents >1.5%, 2091 and 8091 form S + δ', but no T_1
- 8090 (Li content 2.5%) forms S + δ' + T_1
- 1464 (1.7% Li) and 2090 (1.9−2.6% Li) form S + δ' + T_1.
- 2095 (~1.0% Li) forms T_1 + θ'.

Also, some of the second-generation alloys, for example, 8090 and 8091, have been developed with the objective of exploiting the beneficial effects of the S phase on mechanical properties (Gregson and Flower, 1985; Harris et al., 1986). However, in third-generation alloys with relatively low Li and Mg contents and high Cu contents, for example 2095, the main strengthening phases are T_1, θ', and δ'; and also Ω when Ag and Mg are present.

The T_1 and Ω phases and the roles of Ag and Mg: Precipitation of the T_1 phase has attracted a lot of attention since it is an important strengthening phase in the third-generation alloys. In the absence of Mg and Ag, T_1 nucleates heterogeneously on grain boundaries (both low- and high-angle) and on dislocation loops and helices formed by the condensation of quenched-in vacancies. However, in copper-rich alloys (>2.5% Cu) containing sufficient amounts of Mg, the precipitation of T_1 occurs on homogenously formed GP zones (clusters of Cu and Li atoms). Increasing Cu content favors T_1 formation at the expense of S' (Flower and Gregson, 1987; Harris et al., 1987).

Additions of Ag and Mg to Al–Li–Cu alloys result in a uniform dispersion of the plate-like T_1 phase: Mg is particularly effective in promoting a fine and highly dense distribution (Shollock et al., 1990). There are several similarities between T_1 and Ω (which is promoted by Ag). Both have a hexagonal structure (note that for the Ω phase this is not the widely accepted structure), with nearly the same 'a' value, same orientation relationship with the matrix, and similar formation as thin hexagonal plates. Also, it is difficult to distinguish between T_1 and Ω using conventional TEM, since they show similar morphologies and selected area diffraction (SAD) patterns. Only HREM can reveal the exact nature of these precipitates. All these features and similarities have led to the conclusion that the roles of Ag and Mg may be similar in the formation of T_1 and Ω.

Itoh et al. (1996) reported three types of heterogeneous nucleation sites for the T_1 plates, namely dislocation loops, octahedral voids, and Mg-enriched $\{111\}_{\alpha}$ GP zones. Particles nucleating at the octahedral voids do not contain Ag and Mg, while those formed on $\{111\}_{\alpha}$ GP zones contain both elements. It is difficult to determine the exact location of Ag and Mg atoms (i.e. whether or not they are segregated to the T_1/α interface) in the T_1 plates, since they are very thin. Murayama and Hono (2001) have studied the early stages of aging of quenched Al–5.0Li–2.25Cu–0.4Mg–0.1Ag–0.04Zr (at.%) by 3DAP in order to understand the distribution of Ag and Mg atoms. Their results show that unlike the case of Al–Cu–Mg–Ag alloys, only Mg atoms are involved in the formation of clusters in Al–Li–Cu–Mg–Ag alloys. The absence of Ag in the clusters is possibly associated with strong binding between quenched-in vacancies and Li atoms and consequent reduced concentration of Ag-vacancy pairs, leading to slowdown of the migration of Ag atoms. These results suggest that the nucleation mechanism of the T_1 phase is different from that of the Ω phase.

The effect of independent and combined additions of Ag and Mg on precipitation characteristics was studied for the alloy Al–4.01Cu–1.11Li–0.19Zr–0.11Ti (2095 type) by Huang and Zheng (1998):

1. Ag additions retard GP-zone formation, stimulate T_1 precipitation, and increase the growth rates of T_1 and θ' plates. Also, Ag segregates to the $\{111\}T_1$/matrix and $\{100\}\theta'$/matrix interfaces and accommodates the misfit energy between the phases and matrix.
2. Mg additions promote copious GP-zone formation and precipitation of θ' in the early stages of aging, while T_1 formation is facilitated in the later stages. The Mg atoms segregate at the jogs of T_1 and θ' particles, reducing the large misfit energy normal to the habit planes of these phases.
3. Combined additions of Ag and Mg accelerate the nucleation of T_1 but retard its growth.

There are two interrelated reasons for this: (a) the independent role of Mg in promoting precipitation of θ' is somewhat weakened due to a strong interaction between Ag and Mg atoms; and (b) the role of Ag in favoring the growth of θ' plates is diminished in the presence of Mg. It is suggested that some Ag atoms are bound by Mg atoms segregating at the jogs of T_1 plates on the $\{111\}_\alpha$ planes and substituting for Cu in the matrix lattice.

3.6 MINOR ALLOYING ADDITIONS TO Al–Li ALLOYS

Minor alloying additions have been made to improve the mechanical properties of Al–Li alloys. These additions include Co, Ti, Y, Mn, Ni, Zr, Sc, Cr, Cd, Ge, In, Sn, and combinations of elements such as Al–Ti–B and Ti–C–Al in the form of master alloys (Birch, 1986; Birch and Cowell, 1987; Blackburn and Starke, 1989; Cassada et al., 1986; Flower and Gregson, 1987; Gayle and Vandersande, 1984b; Gu et al., 1985; Srivatsan, 1988). Some of these elements form fine, coherent, and non-shearable ordered dispersoids, which contribute to strengthening. However, the strength increases are marginal. (The important contribution of Zr in forming Al_3Zr, and its associated beneficial effects on retardation of recrystallization and homogenization of slip, has been mentioned in section 3.4.4.)

Minor element additions influence grain size and precipitation processes in a number of ways (Polmear, 2006). These include: (a) grain refinement of the alloy solidified from the melt by the addition of a few hundred parts per million of Ti in the form of Al–Ti–B or Al–Ti–C master alloys, (b) grain refinement in the solid state due to retardation of recrystallization and grain growth, as observed with small additions of Zr, Cr, Sc, La, etc., which form fine dispersoids, (c) strong binding between quenched-in vacancies and solute additions leading to reduced nucleation of GP zones and reduced widths of PFZs, (d) raising the GP solvus line, thereby altering the regions of stability (or metastability) of phases, (e) segregation at the nucleating precipitate/matrix interface and reduction of surface energy, thereby facilitating easier nucleation of the phase of interest and contributing to finer dispersion of the precipitates, (f) promoting the formation of a new phase, (g) forming a part of the clusters in the initial stages of the aging process and contributing to heterogeneous nucleation of the existing or formation new phases, and (h) decrease of solubility of the major elements in the matrix, resulting in higher supersaturation in the quenched alloy. Some of these effects have already been discussed. Some more are briefly discussed in this section.

3.6.1 Minor Alloying Additions for Grain Refinement

Grain size control of Al–Li alloys is important for obtaining good mechanical properties. This is because superlattice alloys, such as Al–Li, with coarse grain structures exhibit low ductility and fracture toughness as a result of localized strain hardening and stress concentrations at the grain boundaries. Al–Li alloys were often found to possess coarse recrystallized grain structures despite various thermomechanical processing schedules. Hence the control of grain structure by minor alloying additions has become an important step in these alloys.

1. *Master alloys.* Grain refinement of aluminum alloys during solidification from the melt is normally achieved by introducing Ti and B in the form of Al–Ti–B or Al–Ti–C master alloys. Typical Ti concentrations in the melt are $\sim 0.05\%$. The master alloys contain Al_3Ti, TiB_2, or TiC in addition to the Al-rich solid solution. Small additions of Zr to the Al–Li alloys may introduce some complications in the grain refining process, since it acts as a poisoner and reduces the effect of grain refinement. This problem may be minimized by the addition of more master alloy (increasing the effective concentration of Ti) or replacing Al–Ti–B with Al–Ti–C. Consequently, some borides and carbides are present in the solidified alloy (Birch, 1986; Chakravorty and Chakraborty, 1991; Labarre et al., 1987; Srivatsan, 1988).

2. *Transition elements.* Several transition elements are added to aluminum alloys to retard recrystallization and grain growth: the more common additions are Cr, Mn, and Zr. The relatively low solubility and diffusion rate of these elements result in the formation of dispersoids which exhibit slow coarsening rates at elevated temperatures. However, Cr and Mn additions have not been beneficial in Al–Li alloys (Miller et al., 1984; Niskanen et al., 1981; Vasudevan et al., 1986b). La addition (0.05%) to an Al–2.5Li–2.0Cu–1.3Mg reduces segregation of Fe and Si to the grain boundaries, contributes to grain refinement, and retards the growth of δ' particles. The result is that tensile and yield strengths and elongation (ductility) are higher for La-containing alloys compared with La-free alloys. La also reduces the loss in strength when going from the peak-aged to overaged condition. Reduction in the levels of Fe and Si, and also the tramp element Na, at the grain boundaries contributes to improved fracture toughness (Ziqiao Zhao et al., 1993).

3.6.2 Minor Alloying Additions for Strengthening

1. *Cd, In, Sn*: Minor additions of Cd, In, and Sn to Al–Cu have significant effect on age-hardening, increasing the rate and improving the strength (Noble, 1984). These elements suppress the formation of GP zones and θ'' and promote a fine and uniform dispersion of θ', resulting in a strength increase of 15–20%. This strength benefit is carried over into Al–Li–Cu alloys. Two explanations were offered for the promoted nucleation of θ': absorption of Cd, In, and Sn at the θ'/matrix interface, reducing the interfacial energy and favoring nucleation; and heterogeneous nucleation of θ' on clusters of the trace elements or dislocation loops formed by the

aggregation of vacancies transported by these elements. The results of 1D atom probe and field ion microscopy (FIM) studies are in conformity with the second explanation (Ringer and Hono, 2000).

2. *In*: Additions of In to an Al–Li–Cu alloy were seen to increase the homogeneity of the two major strengthening phases, T_1 and θ', and thereby increase the strength (Blackburn and Starke, 1989). The effect of 0.5% In addition to Al–3.3Cu–0.8 Li at the aging temperature of 175°C has been studied by Zhengrong Pan et al. (2010) using selected area electron diffraction (SAED), HREM, scanning transmission electron microscopy (STEM), and 3DAP. The T_1 phase is largely replaced by the cubic χ phase ($Al_5Cu_6Li_2$, cubic pm$\bar{3}$, a∼0.83 nm) in the T6 condition. However, deformation of the sample before aging (T8) does not favor the formation of the χ phase. This phase has the following orientation relationship with the matrix:

$$\{001\}_\chi \| \{001\}_\alpha \; [010]_\chi \| [010]_\alpha$$

3DAP analysis of the cubic χ phase indicated 60 at.% Al, 30 at.% Cu and 10at.% Li. The phase is considered to be Li-rich γ_1 (Al_4Cu_9) phase. Also, In could not be detected around or within the cubic phase by energy dispersive X-ray analysis (EDX) and 3DAP. We note here that this study is included because it is most interesting from a fundamental viewpoint. However, it concentrated on the nature of phases and did not provide information on possible improvements in mechanical properties.

3. *Yb*: The effect of 0.02 at.% Yb addition to an Al–6.3Li–0.07Sc alloy subjected to a double aging treatment was studied by Monachon et al. (2011) using atom probe tomography and TEM techniques. Aging at 325°C leads to the formation of coherent nano-size precipitates with an Al_3Yb-rich core and Al_3Sc-rich shell, resulting in doubling the microhardness (an indication of considerable strengthening). Both the cores and shells contain a high concentration of Li, up to 50%. These nanoparticles provide excellent resistance to overaging. Also, a second aging at 170°C increases the microhardness by an additional 30% due to δ' precipitating exclusively as second shell on the already formed Al_3Yb–Al_3Sc particles.

3.7 IMPURITIES AND GRAIN BOUNDARY PRECIPITATION

Several impurity elements, Fe, Si, Na, K, and Ca find their way into the alloys from the starting materials used for smelting. Alkali metals are the main impurities in Li, owing to its high reactivity (Starke et al., 1981). Fe and Si are the inevitable impurities in commercial aluminum. Traces of Cl, F, and H are present in the Al–Li alloys due to improper melting and casting processes (Hill et al., 1984; Vijaya Singh and Chakravorty, 1989).

3.7.1 Impurities

As pointed out at the beginning of Section 3.2, Fe and Si form coarse inter-metallic compounds with Al. These constituent particles precipitate on grain boundaries, deleteriously affecting fracture toughness. The alkali metals form low-melting eutectics as thin films on the grain boundaries, with adverse influence on mechanical properties. Small additions of Bi are effective in counteracting the adverse effect of Na.

3.7.2 Precipitates and PFZs

Studies of grain boundary precipitation are relevant to fracture resistance and stress corrosion aspects. Fracture properties can be adversely affected by the presence of PFZs (due to solute depletion) near grain boundaries. PFZs may cause strain localization during deformation and cause premature failure. Also, grain boundary precipitates can be anodic with respect to the grain interior and therefore promote corrosion and stress corrosion.

Tosten et al. (1984) detected the equilibrium δ phase on high-angle grain boundaries in Al−2%Li−3%Cu and Al−2.6%Li−1%Cu alloys and a δ'-PFZ associated with these boundaries. The predominant phase on the low-angle boundaries was T_1, but there was also some θ'. Only one variant of the T_1 and or θ' was observed on any one planar boundary segment, namely the one that minimized the angle between the precipitate habit normal and the boundary plane normal. PFZs were not observed in the vicinity of low-angle grain boundaries in the 2.6%Li−1%Cu alloy but had developed in the 2%Li−3%Cu alloy.

Several studies on grain boundary precipitation were made on the Weldalite-type alloys:

1. *AA 2198 alloy*: This alloy shows grain boundary T_1 and S precipitates that are usually thicker and shorter at the grain boundary than in the matrix. However, the morphology changes to long plates when the habit plane of the precipitate is nearly parallel to the grain boundary plane. The S particles have lath and block morphology at low- and high-angle boundaries, respectively. High aging temperatures and long times lead to nonplate-like particles of T_1 at high-angle boundaries. Fracture studies indicated void nucleation at blocky S particles, long S' laths, and long, thick T_1 plates (Yiwen Mou et al., 1995).

2. *AA 2195 alloy*: Chen and Bhat (2002) established time-temperature-transformation curves for the 2195 alloy. They found that at temperatures $< 315°C$, T_1 was the only phase present at the subgrain boundaries, and between 150°C and 260°C initial nucleation of T_1 occurred preferentially at the subgrain boundaries. At temperatures $> 315°C$, θ', θ, T_2, and T_B formed at the subgrain boundaries and in the matrix. Prolonged exposure

at 425°C resulted in the precipitation of the equilibrium phases T_B and T_2 with T_B growing at the expense of θ' and θ.

A recent study on 2096 alloy by Williams (2010) to correlate the grain boundary microstructure with susceptibility to stress corrosion cracking showed that the T_1 precipitate forms in an underaged temper with continuous dense precipitation on low-angle boundaries and isolated large particles on high-angle boundaries. For short aging times, T_1 PFZs are observed near the small-angle boundaries but not near high-angle boundaries. Longer aging eliminates the PFZs. EDX line scans showed Cu-depleted regions corresponding to the PFZs near low-angle boundaries, but no Cu-depletion near high-angle boundaries. Enhanced stress corrosion in the underaged temper was ascribed to preferential anodic dissolution along the grain boundaries.

3.8 CHAPTER SUMMARY

Intense research activities in the development of Al–Li alloys since the late 1980s have overcome the problems of strongly anisotropic mechanical properties, low short-transverse ductility and fracture toughness, and loss of toughness owing to thermal instability. Combinations of experimental techniques including DSC, electrical resistivity, X-ray diffraction, conventional and high-resolution electron microscopy, and 3D atom probe measurements have been used to study the formation of metastable and equilibrium phases in binary Al–Li, ternary Al–Li–Mg and Al–Li–Cu, and quaternary Al–Cu–Li–Mg alloys:

- *Al–Li*. There is ample experimental evidence for the formation of the precursor phase (congruent ordering) prior to precipitation of δ' in binary Al–Li alloys.
- *Al–Li–Mg*. The δ' phase is the only strengthening phase in Al–Li–Mg alloys. Mg reduces the solubility of Li in Al, thereby increasing the volume fraction of δ' for a given aging treatment. There is also evidence for the substitution of Li by Mg in this phase.
- *Al–Li–Cu*. The characteristics of $\theta'(\theta)$ and T_1 phases in Al–Li–Cu alloys and the composition limits where these phases are formed are well understood. For low Li contents ($<0.6\%$), $\theta'(\theta)$ is formed, while for medium Li content ($<1.4–1.5\%$) the main strengthening phase is T_1. Formation of T_1 is promoted by small additions of Ag and Mg and by cold work prior to artificial aging (T8 treatment).

  In high Li alloys ($>1.4–1.5\%$) δ' also precipitates during artificial aging and contributes to strengthening, but to a lesser extent than T_1. However, δ' can be detrimental to toughness. Zr forms the metastable β' (Al_3Zr) phase, which has an appreciable effect on retarding recrystallization besides providing nucleation sites for composite δ' particles. These composite particles are resistant to shearing, thereby homogenizing slip.

Sc and Yb additions behave in a similar way; the added advantages are excellent resistance to recrystallization and improvement in creep strength.

- *Al−Li−Cu−Mg*. The available information from phase equilibria studies of Al−Li−Cu−Mg alloys is somewhat limited, but sufficient to give an indication of the desirable solution treatment and aging temperatures and the phases formed at these temperatures. Considerable progress has been made in understanding the effects of small additions of Ag and Mg on precipitation in Al−Li−Cu alloys.

  3D atom probe studies suggest a basic difference in the nucleation characteristics of the Ω and T_1 phases. Ω evolves from Ag−Mg co-clusters formed in a short time during artificial aging in Al−Cu−Mg−Ag alloys while only Mg atoms are involved in the formation of clusters in Al−Li−Cu−Mg−Ag alloys. The absence of Ag in the clusters is possibly related to strong binding between quenched-in vacancies and Li atoms and the consequent slow down of diffusion of Ag atoms. Segregation of the Ag and Mg atoms is found in the Ω/α interface in both cases; this contributes to coherency strain relief at this interface.

Trace element additions to the alloys are effective in grain refinement (Ti-B, Ti-C), improving the strength (Ag) and creep properties (Sc, Yb).

Several considerations have gone into addition of alloying elements in the design of third-generation Al−Li alloys. These include the elements themselves, their concentrations, and the heat-treatment schedules. A typical alloy contains Li and Mg for density reduction, Li, Mg, Cu, and Ag for solid solution and precipitation strengthening, Zn for corrosion resistance, and Zr and Mn for homogenizing slip and controlling recrystallization and texture. In addition, impurities such as Fe, Si, Na, and K are minimized to improve fracture toughness.

An important point to note is that some Al−Li alloys, notably second-generation alloys, which have high Li contents, show thermal instability at temperatures considerably lower than the artificial aging temperatures. This thermal instability is characterized by precipitation of new δ', and also— most probably—by lithium segregation to grain boundaries (Pasang et al., 2012). There is then an increase in strength but more importantly a loss of fracture toughness. This problem has been taken into account in the development of the third-generation alloys, which all have reduced Li contents below 2% (Rioja and Liu, 2012).

ACKNOWLEDGMENTS

The authors express their profound gratitude to Professor David E. Laughlin of the Carnegie Mellon University, USA, for his constructive comments on the manuscript. They are indebted to Dr. K. Balasubramanian (Director, Non-Ferrous Materials Technology

Development Centre) and Professors Vikram Jayaram, T.A. Abinandanan (Indian Institute of Science), and Dr. K Tamilmani (CE (A), CEMILAC, Defence Research and Development Organisation) for their help, support, and encouragement.

REFERENCES

Bagaryatsky, Y.A., 1948. Rentgenograficheskoe issledovanie stareniya alyuminievykh splavov 1. Primenenie monokhromaticheskikh rentgenovykh luchei dlya izucheniya struktury ostarennykh splavov. Zhur. Tech. Fiziki 18, 828–830.

Baumann, S.F., Williams, D.B., 1984. The effect of ternary additions on the δ'/α misfit and the δ' solvus line Al–Li alloys. In: Sanders, T.H., Starke, E.A. (Eds.), Proceedings of the Second International Conference on Aluminum–Lithium Alloys II. The Metallurgical Society of AIME, Warrendale, PA, pp. 17–29.

Belov, N.A., Eskin, D.G., Aksenov, A., 2005. Multicomponent Phase Diagrams: Applications for Aluminium Alloys. Elsevier, New York, NY, pp. 217-222.

Birch, M.E.J., 1986. Grain refining of aluminium–lithium based alloys with titanium–boron–aluminium. In: Baker, C., Gregson, P.J., Harris, S.J., Peel, C.J. (Eds.), Proceedings of the Third International Conference on Aluminium–Lithium Alloys III. The Institute of Metals, London, UK, pp. 152–158.

Birch, M.E.J., Cowell, A.J.J., 1987. Titanium–carbon–aluminium: a novel grain refiner for aluminium–lithium alloys. In: Champier, G., Dubost, B., Miannay, D., Sabatay, L. (Eds.), Proceedings of the Fourth International Conference on Aluminium–Lithium Alloys. J. Phys. Colloque, vol. 48, pp. C3.103–C3.108.

Blackburn, L.B., Starke, E.A., 1989. Effect of In additions on microstructure mechanical property relationships for an Al–Cu–Li alloy. In: Sanders, T.H., Starke, E.A. (Eds.), Proceedings of the Fifth International Conference on Aluminium–Lithium Alloys, vol 2. Materials and Component Engineering Publications Ltd., Birmingham, UK, pp. 751–766.

Blankenship, C.P., Starke, E.A., 1992. Fracture behavior of aluminum–lithium–X alloys. In: Peters, M., Winkler, P.J. (Eds.), Aluminum–Lithium, Vol. 1.. Informationsgesellschaft, Verlag, Germany, pp. 187–201.

Brook, J.B., 1963. Precipitation in Metals, Special Report No. 3. Fulmer Research Institute, UK

Butts, D.A., Gale, W.F., 2004. Equilibrium Diagrams, Smithells Metals Reference Book, VIII Edition. In: Gale, W.F., Totemeier, T.C. (Eds.), pp. 1–534.

Cassada, W.A., Shiflet, G.J., Starke, E.A., 1986. The effect of germanium on the precipitation and deformation behavior of Al–2Li alloys. Acta Metall. 34, 367–378.

Cassada, W.A., Shiflet, G.J., Starke, E.A., 1991a. Mechanism of Al_2CuLi (T_1) nucleation and growth. Metall. Trans. 22A, 287–296.

Cassada, W.A., Shiflet, G.J., Starke, E.A., 1991b. The effect of plastic deformation on Al_2CuLi (T_1) precipitation. Metall. Trans. A 22A, 299–306.

Chakravorty, C.R., Chakraborty, M., 1991. Grain refining of aluminium–lithium alloy with Al–Ti–B. Cast Met. 4, 98–100.

Chen, P.S., Bhat, B.N., 2002. Time-temperature-precipitation behavior in Ai–Li alloy 2195. NASA/TM–2002-211548.

Deschamps, A., Sigli, C., Mourey, T., de Geuser, F., Lefebvre, W., Davo, B., 2012. Experimental and modelling assessment of precipitation kinetics in an Al–Li–Mg alloy. Acta Mater. 60, 1917–1928.

Dhers, J., Driver, J., Fourdeux, A., 1986. Cyclic deformation of binary Al–Li alloys. In: Baker, C., Gregson, P.J., Harris, S.J., Peel, C.J. (Eds.), Proceedings of the Third International Conference on Aluminium–Lithium Alloys III. Institute of Metals, London, UK, pp. 233–238.

Dinsdale, K., Harris, S.J., Noble, B., 1981. Relationship between microstructure and mechanical properties of aluminium–lithium–magnesium alloys. In: Sanders, T.H., Starke, E.A. (Eds.), Proceedings of the First International Conference on Aluminum–Lithium Alloys. The Metall. Soc. AIME, Warrendale, PA, pp. 101–118.

Donnadieu, P., Shao, Fe, Y., De Geuser, F., Botton, G.A., Lazar, S., Cheynet, M., et al., 2011. Atomic structure of T_1 precipitates in Al–Cu–Li alloys revisited with HAADF-STEM imaging and small angle X-ray scattering. Acta Mater. 59, 462–472.

Dorward, R.C., 1988. Solidus and solvus isotherms for quaternary Al–Li–Cu–Mg alloys. Metall. Trans. 19A, 1631–1634.

Drits, M.E., Kadaner, E.S., Padezhnova, E.M., Rokhlin, L.L., Sviderskaya, Z.A., Turkina, N.I., 1977. In: Abrokosov, N.Kh. (Ed.), Phase Diagrams of Aluminum- and Magnesium-Based Systems. Nauka, Moscow.

Eswara Prasad, N., 1993. In-Plane Anisotropy in the Fatigue and Fracture Properties of Quaternary Al–Li–Cu–Mg Alloys. Indian Institute of Technology (BHU) [formerly Institute of Technology, Banaras Hindu University], Varanasi, India, Doctoral thesis.

Eswara Prasad, N., Gokhale, A.A., Rama Rao, P., 2003. Mechanical behaviour of aluminium–lithium alloys. Sadhana 28 (Parts 1 & 2), 209–246.

Floriano, M.A., Triolo, A., Caponetti, E., Triolo, R., 1996. On the nature of phase separation in a commercial aluminium–lithium alloy. J. Mol. Struct. 383, 277–282.

Flower, H.M., Gregson, P.J., 1987. Critical assessment: solid state phase transformations in aluminium alloys containing lithium. Mater. Sci. Technol. 3, 81–90.

Fridlyander, I.N., 1965. Phase composition and mechanical properties of aluminium alloys containing magnesium and lithium. Russ. Metall. (Engl. Transl.) 2, 83–90.

Fridlyander, L.N., Rokhlin, L.L., Dobatkina, T.V., Nikitina, N.L., 1993. Metalloved. Term. Obrab. Met. 10, 16–19.

Garg, A., Howe, J.M., 1991. Convergent beam electron diffraction analysis of the Ω phase in an Al–4.0Cu–0.5Mg–0.5Ag alloy. Acta Metall. Mater. 39, 1939–1946.

Gault, B., Cui, X.Y., Moody, M.P., De Geuser, F., Sigli, C., Ringer, S.P., et al., 2012. Atom probe microscopy investigation of Mg site occupancy within δ' precipitates in an Al–Mg–Li alloy. Scr. Mater. 66, 903–906.

Gayle, F.W., Vandersande, J.B., 1984a. The aluminium - lithium system. Bull. Alloy Phase Diagrams 5, 19–20.

Gayle, F.W., Vandersande, J.B., 1984b. Composite precipitates in an Al–Li–Zr alloy. Scr. Metall. 18, 473–478.

Ghosh, G., 1993. A comprehensive compendium of evaluated constitutional data and phase diagrams. In: Petzow, G., Effenberg, G. (Eds.), Ternary Alloys, vol. 6. VCH, pp. 356–375.

Giummarra, C., Thomas, B., Rioja, R.J., 2007. New aluminium–lithium alloys for aerospace applications. Proceedings of the Light Metals Technology Conference.

Gregson, P.J., Flower, H.M., 1985. Microstructural control of toughness in aluminium–lithium alloys. Acta Metall. 33, 527–537.

Gregson, P.J., Flower, H.M., 1986. In: Shepherd, T. (Ed.), Aluminium Technology. Institute of Metals, London, UK, p. 423.

Gu, B.P., Liedl, G.L., Sanders, T.H., Welpmann, K., 1985. The influence of zirconium on the coarsening of δ' (Al_3Li) in an Al–2.8wt%Li–0.14wt%Zr alloy. Mater. Sci. Eng. 76, 147–157.

Hallstedt, B., Kim, O., 2007. Thermodynamic assessment of the Al–Li system. Inter. J. Mater. Res. 98, 961–969.

Harmelin, M., Legendre, B., 1991. In: Petzow, G., Effenberg, G. (Eds.), Ternary Alloys, vol. 4. VCH, Weinheim, pp. 538–545.

Harris, S.J., Noble, B., Dinsdale, K., Peel, C.J., Evans, B., 1986. Mechanical properties of Al–Li–Zn–Mg alloys. In: Baker, C., Gregson, P.J., Harris, S.J., Peel, C.J. (Eds.), Proceedings of the Third Internationl Conference on Aluminium–Lithium Alloys. Institute of Metals, London, UK, pp. 610–620.

Harris, S.J., Noble, B., Dinsdale, B., 1987. The role of magnesium in Al–Li–Cu–Mg–Zr alloys. In: Champier, G., Dubost, B., Miannay, D., Sabatay, L. (Eds.), Proceedings of the Fourth International Conference on Aluminium–Lithium Alloys. J. Phys. Colloque, vol. 48, pp. C415–C423.

Hatch, J.E. (Ed.), 1984. Aluminum: Properties and Physical Metallurgy. American Society for Metals, Materials Park, OH.

Hill, D.P., Williams, D.N., Mobley, C.E., 1984. The effect of hydrogen on the ductility, toughness and yield strength of an Al–Mg–Li alloy. In: Sanders, T.H., Starke, E.A. (Eds.), Proceedings of the Second International Conference on Aluminum–Lithium Alloys II. The Metallurgical Society of AIMIE, Warrendale, PA, pp. 201–218.

Howe, J.M., Laughlin, D.E., Vasudevan, A.K., 1988. A high-resolution transmission electron mciroscopy investigation of the δ'–θ' precipitate structure in an Al–2wt%Li–wt% Cu alloy. Phil. Mag. A 57, 955–969.

Huang, B.P., Zheng, Z.Q., 1998. Independent and combined roles of trace Mg and Ag additions in properties precipitation process and precipitation kinetics of Al–Cu–Li–(Mg)–(Ag)–Zr–Ti alloys. Acta Mater. 46, 381–393.

Huang, J.C., Ardell, A.J., 1986. Microstructural evolution in two Al–Li–Cu alloys. In: Baker, C., Gregson, P.J., Harris, S.J., Peel, C.J. (Eds.), Proceedings of the Third International Conference on Aluminium–Lithium Alloys. Institute of Metals, London, UK, pp. 455–470.

Itoh, G., Cui, Q., Kanno, M., 1996. Effects of small addition of magnesium and silver on the precipitation of T_1phase in an Al–4%Cu–1.1Li–0.2Zr alloy. Mater. Sci. Eng. A211, 128–137.

Jha, S.C., Sanders, T.H., Dayananda, M.A., 1987. Grain boundary precipitate free zones in Al–Li alloys. Acta Metall. 35, 473–482.

Jo, H.H., Hirano, K.I., 1987. Precipitation processes in Al–Cu–Li alloy studied by DSC. Mater. Sci. Forum 13/14, 377–382.

Katsikis, S., 2001. Effect of Copper and Magnesium on the Precipitation Characteristics of Al–Li–Mg, Al–Li–Cu and AI–Li–Cu–Mg Alloys. University of Nottingham, Nottingham, UK, Ph.D. Thesis.

Khachaturyan, A.G., Lindsey, T.F., Morris, J.W., 1988. Theoretical investigation of the precipitation of δ' in Al–Li. Metall. Trans. 19A, 249–258.

Knowles, K.M., Stobbs, W.M., 1988. The structure of {111} age-hardening precipitates in Al–Cu–Mg–Ag alloys. Acta Cryst. B44, 207–227.

Kostrivas, A., Lippold, J.C., 1999. Weldability of Li-bearing aluminium alloys. Int. Mater. Rev. 44, 217–237.

Kulkarni, G.J., Banerjee, D., Ramachandran, T.R., 1989. Physical metallurgy of aluminum–lithium alloys. Bull. Mater. Sci. 12, 325–340.

Laberre, L.C., James, R.S., Witters, J.J., O'Malley, R.J., Emptage, M.R., 1987. Difficulties in grain refining aluminium lithium alloys using commercial Al–Ti and Al–Ti–B master alloys. In: Champier, G., Dubost, B., Miannay, D., Sabatay, L. (Eds.), Proceedings of the

Fourth International Conference on Aluminium–Lithium Alloys. J. Phys. Colloque, vol. 48, pp. C3.93–C3.102.

Levinson, D.W., McPherson, D.J., 1956. Phase relations in magnesium–lithium–aluminium alloys. Trans. ASM 48, 689–705.

Lifshitz, I.M., Slyozov, V.V., 1961. The kinetics of precipitation from supersaturated solid solutions. J. Phys. Chem. Solids 19, 35–50.

Ludwiczak, E.A., Rioja, R.J., 1991. T_B precipitates in an Al–Cu–Li alloy. Scr. Metall. Mater. 25, 1415–1419.

Lynch, S.P., Shekhter, A., Moutsos, S., Winkelman, G.B., 2003. Challenges in developing high performance Al–Li alloys. In: LiMAT 2003, Third International Conference in Light Materials for Transportation Systems. Published on a CD by the Center for Advanced Aerospace Materials, Pohang University of Science and Technology. Pohang, Korea.

Miller, W.S., Cornish, A.J., Titchener, A.O., Bennet, D.A., 1984. Development of lithium containing aluminium alloys for the ingot metallurgy production route. In: Sanders, T.H., Starke, E.A. (Eds.), Proceedings of the Second International Conference on Aluminum–Lithium Alloys II. The Metallurgical Society of AIME, Warrendale, PA, pp. 335–362..

Miura, Y., Horikawa, K., Yamada, K., Nakayama, M., 1994. Precipitation hardening in an Al–2.4Li–0.19Sc alloy. In: Sanders, T.H., Starke, E.A. (Eds.), Proceedings of the Fourth International Conference on Aluminum Alloys: Their Physical and Mechanical Properties, vol. 2. Georgia Institute of Technology, Atlanta, GA, pp. 161–168.

Monachon, C., Krug, M.E., Seidman, D.N., Dunand, D.C., 2011. Chemistry and structure of core/double-shell nanoscale precipitates in Al–6.5Li–0.07Sc–0.02Yb (at.%). Acta Mater. 59, 3396–3409.

Montoya, K.A., Heubaum, F.H., Kumar, K.S., Pickens, J.R., 1991. Compositional effects on the solidus temperature of an Al–Cu–Li–Ag–Mg alloy. Scr. Metall. Mater. 25, 1489–1494.

Mou, Y., Howe, J.M., Starke, E.A., 1995. Grain boundary precipitation and fracture behaviour of an Al–Cu–Li–Mg–Ag alloy. Metall. Mater. Trans. A 26A, 1591–1595.

Murayama, M., Hono, K., 1998. Three-dimensional atom probe analysis of pre-precipitate clustering in Al–Cu–Mg–Ag alloys. Scr. Mater. 38, 1315–1319.

Murayama, M., Hono, K., 2001. Role of Ag and Mg on precipitation of T_1phase in an Al–Cu–Li–Mg–Ag alloy. Scr. Mater. 44, 701–706.

Nes, E., 1972. Precipitation of the metastable cubic Al_3 Zr phase in subperitectic Al–Zr alloys. Acta Metall. 20, 499–506.

Niskanen, P., Sanders, T.H., Marek, M., Rinker, J.G., 1981. The influence of microstructure on the corrosion of Al–Li, Al–Li–Mn, Al–Li–Mg and Al–Li–Cu alloys in 3.5% NaCl solution. In: Sanders, T.H., Starke, E.A. (Eds.), Proceedings of the First International Conference on Aluminum–Lithium Alloys. The Metallurgical Society AIME, Warrendale, PA, pp. 347–376.

Noble, B., Bray, S.E., 1998. On the $\alpha(Al)/\delta'(Al_3Li)$ metastable solvus in Al–Li alloys. Acta Mater. 46, 6163–6171.

Noble, B., Thompson, G.E., 1971. Precipitation characteristics of aluminium–lithium alloys. Met. Sci. J. 5, 114–120.

Noble, B., Thompson, G.E., 1972. $T_1(Al_2CuLi)$ precipitation in aluminium–copper–lithium alloys. Met. Sci. J. 6, 167–174.

Noble, B., Thompson, G.E., 1973. Precipitation characteristics of aluminium–lithium alloys containing magnesium. J. Inst. Met. 101, 111–115.

Noble, B., Trowsdale, A.J., 1995. Precipitation in an aluminium–14 at.% lithium alloy. Phil. Mag., A 71, 1345–1362.

Noble, B., McClaughlin, I.R., Thompson, G.E., 1970. Solute atom clustering processes in aluminium–copper–lithium alloys. Acta Metall. 18, 339–345.

Nozato, R., Nakai, G., 1977. Thermal analysis of precipitation in Al–Li alloys. Trans. Jpn. Inst. Met. 18, 679–689.

Osamura, K., Okuda, H., 1993. Phase decomposition and reversion in Al–Li alloys. In: Champier, G., Dubost, B., Miannay, D., Sabatay, L. (Eds.), Proceedings of the Fourth International Conference on Aluminium–Lithium Alloys. J. Phys. Colloque, vol. 48, pp. C8-311–C8-316.

Pasang, T., Symonds, N., Moutsos, S., Wanhill, R.J.H., Lynch, S.P., 2012. Low-energy intergranular fracture in Al–Li alloys. Eng. Fail. Anal. 22, 166–178.

Pickens, J.R., 1985. The weldability of lithium containing aluminium alloys. J. Mater. Sci. 20, 4247–4258.

Pickens, J.R., 1990. Recent developments in the weldability of Li containing aluminum alloys. J. Mater. Sci. 25, 3035–3047.

Polmear, I.J., 2004. Aluminium alloys—a century of age hardening. In: Nie, J.F., Morton, A.J., Muddle, B.C. (Eds.), Mater. Forum., vol. 28, pp. 1–14.

Polmear, I.J., 2006. Light Alloys, fourth ed. Butterworth-Heinemann, Burlington, MA-01803.

Radmilovic, V., Fox, A.G., Thomas, G., 1989. Spinodal decomposition of Al-rich Al–Li alloys. Acta Metall. 37, 2385–2394.

Raghavan, V., 2010. Phase diagram evaluations: Al–Li–Cu. J. Phase Equilib. Diffus. 31, 288–290.

Reich, l., Murayama, M., Hono, K., 1998. Evolution of Ω phase in an Al–Cu–Mg–Ag alloy—a three-dimensional atom probe study. Acta Mater. 46, 6053–6062.

Ringer, S.P., Hono, K., 2000. Microstructural evolution and age hardening in aluminium alloys: atom probe field-ion microscopy and transmission electron microscopy studies. Mater. Charact. 44, 101–131.

Rioja, R.J., Liu, J., 2012. The evolution of Al–Li base products for aerospace and space applications. Metall. Mater. Trans. A 43, 3325–3337.

Rokhlin, L.L., Dobatkina, T.V., Muratova, E.V., Korol'kova, I.G., 1994. Multicomponent phase diagrams: applications for commercial alloys. Izv. RAN, Met. 1, 113–118.

Röyset, J., 2007. Scandium in aluminium alloys—overview: physical metallurgy, properties and applications. Metall. Sci. Technol. 25, 11–21, Teksid Aluminum.

Sainfort, P., Guyot, P., 1986. Fundamental aspects of hardening in Al–Li and Al–Li–Cu alloys. In: Baker, C., Gregson, P.J., Harris, S.J., Peel, C.J. (Eds.), Proceedings of the Third International Conference on Aluminium–Lithium Alloys III. The Institute of Metals, London, UK, pp. 420–426.

Sanders, T.H., Balmuth, E.S., 1978. Aluminum–lithium alloys: low density. Met. Prog., 32–37.

Sanders Jr., T.H., Niskanen, P.W., 1981. Microstructure, mechanical properties and corrosion resistance of Al–Li–X alloys—an overview. Res. Mech. Lett. 1, 363–370.

Sankaran, K.K., O'Neal, E.A., 1984. Structure-property relationship in Al–Cu–Li alloys. In: Sanders, T.H., Starke, E.A. (Eds.), Proceedings of the Second International Conference on Aluminum–Lithium alloys II. The Metallurgical Society of AIME, Warrendale, PA, pp. 393–405.

Satya Prasad, K., Mukhopadhyay, A.K., Gokhale, A.A., Banerjee, D., Goel, D.B., 1994. δ precipitation in an Al–2.2Li–1.0Cu–0.7Mg–0.04 Zr alloy. Scr. Met. Mater. 30 (10), 1299–1304.

Saunders, N., 1989. Calculated stable and metastable phase equilibria in Al−Li−Zr alloys. Z Metallkde. 80, 894−903.

Schmitz, G., Hono, K., Haasen, P., 1994. High resolution electron microscopy of the early decomposition stage of Al−Li alloys. Acta Metall. Mater. 42, 201−211.

Shchegoleva, T.V., Rybalko, O.F., 1980. Structure of the metastable S'-phase in alloy Al−Mg−Li. Fiz. Metal. Metalloved. 50, 86−90.

Shollock, B.A., Grovenor, C.R.M., Knowles, K.M., 1990. Compositional studies of Ω and θ' precipitates in an Al−Cu−Mg−Ag alloy. Scr. Metall. Mater. 24, 1239−1244.

Sigli, C., Sanchez, J.M., 1986. Calculation of phase equilibrium in Al−Li alloys. Acta Metall. 34, 1021−1028.

Silcock, J.M., 1959. The structural ageing characteristics of Al−Cu−Li alloys. J. Inst. Met. 88, 357−364.

Silcock, J.M., 1960. The structural ageing characteristics of Al−Cu−Mg alloys with copper: magnesium weight ratios of 7:1 and 2.2:1. J. Inst. Met. 89, 203−210.

Soffa, W.A., Laughlin, D.E., 1989. Decomposition and ordering processes involving thermodynamically first-order order→disorder transformations. Acta Metall. 37, 3019−3028.

Soni, K.K., Williams, D.B., Chabala, J.M., Levi-Setti, R., Newbury, D.E., 1992. Electron and ion microscopy studies of Fe-rich second-phase particles in Al−Li alloys. Acta Metall. Mater. 40, 663−671.

Spowage, A.C., Bray, S., 2011. Characterization of nanoprecipitation mechanisms during isochronal aging of a pseudo-binary Al−8.7at.pct Li alloy. Metall. Mater. Trans. A 42, 227−230.

Srivatsan, T.S., 1988. The effect of grain-refining additions to lithium-containing aluminum alloys. J. Mater. Sci. Lett. 7, 940−943.

Starke, E.A., Sanders, T.H., Palmer, I.G., 1981. New approaches to alloy development in the Al−Li system. J. Met. 33, 24−36.

Tiryakioglu, M., Staley, J.T., 2003. In: Totten, G.E., Mackenzie, D.S. (Eds.), Handbook of Aluminum, vol. 1, Physical Metallurgy and Processes. Marcel Dekker, Inc., New York, NY-10016, p. 123.

Tosten, M.H., Vasudevan, A.K., Howell, P.R., 1984. Grain boundary prcepitation in Al-Li-Cu alloys (Al-3%Cu-2%Li and Al-1%Cu-2.6%Li), TMS (The Metallurgical Society) Paper Selection; (USA); 56; Conference: TMS-AIME fall meeting, Detroit, MI (USA), 16−20 Sep 1984.

Tosten, M.H., Vasudevan, A.K., Howell, P.R., 1986. Microstructural development in Al-2%Li-3%Cu alloy. In: Baker, C., Gregson, P.J., Harris, S.J., Peel, C.J. (Eds.), Proceedings of the Third International Conference on AluminiumLithium Alloys. Institute of Metals, London, UK, pp. 483−489.

Tosten, M.H., Vasudevan, A.K., Howell, P.R., 1988. The aging characteristics of an Al-2pct Li-3pct Cu-0.12pct Zr alloy. Metall. Trans. A. 19, 51−66.

The Aluminum Association, February 2009. International Alloy Designations and Chemical Composition Limits for Wrought Aluminum and Wrought Aluminum Alloys − Registration Record Series and Teal Sheets.

Van Smaalen, S., Meetsa, A., DeBoer, J.L., Bronsveld, P.M., 1990. Refinement of the crystal structure of hexagonal Al_2CuLi. J. Solid State Chem. 85, 293−298.

Vasudevan, A.K., Ziman, P.R., Jha, S.C., Sanders, T.H., 1986a. Stress corrosion resistance of Al−Cu−Li−Zr alloys. In: Baker, C., Gregson, P.J., Harris, S.J., Peel, C.J. (Eds.), Proceedings of the Third International Conference on Aluminium−Lithium Alloys III. The Institute of Metals, London, UK, pp. 303−309..

Vasudevan, A.K., Ludwiczak, E.A., Baumann, S.F., Howell, P.R., Doherty, R.D., Kersker, M. M., 1986b. Grain boundary fracture in Al–Li alloys. Mater. Sci. Technol. 2, 1205–1209.

Vecchio, K.S., Williams, D.B., 1987. Convergent electron beam diffraction study of Al₃Zr in Al–Zr and Al–Li–Zr alloys. Acta Metall. 35, 2959–2970.

Venables, D., Christodoulou, L., Pickens, J.R., 1983. On the $\delta' \rightarrow \delta$ transformation in Al–Li alloys. Scr. Metall. 17, 1263–1268.

Vijaya Singh, Chakravorty, C.R., 1989. Melting and casting of Al–Li alloys—A review. Science and Technology of Aluminium—Lithium Alloys, Conference Proceedings. Hindustan Aeronautics Limited, Bangalore, India, pp. 83–91.

Villars, P., Calvert, L.D., 1985. Pearson's Handbook of Crystallographic Data of Intermetallic Phases, vol. 2. ASM, Metals Park, OH.

Wagner, C., 1961. Theory of aging of precipitates by redissolution (Ostwald ripening). Z. Elektrochem. 65, 581–591.

Wang, S.C., Starink, M.J., 2005. Review of precipitation in Al–Cu–Mg(-Li) alloys. Inter. Mater. Rev. 50, 193–215.

Williams, A., 2010. Microstructural Analysis of Aluminium Alloy 2096 as a Function of Heat Treatment. University of Birmingham, Birmingham, UK, Master of Research Thesis.

Williams, D.B., 1981. Aluminum–lithium alloys. In: Sanders, T.H., Starke, E.A. (Eds.), Proceedings of the First International Conference on Aluminum–Lithium Alloys I. The Metallurgical Society of AIME, Warrendale, PA, pp. 89–100.

Williams, D.B., Edington, J.W., 1975. The precipitation of δ' (Al₃Li) in dilute aluminium–lithium alloys. Met. Sci. J. 9, 529–532.

Williams, D.B., Edington, J.W., 1976. The discontinuous precipitation reaction in dilute Al–Li alloys. Acta Metall. 24, 323–332.

Yoshimura, R., Konno, T.J., Abe, E., Hiraga, K., 2003a. Transmission electron microscopy study of the early stage of precipitates in aged Al–Li–Cu alloys. Acta Mater. 51, 2891–2903.

Yoshimura, R., Konno, T.J., Abe, E., Hiraga, K., 2003b. Transmission electron microscopy study of the evolution of precipitates in aged Al–Li–Cu alloys: the θ' and T₁phases. Acta Mater. 51, 4251–4266.

Zhengrong Pan, Zheng. Ziqiao, Lial Zhongquan, Li Shichen, 2010. Effects of indium on precipitation in Al–3.3Cu-0.8Li alloy. Acta Metall. Sin.(Engl. Lett.) 23, 285–292.

Ziqiao Zhao, Z., Liu, Y., Yin, M., Denfeng, 1993. Microstructure and mechanical properties of an Al–Li–Cu–Mg–Zr alloy containing minor lanthanum additions. Trans. NF Soc. 3, 37–42.

Chapter 4

Microstructure and Precipitate Characteristics of Aluminum—Lithium Alloys

K. Satya Prasad*, N. Eswara Prasad†, and Amol A. Gokhale*
*Defence Metallurgical Research Laboratory, Hyderabad-500 058, India, †Regional Centre for
Military Airworthiness (Material), CEMILAC, Hyderabad-500 058, India

Contents

4.1 Introduction 99
4.2 Microstructures in the
Solution-Treated
Condition 101
  4.2.1 Alloys with Zr 102
4.3 Age Hardening Behavior 102
  4.3.1 Age Hardening
Curve 103
  4.3.2 Microstructures at
Different Stages of
Aging 104
  4.3.3 Microstructures After
High-Temperature
Exposure 105
4.4 Characteristics of
Precipitates 106
  4.4.1 Al₃Li (δ′) Phase 107
  4.4.2 Al₂Cu (θ′) Phase 111

4.4.3 Al₂CuLi (T₁) Phase in
Ternary Al—Li—Cu
Alloys 112
4.4.4 Al₂CuLi (T₁) Phase in
Quaternary
Al—Li—Cu—Mg
Alloys 115
4.4.5 Al₂CuMg (S′) Phase and
the Equilibrium S
Phase 116
4.4.6 Al₃Zr (β′) Phase 121
4.4.7 AlLi (δ) Phase 125
4.4.8 Al₆CuLi₃ (T₂)
Phase 127
4.5 Summary 130
Acknowledgments 131
References 131

4.1 INTRODUCTION

The mechanical properties of aluminum—lithium (Al—Li) alloys are determined by the microstructure, particularly the type, size, volume fraction, and distribution of precipitates within the grains and at the grain boundaries.

The precipitation, in turn, depends upon chemistry, grain structure, and total thermomechanical history. In Al−Li alloys, the strengthening from Li additions is due to both solid solution strengthening and precipitation hardening. The precipitation hardening is primarily due to the metastable strengthening phase, δ' (Al_3Li), which forms spherical, coherent, and ordered precipitate particles having a cube-on-cube orientation relationship with the aluminum matrix. At equilibrium, and at its simplest in binary Al−Li alloys, the only phases present are the aluminum-rich solid solution and the δ (AlLi) phase.

Addition of copper and magnesium to binary Al−Li alloys enables more varied precipitation. The addition of copper produces θ' (Al_2Cu) and T_1 (Al_2CuLi) precipitates in artificially aged alloys, although θ' has only been observed in alloys with Cu:Li > 1:3 (on wt% basis). T_1 is commonly present in commercial Al−Li−Cu and Al−Li−Cu−Mg alloys. Magnesium initially affects aging by increasing δ' precipitation, since the solid solubility of Li in aluminum is lowered. Later on, overaging results in the ternary precipitate T (Al_2MgLi).

Further alloying of Al−Li alloys with copper and magnesium results in S′ and/or S (Al_2CuMg) precipitation as well as δ' and T_1. The balance of S′ (or S) and T_1 phases in these alloys depends critically on the relative concentrations of all three elemental additions: high Cu and Li contents relative to Mg result in the predominance of T_1, whereas high Cu and Mg contents relative to Li result in the predominance of S′ (or S).

Besides copper and magnesium, zirconium is an important addition to Al−Li alloys. Zirconium is very effective in inhibiting recrystallization in a wide variety of aluminum alloys, and this effect is associated with the presence of fine (20−30 nm diameter), spherical, and coherent β' (Al_3Zr) precipitates in metastable cubic ($L1_2$) form (Ryum, 1966, 1969; Nes, 1972). These precipitates are stable as a result of (i) low solid solubility of Zr in Al, (ii) small β'/matrix misfit strains, and (iii) sluggish diffusion of Zr in Al (Ryum, 1969). Consequently, the β' precipitates are very effective in pinning grain and subgrain boundaries during thermomechanical processing of commercial aluminum alloys (Ryum, 1969). The presence of Li has no influence on the effectiveness of Zr as a recrystallization inhibitor, but the coherent Al_3Zr precipitates provide heterogeneous nucleation sites for δ', and this results in an apparent acceleration of the δ' aging process (Gayle and Vander Sande, 1984a,b; Gregson and Flower, 1984). Mn is also helpful in retarding recrystallization by forming dispersoids of the type $Al_{20}Cu_2Mn_3$. However, a recent study indicates that the combined presence of Zr and Mn may be undesirable, since Mn reduces the concentration of Al_3Zr due to incorporation of some Zr in the Mn-rich phase (Tsivoulas et al., 2012).

Al−Li alloys developed prior to and in the early 1980s (the first- and second-generation alloys), such as 01420 and AA 2020, 8090, 8091, 2090, and 2091, contain 2−3 wt% each of Li and Cu (Cu in 8XXX alloys is less than 2%) with small amounts of Mg and Zr. These alloys suffered from several disadvantages, including poor corrosion resistance, low toughness, and high anisotropy, and therefore were of limited commercial interest. The introduction of the third-generation alloys, beginning in the 1990s, has largely overcome these problems. These alloys all contain less than 2 wt% Li, see Table 3.1. Also, minor additions of Zn (0.1–0.7 wt%) are present for improving the corrosion resistance. Some of the third-generation alloys have been used in the US space program for the "Super Lightweight Tank" of the Space Shuttle and the F-16 fighter aircraft, replacing the incumbent conventional alloys AA 2124 and 2219, as discussed in Chapters 2 and 15.

Apart from alloy composition, the precipitation of the metastable phases δ', T_1, and S' is primarily governed by preaging and aging parameters. Controlled thermomechanical processing enables these precipitates to contribute to improvements in strength, ductility, and fracture toughness. The various phases that may be present in different Al–Li alloy systems, depending upon the aging conditions, are listed in Table 4.1. The characteristics of the metastable phases δ', T_1, S' (and equilibrium S), and β' are described in Section 4.4, and the equilibrium phases δ and T_2 are considered in Section 4.5.

In addition to precipitate phases, there are insoluble constituent particles that occur in the matrix or at the grain boundaries of aluminum alloys, including Al–Li alloys (Kulkarni et al., 1989). Impurities like iron and silicon give rise to coarse intermetallic compounds, which can adversely affect the properties, especially the fracture toughness. Details of these intermetallic compounds are given in Table 4.2 (Kulkarni et al., 1989).

4.2 MICROSTRUCTURES IN THE SOLUTION-TREATED CONDITION

In the solution heat-treated condition, the microstructures of Al–Li alloys generally consist of grain structure, insoluble impurity particles (Table 4.2), and Al_3Zr (β') precipitates in zirconium-containing alloys (all modern alloys). The β' phase precipitates during homogenization, and the amount of this precipitate determines its effectiveness in inhibiting recrystallization.

TABLE 4.1 Different Precipitate Phases in Al–Li Alloy Systems

Alloy Systems	Precipitate Phases
Al–Cu–Li–Zr	β' (Al_3Zr), δ' (Al_3Li), δ ($AlLi$), θ' (Al_2Cu), T_1 (Al_2CuLi), T_2 (Al_6CuLi_3)
Al–Mg–Li–Zr	β' (Al_3Zr), δ' (Al_3Li), δ ($AlLi$), T (Al_2MgLi)
Al–Li–Cu–Mg–Zr	β' (Al_3Zr), δ' (Al_3Li), δ ($AlLi$), T_1 (Al_2CuLi), S' and S (Al_2CuMg), T_2 (Al_6CuLi_3)

TABLE 4.2 Insoluble Particles in Al–Li Alloys

Phase	Crystal Structure	Lattice Parameters (nm)
Al_7Cu_2Fe	Tetragonal	$a = 0.6336, c = 1.487$
$Al_{23}CuFe_4$	Orthorhombic	$a = 0.7664, b = 0.6441, c = 0.8778$
AlFeCuNi	Monoclinic	$a = 1.51, b = 0.83, c = 1.25, \beta = 107.7°$
$Al_6(CuFeMn)$	Orthorhombic	$a = 0.646, b = 0.746, c = 0.879$
$Al_{20}Cu_2Mn_3$	Orthorhombic	$a = 2.411, b = 1.25, c = 0.72$
Al_5FeSi	Monoclinic	$a = 0.612, b = 0.612, c = 4.15, \beta = 91°$
Al_8Fe_2Si	Hexagonal	$a = 1.23, c = 2.63$
$Al_{12}(FeMn)_3Si$	Cubic	$a = 1.265$

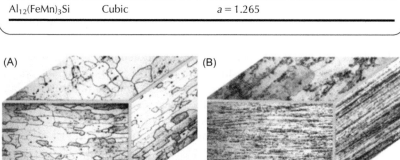

FIGURE 4.1 Optical microstructures of Al–Li–Cu–Mg–Zr alloys: (A) recrystallized low-Zr (0.08 wt%) Lital C alloy and (B) nearly unrecrystallized high-Zr (0.12 wt%) AA 8090 alloy (Eswara Prasad, 1993).

4.2.1 Alloys with Zr

Low Zr content, less than about 0.08 wt%, permits recrystallization of α-aluminum, resulting in an equiaxed grain structure as shown in Figure 4.1A. Modern Al–Li alloys contain about 0.1 wt% Zr, which is sufficient to inhibit recrystallization such that the worked grain structure (also called a pancake structure for rolled products) is retained after solution heat treatment (Figure 4.1B). All modern Al–Li alloys are subjected to multistage thermomechanical processing to achieve the desired degrees of recrystallization, which have important and basic influences on the engineering properties.

4.3 AGE HARDENING BEHAVIOR

On the aluminum-rich side of the Al–Li phase diagram, the solid solubility of Li in Al decreases with decreasing temperature (Figure 4.2). Thus, Li in solution

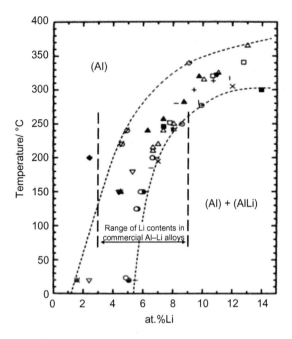

FIGURE 4.2   The Al–Li binary system with $\alpha–\delta'$ solvus line. *Source: Data from several investigators summarized by Noble and Bray (1998).*

at elevated temperatures (4 wt% at 600°C) will precipitate out during (i) slow cooling or (ii) natural or artificial aging after rapid cooling or quenching. During aging, the metastable δ' phase precipitates out first, and continued aging results in the equilibrium δ (AlLi) phase. Since δ' is the primary strengthening phase for Al–Li alloys, the position of the $\alpha–\delta'$ solvus line is important for the practical use of age hardening. This line (indicated as a band in Figure 4.3) lies roughly in the temperature range 150–250°C for commercial alloys.

4.3.1 Age Hardening Curve

Figure 4.3 shows a typical age hardening curve, obtained by isothermal aging at 190°C, for the second-generation Al–Li alloy AA 8090 (Prasad, 1999). The peak hardness of 142 VHN was reached after aging for 40 h. Also studied was the evolution of precipitate structure during aging through the various stages of underaging (UA), peak aging (PA), and overaging (OA). Four representative aging times of 1.5 and 10 (UA), 40 (PA), and 100 h (OA) were chosen, and some of the results are presented in the next section.

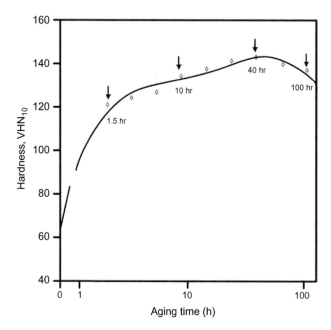

FIGURE 4.3 Typical isothermal age hardening curve at 190°C for the second-generation Al–Li–Cu–Mg–Zr alloy AA 8090 (Prasad, 1999).

4.3.2 Microstructures at Different Stages of Aging

δ' is the most abundant metastable phase in artificially aged Al–Li alloys and can be imaged in dark field mode with {001} reflections (Prasad, 1999). Figure 4.4 shows the δ' precipitates after aging Al–Li alloy AA 8090 at 190°C for different times. The sizes and spacings of the δ' precipitates increase with aging time, and the sizes change from 14 to 45 nm over the range of 1.5–100 h aging. Other important microstructural features are the regions devoid of δ' precipitates, the precipitate-free zones (PFZs), which are adjacent to the grain boundaries (Figure 4.5). The PFZ widths increase from 32 to 230 nm over the same period of aging from 1.5 to 100 h.

Figures 4.6 and 4.7 show that the metastable phases T_1, S, and β' and equilibrium phases δ and T_2 can also be found in the AA 8090 alloy for longer aging times. The S (Al_2CuMg) precipitates are characteristically lath-shaped, and the T_1 (Al_2CuLi) and T_2 (Al_6CuLi_3) precipitates are plate-like. Figure 4.7 also gives examples of δ precipitated on T_2, and T_2 precipitated mostly on the β' particles, though T_2 can precipitate independently within the grains. β' is generally described as spherical, but other morphologies are also found: faceted, rod-shaped, and cauliflower-shaped (Tsivoulas, 2010). The spherical form of β' provides nucleation sites for δ' (Figure 4.7A), resulting

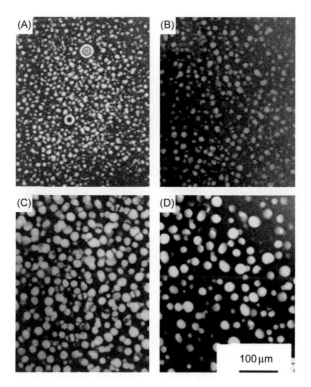

FIGURE 4.4 TEM dark-field images showing δ' precipitates for (A) 1.5 h, (B) 10 h, (C) 40 h, and (D) 100 h aging of the Al–Li alloy AA 8090 at 190°C (Prasad, 1999).

inacharacteristic "bull's-eye" structure; and the faceted β' appears to aid nucleation of T_2 (Figure 4.7B).

4.3.3 Microstructures After High-Temperature Exposure

High-temperature exposure (above the age hardening temperatures but below the solvus temperature) is important from the hot working point of view, since coarse precipitates forming at high temperatures can influence the micromechanisms of deformation and therefore the grain structure and texture of the wrought semiproduct, see Chapter 5. Thus, it is important to obtain a detailed characterization of Al–Li alloys in the hot working temperature range between 250°C and 450°C.

Examples of microstructures resulting from high-temperature exposure of Al–Li alloy AA 8090 are shown in Figures 4.8 and 4.9 (Prasad, 1999):

1. Figure 4.8 shows scanning electron microscope (SEM) micrographs for 16 h exposure over the range 250–350°C. At 250°C and 300°C, the precipitates forming at the grain boundaries and within the grains are of the

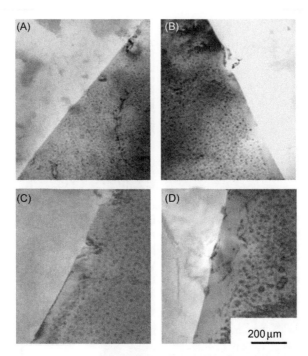

FIGURE 4.5   TEM bright-field images showing δ' PFZs at the grain boundaries after (A) 1.5 h, (B) 10 h, (C) 40 h, and (D) 100 h aging of the Al–Li alloy AA 8090 at 190°C (Prasad, 1999).

same size. At 350°C, the grain boundary precipitates are much coarser than those in the grains, and increased exposure temperatures generally increase the precipitate sizes and decrease their numbers.

2. Figure 4.9 presents transmission electron microscope (TEM) micrographs for 16 h exposure at 250°C. These show precipitates (a) at grain boundaries and (b) within a grain. Besides β' (mainly spherical), there were two other types of precipitate within the grains. Most were T_2 (icosahedral), and the rest were identified as δ. An example of a δ precipitate is indicated in Figure 4.9B.

4.4 CHARACTERISTICS OF PRECIPITATES

As mentioned in the introduction to this chapter, there are several metastable and stable phases that may be present in Al–Li alloys, depending on the alloying elements. The metastable precipitates to be discussed in this section are δ' (Al_3Li), θ' (Al_2Cu), T_1 (Al_2CuLi), S (Al_2CuMg), and β' (Al_3Zr), while the stable phases discussed are δ (AlLi) and T_2 (Al_6CuLi_3).

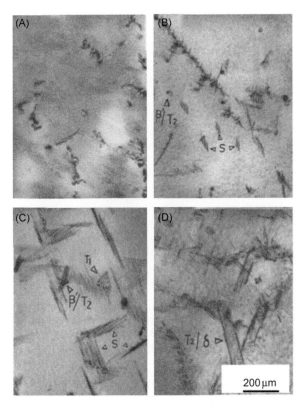

FIGURE 4.6   TEM bright-field images showing various other precipitates after (A) 1.5, (B) 10, (C) 40, and (D) 100 h aging of the Al−Li alloy AA 8090 at 190°C (Prasad, 1999).

4.4.1 Al₃Li (δ′) Phase

The Al_3Li (δ') phase precipitates homogeneously during the aging of a super-saturated Al−Li solid solution that has been obtained by quenching into the two-phase field (Figure 4.2). The δ' phase has an $L1_2$ (ordered fcc) superlattice crystal structure. These precipitates are spherical, except when precipitated discontinuously, see the micrographs in Figures 4.4 and 4.7. They possess a cube-on-cube orientation relationship with respect to the aluminum matrix (Noble and Thompson, 1971; Silcock, 1959−1960; Williams and Edington, 1975), and they are coherent with the matrix even when coarsened to large diameters (>300 nm) (Williams, 1981). The lattice mismatch with α-Al is small, and reported to be in the range 0.08−0.3% (Noble and Thompson, 1971; Tamura et al., 1970; Williams and Edington, 1975).

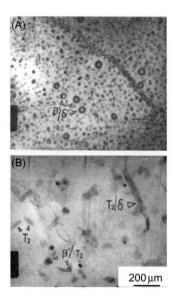

FIGURE 4.7 TEM bright-field images showing (A) spherical β' precipitates enveloped by δ' (bull's-eye structure) and (B) T_2 nucleated on faceted β' precipitates and also independently in the matrix, for 100 h aging of the Al–Li alloy AA 8090 at 190°C (Prasad, 1999).

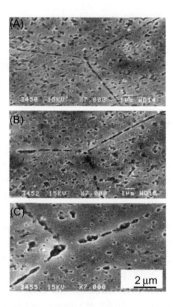

FIGURE 4.8 SEM micrographs of precipitate distributions and sizes after exposing the Al–Li alloy AA 8090 for 16 h at (A) 250°C, (B) 300°C, and (C) 350°C (Prasad, 1999).

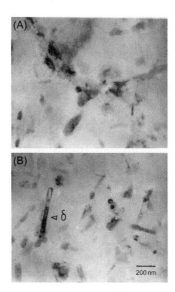

FIGURE 4.9   TEM microstructure after aging the Al–Li alloy AA 8090 at 250°C for 16 h, showing precipitates (A) at the grain boundaries and (B) within the grains (Prasad, 1999).

δ' Precursors

Several researchers have predicted the presence of one or two precursor Guinier–Preston (GP) zones and another precursor δ'' phase. However, GP zones have never been observed directly in the TEM or field ion microscope (FIM). TEM diffraction contrast images will never reveal GP zones, because the inevitable presence of strong contrast images from δ' will mask the small strain contrast resulting from GP zones. Therefore, it is necessary to use techniques such as high-resolution phase contrast imaging, which should readily distinguish GP zones and the δ' phase.

δ' Nucleation

In directly observing the initial stages of δ' precipitation by TEM of thin foils, Sato and Kamio (1990) stated that the interface contrast is "unclear" at the beginning and then becomes "clearer" as the particle size increases. However, if δ' particles overlap in the thin foil, the nature of the interface will be obscured. Overlaps are quite likely, owing to the small size and relatively large numbers of particles in the early stages of precipitation (Figure 4.4A).

In zirconium-containing Al–Li alloys, the Al_3Zr (β') particles act as heterogeneous nucleation sites for δ' precipitation. This was first observed in rapidly solidified Al–Li–Zr alloys containing 1 wt% Zr (Gayle and Vander Sande, 1984a,b) and subsequently in conventionally cast alloys (Gregson and

Flower, 1984; Makin and Ralph, 1984). Flower and Gregson (1987) stated that the reduction in both strain and surface energy terms associated with δ' nucleation are responsible for the effectiveness of the β' particles as nucleation sites. Precipitation of δ' on the β' particles results in coarsening, and the δ'-coated Al_3Zr behave as a population of larger δ' particles. This has been studied in detail by Gu et al. (1975).

Gayle and Vander Sande (1984a,b) reported that there is an intersolubility of β' and δ', basing this observation on electron diffraction and electron energy loss spectroscopy (EELS) results. However, Makin and Ralph (1984) showed that there is negligible chemical interaction between Li and Zr in either the precipitates or the matrix. Also, Stimson et al. (1986) concluded from their EELS analysis of semi-coherent β' particles in the matrix that there is no evidence for diffusion of Li into Al_3Zr. They also observed that Al_3Li can form continuous shells around coherent Al_3Zr precipitates, and the shells coarsen at the expense of matrix Al_3Li. Atomic resolution microscopy (ARM) images studied by Radmilovic and Thomas (1987) confirmed that δ' forms a complete shell around β' particles, but the shell is not spherical as was reported previously (Galbraith et al., 1987). ARM imaging revealed that the δ' shell is truncated by facets, leading to polygonal shapes. Also, δ' does not increase symmetrically in thickness around β' particles during further aging.

δ' Growth (Coarsening)

There have been numerous studies of δ' coarsening in binary and more complex alloys (Baumann and Williams, 1985; Broussaud and Thomas, 1986; Gu et al., 1985,1986; Kulwicki and Sanders, 1983; Mahalingam et al., 1987; Noble and Thompson, 1971; Sanders et al., 1980; Williams, 1974; Williams and Edington, 1975). δ' coarsening is governed by the classical Lifshitz, Slyozov, and Wagner (LSW) kinetics (Lifshitz and Slyozov, 1961; Wagner, 1961). For diffusion-controlled growth, the precipitate radius, $r \propto t$ (time)$^{1/3}$. However, Berezina et al. (1983a) found that although particle coarsening in an Al−2.8Li alloy at 250°C obeyed this theory, the coarsening resulted in a non-predicted asymmetrical particle size distribution, the shape of which was stable with time. Gu et al. (1986) showed that this asymmetry developed in single-step aging at temperatures between 200°C and 225°C and is a function of alloy content: they used the Weibull statistical distribution function to describe the observed precipitate size distributions. In addition, it has been observed that two-step aging treatments affect the size distribution and effective supersaturation and can thus result in significant deviations from LSW conditions (Berezina et al., 1983b).

δ' coarsening has also been modeled with the incorporation of PFZ formation at high-angle boundaries, where equilibrium δ phase is formed, and good agreement was achieved between the model and the experimental results.

Finally, one should note that a comparison of the coarsening rate constants for the two different types of δ' precipitates (independently nucleated and nucleated on β') showed that the rate constant for coarsening of the composite $\delta' + \beta'$ precipitates is significantly greater (Valentine and Sanders, 1989).

4.4.2 Al$_2$Cu (θ') Phase

The most widely acknowledged precipitation sequence of Al$_2$Cu precipitates, either equilibrium θ or metastable θ', has been given by Wang and Starink (2005):

$$\text{SSS} \quad \rightarrow \quad \text{GPI (Al}_9\text{Cu, Al}_7\text{Cu, Al}_5\text{Cu, Al}_3\text{Cu)} \quad \rightarrow \quad \text{GPII}/\theta''(\text{Al}_3\text{Cu})$$
$$\rightarrow \quad \theta' \ (\text{Al}_2\text{Cu}) \qquad\qquad\qquad\qquad\qquad\quad \rightarrow \quad \theta \ (\text{Al}_2\text{Cu})$$

The θ phase is in coherent with matrix aluminum and has $I4/mcm$ crystal structure with $a = 0.607$ nm and $c = 0.488$ nm. However, most strengthening in Al−Cu alloys, with or without Li, occurs due to metastable Al$_2$Cu (θ'), which has a tetragonal structure, $a = 0.82$ nm, $c = 1.16$ nm or $a = 0.57$ nm, $c = 0.58$ nm (Preston, 1938; Porter et al., 2009; Wassermann and Weerts, 1935).

Figure 4.10 shows the Silcock model (Silcock et al., 1953−1954) for the tetragonal θ' phase with $a = 0.404$ nm, $c = 0.58$ nm, and space group $I4m2$ (rather than the above-mentioned $I4/mcm$ crystal structure). The θ' precipitates are either rectangular or octagonal on {100} planes and have an orientation relationship $(100)_{Al}//(100)_\theta, [001]_{Al}//[001]_\theta$. Figure 4.10B shows the experimentally obtained θ' selected area diffraction (SAD) pattern.

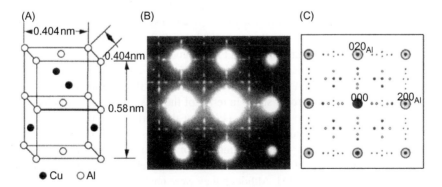

FIGURE 4.10   (A) Structural model of θ', (B) experimentally obtained [001] SAD pattern, and (C) simulated [001] Al diffraction patterns with reflections from θ' (Wang and Starink, 2005).

4.4.3 Al₂CuLi (T₁) Phase in Ternary Al–Li–Cu Alloys

The Al_2CuLi (T_1) phase was first identified as an important constituent in the Al–Li–Cu alloy system (Hardy and Silcock, 1955–1956; Silcock, 1959–1960). The phase boundaries after 16 h of aging at 165°C and the 350°C isothermal sections of the ternary phase diagram were determined by Hardy and Silcock (1955–1956). They also established the crystal structure of T_1 as hexagonal, with lattice parameters $a_0 = 0.4965$ nm and $c_0 = 0.9345$ nm; and also suggested the possible space groups, namely $P622$, $P6mm$, $P\bar{6}m2$, and $P6/mmm$. Hardy and Silcock (1955–1956) also determined the orientation relationship of T_1 with matrix aluminum as follows:

$$(0001)T_1//\{111\}Al \quad \text{and} \quad <10\bar{1}0>T_1//<\bar{1}10>Al$$

Noble and Thompson (1973) subsequently observed the formation of plate-shaped T_1 precipitates on $\{111\}$ in ternary Al–Li–Cu alloys with an orientation relationship $(0001)T_1//(111)Al$, $(10\bar{1}0)T_1//(110)Al$, and $(11\bar{2}0)$ $T_1//(2\bar{1}\bar{1})Al$, which agrees with the Hardy and Silcock relationship. They proposed that in alloys of low supersaturation the T_1 phase forms by dissociation of $1/2 <110>$ dislocations into $1/6 <211>$ Shockley partials bounding a region of intrinsic stacking fault. Copper and lithium enrichment of the fault produces a thin layer of the T_1 phase, and growth of this thin nucleus proceeds easily since it involves only further separation of the Shockley partials. Noble and Thompson (1973) also observed GP zones and θ' (Al_2Cu) precipitates after aging highly supersaturated alloys for various times but did not report the presence of a precursor to T_1 formation.

T₁ Precursors

Suzuki et al. (1982) were the first to suggest that precipitation of T_1 plates might be preceded by another metastable precipitate, T_1', in an alloy containing $\sim 4.2\%$ Cu and $\sim 0.9\%$ Li, but the evidence was insufficient to support this claim. More substantial evidence for metastable T_1' precipitate was presented by Rioja and Ludwiczak (1986). They studied selected area electron diffraction (SAED) patterns from Al–Li alloy AA 2090 thin foils prepared from samples subjected to different aging conditions. From some patterns they observed extra reflections at the positions of $1/3\{311\}$ and $2/3\{311\}$ in the matrix $<112>$ zone axis. They could not rationalize the existence of these reflections or the intensity maxima in reciprocal space in terms of the T_1 structure and concluded from these results and the dark-field TEM that they originated from a transitional T_1' phase. Rioja and Ludwiczak (1986) also postulated that the crystal structure of the T_1' phase was orthorhombic, of Pt_2Mo type, with $a = 0.2876$ nm, $b = 0.86$ nm, and $c = 0.406$ nm and an orientation relationship of $(010)T_1'//$ $(011)Al$ and $[001]T_1'//[100]Al$. Although they were unable to determine the precise chemistry of the T_1' phase and the positions of the Li atoms in it, they nevertheless determined interplanar spacings and relative intensities of the diffraction peaks, which were in fair agreement with the results of their own Guinier-de Wolff X-ray experiments. Huang and Ardell (1986, 1987) investigated the aging sequence at 160°C and 190°C in Al–Li–Cu alloys. Diffraction effects were observed similar to those reported by Rioja and Ludwiczak (1986) in the <112> zone axis. However, the dark-field TEM images taken using the variety of reflections in the <112> zone axis were inconsistent in suggesting the coexistence of T_1 and T_1' in the aged microstructure.

T_1 Crystal Structure

Huang and Ardell (1987) studied the crystal structure of T_1 in detail using electron diffraction and X-ray diffraction in conjunction with stereographic projections. They proposed a crystal structure for T_1 (Al_2CuLi) as shown in Figure 4.11. This crystal structure possesses the same hexagonal cell parameters determined by Hardy and Silcock (1955–1956) and belongs to the *P6/mmm* space group. Huang and Ardell stated that whatever the aging time, all the electron diffraction patterns (EDPs) can be explained using T_1 reflections and the streaking associated with them, and so there is no need to invoke the presence of T_1' at short aging times. Hence, they concluded that the orthorhombic T_1' phase does not exist.

Huang (1992) also studied the crystal structure of T_1 with the help of simulated diffraction patterns using a computer program "DIFFRACT." He compared the experimental diffraction patterns with the simulated ones and confirmed the hexagonal structure model, including the high symmetry requirement *P6/mmm*. Alternative models have been proposed by Cassada et al. (1987) and on the other hand, Howe et al. (1988) employed image simulation of high-resolution TEM (HRTEM) images to suggest a different T_1

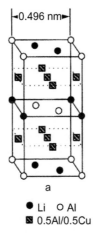

FIGURE 4.11   Huang and Ardell (1986, 1987) model for T_1 structure with space group *P6/mmm* and lattice parameters $a = 0.496$ and $c = 0.935$ nm.

structure. They proposed a four-layer cell consisting of stacks of A_1BA_2C layers, where all the layers are closely packed. The B and C layers are compositionally different (varying from 50%Cu/50%Al to 35%Cu/65%Al, and vice versa); A_1 is mostly aluminum and A_2 is mostly lithium. This four layer cell produces the best agreement with experimental EDPs and HRTEM images but has space group symmetry $P6/mm$ instead of the more commonly found $P6/mmm$. Convergent beam electron diffraction (CBED) studies of Vecchio and Williams (1988a) support the Huang–Ardell model (Cassada et al., 1987; Howe et al., 1988).

T_1 Crystal Structure and Nucleation

The precipitation of T_1 in quaternary Al–Li–Cu–Mg alloys is essentially similar to that occurring in Al–Li–Cu alloys but is influenced by competition with the S phase for both heterogeneous nucleation sites and available copper atoms in the matrix. It has also been reported that Mg can become incorporated into the T_1 precipitate and modify the usual plate-like morphology to form rounded laths (Meyer and Dubost, 1986). The addition of small amounts (0.5–1.0 wt%) of Mg to higher-Cu-containing alloys, such as 2090, suppresses the formation of θ' and introduces the S' (S) phase (Figure 4.6; Crookes and Starke, 1984). In higher-Li and lower-Cu alloys such as 8090, S', which precipitates with δ' and small amounts of T_1, becomes the dominant Cu-bearing phase (Gregson and Flower, 1985; Livet and Bloch, 1985). When the Cu:Mg ratio is progressively decreased below the value for 8090, the precipitation of T_1 is fully suppressed and S' becomes the primary Cu-bearing phase, as in the case of 2091 (Al–2Li–2.2Cu–1.5Mg–0.08Zr) (Hautefeuille et al., 1987; Sainfort and Dubost, 1987).

Three different groups have investigated the structure and nucleation mechanisms of T_1 using HRTEM (Cassada et al., 1988; Howe et al., 1988; Radmilovic and Thomas, 1988). Cassada et al. (1988) employed a modified version of the Huang and Ardell (1986) structure as the basis for interpreting high-resolution images. This arrangement maintains the sequence of compositional layers (ABCB...), but treats the stacking of T_1 layers as ABAB ..., that is the stacking of close-packed planes. Radmilovic and Thomas (1988) also assumed a close-packed modification of the Huang and Ardell (1986) structure in the interpretation of their high-resolution images. From their studies it was clear that in underaged alloys T_1 nucleates preferentially inside the ordered δ' phase. They stated that this must have been caused by plastic flow due to glide on {111} planes, thereby emphasizing that dislocation–solute atom or dislocation–particle interactions are needed for T_1 nucleation. This evidence-based conclusion provides support for the model proposed by Noble and Thompson (1973) to describe growth of T_1 from split unit dislocations.

In another study, Cassada et al. (1991) observed that the nucleation and growth of T_1 plates occur by the dissociation of matrix dislocations into partials ($1/6 < 112>$) which form the growth interfaces of the plates and the plate edges, as proposed by Noble and Thompson (1973). Cassada et al. (1991) modeled T_1 plate nucleation from the dissociation of a dislocation on opposite sides of a jog (or cross-slipped screw segment) one or two {111} planes high. Such dissociation can account for all of the observed T_1 plate characteristics in high-resolution images. The transformation produces a four-layer hexagonal plate by the coupling of partials on these layers, and the model can produce either perfect or imperfect stacking with displacement faults. Furthermore, Cassada et al. (1991) stated that the nucleation of growth ledges on existing plates can also be modeled by the same mechanism.

4.4.4 Al_2CuLi (T_1) Phase in Quaternary Al–Li–Cu–Mg Alloys

The precipitation of T_1 in quaternary Al–Li–Cu–Mg alloys is essentially similar to that occurring in Al–Li–Cu alloys but is influenced by competition with the S phase for both heterogeneous nucleation sites and available copper atoms in the matrix. It has also been reported that Mg can become incorporated into the T_1 precipitate and modify the usual plate-like morphology to form rounded laths (Meyer and Dubost, 1986). Several mechanisms have been proposed for the nucleation of the T_1 phase but no definite conclusions are drawn. The various possibilities are as follows (Tsivoulas, 2010):

- Heterogeneous nucleation on dislocations (e.g., during a T8 temper) which involves cold work before artificial aging. This nucleation mechanism is associated with enhanced aging kinetics, improved strength, and elongation. There is some controversy whether dislocation loops surrounding the Al_3Zr dispersoids promote the formation of T_1.
- Reduction in stacking fault energy due to Mg and Ag additions, leading to the formation of {111}$_{Al}$ stacking faults, which favor nucleation.
- Low-angle grain boundaries favor the formation of short and thick plates; longer plates are observed when the habit plane of the boundary is parallel to the boundary plane.
- High-angle grain boundaries favor the formation of short and thick plates of T_1 with associated PFZs.
- Vacancies or clusters of vacancies (octahedral voids or dislocations loops formed by vacancy condensation) provide nucleation sites for T_1. Enhanced diffusion of vacancies (e.g., direct aging after solution treatment) reduces nucleation rates within the grains as compared to slow heating to the aging temperature or use of two-stage aging).
- Co-clusters of Mg–Ag forming on {111}Al planes minimize the elastic energy and act as heterogeneous nucleation sites. Their composition is

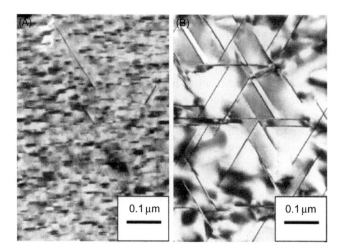

FIGURE 4.12 Electron micrographs of T_1 phase in the Al–5.3%Cu–1.3%Li–0.4%Mg–0.4% Ag–0.16%Zr alloy: (A) quenched and T6 aged at 160°C for 8 h, showing GP zones and a few plates of T_1 and (B) quenched and T8 cold worked 6% and aged at 160°C for 8 h, showing a uniform dispersion of T_1 plates with high aspect ratio (Polmear, 2004).

reported to be Mg₃Ag. These clusters probably favor Cu–Li co-clustering, which would increase the T_1 precipitation kinetics. GP zones, which are another type of clustering, also facilitate the nucleation of T_1.

● Dispersoids such as Al₃Zr act as nucleation.

Precipitation of the T_1 phase is facilitated by deformation of the quenched alloy before artificial aging. This is illustrated in Figure 4.12 by the differences in the distribution of T_1 after T6 and T8 aging treatments.

4.4.5 Al₂CuMg (S′) Phase and the Equilibrium S Phase

Al–Cu–Mg Alloys

The addition of small amounts (0.5–1.0 wt%) of Mg to higher-Cu-containing alloys, such as AA 2090, suppresses the formation of θ' and introduces the S′ phase (Crookes and Starke, 1984). In higher-Li and lower-Cu alloys such as AA 8090, the S′ phase (which precipitates with δ' and small amounts of T_1) becomes the dominant Cu-bearing phase (Gregson and Flower, 1985; Livet and Bloch, 1985). As the Cu:Mg ratio progressively decreases below the value for AA 8090, the precipitation of T_1 is fully suppressed and S′ becomes the primary Cu-bearing phase, as in the case of 2091(Al-2Li-2.2Cu-1.5Mg–0.08Zr) (Sainfort and Dubost, 1987; Hautefeuille et al., 1987) In Al–Cu alloys, the replacement of some copper by magnesium is advantageous to the alloy density and also offers the possibility of utilizing additional precipitation of the S′ and S (Al₂CuMg) phases in addition to θ' and θ phases. In the

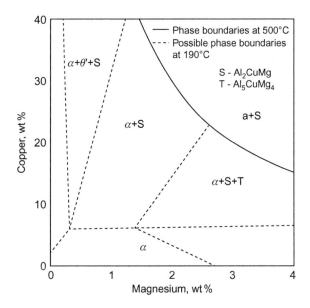

FIGURE 4.13 Phase equilibria in the Al−Cu−Mg system.

Al−Cu−Mg ternary alloy system, the precipitation reactions are complex but may be described in terms of the isothermal sections shown in Figure 4.13 (Weatherly, 1966). Precipitation depends on the Cu:Mg ratio, and three equilibrium precipitates are found: θ (Al$_2$Cu) in high Cu alloys; S (Al$_2$CuMg) when the Cu:Mg ratio is less than 3:1; and T$_1$ (Al$_2$CuLi) when the Cu:Mg ratio is higher than 3:1.

We note here that in commercial second-generation Al−Li alloys, the aging reactions involve principally the $\alpha + S$ field of the Al−Cu−Mg phase diagram. The precipitation sequence has been summarized by Silcock (1961):

$$\alpha_{ss} \rightarrow GP \text{ zones} \rightarrow S' \rightarrow S(Al_2CuMg)$$

Al−Li−Cu−Mg Alloys

The precipitation behavior and hardening of a wide range of quaternary Al−Li−Cu−Mg alloys have been investigated by many workers and also reviewed (Flower and Gregson, 1987; Sanders and Starke, 1989). Cu or Mg in solution does not affect the general characteristics of δ' precipitation (Baumann and Williams, 1983; Dinsdale et al., 1981; Thompson and Noble, 1973). There is evidence to suggest that magnesium decreases the solid solubility of lithium in aluminum and may be incorporated into δ', thereby increasing its volume fraction (Dinsdale et al., 1981).

The numerous matrix precipitate phases observed in the quaternary alloy systems are discussed in Section 3.5.2. In the rest of this section, we shall

TABLE 4.3 Different Variants of S′ Laths and Their Associated Crystallographic Relationships

$[100]S′//$	$[010]S′//$			
$[100]m$	$[021]mS_1$	$[012]mS_2$	$[0\bar{1}2]mS_3$	$[0\bar{2}1]mS_4$
$[010]m$	$[102]mS_5$	$[201]mS_6$	$[20\bar{1}]mS_7$	$[10\bar{2}]mS_8$
$[001]m$	$[210]mS_9$	$[120]mS_{10}$	$[\bar{1}20]mS_{11}$	$[\bar{2}10]mS_{12}$

concentrate on the precipitation reactions for the S′ and S (Al_2CuMg) phases which, besides T_1, are the most important in commercial Al—Li alloys.

S′ and S Crystal Structure

The S′ phase has an orthorhombic crystal structure with the lattice parameters, $a = 0.405$ nm, $b = 0.916$ nm, and $c = 0.720$ nm (Gupta et al., 1987). This phase grows as laths on $\{012\}$ matrix planes and along $<100>$ matrix directions, with the following orientation relationship (Mondolfo, 1976):

$$[100]S′ // <100> \text{matrix}; \quad [010]S′ // <012> \text{matrix}$$

This precipitate/matrix orientation relationship gives rise to 12 possible orientations of S′ laths in the aluminum matrix: four variants corresponding to each of the three $<100>$ matrix directions (Table 4.3; Khireddine et al., 1988). The S′ phase is very similar to the equilibrium S, which has the same morphology and orientation relationship, and only slightly different lattice parameters: $a = 0.40$ nm, $b = 0.923$ nm, and $c = 0.714$ nm. In fact, S′ can be considered as a distorted epitaxial form of S with a lath morphology (Gupta et al., 1987).

S′ and S Nucleation and Growth

The low-temperature decomposition sequence of Al—Cu—Mg alloys with atomic ratio Cu/Mg close to 1 is not clearly established, and contradictory interpretations may result from investigation of different compositions (Gupta et al., 1987). The elastic strains favor heterogeneous nucleation of S′ on edge dislocations. However, these precipitates are also partially coherent with the matrix and show a strong tendency for heterogeneous nucleation along matrix dislocations, low-angle grain boundaries and other structural inhomogeneities (Sen and West, 1969). These various nucleation sites are discussed in the following sections:

1. *GPBs*: The GPB nomenclature was used by Bagaryatsky (1952) to differentiate GPBs from the AlCu GP zones. The existence of two types,

GPB(I) and GPB(II), was proposed. GPB(I) diffraction displays no streak, but diffuse spots appear near {100} planes in reciprocal space. According to Silcock (1961), these zones are cylindrical, 1–2 nm in diameter and 4 nm long. They are also formed during the early stage of aging at 190°C and are followed by nucleation of S′ precipitates with a {120} habit plane.

Bagaryatsky (1948, 1952a) stated that GPB zones do not act as nuclei for the S′ phase in ternary Al−Cu−Mg alloys. In another study, Bagaryatsky (1952b) proposed that a further metastable phase S″ precedes S′ formation, and the existence of such a precursor has received support from Cuisiat et al. (1984). However, Weatherly (1966) found strong evidence for GPB zones transforming to S″ precipitates at 190°C in an Al−2.7Cu−1.5Mg−0.2Si alloy.

Yet another complication comes from Wilson (1969), who suggested that Si increases the effective binding energy between solute atoms, vacancies, and GPB zones, thereby enhancing the stability of the zones and raising the temperature at which S′ could form by transformation from GPB zones.

The situation with regard to GPBs acting as nucleation sites for S′ is far from clear, and in the absence of convincing arguments that S″ and even GPB(II) are different phases, it seems more reasonable to assimilate these "precursors" into the initial state of S′ precipitation. However, the GPB concept persists.

2. *Dislocations, boundaries, and vacancy clusters*: In ternary Al−Cu−Mg alloys, heterogeneous nucleation of S occurs on the high density of dislocation loops and helices which are present in the as-quenched condition (Figure 4.14). However, in lithium-containing alloys, the Li atom/vacancy binding is very strong (Cerasara et al., 1977; Ranian et al., 1970). The Li−vacancy binding energy, 0.25−0.26 eV, is high enough to allow Li atoms to trap vacancies during quenching and prevent the formation of loops and helices (Cerasara et al., 1977). This leaves only grain and subgrain boundaries available as heterogeneous nucleation sites (Flower and Gregson, 1987). Nucleation occurs at these sites and eventually homogeneous nucleation of S takes place, even in fairly (Mg + Cu) dilute alloys (Gregson et al., 1986). A widespread distribution of vacancy clusters may provide sites for the precipitation of S′ laths. Alternatively, the clusters may become enriched with Cu and Mg and then develop into S′ precipitates via a model of the classic GP zone type (Gregson et al., 1986).

Radmilovic et al. (1989) studied the nucleation and growth mechanisms of S′ by using high-resolution electron microscopy (HREM) for both non-Li (Al−2.01Cu−1.06Mg−0.14Zr) and Li-containing (Al-2.5Li-1.3Cu-1Mg−0.9Zr) alloys. For the non-Li alloy, the HREM images suggested two primary nucleation mechanisms. The first was clustering of

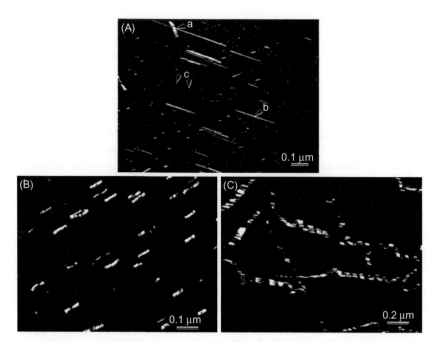

FIGURE 4.14 S′ precipitates in AA 8090 alloy aged at 190°C for 16 h: (A) different morphologies—sheets (a), laths (b), and finely dispersed rods (c); (B) S′ nucleated at dislocation loops; and (C) S′ nucleation on helical dislocations (Tang, 1995).

Cu- and Mg-rich regions between 2 nm and 4 nm in size, slightly larger than the GPB size discussed by Silcock (1961). Radmilovic et al. (1985) indicated that these clusters can develop into S′ according to the GP zone type model suggested by Flower and Gregson (1987). The second mechanism was S′ nucleation at subgrain or grain boundaries.

For the Li-containing alloy, the HREM images also indicated two types of nucleation mechanisms, but significantly different from those in the non-Li-containing alloy:

- *Nucleation on the δ/matrix interface*: During δ′ growth excess, Cu and Mg concentrations occur at the growth front. These excesses, combined with the excess vacancies released when Li adds to the δ′ precipitate, result in favorable condition for S′ nucleation.
- *Nucleation on matrix dislocations and on subgrain boundary dislocations*: It was concluded that in Li-containing alloys, the dominant nucleation sites were the δ′/Al-matrix interface boundary and dislocations.

In both alloys, the growth mechanism was controlled by ledge migration and ordered periodicities along $<102>$ matrix directions. One important conclusion made by Radmilovic et al. (1989) is that S′ and S are the same, i.e., there is actually only one S precipitate.

Heterogeneous and Homogeneous Precipitation of S

Gomiero et al. (1991, 1992) used small-angle X-ray scattering (SAXS) and TEM to investigate the precipitation behavior of the S phase in ternary alloys with and without lithium. They observed GPB zones in both alloys up to 48 h of aging at 150°C, S′ precipitation after only 6 h on dislocations and quenched-in defects, and heterogeneous S precipitation on grain boundaries. They also established the precipitation sequence in an Al–Li–Cu–Mg alloy aged at low temperatures as: $\alpha \rightarrow \alpha + GPB + \delta' \rightarrow \alpha + S' + \delta'$. An important feature of this sequence is that the δ' precipitation is not markedly influenced by the precipitation of S, since S has the composition Al_2CuMg and does not contain lithium. In addition, S precipitation does not cause δ'-free PFZs along either high- or low-angle grain boundaries, unlike the situation when T_1 (Al_2CuLi) precipitates.

Homogeneous precipitation of S in commercially feasible age hardening treatments cannot be achieved because there is a low density of quenched-in defects to act as nucleation sites, and this is due to the high lithium–vacancy binding energy (Cerasara et al., 1977). Thus, reliance must be made on heterogeneous precipitation. The density of heterogeneous nucleation sites is conventionally increased by giving an alloy a preaging stretch, which encourages more widespread heterogeneous precipitation of S (Ahmad and Ericsson, 1986; Gregson and Flower, 1985; Peel et al., 1983; Sainfort and Dubost, 1987; Welpmann et al., 1986). Furthermore, combining a preaging stretch with a two-step aging treatment (170°C/1.5 h + 190°C/24 h) results in a relatively uniform precipitation of fine S laths (Gregson and Flower, 1985).

4.4.6 Al₃Zr (β′) Phase

Zirconium additions to aluminum alloys are very effective in inhibiting recrystallization (Nes, 1972; Nes and Ryum, 1971; Ryum, 1966, 1969) and improving the toughness and resistance to stress corrosion cracking (Di Russo, 1964). The property improvements are due to the formation of spherical and coherent β' (Al_3Zr) precipitates and the way they interact with grain boundaries and affect δ' precipitation in Al–Li alloys. As mentioned in the introduction to this chapter, the β' precipitates are stable as a result of (i) low solid solubility of Zr in Al, (ii) small β'/matrix misfit strains, approximately 0.8% (Nes, 1972; Ryum, 1969), and (iii) sluggish diffusion of Zr in Al (Ryum, 1969). Consequently, β' precipitates are very effective in pinning grain and subgrain boundaries during thermomechanical processing of commercial aluminum alloys (Ryum, 1969). Also, for Al–Li alloys the β' precipitates provide heterogeneous nucleation sites for δ' (Gayle and Vander Sande, 1984a,b; Gregson and Flower, 1984), see Section 4.4.1. The reductions in both strain and surface energy associated with δ' nucleation on β' are considered to be responsible for the effectiveness of β' particles as nucleation sites.

β' Crystal Structure

β' has a metastable cubic (L1$_2$) structure with a lattice parameter $a_0 = 0.408$ nm. The precipitates are typically 20–30 nm in diameter and have a cube-on-cube orientation relationship with the matrix (Figure 4.15). β' is normally spherical and is reported to nucleate heterogeneously on dislocations and grain boundaries (Nes and Ryum, 1971). The ordered L1$_2$ Al$_3$Zr phase has been observed to be stable up to temperatures as high as 600°C (Lederich and Sastry, 1984;

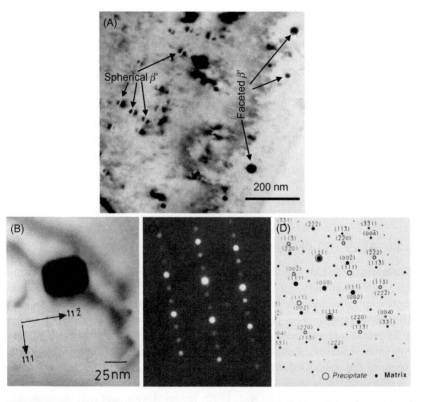

FIGURE 4.15 (A) TEM bright-field image showing spherical (Ashby–Brown contrast) and faceted Al$_3$Zr (β') precipitates; (B) faceted β'; (C) micro-diffraction pattern obtained from the β' precipitate shown in (B), representing the twin orientation relationship with the matrix α-Al; and (D) corresponding indexing of the diffraction pattern in (C).

Wadsworth and Pelton, 1984). The metastable $L1_2$ and equilibrium $D0_{23}$ structures are very similar. They have an identical first-nearest neighbor configuration, which accounts for a large fraction of the ordering energy (Flinn, 1960). The $L1_2$ ordered β'/matrix interfacial energy has been reported to be 66 mJ/m^2, which value is more than two times the interfacial energy value (<30 mJ/m^2) of the δ' precipitates (Baumann and Williams, 1984; Sainfort and Guyot, 1986). The precipitates having higher interfacial energy contribute more effectively by means of chemical strengthening to the overall strengthening. Collectively, interfacial energies of both δ' and β' are smaller than those of precipitates such as S, so that they remain coherent with α-Al. However, unlike higher δ' volume fractions, the β' volume fractions are far less in Al−Li alloys as Zr is a minor alloying addition, essentially meant for grain refining rather than for strengthening. These aspects are covered in detail in Sections 3.3 and 10.2.

Significance of β' Precipitation Behavior

1. *Recrystallization*: Gayle and Vander Sande (1989) studied β' precipitation in an Al−Li−Zr alloy by using TEM techniques. According to them, β' is a stable precipitate at temperatures below the solidus and consists of $Al_3(Zr,Li)$ with varying Zr/Li ratios, depending upon the precipitation mechanism. It was also stated that β' maintains an extremely stable fine distribution with little coarsening for extended periods at 450°C, although transformation to an equilibrium tetragonal phase appeared to be initiated during prolonged heat treatment.

   The distribution of β' precipitates is usually inhomogeneous in Al−Li alloys. This is attributed to retained as-cast Zr segregation because of the low diffusivity of Zr in Al (Hatch, 1983). Consequently, Al−Li alloys tend to have partially recrystallized regions. The β' precipitates produce a drag effect on grain boundaries, and TEM studies (Gonçalves, 1991) have revealed that β' impedes the movement of high-angle boundaries during recrystallization. It was also observed that the moving boundaries do not cut through the precipitates but completely envelop them.

2. *Quench sensitivity*: A major reason why Zr (rather than Cr and Mn) has become the preferred choice as the recrystallization-inhibiting element in Al−Li alloys is that Zr has the least effect of these three elements on the quench sensitivity of commercial Al alloys (Holl, 1969). The reason for this difference lies in the coherency of particle/matrix interfaces: the incoherent interfaces between Cr- and Mn-bearing dispersoids and the aluminum matrix act as heterogeneous nucleation sites for undesirable phases when the alloys are slowly cooled from the solution treatment temperature, thus leading to reduced age hardening of the alloys. These incoherent interfaces also act as sinks for vacancies during thermal treatments, leading to undesirable PFZs around the dispersoids in the fully heat-treated microstructure.

The fully coherent β'/matrix interfaces, on the other hand, act as heterogeneous nucleation sites for only the desirable metastable strengthening precipitates, which have coherent interfaces during aging (Galbraith et al., 1987; Gregson and Flower, 1984; Makin and Ralph, 1984). Also, these interfaces do not act as sinks for vacancies, although vacancies can be attracted toward the interfaces, owing to the positive ($+$) δ'/matrix volume misfit strain and can condense to form dislocation loops (Mukhopadhyay et al., 1990).

Additional Aspects of β' Precipitation Behavior

Several studies on other types of aluminum alloys have shown that there can be a change in the coherency state of β' precipitates during subgrain boundary migration or recrystallization (Nes, 1979; Kanno and Ou, 1991; Kikuchi et al., 1993; Prangnell et al., 1994) but without a change in crystal structure. This transition occurs when a migrating grain boundary bypasses β' particles without dissolving them, so that the particles are reoriented to a new relationship with the (new) grain. However, it is also possible for β' precipitates to dissolve at migrating grain boundaries and reprecipitate behind them in either (i) a coherent spherical form with the usual cube-on-cube orientation relationship (OR) or (ii) a faceted morphology with the same crystal structure (Figure 4.15B), i.e., ordered fcc (L1$_2$), in a twin relationship with the matrix (Figures 4.15C,D) but not with the cube-on-cube orientation relationship.

The change in OR has important implications for quench sensitivity and subsequent aging behavior. Reoriented β' does not act as nucleation sites for δ' (Prangnell et al., 1994), but it has been shown to do so for the T$_2$ phase in an Al—Li—Cu—Mg—Zr alloy (Prasad et al., 1994, 1999) and for stable MgZn$_2$ precipitates in an Al—Zn—Mg—Cu—Zr alloy (Kikuchi et al., 1993).

There is some recent interest in the study of Al—Li alloys with small additions of Zr, Sc, and Yb. In common with Zr, the elements Sc and Yb also form Al$_3$(Sc,Yb,Zr)-type dispersoids that offer sites for formation of δ' (Al$_3$Li). The composite phase in the Al—Li—Sc system consists of a core of Al$_3$(Sc,Li) surrounded by a shell of pure Al$_3$Li. In order to develop the optimum precipitate distribution, it is necessary to carry out a double-aging treatment of the quenched alloy: a 325°C anneal to precipitate out Al$_3$(Sc,Li) followed by an anneal at 170°C when Al$_3$Li forms as a shell surrounding the Al$_3$Sc core. The elemental distribution in the core and shell is confirmed by using the aberration-corrected transmission electron microscope (TEAM 0.5) in the TEM and STEM modes (Dahmen et al., 2009). Yb additions lead to the formation of a double shell-core structure in which Al$_3$Yb forms the core, which is surrounded by an inner shell of Al$_3$Sc and an outer one of Al$_3$Li. Two important features of these composite structures are a nearly-uniform size dispersion of nano particles and considerably improved strength (Krug et al., 2008; Monachon et al., 2011).

4.4.7 AlLi (δ) Phase

In binary Al–Li alloys, annealing above the δ' solvus and overaging below the δ solvus lead to the formation of equilibrium δ (AlLi) phase at the grain boundaries and within the grains (Figures 4.8 and 4.10; Nozato and Nakai, 1977; Silcock, 1959–1960). The optical microstructure of Al–3.02 wt% Li aged at 190°C for 613 h showed that most of the δ phase was confined to the grain boundary regions (Jones and Das, 1958–1959). For other alloys, Noble and Thompson (1971) observed plate-like δ within the grains; and Williams and Edington (1975) observed the δ phase in association with a high density of misfit dislocations and a δ'-free PFZ surrounding it.

δ Crystal Structure

Silcock (1959–1960) has shown by the X-ray technique that the δ phase has a B32 (NaTl) cubic structure with a lattice parameter of 0.637 nm. The orientation relationships between the δ phase and the matrix have been determined by electron diffraction (Noble and Thompson, 1971) as: (100) $\delta//(110)\alpha$ and (011) $\delta//(\bar{1}11)\alpha$, with {111} habit planes.

δ Formation

Different mechanisms for δ formation have been reported. Niskanen et al. (1982) suggested that the formation of δ phase at grain boundaries is a result of preferential coarsening of δ'. They added that the grain boundaries provide the energy required to overcome the interfacial energy barrier for the $\delta' \to \delta$ transformation. Williams (1981) rejected this suggestion and stated that spherical δ precipitates of size 0.3 μm were surrounded by misfit dislocations at the grain boundaries (Williams, 1974). However, δ' has been observed to grow to a size larger than 0.3 μm, maintaining its spherical shape and without interface dislocations. This observation contradicts Williams (1981) and also the idea that misfit dislocations are required for δ' to transform to δ (Sanders and Starke, 1983).

Yet another suggestion was made by Williams (1981), who proposed a mechanism in which δ nucleates independently of δ' on large heterogeneous nucleation sites such as grain boundaries or matrix dislocations. This proposal was supported by an investigation which showed that at lower aging temperatures δ does indeed nucleate on grain boundaries and matrix dislocations, although at higher aging temperatures it nucleates at more highly strained δ'/matrix interfaces (Venables et al., 1983). To explain this latter observation it was proposed that δ must form as a coherent precipitate and transform to an incoherent precipitate later. However, in view of the enormous difference in the lattice parameters of δ and δ' (about 50%), the difference in crystal structure, and the very low value of the interfacial energy (γ_{ls})

for the δ'/matrix interface, even at high temperatures, this mechanism must be considered doubtful (Kaufman et al., 1992).

The above-mentioned investigations of δ formation were all on experimental alloys. Prasad et al. (1994) have, however, studied the precipitation of δ in Al–Li AA 8090 alloy, heat treated in the temperature range of 170–375°C. Their TEM studies showed that δ nucleated both at the grain boundaries and within the grains, at T_2/matrix interfaces (Figure 4.16). This latter observation illustrates the importance of studying commercial alloys as well as experimental ones.

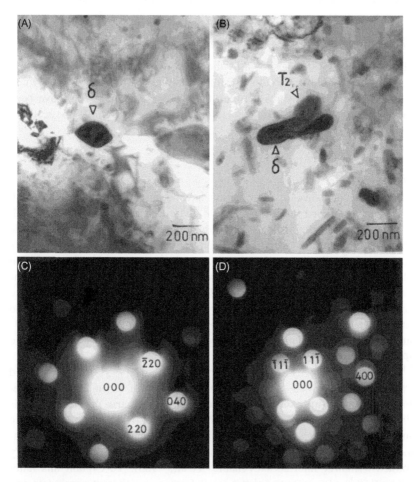

FIGURE 4.16 TEM micrographs showing (A) δ at the grain boundary, (B) δ on T_2 within the grain, and (C) and (D) micro-diffraction patterns from δ phase corresponding to $<100>$ and $<110>$ zone axes, respectively in a commercial Al-Li alloy AA 8090 (Prasad et al., 1994).

4.4.8 Al_6CuLi_3 (T_2) Phase

Besides the metastable phases θ', T_1, and S', the addition of Cu to Al–Li based alloys introduces another Cu-containing phase, the equilibrium T_2 phase, with composition (Al_6CuLi_3) (Silcock, 1959–1960). This phase has been shown to display icosahedral symmetry (Ball and Lloyd, 1985; Ball and Lagace, 1986; Cassada et al., 1986a,b; Crookes and Starke, 1984; Loiseau and Lapasset, 1986) and it nucleates predominately on high-angle boundaries (Figure 4.7).

T_2 Crystal Structure

It is now known that the T_2 phase shows fivefold diffraction symmetry even after extended heat treatment at temperatures close to its melting point. Also, T_2 can be grown to rather large ($\sim$2 mm) crystals in the stoichiometric or near-stoichiometric compositions (Bartges et al., 1987; Gayle, 1987; Sainfort et al., 1985).

Hardy and Silcock (1955–1956) first studied the T_2 structure by using powder X-ray diffraction and were unable to determine it. However, they did suggest that the observed very weak reflections did not correspond to a cubic structure. After Hardy and Silcock (1955–1956), a number of other investigators (Ball and Lloyd, 1985; Cassada et al., 1986a,b; Gayle, 1987) examined the T_2 phase using a variety of analytical techniques. Ball and Lloyd (1985) reported that HREM images of T_2 precipitates displayed Moiré fringes, which indicated that micro-twinning was responsible for the fivefold diffraction symmetry. Cassada et al. (1986a,b) interpreted their CBED patterns of T_2 in terms of a fivefold symmetry. Gayle (1987) reported that T_2 crystals which grew with a faceted external morphology possessed shapes consistent with both $m35$ and 235 fivefold point group symmetries.

Vecchio and Williams (1988a,b) studied T_2 precipitates in dilute Al–Li–Cu alloys and showed that diffraction maxima in the apparent fivefold SAD patterns were composed of multiple spots rather than discrete single spots; and that by preparing very thin specimens the T_2 precipitates appeared to be composed of cubic microcrystals less than $\sim$60 nm in size. CBED patterns taken along the apparent fivefold axis of the precipitates revealed that no rotational symmetry was present in the patterns. Also the T_2 crystals displayed an apparent fivefold symmetry in the alloys of stoichiometric composition. However, asymmetries observed in both the SAD and CBED patterns indicated that the symmetry of the structure is not truly fivefold. Vecchio and Williams (1988a,b) also conducted diffraction analysis on very thin crystals and stated that the T_2 phase was composed of microcrystals or microtwins which produce the *apparent* fivefold diffraction observed. Furthermore, the individual microcrystals produced diffraction patterns consistent with a twinned cubic structure. Nevertheless, whether the fivefold symmetry observed is the result of diffraction from a quasicrystal or the result of multiple diffraction from microtwins is still under debate.

Tosten et al. (1989) and Howell et al. (1989) used TEM of thin foils to study the origin and nature of microcrystalline regions in an Al–Li–Cu alloy. They found that the microcrystalline regions were developed as a result of selective leaching of Li and Cu from the T_2 particles, either during foil preparation or when T_2 precipitates were in contact with one or both of the foil surfaces. The transformation products appeared to be either the aluminum matrix solid solution or $\alpha + \delta'$ (Al_3Li). It was also concluded that the existence of these microcrystalline regions could not be used as evidence for a non-icosahedral T_2 phase. During TEM *in situ* heating, Howell et al. (1989) found that the T_2 precipitates transformed to duplex microcrystalline regions which consisted of a matrix solid solution and the T_B ($Al_{7.5}Cu_4Li$) phase. They stated that this transformation was most probably nonequilibrium in nature since it involved selective leaching of Li.

A similar kind of T_2 phase (Figure 4.17) that showed a fivefold symmetry was also observed in Al–Li–Cu–Mg alloys (Kulkarni et al., 1989; Prasad, 1999; Sainfort et al., 1985). Mukhopadhyay et al. (1992) studied the effect of varying Cu:Mg ratios on the formation of T_2 phase in AA 8090 alloy and found that decreasing the Cu:Mg ratio had the effect of replacing the T_2 phase by the tetragonal C phase: such changes were observed in as-cast materials as well as heat-treated materials.

Lapasset and Loiseau (1987) established the orientation relationships between the T_2 phase and matrix in an AA 8090 alloy. They found four different orientation relationships and represented them in a stereographic projection. These relationships are described as follows:

OR1: the twofold axes are parallel to the $[11\bar{1}]$ and $[\bar{2}1\bar{1}]$ Al directions.
OR2: the twofold and fivefold axes roughly coincide with the [100], $[01\bar{1}]$, $[11\bar{1}]$, and $[\bar{1}1\bar{1}]$ Al directions.
OR3: one of the fivefold axes is parallel to the $[\bar{1}1\bar{1}]$ Al direction and a threefold axis is close to the $[01\bar{1}]$ Al direction.
OR4: as in OR1, a twofold axis coincides with the $[11\bar{1}]$ Al direction and a fivefold axis is nearly parallel to the $[01\bar{1}]$ Al direction.

In this detailed study, the growth direction was found to be parallel to the $<110>$ Al directions, and the precipitate morphology was "pencil-like," i.e., rod-shaped. Similar orientation relationships (one of which was different) were observed by Kim and Cantor (1992) for a rapidly solidified Al–Cu–Li–Mg alloy in both the as-spun and heat-treated conditions. These orientation relationships were described as follows:

OR-I: [twofold]//[011], $[1\bar{1}1]$ and $[21\bar{1}]$ of Al.
OR-II: [twofold]//[011], $[11\bar{1}]$ and $[2\bar{1}1]$ of Al.

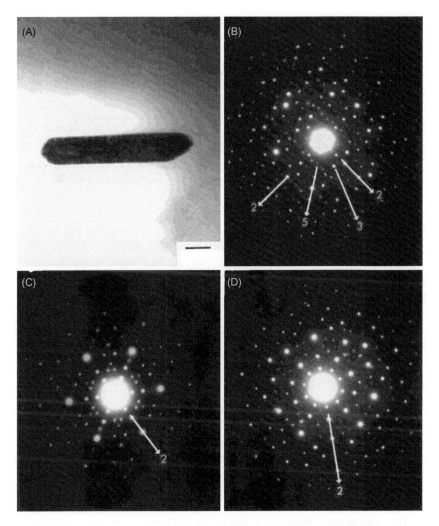

FIGURE 4.17 (A) TEM bright-field image showing a T_2 precipitate (at $350°C$) within a grain, and (B), (C), and (D) selected area EDPs of T_2 phase along two-, three-, and fivefold symmetry axes, respectively (Prasad, 1999).

OR-III: [twofold]//[011], [0$\overline{1}$1] and [100] of Al.
OR-IV: [twofold]//[010], [100] and [001] of Al.

To conclude this section, we present schematics of typical microstructural features in second- and third-generation alloys in Figure 4.18. It should be noted that that the T_1 phase is the most important phase for improving the strength. T_1 formation is facilitated by deformation of the quenched alloy before aging (T8 treatment) and by the addition of small amounts of Mg and Ag. The T_1 phase is commonly found in third-generation Al−Li alloys, e.g.,

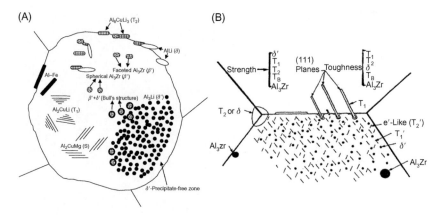

FIGURE 4.18 Schematics of typical microstructural features in (A) second- and (B) third-generation Al−Li alloys (Prasad unpublished work; Liu and Rioja, 2012).

AA 2050 and AA 2198. S phase is encountered only in alloys with relatively low Cu content and reasonable additions ($\sim 0.8\%$) of Mg, and there is ample evidence for heterogeneous nucleation of this phase. The δ' and θ' phases are also found in these alloys: the relative amounts of these phases depend on the alloy composition.

4.5 SUMMARY

Extensive studies have been made of the precipitation of metastable and equilibrium phases in the second- and third-generation Al−Li alloys. The studies include various stages of precipitation, the structures of the phases, and nucleation mechanisms. Besides precipitates, there are dispersoids such as Al_3Zr and $Al_{20}Cu_2Mn_3$ which minimize recrystallization, and also constituent intermetallic particles, which are mainly compounds of Al with Fe, Mn, Si, and Cu and sometimes Mg.

The precipitates have very important influences on the mechanical properties. There are several types of hardening (strengthening) precipitates, whose formation depends on the Li and Cu contents and the presence of Mg and Ag:

- Nonshearable θ' (Al_2Cu) phase is found in alloys with low Li content ($<0.6\%$).
- T_1 (Al_2CuLi) is the most prominent phase in alloys with medium Li content ($<1.4-1.5\%$). T_1 has a significant effect on strength. Formation of this phase is favored by the presence of Cu and small amounts of Mg and Ag. Also, cold working the quenched alloy before artificial aging (i.e., a T8-type temper) promotes copious formation of this phase.

- δ' (Al_3Li) phase is formed in high Li-containing alloys ($>1.4-1.5\%$). Its contribution to strengthening is less than that of T_1, and it can also be detrimental to toughness.
- Besides offering resistance to recrystallization, the metastable β' (Al_3Zr) phase offers sites for nucleation of the δ' phase. This facilitates homogenization of slip and hence better mechanical properties.
- Recent studies on the $L1_2$ phases containing Sc and Yb in addition to Li indicate that a uniform distribution of nanoparticles of these phases can be obtained by a double-aging treatment, typically aging at 325°C followed by aging at 170°C. The particles have a double shell-core structure, with core of Al_3Yb, surrounded by a shell enriched with Sc and another with Li. This structure is characterized by high strength and marked resistance to coarsening.

The dispersoids form during homogenization treatments of the cast ingots. A two-step homogenization treatment, starting at a lower temperature and ending at a higher temperature, leads to a higher concentration of dispersoids and hence to more resistance to recrystallization. Recent studies indicate that the combined presence of Zr ($\sim 0.1\%$) and Mn (0.3%) may reduce the beneficial effect of Zr alone in controlling recrystallization.

The constituent particles are undesirable because they sequester Cu and Mg, reducing the solute available for formation of strengthening precipitates. They also reduce the fracture toughness.

ACKNOWLEDGMENTS

The authors thank Professor TR Ramachandran and Dr. RJH Wanhill profoundly for their expert comments on the book contents and also their meticulous technical and editorial corrections. They also thank the Defence Research and Development Organisation (DRDO) of the Government of India for sponsoring a number of studies that form the basis of this chapter.

REFERENCES

Ahmad, M., Ericsson, T., 1986. Coarsening of δ', T_1, S' phases and mechanical properties of two Al–Li–Cu–Mg alloys. In: Baker, C., Gregson, P.J., Harris, S.J., Peel, C.J. (Eds.), "Aluminium–Lithium Alloys III," Proceedings 3rd International Conference on Aluminum–Lithium Alloys. The Institute of Metals, London, UK, pp. 509–515.

Bagaryatsky, Y.A., 1948. Rentgenograficheskoe issledovanie stareniya alyuminievykh splavov 1. Primenenie monokhromaticheskikh rentgenovykh luchei dlya izucheniya struktury ostarennykh splavov. Zhur. Tech. Fiziki 18, 827–830.

Bagaryatsky, Y.A., 1952. Precipitation behaviours in Al–Cu–Mg and 2024 aluminium alloys. Dokl. Akad. 87, 397.

Bagaryatsky, Y.A., 1952a. Deformation behaviour of the AlMg4.5Cu0.5 type alloy sheet. Dokl. Akad. 87, 559–562.

Bagaryatsky, Y.A., 1952b. Dokl. Akad. 87, 557−560.

Ball, M.D., Lagace, H., 1986. Characterisation of coarse precipitates in an overaged Al−Li−Cu−Mg alloy. In: Baker, C., Gregson, P.J., Harris, S.J., Peel, C.J. (Eds.), Proceedings 3rd International Conference "Aluminium−Lithium Alloys III". The Institute of Metals, London, UK, pp. 555−564.

Ball, M.D., Lloyd, D.J., 1985. Particles apparently exhibiting five-fold symmetry in Al−Li−Cu−Mg alloys. Scr. Metall. 19, 1065−1068.

Bartges, C., Tosten, M.H., Howell, P.R., Ryba, E.R., 1987. A combined single crystal X-ray diffraction and electron diffraction study of the T_2phase in Al−Li−Cu alloys. J. Mater. Sci. 22, 1663−1669.

Baumann, S.F., Williams, D.B., 1983. The effect of ternary additions on the δ'/δ misfit and the δ'solvus line Al−Li alloys. In: Sanders Jr., T.H., Starke Jr., E.A. (Eds.), Proceedings 2nd International Conference "Aluminum−Lithium alloys II". The Metallurgical Society of AIME, Warrendale, PA, pp. 17−29.

Baumann, S.F., Williams, D.B., 1984. Determining the effect of microstructure and heat treatment on the mechanical strengthening behaviour of an aluminium alloy containing lithium precipitation hardening with the δ' Al_3Li intermetallic phase. Scr. Metall. 18, 611−616.

Baumann, S.F., Williams, D.B., 1985. Experimental observations of the nucleation and growth of δ' (Al_3Li) in dilute Al−Li alloys. Metall. Trans. A 16A, 1203−1211.

Berezina, A.L., Trofimova, L.N., Chivustov, K.V., 1983a. Investigation of decomposition kinetics of alloy Al-2.8wt% Li after two stage ageing. Phys. Met. Maetallogr. 55, 111.

Berezina, A.L., Trofimova, L.N., Chivustov, K.V., 1983b. Phys. Met. Maetallogr. 55, 103−107.

Broussaud, F., Thomas, M., 1986. Influence of δ' phase coalescence on Young's modulus in an Al-2.5 wt%Li alloy. In: Baker, C., Gregson, P.J., Harris, S.J., Peel, C.J. (Eds.), Proceedings 3rd International Conference "Aluminium−Lithium Alloys III". The Institute of Metals, London, UK, pp. 442−447.

Cassada, W.A., Shiflet, G.J., Starke Jr., E.A., 1986a. Grain boundary precipitates with five-fold diffraction symmetry in an Al−Li−Cu alloy. Scr. Metall. 20, 751−756.

Cassada, W.A., Shiflet, G.J., Starke Jr., E.A., 1986b. A high resolution electron microscopic study of the microstructure in rapidly solidified Al−Li−Cu−Mg−Zr alloys. Phys. Rev. Lett. 56, 2276−2279.

Cassada, W.A., Shiflet, G.J., Starke Jr., E.A., 1988. Mechanism of Al_2CuLi (T_1) nucleation and growth. J. Phys. 48 (Suppl. Ce (9)), 397−406.

Cassada, W.A., Shiflet, G.J., Starke Jr., E.A., 1991. Mechanism of Al_2CuLi (T_1) nucleation and growth. Metall. Trans. A 22A, 287−296.

Cerasara, S., Girada, A., Sanchez, A., 1977. Annealing of vacancies and ageing in Al−Li alloys. Philos. Mag. 35, 97.

Crookes, R.E., Starke Jr., E.A., 1984. The microstructure and tensile properties of a splat quenched Al−Cu−Li−Mg−Zr alloy. Metall. Trans. A 15A, 1367.

Cuisiat, F., Duval, P., Graf, R., 1984. On the crystal structure of S' phase in Al−Cu−Mg alloy. Scr. Metall. 18, 1051−1056.

Di Russo, E., 1964. Microstructures, tensile properties, fatigue crack growth behaviour of the zirconium modified 2024 alloys processed by liquid dynamic compaction. Alluminio e Nuova Metall. 33, 505.

Dinsdale, K., Harris, S.J., Noble, B., 1981. Relationship between microstructure and mechanical properties of aluminium−lithium−magnesium alloys. In: Sanders, T.H., Starke, E.A. (Eds.),

Proceedings 1st International Conference "Aluminium–Lithium Alloys". The Metallurgical Society of AIME, Warrendale, PA, pp. 101–118.

Eswara Prasad, N., 1993. In-plane anisotropy in the fatigue and fracture properties of quaternary Al-Li-Cu-Mg alloys. Indian Institute of Technology (BHU) (formerly Institute of Technology, Banaras Hindu University), Varanasi, India. Doctoral Thesis.

Flinn, P.A., 1960. High temperature creep in a semi-coherent NiAl-Ni2AlTi alloy. Trans. AIME 218, 145–154.

Flower, H.M., Gregson, P.J., 1987. Solid state phase transformations in aluminium alloys containing lithium. Mater. Sci. Technol. 3, 81–90.

Galbraith, J.M., Tosten, M.H., Howell, P.R., 1987. On the nucleation of δ' and T_1 on Al_3Zr precipitates in Al–Li–Cu–Zr alloys. J. Mater. Sci. 22, 27–36.

Gayle, F.M., 1987. Study of transformation from T2 to R phase in Al–Cu–Mg alloy. J. Mater. Res. 2 (1), 1–4.

Gayle, F.W., Vander Sande, J.B., 1984a. Composite precipitates in an Al–Li–Zr alloy. Scr. Metall. 18, 473–478.

Gayle, F.W., Vander Sande, J.B., 1984b. Fatigue crack growth and fatigue toughness behavior of an Al–Li–Cu alloy. In: Proceedings of ASTM Conference on "Rapidly Solidified Powder Aluminium Alloy".

Gayle, F.W., Vander Sande, J.B., 1989. Phase transformations in the Al–Li–Zr system. Acta Metall. 37, 1033–1046.

Gomiero, P., Livet, F., Brechet, Y., Louchet, F., 1992. Microstructure and mechanical properties of a 2091 Al–Li alloy–I. Microstructure investigated by SAXS and TEM. Acta Metall. 40, 847–855.

Gomiero, P., Livet, F., Lyon, O., Simon, J.P., 1991. Double structural hardening in an Al–Li–Cu–Mg alloy studied by anomalous small angle X-ray scattering. Acta Metall. 39, 3007–3014.

Gonçalves, M., 1991. Grain size control in Ti-48Al-2Cr-2Nb with yttrium additions. Scr. Metall. 24, 835–840.

Gregson, P.J., Flower, H.M., 1984. δ' precipitation in Al–Li–Mg–Cu–Zr alloys. J. Mater. Sci. Lett. 3, 829–834.

Gregson, P.J., Flower, H.M., 1985. Microstructural control of toughness in aluminium–lithium alloys. Acta Metall. 33, 527–537.

Gregson, P.J., Flower, H.M., Tite, C.N.J., Mukhopadhyay, A.K., 1986. Role of vacancies in precipitation of δ'- and S-phases in Al–Li–Cu–Mg alloys. Mater. Sci. Technol. 2, 349–353.

Gu, B.L., Liedl, G.L., Kululichi, J.H., Sanders Jr., T.H., 1985. Coarsening of δ' (Al_3Li) precipitates in an Al-2.8Li-0.3Mn alloy. Mater. Sci. Eng. 70, 217–228.

Gu, B.P., Liedl, G.L., Sanders Jr., T.H., Welpmann, K., 1975. The influence of zirconium on the coarsening of δ' (Al_3Li) in an Al-2.8wt%Li-0.14wt%Zr alloy. Mater. Sci. Eng. 76, 147–157.

Gu, B.P., Mahalingam, K., Liedl, G.L., Sanders Jr., T.H., 1986. The δ'(Al_3Li) particle size distributions in a variety of Al–Li alloys. In: Baker, C., Gregson, P.J., Harris, S.J., Peel, C.J. (Eds.), Proceeding 3rd International Conference "Aluminum–Lithium Alloys". The Institute of Metals, London, UK, pp. 360–368.

Gupta, A.K., Gaunt, P., Chaturvedi, M.C., 1987. Quantitative analysis of interfacial chemistry in TiC/Ti using electron energy loss spectroscopy. Philos. Mag. A 55 (3), 375–387.

Hardy, H.K., Silcock, J.M., 1955–1956. The phase sections at 500°C and 350°C of aluminium-rich aluminium–copper–lithium alloys. J. Inst. Metall. 84, 423–428.

Hatch, J.E., 1983. Scandium in aluminium alloys overview: physical metallurgy properties and applications. Phys. Met. Metallogr., 144.

Hautefeuille, L., Rahouadj, R., Barbaux, Y., Clavel, M., 1987. Toughness and heat treatment relationship in a 2091 aluminum alloy. In: Champier, G., Dubost, B., Miannay, D., Sabtay, L. (Eds.), Proceedings of 4th International Conference "Aluminium–Lithium Alloys". J De Physique, vol. 48, pp. C3 669–C3 676.

Holl, H.A., 1969. Effect of manganese on the precipitation in an Al–Zn–Mg alloy. J. Inst. Metals 97, 200–205.

Howe, J.M., Lee, J., Vasudevan, A.K., 1988. Structure and deformation behavior of T_1precipitate plates in an Al-2Li-1Cu alloy. Metall. Trans. A 19A, 2911–2920.

Howell, P.R., Michel, D.J., Ryba, E., 1989. The nature of micro-crystalline regions produced by an in situ transformation of T_2particles in a ternary Al-2.5%Li-2.5%Cu alloy. Scr. Metall. 23, 825–828.

Huang, J.C., 1992. On the crystal structure of the T_1phase in Al–Li–Cu alloys. Scr. Metall. 27, 755–760.

Huang, J.C., Ardell, A.J., 1986. Microstructure evolution in two Al–Li–Cu alloys. In: Baker, C., Gregson, P.J., Harris, S.J., Peel, C.J. (Eds.), Proceedings 3rd International Conference "Aluminum–Lithium Alloys". The Institute of Metals, London, UK, pp. 455–470.

Huang, J.C., Ardell, A.J., 1987. On the crystal structure and stability of the T1 precipitates in aged Al–Li–Cu alloys. Mater. Sci. Technol. 3, 176–188.

Jones, W.R.D., Das, P.P., 1958–1959. The mechanical properties of aluminium–lithium alloys. J. Inst. Metals 87, 133–160.

Kanno, M., Ou, B.L., 1991. New composite material for lithium ion batteries. Metall. Mater. Trans. A 32, 450–455.

Kaufman, M.J., Morrone, A.A., Lewis, R.E., 1992. Complications concerning TEM analysis of the δ-AlLi phase in aluminum–lithium alloys. Scr. Metall. Mater. 27, 1265–1270.

Khireddine, D., Rahouadj, R., Clavel, M., 1988. Evidence of S′ phase shearing in an aluminium–lithium alloy. Scr. Metall. 22, 167–172.

Kikuchi, S., Yamazaki, H., Otsuka, T., 1993. Grain structure and quench rate effects on strength and toughness of AA 7050 Al–Zn–Mg–Cu–Zr alloy plate. J. Mater. Process. Technol. 38, 689–701.

Kim, D.H., Cantor, B., 1992. Microstructure of rapidly solidified eutectic and micro eutectic Al–Cu alloys. Int. J. Rapid Solidification 7, 67–81.

Krug, M.E., Dunand, D.C., Seidman, D.N., 2008. Composition profiles within Al_3Li and Al_3Sc/Al_3Li nanoscale precipitates in aluminum. Appl. Phys. Lett. 124, 107.

Kulkarni, G.J., Banerjee, D., Ramachandran, T.R., 1989. Physical metallurgy of aluminium–lithium alloys. Bull. Mater. Sci. 12, 325–340.

Kulwicki, J.H., Sanders Jr., T.H., 1983. Coarsening of δ'(Al3Li) precipitates in a Al-2.7Li-0.3 Mn alloy. In: Sanders Jr., T.H., Starke Jr., E.A. (Eds.), Proceeding 2nd International Conference "Aluminum–Lithium alloys II". The Metallurgical Society of AIME, Warrendale, PA, pp. 31–51.

Lapasset, G., Loiseau, A., 1987. A TEM study of icosahedral and near icosahedral phases in 8090 alloy. In: Champier, G., Dubost, B., Miannay, D., Sabatay, L. (Eds.), Proceedings 4th International Aluminium–Lithium Conference. J. De Physique, vol. 48. pp. C3.489–C3.495.

Lederich, R.J., Sastry, S.M.L., 1984. Superplastic deformation of P/M and I/M Al–Li based alloys. In: Sanders Jr., T.H., Starke Jr., E.A. (Eds.), Proceedings 2nd International Conference 'Aluminum–Lithium Alloys II. The Metallurgical Society of AIME, Warrendale, PA, pp. 137–152.

Lifshitz, I.M., Slyozov, V.V., 1961. Kinetics of precipitation from supersaturated solid solutions. J. Phys. Chem. Solids 19, 35–50.

Livet, A., Bloch, D., 1985. A kinetic analysis of Al—Al$_3$Li unmixing. Scr. Metall. 19, 1147—1151.

Loiseau, A., Lapasset, G., 1986. Organization of defects in first cubic approximant of the quasi crystal Al$_6$Li$_3$Cu. J. Phys. 47, C3.331.

Mahalingam, K., Gu, B.P., Liedl, G.L., Sanders Jr., T.H., 1987. Coarsening of δ′ precipitates in binary Al—Li alloys. Acta Metall. 35, 483—498.

Makin, P.L., Ralph, B., 1984. On the ageing of an aluminium—lithium—zirconium alloy. J. Mater. Sci. 19, 3835—3843.

Meyer, P., Dubost, B., 1986. Production of aluminium—lithium alloy with high specific properties. In: Baker, C., Gregson, P.J., Harris, S.J., Peel, C.J. (Eds.), Proceedings 3rd International Conference "Aluminum—Lithium Alloys". The Institute of Metals, London, UK, pp. 37—46.

Monachon, C., Krug, M.E., Seidman, D.N., Dunand, D.C., 2011. Chemistry and structure of core/double-shell nanoscale precipitates in Al-6.5Li-0.07Sc-0.02Yb (at.%). Acta Mater. 59, 3396—3409.

Mondolfo, L.F., 1976. Microstructures in overaged Al—Li—Cu—Mg—Ag alloys. Aluminium Alloys: Structure and Properties. Butterworth & Co. Ltd., London, UK, pp. 497—504.

Mukhopadhyay, A.K., Shiflet, G.J., Starke Jr., E.A., 1990. Influence of zirconium and copper on the early stages of aging in Al—Zn—Mg alloys. Scr. Metall. 24, 307.

Mukhopadhyay, A.K., Zhou, D.S., Yang, Q.B., 1992. Effect of variation in the Cu: Mg ratios on the formation of T$_2$ and C phases in AA 8090 alloys. Scr. Metall. Mater. 26, 237—242.

Nes, E., 1972. Study of the high-temperature behaviour of aluminium alloy EN AW 2014. Acta Metall. 20, 499—506.

Nes, E., 1979. Superplastic deformation of Al—Cu—Mg alloy sheet. Mater. Sci. 13, 211—215.

Nes, E., Ryum, N., 1971. The ageing characteristics of an Al2 Pct Li-3 Pct Cu-0.12 Pct Zr alloy at 190°C. Scr. Metall. 5, 987—990.

Niskanen, P., Sanders Jr., T.H., Rinker, J.G., Marek, M., 1982. Corrosion of aluminium alloys containing lithium. Corros. Sci. 22, 283—304.

Noble, B., Thompson, G.E., 1971. Precipitation characteristics of aluminium—lithium alloys. Metals Sci. J. 5, 114—120.

Noble, B., Thompson, G.E., 1973. Precipitation characteristics of aluminum—lithium alloys containing magnesium. J. Inst. Metals 101, 111—115.

Nozato, R., Nakai, G.R., 1977. Corrosion of aluminium alloys containing lithium. Japan Inst. Metals 18, 357.

Peel, C.J., Evans, B., Baker, C.A., Bannet, D.A., Gregson, P.J., Flower, H.M., 1983. The development and application of improved aluminium—lithium alloys. In: Proceedings 2nd international Conference "Aluminum—Lithium Alloys", pp. 363—392.

Polmear, I.J., 2004. Aluminium Alloys-A Century of Age Hardening Materials, vol. 28. Institute of Materials Engineering Australasia Ltd., p. 1.

Porter, D.A., Easterling, K.E., Sherif, M., 2009. Phase Transformations in Metals and Alloys, third ed. CRC Press, New York.

Prangnell, P.B., Ozkaya, D., Stobbs, W., 1994. Discontinuous precipitation in high lithium content Al—Li—Zr alloys. Acta Metall. 42, 419—433.

Prasad, K.S., 1999. Solid State Phase Transformations in AA 8090 Al—Li Alloys. Doctoral Thesis, University of Roorkee, Roorkee, India.

Prasad, K.S., Gokhale, A.A., Banerjee, D., Goel, D.B., 1992. New nanocrystalline regions in 1420 Al—Li alloy. Scr. Metall. Mater. 26, 1803—1808.

Prasad, K.S., Mukhopadhyay, A.K., Gokhale, A.A., Banerjee, D., Goel, D.B., 1994. δ precipitation in an Al—Li—Cu—Mg—Zr alloy. Scr. Metall. 30, 1299—1304.

Prasad, K.S., Mukhopadhyay, A.K., Gokhale, A.A., Banerjee, D., Goel, D.B., 1999. On the formation of facetted Al₃Zr(β') precipitates in Al—Li—Cu—Mg—Zr alloys. Acta Metall. 47, 2581—2592.

Preston, G.C., 1938. Effect of stress during ageing on the precipitation of Φ' Al4 wt pct Cu. Philos. Mag. 26, 855—871.

Radmilovic, V., Thomas, G., 1987. Atomic resolution imaging in Al—Li—Cu alloy. In: Champier, G., Dubost, B., Miannay, D., Sabetay, L. (Eds.), Proceedings 4th International Conference "Aluminium—Lithium Alloys," J. de Physique, vol. 48. pp. C3.385—C3.396.

Radmilovic, V., Thomas, G., Shiflet, G.J., Starke Jr., E.A., 1989. On the nucleation and growth of Al₂CuMg(S′) in Al—Li—Cu—Mg and Al—Cu—Mg alloys. Scr. Metall. 23, 1141—1146.

Rioja, R.J., Ludwiczak, E.A., 1986. The effect of alloy composition and the processing of structure and properties of Al—Li—X alloys. Proceeding 3rd International Conference "Aluminum—Lithium Alloys-III". Institute of Metals, pp. 471—478.

Ryum, N., 1966. Precipitation in an Al-1.78 wt% Hf alloy after rapid solidification. J. Inst. Metals 94, 191.

Ryum, N., 1969. Precipitation and recrystallization in an 0.5%wt Zr alloy. Acta Metall. 17, 269—278.

Sainfort, P., Dubost, B., 1987. Coprecipitation hardening in Al—Li—Cu—Mg alloys. In: Champier, G., Dubost, B., Miannay, D., Sabetay, L. (Eds.), Proceedings 4th International Conference "Aluminum—Lithium Alloys", J. de Physique, vol. 48. pp. C3.407—413.

Sainfort, P., Dubost, B., Dubus, A., 1985. Precipitation of "quasicrystals" by decomposition of solid solutions of the Al—Li—Cu—Mg system. Comptes Rendus 301 (II), 689—692.

Sainfort, P., Guyot, P., 1986. Fundamental aspects of hardening in Al—Li and Al—Li—Cu alloys. In: Baker, C., Gregson, P.J., Harris, S.J., Peel, C.J. (Eds.), "Aluminium—Lithium Alloys III," Proceedings 3rd International Conference. The Institute of Metals, London, UK, pp. 420—426.

Sanders Jr., T.H., Ludwiczak, E.A., Sawtell, R.R., 1980. The fracture behaviour of recrystallized Al-2.8% Li-0.3% Mn sheet. Mater. Sci. Eng. 43, 243—260.

Sanders Jr., T.H., Starke Jr., E.A., 1989. Microstructure and mechanical behaviour of Al—Li—Cu—Mg alloy 8090 microalloyed with V and Be. Proceedings 5th International Conference "Aluminium—Lithium Alloys V". MCEP Publications, Birmingham, UK, pp. 1—37.

Sato, T., Kamio, A., 1990. Computer simulation and micrographical evaluation coarsening δ' (Al₃Li) precipitates in Al—Li alloys. Mater. Trans. Japan Inst. Metals 31, 25—38.

Silcock, J.M., 1959—1960. The structural ageing characteristics of Al—Cu—Li alloys. J. Inst. Metals 88, 357—364.

Silcock, J.M., 1961. Effect of reinforcement size on age hardening of Al-SiC 20 vol % particulate composites. Japan Inst. Metals 89, 203—210.

Silcock, J.M., Heal, T.J., Hardy, H.K., 1953—1954. A study of age hardening Al-3.85% Cu by the divergent X-ray beam method. J. Inst. Metals, 82, 239—248.

Stimson, W., Tosten, M.H., Howell, P.R., Williams, D.B., 1986. Precipitation and lithium segregation studies in Al-2wt%Li-0.1wt%Zr. In: Baker, C., Gregson, P.J., Harris, S.J., Peel, C.J. (Eds.), Proceedings 3rd International Conference "Aluminium—Lithium Alloys". The Institute of Metals, London, UK, pp. 386—391.

Suzuki, H., Kanno, M., Hayashi, N., 1982. High strength of Al—Cu—Li—Zn alloys. Japan Inst. Metals 32, 88.

Tamura, M., Mori, T., Nakamura, T., 1970. Experimental observations on the nucleation and growth of δ' (Al$_3$Li) in dilute Al–Li alloys. Japan Inst. Metals 34, 919–925.

Tang, X., 1995. Microstructural Development in Al–Li–Cu–Mg Alloys and Metal Matrix Composites. PhD Thesis, Pennsylvania State University, USA.

Thompson, G.E., Noble, B., 1973. Precipitation characteristics of Al–Li alloys containing Mg. J. Inst. Metals 101, 111–115.

Tosten, M.H., Ramani, A., Bartges, C.W., Michel, D.J., Ryba, E., Howell, P.R., 1989. "On the origin and nature of microcrystalline regions in an Al–Li–Cu–Zr alloy. Scr. Metall. 23, 829–834.

Tsivoulas, D., 2010. Effects of Combined Zr and Mn Additions on the Microstructure and Properties of AA2198 Sheet. PhD Thesis, University of Manchester, 2010.

Tsivoulas, D., Robson, J.D., Sigli, C., Prangnell, P.B., 2012. Interactions between zirconium and manganese dispersoid-forming elements on their combined addition in Al–Cu–Li alloys. Acta Mater. 60, 5245–5259.

Valentine, M.G., Sanders Jr., T.H., 1989. The influence of temperature and composition on the distribution of δ' (Al$_3$Li). In: Sanders, T.H., Starke, E.A. (Eds.), "Aluminium–Lithium Alloys," Proceedings 5th International Conference, vol. 2. Materials and Component Engineering Publications Ltd., Birmingham, UK, pp. 575–584.

Vecchio, K.S., Williams, D.B., 1988a. Convergent beam electron diffraction analysis of the (Ai$_2$CuLi) phase in Al–Li–Cu alloys. Metall. Trans. A 19A, 2885–2891.

Vecchio, K.S., Williams, D.B., 1988b. A non-icosahedral T$_2$ (Al$_6$Li$_3$Cu) crystals. Philos. Mag. B. 57, 535B.

Vecchio, K.S., Williams, D.B., 1988c. The apparent 'Five-Fold' nature of large T$_2$ (Al6Li3Cu) crystals. Metall. Trans. A 19A, 2875–2884.

Venables, D., Christodoulou, L., Pickens, J.R., 1983. On the $\delta' \rightarrow \delta$ transformation in Al–Li alloys. Scr. Metall. 17, 1263–1268.

Wadsworth, J., Pelton, A.R., 1984. Superplastic behavior of a powder-source Aluminium–Lithium based alloy. Scr. Metall. 18, 473–478.

Wagner, C., 1961. Self consistent forms of the chemical rate theory of Ostwald ripening. Z. Electrochem. 65, 581–591.

Wang, S.C., Starnik, M.J., 2005. Precipitates and intermetallic phases in precipitation hardening Al–Cu–Mg–(Li) based alloys. Int. Mater. Rev. 50, 193–215.

Wassermann, G., Weerts, J., 1935. Precipitates and Intermetallic phases in precipitation hardening of Al–Cu–Mg alloy. Metallwirtschaft 14, 605.

Weatherly, G.C., 1966. Conditions of S'-formation in Al–Cu–Mg alloy. PhD Thesis, University of Cambridge.

Welpmann, K., Peters, M., Sanders Jr., T.H., 1986. Age hardening behaviour of DTD XXXA. In: Baker, C., Gregson, P.J., Harris, S.J., Peel, C.J. (Eds.), Proceedings 3rd International Conference, "Aluminium–Lithium Alloys". The Institute of Metals, London, UK, pp. 524–529.

Williams, D.B., 1974. On the Ageing of Aluminium–Lithium–Zirconium Alloy. Ph.D Thesis, University of Cambridge, Cambridge, UK.

Williams, D.B., 1981. Aluminum–lithium alloys. In: Sanders, T.H., Starke, E.A. (Eds.), "Aluminum–Lithium Alloys I," Proceedings 1st International Conference. The Metallurgical Society of AIME, Warrendale, PA, pp. 89–100.

Williams, D.B., Edington, J.W., 1975. The precipitation of δ' (Al$_3$Li) in dilute aluminium–lithium alloys. Metal Sci. J. 189, 529–532.

Wilson, R.N., 1969. The ageing response of Al–Cu and Al–Cu–Mg directly solidified eutectis. J. Inst. Metals 97, 80–87.

Chapter 5

Texture and Its Effects on Properties in Aluminum—Lithium Alloys

K.V. Jata* and A.K. Singh[†]

*MMD/AFRL, Wright Patterson Air Force Base, OH, USA, [†]Defence Metallurgical Research Laboratory, Hyderabad, India

Contents

5.1 Introduction	140	
5.2 Texture in Al—Li Alloys	142	
5.3 Texture Evolution During Primary Processing	145	
5.3.1 Beta Fiber Components (Bs, S, and Cu) in Binary and Ternary Al—Li Alloys	146	
5.3.2 Alpha Fiber Components (Goss and Cube/Rotated Cube)	149	
5.3.3 Role of Precipitates and Slip Character During Hot and Cold Forming	151	
5.3.4 Influence of Non-octahedral Slip	152	
5.4 Macroscopic Anisotropy of Yield Strength	152	
5.5 Practical Methods to Reduce Texture in Al—Li Alloys	154	
5.5.1 Processing Methods	154	
5.5.2 Thick Plate Anisotropy	156	
5.5.3 Sheet Product Fatigue Crack Deviations	157	
5.5.4 Example Microstructures for Third-Generation Al—Li Alloys	159	
5.6 Summary	160	
Acknowledgments	160	
References	160	

Aluminum—Lithium Alloys.

5.1 INTRODUCTION

Crystallographic texture evolves during forming wrought products from the cast ingots. The processing conditions discussed in this chapter are (i) primary processing, where the cast material is hot or cold worked to a wrought product (covered in detail in Chapters 6−8) and (ii) secondary processing, followed by solution heat treatment, stretching, and aging that are applied to the wrought product to obtain the desired mechanical properties.

The time and temperature at which the primary processing operations are performed play a key role because precipitates that are stable at these temperatures evolve during this stage. Precipitate/particle interactions with dislocations during primary processing control the development of texture in any wrought product, including aluminum−lithium (Al−Li) alloys.

After primary processing, secondary processing including solution heat treatment and quenching is performed on the wrought product, followed invariably by a 1−2% plastic strain (stretch) before ageing. The dislocations generated during stretching act as sites for nucleation of semicoherent precipitates during ageing and enhance the uniformity of distribution of precipitates. The angle between the stretch axis and the forming direction (rolling or extrusion direction) also plays a role in determining the anisotropy of the mechanical properties.

One of the primary concerns in wrought product development is the anisotropy of mechanical properties because it can affect the design allowables (Eswara Prasad et al., 1993a). As discussed later, not all the mechanical property anisotropy can be attributed to the crystallographic texture developed during primary processing. The distribution of precipitates, specifically T_1 (Al_2CuLi), plays a major role in determining property anisotropy in 2XXX series Al−Li alloys. These particles precipitate on the {111} planes, which are also the slip planes for face-centered cubic (FCC) materials like Al−Li alloys.

The main subjects of this chapter are the crystallographic texture development in Al−Li alloys and the associated anisotropy of mechanical properties, particularly the yield strength (Jata et al., 1999; Panchanadeeswaran and Field, 1995; Przystupa et al., 1994; Singh et al., 1999; Spriano et al., 1988; Zeng and Barlat, 1994). Therefore, besides understanding the evolution of texture during primary processing in Al−Li alloys, there has also been a strong desire to relate the yield strength anisotropy to the crystallographic texture (e.g., see the proceedings of aluminum−lithium alloys conference volumes, particularly volume III; Cho et al., 1999; Jata et al., 1998; Lyttle and Wert, 1996; Vasudevan et al., 1988a,b, 1990, 1995).

In plates, sheets, and extruded products, anisotropy has been attributed to crystallographic texture as well as strengthening precipitates. Studies have also been conducted by the NASA Langley group (Tayon et al., 2010) to understand the effect of microtexture on delamination fracture common to some of the Al−Li alloys, particularly second-generation alloys such as AA

2090 and AA 8090. Tortuous crack profiles under fatigue conditions have been attributed partly to crystallographic texture (Pao et al., 1989; Roven et al., 1990; see also Chapter 12). The periodic large and small misorientations between grains have been determined through microtexture measurements and correlated to the spacing of the fine delamination. Cracks turning parallel to the loading direction along grain boundaries under sustained loads (creep crack growth) have been observed by Sadananda and Jata (1998).

Texture "tailoring" is critical to the manufacturing of structural components because the presence of intense (or sharp) textures can limit aerospace applications. Thin sheet Al−Li products usually show in-plane anisotropy, with a 45° angle to the rolling direction giving the lowest strength properties. Lower yield strengths along the 45° axis negatively impact the benefits of these sheet products since the minimum yield strength has to be included in structural design. Thick plate products have through-thickness anisotropy: the strength properties in the centers of thick plate are usually higher than those at the surfaces. Again, these variations must be taken into account when designing and making components.

In the past 30 years, numerous experiments and theoretical work have been conducted to answer many questions regarding texture. Some of the questions raised by investigators in this field are listed here:

1. How do the primary processing temperature and the dwell time affect the texture (De et al., 2011)? How do the alpha fiber, the beta fiber, and the various brass (Bs) texture components in the texture evolve as a function of hot and cold deformation conditions?
2. What are the influences of the precipitates that form during hot working on the texture components? Similarly, what are the consequences of the precipitate−dislocation interactions on texture during cold rolling (Contrepois et al., 2010; Engler and Lücke, 1991).
3. Recently, there has been a strong interest in understanding texture evolution in processes such as friction-stir welding and equal-channel angular pressing that belong to the class of severe plastic deformation (SPD) processes. How does Al−Li alloy microtexture evolve in such processes (Field et al., 2001; Jata and Semiatin, 2000)?
4. What are the major mechanisms in the evolution of Al−Li alloy textures, and how do they compare to the traditional aluminum alloys? In particular, what is the role of
   a. composition;
   b. stacking fault energy (SFE): Al−Li alloys are considered as low SFE materials compared to traditional aluminum alloys (considered as high SFE materials);
   c. deformation modes (planar slip versus homogeneous deformation)?
5. What are the ageing effects on texture of Al−Li alloy sheets and thick plate (Jata et al., 1996; Rioja, 1998)?

6. What parameters affect the texture evolution during thin sheet development (Barlat et al., 1992; Choi and Barlat, 1999)?
7. How can texture be controlled in these alloys so that desired product forms with controlled texture can be produced (Rioja, 1998; Rioja and Liu, 2012)?

The products obtained from second-generation Al—Li alloys such as 2090 and 8090 have suffered from severe anisotropy of mechanical properties compared to the newer/third-generation Al—Li alloys (Chen et al., 1999; Chung et al., 2000; Gokhale and Singh, 2005). In fact, texture control and limitation has been a prime goal in the development of third-generation alloys (Rioja and Liu, 2012). Many of the challenges in reducing sharp texture and property anisotropy have been overcome and are discussed in Section 5.5.

Table 5.1 lists the nominal compositions and densities of most of the first-, second-, and third-generation commercial Al—Li alloys and their year of introduction. There are many third-generation alloys now commercially available. This emphasizes that a detailed knowledge of texture and its effect on the mechanical properties of Al—Li alloys is very important.

5.2 TEXTURE IN Al—Li ALLOYS

The beta fiber comprising brass (Bs), S, and copper (Cu) texture components has been observed during thermomechanical processing of aluminum alloys. Al—Li alloys also exhibit the same texture components except that the Bs component is high. If Al—Li alloys undergo recrystallization, the usual recrystallized texture components cube, Cu, and Goss (G) are exhibited. It was stated in Chapter 4 that most Al—Li alloys contain approximately 0.12 wt% of Zr to inhibit recrystallization through Al_3Zr dispersoids. Unrecrystallized microstructures with pancake-shaped grains are preferred, as they provide superior fracture toughness and fatigue crack growth resistance. Therefore Zr, a microalloying element, is preferred as a recrystallization inhibitor rather than Mn or Cr.

Many studies have been performed to understand the evolution of texture in aerospace aluminum alloys and aluminum—lithium alloys. Typically, these alloys show a string of texture orientations ranging from Bs {110}⟨112⟩ component through the S {123}⟨634⟩ component to the Cu {112}⟨111⟩ component. A distinctive aspect of hot rolled or hot extruded aluminum alloys is that a strong Bs texture component is present and is responsible for property anisotropy in these alloys.

Figure 5.1 schematically illustrates the positions of the various texture fibers located in Euler space as represented in the Bunge notation (Hales and Hafley, 1998, 2001). This illustration is helpful in identifying the presence of texture components in orientation distribution function (ODF) plots. Intensities of the texture from the ODF plots are normally shown as 2-dsections for constant φ_2 sections of 0°, 45°, 65°, 90°, or $45° < \varphi_2 < -45°$ designation, according to choice.

TABLE 5.1 Compositions and Densities of Al–Li Alloys

Alloys	Li	Cu	Mg	Ag	Zr	Sc	Mn	Zn	Density ρ (g/cm³)	Introduction
First Generation										
2020	1.2	4.5					0.5		2.71	Alcoa 1958
1420	2.1		5.2		0.11				2.47	Soviet 1965
1421	2.1		5.2		0.11	0.17			2.47	Soviet 1965
Second Generation (Li ≥ 2wt%)										
2090	2.1	2.7			0.11				2.59	Alcoa 1984
2091	2.0	2.0	1.3		0.11				2.58	Pechiney 1985
8090	2.4	1.2	0.8		0.11				2.54	EAA 1984
1430	1.7	1.6	2.7		0.11	0.17			2.57	Soviet 1980s
1440	2.4	1.5	0.8		0.11				2.55	Soviet 1980s
1441	1.95	1.65	0.9		0.11				2.59	Soviet 1980s
1450	2.1	2.9			0.11				2.60	Soviet 1980s
1460	2.25	2.9			0.11	0.09			2.60	Soviet 1980s
Third Generation (Li < 2wt%)										
2195	1.0	4.0	0.4	0.4	0.11				2.71	LM/Reynolds 1992
2196	1.75	2.9	0.5	0.4	0.11		0.35 max	0.35 max	2.63	LM/Reynolds/ McCook Metals 2000
2297	1.4	2.8	0.25 max		0.11		0.3	0.5 max	2.65	LM/Reynolds 1997
2397	1.4	2.8	0.25 max		0.11		0.3	0.10	2.65	Alcoa 2002
2098	1.05	3.5	0.53	0.43	0.11		0.35 max	0.35	2.70	McCook Metals 2000
2198	1.0	3.2	0.5	0.4	0.11		0.5 max	0.35 max	2.69	Reynolds/ McCook Metals/Alcan 2005
2099	1.8	2.7	0.3		0.09		0.3	0.7	2.63	Alcoa 2003
2199	1.6	2.6	0.2		0.09		0.3	0.6	2.64	Alcoa 2005
2050	1.0	3.6	0.4	0.4	0.11		0.35	0.25 max	2.70	Pechiney/ Alcan 2004[a]

(Continued)

TABLE 5.1 (Continued)

Alloys	Li	Cu	Mg	Ag	Zr	Sc	Mn	Zn	Density ρ (g/cm³)	Introduction
2296	1.6	2.45	0.6	0.43	0.11		0.28	0.25 max	2.63	Constellium Alcan 2010[a]
2060	0.75	3.95	0.85	0.25	0.11		0.3	0.4	2.72	Alcoa 2011
2055	1.15	3.7	0.4	0.4	0.11		0.3	0.5	2.70	Alcoa 2011
2065	1.2	4.2	0.50	0.30	0.11		0.40	0.2	2.70	Constellium 2012
2076	1.5	2.35	0.5	0.28	0.11		0.33	0.30 max	2.64	Constellium 2012

[a]Pechiney acquired by Alcan 2003; Constellium, formerly Alcan Aerospace.
Source: After Rioja and Liu (2012); see also Chapter 2.

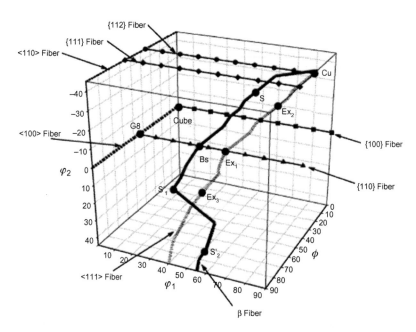

FIGURE 5.1 Schematic illustration of relative positions of $\{hkl\}\langle uvw \rangle$ texture—fiber in aluminum alloys using the Bunge notation (Hales and Hafley, 2001).

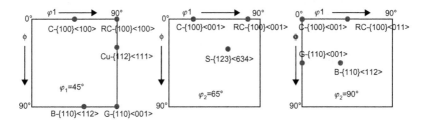

FIGURE 5.2   Diagram showing locations of important texture components using the Bunge notation in $\varphi_2 = 45°$, $65°$, and $90°$ sections.

Figure 5.2 depicts the locations of important texture components using the Bunge notation in $\varphi_2 = 45$, 65, and 90° sections. Roe's notation is also alternatively used, and examples of ODF plots in Roe's Euler angles ϕ, ψ, and θ in the Al–Li alloy literature have been shown elsewhere (Vasudevan et al., 1995).

The development of texture and microstructure for a quaternary Al–Li–Cu–Mg alloy sheet have been investigated in detail at different stages of processing (Singh et al., 1996). The processing schedule involved hot cross-rolling followed by several stages of cold rolling with intermediate solution treatment. They observed marked through-thickness variation in the processed sheet during the course of thermomechanical treatment. The overall intensity of texture at the mid-thickness level was 2–3 times sharper than the texture at surface. In more detail, after hot cross-rolling the alloy had equally strong Bs and Bs/S $(168)\langle 211 \rangle$ components at mid-thickness, and during three subsequent cycles of cold rolling the overall texture intensity was reduced by a factor of nearly 2. The final processed sheet material had a moderately strong Bs and a predominant S $(123)\langle 634 \rangle$ component. It was also shown that solution treatment of the cold rolled material caused extensive recovery but only partial recrystallization. It appeared that the second phase particles did not affect the texture development during processing.

5.3 TEXTURE EVOLUTION DURING PRIMARY PROCESSING

One of the major concerns about texture development in the primary processing stage (where the cast structure is broken down into wrought form either by rolling, extrusion, or forging) is the role of impurity inclusions, constituent particles, and strengthening precipitates. A number of precipitates that are stable during high-temperature extrusion processing have been shown to form in Al–Li alloys (Jata et al., 1998), and these precipitates strongly influence the texture. Inclusions usually contain elements such as Si and Fe. The secondary phase particles include Mn-containing dispersoids, Al_6Mn or $Al_{20}Cu_2Mn_3$, and Al_3Zr. These particles principally control the grain size in Al/Al–Li alloys, although they also affect the fracture properties, as do the constituent particles (see Chapter 13).

5.3.1 Beta Fiber Components (Bs, S, and Cu) in Binary and Ternary Al—Li Alloys

Alloy composition as well as the time and temperature during mechanical working determine the size and volume fraction of the precipitates that can influence the slip processes and texture in Al—Li alloys. Spherical incoherent δ as well as platelet-shaped $T_1(Al_2CuLi)$, $T_b(Al_{7.5}Cu_4Li)$, and $T_2(Al_6CuLi_3)$ precipitates of varying volume fractions have been observed to form at the hot forming temperatures. High-temperature Mn (Al_6Mn, $Al_{20}Cu_2Mn_3$) and Zr (Al_3Zr) bearing dispersoids are also present during these high-temperature forming operations, and these can also influence the slip processes and the resultant texture.

In the case of high-temperature forming, the starting temperatures for rolling binary Al—Li and ternary Al—Li—Cu alloys have generally been in the vicinity of 550°C, and the final temperatures are around 425°C. Texture development during hot forming of Al—Li binary and Al—Li—Cu ternary alloys has been studied in depth by Vasudevan et al. (1988a,b, 1990, 1995). The binary alloys with lithium contents < 1.4 wt% consist of the solid solution and homogeneously dispersed δ precipitates during hot rolling in the temperature range 425—450°C (Vasudevan et al., 1995). Thus dislocations interact with the homogeneously dispersed δ precipitates, and the deformation is essentially homogeneous. As the lithium content increases the volume fraction of the Li-containing precipitates increases, and at high hot-rolling temperatures Li-containing precipitates impart homogeneous slip and reduce the overall ODF intensities. For example, for the 2.9 wt% Li binary alloy the ODF intensity was 20 times random, while for the 4.6 wt% Li alloy the corresponding ODF value was only 5 times random. The hot rolled binary Al—Li alloys with high Li contents have a distinct beta fiber texture with a maximum intensity near the Bs—G position $(101)\langle141\rangle$ (Vasudevan et al., 1995).

Interesting results have also been obtained for ternary Al—Li—Cu alloys. A systematic variation of the Li:Cu ratio was investigated for texture evolution with processing (Vasudevan et al., 1988a,b, 1990). As shown in Figure 5.3, the maximum ODF intensity for the Bs components was observed for the alloy with 2.1 wt% Li and 2.9 wt% Cu. The development and variation of these textures is consistent with hot deformation theories. During high-temperature rolling the incoherent δ precipitates homogenize the deformation and reduce the sharpness of the texture.

In a comparison between experimental Al—Cu—Li and Al—Zn—Mg—Cu alloys, Contrepois et al. (2010) have concluded that the textures in both alloys are very similar, and the fact that the Bs texture $\{110\}\langle112\rangle$ in commercial Al—Cu—Li alloys is higher than that in commercial Al—Zn—Mg—Cu products was attributed to the higher rolling temperature. This conclusion was derived based on plane strain compression tests and observed to be consistent with non-octahedral slip at high temperatures.

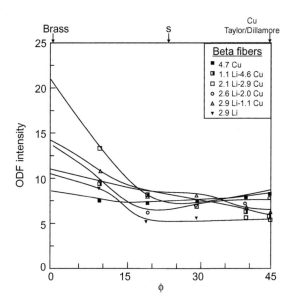

FIGURE 5.3   ODF texture intensities along the β fibers for Al—Li—Cu alloys (Vasudevan et al., 1988a,b, 1990).

Jata et al. (1996, 1998) have investigated the evolution of microstructure and texture in an experimental alloy, AF/C489 (2.06Li, 2.7Cu, 0.3Mg, 0.3Mn, 0.05Zr), during high-temperature extrusion. They conducted systematic extrusion experiments in an effort to understand the influence of primary processing condition on the evolution of texture. The processing windows in terms of temperature, strain rate, and microstructure were provided by Jain et al. (1998) from hot compression testing, whose processing variables are applicable to the extrusion process. Extrusion samples were made at 55°C intervals in the range of 260—538°C, with a constant extrusion ratio of 7:1, and a strain rate of 1 s^{-1}. The samples were quenched as soon as they came out of the extrusion press in order to "freeze" the microstructures and textures. In another set of experiments, the as-extruded specimens were solution heat treated and the associated texture components measured:

1. Figure 5.4 shows significant amounts of texture components for all extrusion temperatures in the as-extruded condition.
2. Figure 5.5 shows that the Bs texture intensity is maintained at the two highest extrusion temperatures (>480°C) in the solution treated condition.

Jata et al. (1999) have used extensive thin-foil transmission electron microscopy (TEM) to measure the sizes of subgrains and interplatelet

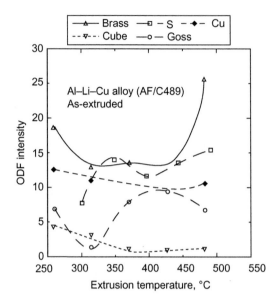

FIGURE 5.4 Variation of major texture components in the as-extruded AF/C489 alloy. The Bs component variation is relatively high (Jata et al., 1998).

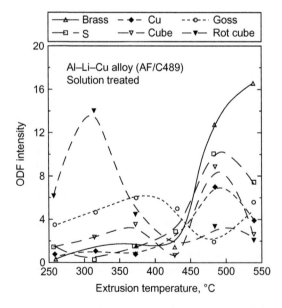

FIGURE 5.5 Variation of major texture components in the extruded AF/C489 alloy after solution heat treatment.

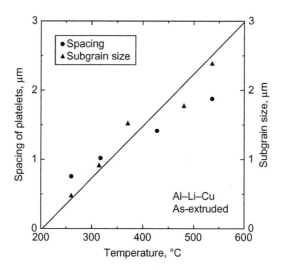

FIGURE 5.6 TEM micrograph data showing the relationship between subgrain size and spacing of platelet particles at various extrusion temperatures (Jata et al., 1999).

spacing of the second phase platelet precipitates. The variation of subgrain size and spacing of platelets with extrusion temperature is shown in Figure 5.6. The subgrains were absent in the solution treated samples that were extruded at 538°C and rarely present in the materials extruded at 482°C. The materials extruded below 482°C and then solution treated did not show the presence of subgrains. This suggests that the material extruded at lower temperatures recrystallized upon solution treatment.

It is known that the increase in dynamic recovery at high extrusion temperatures enhances the formation of shear bands and consequently increases the Bs texture formation (Lyttle and Wert, 1996). This is also consistent with negligible recrystallization during solution heat treatment after extrusion (Jata et al., 1998, 1999). It is to be noted that Lee et al. (1993) have suggested that when FCC materials with high SFE energy are rolled at temperatures where significant dynamic recovery is possible, the volume fraction of Bs intensity is high. For AF/C489 in the extruded condition a high Bs texture is consistent with high SFE. (As mentioned in Section 5.1, the SFEs of Al–Li alloys are less than those of traditional aluminum alloys. However, the SFEs of Al–Li alloys are still higher than for many other FCC materials.)

5.3.2 Alpha Fiber Components (Goss and Cube/Rotated Cube)

In this section the recrystallization texture components are discussed first with reference to Figure 5.5, concerning the AF/C489 extruded and solution

TABLE 5.2 Phases Detected by X-Rays in the Experimental AF/C489 Alloy Extruded at Various Temperatures and Then Quenched to Retain the Microstructure and Texture for Further Analysis (Csontos and Starke, 2000)

X-Ray Guinier Analysis of the AF/C489 Extrusions in the As-Extruded Condition

Extrusion Temperature (°C)	T_1	T_B	T_2	$Al_{20}Cu_2Mn_3$
260	Small	Trace	Medium +	Small
315	Small	Trace	Large	
371	Very small	Possible trace	Medium +	
426				
482				
538				Very small

treated alloy. The rotated cube texture component (100)⟨130⟩ is highest for extrusion at the lower temperatures. Next highest, up to the mid-range of extrusion temperatures, is the Goss texture component intensity. The strong presence of the rotated cube component is consistent with the fact that second phase particles are present at low temperatures and provide nucleation sites for recrystallization (Table 5.2). It is to be noted that the strong Bs component for high-temperature extrusion followed by solution heat treatment is obviously resistant to recrystallization (see Mondal et al., 2011).

Secondly, binary alloys with <1.4 wt% Li behave as a solid solution during hot rolling, and Vasudevan et al. (1995) in their texture evolution studies on binary Al–Li alloys have observed recrystallization texture components typical of aluminum alloys, namely the alpha fiber with the strong Goss {101}⟨010⟩ and weak cube {001}⟨100⟩ orientations (Figure 5.7).

Models describing the development and predominance of recrystallization texture (cube {001}⟨100⟩) have been discussed by several investigators (Bacroix and Jonas, 1988; Dillamore and Roberts, 1964; Hirsch and Lücke, 1985). Dillamore and Roberts (1964) have suggested that the presence of {112}⟨111⟩ orientation in the deformed material has a strong influence on the subsequent recrystallization behavior and formation of the cube component. Hirsch and Lücke (1985) observed a minimum cube component in dilute Al–Fe alloys at an intermediate annealing temperature (360°C) and termed it continuous recrystallization. This minimum cube component has been attributed to precipitation, which is known to efficiently obstruct grain boundary motion during recrystallization. The strength of the cube component increases at lower and higher annealing temperatures owing to discontinuous recrystallization.

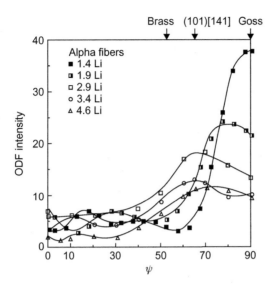

FIGURE 5.7 ODF texture intensities along the alpha fiber for hot rolled binary Al–Li alloys (Vasudevan et al., 1995).

5.3.3 Role of Precipitates and Slip Character During Hot and Cold Forming

Engler et al. (1989) have investigated the development of texture and structure of an Al–1.8% Cu alloy (without Li) as a function of precipitation state, which was extensively varied by a series of annealing treatments. For a solution heat-treated condition and homogeneous deformation (cold rolling), the texture corresponded to a slightly increased Bs orientation as compared to that of pure aluminum. The homogeneous deformation was simulated by the relaxed constrained model (Van Houtte, 1988). As aging progressed to the peak-aged temper, the Bs texture component further increased owing to the cutting of coherent Al_2Cu precipitates and the predominant occurrence of slip localization. Because this localized slip is very similar to that of a deforming single crystal, it was modeled using the Sachs approach. With further increase in ageing, that is, overageing, the slip was controlled by incoherent Al_2Cu precipitates. As expected, the slip was much more homogeneous due to dislocation bypassing and looping. This transition from localized slip to homogeneous slip resulted in a decrease of the Bs texture component and an increase in the Cu component. Since slip in the overaged condition has more influence on neighboring grains, it was simulated using the fully constrained model (Taylor, 1938).

Engler and Lücke (1991) performed experiments to understand texture evolution during cold rolling of an already hot rolled AA 8090 alloy plate

(2.4Li−1.2Cu−0.7Mg, 0.1Zr in wt%). The first set of experiments consisted of quenching the hot rolled as-received plate and cold rolling it. The behavior of the quenched alloy after solution heat treatment was consistent with the behavior of a solid solution, with a possibility of δ' precipitation if left to age at room temperature. These investigators observed strong texture components at the S orientation and in the later stages of deforming a Bs {011}⟨211⟩ component. They concluded that the solid solution alloy behaved very similarly to a traditional aluminum alloy. When cold rolling was performed on alloys that underwent further ageing, slip localization due to shear of δ' precipitates resulted in a higher fraction of the Bs texture component. After extreme overageing, the contribution of dispersed slip from S' and T_1 phases resulted in Cu {112}⟨111⟩ and S {123}⟨634⟩ orientations. Note that in their earlier work, Engler et al. (1989) attributed evolution of a Bs texture component to planar slip, and evolution of the other texture components (S and Cu) to homogeneous slip.

5.3.4 Influence of Non-octahedral Slip

The influence of planar/localized slip on formation of a Bs texture component has been discussed earlier (Engler et al., 1989). The other theory, that is, the influence of non-octahedral slip on the evolution of Bs texture in aluminum alloys, has been studied by Bacroix and Jonas (1988) and Contrepois et al. (2010).

Contrepois et al. (2010) compared texture evolution in a Li-containing alloy and a non-lithium alloy under a variety of stress-state conditions, a wide range of temperatures (350−500°C) and strain rates. Tests were done using plane-strain channel-die compression, which may be considered comparable to industrial rolling conditions. One of the important results was to show that the Bs texture component in hot rolled products is not necessarily due to the presence of Li (and Li-containing precipitates) in aluminum alloys and therefore cannot be attributed to planar slip deformation. Instead, the Bs texture component is due to the participation of non-octahedral slip systems: slip traces from the non-octahedral slip planes {110}, {100}, and {112} were observed. The presence of such slip traces has not been confirmed in Al−Li alloys, but it has been suggested that non-octahedral slip could play a role in the evolution of a Bs texture component during high-temperature deformation/processing of these alloys. There is a need for experiments to validate this suggestion.

5.4 MACROSCOPIC ANISOTROPY OF YIELD STRENGTH

From a manufacturing and process design point of view, the influence of macroscopic anisotropy and texture on mechanical properties is important.

Thus there has been an extensive effort to evaluate and understand the effect of anisotropy on mechanical properties. In particular, the yield strength anisotropy has to be minimized for most applications, and furthermore the combination of yield strength and fracture toughness has also to be optimized.

The yield strength anisotropy has been attributed to slip planarity. Hence the earlier efforts to avoid slip planarity have concentrated on reducing the misfit between the crystal lattice and the δ' precipitates. This approach was soon abandoned because the δ' precipitates maintain their coherency, and hence shearability by dislocations, well beyond peak ageing. An alternative, and a more successful, approach has been to disperse the planar slip by introducing Mg and precipitating the semicoherent Al_2CuMg (S' phase). The AA 2091 and AA 8090 alloys are second-generation examples of this approach. The third-generation alloys (Table 5.1) generally include less Mg than 2091 and 8090, but most (especially the newer alloys) contain Ag, which has a similar function.

Both hot and cold deformation processing have been found to contribute to anisotropy. In addition, the subsequent solution heat treatment and ageing processes can increase this anisotropy. In sheet alloys the texture and the associated anisotropy in properties are mostly in-plane, while in thick plate products the anisotropy extends to the through-thickness direction.

Unfortunately, the anisotropic properties in commercial Al—Li alloys are severely complicated by the fact that they are influenced not only by crystallographic texture but also by numerous other parameters such as grain shape morphology (pancake-shaped grains ranging from 0.1 to 0.25 mm in unrecrystallized microstructures), subgrains of the order of a few micrometers ($10-20\,\mu m$), grain and subgrain orientations, gradients in grain size and morphology, directional precipitates (e.g., T_1, S'), and slip nature or dislocation structure.

1. The texture in Al—Li is primarily crystallographic in origin. However, the yield strength anisotropy has been related not only to the crystallographic texture (and hence Taylor factor for a polycrystalline material) but also to the texture or anisotropy that arises due to the precipitate—dislocation interactions. Bearing in mind this important limitation, there have been attempts to model yield strength anisotropy from purely crystallographic considerations.

   The variation in yield strength anisotropy was initially attributed to the variation in Schmid and Taylor factors (Vasudevan et al., 1990). The relaxed constraint models were employed to include the grain morphology, but the predictions did not fully agree with the observations. Choi and Barlat (1999) extended the modeling using a viscoplastic self-consistent (VPSC) polycrystal model: they chose *solution heat treated* AA 2090 alloy to avoid precipitation effects. The VPSC model gave a better correlation than the Taylor model, but of course the modeling should incorporate microstructural parameters. The Taylor—Bishop—Hill (TBH) polycrystal model theory is an alternative, but it has been found to be insufficient when precipitates are present in the alloy (Fricke et al., 1986).

Determining and correlating the anisotropy of basic mechanical properties under monotonic and fatigue conditions have been performed in non-Li—Al alloys, notably by Hosford and Zeisloft (1972) and Bate et al. (1981, 1982). Many of the early findings in these research endeavors have been applied to understand anisotropy due to crystallographic texture in Al—Li alloys (Barlat et al., 1992; Garmestani et al., 2002) and to understand the effects of various precipitation states and type of slip on the rolling texture evolution (Barlat and Liu, 1998; Hargarter et al., 1998; Lyttle and Wert, 1996).

2. Contributions of precipitates to mechanical property anisotropy are due to their formation on specific crystallographic planes, followed by their interactions with dislocations. As a result, apart from the sizes, shapes, and distributions of precipitates, their orientation relationships with the matrix are also important.

As discussed earlier in this chapter, as well as in Chapters 3 and 4, the precipitate microstructures in Al—Li alloys can be complex and depend on the Li:Cu ratio and on the Mg content in alloys containing Mg. The choice of grain-refining agent is also important: recent alloy development programs have resulted in Mn being added as well as Zr (see Table 5.1). Mn combines to form Al_6Mn and $Al_{20}Cu_2Mn_3$ dispersoids that not only control grain size but break up and disperse the planar slip resulting from shearing of the ordered and coherent δ' precipitates.

Another effect of precipitation concerns the very thin T_1 (Al_2CuLi) platelets that (depending on the Li:Cu ratio) precipitate on the {111} slip planes of the aluminum alloy matrix. These precipitates can interact with dislocations during deformation and therefore influence the yield strength.

5.5 PRACTICAL METHODS TO REDUCE TEXTURE IN Al—Li ALLOYS

5.5.1 Processing Methods

Processing methods have had to be optimized to reduce the anisotropy in Al—Li alloy products while maintaining yield strength, fracture toughness, and crack growth resistance. Rioja (1998) has pointed out that at least five approaches have been used to reduce anisotropy in Al—Li alloys:

1. *Overage the material prior to performing the processing steps*: The idea behind this is that overageing the material and then performing a cold or hot forming operation will lead to homogeneous slip during processing.

This decreases the intensity of the Bs texture component and in turn reduces anisotropy in sheet products (Gregson and Flower, 1985).

2. *Adding grain refiners other than Zr*: This has been suggested to reduce the effectiveness of Al_3Zr in pinning the grain boundaries and preventing recrystallization. Replacement (or reduction) of Zr-containing dispersoids by less effective (in terms of pinning effect) dispersoids containing Mn and Cr is the most obvious possibility. It has been shown that the Zr-containing alloy exhibits a strong rolling texture while the (Mn + Cr)-containing alloy displays a reasonably weak texture (Dorward, 1987).

3. *Stretch in directions at an angle to the rolling direction after solution heat treatment (ST)*: Stretching or cold rolling in different directions, as suggested by Lee et al. (1999), could be done to align the precipitates in directions other than the rolling direction.

4. *Magnitude of deformation processing*: The amount of deformation during hot working can be reduced to avoid the sharpness of texture.

5. *Recrystallization as an intermediate step*: Introduction of recrystallization during an intermediate step will result in overall reduction in texture intensity.

Most approaches to reduce texture and decrease anisotropy that work well on laboratory-scale ingots are also successful on an industrial scale. This is also true for the five approaches listed earlier. Some important observations that have been made are as follows:

1. Overageing Al–Li alloys reduces the fracture toughness. This behavior is unlike that of conventional alloys, where toughness improves with overageing. Thus, even though overageing reduces anisotropy in Al–Li alloys, it is impractical for structural components that need high toughness.

2. For laboratory-scale processing of Al–Li–Cu–Mg–Mn–Zr alloys it has been shown by Cho and Sawtell (1991) and Jata et al. (1996) that unrecrystallized alloys subjected to a recrystallization anneal as an intermediate step exhibit lower anisotropy. This observation is particularly important for plate products. The laboratory process was subsequently scaled up to an industrial level by Rioja (1998).

Figure 5.8 shows details of the laboratory-scale Alcoa/AFRL processing steps and a generic industrial-scale fabrication map for reducing the Bs texture intensities of plate products. Figure 5.8A shows specific examples of the change in the Bs texture intensity associated with each processing step. Some information about the effects of this modern processing on the yield strength of a third-generation Al–Li plate alloy, AA 2199, is given in Section 5.5.2.

Texture control is also important for sheet products, but other factors also play a significant role in determining the anisotropy (Rioja and Liu, 2012). These are discussed briefly in Section 5.5.3.

(A)

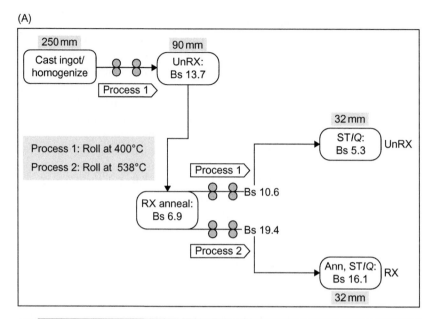

(B)

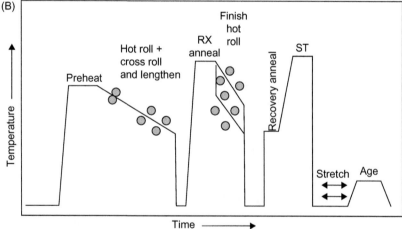

FIGURE 5.8 (A) Alcoa/AFRL process flow chart for reducing texture in an Al–Li–Cu–Mn–Zr thick plate. (B) Generic fabrication map for low Bs texture intensity in unrecrystallized Al–Li plates. UnRX, unrecrystallized; RX, recrystallized. *Source: (A) After Rioja (1998) and (B) After Rioja and Liu (2012).*

5.5.2 Thick Plate Anisotropy

Rioja and Liu (2012) have given an example of texture control for plate products of the third-generation Al–Li alloy AA 2199. Figure 5.9A

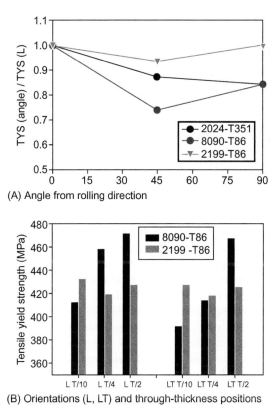

FIGURE 5.9 Comparisons of (A) in-plane yield strengths for 2199-T86, 8090-T86, and 2024-T351 plate and (B) through-thickness yield strength anisotropy for 2199-T86 and 8090-T86 plate (Rioja and Liu, 2012).

compares the in-plane yield strengths for unrecrystallized plates of 2199-T86, the second-generation Al–Li alloy AA 8090-T86, and the conventional alloy AA 2024-T351. It is evident that the 2199 alloy is almost isotropic in yield strength, demonstrating the success of the modern processing methods illustrated in Figure 5.8. This success is emphasized in Figure 5.9B, which compares the through-thickness anisotropy of unrecrystallized 2199-T86 and 8090-T86 plate for both the longitudinal (L) and long-transverse (LT) orientations. The 2199 plate shows minimal through-thickness and in-plane anisotropy, while the 8090 plate is highly (and unacceptably) anisotropic.

5.5.3 Sheet Product Fatigue Crack Deviations

During the (unsuccessful) development of second-generation Al–Li alloys, some "damage tolerant" AA 8090-T81 sheets showed macroscopic deviations of fatigue cracks when tested in the nominal T–L crack plane orientation (Grimes et al., 1990; McDarmaid and Peel, 1989, 1990); and also sometimes when tested in the L–T orientation (Eswara Prasad et al., 1993b, 1994). Some examples for the T–L orientation are shown in Figure 5.10. This crack plane orientation is especially important because it corresponds to the longitudinal skin crack orientation in aircraft fuselage panels (see Figure 13.1 in Chapter 13).

As may be seen from Figure 5.10, the crack deviations varied. Large deviations, as in sheet 1, would actually have a beneficial effect on in-service fatigue crack growth rates (Eswara Prasad et al., 1993b; Peel and McDarmaid, 1989). However, the whole phenomenon of variable crack deviations was considered unacceptable for the design of fail-safe stiffened panel structures, notably fuselage panels in transport aircraft pressure cabins (Grimes et al., 1990).

This issue was addressed by a consortium consisting of Deutsche Airbus, Alcan, Hoogovens (now Tata Steel), the UK Defence Research Agency, and Alcoa (Rioja and Liu, 2012). The crack deviations were recognized to depend on crystallographic texturing (Peel and McDarmaid, 1989), and modification of the sheet rolling practice reduced the variability (Grimes et al., 1990; McDarmaid and Peel, 1990). Nevertheless, the problem was not eliminated, and it was subsequently found that slip planarity had to be reduced by decreasing the amount of δ', which was possible in most cases by lowering the lithium content to a maximum of 1.8 wt% (Rioja and Liu, 2012). This has been one of the key measures in developing the third-generation Al–Li alloys (see Sections 2.3.1 and 13.3.8 in Chapters 2 and 13, respectively).

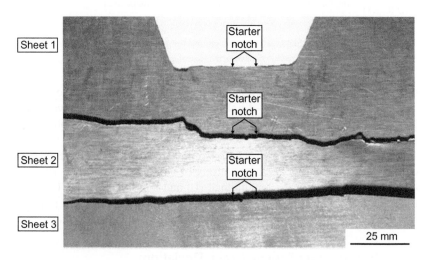

FIGURE 5.10 Variable amounts of fatigue crack deviation in second-generation Al–Li AA 8090-T81 sheet: the starter notches are in the T–L crack plane orientation. *Source: Photograph courtesy of C.J. Peel and R.J.H. Wanhill.*

Rioja and Liu (2012) have stated, "successful mitigation of mechanical property anisotropy by composition, optimization and control of crystallographic texture, grain size and shape, cold deformation, and amount and type of precipitates solved the crack deviation problem." This is exemplified by the crack path in the third-generation Al—Li alloy 2199-T8X (Figure 5.11).

5.5.4 Example Microstructures for Third-Generation Al—Li Alloys

Examples of microstructures obtained by reduced-texture processing, along the lines of the generic fabrication map in Figure 5.8B, are shown in Figure 5.12. For optimum mechanical properties, the extrusions and plate products are typically controlled to be unrecrystallized, whereas sheet products are typically recrystallized (Rioja and Liu, 2012).

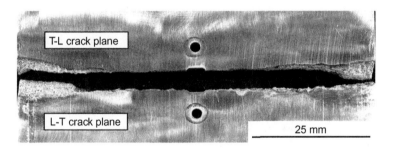

FIGURE 5.11   Fatigue crack growth test specimens of the third-generation Al—Li alloy AA 2199-T8X showing straight crack paths perpendicular to the loading axis. *Source: Photograph courtesy of G.H. Bray.*

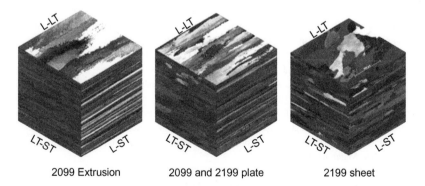

FIGURE 5.12   Optical micrographs of third-generation Al—Li alloys showing thin, elongated unrecrystallized grains in an extrusion; elongated unrecrystallized grains in plate; and recrystallized grains in sheet (Rioja and Liu, 2012). *Source: Photograph courtesy of G.H. Bray.*

5.6 SUMMARY

There is now a sound understanding of the evolution of various texture components in complex aluminum–lithium (Al–Li) alloy compositions. Practical approaches to control texture have been developed, and many of these have been successfully used to tailor texture in products obtained from large scale ingots.

There are a number of theoretical approaches to modeling the anisotropy in yield strength of aluminum alloys in the presence of complex precipitates, but such approaches have not been totally validated for predicting the dependence of in-plane anisotropy in thin sheets and through-thickness anisotropy in thick plates on the crystallographic texture and precipitate properties.

ACKNOWLEDGMENTS

The authors would like to thank the editors profoundly for a thorough review of the book chapter and also for many valuable suggestions. They would like to particularly thank Dr. Russell J.H. Wanhill for many valuable inputs, which made this book chapter one of the most comprehensive documents on the texture of Al–Li alloys. One of the authors (AKS) gratefully acknowledges the support and provision of facilities by the Defence Research and Development Organization, New Delhi, India.

REFERENCES

Bacroix, B., Jonas, J.J., 1988. The influence of non octahedral slip on texture in FCC metals. Textures Microstruct. 8 and 9, 267–311.

Barlat, F., Liu, J., 1998. Precipitation induced anisotropy in binary Al–Cu alloys. Mater. Sci. Eng. A257, 47–61.

Barlat, F., Bren, J.C., Liu, J., 1992. On crystallographic texture gradient and its mechanical consequence in rolled aluminum–lithium sheet. Scr. Metall. Mater. 27, 1121–1126.

Bate, P., Roberts, W.T., Wilson, D.V., 1981. The plastic anisotropy of two phase aluminum alloys—I. Anisotropy in unidirectional deformation. Acta Metall. 29, 1797–1814.

Bate, P., Roberts, W.T., Wilson, D.V., 1982. The plastic anisotropy of two phase aluminum alloys—II. Anisotropy in load-reversal tests. Acta Metall. 30, 725–732.

Chen, D.L., Chaturvedi, M.C., Goel, N., Richards, N.L., 1999. Fatigue crack growth behaviour of X2095 Al–Li alloy. Int. J. Fatigue 21, 1079–1086.

Cho, C.W., Sawtell, R.R., 1991. Al–Li alloys and method of making the same, U.S. Patent No. 5, 066, 342, pp. 11–19.

Cho, K.K., Chung, Y.H., Lee, C.W., Kwun, S.I., Shin, M.C., 1999. Effects of grain shape and texture on the yield strength anisotropy of Al–Li alloy sheet. Scr. Mater. 40, 651–657.

Choi, S.H., Barlat, F., 1999. Prediction of macroscopic anisotropy in rolled aluminum–lithium sheet. Scr. Mater. 41, 981–987.

Chung, Y.H., Cho, K.K., Han, J.H., Shin, M.C., 2000. Effect of grain shape and texture on the earings in an Al–Li alloy. Scr. Mater. 43, 759–764.

Contrepois, Q., Maurice, C., Driver, J.H., 2010. Hot rolling textures of Al–Cu–Li and Al–Zn–Mg–Cu aeronautical alloys: experiments and simulations to high strains. Mater. Sci. Eng. A527, 7305–7312.

Csontos, A.A., Starke E.A., Jr., 2000. The effect of processing and microstructure development on the slip and fracture behavior of the 2.1 wt. pct. Li AF/C-489 and 1.8 wt. pct. Li AF/C-458 Al—Li—Cu—X alloys. Metall. Mater. Trans. A 31A, 1965—1976.

De, P.S., Mishra, R.S., Baumann, J.A., 2011. Characterization of high cycle fatigue behavior of a new generation aluminum lithium alloy. Acta Mater. 59, 5946—5960.

Dillamore, I.L., Roberts, W.T., 1964. Rolling textures in FCC and BCC materials. Acta Metall. 12, 281—293.

Dorward, R.C., 1987. Zirconium vs manganese—chromium for grain structure control in an Al—Cu—Li alloy. Metall. Trans. 18A, 1820—1823.

Engler, O., Lücke, K., 1991. Influence of precipitation state on the cold rolling texture in 8090 Al—Li materials. Mater. Sci. Eng. A148, 15—23.

Engler, O., Hirsch, J., Lücke, K., 1989. Texture development in an Al—1.8% Cu depending on the precipitation state-1 rolling textures. Acta Metall. 37, 2743—2753.

Eswara Prasad, N., Kamat, S.V., Prasad, K.S., Malakondaiah, G., Kutumbarao, V.V., 1993a. In-plane anisotropy in the fracture toughness of an Al—Li alloy 8090 alloy plate. Eng. Fract. Mech. 46, 209—223.

Eswara Prasad, N., Kamat, S.V., Malakondaiah, G., 1993b. Effect of crack deflection and branching on the R-curve behaviour of an Al—Li alloy 2090 sheet. Int. J. Fract. 61, 55—69.

Eswara Prasad, N., Kamat, S.V., Prasad, K.S., Malakondaiah, G., Kutumbarao, V.V., 1994. Fracture toughness of quaternary Al—Li—Cu—Mg alloy under mode I, mode II and mode III loading conditions. Metall. Mater. Trans. A25, 2439—2452.

Field, D.P., Nelson, T.W., Hovanski, Y., Jata, K.V., 2001. Heterogeneity of crystallographic texture in friction stir welds of aluminum. Metall. Mater. Trans. A 32A, 2869—2877.

Fricke W.G., Jr., Przystupa, M.A., Barlat, F., 1986. Modelling mechanical properties from crystallographictexture (ODF) of aluminium alloys. In: Sheppard, T. (Ed.), Aluminium Technology. The Institute of Metals, London, pp. 310—316.

Garmestani, H., Kaldindi, S.R., Williams, L., Bacaltchuk, C.M., Fountain, C., Lee, E.W., et al., 2002. Modeling the evolution of anisotropy in Al—Li alloys: application to Al—Li 2090-T8E41. Int. J. Plast. 18, 1373—1393.

Gokhale, A.A., Singh, V., 2005. Effect of Zr content and mechanical working on the structure and tensile properties of AA 8090 alloy plates. J. Mater. Process. Technol. 159, 369—376.

Gregson, P.J., Flower, H.M., 1985. Microstructural control of toughness in aluminium—lithium alloys. Acta Metall. 33, 527—537.

Grimes, R., Gatenby, K.M., Reynolds, M.A., Gray, A., Palmer, I.G., 1990. The evolution of damage tolerant aluminium—lithium alloys. In: Khan, T., Effenberg, G. (Eds.), Advanced Aluminium and Magnesium Alloys: Proceedings of the International Conference on Light Metals. ASM International European Council, Brussels, pp. 131—138.

Hales, S.J., Hafley, R.A., 1998. Texture and anisotropy in Al—Li alloy 2195 plate and near net shaped extrusions. Mater. Sci. Eng. A257, 153—164.

Hales, S.J., Hafley, R.A., 2001. Structure—property correlations in Al—Li alloy integrally stiff-ened extrusions, NASA Technical Publication TP-2001-210839.

Hargarter, H., Lyttle, M.T., Starke Jr., E.A., 1998. Effect of preferentially aligned precipitates on plastic anisotropy in Al—Cu—Mg—Ag and Al-Cu alloys. Mater. Sci. Eng. A257, 87—99.

Hirsch, J., Lücke, K., 1985. The application of quantitative texture analysis for investigating continuous and discontinuous recrystallization processes of Al—0.01Fe. Acta Metall. 33, 1927—1938.

Hosford, W.F., Zeisloft, R.H., 1972. The anisotropy of age hardened Al—4pct Cu single crystals during plane strain compression. Metall. Trans. 3, 113—121.

Jain, V.K., Jata, K.V., Rioja, R.J., Morgan, J.T., Hopkins, A.K., 1998. Processing of an experimental aluminum−lithium alloy for controlled microstructure. J. Mater. Process. Technol. 73, 108−118.

Jata, K.V., Semiatin, S.L., 2000. Continuous dynamic recrystallization during friction stir welding of high strength aluminum alloys. Scr. Mater. 43, 743−749.

Jata, K.V., Hopkins, A.K., Rioja, R.J., 1996. The anisotropy and texture of Al−Li alloys. In: Driver, J.H., Dubost, B., Durand, R., Fougeres, R., Guyot, P., Sainfort, P., Surrey, M. (Eds.), Aluminium Alloys, Their Physical and Mechanical Properties: Proceedings, ICAA5, Grenoble France, Part 1. Trans Tech Publications, Basle, Switzerland, p. 647.

Jata, K.V., Panchanadeeswaran, S., Vasudevan, A.K., 1998. Evolution of texture, micro structure and mechanical property anisotropy in an Al−Li−Cu alloy. Mater. Sci. Eng. A257, 37−46.

Jata, K.V., Panchanadeeswaran, S., Vasudevan, A.K., 1999. Evolution of texture, microstructure and mechanical property anisotropy in an Al−Li−Cu alloy. In: Singh, A.K., Ray, R.K. (Eds.), Textures in Materials Research. Oxford and IBH Publishing Co. Pvt. Ltd., New Delhi, pp. 161−181.

Lee, C.S., Duggan, B.J., Smallman, R.E., 1993. A theory of deformation banding in cold rolling. Acta Metall. 41, 2265−2270.

Lee, E.W., Kalu, P.N., Brandao, L., Es-Said, O.S., Foyos, J., Garmestani, H., 1999. The effect of off-axis thermo mechanical processing on the mechanical behavior of textured 2095 Al−Li alloy. Mater. Sci. Eng. A265, 100−109.

Lyttle, M.T., Wert, J.A., 1996. The plastic anisotropy of an Al−Li−Cu−Zr alloy extrusion in unidirectional deformation. Metall. Mater. Trans. A 27A, 3503−3513.

McDarmaid, D.S., Peel, C.J., 1989. Aspects of damage tolerance in 8090 sheet. In: Sanders Jr., T.H., Starke Jr., E.A. (Eds.), Aluminium−Lithium Alloys: Proceedings of the Fifth International Aluminium−Lithium Conference, vol. 2. Materials and Component Engineering Publications, Birmingham, pp. 993−1002.

McDarmaid, D.S., Peel, C.J., 1990. Damage tolerance in recrystallized 8090 aluminium−lithium sheet. In: Khan, T., Effenberg, G. (Eds.), Advanced Aluminium and Magnesium Alloys: Proceedings of the International Conference on Light Metals. ASM International European Council, Brussels, pp. 139−146.

Mondal, C., Singh, A.K., Mukhopadhyay, A.K., Chattopadhyay, K., 2011. Formation of a single rotated brass-{110}⟨556⟩ texture by hot cross rolling of an Al−Zn−Mg−Cu−Zr alloy. Scr. Mater. 64, 446−449.

Panchanadeeswaran, S., Field, D.P., 1995. Texture evolution during plane strain deformation of aluminum. Acta Metall. Mater. 43, 1683−1692.

Pao, P., Cooley, L.A., Imam, M.A., Yoder, G.R., 1989. Fatigue crack growth in 2090 Al−Li alloy. Scr. Metall. 23, 1455−1460.

Peel, C.J., McDarmaid, D.S., 1989. The present status of the development and application of aluminium−lithium alloys 8090 and 8091. Aerospace 16 (5), 18−23.

Przystupa, M.A., Vasudevan, A.K., Rollet, A.D., 1994. Crystallographic texture gradients in the aluminum 8090 matrix alloy and 8090 particulate composites. Mater. Sci. Eng. A186, 35−44.

Rioja, R.J., 1998. Fabrication methods to manufacture isotropic Al−Li alloys and products for space and aerospace applications. Mater. Sci. Eng. A257, 100−107.

Rioja, R.J., Liu, J., 2012. The evolution of Al−Li base products for aerospace and space applications. Metall. Mater. Trans. A 43A, 3325−3337.

Roven, H.J., Starke Jr., E.A., Sodahl, O., Hjelen, J., 1990. Effect of texture on delamination behavior of a 8090 type Al–Li alloy at cryogenic and room temperature. Scr. Metall. 24, 421–426.

Sadananda, K., Jata, K.V., 1998. Creep crack growth behavior of two Al–Li alloys. Metall. Mater. Trans. A 19A, 846–853.

Singh, A.K., Saha, G.G., Gokhale, A.A., Ray, R.K., 1996. Evolution of texture and microstructure in a thermomechanically processed Al–Li–Cu–Mg Alloy. Metall. Mater. Trans. A 29A, 665–675.

Singh, A.K., Gokhale, A.A., Saha, G.G., Ray, R.K., 1999. Texture evolution and anisotropy in Al–Li–Cu–Mg alloys. In: Singh, A.K., Ray, R.K. (Eds.), Textures in Materials Research. Oxford and IBH Publishing Co. Pvt. Ltd., New Delhi, pp. 219–232.

Spriano, S., Dogoline, R., Baricco, M., 1988. Texture hardening and mechanical anisotropy in AA 8090-T851 plate. Mater. Sci. Eng. A257, 134–138.

Taylor, G.I., 1938. Plastic strain in metals. J. Inst. Met. 62, 307–324.

Tayon, W., Crooks, R., Domack, M., Wagner, J., Elmustafa, 2010. EBSD study of delaminations fracture in Al–Li 2090. Exp. Mech. 50, 135–143.

Van Houtte, 1988. A comprehensive mathematical formulation of an extended Taylor–Bishop–Hill model featuring relaxed constraints, the Renouard–Wintenberger theory and a strain rate sensitivity model. Textures Microstruct. 8 and 9, 313–350.

Vasudevan, A.K., Fricke, W.G., Malcolm, R.C., Bucci, R.J., Przystupa, M.A., Barlat, F., 1988a. On through thickness crystallographic texture gradient in Al–Li–Cu–Zr alloy. Metall. Trans. A 19A, 731–732.

Vasudevan, A.K., Fricke, W.G., Przystupa, M.A., Panchanadeeswaran, S., 1988b. Synergistic effects of crystallographic texture and precipitation on the yield stress anisotropy in Al–Li–Cu–Zr alloy. In: Proceedings of the Eighth International Conference on Texture of Materials (ICOTOM-8), The Metallurgical Society, Warrendale, PA, pp. 1071–1077.

Vasudevan, A.K., Przystupa, M.A., Fricke Jr., W.G., 1990. Texture-microstructure effects in yield strength anisotropy of 2090 sheet alloy. Scr. Metall. Mater. 24, pp. 1429–1234.

Vasudevan, A.K., Przystupa, M.A., Fricke Jr., W.G., 1995. Effect of composition on crystallographic texture of hot rolled Al–Li binary alloys. Mater. Sci. Eng. A196, 1–8.

Zeng, X.-H., Barlat, F., 1994. Effects of texture gradients on yield loci and forming limit diagrams in various aluminum-lithium sheet alloys. Metall. Trans. A 25A, 2783–2795.

Processing Technologies

Chapter 6

Melting and Casting of Aluminum−Lithium Alloys

Vijaya Singh and Amol A. Gokhale
Defence Metallurgical Research Laboratory, Hyderabad-58, India

Contents

6.1 Introduction	167	6.3 Crucible Materials	172
6.2 Melt Protection from the		6.4 Hydrogen Pickup and Melt	
Atmosphere	168	Degassing	174
6.2.1 Lithium Reactivity	168	6.5 Grain Refinement	174
6.2.2 Addition of Lithium		6.6 Casting Practices	178
via an Al−Li		6.6.1 Metal/Mold Reactions,	
Master Alloy	170	Castability, and Shape	
6.2.3 Li Addition and Further		Casting	178
Melt Processing Under		6.6.2 DC Casting	180
Flux Cover	170	6.7 Summary	181
6.2.4 Melting Under Inert		References	182
Atmosphere	171		

6.1 INTRODUCTION

Wrought aluminum alloys are produced mainly by melting in fuel-fired reverberatory furnaces in an air atmosphere, whereby fluxes are added to the melt to reduce atmospheric oxidation. The alloy melt is usually held at an appropriate temperature in larger capacity holding furnaces, where it undergoes melt degassing and filtration operations. Finally, the melt is Direct Chill (DC) cast into slabs and billets that are thermomechanically processed to various semi-finished product forms such as plates, sheets, extrusions, and forging stock (Polmear, 2006). Aluminum−lithium (Al−Li) alloys are produced by the same route, except that the casting production technology is much more challenging owing to the high reactivity of lithium in the molten alloy.

Though initial work on Al—Li melting and casting technologies was carried out in the 1960s and 1970s as part of the development of first-generation Al—Li alloys, these technologies were improved and made safer in the 1980s and 1990s during the development of second-generation Al—Li alloys. These improvements have largely been in the areas of melt protection, refining, and DC casting in order to obtain melts with low contents of inclusions and good chemical composition with low alkali metal impurities and hydrogen contents. There are no reports published in recent years on any further modifications or improvements to the melting and casting technology, perhaps because the third-generation alloys, with their lower Li contents, are easier to produce than the second-generation alloys. In this chapter, various aspects of melting and casting of Al—Li alloys are covered, based mainly on the experience with first- and second-generation alloys.

6.2 MELT PROTECTION FROM THE ATMOSPHERE

6.2.1 Lithium Reactivity

Melting and casting of Al- and Li-based alloys is difficult due to the high reactivity of lithium with oxygen and nitrogen. Table 6.1 lists the free energies (ΔG) of some oxidation reactions which are possible in commercial Al—Li alloys at 800 (Partridge, 1990) and 1000 K (Thermo-Calc®). The reactions are ranked in the order of decreasing driving forces (lower negative ΔG) for oxidation.

Although MgO is thermodynamically more stable than Li_2O, the higher mobility and higher atomic fraction of Li favor early formation of Li_2O rather than other compounds in Al—Li—Mg alloys (Tarasenko et al., 1980). Li reacts with dry oxygen already at temperatures above 100°C to form Li_2O. Similarly, Li reacts with water vapor (present in the furnace environment, crucible, etc.) and forms Li_2O and H_2. The presence of moisture and CO_2 in the furnace atmosphere also results in the formation of minor quantities of LiOH and $LiOH.H_2O$. Furthermore, molten lithium quickly reacts with nitrogen and forms black-colored hygroscopic nitride, Li_3N.

Lithium oxide and nitride are solid, and lithium hydroxide is molten, during melting and casting of Al—Li alloys. In the temperature range of 500—800°C, lithium can also react with hydrogen and form lithium hydride, LiH, which is relatively stable (Foote Mineral Co. Bulletin, 1984). LiH with melting point of 690.8°C and density 0.772 g/cm^3 has mutual solubility with LiF (Mackay, 1966). Thus, LiH is likely to float to the surface and dissolve in a flux cover containing LiF. The compound Li_3AlH_6 can also form in the presence of molten aluminum (Hill et al., 1984).

The various reaction products formed by reactions of Li with O_2, N_2, H_2, H_2O, etc., must be removed from the metal through degassing and filtration before casting in order to get sound cast billets suitable for further processing.

TABLE 6.1 Free Energy Change ΔG per Mole of Gaseous Reactants for Surface Reactions at 800 (Partridge, 1990) and 1000 K (Thermo-Calc® Calculations)

S. No.	Reaction	ΔG (kJ) at	
		800 K	1000 K
1	$2Mg + O_2 = 2MgO$	-1029	-986.9
2	$Li + Al + O_2 = LiAlO_2$	-1018.4	-974.5
3	$4Li + O_2 = 2Li_2O$	-988.4	-930.2
4	$Li + 5Al + 4O_2 = LiAl_5O_8$	-962.0	$-$
5	$1.33Al + O_2 = 0.66Al_2O_3$	-950.0	-907.6
6	$4LiH + O_2 = 2Li_2O + 2H_2$	-879.5	-880.2
7	$Li + Al + CO_2 = LiAlO_2 + C$	-622.8	-578.6
8	$4Li + CO_2 = 2Li_2O + C$	-592.8	-534.3
9	$4Li + Li_2CO_3 = 3Li_2O + C$	-494.4	-466.8
10	$2.5Li + 0.5Al + H_2O = 0.5LiAlO_2 + 2LiH$	-360.2	-318.3
11	$4Li + H_2O = Li_2O + 2LiH$	-345.2	-297.5
12	$0.5Li + 0.5Al + H_2O = 0.5LiAlO_2 + H_2$	-305.7	-294.6
13	$2Li + H_2O = Li_2O + H_2$	-290.7	-273.7
14	$2LiH + H_2O = Li_2O + 2H_2$	-290.7	-249.9
15	$0.66Al + H_2O = 0.33Al_2O_3 + H_2$	-271.5	-255.4
16	$1.32Li + CO_2 = 0.66Li_2CO_3 + 0.33C$	-263.2	-225.8
17	$Cu + O = CuO$	-223.0	-233.9
18	$2Li + H_2O = LiOH + LiH$	-185.1	-155.7
19	$Li_2O + CO_2 = Li_2CO_3$	-98.4	-71.5
20	$Li_2O + H_2O = 2LiOH$	-25.1	-13.8

Thermo-Calc® is copyrighted software of Thermo-Calc Software AB and Foundation of Computational Thermodynamics, Stockholm, Sweden.

As mentioned earlier, lithium's high reactivity makes it difficult to add to aluminum alloys. Three methods are reported in the literature. These are (i) Li addition in the form of an Al–Li master alloy, (ii) Li addition and further melt processing under a flux cover, and (iii) lithium addition under vacuum or inert gas atmosphere. These methods will be discussed in the next three sections.

6.2.2 Addition of Lithium via an Al–Li Master Alloy

Owing to the high reactivity of Li in the pure form, some researchers have tried to prepare Al–Li master alloys and add them to aluminum melts. Al–Li master alloys have generally been prepared by electrochemical reactions. The advantage of direct production of Al–Li alloys by electrolysis is that alloys with very low contents of other alkali metal impurities like Na, K, and Ca can be made. For example, an Al–20% Li master alloy with Na, K, and Ca contents each below 5 ppm was prepared by LiCl–KCl molten salt electrolysis using Al as the cathode (Watanabe et al., 1987). Kamaludeen et al. (1987) prepared an Al–11% Li master alloy by fused salt electrolysis of an LiCl, KCl, and LiF mixture using molten Al as the cathode. Sumitomo Light Metal Industry, Japan, was reported to be commissioning pilot plant production of Al–Li alloys by electrolysis (Mahi et al., 1986). However, there were no further reports on the plant being operational.

The Al–Li phase diagram (Figure 3.1) shows that the liquidus temperatures of Al-based master alloys containing 10–20 wt% Li are in the temperature range of 650–700°C. This is below the temperature range at which aluminum melts are treated before casting, and so adding Al–Li master alloys should be easy. However, preparation of Al–Li alloys containing greater than 2 wt% Li requires relatively large amounts of master alloy additions (especially if the master alloy contains less than 10 wt% Li) that can cause sudden drops in melt temperature and disturb the melt cycle. In fact, if the master alloy is added all at once, it can lead to solidification of the aluminum melt in the crucible, requiring reheating for further processing. To avoid this situation, the master alloy should be either preheated sufficiently or added in several batches with intermittent superheating.

The use of Al–Li master alloys may be justified for making third-generation lower-lithium content alloys such as 2195, 2050, 2198, and 2060 (0.75–1.0% Li). However, since master alloys typically contain large atomic percentages of Li (e.g., 10–20 wt% Li corresponds to ~30–50 at.% of Li in binary Al–Li alloys), they are prone to atmospheric oxidation and require special storage even in the solid state. This partly reduces their processing advantage over using pure Li for making commercial Al–Li alloys.

6.2.3 Li Addition and Further Melt Processing Under Flux Cover

Fluxes used in alloy melting practices have two distinct functions: melt protection from the atmosphere and melt refining. For Al–Li alloys, the melting fluxes should be capable of protecting the reactive melts from atmospheric gases and should also refine the melts with respect to unwanted alkali metal impurities.

Conventional Al-alloy melting fluxes, which are normally based on NaCl–KCl–NaF formulations, lead to high pickup of Na in Al–Li melts. For example, sodium-containing carnallite-based fluxes or impure KCl-based fluxes resulted in as high as 50 ppm Na in the melt (Komarov et al., 1995). Instead, sodium-free flux (containing halides of Ca, Mg, and K) can be used to melt Al–Li alloys with some degree of success, since Na pickup is avoided and K pickup is negligible. However, Ca is picked up by the melt when $CaCl_2$ is present in the flux (Singh, 1997). Owing to the negligible pickup of K from KCl, it is possible to melt Al–Li alloys under an LiCl–KCl cover (Birch, 1986) and to cast by DC casting (US Patent No. 5415220, May 1995). High-purity LiCl, which has a melting point of 613°C, has also been used for melting Al–Mg–Li alloy 1420, resulting in less than 10 ppm Na pickup.

Successful melting of Li separately under high-purity LiCl–LiF flux cover has been reported by Starke et al. (1981) and Chakravorty (1988). Separate melting of Li under an LiCl–KCl cover and mixing into molten Al have been successfully practiced in the authors' laboratory for melts up to 50 kg. However, there is a possibility of flux entrapment in the final alloy, and the process is not amenable to automation and control.

6.2.4 Melting Under Inert Atmosphere

Owing to the difficulties associated with scale-up, flux cover melting is not used for industrial production. Like any other reactive metal, lithium can also be melted under argon/helium cover on an industrial scale. Good quality ingots of Al–Li alloys have been produced by melting in a vacuum-induction or resistance-melting furnace filled with high-purity argon or helium (Ashton et al., 1986; Cassada et al., 1986; Divecha and Karmarkar, 1981; Jones and Das, 1959; Lin et al., 1982; Suzuki et al., 1982). Jones and Das (1959) melted super-pure aluminum in a graphite crucible and plunged distilled lithium into it under argon cover. The alloy was poured into a cast iron mold under a rapid stream of argon. Dinsdale et al. (1981) used an argon atmosphere for melting Al–Li alloys in an induction furnace: lithium was added either in pure ingot form or as an Al–19.5 wt% Li master alloy. Similar observations were made by Hill et al. (1984), showing that an arc furnace could also be employed to melt the alloys. Cassada et al. (1986) reported melting of Al–Li alloys in a graphite crucible under argon atmosphere.

Divecha and Karmarkar (1981) reported resistance melting of Al–Li alloys in a dry box under a helium atmosphere. Melts of Al and Li were prepared in separate crucibles under protective atmosphere, and molten aluminum was poured into the crucible containing liquid lithium. High-purity helium had been previously bubbled through the molten lithium to remove impurities like Na, K, and Ca, although this process was not very effective.

For commercial alloys, flux-free melting and pouring under argon were employed to get improved quality Russian grade Al–Li alloy 1420 (Fridlyander et al., 1992). Adding solid Li into molten aluminum under argon flushing is also possible. Use of dichlorodifluoromethane gas with an inert gas for Al–Li melt blanketing has been found to result in a passive, self-healing viscous liquid layer that protects the melt from oxidation, lithium loss and hydrogen pickup (US Patent No. 4770697, September 1988). A vacuum induction degassing pouring furnace system developed by Leybold AG relied on melting and casting Al–Li alloys in a sealed chamber under argon atmosphere (Lorke and Stenzel, 1989). However, operational costs of all these practices would be expected to be high due to the large volume of inert gas required.

To overcome the high cost of inert gas, there are reports on melting Al–Li alloys under vacuum which promise to result in lower levels of alkali impurities, though higher loss of lithium would be expected compared to inert gas melting and casting. A vacuum refining process for reducing alkali metal impurities to very low levels (less than 1 ppm) was patented (US Patent No. 5320803, June 1994), promising to improve alloy toughness owing to the reduced levels of alkali elements. Among various options for melting, an inert gas flushing system has reached industrial-scale production due to its techno-economic viability compared to either completely sealed inert gas or vacuum melting.

6.3 CRUCIBLE MATERIALS

Since Li is a reactive metal, crucibles which are used to melt conventional aluminum alloys cannot be directly used to melt Li-bearing alloys. Special grades of stainless steel like 304L, 321, and 347 are reported to have very good resistance to lithium attack. Considering the various grades of stainless steels, Averill et al. (1981) pointed out that excellent service life can be obtained from 304L grade if the nitrogen concentration in liquid lithium is less than 100 ppm. In certain cases, 2.5Cr–1.0Mo steel has been found to have better corrosion resistance than 304L- and 321-type stainless steels. HSLA steels have also been suggested as possible materials to melt lithium. Pure iron has an excellent resistance to lithium attack, but this resistance decreases under stress, thus limiting its practical application.

If aluminum is melted in the same crucible as lithium, it will result in pickup of Fe (and Ni in the case of stainless steel) from the crucible material, owing to the high solubility of these elements in liquid Al. For example, Divecha and Karmarkar (1981) used pure iron as the crucible material for preparation of Al–Li alloys but reported an iron pickup problem. They further suggested the use of tantalum as crucible material because of its very good resistance to both molten lithium and aluminum. However, tantalum is very expensive, making it less feasible commercially. Also, tantalum is highly susceptible to lithium attack when the oxygen concentration in tantalum exceeds a threshold value of 100 ppm (Klueh, 1974).

Graphite is a widely used material for making Al-melting crucibles, as it does not add metallic impurities to the melt (Aluminum Casting Technology, 1986). Graphite has been used by some researchers (Cassada et al., 1986; Lin et al., 1982; Jones and Das, 1959) for preparing Al–Li alloys under inert atmosphere. However, graphite's suitability for melting Al–Li alloys in air is questionable since lithium reacts with carbonaceous materials in air. Clay-bonded graphite crucibles have been found compatible with melts containing up to 1.5 wt% lithium (Divecha and Karmarkar, 1986).

From a thermodynamic stability point of view, MgO-based refractories may be suitable for use as ramming mass in induction-melting furnaces, but there is no information about this in the open literature. A silicon carbide crucible has been used by Birch (1986) for holding Al–Li alloy melts under a flux cover added to avoid attack on the crucible. However, the use of alumina, magnesia, and silicon carbide as crucible materials is not recommended by Divecha and Karmarkar (1986) because of the risks of molten metal attack and crucible shattering.

Besides metallic and ceramic materials, composite materials have been reported to be used for Al–Li alloy melting. Oka et al. (1992) have evaluated different refractories for melting Al–Li alloys and found that high-purity alumina + spinel and composites of alumina + silicon nitride and graphite + alumina have good resistance to corrosion from alloy melts. Use of nitride (silicon nitride, silicon oxynitride)-bonded silicon carbide refractory bricks with mortar mix containing particulate nitride filler and a colloidal sol binder has been patented for melting Al–Li alloys by Alcoa (US Patent No. 4581295 April 8, 1986). A refractory material based on graphite—silicon carbide—alumina, and with good infiltration and penetration resistance, has been patented in Japan (Japan Patent No. 1992–285065 October 9, 1992).

Since there is no single crucible material compatible with both aluminum and lithium, it is necessary to adopt a melting and casting sequence which avoids long holding times after mixing Li and Al, in order to reduce the crucible attack and melt contamination. Similarly, melt transfer to the casting station is another area of concern, due to the requirement of controlling atmospheric oxidation and reaction between the melt and the container materials. Developments in this area are generally proprietary in nature (Miller et al., 1984), although some information is available in the open literature. In one case, lithium was encapsulated in super-purity aluminum foil and then plunged into molten aluminum using a graphite rod (Jones and Das, 1959). Another method used by some researchers is to transfer the aluminum melt into the lithium melt (Chakravorty, 1988; Divecha and Karmarkar, 1981). A technique of alloy mixing reported in a patent filed by Alcoa involves continuous pouring of two separate streams of molten aluminum and lithium into a small mixing vessel, which then allows an outflow of an Al–Li alloy stream (US Patent No. 4556535 December 3, 1985). Since the mixing vessel is small, costly refractories or tantalum metal can be used as linings for this vessel.

6.4 HYDROGEN PICKUP AND MELT DEGASSING

Lithium-containing alloys are more prone to hydrogen pickup than conventional aluminum alloys. Higher solubility of hydrogen in liquid Al–Li alloys than in pure aluminum or other alloys poses greater difficulty in degassing them. Despite this, the incidence of porosity in Al–Li alloy ingots is much less than for ingots of other alloys cast from liquid metals having the same hydrogen content. This lower porosity is due to the relatively low solubility of hydrogen in the solid compared to the solubility in the liquid at temperatures near the liquidus and solidus temperatures (Talbot, 1988). Higher solubility of hydrogen in liquid aluminum alloy depends on alloying and decreases with increase in Cu, Si, Zn, and Fe levels but increases with rising levels of Mg, Li, and Ti (Anyalebechi, 1995). Also, hydrogen content in the Al–Li melt increased from 0.96 to $3.14\,cm^3/100\,g$ when the Li content increased from 0.5 to 5 wt% (Fedosov et al., 1992).

Control of hydrogen content in the alloy is possible through the use of suitable fluxes, inert gas shrouding, and degassing using inert/ reactive gases. Use of an LiCl–LiF flux cover is reported to reduce the hydrogen pickup (Starke et al., 1981). LiF can also help in removing hydrogen, since LiH has good solubility in lithium fluoride (Mackay, 1966; Messer and Mellor, 1960). Al–Li alloys have been produced with hydrogen levels as low as 0.3 ml/100 g by a proprietary technique employing fluxes and inert gas treatment (Miller et al., 1984). Hydrogen gas nucleates at the oxide particles suspended in the melt and, coupled with inert gas treatment, can effectively degas the Al–Li melts (Tiwari and Beech, 1978). Laboratory-scale melts have been effectively degassed in the authors' laboratory using hexachloroethane tablets (Chakravorty et al., 1987; Singh, 1997). However, this technique is not suitable for use on an industrial scale owing to environmental concerns. On an industrial scale, melt holding under vacuum and/or argon purging through the melt have been used for effective degassing (Ji et al., 1992; Semenchenkov et al., 1992), with the best results from combined vacuum degassing and argon purging (Komarov, 1995; Makarov and Komarov, 1994).

6.5 GRAIN REFINEMENT

Like any other aluminum alloy, control of final grain size in Al–Li alloys is important to achieve high strength, ductility, toughness, fatigue life, and fatigue crack growth resistance. A fine cast grain size is also desired for improving hot workability of the alloy. The final grain size in wrought products depends on two factors: the fine as-cast grain size and grain growth control during thermomechanical processing. Aluminum is one of the few metals in which a small addition of one or more of many elements, including boron, niobium, tantalum, titanium, zirconium, vanadium, molybdenum, tungsten, hafnium, and chromium, cause grain refinement during casting without substantially affecting the properties. Combinations of titanium and boron in TiBAl master alloys are most commonly used (De Ross and Mondolfo, 1980).

Grain refinement of the second-generation Al—Li alloy AA8090 has been studied in detail by Birch (1986) using 0.2 and 0.4 wt% of 3/1 TiBAl and 5/1 TiBAl master alloys, respectively. Some of the results of their work, shown in Figure 6.1, bring out two important observations: (i) in all cases, there is an optimum holding time with respect to grain refinement and (ii) higher amounts of 0.4 wt% 3/1 TiBAl are required to give maximum grain refinement.

Al—Li alloys contain Zr to control grain size during high-temperature processing and to promote the formation of composite $Al_3Zr—Al_3Li$ dispersoids, which improve alloy ductility by minimizing the planar slip. The presence of Zr in the alloy makes grain refining difficult due to a poisoning effect: poisoning by Zr is supposed to be due to the presence of a zirconium boride layer on the surface of the TiB_2 nuclei when the ratio of zirconium to

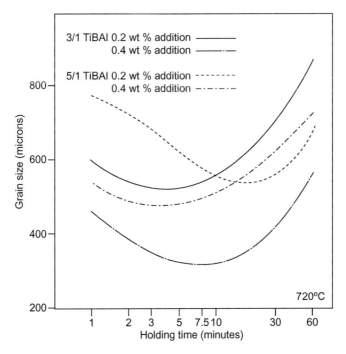

FIGURE 6.1   Effect of titanium content and addition rate of TiBAl alloy on grain refinement of AA8090 (Birch, 1986).

titanium contents is very high (Jones and Pearson, 1976). However, work by Spittle and Sadli (1994) on grain refinement of 99.99% Al by inoculation with Ti + B has shown that Zr does not result in the poisoning of Ti-containing nucleant substrate, and that the coarsening of as-cast Al (-Zr) alloys, grain refined with Ti + B, is possibly due to intermediate phases formed as a result of interaction between Zr and other solute elements present in the alloy. Chakravorty (1988) has also reported the use of 0.8 wt% of 3/1 TiBAl for effective grain refinement of binary Al–Li alloys as against 0.2 wt% of grain refiner required for most commercial Al-alloys. Birch (1986) has also shown that copper additions up to 1.5% and magnesium additions up to 0.5% reduce the grain size of binary Al-alloys, but increasing lithium content did not change the grain refining behavior.

Besides TiBAl, use of carbide for grain refinement of Al–Li alloys has been tried for Al–Li alloys by Birch and Cowell (1987). In their experiments a TiCAl (8 wt% Ti and 1 wt% C) master alloy was used for grain refinement of 8090. The results of this grain refiner addition were compared with the results for 0.2 and 0.4 wt.% 3/1 TiBAl additions, respectively (Figure 6.2). This shows that the test-to-test variations in grain refining response were less using TiCAl. It is also seen that for a given wt% addition the grain size was finer with TiCAl than with TiBAl: the unrefined grain size was ~1400 μm, which refined to ~210 and 275 μm with 0.4 wt% addition of 8/1 TiCAl and 3/1 TiBAl, respectively. The improvement in grain refinement by TiCAl

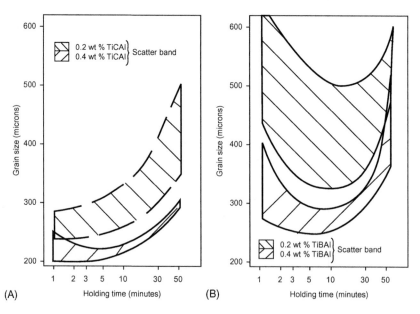

FIGURE 6.2   Effect of grain refining 8090 alloy with (A) 8/1TiCAl and (B) 3/1 TiBAl (Birch and Cowell, 1987).

inoculation is attributed due to the nonpoisoning of TiC nuclei by Zr. Using an Al–10% Ti master alloy, Birch and Cowell (1987) have also confirmed that grain refinement due to titanium alone is considerably enhanced by the addition of carbon to form titanium carbide. However, the use of TiCAl as a grain refiner has not been reported commercially.

The work of Labarre et al. (1987) on grain refinement of 2090 and 2091 alloys by Al–Ti and Al–Ti–B master alloys has revealed that the effect of Ti on grain refinement is very large, and therefore it should be controlled very carefully. They also showed a strong positive effect of magnesium on grain refinement, which indicated that the practice for the grain refinement of magnesium-containing Al–Li alloys would have to be different from that of other Al–Li alloys. However, the boron content and temperature of the melt were found to be less critical.

Vives et al. (1987) have studied the effect of electromagnetic stirring on grain refinement of Al–Li alloys and found a significant reduction in the diameter of columnar crystals and finer equiaxed grains. Their study also showed better temperature uniformity for the bulk liquid and a faster decrease of superheating because of the vigorous forced convection induced by the electromagnetic forces. This technique also has the advantage of avoiding melt contamination, since there is no direct contact between the melt and a mechanical stirrer.

In the authors' laboratory, an Al–Li alloy equivalent to the 8090 composition and inoculated with 0.2 wt% 5/1 TiBAl was cast under LiCl–LiF flux cover and held and poured under argon cover into a rectangular metal mold, resulting in a rectangular alloy plate (Figure 6.3). The mold cavity size was approximately $300 \times 300 \times 33$ mm, and the average cooling rate at the center

FIGURE 6.3 Al–Li alloy being poured into the metal mold (with argon gas flushing in the crucible as well as in the metal mold).

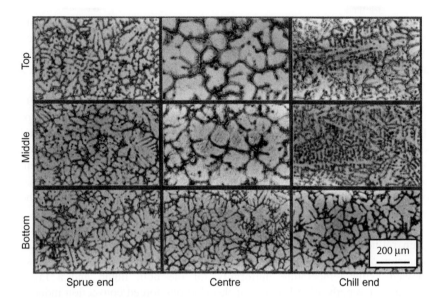

FIGURE 6.4 Microstructures of TiBAl grain-refined 8090 alloy in the as-cast condition for different locations in the metal mold.

of the mold was measured as 0.7°C/s. The alloy microstructures from different locations of the alloy plate are shown in Figure 6.4. These results show a cellular dendritic solidification structure with equiaxed grains in the center of the plate and elongated grains near the bottom. This grain shape variation is most probably due to low thermal gradients and slow cooling in the center of the plate and high cooling rates and high thermal gradients near the bottom of the plate.

Further grain refinement of this nominally 8090 alloy was possible by combined addition of zirconium and scandium (Singh and Gokhale, 2000; Singh et al., 2004). Addition of 0.11 and 0.22 wt% Sc did not change the grain size, which was found to be in excess of 100 μm. Higher levels of Sc (0.43 and 0.84 wt%) resulted in a fine nondendritic structure with the grain size reduced to less than 30 μm (Figure 6.5). However, high Sc additions are not acceptable in wrought alloys owing to the formation of coarse particles in the alloy, which may adversely affect the fatigue properties. The high cost of Sc is another reason for restricting its addition to aluminum alloys.

6.6 CASTING PRACTICES

6.6.1 Metal/Mold Reactions, Castability, and Shape Casting

All commercial Al–Li alloys are wrought alloys, hence the importance of castability evaluation is limited. In the late 1980s, when second-generation Al–Li alloys development work was at its peak, there were reports on evaluation of

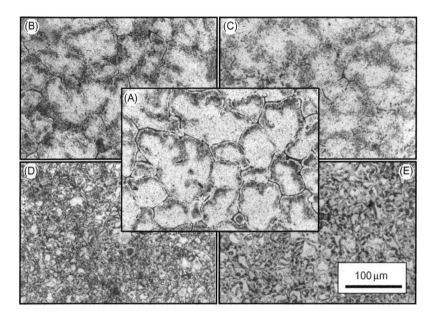

FIGURE 6.5 Optical micrographs of as-cast ingots having 0.12% Zr with (A) 0% Sc, (B) 0.11% Sc, (C) 0.22% Sc, (D) 0.43% Sc, and (E) 0.84% Sc.

metal/mold reactions and castability of these alloys. Use of CO_2—sodium silicate and sodium aluminate-bonded silica sand systems (Chakravorty, 1988)—for shape casting of Al–Li alloys was shown to result in severe gas porosity due to a metal/mold reaction: graphite wash was found useful to prevent this reaction. On the other hand, Al–Li alloys could be successfully cast in metal and graphite molds or no-bake organic binder-based sand molds.

The effect of superheating and lithium content on the fluidity of binary Al–Li alloys was also studied from the viewpoint of casting design. It was found that fluidity gradually increases with superheating and decreases with increasing lithium content of the alloy. Tong et al. (1987) studied the castability of Al–Li–Mg and Al–Li–Cu–Mg alloys containing 2.5 wt% Li. They found that cast iron molds and resin-bonded sand molds were suitable for Al–Li castings, while a severe metal-mold reaction occurred with a sodium silicate-bonded sand mold. Furthermore, mold coatings of alumina, zirconia, and silicon carbide were found to reduce the surface scaling on the castings. With respect to alloying, the Al–Li–Mg alloys had better combinations of fluidity and strength. Tong et al. also claimed that Al–Li–Mg investment castings could be used in place of forgings, since the castings had about 80% of the strength of their forged counterparts.

The work of Webster et al. (1988) on thin-walled complex-shaped Al–Li investment castings with 1.1–3.6 wt% Li additions showed that these had mechanical properties well above the baseline properties of the conventional aluminum casting alloy A356 and with lower densities. Colvin et al. (1988) developed Al−Li−Cu−Mg alloy premium quality investment castings with 5% higher specific strength and 11% higher modulus than A357 alloy. Haynes et al. (1986, 1987) have also explored the possibility of using Al−Li alloys for producing near-net-shaped components by investment casting. However, so far no commercial application of any cast Al−Li alloy is reported, possibly because of low ductility and/or proneness to brittle intergranular fracture.

6.6.2 DC Casting

Like any other aerospace aluminum alloy, the production of Al−Li alloys involves the process of semicontinuous, vertical, DC casting to convert the molten alloy into a solid ingot for further processing. In comparison to conventional aluminum alloys, the problem of microsegregation is less severe in Al−Li alloys owing to the lower partition coefficient (0.5) of Li in Al (Adam, 1981). The microsegregation problem is further reduced when the DC casting method is used to produce Al−Li alloy ingots. The most important considerations in the DC casting of Al−Li alloys are the increased risk of cracking (Bretz, 1987) and the explosion hazard when molten alloy comes in contact with water (Bretz and Gilliland, 1987; Binczeveki, 1986; Long, 1957). The explosion hazard of molten Al−Li alloys with various coolants like water, ethyl alcohol, propyl alcohol, propylene, and ethylene glycol has been studied (Page et al., 1987), and it was reported that the energy release in an explosion increases exponentially with lithium content (Figure 6.6).

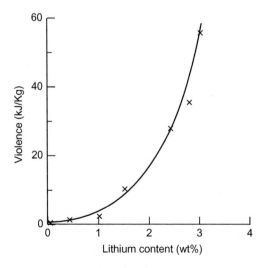

FIGURE 6.6 Energy released versus lithium content in an explosion with water coolant (Page et al., 1987).

Thus the casting problems have been potentially more severe for the second-generation high-Li 8XXX series alloys than for the lower-Li content 2XXX series alloys (Bretz and Gilliland, 1987).

Page et al. (1987) also showed that when the lithium content is 0.6–2.5 wt%, the energy released in an explosion with ethylene glycol as coolant is very low. Therefore the advantages of using ethylene glycol over water are most important, and patents were obtained by Alcoa in this regard (US Patent Nos. 4610295, September 1986, and 4709740, December 1987): their preferred coolant contains ethylene glycol with a maximum of 25 vol.% water. Another technique to prevent explosion hazard involves *in situ* cladding of DC cast Al–Li ingots with pure aluminum (US Patent No. 4567936, February 1986). From a practical point of view, DC casting pits are designed such that there is no water accumulation, so that an explosion is avoided if melt bleed out/run out occurs. Such dry pit technology has been patented (EP 0150922 A2, August 1985). In this technique, casting is commenced without a pool of water within the pit, supplying cooling water to the emerging ingot at a predetermined rate, and continuously removing water from the pit to ensure no buildup of a water pool as casting continues.

The advent of third-generation Al–Li alloys has resulted in increased industrial investment to set up dedicated melting and casting facilities. In July 2010, Alcan Global ATI unveiled plans for setting up a production site at Issoire and an R&D center at Voreppe, both located in France. These facilities are now part of the Constellium consortium. Aleris Aluminum Koblenz GmbH, Germany, announced in December 2011 that a specialized casting facility for Al–Li plate and sheet products would be built, and that this would be able to cast full-scale ingots for trials as well as for serial production of Al–Li by the first quarter of 2013. Similarly, in May 2012, Alcoa celebrated launch of its "green field" state-of-the-art 20,000 metric tons aluminum–lithium plant. This facility is located next to its existing plant at Lafayette and, via a US$90 million, 115,000 square feet expansion, will start producing its third-generation Al–Li alloys by the end of 2014.

Finally, it is worth noting that owing to their lower Li contents, less than 2 wt% (see Table 2.1), the DC casting of third-generation Al–Li alloys is expected to be easier than that of the earlier alloys.

6.7 SUMMARY

Melting and casting practices for Al–Li alloys differ from those for conventional Al-alloys because of the high reactivity of molten lithium with the atmosphere and the materials with which it comes into contact during processing. To overcome reactivity problems, options such as melting under Li-halide-containing specialized fluxes, addition of Li in the form of master alloys, and vacuum or inert gas melting have been suggested. Melting under inert gas (argon) has been commonly practiced on an industrial scale.

Stainless steel 304L, pure iron, tantalum metal, and certain ceramics have been suggested for containment of Al–Li melts. The investigated ceramics include compositions based on alumina, silicon carbide, and silicon carbide + silicon oxynitride.

Al–Li alloys have a higher solubility for hydrogen than other aluminum alloys when molten. However, the occurrence of porosity in the solid Al–Li ingots is inhibited owing to the much lower solubility of hydrogen in the solid phase at temperatures near the liquidus and solidus temperatures. Although the use of an LiF flux for hydrogen removal from the melt has been reported, vacuum degassing and argon purging, individually or in combination, have been adopted for effective degassing of industrial-scale Al–Li melts.

Grain refinement of Al–Li alloys is difficult due to the presence of Li and Zr in the alloys. Higher than normal amounts of grain refiner (TiBAl and TiCAl) are needed for effective refinement. Addition of scandium with zirconium also results in grain refinement of the alloy, but the possibility of coarse scandium-containing particles being present in such alloys is undesirable from the point of view of fatigue properties. This is in addition to the high cost of scandium.

There are no reported applications for cast Al–Li alloys. Only wrought Al–Li alloys are being produced for aerospace applications. However, like conventional Al-alloys, these wrought Al–Li alloys are also DC cast into solid ingots for further processing. The explosion hazard during DC casting Al–Li alloys is higher than that for conventional alloys, and the risk increases exponentially with increasing lithium contents in the alloys. To avoid this hazard, it is recommended to use ethylene glycol as coolant and the dry pit technology, whereby the accumulation of cooling water in the casting pit is avoided. Since most of the melting and casting problems are due to lithium's presence and reactive nature, third-generation alloys, which contain less than 2 wt% Li, should be less difficult to melt and cast.

REFERENCES

Adam, C.M., 1981. Overview of D.C. casting. In: Sanders, T.H., Starke, E.A. (Eds.), Aluminium–Lithium Alloys. AIME, New York, NY, pp. 37–48.

Aluminum Casting Technology, 1986. American Foundrymen's Society, Inc. Des Plaines, Illinois, 18.

Anyalebechi, P.N., 1995. Analysis of the effects of alloying elements on hydrogen solubility in liquid aluminum alloys. Scripta Met. Mater. 33 (**8**), 1209–1216.

Ashton, R.F., Thompson, D.S., Starke Jr., E.A., Lin, F.S., 1986. Processing Al–Li–Cu–(Mg) alloys. In: Baker, C., Gregson, P.J., Harris, S.J., Peel, C.J. (Eds.), Aluminium–Lithium Alloys III. The Institute of Metals, London, UK, pp. 66–77.

Averill, W.A., Olsen, D.L., Matlock, D.K., Edwards, G.R., 1981. Lithium reactivity and containment. In: Sanders, T.H., Starke, E.A. (Eds.), Proceedings of the First International Aluminium–Lithium Conference. The Metallurgical Society of AIME, Warrandale, PA, pp. 9–28.

Binczeveki, G., 1986. Producing aluminium—lithium ingot encased in outer metal cladding. Light Met. Age June, 31—36.

Birch, M.E.J., 1986. Grain refining of aluminium—lithium based alloys with titanium boron aluminium. In: Baker, C., Gregson, P.J., Harris, S.J., Peel, C.J (Eds.), Aluminium—Lithium Alloys III, The Institute of Metals, London, UK, pp. 152—158.

Birch, M.E.J., Cowell, A.J., 1987. Titanium—carbon—aluminium: a novel grain refiner for aluminium—lithium alloys. In: Champier, G., Dubost, B., Miannay, D., Sabetay, L. (Eds.), Proceedings of the Fourth International Conference on Aluminium—Lithium Alloys. J. Phys., Colloque, 48, C3.103—C3.108.

Bretz, P.E., 1987. Alithalite® alloy development and production. In: Champier, G., Dubost, B., Miannay, D., Sabetay, L. (Eds.), Proceedings of the Fourth International Conference on Aluminium—Lithium Alloys. J. Phys., 48, C3.25—C3.31.

Bretz, P.E., Gilliland, R.G., 1987. The intensive development programme that produced aluminium—lithium alloys. Light Met. Age April, 5—12.

Cassada, W.A., Shiflet, G.J., Starke, E.A., 1986. The effect of Germanium on the precipitation and deformation behaviour of Al—2Li alloys. Acta Metall. 34 (3), 367—378.

Chakravorty, C.R., 1988. Studies on the Characteristics of Cast Binary Aluminium—Lithium Alloys, Doctoral thesis. Indian Institute of Technology, Kharagpur, India.

Chakravorty, C.R., Singh, V., Gokhale, A.A., 1987. Melting and casting of aluminium—lithium alloys. Proceedings of the 36th Annual Convention of Institute of Indian Foundrymen (IIF), The Institute of Indian Foundrymen, Calcutta, India, 135—140.

Colvin, G.N., Tak, J.H., Veeck, S.J., 1988. An investment cast Al—Li alloy HTC 321. In: Kar, R.J., Agarwal, S.P., Quist, W.E. (Eds.), Aluminium—Lithium Alloys: Design, Development and Application Update. ASM International, Metals Park, OH, pp. 453—465.

De Ross, A.B., Mondolfo, L.F., 1980. Metallurgical aspects of casting aluminium alloys. In: Pampilo, C.A., Biloni, H., Embury, D.E. (Eds.), Aluminium Transformation Technology. ASM, Metals Park, OH, pp. 81—140.

Dinsdale, K., Harris, S.J., Noble, B., 1981. Relationship between microstructure and mechanical properties of aluminium—lithium—magnesium alloys. In: Sanders, T.H., Starke, E.A. (Eds.), Aluminium—Lithium Alloys, Proceedings of the First International Aluminium—Lithium Conference. The Metallurgical Society of AIME, Warrendale, PA, pp. 101—118.

Divecha, A.P., Karmarkar, S.D., 1981. Casting problems specific to aluminium-lithium alloys. In: Sanders, T.H., Starke, E.A. (Eds.), Aluminium—Lithium Alloys I. AIME, New York, NY, pp. 49—62.

Divecha, A.P., Karmarkar, S.D., 1986. The search for aluminium-lithium alloys. Adv. Mater. Processes Inc. Met. Prog. 10, 75—79.

Fedosov, A.S., Danilkin, V.A., Grigoryeva, A.A., 1992. Influence of lithium on balance between hydrogen and aluminium melts. Tsvetn. Met. 2, 61—63.

Foote Mineral Co. Bulletin, 1984. Technical Data Bulletin 101.

Fridlyander, I.N., Bratukhin, A.G., Davydov, V.G., 1992. Soviet Al—Li alloys of aerospaceapplication. In: Peters, M., Winkler, P.J. (Eds.), 'Aluminium—Lithium', Volume 2, Conf. Proc. Sixth International Aluminium-Lithium Conference, 1991. DGM Informationgesellschaft, Verlag, Germany, pp. 35—42.

Haynes, T.G., Tesar, A.M., Webster, D., 1986. Developing aluminium—lithium alloys for investment casting, Modern Casting. 76 (10), 26—28.

Haynes, T.G., Wyte, M., Webster, D., 1987. Mechanical properties and microstructure of Al—Li investment castings. In: Champier, G., Dubost, B., Miannay, D., Sabatay, L. (Eds.),

Proceedings of the Fourth International Conference on Aluminum–Lithium Alloys. J. Phys., Colloque, 48, C3.123–C3.128.

Hill, D.P., Williams, D.N., Mobley, C.E., 1984. The effect of hydrogen on the ductility, toughness and yield strength of an Al–Mg–Li alloy. In: Sanders, T.H., Starke, E.A. (Eds.), Proceedings of the Second International Conference on Aluminum–Lithium Alloys. The Metallurgical Society of AIME, Warrendale, PA, pp. 201–218.

Ji, D.X., Tian, S.X., Chen, C.Q., 1992. A promising melting processing for Al–Li alloys—vacuum melting. In: Peters, M., Winkler, P.J. (Eds.), Aluminium–Lithium Alloys VI. Deutsche Gesellschaft fur Materiakundi, Germany, pp. 863–876.

Jones, G.P., Pearson, J., 1976. Factors affecting the grain refinement of aluminium using titanium and boron additives. Metall. Trans. 7B, 223–234.

Jones, W.R.D., Das, P.P., 1959. The mechanical properties of aluminium–lithium alloys. J. Inst. Met. 80, 435–443.

Kamaludeen, M., Ranganathan, N.G., Sundaram, M., Vasu, K.I., 1987. Bull. Electrochem. 3 (2), 143–145.

Klueh, R.L., 1974. Oxygen effects on the corrosion of niobium and tantalum by liquid lithium. Metall. Trans. 5B, 875–879.

Komarov, S.B., 1995. Improvement of the process for vacuum treatment of Al–Li alloys. Tekhnologiya Legkikh Splavov 5, 43–45.

Komarov, S.B., Mozharovsky, S.M., Ovsyannikov, B.V., Makarov, G.S., Grushko, O.E., 1995. Protection and refining of aluminium–lithium alloys by fluxes. NonFerrous Met. (in Russian) 10, 57–60.

Labarre, L.C., James, R.S., Witters, J.J., O'Malley, R.J., Emptage, M.R., 1987. Difficulties in grain refining aluminium lithium alloys using commercial Al–Ti and Al–Ti–Br master alloys. In: Champier, G., Dubost, B., Miannay, D., Sabetay, L. (Eds.), Proceedings of the Fourth International Conference on Aluminium–Lithium Alloys. J. Phys., 48, C3. 93–C3.102.

Lin, F.S., Chakraborty, S.B., Starke Jr., E.A., 1982. Microstructure—property relationships of two Al-3Li-2Cu-0.2Zr-xCd alloys. Metall. Trans. 13A, 401–410.

Long, G., 1957. Explosion of molten metal in water—causes and prevention. Met. Prog. 71, 107–112.

Lorke, M., Stenzel, O.W., 1989. Melting of Al–Li alloys-the equipment manufacturer's view. In: Sanders Jr., T.H., Starke Jr., E.A. (Eds.), Proceedings of the Fifth International Conference on Aluminum–Lithium Alloys, vol. 1. MCEP, Birmingham, UK, pp. 41–53.

Mackay, K.M., 1966. Ionic Hydrides in Hydrogen Compounds of the Metallic Elements. F.N. Spon Ltd, London, UK, pp. 18–35.

Mahi, P., Smeets, A.A.J., Fray, D.J., Charles, J.A., 1986. Lithium metal of the future. J. Met. 38, 20–26.

Makarov, G.S., Komarov, S.B., 1994. Effective methods of Al–Li melt degassing. Tekhnologiya Legkikh Splavov, 9–12, 94–98.

Messer, C.E., Mellor, J., 1960. The system lithium hydride–lithium fluoride. J. Phys. Chem. 64 (4), 503–505.

Miller, W.S., Cornish, A.J., Titchener, A.O., Bennet, D.A., 1984. Development of lithium containing aluminium alloys for the ingot metallurgy production route. In: Sanders, T.H., Starke, E.A. (Eds.), Aluminum–Lithium Alloys II. The Metallurgical Society of AIME, Warrandale, PA, pp. 335–362.

Oka, K., Wakasaki, O., Ohzono, T., Hayashi, Y., Washio, S., Yamamoto, K., 1992. Behavior of refractory for molten Al–Li alloy. In: Peters, M., Winkler, P.J. (Eds.), Proceedings of the

Sixth International Conference on Aluminum—Lithium Alloys. Deutsche Gesellschaft fur Materiakundi, Germany, pp. 877—882.

Page, F.M., Chamberlain, A.T., Grimes, R., 1987. The safety of molten aluminium—lithium alloys in the presence of coolants. In: Champier, G., Dubost, B., Miannay, D., Sabetay, L. (Eds.), Proceedings of the Fourth international Conference on Aluminum—Lithium Alloys. J. Phys., Colloque, 48, C3.63—C373.

Partridge, P.G., 1990. Oxidation of aluminium-lithium alloys in the solid and liquid states. Int. Mater. Rev. 35, 37—58.

Polmear, I.J., 2006. Light Alloys, fourth ed. Butterworth—Heinemann, pp. 97—99.

Semenchenkov, A.A., Fedosov, A.S., Davydov, V.G., 1992. Degassing of aluminium—lithium alloys. In: Peters, M., Winkler, P.J. (Eds.), Proceedings of the Sixth International Conference on Aluminium—Lithium Alloys. Deutsche Gesellschaft fur Materiakundi, pp. 873—876.

Singh, V., 1997. Preparation and Characterisation of Al—Li—Cu—Mg—Zr Based Alloys, Doctoral thesis. Indian Institute of Technology (BHU) [formerly Institute of Technology, Banaras Hindu University], Varanasi, India.

Singh, V., Gokhale, A.A., 2000. Control of grain structure through transition elemental additions in an Al—Li base alloy. Mater. Sci. Forum, 331—337, 447—482.

Singh, V., Prasad, K.S., Gokhale, A.A., 2004. Effect of minor Sc additions on structure, age hardening and tensile properties of aluminium alloy AA8090 plate. Scripta Mater. 50, 903—908.

Spittle, J.A., Sadli, S., 1994. The influence of zirconium and chromium on the grain refining efficiency of Al—Ti—B inoculants. Cast Met. 7 (4), 247—253.

Starke, E.A., Sanders, T.H., Palmer, I.G., 1981. New approaches to alloy development in the Al—Li system. J. Met. 33, 24—32.

Suzuki, H., Kanno, M., Hayashi, N., 1982. Aging phenomena in aluminium—2.26% lithium alloys. Aluminium 58 (2), 120—122.

Talbot, D.E.J., 1988. Influence of hydrogen on quality of aluminium and aluminium alloy ingots. In: Beech, J., Jones, H. (Eds.), Solidification Processing 1987. The Institute of Metals, pp. 29—36.

Tarasenko, L.V., Grushko, O.E., Zasypkin, V.A., Ivanova, L.A., 1980. Phase composition of surface films on aluminium alloys with lithium. Russ. Metall. 2, 174—177.

Tiwari, S.N., Beech, J., 1978. Origin of gas bubbles in aluminium. Met. Sci. 12, 356—362.

Tong, C.H., Yao, I.G., Nieh, C.Y., Chang, C.P., Hsu, S.E., 1987. Castability of Al—Li—Mg and Al—Li—Cu—Mg alloys. In: Champier, G., Dubost, B., Miannay, D., Sabetay, L. (Eds.), Proceedings of the Fourth International Conference on Aluminium—Lithium Alloys. J. Phys., Colloque, 48, C3.117—C3.122.

Vives, C., Bas, J., Cans, Y., 1987. Grain refining in Al—Li alloys by electromagnetic stirring. In: Proceedings of the Fourth International Conference on Aluminium—Lithium Alloys. J. Phys., 48, C3.109—C3.115.

Watanabe, Y., Toyoshima, M., Itah, K., September 1987. High purity aluminium—lithium master alloy by molten salt electrolysis. In: Proceedings of the Fourth International Conference on Aluminium—Lithium Alloys. J. Phys., 48, C3.85—C3.91.

Webster, D., Haynes, T.G., Flemings, R.H., 1988. Al—Li investment castings coming of age. Adv. Mater. Processes Inc. Met. Prog. J133 (6), 25—30.

Chapter 7

Mechanical Working of Aluminum–Lithium Alloys

G. Jagan Reddy*, R.J.H. Wanhill**, and Amol A. Gokhale*

*Defence Metallurgical Research Laboratory, Hyderabad-500058, India, **NLR, Emmeloord, the Netherlands*

Contents

7.1 Introduction	**187**	7.4.1 Process Maps and	
7.1.1 Overview	189	EBSD–OIM	196
Part 1: Workability	**191**	**7.5 Summary of the Hot**	
7.2 Introduction	**191**	**Deformation**	
7.3 General Review of Hot		**Investigations**	**203**
Deformation		**Part 2: Processing of Al–Li**	
Characteristics of Al–Li		**Alloys**	**203**
Alloys	**192**	**7.6 General Considerations**	**203**
7.3.1 Hot Workability		7.6.1 Rolled Products	205
of a Binary		7.6.2 Extrusions and	
Al–Li Alloy	192	Forgings	209
7.3.2 Hot Workability of		**7.7 Industrial-Scale**	
Multicomponent		**Processing**	**211**
Al–Li Alloys	193	7.7.1 Introduction	211
7.3.3 Hot Deformation		7.7.2 Review and	
Regions Favoring		Discussion of Available	
Superplasticity	195	Information for Third-	
7.4 Hot Deformation and		Generation Alloys	212
Process Mapping of		**7.8 Summary**	**216**
Al–Li Alloy UL40	**196**	**References**	**216**

7.1 INTRODUCTION

Most aerospace aluminum alloy components and structures are made from ingot metallurgy (IM) wrought products. These products are obtained by hot working, either by rolling, extrusion, or forging. The starting stock is DC cast aluminum alloy ingots, either round or slabs. Round ingots nearly always undergo initial mechanical processing (e.g., extrusion) before further working, and slabs are given a homogenization heat treatment before rolling.

Rolled products are classified either as sheet (from 0.15 to <6.3 mm) or plate (>6.3 mm) and are obtainable in widths up to about 2.5 m (Starke and Staley, 1996). For aircraft applications, sheet is used for fuselage skins and stringers; thin plate is used for frames; and thicker plate is used for wing covers, bulkheads, wing spars, and other supporting structures. Plate gauges ranging from thin to thick are also used for spacecraft liquid propellant tanks.

Extrusions have several applications covering a wide range of product thicknesses and profiles. Aircraft applications include fuselage stringers and frames, upper and lower wing stringers, and integrally stiffened floor beams and seat rails.

Forgings in near-net shapes are supplied as hand forgings or closed-die forgings. Hand forgings are generally produced in thicknesses from 75 to about 300 mm; die forgings have thicknesses up to about 250 mm (Starke and Staley, 1996). Both types of forgings provide alternatives to thick plate for bulkheads and other supporting structures, since forging to near-net shapes allows thinner cross-sectional products to be made before heat treatment, and also creates a favorable grain flow pattern that can improve the properties and avoid "end-grain" exposure in the finished product: end-grain exposure can be an important consideration for avoiding intergranular corrosion and stress corrosion. Additional benefits from forgings are that they require less machining, and they are likely to have less porosity than thick plate.

As an illustration of the use of various aluminum alloy product forms, Figure 7.1 shows the construction of a large aircraft wing. Apart from stringers, which are extrusions, all of the components are usually made from plates of varying thickness. However, forgings may be considered for heavy-section ribs if thick plate is deemed unsuitable, e.g. because the required thicknesses limit the properties that can be achieved by heat treatment of the plates before they are machined to the finished products.

Figure 7.2 shows a machined wing rib, illustrating the very large differences in thickness that may be required: the challenge for the alloy developers and producers is to provide a material and processing schedule that results in near-uniform properties. Figure 7.3 shows some as-forged bulkheads, illustrating the achievement of near-net shapes by this processing method.

Al—Li alloy wrought products

The wrought product forms developed for aluminum—lithium (Al—Li) alloys are listed in Table 7.1, which is compiled from various sources, including Alcoa and Constellium. The majority are rolled products (sheet and plate). More details, notably on the third-generation alloy specific applications and heat treatment conditions, are given in Chapters 2 and 15.

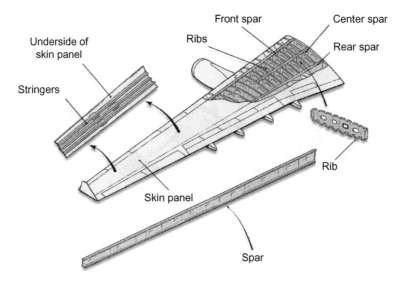

FIGURE 7.1 Schematic construction of a transport aircraft wing.

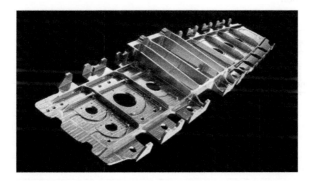

FIGURE 7.2 Aluminum alloy wing rib machined from thick plate, illustrating the large variations in thickness of the finished product.

7.1.1 Overview

This chapter is in two main parts. In Part 1 we first discuss the "workability" of metals and alloys, concentrating on thermomechanical (hot) working and introducing the concept of Process Maps, based on process modeling (Prasad et al., 1984). This is followed by a general review of the hot deformation characteristics of Al–Li alloys, and then an illustration of Process Mapping for a binary Al–Li alloy, UL40, subjected to hot compression and having different starting conditions (Jagan Reddy, 2010).

FIGURE 7.3   Aluminum alloy bulkheads (foreground) after pressing and before finish machining.

TABLE 7.1 Wrought Product Forms Developed for Al–Li Alloys

Product Forms	Alloys
First Generation	
Sheet, plate	1420; 2020
Extrusions	1420
Forgings	1420, 1421
Second Generation (Li ≥ 2wt%)	
Sheet	2090
Sheet, plate	1430, 1440, 1441, 1450, 1460; 2090, 2091, 8090
Extrusions	2090
Extrusions, forgings	1450; 2091, 8090
Third Generation (Li < 2wt%)	
Sheet, plate	2198, 2060, 2098, 2199
Plate	2050, 2055, 2099, 2195, 2297, 2397
Extrusions	2055, 2065, 2076, 2099, 2196, 2296
Forgings	2050, 2060

Part 2 gives an overview of the main processes of rolling, extrusion and forging of Al–Li alloys in Section 7.6. This is a general treatment, since the processes used by commercial manufacturers are essentially proprietary. However, some information about industrial processing is available from the open literature, and this is discussed specifically for the third-generation alloys in Section 7.7.

A summary of the entire chapter is given in Section 7.8.

PART 1: WORKABILITY

7.2 INTRODUCTION

Mechanical processing of engineering alloys is used not only to achieve the required shape but also to obtain microstructural changes that will lead to optimum properties after any subsequent heat treatments. In designing the working processes for products and components to be made from complex materials such as Al—Li alloys, the most important task is selection of the controlling process parameters that will ensure the required properties and an acceptable/desired final component quality. The controlling process parameters are the sequence and number of mechanical and thermomechanical working operations, the heat treatment conditions, and the associated quality assurance tests.

Hot working of metals and their alloys is performed at temperatures greater than about $0.5T_M$, where T_M is the melting point of the alloy measured on the Kelvin scale. The main hot working processes are rolling, extrusion, and to a lesser extent forging. During rolling, and also particularly during extrusion, the strain rate varies. Moreover, the plastic strain is not uniform throughout the material cross-section. Because of these processing features, and also their relatively complicated stress states, rolling and extrusion are seldom used to characterize the hot deformation characteristics (workability) of materials. Workability is an important engineering parameter: it refers to the ease with which a material can be shaped by plastic flow without the onset of fracture.

The workability of an alloy is influenced by its microstructure, the work piece temperature, applied strain rate, and stress—strain state in the deformation zone. It is therefore convenient to consider workability as composed of two independent parts: (i) the state of stress (SOS) workability and (ii) the intrinsic workability. The SOS workability depends upon the geometry of the deformation zone in which the work piece is subjected to a three-dimensional stress state, and the frictional conditions are based on the types of metal working processes (Dieter, 1988). The SOS workability is thus independent of the material behavior and will not be considered further.

The intrinsic workability, on the other hand, depends upon the initial microstructure (as influenced by the alloy composition and prior processing history) and its response to applied temperature, strain rate, and strain during processing. This response is embedded implicitly in the flow stress variation with temperature, strain rate, and strain, which can be represented mathematically as a constitutive equation. However, the explicit response of the material occurs in two complementary ways: initially as a temperature rise and later, more importantly, as a change in the microstructure.

The microstructural changes decide the intrinsic workability. Hence modeling techniques (Jonas and Sakai, 1982; Venugopal et al., 1997; Wellstead, 1979) have been developed for predicting the microstructural changes during hot working and for optimizing intrinsic workability. The resulting Process Maps are generated/developed on the basis of dynamic materials modeling (DMM). A detailed description of the DMM approach, as well as interpretation of the Process Maps, is given by Prasad et al. (1984). Process Maps have been widely used to establish the microstructural mechanisms operating during hot deformation of various materials (Chakraborthy et al., 1992; Kang et al., 2008; Prasad and Ravichandran, 1991), including Al–Li alloys (Gokhale et al., 1994; Prasad and Sashidhara, 1997).

A Process Map is generated using flow stress data for a wide range of temperature and strain rate. These data are best obtained using hot compression tests. The flow stress variation with strain rate at a given temperature is curve-fitted using a spline function, and the strain rate sensitivity is calculated as a function of strain rate ($\dot{\varepsilon}$). The efficiency (η) given by Eq. (7.1) is then calculated and plotted as a contour map in a frame of temperature and strain rate, also called a Power Dissipation Map.

$$\eta = J/J_{max} = \frac{2m}{(m+1)} \tag{7.1}$$

J is the co-content of power dissipation, J_{max} is the maximum value of J, and m is the strain rate sensitivity. The instability parameter $\xi(\dot{\varepsilon})$ given by Eq. (7.2) is calculated and plotted as an Instability Map, which is superimposed on the Power Dissipation Map to obtain a Process Map.

$$\xi(\dot{\varepsilon}) = \partial \ln[m/(m+1)]/\partial \ln(\dot{\varepsilon}) + m < 0 \tag{7.2}$$

Owing to the importance of Al–Li alloys, a large amount of research has been carried out to characterize their workability, including the use of Process Maps. A summary of the results of previously reported work pertinent to the deformation behavior of Al–Li alloys is presented in the following section.

7.3 GENERAL REVIEW OF HOT DEFORMATION CHARACTERISTICS OF Al–Li ALLOYS

7.3.1 Hot Workability of a Binary Al–Li Alloy

Menon and Rack (1994) used the DMM approach to obtain Process Maps for an Al–2 wt% Li alloy. Based on the maps, dynamic recovery (DRV) was expected at 400°C and $\dot{\varepsilon} = 1 \times 10^{-2}/s$, while dynamic recrystallization (DRX) was expected at 500°C and $\dot{\varepsilon} = 10^{-3}/s$.

In the temperature range 400—450°C, increasing the strain rate beyond 10^{-1}/s resulted in unstable flow that was attributed to macroscopic shear band formation. At temperatures between 500°C and 550°C instability was observed at low strain rates (10^{-3}/s to 5×10^{-4}/s), and this unstable flow was attributed to dynamic grain growth.

Low temperature conditions that promote DRV result in pancake grain structures, while the higher temperature conditions favoring DRX result in equiaxed grain structures. Either of these deformation regimes (domains) may be selected for processing, depending on the desired grain structure and properties. However, the process parameters resulting in unstable deformation, due to either shear band formation or grain growth, are obviously to be avoided.

7.3.2 Hot Workability of Multicomponent Al—Li Alloys

Experimental Alloys

Niikura et al. (1985) made a very detailed study of the hot deformation behavior of Al—Li—Cu—Mg—Zr alloys with Li contents ranging from effectively 0 to 4 wt% (Table 7.2).

They reported that DRX was inhibited under conditions where δ phase remained undissolved or precipitated dynamically, resulting in a pancake grain structure. This effect of δ phase agrees with the observation of only DRV at 400°C by Menon and Rack (1994) (Section 7.3.1) since this temperature is below the δ solvus for binary Al—Li alloys (419°C), and hence the presence of δ during deformation at 400°C would be expected.

Near-Commercial and Commercial Alloys

Gokhale et al. (1994) studied the hot deformation characteristics of compositional variations of the second-generation Al—Li—Cu—Mg—Zr alloy AA 8090 by generating Process Maps over the temperature range of 300—550°C and strain rate range of 1×10^{-3} to 1×10^{-2}/s. This study included the effects of

TABLE 7.2 Compositions (wt%) of the Al—Li Alloys (Niikura et al., 1985)

Alloy	Li	Cu	Mg	Zr	Fe	Si	Na	K
Li-free	trace	1.18	0.92	0.14	0.03	<0.01	<0.005	0.019
1 Li	1.24	1.20	1.25	0.14	0.03	<0.01	<0.005	0.024
2 Li	2.16	1.28	1.24	0.14	0.03	<0.01	<0.005	0.027
3 Li	3.13	1.30	1.32	0.13	0.03	<0.01	<0.005	0.022
4 Li	3.92	1.24	1.18	0.14	0.03	<0.01	<0.005	0.038

TABLE 7.3 Compositions (wt%) of 8090-Type Alloys (Gokhale et al., 1994) and AA 8090

Alloy	Li	Cu	Mg	Zr	Be	Fe	Si
A	2.12	1.10	0.87	0.09	0.026	0.08	0.04
B	2.32	1.18	0.92	0.08	0.033	0.08	0.04
C	1.85	1.15	0.91	0.09		0.07	0.04
D	2.32	1.18	0.97	0.07		0.08	0.04
AA 8090	2.4	1.2	0.8	0.11			

trace additions of Be (0.02—0.03 wt%) and variations in Li content (above or below 2 wt%) on the Process Maps, and the properties of extrusions processed near optimum conditions. The alloy compositions are given in Table 7.3, which includes the nominal composition of the commercial 8090 alloy.

Gokhale et al. observed that the efficiency parameter (η) increased with increasing temperature and decreasing strain rate, while the process instability parameter $\xi(\dot{\varepsilon})$ became increasingly negative with decreasing temperature and increasing strain rate. Furthermore, they reported the occurrence of DRX at about 550°C and $\dot{\varepsilon} = 1 \times 10^{-3}/s$ with an efficiency of 40%; and it was also found that lower Li contents, higher temperatures, and low strain rates favored DRX. The trace additions of Be and (to a lesser extent) reduction in Li content below 2 wt% expanded the region of stable deformation and also improved the tensile strength and notch strength ratio. These effects are worth noting in the light of development of the third-generation alloys, which have Li contents below 2 wt% (Rioja and Liu, 2012).

Another study of 8090-type alloys was carried out by Prasad and Sashidhara (1997). They presented Process Maps resulting from three initial material conditions: as-cast, homogenized, and extruded. In all cases the maximum process efficiency (η_{max}) and DRX were observed at 550°C, but at different strain rates of 0.001, 0.01, and 100/s for the as-cast, homogenized, and extruded conditions respectively. These results have important implications for deciding upon the processing parameters for billet forging and plate rolling. For example, forging, which is normally done using extruded billets, can be done at higher strain rates than rolling, for which homogenized slabs are used. As-cast ingots are rarely worked directly, as mentioned in the introduction to this chapter.

Kridli et al. (1998) developed Process Maps for Weldalite™ 049 (a low Li content forerunner of the third-generation alloy AA 2195; Table 7.4). The weldable 2195 alloy has been much used for spacecraft liquid propellant tanks.

TABLE 7.4 Compositions (wt%) of Weldalite™ 049 (Kridli et al., 1998) and AA 2195

Alloy	Li	Cu	Mg	Ag	Zr	Fe	Si
Weldalite™ 049	1.0–1.3	3.9–4.6	0.25–0.80	0.25–0.60	0.08–0.16	0.15 max	0.12 max
AA 2195	1.0	4.0	0.4	0.4	0.11		

In their study, Kridli et al. (1998) found that Weldalite™ 049 exhibits maximum formability in a Process Map region bounded by a temperature of 485°C and a strain rate of 5×10^{-4}/s. Slower strain rates resulted in large decreases in formability, and it was also found that below a strain rate of 5×10^{-5}/s the stress–strain behavior showed oscillatory hardening and softening. This behavior became more evident at higher temperatures (over the range 463–496°C) and lower strain rates and is due to DRX (Frost and Ashby, 1982; Ghosh and Raj, 1996; Eddahbi et al., 2000).

These observations of DRX by Kridli et al. (1998) confirm the above-mentioned findings of Gokhale et al. (1994) that lower Li contents, high temperatures, and low strain rates favor DRX in Al–Li alloys. These findings have significant implications for the processing of third-generation Al–Li alloys, though no details are available from the commercial producers.

7.3.3 Hot Deformation Regions Favoring Superplasticity

Superplastic behavior appears to be a general feature of Al–Li alloys. A detailed survey and discussion of superplasticity in Al–Li alloys is given in Chapter 8. Here we mention a few investigations examining the transitions between superplastic and nonsuperplastic behaviors. Also, it is noteworthy that detailed Process Mapping of the alloy discussed in Section 7.4 showed *two* regions of superplasticity.

The earliest mention of superplasticity in Al–Li alloys was made by Wadsworth et al. (1983). They prepared two experimental alloys, Al–3Li–0.5Zr and Al–3Li–4Cu–0.5Zr, by DC casting and three thermomechanical processing (TMP) schedules (isothermal rolling + heat treatment). Tensile tests revealed superplasticity for all three conditions at 450°C and over a wide strain rate range.

Superplastic behavior in the temperature range 450–525°C for a rapidly solidified Al–3Li–1Cu–0.5Mg–0.5Zr alloy was reported by Koh et al. (1998). The occurrence of superplasticity was based on tensile test data generated over a range of temperatures (430–570°C) and strain rates (2×10^{-4}

to 6×10^{-3}/s). The superplastic characteristics depended markedly upon the strain, strain rate, and temperature, as might be expected.

Another study of superplasticity, this time for the second-generation alloy AA 8090, was made by Fan et al. (2001). They investigated the deformation behavior in the temperature range 250–573°C and for a wide strain rate range of 1×10^{-5} to 1×10^{-2}/s. During early deformation the flow curves exhibited strain hardening, whose rate increased with increasing strain rate and decreasing temperature; and there were significant slip-dislocation contributions to both strain hardening and the mechanism of superplastic flow. Superplastic deformation was characterized by grain growth, cavitation, dislocation interactions, and texture weakening; and this microstructural evolution facilitated grain boundary sliding (GBS) and its accommodation by a diffusion process.

7.4 HOT DEFORMATION AND PROCESS MAPPING OF AL–LI ALLOY UL40

Osprey Alcan International used the spray-casting technique to develop the Al–Li alloy UL40 (Prangnell et al., 1994; Singer, 1982). This is a medium strength alloy containing approximately 4 wt% Li and 0.2 wt% Zr and has a very low density ($\rho = 2.43$ gm/cm^3). The deformation characteristics of this alloy were studied extensively by Jagan Reddy (2010), using Process Maps for four different initial conditions:

1. spray cast,
2. spray cast and hot isostatically pressed (termed hereafter as "HIPed"),
3. spray cast, HIPed and homogenized (termed hereafter as "homogenized"),
4. spray cast and extruded (termed hereafter as "extruded").

The results from these studies are reviewed in this section, which includes the Process Maps and examples of electron backscatter diffraction orientation imaging microscopy (EBSD–OIM).

7.4.1 Process Maps and EBSD–OIM

The Process Maps for all initial conditions are presented in Figure 7.4. The features (domains) indicated on these maps are given in Table 7.5, which includes the temperature–strain rate ranges of the different domains, the efficiency of power dissipation (η), instability regions ($\zeta < 0$), tensile ductility, microstructural characteristics, and the rate controlling mechanism (cross-slip in the safe domains). In more detail:

1. *Safe domains*: All four initial conditions enabled similarly-located safe domains of hot working characterized by DRV. The HIPed and extruded materials had extended safe windows (Figure 7.4A and B). Also, the

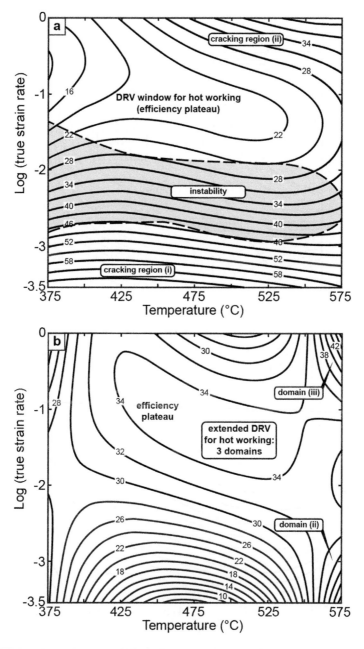

FIGURE 7.4   Processing maps of UL40 alloy starting from four different initial conditions: (A) spray cast, (B) HIPed, (C) homogenized, and (D) extruded. The different regions on the maps are described in Table 7.5. DRV, dynamic recovery; CDRX, continuous dynamic recrystallization; DDRX, discontinuous dynamic recrystallization.

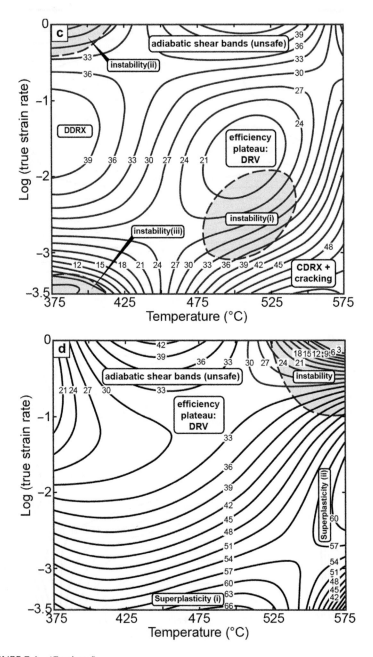

FIGURE 7.4 (Continued).

TABLE 7.5 Summary of Key Features of Process Maps for UL40 Alloy in Different Initial Conditions

Initial Conditions	Safe Windows for Hot Deformation	Unsafe Windows for Hot Deformation		Remarks
		Cracking Regions	Instability Regions	
(a) Spray cast	DRV: 425–525°C 5×10^{-2} to 3×10^{-1}/s $\eta = 22\%$	(i) 375–575°C 1×10^{-3} to 3×10^{-4}/s $\eta = >52\%$ (ii) 475–575°C 3×10^{-1} to 1×10^{0}/s $\eta = 34\%$ No DRX domain	375–575°C 1×10^{-3} to 5×10^{-2}/s $\zeta < 0$	• Pancake grains in sample deformed at 475°C/10^{-1}/s and also subgrains • Ductility maximum (34%) at 425°C/10^{-1}/s but low at 3×10^{-4}/s and decreasing with temperature • DRV is microstructural mechanism in safe domain • Rate controlling mechanism in DRV regime is cross-slip • Cracking and instability due to coarse δ (AlLi) precipitates and porosity
(b) HIPed at 500°C for 1 h	Extended DRV: 425–575°C 3×10^{-3} to 3×10^{-1}/s $\eta = 34\%$	DRX + cracking at 550–575°C 10^{-3} to 3×10^{-4}/s $\eta = 46\%$	No instability	• Like spray-formed samples, pancake grains in samples deformed at 475°C/10^{-1}/s, but more subgrains • Ductility maximum (36%) at 425°C/10^{-1}/s but low at 3×10^{-4}/s and decreasing with temperature • Extended DRV is microstructural mechanism in safe domain • Rate controlling mechanism in DRV regime is cross-slip • Absence of instability attributed to pore closure during HIPing

(Continued)

TABLE 7.5 (Continued)

Initial Conditions	Safe Windows for Hot Deformation	Unsafe Windows for Hot Deformation		Remarks
		Cracking Regions	Instability Regions	
(c) HIPed and homogenized at 565°C for 4 h	(i) DDRX: 375–425°C 3×10^{-3} to 3×10^{-1}/s $\eta = 39\%$ (ii) Efficiency plateau: DRV 475–525°C 1×10^{-2} to 5×10^{-1}/s $\eta = 21\%$	(i) CDRX + cracking: 525–575°C 10^{-3} to 3×10^{-4}/s $\eta = 57\%$ (ii) Adiabatic shear bands at 425–575°C 3×10^{-1} to 1×10^{0}/s	(i) 475–500°C 10^{-3} to 10^{-2}/s (ii) 375–400°C 3×10^{-1} to 10^{0}/s and also (iii) 10^{-3} to 3×10^{-4}/s	• Pancake grains in sample deformed at 475°C/10^{-1}/s but less subgrains than for spray formed and HIPed alloys • Ductility maximum (26%) at high strain rate (525°C/10^{-1}/s) but low at 3×10^{-4}/s and decreasing with temperature • Rate controlling mechanism in DRV regime is cross-slip • Instabilities attributed to (i) solute drag, (ii) adiabatic shear banding, and (iii) DSA • DDRX due to presence of β' precipitates • Uniform grain size in CDRX sample but ductility low due to cavitation
(d) Extruded at 475°C/1×10^{-1}/s	(i) Efficiency plateau: DRV 425–525°C 5.5×10^{-2} to 3×10^{-1}/s (ii) Superplasticity-I 425–525°C 3×10^{-4} to 10^{-3}/s (iii) Superplasticity-II 550–575°C 1.8×10^{-3} to 5.5×10^{-2}/s	Adiabatic shear bands at 425–475°C 3×10^{-1} to 10^{0} s	525–575°C 3×10^{-1} to 10^{0}/s	• No pancake grains under any deformation conditions, unlike in spray formed, HIPed, and homogenized alloys • Maximum ductility (180%) occurred at 525°C/3×10^{-4}/s owing to superplasticity and DRX • At high strain rate (10^{-1}/s) the maximum ductility (~50%) was at 475°C • Grain boundary diffusion is rate controlling mechanism for GBS in superplasticity domain-I at 425–475°C/3×10^{-4}/s, while it is lattice diffusion in superplasticity domain–II at 550–575°C/10^{-2}/s • Rate controlling mechanism in DRV regime is cross-slip • Instability due to wide grain size distribution

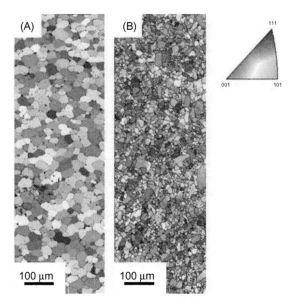

FIGURE 7.5 EBSD–OIM images of a homogenized sample: (A) undeformed and (B) deformed at 575°C/3 × 10⁻⁴/s.

homogenized material had another safe domain characterized by discontinuous dynamic recrystallization (DDRX) (Figure 7.4C).

Figure 7.5 shows a typical EBSD–OIM example for an undeformed and deformed microstructure when processing is done in a safe region. In this example the safe region was in the bottom RH corner of Figure 7.4C (CDRX and avoiding cracking). The as-processed microstructure is completely recrystallized, with some grain growth owing to the high temperature (575°C).

2. *Peak efficiency*: For all four initial conditions the peak efficiency occurred at the same temperature of 475°C and the same strain rate of 1×10^{-1}/s. However, the peak efficiency values differed: 22% for spray cast, 34% for HIPed, 21% for homogenized, and 33% for extruded. The higher efficiencies for the HIPed condition are attributable to the development of more subgrains and significant grain refinement (Figure 7.6). The high efficiency for the extruded material was also due to grain refinement.

3. *Adverse effects on DRV*: The coarse δ precipitates and porosity in the as-spray formed alloy were detrimental to the DRV process, resulting in instability and cracking, and lower maximum efficiency (22%) (Figure 7.4A). The lowest maximum efficiency (21%) for the homogenized alloy is attributable to a higher Li content in solid solution and a coarse initial grain size.

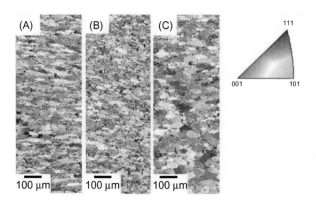

FIGURE 7.6 EBSD–OIM images of UL40 alloys deformed on the "efficiency plateaus" under the same conditions ($475°C/1 \times 10^{-1}$/s): (A) spray cast, (B) HIPed, and (C) homogenized.

4. *Instability regions*: The spray cast, homogenized, and extruded materials had evident instability regions, shown in Figure 7.4 and summarized in Table 7.5. The characteristics of these regions were, however, very different:

   a. Spray cast material instability occurred over the entire temperature range ($375–575°C$) at strain rates between $\sim 10^{-3}$ to 5×10^{-2}/s (Figure 7.4A). Instability was caused by intergranular cracking that was probably due to decohesion of grain boundary particles (Jagan Reddy et al., 2008).

   b. The homogenized material showed three instability regions (Figure 7.4C). Instability (i) is in an intermediate temperature–strain rate range and may be attributed to the effect of solute drag; instability (ii) was associated with adiabatic shear band formation; and instability (iii) may have been due to dynamic strain ageing (DSA).

   c. The extruded material exhibited instability at high temperatures ($\sim 525–575°C$) and strain rates (1×10^{-1} to 1×10^{0}/s) (Figure 7.4D). Microstructural examination suggested that instability was due to formation of adiabatic shear bands.

   The HIPed material did not show instability under the investigated temperature and strain rate conditions. Absence of instability was attributed to the uniform microstructure and low porosity in the initial condition (Jagan Reddy, 2010).

5. *Superplasticity*: Uniquely, in this investigation, the homogenized material showed *two* superplasticity regions (Figure 7.4D). As mentioned in Table 7.5, these regions were defined by different temperature and strain rate ranges, and the rate controlling mechanisms were different.

This extensive Process Mapping investigation illustrates the importance of the initial condition for subsequent hot deformation behavior and also the complexity of Al–Li alloy hot deformation.

7.5 SUMMARY OF THE HOT DEFORMATION INVESTIGATIONS

The investigations selected and reviewed in Sections 7.3. and 7.4 are listed in Table 7.6 with the alloys separated into two groups: experimental and near- or fully-commercial alloys. Main aspects of the investigations are indicated, and it is seen that several investigations used Process Mapping, which has greatly increased the systematic knowledge about hot deformation of Al–Li alloys.

Process Maps for both experimental and commercially based Al–Li alloys have shown that although their basic features are similar (Figure 7.4; Gokhale et al., 1994), there are positional and size or spatial changes in the safe working domains and instability regions. There can also be large differences in the optimum strain rates and efficiency. These changes and differences depend on alloy composition (Gokhale et al., 1994) and also the initial condition before deformation (Jagan Reddy, 2010; Prasad and Sashidhara, 1997).

Limitations to Process Maps are that they predict restoration mechanisms and resultant microstructures only under isothermal and constant true strain rate deformation. However, real metal working processes usually involve decreases in temperature and increases in strain rate with time and are usually accompanied by complex stress states and friction, contributing to the so-called "extrinsic workability," which controls microstructural changes.

It follows that Process Map predictions are only guides, albeit excellent ones, to the choice of process parameters. More specifically, Process Maps are useful for (i) establishing safe and efficient windows for hot working, (ii) determining the optimum strain rates, and (iii) providing information about controlling the amounts of recrystallization and recovery. For example, as discussed earlier, the optimum strain rates for billet forging and plate rolling can be very different (Prasad and Sashidhara, 1997); and, on an *apparently* very different topic, a Process Map is essential to determine the best temperature–strain rate conditions for superplastic forming, which is a time-consuming process.

PART 2: PROCESSING OF Al–Li ALLOYS

7.6 GENERAL CONSIDERATIONS

As shown in Table 7.1, the wrought product forms developed for Al–Li alloys are sheet, plate, extrusions, and forgings, the majority being rolled products (sheet and plate). However, it is worth pointing out that aluminum alloy sheet, plate and extrusions are all widely used for aerospace components and structures, while forgings—though important—are generally less used. More information on this topic, albeit specifically for Al–Li alloys, is given in Chapters 2 and 15.

TABLE 7.6 Selected and Reviewed Hot Deformation Investigations on Al–Li Alloys

References	Alloys		Main Aspects
	Alloying Elements/Designations	Initial Conditions[a]	
Wadsworth et al. (1983)	Al–Li–Zr, Al–Li–Cu–Zr	R, A + R, STA + R	Superplasticity
Niikura et al. (1985)	Al–Li–Cu–Mg–Zr	H + R + ST	Influence of δ on DRX
Menon and Rack (1994)	Al–Li	H	Process Maps: DRV, DRX
Jagan Reddy (2010)	Al–Li–Zr	C, HIP, HIP + H, E	Process Maps: Figure 7.1; Table 7.5
Gokhale et al. (1994)	Al–Li–Cu–Mg–Zr (8090-type)	H + E	Process Maps: DRX
Prasad and Sashidhara (1997)	Al–Li–Cu–Mg–Zr (8090-type)	C, H, E	DRX at different strain rates
Kridli et al. (1998)	Al–Li–Cu–Mg–Zr (Weldalite™ 049)	R (unrecrystallized)	Process Maps: superplasticity
Koh et al. (1998)	Al–Li–Cu–Mg–Zr	Rapidly solidified	Superplasticity
Fan et al. (2001)	Al–Li–Cu–Mg–Zr (8090)	Partly recrystallized	Superplasticity

[a]A, aged; C, cast; DRV, dynamic recovery; DRX, dynamic recrystallization; E, extruded; H, homogenized; HIP, hot isostatically pressed; R, hot rolled; ST, solution treated; STA, solution treated and aged.

7.6.1 Rolled Products

Al–Li rolled products are sheets (as low as 0.6 mm thick) and plates (greater than 6 mm thick). The third-generation Al–Li rolled products are made from DC cast slabs using TMP combinations of hot rolling and heat treatment (Rioja, 1998; Rioja and Liu, 2012). The exclusive use of hot rolling during primary working (as opposed to combinations of hot and cold rolling for some conventional aluminum alloys) is necessary to control the texture and the texture-related mechanical property anisotropy of these alloys. This is discussed in detail in Chapter 5, Section 5.5.

Plates up to 50 mm thickness are more usual, but the alloys AA 2050 and AA 2055 are being produced in thicknesses up to 125–150 mm (Lequeu et al., 2010; Magnusen et al., 2012). Plate products are characterized by (i) grain structure (pancake versus equiaxed) and its through-thickness variation, (ii) texture, (iii) property anisotropy in the rolling plane (in-plane anisotropy) and through-thickness (short-transverse) direction, and (iv) quench sensitivity. Component manufacturing and assembly issues are no less important, owing to problems arising from the typical pancake grain structures. These problems include property variations in different directions and "end-grain" exposure, which can facilitate intergranular corrosion and stress corrosion, as mentioned in the introduction to this chapter.

Aerospace aluminum alloy sheets have usually been produced as clad sheets for improved corrosion resistance. However, addition of zinc to third-generation Al–Li alloys (Magnusen et al., 2012; Rioja and Liu, 2012) may make cladding unnecessary for some applications (Giummarra et al., 2007). Thin Al–Li alloy sheets are often heat-treated to produce recrystallized microstructures, which reduce in-plane anisotropy and give a better balance of mechanical properties (Rioja and Liu, 2012). In such cases, heating rates to the solution treatment (ST) temperature are important to minimize the surface depletion of lithium, which is fast diffusing and highly reactive with the environment. Lithium diffusion into cladding, if cladding is used, is also important (Mallesham et al., 2003).

Grain Structure and Texture

Plates usually receive less total deformation than sheet. The thicker the final plate product, the less strain that is imparted, and the coarser is the overall grain structure, which is typically pancake-shaped. If the rolling reduction is low, the strain varies through the plate thickness, resulting in a (very) coarse grain structure in the plate center and a relatively fine structure in the surface layers. When the rolling reduction is very low, the cast structure, including porosity, may have been retained in the plate center, adversely affecting the mechanical properties.

As is well known, both plate and sheet Al−Li alloy products have a tendency to be strongly textured. The texturing interacts with grain size and shape, cold deformation (stretching before ageing) and the precipitates developed during ageing, and this interaction can result in high degrees of anisotropy. In second-generation plate products the strong texture, combined with planar slip induced by the shearing of δ' precipitates, resulted in wide variations in through-thickness properties and poor short-transverse ductility and fracture toughness. The persistence of these texture- and microstructure-related problems was primarily responsible for preventing second-generation alloys from achieving widespread application.

The solution to the anisotropy problem has been to (i) reduce slip planarity by decreasing the amount of δ', usually by lowering the lithium content of the alloys to a maximum of 1.8 wt%; (ii) rely on other types of precipitates, besides δ', for achieving high mechanical properties; and (iii) employ multistep TMP to control the texture, grain size and shape, and precipitate distributions (Rioja and Liu, 2012). All of these modifications have been implemented for the third-generation Al−Li alloys.

Another important aspect relating to the multistep TMPs used for third-generation Al−Li alloys is the probable difficulty in controlling the grain boundary microstructures when plate thicknesses exceed about 30 mm (see Chapter 14). This may have repercussions on the fracture toughness and stress corrosion cracking (SCC) resistance (see Section 7.6.2).

Quench Sensitivity (plates)

Some information on quench sensitivity obtained from AA 8090 alloy variants is of interest for third-generation alloys. Colvin and Starke (1988) systematically studied the quench sensitivity of two versions of AA 8090, one containing 2.28 wt% Li and 0.86 wt% Cu, and the other with higher Li and Cu contents, 2.58 wt% Li and 1.36 wt% Cu, respectively. They cooled 25 mm thick plates at 150°C/s (water quench), 30°C/s (polymer quench), and 0.2°C/s (air cooling) after ST at 550°C. The highest cooling rate resulted in both alloys being free of detrimental coarse constituents; and the lowest cooling rate resulted in both alloys containing coarse T_2 and S precipitates. However, cooling at the intermediate rate resulted in coarse T_2 and S only in the higher Li + Cu alloy. The beneficial effect of lower Li may be a pointer to less quench sensitivity in third-generation alloys.

More definitively, a study of the "transitional" alloy AF/C-458, which is similar in composition to the third-generation alloys AA 2099 and AA 2199 (Table 7.7), showed it to be relatively quench-insensitive, "similar to other Al−Li−Cu−X alloys" (Csontos et al., 2004).

TABLE 7.7 Compositions and Densities of Second-, Transitional, and Third-Generation Western Al–Li Alloys

Alloys	Li	Cu	Mg	Ag	Zr	Sc	Mn	Zn	Density ρ (g/cm^3)	Introduction
Second Generation (Li ≥ 2wt%)										
2090	2.1	2.7			0.11				2.59	Alcoa 1984
2091	2.0	2.0	1.3		0.11				2.58	Pechiney 1985
8090	2.4	1.2	0.8		0.11				2.54	EAA 1984
Transitional										
AF/C-489	2.05	2.7	0.3		0.04		0.3	0.6	2.59	Investigation from 1992–2000
AF/C-458	1.73	2.58	0.26		0.09		0.25	0.6	2.64	
Third Generation (Li < 2wt%)										
2195	1.0	4.0	0.4	0.4	0.11				2.71	LM/Reynolds 1992
2196	1.75	2.9	0.5	0.4	0.11		0.35 max	0.35 max	2.63	LM/Reynolds/McCook Metals 2000
2297	1.4	2.8	0.25 max		0.11		0.3	0.5 max	2.65	LM/Reynolds 1997
2397	1.4	2.8	0.25 max		0.11		0.3	0.10	2.65	Alcoa 2002

(Continued)

TABLE 7.7 (Continued)

Alloys	Li	Cu	Mg	Ag	Zr	Sc	Mn	Zn	Density ρ (g/cm^3)	Introduction
2098	1.05	3.5	0.53	0.43	0.11		0.35 max	0.35	2.70	McCook Metals 2000
2198	1.0	3.2	0.5	0.4	0.11		0.5 max	0.35 max	2.69	Reynolds/McCook Metals/Alcan 2005
2099	1.8	2.7	0.3		0.09		0.3	0.7	2.63	Alcoa 2003
2199	1.6	2.6	0.2		0.09		0.3	0.6	2.64	Alcoa 2005
2050	1.0	3.6	0.4	0.4	0.11		0.35	0.25 max	2.70	Pechiney/Alcan 2004
2296	1.6	2.45	0.6	0.43	0.11		0.28	0.25 max	2.63	Constellium Alcan 2010
2060	0.75	3.95	0.85	0.25	0.11		0.3	0.4	2.72	Alcoa 2011
2055	1.15	3.7	0.4	0.4	0.11		0.3	0.5	2.70	Alcoa 2012
2065	1.2	4.2	0.50	0.30	0.11		0.40	0.2	2.70	Constellium 2012
2076	1.5	2.35	0.5	0.28	0.11		0.33	0.30 max	2.64	Constellium 2012

Source: Various sources, including Chapter 2.

7.6.2 Extrusions and Forgings

Although rolled products are the main forms used for aerospace structures, extrusions are widely used since the processing is straightforward and they can be produced in a variety of shapes (profiles). Forgings are mainly used for heavy section components like bulkheads, as illustrated in Figure 7.3. A recent "advanced aluminum alloy extrusions and forgings map" is shown in Figure 7.7. Extrusions are used principally for wing and fuselage stringers and fuselage frames and also find use in floor beams and seat tracks. An example of a recent floor beam application for third-generation Al–Li alloys is shown in Figure 7.8.

Extrusion Characteristics

The processing parameters for extrusion are temperature, extrusion ratio, and ram speed, all of which influence the microstructure and properties. Both DRV and DRX occur during the hot deformation. Investigation of the second-generation alloy AA 8090 showed that the extrusion surfaces are typically recrystallized and the cores unrecrystallized, and increasing the three processing parameters increases the depth of the recrystallized layers, especially the extrusion ratio (Mukhopadhyay et al., 1990).

A more extensive investigation of the second-generation Al–Li–Cu alloy AA 2090 and the Al–Li–Cu–Mg alloys AA 8090 and AA 8091 showed several significant trends (Skillingberg and Ashton, 1987):

1. The inclusion of Mg in the alloy composition reduces the hot working range by increasing the flow stress and enhancing the onset of incipient

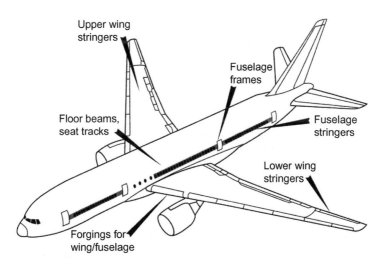

FIGURE 7.7 Extrusion and forging applications in a modern transport aircraft. *Source: Original source Alcoa.*

FIGURE 7.8 Third-generation Al—Li alloy floor beam extrusions and carbon fiber reinforced plastic (CFRP) floor beams in the Airbus A380. *Source: Image source Airbus (public domain). More details are given in Section 15.4.1.*

  melting and surface tearing. This is also relevant for third-generation alloys, all of which contain magnesium.

2. As observed by Mukhopadhyay et al. (1990), increased temperature and extrusion ratio tended to cause more recrystallization, as would be expected.

3. The mechanical properties, notably strength and toughness, were strongly influenced by recrystallization and grain shape. Elongated and unrecrystallized grains resulted in high longitudinal strengths and toughness but poor transverse properties. Recrystallization and retention of pancake grain shapes resulted in less anisotropy but lower longitudinal strength and toughness.

  As discussed in Section 7.7.2, third-generation Al—Li alloy extrusions are typically unrecrystallized (or mainly unrecrystallized) since this provides optimum combinations of properties (Rioja and Liu, 2012; Tchitembo Goma et al., 2012).

Forging Characteristics

Closed-die forgings are most commonly used for aluminum alloys, which are easily formed into near-net shapes. There are several reasons for this: (i) high ductility, (ii) low forging temperatures, typically 350—450°C for high-strength aerospace alloys, (iii) low forging pressures, and (iv) no scale development, so preheating furnace atmospheres are not important provided the furnace is initially dry (this prevents hydrogen pickup and blistering). Open-die, drop-, press-, and hand forgings can also be made. Any of these may be convenient for forging simple shapes.

Open literature information on Al–Li alloy forgings is mostly limited to results for simple shapes of the second-generation AA 8090 and AA 8091 alloys (Lewis et al., 1987; Pitcher, 1988a,b; Terlinde et al., 1992). There is also some information on small closed-die forgings of the Weldalite™ 049 developmental alloys X2094 and X2095 (McNamara et al., 1992); a paper observing that hand, die, and ring forgings are readily made with Al–Li alloys (Babel and Parrish, 2004); and a nonarchival mention of second-generation AA 2090 forgings for landing gear parts.

The results for the second-generation alloys were discouraging, since the forgings had low short-transverse fracture toughness and ductility. Despite this, AA 8090 forgings were used in the AgustaWestland EH101 helicopter (Chapter 2)—with spectacular results owing to a "hard landing" (see Figure 13.3 in Chapter 13).

The Weldalite™ 049 alloys had excellent forging characteristics, but tended to have low *longitudinal* ductility, especially in a high-strength condition (McNamara et al., 1992). These results might be considered discouraging since the ductility would be expected to be highest in the longitudinal direction. However, the authors remained optimistic.

Future investigations of forgings for third-generation alloys could consider more sophisticated shapes for property determinations. An example used for several high-strength 7XXX forgings in the 1970s and 1980s is schematically illustrated in Figure 7.9, showing the variety of properties that may be evaluated (Wanhill, 1994).

7.7 INDUSTRIAL-SCALE PROCESSING

7.7.1 Introduction

As mentioned at the beginning of this chapter, in Section 7.1.1, the commercial manufacturing processes are essentially proprietary. However, there is some recently published information for third-generation Al–Li alloys (Rioja and Liu, 2012); and also for two "transitional" alloys, AF/C-458 and AF/C-489, that aided development of the third generation (Rioja, 1998).

Table 7.7 indicates (by box shading) how the transitional alloys relate to the second- and third-generation alloys from the main Western manufacturers. The earlier alloy, AF/C-489, has a composition and density closely matching those of the second-generation alloy AA 2090. However, owing to ductility problems in the peak strength condition, AF/C-489 was superseded by the derivative lower Li content alloy AF/C-458 (Csontos and Starke,

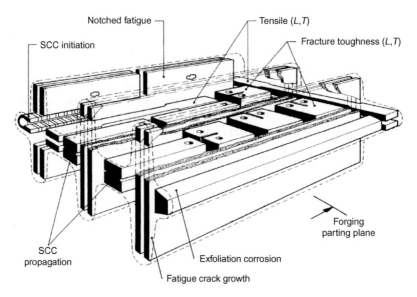

Notched fatigue

SCC initiation

Tensile (L,T)

Fracture toughness (L,T)

Forging
parting plane

SCC
propagation

Exfoliation corrosion

Fatigue crack growth

FIGURE 7.9 Schematic of a closed-die forging shape and test specimens for property evaluations in high-strength aluminum alloys (Wanhill, 1994); approximate forging size $L = 355$ mm,
$B = 215$ mm, $H = 125$ mm.

2000). It may be seen from Table 7.7 that AF/C-458 has essentially the
same composition and density as the third-generation alloys AA 2099 and
AA 2199.

7.7.2 Review and Discussion of Available Information for Third-Generation Alloys

The available processing information for the transitional and third-generation
alloys is summarized in Table 7.8. The main sources of information are the
surveys by Rioja (1998) and Rioja and Liu (2012), specifically concerning
texture control. There are several points to be noted:

1. *Grain structure*: Both unrecrystallized (UnRX) and recrystallized (RX)
   sheet and plate are produced, depending on the alloys and property
   requirements. Rioja and Liu (2012) state that for optimum mechanical
   properties plates and extrusions are typically unrecrystallized, whereas
   sheet products are typically recrystallized.

   Highly elongated grains in recrystallized AA 2199 sheet and partially
   recrystallized AA 2050 plate are mentioned by Rioja and Liu (2012) and
   Richard et al. (2012), respectively, and appear to be regarded as typical
   or normal.

TABLE 7.8 Transitional (AF/C-458, AF/C-489) and Third-Generation Al–Li Alloy Processing Information from the Open Literature

References	Materials	Products	Grain Structures	Remarks
Rioja (1998)	AF/C-458,489	Plate	UnRX	• TMP to obtain low "Brass" texture
Giummarra et al. (2007)	2099, 2199 2099	Sheet, plate, extrusion	UnRX UnRX	• TMP to obtain low "Brass" texture • At least partially UnRX
Lequeu et al. (2010)	2050	Plate	UnRX	• Strengthening mainly by T_1; some θ'
Boselli et al. (2012)	2060	Plate	Not stated	• Strengthening mainly by T_1; some θ' (inferred from precipitation data)
Daniélou et al. (2012)	2050	Plate	Partially RX	• Strengthening mainly by T_1 • Standard TMP: H + HR + ST + Q + stretch + A
Denzer et al. (2012)	2055	Plate, Extrusion	UnRX UnRX	• Strengthening mainly by T_1 and θ' • TMP to obtain precipitate-free grain boundaries (toughness benefit)
Karabin et al. (2012)	2060, 2199	Plate	UnRX	• UnRX good for strength and toughness • 2060 strengthened by T_1 • 2199 strengthened by T_1, δ', θ'
Magnusen et al. (2012)	2060, 2199	Sheet	RX	• RX and grain size controlled by β' and $Al_{20}Cu_2Mn_3$ dispersoids • 2060 strengthened by T_1 • 2199 strengthened by T_1, δ', θ'

(Continued)

TABLE 7.8 (Continued)

References	Materials	Products	Grain Structures	Remarks
Richard et al. (2012)	2050	Plate	Partially RX	• Strengthening mainly by T_1 • Grains strongly elongated in rolling direction
Rioja and Liu (2012)	2199 2099, 2199 2099	Sheet Plate Extrusion	RX UnRX UnRX	• Elongated grains • UnRX for optimum mechanical properties: low "Brass" texture • UnRX for optimum mechanical properties
Tchitembo Goma et al. (2012)	2099	Extrusion	Mainly UnRX	• UnRX better for FCG properties (crack deflection and deviation)
Holroyd et al. (Chapter 14)		Thick sections	Partially RX	• SCC resistance for thicknesses >30 mm may be a problem since TMP to control the grain boundary microstructure will become more difficult

*Key to Abbreviations: A, aged; FCG, fatigue crack growth; H, homogenized; HR = hot rolled; Q, quenched; RX, recrystallized; SCC, stress corrosion cracking; ST, solution treated; TMP, thermomechanical processing; T4, solution treated + naturally aged; UnRX, unrecrystallized.
Note: The final tempers are all T8-type, whereby the product is solution treated, cold worked (e.g., by a controlled amount of stretching), and artificially aged.

2. *Texture*: TMP schedules for unrecrystallized sheet and plate are designed to avoid the "Brass" texture in order to obtain optimum mechanical properties. Avoidance of the "Brass" texture is particularly important for minimizing in-plane and through-thickness anisotropy of plate products (Rioja, 1998; Rioja and Liu, 2012) (see Section 5.5 in Chapter 5).

3. *Precipitates*: Alloy strengthening relies on T_1 precipitation, with additional contributions from δ' and θ', depending on the alloy. The low-lithium content AA 2060 alloy (0.75 wt% Li; Table 7.7) relies mainly or entirely on T_1, while the higher-lithium content alloy 2199 (and presumably 2099) is strengthened by all three types of precipitate.

   (Here it must be stated that the list of precipitates contributing to strengthening is not necessarily confined to T_1, δ', and θ', see Chapters 3 and 4.)

4. *Dispersoids*: Grain size is controlled by Zr-containing β' (Al_3Zr) and Mn-containing $Al_{20}Cu_2Mn_3$ dispersoids (Magnusen et al., 2012; Rioja and Liu, 2012).

5. *Grain boundaries*: Fracture toughness in AA 2055 is benefited by precipitate-free grain boundaries (Denzer et al., 2012). This is also likely for other alloys (see Lynch et al. (2008) and Section 13.3.8 in Chapter 13).

   The situation with regard to SCC resistance is more complex. In Chapter 14, Holroyd et al. state that the processes controlling Al—Li alloy SCC may involve grain boundary precipitates, Li or trace element segregation to the boundaries, or copper-depleted zones along the boundaries. Whatever the mechanism(s), the grain boundary microstructures must be important.

Point (5) is particularly important for thicker section products. In Section 7.6.1 we mentioned that the multistep TMP schedules employed for third-generation Al—Li alloys may encounter difficulties in controlling the grain boundary microstructures when plate thicknesses exceed about 30 mm. To illustrate this point, Figure 7.10 shows a generic TMP schedule for finally unrecrystallised plate (Rioja and Liu, 2012). The process is evidently complex, even without details of the temperatures and strain rates involved.

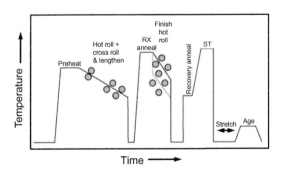

FIGURE 7.10 Generic TMP schedule for third-generation Al—Li plate products (Rioja and Liu, 2012). Note that the processing includes recrystallization (RX) at an intermediate step.

7.8 SUMMARY

Mechanical working of Al–Li alloys is primarily concerned with aerospace alloy rolled products (sheet and plate), extrusions, and to a lesser extent forgings. These product forms are fabricated by hot working with intermittent and final heat treatments. This thermomechanical processing (TMP) can be rather complex for the modern third-generation Al–Li alloys, but it is necessary to obtain optimum combinations of properties.

In this chapter we have first discussed the "workability" of metals and alloys and the hot deformation characteristics of Al–Li alloys, leading to the concept of Process Maps. A comprehensive example of Process Mapping for a binary Al–Li alloy, UL40, illustrates the usefulness of these maps for defining temperature–strain rate regions for safe and unsafe hot working, recrystallization and recovery, and superplastic behavior. Although these maps are not directly applicable to industrial-scale production, their basic features provide excellent guides to establishing the required processing parameters.

The second part of the chapter provides some general considerations about processing Al–Li alloy products, and this is followed by a review and discussion of the available information for third-generation alloys. Owing to the complex TMP schedules, it may be difficult to obtain optimum combinations of properties for thicker products (e.g., heavy-section plate and forgings).

REFERENCES

Babel, H., Parrish, C., 2004. Manufacturing considerations for aluminum–lithium alloys. Paper 03H0108, WESTEC 2004, March 22–25, 2004, Los Angeles, CA.

Boselli, J., Bray, G., Rioja, R.J., Mooy, D., Venema, G., Feyen, G., et al., 2012. The metallurgy of high fracture toughness aluminum-based plate products for aircraft internal structure. In: Weiland, H., Rollett, A.D., Cassada, W.A. (Eds.), Proceedings of the 13th International Conference on Aluminum Alloys (ICAA13). The Minerals, Metals and Materials Society (TMS) and John Wiley & Sons, Hoboken, NJ, pp. 581–586.

Chakraborthy, J.K., Banerjee, S., Prasad, Y.V.R.K., 1992. Superplasticity in β-zirconium: a study using a processing map. Scr. Metall. Mater. 26, 75–78.

Colvin, G.N., Starke Jr., E.A., 1988. Quench sensitivity of the Al–Li–Cu–Mg alloy 8090. SAMPE Q. 19, 10–21.

Csontos, A.A., Starke Jr., E.A., 2000. The effect of processing and microstructure development on the slip and fracture behavior of the 2.1 wt pct Li AF/C-489 and 1.8 wt pct Li AF/C-458 Al–Li–Cu–X alloys. Metall. Mater. Trans. A 31A, 1965–1976.

Csontos, A.A., Gable, B.M., Starke Jr., E.A., 2004. Effect of quench rate, pre-age stretch, and artificial aging on the Al–Li–Cu–X alloy AF/C-458. In: Tabereaux, A.T. (Ed.), Light Metals 2004. The Minerals, Metals and Materials Society, Warrendale, PA, pp. 891–896.

Daniélou, A., Ronxin, J.P., Nardin, C., Ehrström, J.C., 2012. Fatigue resistance of Al–Cu–Li and comparison with 7XXX aerospace alloys. In: Weiland, H., Rollett, A.D., Cassada, W.A. (Eds.), Proceedings of the 13th International Conference on Aluminum Alloys (ICAA13).

The Minerals, Metals and Materials Society (TMS) and John Wiley & Sons, Hoboken, NJ, pp. 511—516.

Denzer, D.K., Rioja, R.J., Bray, G.H., Venema, G.B., Colvin, E.L., 2012. The evolution of plate and extruded products with high strength and toughness. In: Weiland, H., Rollett, A.D., Cassada, W.A. (Eds.), Proceedings of the 13th International Conference on Aluminum Alloys (ICAA13). The Minerals, Metals and Materials Society (TMS) and John Wiley & Sons, Hoboken, NJ, pp. 587—592.

Dieter, G.E., 1988. Mechanical Metallurgy. SI Metric Edition. McGraw-Hill Book Company, London, UK.

Eddahbi, M., Thomson, C.B., Carreño, F., Ruano, O.A., 2000. Grain structure and microtexture after high temperature deformation of an Al—Li (8090) alloy. Mater. Sci. Eng. A A284, 292—300.

Fan, W., Kashyap, B.P., Chaturvedi, M.C., 2001. Effects of strain rate and test temperature on flow behaviour and microstructural evolution in AA 8090 Al—Li alloy. Mater. Sci. Technol. 17, 431—438.

Frost, H.J., Ashby, M.F., 1982. Deformation-Mechanism Maps, The Plasticity and Creep of Metals and Ceramics. Pergamon Press, New York, NY.

Gokhale, A.A., Singh, V., Eswara Prasad, N., Chakravorty, C.R., Prasad, Y.V.R.K., 1994. Process maps for aluminium—lithium alloys. In: Sanders Jr., T.H., Starke Jr., E.A. (Eds.), "Aluminum Alloys: Their Physical and Mechanical Properties," Proceedings of the Fourth International Conference, vol. 2. Georgia Institute of Technology, Atlanta, GA, pp. 242—249.

Ghosh, A.K., Raj, R., 1996. A model for the evolution of grain size distribution during superplastic deformation. Acta Metall. 34, 447—456.

Giummarra, C., Thomas, B., Rioja, R.J., 2007. New aluminum lithium alloys for aerospace applications. In: Sadayappan, K., Sahoo, M. (Eds.), Proceedings of the Third International Conference on Light Metals Technology. CANMET, Ottawa, ON, pp. 41—46.

Jagan Reddy, G., 2010. Study on High Temperature Flow Properties of Al—Li Alloy UL40 and Development of Processing Maps, Doctoral Thesis. Indian Institute of Technology, Bombay, India.

Jagan Reddy, G., Srinivasan, N., Gokhale, A.A., Kashyap, B.P., 2008. Characterization of dynamic recovery during hot deformation of spray cast Al—Li alloy UL40 alloy. Mater. Sci. Technol. 24, 725—733.

Jagan Reddy, G., Srinivasan, N., Gokhale, A.A., Kashyap, B.P., 2009. Processing map for hot working of spray formed and hot isostatically pressed Al—Li alloy (UL40). J. Mater. Process. Technol. 209, 5964—5972.

Jonas, J.J., Sakai, T., 1982. A new approach to dynamic recrystallization. In: George, K. (Ed.), Deformation, Processing and Structures. ASM International, Metals Park, OH, pp. 185—243.

Kang, F.W., Zhang, G.Q., Sun, J.F., Li, Z., Shen, J., 2008. Hot deformation behavior of a spray formed superalloy. J. Mater. Process. Technol. 204 (1—3), 147—151.

Karabin, L.M., Bray, G.H., Rioja, R.J., Venema, G.B., 2012. Al—Li—Cu—Mg—(Ag) products for lower wing skin applications". In: Weiland, H., Rollett, A.D., Cassada, W.A. (Eds.), Proceedings of the 13th International Conference on Aluminum Alloys (ICAA13). The Minerals, Metals and Materials Society (TMS) and John Wiley & Sons, Hoboken, NJ, pp. 529—534.

Koh, H.J., Kim, N.J., Lee, S., Lee, E.W., 1998. Superplastic deformation-behavior of a rapidly solidified Al—Li alloy. Mater. Sci. Eng. A A256, 208—213.

Kridli, G.T., El-Gizawy, A.S., Lederich, R., 1998. Development of process maps for superplastic forming of Weldalite™ 049. Mater. Sci. Eng. A A244, 224−232.

Lequeu, P., Smith, K.P., Daniélou, A., 2010. Aluminum−copper−lithium alloy 2050 developed for medium to thick plate. J. Mater. Eng. Perform. 19 (6), 841−847.

Lewis, R.E., Starke Jr., E.A., Coons, W.C., Shiflet, G.J., Willner, E., Bjeletich, J.G., et al., 1987. Microstructure and properties of Al−Li−Cu−Mg−Zr (8090) heavy section forgings. In: Champier, G., Dubost, B., Miannay, D., Sabetay, L. (Eds.), Proceedings of the Fourth International Conference on Aluminium−Lithium Alloys. J. Phys., vol. 48, pp. C3.643−C3.652.

Lynch, S.P., Knight, S.P., Birbilis, N., Muddle, B.C., 2008. Heat-treatment, grain-boundary characteristics and fracture resistance of some aluminium alloys. In: Hirsch, J., Skrotski, B., Gottstein, G. (Eds.), "Aluminium Alloys—Their Physical and Mechanical Properties," Proceedings of the 11th International Conference on Aluminum Alloys (ICAA11). Wiley-VCH Verlag GmbH & Co. KGaA, Weinheim, Germany, pp. 1409−1415.

Magnusen, P.E., Mooy, D.C., Yocum, L.A., Rioja, R.J., 2012. Development of high toughness sheet and extruded products for airplane fuselage structures. In: Weiland, H., Rollett, A.D., Cassada, W.A. (Eds.), Proceedings of the 13th International Conference on Aluminum Alloys (ICAA13). The Minerals, Metals and Materials Society (TMS) and John Wiley & Sons, Hoboken, NJ, pp. 535−540.

Mallesham, P., Gokhale, A.A., Murti, V.S.R., 2003. Interface microstructure and bond shear strength of aluminium alloy AA8090/AA7072 roll clad sheets. J. Mater. Sci. Lett. 22, 1793−1795.

McNamara, D.K., J. R. Pickens, J.R., Heubaum, F.H., 1992. Forgings of Weldalite (TM) 049 Alloys X2094 and X2095. In: Peters, M., Winkler, P.J. (Eds.), Aluminium−Lithium Alloys VI, vol. 2. Deutsche Gesellschaft für Metallkunde, Frankfurt, Germany, pp. 921−926.

Menon, S.S., Rack, H.J., 1994. Flow characterization of cast Al−2Li. In: Sanders Jr., T.H., Starke Jr., E.A. (Eds.), "Aluminum Alloys: Their Physical and Mechanical Properties," Proceedings of the Fourth International Conference, vol. 2. Georgia Institute of Technology, Atlanta, GA, pp. 230−237.

Mukhopadhyay, A.K., Flower, H.M., Sheppard, T., 1990. Development of microstructure in an AA 8090 alloy produced by extrusion processing. Mater. Sci. Technol. 6, 461−468.

Niikura, M., Takahashi, K., Couchi, C., 1985. Hot deformation behaviour in Al−Li−Cu−Mg−Zr alloys. In: Baker, C., Gregson, P.J., Harris, S.J., Peel, C.J. (Eds.), "Aluminium−Lithium Alloys III," Proceedings of the Third International Conference on Aluminium−Lithium Alloys. The Institute of Metals, London, UK, pp. 213−221.

Pitcher, P.D., 1988a. Ageing of forged aluminium−lithium 8091 alloy. Scr. Metall. 22, 1301−1306.

Pitcher, P.D., 1988b. Ageing of forged aluminium−lithium 8091 alloy. Royal Aircraft Establishment Technical Memorandum MAT/STR 1120, Farnborough, UK.

Prangnell, P.B., Hulley, S.I., Palmer, I.G., 1994. Microstructure and mechanical properties of hot rolled high Li content aluminum alloys processed by spray casting. In: Sanders Jr., T.H., Starke Jr., E.A. (Eds.), "Aluminum Alloys: Their Physical and Mechanical Properties," Proceedings of the Fourth International Conference, vol. 2. Georgia Institute of Technology, Atlanta, GA, pp. 106−113.

Prasad, Y.V.R.K., Ravichandran, N., 1991. Effect of stacking fault energy on the dynamic recrystallization during hot working of FCC metals: a study using processing maps. Bull. Mater. Sci. 14, 1241−1248.

Prasad, Y.V.R.K., Sashidhara, S., 1997. Aluminum alloys. In: Prasad, Y.V.R.K., Sashidhara, S. (Eds.), Hot Working Guide: A Compendium of Processing Maps. ASM International, Materials Park, OH, pp. 160–177.

Prasad, Y.V.R.K., Gegel, H.L., Doraivelu, S.M., Malas, J.C., Morgan, J.T., Lark, K.A., et al., 1984. Modeling of dynamic material behavior in hot deformation: forging of Ti-6242. Metall. Trans. A 15A, 1883–1892.

Richard, S., Sarazzin-Baudoux, C., Petit, J., 2012. Fatigue crack propagation in new generation aluminium alloys. Key Eng. Mater. 476–479, 488–489.

Rioja, R.J., 1998. Fabrication methods to manufacture isotropic Al–Li alloys and products for space and aerospace applications. Mater. Sci. Eng. A A257, 100–107.

Rioja, R.J., Liu, J., 2012. The evolution of Al–Li base products for aerospace and space applications. Metall. Mater. Trans. A 43A, 3325–3337.

Singer, A.R.E., 1982. The challenge of spray forming. Powder Metall. 25, 195–200.

Skillingberg, M.H., Ashton, R.F., 1987. Processing and performance of Al–Li–Cu–X extrusions. In: Champier, G., Dubost, B., Miannay, D., Sabetay, L. (Eds.), Proceedings of the Fourth International Conference on Aluminium–Lithium Alloys. J. Phys., vol. 48, pp. C3.179–C3.186.

Starke Jr., E.A., Staley, J.T., 1996. Application of modern aluminum alloys to aircraft. Prog. Aerosp. Sci. 32, 131–172.

Tchitembo Goma, F.A., Larouche, D., Bois-Brochu, A., Blais, C., Boselli, J., Brochu, M., 2012. Fatigue crack growth behavior of 2099-T83 extrusions in two different environments. In: Weiland, H., Rollett, A.D., Cassada, W.A. (Eds.), Proceedings of the 13th International Conference on Aluminum Alloys (ICAA13). The Minerals, Metals and Materials Society (TMS) and John Wiley & Sons, Hoboken, NJ, pp. 517–522.

Terlinde, G., Sauer, D., Fischer, G., Smith, A., 1992. Development of Al–Li forgings and extrusions for aerospace applications. In: Peters, M., Winkler, P.J. (Eds.), Aluminium–Lithium Alloys VI, vol. 2. Deutsche Gesellschaft für Metallkunde, Frankfurt, Germany, pp. 927–932.

Venugopal, S., Mannan, S.L., Prasad, Y.V.R.K., 1997. Instability map for cold and warm working of as-cast 304 stainless steel. J. Mater. Process. Technol. 65 (1–3), 107–115.

Wadsworth, J., Palmer, I.G., Crooks, D.D., 1983. Superplasticity in Al–Li based alloys. Scr. Metall. 17, 347–352.

Wanhill, R.J.H., 1994. Damage tolerance engineering property evaluations of aerospace aluminium alloys with emphasis on fatigue crack growth. National Aerospace Laboratory NLR Technical Publication NLR-TP-94177, Amsterdam, the Netherlands.

Wellstead, P.E., 1979. Introduction to Physical System Modelling. Academic Press Ltd., London, UK.

Superplasticity in and Superplastic Forming of Aluminum−Lithium Alloys

S. Balasivanandha Prabu* and K.A. Padmanabhan†

*Associate Professor, Department of Mechanical Engineering, Anna University, Chennai, India,
†University Chair Professor, School of Engineering Sciences and Technology and Centre for
Nanotechnology, University of Hyderabad, Hyderabad, India

Contents

8.1 Introduction	221	8.3.1 Cavity Nucleation
8.2 Superplasticity	225	and Growth	240
8.2.1 Phenomenon of		8.4 Role of Friction Stir Processing
Superplasticity	225	on Superplastic Forming	245
8.2.2 Experimental		8.4.1 Superplasticity in
Investigations	230	FSP Materials	245
8.2.3 Low-Temperature		8.4.2 Superplastic Behavior
Superplasticity	234	and Deformation
8.2.4 Effects of Strain Rate		Mechanism	246
and Strain Rate		8.4.3 Cavity Density and
Sensitivity (*m* value)		Size Distribution	250
on Superplasticity	237	8.5 Applications	251
8.3 Superplastic Forming	239	8.6 Concluding Remarks	253
		References	253

8.1 INTRODUCTION

Since the late 1950s, lithium has been used as an alloying element in aluminum (Ivanov et al., 2012). The development of Al−Li alloys offers significant improvements in structural performance of aerospace components through density reduction, stiffness increase, increases in fracture toughness and fatigue crack growth resistance, and enhanced corrosion resistance. The first- and second-generation Al-Li alloys have found limited applications in aerospace structures and components. The first-generation Al—Li alloy for example, AA 2020 (Li—1.2%, Cu—4.5%, Mn—0.5%), was initially used for the wings of the US Navy's RA-5C Vigilante aircraft. The other alloys, 1420 and 1421, were developed by the Soviet Union in 1965 (Rioja and Liu, 2012).

In the second-generation Al—Li alloys the Li concentration is above 2 wt%. Although it led to 7—10% density reduction, these alloys were considered unfit for airframe designs due to lower short-transverse fracture toughness and higher anisotropy in the tensile properties (Rioja and Liu, 2012). These problems were overcome in the third-generation alloys with reduced Li concentration (from 0.75 to 1.8 wt%, along with the minor additions of Zr of the order of ~ 0.1). Alloys such as 2195, 2196, 2297, 2397, 2198, 2099, 2199, 2050, and 2060 were developed for aerospace applications. The chemical compositions of many of the different Al—Li alloys are listed in Table 8.1.

The present authors analyzed the contents of more than 500 papers for this review. However, only a significantly reduced number of papers, which were regarded as essential, are cited here. This chapter discusses the superplastic behavior of Al—Li alloys and also their applications in forming.

The phenomenon of structural superplasticity has been demonstrated in metals and alloys, intermetallics, ceramics, dispersion-strengthened alloys, metal matrix and ceramic matrix composites, and geological materials. This phenomenon is present when the grain size is in the micrometer-, submicrometer-, or nanometer range. With decreasing grain size, however, the phenomenon is manifest at lower temperatures and/or higher strain rates compared with the situation present in coarser-grained materials. The similarity between the deformation behavior of superplastic alloys and that of tar, pitch, "silly putty," plasticine, and "chewing gum" is well known. In the case of metals and alloys, crystal structure has insignificant effect on superplastic behavior; alloys that have Face Centered Cubic (FCC), Body Centered Cubic (BCC) and Hexagonal Close Packed (HCP) structures are all significantly superplastic. Intermetallics, in which movement of dislocations is quite difficult, also display a superplastic response very similar to that of normal metallic materials. Therefore, it is reasonable to suggest that superplastic flow critically depends only on the microstructural features that are common to all the above-mentioned classes of materials (Padmanabhan, 2009; Padmanabhan and Gleiter, 2012).

Many commercial superplastic alloys were successfully developed, and the distinctive properties obtained during superplastic flow are being exploited in forming for quite a while now (Giuliano, 2008; Padmanabhan and Davies, 1980; Tang and Robbins, 1974; Thomsen et al., 1970; Wu et al., 2009; Yuwei et al., 1997). These alloys display behavior similar to that of thermoplastics above their glass transition temperatures, which enable the use of polymer

TABLE 8.1 Nominal Compositions of Key Al–Li Alloys (Wt%)

Alloys	Li	Cu	Mg	Ag	Zr	Sc	Mn	Zn	Year of Induction
First Generation									
2020	1.2	4.5					0.5		Alcoa 1958
1420	2.1		5.2		0.11				Soviet 1965
1421	2.1		5.2		0.11	0.17			Soviet 1965
Second Generation (Li ≥ 2%)									
2090	2.1	2.7			0.11				Alcoa 1984
2091	2.0	2.0	1.3		0.11				Pechiney 1985
8090	2.4	1.2	0.8		0.11				EAA 1984
1430	1.7	1.6	2.7		0.11	0.17			Soviet 1980s
1440	2.4	1.5	0.8		0.11				Soviet 1980s
1450	2.1	2.9			0.11				Soviet 1980s
1460	2.25	2.9			0.11	0.09			Soviet 1980s
Third Generation (Li < 2%)									
2195	1.0	4.0	0.4	0.4	0.11				LM/Reynolds 1992
2196	1.75	2.9	0.5	0.4	0.11		0.35 max	0.35 max	LM/Reynolds/McCook Metals 2000
2297	1.4	2.8	0.25 max		0.11		0.3	0.5 max	LM/Reynolds 1997
2397	1.4	2.8	0.25 max		0.11		0.3	0.10	Alcoa 2002
2198	1.0	3.2	0.5	0.4	0.11		0.5 max	0.35 max	Reynolds/McCook Metals/Alcan 2005
2098	1.0	3.5	0.5	0.4	0.11		0.35 max	0.35 max	McCook Metals 2000
2099	1.8	2.7	0.3		0.09		0.3	0.7	Alcoa 2003
2199	1.6	2.6	0.2		0.09		0.3	0.6	Alcoa 2005
2050	1.0	3.6	0.4	0.4	0.11		0.35	0.25 max	Pechiney/Alcan 2004[a]
2296	1.6	2.45	0.6	0.43	0.11		0.28	0.25 max	Constellium Alcan 2010[a]
2060	0.75	3.95	0.85	0.25	0.11		0.3	0.4	Alcoa 2011
2055	1.15	3.7	0.4	0.4	0.11		0.3	0.5	Alcoa 2012

[a]*Pechiney acquired by Alcan 2003; Constellium formerly Alcan Aerospace.*
Source: After Rioja and Liu (2012).

processing techniques for superplastic forming of metals/alloys (e.g., pressure forming) (Chockalingam et al., 1985; Padmanabhan and Davies, 1980). Aluminum-based alloys could be superplastically deformed when the microstructure satisfies any one of the following conditions: (i) a fully recrystallized condition or (ii) a cold- or warm-rolled condition (Qing et al., 1992; Wadsworth et al., 1983). In an industrial perspective based on mass production, the second route is preferred to the first one since the Al alloys of the second kind often become superplastic at higher strain rates than those of the first type (Qing et al., 1992).

Aluminum alloys containing lithium reduce the density and increase the specific strength and the Young's modulus in direct proportion to the Li content (Lequeu et al., 2010). For lithium additions up to 4 wt%, every weight percent lithium addition to an aluminum alloy reduces the density by $\sim 3\%$ and increases the elastic modulus by $\sim 6\%$ (Chaudhuri et al., 1990). However, with the addition of Li, Al alloys show decreases in ductility and fracture toughness, delamination problems, and poor stress corrosion cracking resistance. Increased strength with only minimal or no decrease in toughness is, therefore, a major desired goal (Moreira et al., 2012). Their potential ability for superplastic forming makes Al–Li alloys the primary choice for many structural applications in the aerospace industry. The superplastic forming route will cut down the number of stages required to form complicated parts, while at the same time bringing down tooling and assembly costs (Bairwa et al., 2005; Padmanabhan and Davies, 1980).

Approximately 10% of weight reductions were reported when 2000 and 7000 series aluminum alloys containing $\sim 2.5\%$ Li were used to fabricate the components. Even greater weight savings might have been possible if these alloys had been superplastically formed (Shakesheff and Partridge, 1986). The high stiffness to density ratio and excellent precipitation-hardening response of Al—Li alloys make them potential replacements for heavier wrought Al alloys in structural components for aerospace applications (Gregson and Flower, 1985). Commercial Al alloys with such high Li contents (4.2 wt%) are generally produced by spray forming, as the problems associated with conventional DC casting (hot tearing) or melt spinning (oxidation) restrict Li additions to a maximum of only 2.5 wt% in commercial production. However, spray-formed materials contain a moderate volume fraction of porosity (4—6 vol.%) and are, therefore, hot isostatically pressed or extruded for porosity closure prior to forging or rolling (Jagan Reddy et al., 2009; Prangnell et al., 1994). Evidently this additional step will add to the cost of manufacture.

Considerable interest in the superplastic forming of aluminum—lithium alloys has followed in the wake of superplastic titanium alloys, which have shown significant payoffs in aircraft structures. Superplasticity of structural Al and Al—Li alloys has been investigated in detail in earlier works. This chapter discusses superplasticity in Al—Li alloy systems and the use of this phenomenon in the forming processes.

8.2 SUPERPLASTICITY

8.2.1 Phenomenon of Superplasticity

The term superplasticity defines very large elongations (of several hundred percent) in tensile deformation, shown by certain fine-grained polycrystalline materials (grain size $< \sim 10 \, \mu m$)—especially metallic materials, which exhibit an unusually high-strain rate sensitivity of flow stress. Usually, the strain rate range is around $10^{-3} - 10^{-5} \, s^{-1}$, and the forming temperatures are above $0.5 T_m$ (where T_m is the absolute melting point). (In recent times in several aluminum alloys, through the use of grain boundary pinning agents, dispersions, etc., grain sizes of $\sim 1-2 \, \mu m$ or even less have been produced; and in these alloys, superplastic deformation has been observed at strain rates of the order of $10^{-2} \, s^{-1}$ or more—the so-called high-strain rate superplasticity (HSRSP).) Despite the understanding of the phenomenon, according to some, the fundamental mechanism responsible for the high rate sensitivity is unclear (Sotoudeh and Bate, 2010). Others contest this view (Padmanabhan, 2009; Padmanabhan and Gleiter, 2004, 2012). This phenomenon is used for producing intricate and large components at very low-applied stresses relative to those required in conventional deformation processing. As pointed out above, the phenomenon imparts a resistance to strain localization that is due to the high sensitivity of the flow stress to the strain rate, and superplastic forming is usually achieved at high homologous temperatures and low strain rates, which normally result when the flow stresses are low (Es-Said et al., 2011).

The stress−strain rate relationship in the superplastic regime can be expressed as $\sigma = K \dot{\varepsilon}^m$, where σ is the stress at a strain rate of $\dot{\varepsilon}$; m is the strain rate sensitivity index of value ≥ 0.3 under uniaxial loading (treated as a constant; in reality it is not), and K is a material constant (Padmanabhan et al., 2001).

One can also rewrite the above equation as follows (Padmanabhan et al., 2001), i.e., $\dot{\varepsilon} = C \sigma^n$, where $C = 1/K^n$ and $n = 1/m$. Then, on logarithmic coordinates $\log \sigma - \log \dot{\varepsilon}$, would be a straight line, the slope of which is equal to m. However, experimental data cannot be normally fitted by a straight line. The actual shape is typically sigmoidal, as shown in Figure 8.1A. Therefore, Eq. (8.1) should be treated as a local approximation of the sigmoidal plot which is valid within a sufficiently narrow strain rate interval where the hypothesis $m \cong \text{const}$ (Vasin et al., 2000) is valid.

Let

$$M = \partial \ln \sigma / \partial \ln \dot{\varepsilon} \tag{8.1}$$

The value of M depends upon strain rate so that the M and $\dot{\varepsilon}$ dependency has a dome-like shape as shown in Figure 8.1B. The maximum value M_{max} corresponds to the optimum strain rate $\dot{\varepsilon}_{opt}$ for superplastic deformation for a given average grain size and temperature of deformation. The boundaries of the optimum strain rate interval for superplastic flow are conventionally found from the

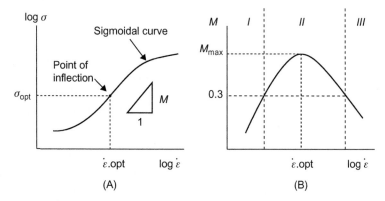

FIGURE 8.1   (A) Sigmoidal log σ–log $\dot{\varepsilon}$ relationship and (B) M variation for superplastic flow (Padmanabhan et al., 2001).

empirical condition $M > 0.3$ (Padmanabhan and Davies, 1980; Padmanabhan et al., 2001; Vasin et al., 2000). It should be noted that region II could be divided into region IIa (left of the maximum) and IIb (right of the maximum). The physical mechanisms operating in the two regions are different. Moreover, in region IIa, an increase in strain rate increases the M value, while in region IIb, a strain rate increase will decrease the M value. Evidently, the necking resistance in the two regions resulting from such strain rate changes would be different (Padmanabhan and Davies, 1980).

The value m depends upon a number of factors: strain, strain rate, structure evolution, deformation mode, type of loading, etc. Therefore, this parameter cannot be considered as a material constant. Hence, m and M are different in nature. It is to be noted that the dependencies $m\,(\dot{\varepsilon})$ and $M\,(\dot{\varepsilon})$ are not characterized by the same function (Padmanabhan et al., 2001).

Many physical models of superplasticity have been suggested. It is sometimes suggested that the superplastic deformation is the result of grain boundary sliding (GBS), diffusion creep (DC), and intragranular slip (IS), added in a linear fashion. In the same way, the total plastic strain rate $\dot{\varepsilon}_p$ is assumed to result as follows (for a summary, see Padmanabhan et al., 2001):

$$\dot{\varepsilon}_p = \dot{\varepsilon}_{GBS} + \dot{\varepsilon}_{DC} + \dot{\varepsilon}_{IS} \tag{8.2}$$

It is further assumed that every mechanism of superplastic deformation is characterized by its own value of the strain rate sensitivity index, m, for example, for GBS it is assumed $m_{GBS} \approx 0.5$, for DC it is assumed $m_{DC} \approx 1$, while for IS it is assumed $m_{IS} = 0.1$. The resulting value of the strain rate sensitivity index, m, is believed to be determined by the interaction of all these mechanisms. Padmanabhan et al. (2001) have shown that such a combination, regardless of the individual percentage contribution, will not lead to a sigmoidal shape. Therefore, if mechanisms are to be combined, one may have to do so by taking two at a time, making sure that the choice is physically realistic.

The maximum elongation obtainable in a superplastic material is strongly dependent on the initial microstructure of the material. Superplasticity also depends on the changes in the microstructure that take place during deformation. The changes in the microstructure occur during superplastic deformation due to dynamic grain growth, dynamic recrystallization, cavitation, etc., in addition to static recrystallization in some cases prior to the onset of deformation. Dynamic grain growth and cavitation (Chokshi and Mukherjee, 1988; Chung et al., 2004; Hirai et al., 2007; Jiang and Bate, 1996; Padmanabhan and Raviathul Basariya, 2009; Rabinovich and Trifonov, 1996) affect the extent of achievable superplastic flow in a material. However, dynamic recrystallization (Maehara and Ohmori, 1987; Qing et al., 1992), when present in a continual manner, could favor superplastic deformation and can enhance ductility (Pancholi and Kashyap, 2007).

It is well known that an equiaxed fine-grained microstructure is essential for obtaining superplastic elongations. A startling observation of the microstructural development is that equiaxed grains remain nearly equiaxed even after large deformations. Further, GBS is assumed by most of the authors as the fundamental mechanism of superplastic flow. Another way of describing the mechanism responsible for superplasticity, not so popular, is Lifshitz sliding, where the grains do not move relative to each other, with intragranular deformation or diffusion being the rate-controlling process (Sotoudeh and Bate, 2010).

Superplasticity in Al−Li alloys is usually induced through thermomechanical treatment (TMT). It is carried out to obtain a very fine microstructure either by recrystallization after large plastic deformation (hot or cold working) or by dynamic recrystallization or "continuous" recrystallization during superplastic deformation (Koh et al., 1998; Qing et al., 1991; Wadsworth et al., 1985). In two-phase alloys, with only a small volume fraction of second phase, the control of grain growth is via the Zener pinning mechanism (Sotoudeh and Bate, 2010). The other noticeable phenomenon in superplasticity is diffusional creep, advocated in particular for ceramic materials. The DC mechanism has been considered as a cause of superplasticity when IS and Rachinger sliding are absent. Some grain elongation does occur, without boundary migration, but simply changes the grain shape (Sotoudeh and Bate, 2010). Chokshi (2000) presented an explanation based on DC and dynamic grain growth mechanisms for superplasticity in oxide ceramics.

Padmanabhan and Raviathul Basariya (2009) have further considered a model due to Padmanabhan and Schlipf (1996) and Padmanabhan and Gleiter (2004, 2012), which is based on the assumption that GBS, which develops to a mesoscopic scale (of the order of a grain diameter or more), is the rate-controlling process and that it is accommodated by faster/nonrate-controlling dislocation activity and diffusion that take place to a limited extent. In fact, Gifkins and Snowden noted as early as in 1966 that at moderately high temperatures ($>0.5T_{\mathrm{m}}$, where T_{m} is the melting point on the absolute scale) and low stresses, Nabarro-Herring and Coble creep mechanisms are too slow to account for the experimentally observed strain rates (Gifkins and Snowden, 1966). Yet this suggestion, based on experimental results, was not invoked in a theoretical model in which GBS is rate controlling until this model was proposed. This model is concerned with the identification of the unit processes that eventually lead to the macroscopic phenomenon of GBS. A 2D section of the 3D microstructure would appear as shown in Figure 8.2A. The shaded portions identify the 2D sections of the peaks and valleys that should be leveled by local boundary migration for plane interface formation (mesoscopic boundary sliding) to facilitate large-scale boundary sliding. Figure 8.2B shows the external shear stress-driven movement of a boundary triple junction. The internal stresses in the vicinity of the triple junction are reduced by the migration of point A to point B, that is, when the semi-angle of the triple junction, α, increases with increasing external shear stress. It can be shown that in the limit $\alpha \to 90°$, the plane interface is formed (Padmanabhan, 2009; Padmanabhan and Gleiter, 2004).

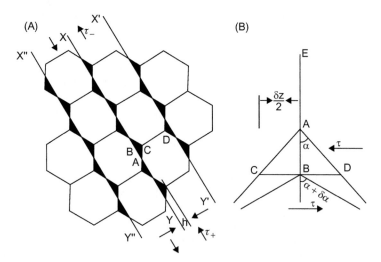

FIGURE 8.2 (A) 2D Section of grains of equal size. Planar interfaces along XY, $X'Y'$, $X''\ Y''$, etc., are obtained if the atoms located in the dark regions, e.g., in region ABC, move by nonrate-controlling dislocation emission/diffusion in such a way as to extend the boundary DC. (B) Shear stress-driven movement of a boundary triple junction, A (formed by three boundaries: AC, AD, and AE). If an external stress, τ, is applied, as indicated, the total free energy of the system will be reduced when the triple junction moves from position A to position B. *Source: After Padmanabhan and Gleiter (2004).*

A detailed analysis has led to the following equations (von Mises criterion is assumed):

$$\Delta F_0 = \frac{1}{2}(\beta_1 \gamma_0^2 + \beta_2 \varepsilon_0^2)GV_0 \tag{8.3}$$

where

$$V_0 = \frac{2}{3}\pi W^3; \beta_1 = 0.994\left(\frac{1.590 - p}{1 - p}\right); \beta_2 = \frac{4(1 + p)}{9(1 - p)}; \varepsilon_0 = \frac{\gamma_0}{\sqrt{3}}$$

and

$$\sigma_0 = \frac{\sqrt{3}C_1}{d} \text{ and } C_1 = \left(\frac{8G\Gamma_B r}{3^{0.25}}\right) \tag{8.4}$$

$$\dot{\varepsilon} = \left(\frac{0.6981 W^4 \gamma_0^2 v}{kTd}\right)(\sigma - \sigma_0)\exp\left(\frac{-\Delta F_0}{kT}\right) \tag{8.5}$$

The model can also estimate the average grain boundary misfit removed by diffusion, r, as the GBS process develops from the microscopic to mesoscopic scale. In addition, this analysis is able to describe superplasticity in metals and alloys, ceramics and intermetallics of micrometer-, submicrometer-, and nanometer-grain size ranges and HSRSP using a single rate-controlling mechanism, namely, mesoscopic GBS. Nonrate-controlling dislocation emission from one grain boundary gets absorbed at the opposite boundary and facilitates local boundary migration. This leads to plane interface formation (mesoscopic sliding), which is the accommodation process. In these equations, ΔF_0 is the free energy of activation, p the Poisson ratio, γ_0 the mean shear strain associated with unit strain in an atomistic scale shear event, $\varepsilon_0 = (1/\sqrt{3})\gamma_0$ (von Mises), G the shear modulus, V_0 the volume of the basic unit of sliding ($= (2/3)\pi W^3$, with W the grain boundary width). σ_0 is the threshold stress necessary to be overcome to lead to mesoscopic boundary sliding, d the average grain size, Γ_B the specific grain boundary energy, $\dot{\varepsilon}$ the external strain rate, v the thermal vibration frequency, k the Boltzmann constant, and T is the temperature of deformation on the absolute scale,

An alternative approach is based on a generic rate equation proposed for high-temperature creep more than 40 years ago, assuming that this "largely empirical" equation (Mukherjee et al., 1969), namely,

$$\dot{\varepsilon} = A\left(\frac{DGb}{kT}\right)\left(\frac{b}{d}\right)^p\left(\frac{\sigma}{G}\right)^n \tag{8.6a}$$

or

$$\dot{\varepsilon} = A'\left(\frac{DGb}{kT}\right)\left(\frac{b}{d}\right)^2\left(\frac{\sigma - \sigma_0}{G}\right)^2 \tag{8.6b}$$

with A and A' constants, D the relevant (lattice, grain boundary, or mixed) diffusivity and b, the Burgers vector, has physical significance and that the activation energy (obtained from D), p, and the n ($=1/m$) values estimated using Eq. (8.6a) or Eq. (8.6b) for a given set of experimental data can help identify the rate-controlling mechanism. Apart from the lack of proof for the physical significance of this equation in the context of superplasticity, this analysis also suggests different rate-controlling mechanisms for different materials, even in a single class, and hence lacks universal applicability, unlike the former analysis.

8.2.2 Experimental Investigations

Many Al–Li systems were investigated for superplastic deformation potential. The work of Mogucheva and Kaibyshev (2008) on alloy 1421 established elongations $>1000\%$ in a wide range of temperatures and strain rates (Figures 8.3 and 8.4). The maximum elongation of 2700% was observed at 450°C and $\dot{\varepsilon} = 1.4 \times 10^{-2}\,\text{s}^{-1}$. The elongation versus strain rate plots at various temperatures shown in Figure 8.3 clearly distinguishes the three characteristic regions of superplasticity. The strain rate of maximum m increases with an increase in temperature, and the absolute value of m increases from 0.32 at 300°C to 0.57 at 450°C. As high elongations and values of $m \geq 0.33$ are observed at strain rates $>10^{-2}\,\text{s}^{-1}$, it is noted that this alloy exhibits "HSRSP."

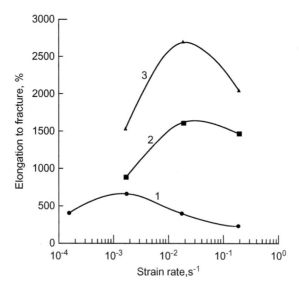

FIGURE 8.3 Effect of strain rate on the elongation to fracture in alloy 1421 at (1) 300°C, (2) 400°C, and (3) 450°C (Mogucheva and Kaibyshev, 2008).

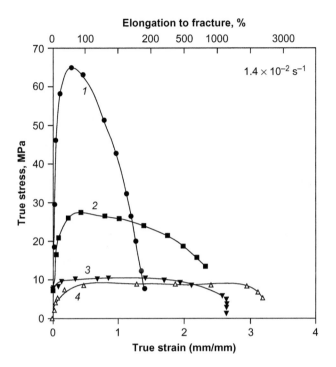

FIGURE 8.4 Plots of the true stress versus imposed strain for alloy 1421 at (1) 300°C, (2) 400°C, (3) 450°C, and (4) 450°C (Mogucheva and Kaibyshev, 2008).

For optimal superplastic effects, microstructural uniformity is of importance and the grain size should be of the order of 0.5−2.5 μm or even less. However, the fracture onset and the attainable relative elongation values depend on the formation of pores. Pore formation is suppressed when the microstructure is homogeneous and the grains are equiaxed (Mogucheva and Kaibyshev, 2008).

Figure 8.4 shows the superplastic flow behavior of alloy 1421 at different temperatures for a constant strain rate of $1.4 \times 10^{-2}\,s^{-1}$. The strain rate sensitivity index m at 450°C increases with the degree of straining from 0.5 for $e = 0.5$, to 0.6 for $e = 2$. Further deformation leads to a decrease in m. However, this value does not fall below 0.4. Thus, evidently the high plasticity of the material is caused by the high-strain rate sensitivity of the flow stress, which prevents strain localization leading to the formation of neck and pore formation.

Grain refinement through thermomechanical processing enhances superplasticity. Several investigators have reported that a submicrocrystalline structure was obtained using the severe plastic deformation (SPD) processes such as equal channel angular pressing (ECAP) (Ehab and El-Danaf, 2012), High Pressure Torsion (HPT) (Lee et al., 2012), and multidirectional extrusion

(Gokhale and Singh, 2005), in which considerable grain refinement was reported and which resulted in a maximum superplastic elongation of up to $\sim 1850\%$.

Superplastic properties were studied using the 2098 (2.2% Li, 1.3% Cu, 0.73% Mg, 0.05% Zr, balance Al) material after it had been subjected to an eight-step equal channel angular extrusion (ECAE) process. The results are shown in Figure 8.5. At 500°C and 2×10^{-3} s^{-1}, a peak flow stress of about 125 MPa was measured that remained nearly constant between true strains of 0.6 and 0.9. The elongation to fracture for this test was 270% (true strain ~ 1.2). The flow stress at 530°C and strain rate 2×10^{-3} s^{-1} was 110 MPa and remained relatively constant between true strains of 0.6 and 1.2, and the elongation to fracture was about 330% (true strain ~ 1.5). In all cases, the flow stress decreased beyond its peak value. This is due to (i) decrease in cross-section with increasing strain and (ii) cavity formation accompanying superplastic flow (Salem and Lyons, 2002).

It is also understood from the annealing results reported by Salem and Lyons (2002) that at higher temperatures, for example, 530°C, the grain size

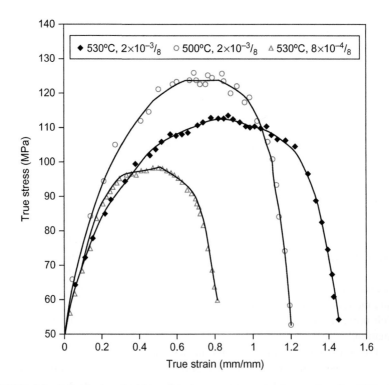

FIGURE 8.5 Superplastic uniaxial tensile-true stress-true strain diagram for alloy 2098 after ECAE processing for four passes at 350°C plus four passes at 200°C with a cumulative effective strain of 9.3 mm/mm (Salem and Lyons, 2002).

increased to about $2\,\mu m$, with dislocations within the grains in the form of dislocation helices and Orowan loops surrounding $\delta'(Al_3Li)$ coherent precipitates, as shown in Figure 8.6. Furthermore, extinction contours with dislocations were observed along the coarsened grain boundaries (Salem and Lyons, 2002). The lack of grain growth inhibitors in the microstructure resulted in low elongation to fracture (premature fracture) at the elevated temperatures during superplastic forming. Therefore, to retain the submicron grain size microstructure developed by ECAE to achieve large superplastic elongations, the presence of a high volume fraction of fine, coherent, and uniformly distributed precipitates which have high stability against dissolution at elevated temperatures is necessary (Salem and Lyons, 2002).

The alloy 2098 billets obtained after ECAE are likely to have subgrains due to dynamic recovery. This is different from what is found in the Al−Cu−Li alloys such as 2095 and 2195 processed under a similar route, which exhibit evidence for particle-stimulated grain nucleation. On the one hand, 2095 and 2195 alloys, at a processing temperature of $350°C$, fall in the Al_{ss} + TB (Al_7CuLi) + T1 (Al_2CuLi) and Al_{ss} + TB phase regions, respectively, where Al_{ss} denotes the phases in solid solution. On the other hand, the 2098 Al−Li−Cu alloy system at $350°C$ falls in the Al_{ss} + T2 + δ (Al−Li) phase region. The low volume fraction and small size ($\leq 100\,nm$) of the particles within the matrix of the 2098 alloy do not effectively promote a fine, well-defined structure by particle-stimulated nucleation (Salem and Lyons, 2002).

Figure 8.7 shows tensile elongation reported by Koh et al. (1998) for a rapidly solidified Al−Li alloy (Al−3Li−1Cu−0.5Mg−0.5Zr) tested at $500°C$ as a function of strain rate and compared with 8090 Al−Li alloy. This experiment was carried out to confirm the importance of the microstructure to superplasticity. In the strain rate range of $10^{-4}\,s^{-1}$ to $\sim 10^{-3}\,s^{-1}$, tensile elongation of the 8090 Al−Li alloy shows a higher value than that of the rapidly solidified Al−Li alloy. But, in the rapidly solidified Al−Li alloy as the strain rate increases, elongation also increases and reaches above 400% at $500°C$, which exceeds the maximum elongation of the 8090 alloy.

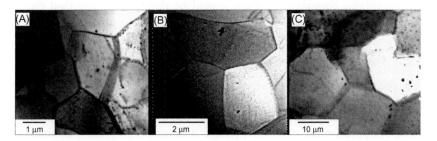

FIGURE 8.6 TEM bright-field micrographs for four passes at $350°C$ plus four passes at $200°C$ ECAE processed alloy 2098. Static annealing at $530°C$ for (A) 10 min, (B) 20 min, and (C) 30 min (Salem and Lyons, 2002).

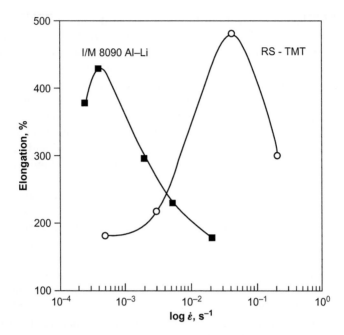

FIGURE 8.7 The effect of strain rate on tensile elongation of the rapidly solidified Al–Li alloy tested at 500°C and compared with 8090 Al–Li alloy (RS-TMT: rapidly solidified thermomechanical treatment) (Koh et al., 1998).

The data in Figure 8.8A show that elongation increases with increasing strain rate at 520°C and reaches the maximum elongation of about 530%. However, a higher stress is required to superplastically deform the materials at the higher strain rates (Figure 8.8A). Figure 8.8B reveals that the strain rate sensitivity index (m) is about 0.4 in the strain rate range of $10^{-2} \, \text{s}^{-1}$ to $4 \times 10^{-2} \, \text{s}^{-1}$ in which range the elongation also is a maximum. Therefore, the superplastic forming equipment should be designed to withstand these high stresses (Koh et al., 1998).

8.2.3 Low-Temperature Superplasticity

The optimum temperature range for superplastic deformation of Al–Li alloys is typically 510–530°C. At such high temperatures, Li or Mg depletion takes place near the surface of a superplastic thin sheet, which is detrimental to post-formed room temperature mechanical properties. Cavitation and grain growth also occur at high temperatures. Therefore, the inducement of low-temperature superplasticity (LTSP) in Al–Li alloys becomes important (Pu et al., 1995).

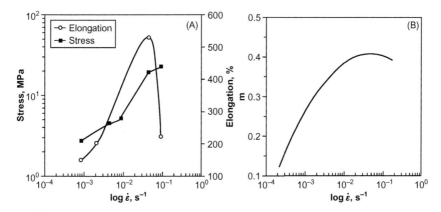

FIGURE 8.8 The effect of strain rate on (A) stress and elongation and (B) m value in case of a rapidly solidified Al−Li alloy tested at 520°C (Koh et al., 1998).

Improved grain refinement facilitates LTSP ($\sim 0.5 T_m$) in Al−Li alloys, which minimizes grain growth and creates new opportunities for superplastic forming technology. In recent years, many SPD techniques were developed to achieve grain refinement in these alloys. The ultrafine-grained (UFG) materials that result from SPD treatments are helpful in achieving LTSP. UFG materials are affected by grain coarsening during the heating before superplastic deformation. Therefore, it is necessary to carry out the experimental study below a temperature of $0.5 T_m$, which eliminates significant grain growth (Ma and Mishra, 2003; Noda et al., 2003).

Many experiments indicate that, as in conventional superplastic deformation, GBS is the dominant mechanism in the metallic alloys that exhibit LTSP or HSRSP at somewhat greater temperatures (Wang and Huang, 2003).

Islamgaliev et al. (2009) established superplasticity in Al 1421 alloy at lower temperatures after the alloy was subjected to the ECAP process. In the temperature range of 300−400°C, 660−780% elongation at deformation rates of $10^{-2}-10^{-1}\,s^{-1}$ is obtained. This was due to the presence of small equiaxed grains ($d = 0.7-0.8\,\mu m$) with a higher volume fraction of grain boundaries and the presence of second-phase grains/particles, which help in the prevention of grain growth by grain boundary pinning.

The superplastic conditions, i.e., optimum superplastic temperature, strain rate, and maximum elongation, of some of the Al−Li alloys are summarized in Table 8.2. Most powder metallurgy alloys display superplastic behavior similar to that of the ingot metallurgy alloys because both types have similar grain sizes after TMT. Many subgrain boundaries are also present.

TABLE 8.2 Summary of Superplastic Deformation in Different Al–Li Alloys

Alloy Designation	Temperature Range (°C)	Strain Rate Range	Average Grain Size (μm)	Elongation (Long) (%)	Reference
2091 (Al–Cu–Li–Mg–Zr)	510–520	8×10^{-4}	–	<470	Yuwei et al. (1997)
Al–3Li–1Cu–0.5Mg–0.5Zr	520	4×10^{-2}	2S	530	Koh et al. (1998)
Al–3Li–1Cu–0.5Mg–0.5Zr (PM)	793	4×10^{-2}	1	530	Pandey et al. (1986)
Al–3Cu–2Li–1Mg–0.15Zr (IM)	723	5.0×10^{-4}	1.5	677	Pandey et al. (1986)
Al–3Cu–2Li–1Mg–0.2Zr (PM)	723	5.0×10^{-4}	1	477	Wadsworth et al. (1985)
Al–3Cu–2Li–1Mg–0.15Zr (IM-Ext.)	773	3.3×10^{-3}	3.0	900	Wadsworth et al. (1985)
Al–3Cu–2Li–1Mg–0.2Zr (PM)	763	3.3×10^{-3}	2.0	680	Cui et al. (1994)
Al–3Li–0.5Zr (PM)	723	3.3×10^{-3}	–	1000	Wadsworth et al. (1985)
8090	350	8×10^{-4}	3.7	700	Pu et al. (1995)
8090 (rolled)	530	1×10^{-3}	–	660	Xun et al. (2005)
8090	520	5×10^{-4}	–	660	Garmestani et al. (1998)
1421 (ECAP)	450	1.4×10^{-2}	1.6	2700	Mogucheva and Kaoebyshev (2008)
1420 (Al–Mg–Li alloy)	525	1×10^{-3}	6.3	915	Ye et al. (2009)

IM, PM, and Ext. refer to ingot metallurgy, powder metallurgy, and extrusion, respectively.

8.2.4 Effects of Strain Rate and Strain Rate Sensitivity (*m* value) on Superplasticity

There is ample evidence in research reports that limited amounts of diffusion and dislocation activity are present during superplastic deformation. Therefore, three different views are available in the literature as to the rate-controlling mechanism during superplastic flow, namely, GBS, diffusion, or dislocation motion. However, the first mentioned mechanism is conceded by most of the researchers to be the dominant process during superplastic deformation. With the assumption that GBS, which develops to a mesoscopic scale, is the rate-controlling mechanism, a quantitative theory for superplastic flow has been worked out (see Padmanabhan and Gleiter, 2012 for a detailed description). This mechanism describes superplastic flow in different classes of materials like metallic alloys, ceramics, metal matrix and ceramic matrix composites, dispersion-strengthened materials, intermetallics, metallic glasses, and geological materials on a common basis. As GBS is most effective only in intermediate ranges, there is a lower and an upper bound to the strain rate range within which significant superplasticity is observed.

The strain rate sensitivity index does not remain constant even over the optimum strain rate range but often varies with strain rate. The degree of thickness variation during forming is a function of the peak strain rate sensitivity index (*m*) of a material and an increase in *m* value reduces the thickness variation in the formed component (Padmanabhan and Davies, 1980; Pancholi and Kashyap, 2007).

The dependence of *m* value in an Al–Li alloy on the superplastic deformation temperature was demonstrated by Mogucheva and Kaibyshev (2008). The study performed on Al alloy 1421 (Al–Li–Mg–Sc), which was first subjected to isothermal rolling after ECA pressing, displays HSRSP. After the ECAP, a uniform microstructure with an average grain size of $1.6\,\mu m$ was obtained in the alloy.

The values of the flow stress, relative elongation, and coefficient of strain rate sensitivity at different temperatures are shown in Figure 8.9. The $\sigma - \dot{\varepsilon}$ curve shows a sigmoid shape at all deformation temperatures (Figure 8.9A). An increase in temperature leads to a displacement of the optimum strain rate to the side of the higher strain rates and to an increase in the value of *m* from 0.32 at 300°C to 0.57 at 450°C (Figure 8.9B and D).

The relative elongation is $\geq 1000\%$ for all the temperatures except at low temperature, i.e., at 300°C, and is observed over a wide range of strain rates. The maximum elongation at fracture (2700%) was observed at a temperature of 450°C and strain rate $= 1.4 \times 10^{-2}\,s^{-1}$. A relative elongation of 1000% was observed at the same temperature and at the same deformation rate in the alloy 1421 subjected to ECAP. Figure 8.9D represents the dependence on the true strain of the coefficient of strain rate sensitivity at a constant strain rate $= 1.4 \times 10^{-2}\,s^{-1}$. At the temperature interval 400–450°C,

Aluminum–Lithium Alloys

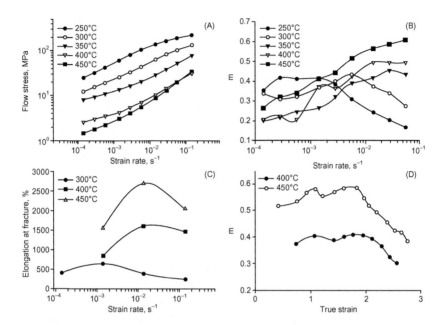

FIGURE 8.9 Effect of deformation rate on (A) the flow stress, (B) the coefficient of strain rate sensitivity, (C) elongation at fracture, and (D) the effect of strain on the coefficient of strain rate sensitivity (Al alloy 1421 (Al–Li–Mg–Sc)) (Mogucheva and Kaibyshev, 2008).

up to a true strain $\varepsilon \sim 2.0$, the value of m exhibits an increase. However, at still higher degrees of deformation, a decrease in m is observed. The high coefficient of strain rate sensitivity ensures high resistance of the material to necking and, thus, extraordinary relative elongations.

Another study on alloy 1420 by Ye et al. (2009) measured the flow stress, at a strain of 0.4, as a function of strain rate. This is plotted on a double-logarithmic scale in Figure 8.10A. A novel thermomechanical processing method was used for producing fine-grained Al–Mg–Li alloy 1420 sheet for superplasticity. The average grain diameter was 6.3 μm with a grain aspect ratio of 1.2 and 7.7 μm with a grain aspect ratio of 1.5 in the surface layer and the center layer, respectively. The values of m are plotted as a function of initial strain rate in Figure 8.10B. When deformed at 510°C and above, the maximum m value at each temperature is found at an initial strain rate of 1×10^{-3} s^{-1}, and with deformation temperature decreasing to 480°C and below, it is found at an initial strain rate of 5×10^{-4} s^{-1}. The m value increases with increasing deformation temperature and the maximum m value of about 0.79 is found at 540°C and an initial strain rate of 1×10^{-3} s^{-1}.

However, a maximum elongation to failure of 915% is observed at 1×10^{-3} s^{-1} initial strain rate, but a lower deformation temperature of

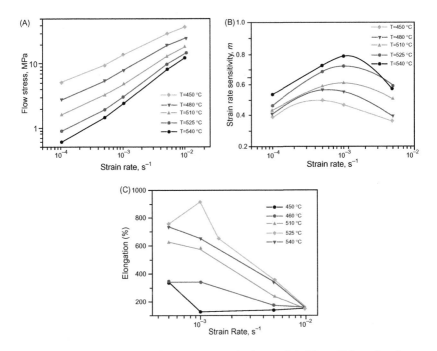

FIGURE 8.10   Variation of (A) flow stress determined at $\varepsilon = 0.4$, (B) coefficient of strain rate sensitivity, m, at $\varepsilon = 0.4$, and (C) elongation to failure with strain rate (Ye et al., 2009).

525°C with a corresponding m value of about 0.72. It is seen that the alloy exhibited considerable elongations of 580–915% when deformed at 510–540°C and initial strain rates ranging from 5×10^{-4} to $1 \times 10^{-3}\ \mathrm{s}^{-1}$ (Ye et al., 2009).

8.3 SUPERPLASTIC FORMING

Superplastic forming is an attractive method for the production of complex components from materials in the superplastic state. The use of superplastically formed aluminum alloy products for aerospace applications has increased in the recent decades (Pancholi and Kashyap, 2007; Shakesheff and Partridge, 1986; Wu et al., 2009).

The intrinsic problem with the present methods of superplastic forming is the nonuniform thickness observed in the formed components: usually the edge region is thicker than the center. Cavity formation is another disadvantage associated with superplasticity. Thickness variation and cavitation affect the mechanical properties of the superplastically formed components. Cavitation can be reduced by using back pressure, whereas the thickness variation after superplastic forming still remains an unsolved problem (Pancholi and Kashyap, 2007).

The thickness variation during superplastic forming strongly depends on the peak strain rate sensitivity index (m) of the material, and an increase in the m value reduces the thickness variation in the formed component. Therefore, most of the reported research works on superplastic forming are focused on the development of variable pressure cycles for blow forming in order to deform the sheet at the strain rate associated with the peak m value.

8.3.1 Cavity Nucleation and Growth

Internal cavity formation during superplastic deformation in aluminum and other alloys has been reported by Cocks and Ashby (1982), Chokshi and Mukherjee (1989), Wu (2000), Gouthama and Padmanabhan (2003), and Chokshi (2005). It was demonstrated that the stress concentrations induced at boundary irregularities such as second-phase particles, boundary ledges, and triple junctions were responsible for the development of cavitation at the high-angle boundary between two grains and this becomes prominent when significant GBS is present (Gouthama and Padmanabhan, 2003; Xun et al., 2005).

Cavitation occurs during the superplastic deformation of high-strength aluminum alloys, resulting in the development of small intergranular voids during straining. The cavities formed during superplastic forming may induce premature failure of superplastic materials during processing, and the presence of these cavities may impose significant limitations on the commercial use of superplastically formed components. Therefore, it is necessary to control cavitation during deformation. There are many studies that describe the mechanisms of cavitation during superplastic deformation (Ohishi et al., 2004; Bae and Ghosh, 2002a,b,c; Stowell et al., 1984; Ghosh and Bae, 1997), where cavitation has been experimentally and theoretically studied. It was reported that the superimposition of a hydrostatic pressure can substantially reduce cavitation. This improves the formability of superplastic metals (Lee and Mukherjee, 1991; Xun et al., 2005).

Several studies have been carried out to investigate the characteristics of cavitation during superplastic deformation. Vacancy diffusion and strain-controlled growth have been reported as the two main mechanisms for cavity growth. The growth rates of these two mechanisms depend on the superplastic conditions; diffusion growth is only dominant at low strain rates and for very small cavities, and cavity growth is essentially strain-controlled during superplastic deformation. Pandey et al. (1986), Chokshi and Mukherjee (1989), Zelin et al. (1993), and Wu (2000) have experimentally demonstrated the cavitation behavior of superplastically formed domes.

The experiments performed by Wu (2000) used different strain rates and strains during the superplastic forming of a fine-grained 8090 Al–Li alloy with an average grain size of 6.87 μm to form rectangular pans, and hemispherical and cylindrical capped parts to study the influence of strain state

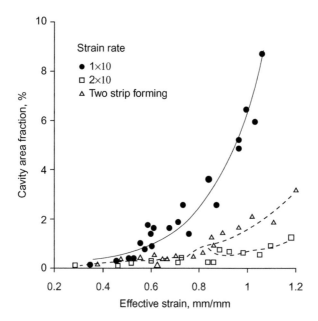

FIGURE 8.11   Cavity area fraction in a rectangular pan as a function of effective strain for various strain rates (Wu, 2000).

on the nucleation and growth of cavities. Figure 8.11 shows cavity formation as a function of effective strain and also shows that cavitation can be effectively reduced by using two-step forming rates. A greater strain rate imposed at the early stage of forming may increase the number of cavity nucleation sites, but grain growth would be reduced at a higher strain rate. Lower strain rate could decrease the cavity growth rate. Hence, using a two-step strain rate during superplastic forming could reduce the degree of cavitation.

Figure 8.12, taken from Wu (2000), explains the effect of post-sintering time (the extended forming time after the part is completely formed) with an imposed pressure of 0.889 MPa on the distribution of cavity volume fraction along the centerline of the cylindrical cup formed at a strain rate of $8 \times 10^{-4}\,\mathrm{s}^{-1}$. The post-sintering treatment reduces the cavitation level. The cavity volume fractions for the cup formed at a strain rate of $8 \times 10^{-4}\,\mathrm{s}^{-1}$ were significantly reduced.

The topography of GBS offsets, grain rotation angles, formation of striated bands or fibers, cavity distribution, cavity formation mechanisms, and contribution of GBS or grain separation to the overall strain were traced using scanning (SEM) or transmission electron microscopy (TEM) (Figure 8.13).

Yuwei et al. (1997) reported that the cavitation can be eliminated when back pressure is used, but the usage of back pressure can result in much higher forming pressure for the production parts of large size. The cavitation

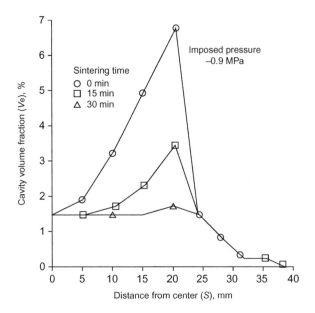

FIGURE 8.12 Distribution of cavity volume fraction along the centerline of the cylindrical cup showing the effect of post-sintering time with an imposed pressure of 0.889 MPa (Wu, 2000).

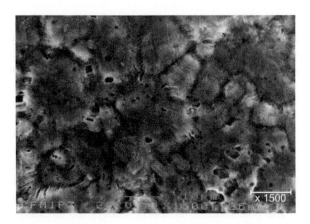

FIGURE 8.13 SEM micrograph of the surface topography of a superplastically formed 8090 Al–Li–Cu–Mg thin sheet following a thickness strain of 1.0. GBS, grain rotation, grain boundary sliding offsets formation of striated bands and fibers, and cavitation distribution were examined (Huang and Chuang, 1999).

percentage is plotted as a function of the strain for 2091 in Figure 8.14. It could be concluded that in 2091 sheets, the effect of cavitation is minimized, or even eliminated, when the SPF strain is <0.8 and a hydrostatic pressure of 2.0 MPa is applied.

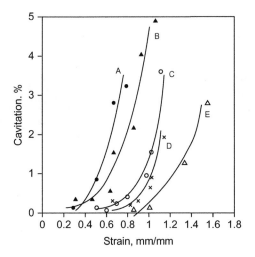

FIGURE 8.14 Cavitation versus superplastic strain curve for 2091 alloy sheets for $\dot{\varepsilon} = 8 \times 10^{-4}\,\text{s}^{-1}$. (A) $P_b = 0.0$, (B) $P_b = 0.5$, (C) $P_b = 1.0$, (D) $P_b = 1.5$, and (E) $P_b = 1.5$ MPa (Yuwei et al., 1997).

It is believed that cavity nucleation during SPD is caused by stress concentration at a particle or ledge in the grain boundary produced by GBS. Based on the existing models, it appears that cavity nucleation is as a result of the stress concentration arising from incomplete accommodations of GBS. However, this concept suffers from the absence of direct experimental evidence showing the formation of cavities at sliding boundaries (Xun et al., 2005), although postmortem studies have provided indirect support to this view.

Xun et al. (2005) characterized the cavitation microstructure of an Al–Li alloy (8090) using electron backscatter diffraction. The microstructure shows equiaxed grains with sizes of $10-20\,\mu\text{m}$ (Figure 8.15A). A cavity of dimension $\sim 15\,\mu\text{m}$ in size can be seen slightly elongated along 45° to the tensile direction. Figure 8.15B shows the orientation map, with superimposed SEM micrographs.

The grains grouped into I and II are separated by a high-angle boundary (dotted line, as shown in Figure 8.15A) along 45° to the tensile direction. GBS takes place preferentially at high-angle boundaries under a shear stress, and in a tensile test on a sheet specimen, the maximum shear stress acts along 45° to the tensile axis (2D). Therefore, the sliding of a group of grains in the direction noted by the arrows along the boundary will take place.

Experimental observations of cavitation in SPD have shown a correlation between the presence of hard second-phase particles and cavitation. From Figure 8.16, it is clear that small second-phase particles $1-4\,\mu\text{m}$ size can serve as potential cavity nucleation sites during deformation. Needleman and Rice (1980) and Chokshi and Mukherjee (1989) proposed a critical particle

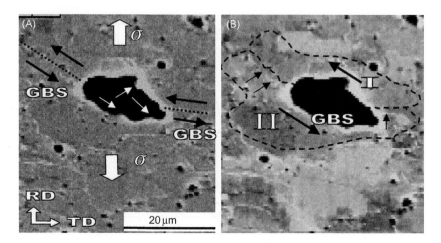

FIGURE 8.15 Orientation imaging maps of the microstructure around a cavity in a specimen deformed to $\varepsilon = 4.0$ at 530°C and initial strain rate of 10^{-3} s^{-1}: (A) SEM micrograph and (B) orientation map with superimposed SEM micrograph (Xun et al., 2005).

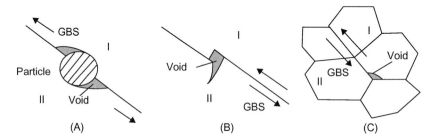

FIGURE 8.16 Models for cavity nucleation at a sliding grain boundary (A) at a second-phase particle, (B) at a grain boundary ledge, and (C) at a triple junction (Xun et al., 2005).

radius, Λ, above which localized Coble DC will not relax the large stress concentrations caused by GBS.

$$\Lambda = \left(\frac{\Omega D_1 \sigma}{\pi k T \dot{\varepsilon}} \right)^{1/2} \qquad (8.7)$$

where Λ is the critical particle radius, D_1 the coefficient of lattice diffusion, σ the applied stress, Ω the atomic volume, k the Boltzmann constant, and T is the absolute temperature of deformation. From this equation, the critical particle radius below which cavities will not be nucleated is 0.19 μm for Al-based superplastic alloys at $\dot{\varepsilon} = 10^{-3}$ s^{-1}.

In order to meet the needs of industrial applications, it is necessary to understand the metal and cavitation behavior under multiaxial deformation during superplastic forming (Wu et al., 2009). Studies of this kind are rather limited in number.

8.4 ROLE OF FRICTION STIR PROCESSING ON SUPERPLASTIC FORMING

In recent years, many grain refinement techniques have been developed to produce UFG materials for structural applications. The industry aims at mass production that can result from HSRSP. It is already seen that SPD techniques are effective ways of producing nonporous and bulk metallic samples with UFG microstructures. Friction stir processing (FSP) has been demonstrated as a potentially useful technique for producing superplastic aluminum alloys due to its capacity to give rise to very fine grain sizes and high grain boundary misorientations (Charit and Mishra, 2003). FSP has drawn much attention as it can improve strength, ductility, corrosion resistance, formability, and other properties and simultaneously eliminate casting defects and refine the microstructure. However, grain growth at elevated temperatures limits the extent of superplasticity in FSP aluminum alloys (Mishra and Ma, 2005). Most of the previous studies (Mao et al., 2005; Zhao et al., 2010) show that the FSP method allows the achievement of grain sizes below 0.3 μm. However, these UFG materials may degrade in properties at elevated temperatures as they possess poor thermal stability, which may be due to the accumulation of high dislocation densities and high residual stresses caused by FSP. These features tend to make the microstructures thermally unstable. This may impose limits on their commercial applications (Adamczyk-Cieślak et al., 2010).

It is possible to obtain a small grain size even in single-phase aluminum alloys (e.g., in Al−Li alloys). This could be due to reduction in dynamic recovery because of the presence of the substitutional Li atoms (Adamczyk-Cieślak et al., 2010). Jata and Semiatin (2000) examined the microstructure of friction stir-processed Al−Cu−Li alloy and reported that the grain refinement was due to continuous dynamic recrystallization through a dislocation glide-assisted subgrain rotation. There are no reports available for binary Al−Li alloys. However, to explain the importance of FSP as a potential technique to prepare the raw materials for superplastic forming, the key improvements in the formability aspects are demonstrated with other Al alloys. There is plenty of scope for the researchers to work in this area as, to the best of our knowledge, there is no report so far. As a result, all the materials considered in the next section to demonstrate the potential of FSP in inducing superplastic forming in Al−Li alloys are alloys that do not contain Li.

8.4.1 Superplasticity in FSP Materials

A recent paper reports grain sizes ranging from 0.8 to 12 μm in aluminum alloys (Mishra et al., 2000). As the grain size in FSP material is smaller, the optimum superplastic response shifts toward higher strain rates and/or lower temperatures (Commin et al., 2009).

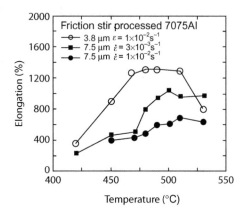

FIGURE 8.17 Elongation versus test temperature results at initial strain rates of 3×10^{-3} s^{-1} and 1×10^{-2} s^{-1} for two FSP 7075 Al alloys (Ma et al., 2002).

Mishra et al. (2000) and Ma et al. (2002) have demonstrated that FSP 7075 Al with a fine-grained microstructure exhibits HSRSP. At 1×10^{-2} s^{-1} strain rate and 490°C, 1000% elongation was noticed, as shown in Figure 8.17. This clearly shows the effectiveness of FSP for HSRSP. Similar effect of FSP on the superplastic ductility of 7075 Al at initial strain rates of 3×10^{-3} s^{-1} and 1×10^{-2} s^{-1} has also been demonstrated. The enhanced superplasticity was noticed in a 3.8 μm-FSP 7075 Al over a wide temperature range of 420–510°C. Even at a higher temperature of 530°C, the 3.8 μm-7075 Al alloy exhibited a large superplastic elongation of 975% (Ma et al., 2002).

Figure 8.18 displays SEM micrographs showing surface morphologies of FSP 3.5 μm-7075 Al which was superplastically deformed to failure at various strain rates and different temperatures. GBS and cavities induced as a result of GBS are clearly seen in Figure 8.18A, which was the result of deformation at a high strain rate of 1×10^{-1} s^{-1}. Figure 8.18B shows elongated grains in the deformation direction, which happened when the strain rate was decreased to 3×10^{-2} s^{-1}. When the strain rate is 1×10^{-2} s^{-1}, GBS, grain elongation, and the development of fibers between slid grains are evident from the micrograph Figure 8.18C. At a still lower strain rate of 3×10^{-3} s^{-1}, extensive GBS and significant grain elongation and strip or ribbon-like surface topography (Figure 8.18D) are seen. At low temperature, no fibers were developed, as shown in Figure 8.18E and F (Ma et al., 2002).

8.4.2 Superplastic Behavior and Deformation Mechanism

Charit and Mishra (2003) have reported that an optimum ductility of ~525% was obtained at a strain rate of 10^{-2} s^{-1} in the FSP 2024 Al alloy at a lower temperature of 430°C. It is also important to note that the ductility was ~280% even at a high strain rate of 10^{-1} s^{-1}.

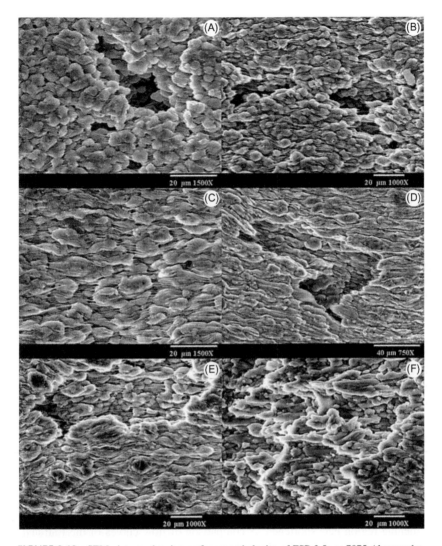

FIGURE 8.18   SEM pictures showing surface morphologies of FSP 3.5 μm-7075 Al superplastically deformed alloy tested to failure at (A) 480°C and $1 \times 10^{-1}\,\text{s}^{-1}$, (B) 480°C and $3 \times 10^{-2}\,\text{s}^{-1}$, (C) 480°C and $1 \times 10^{-2}\,\text{s}^{-1}$, (D) 480°C and $3 \times 10^{-3}\,\text{s}^{-1}$, and (E) 470°C and $1 \times 10^{-2}\,\text{s}^{-1}$, and fracture surface at (F) 470°C and $1 \times 10^{-2}\,\text{s}^{-1}$ (tensile axis is horizontal) (Ma et al., 2002).

Figure 8.19 illustrates in the strain rate/temperature plot for Al 2024 the superplastic domains for different processes, including FSP. Each process is outlined with a rectangular box to show the range of temperature and strain rate where the alloy exhibits superplasticity. A dashed line indicates the lower strain rate limit to HSRSP. It can be observed from Figure 8.19 that

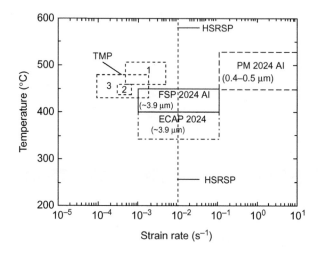

FIGURE 8.19 Temperature–strain rate map depicting the superplastic domains for 2024 Al alloy processed by different routes. For superplasticity of TMP 2024 Al alloys, region 1: grain size 2.5 μm, region 2: grain size 9.2 μm, and region 3: grain size 5–8 μm (Commin et al., 2009).

conventional thermomechanically-processed Al 2024 alloys do not exhibit HSRSP. Superplasticity in aluminum alloys is typically evident at ∼475°C ($0.8T_m$) or higher. The optimum temperature of 430°C ($0.75T_m$) for the FSP Al alloy is ∼50°C lower than the usual temperature range for superplasticity in aluminum alloys. Optimum superplasticity was achieved at a lower temperature in the ECAP Al 2024 alloy due to much finer grain size, since the specimen is subjected to 8–10 passes to obtain the fine-grained microstructure. But in the case of FSP, grain refinement was achieved in a single pass (Commin et al., 2009).

The true stress versus true plastic strain behavior of the FSP 2024 alloy is shown in Figure 8.20A for various strain rates at 430°C. The flow curves exhibit a significant strain-hardening region. As the strain rate increases, the strain-hardening region gets shortened. A comparison of flow curves of the FSP and the parent 2024 Al at 430°C and strain rate of $10^{-2}\,s^{-1}$ is shown in Figure 8.20B. The flow stress for the parent material is much higher than that for the FSP material, while the ductility of the parent material is much less (Charit and Mishra, 2003). The plastic flow of materials is enhanced by the fine-grained microstructure. However, the reduction in strength may have also been due to cavity formation, in addition to the dominance of the creep effects.

Figure 8.21 presents the ductility data obtained for different strain rates (10^{-3}–$10^{-1}\,s^{-1}$) at two temperatures (430° and 450°C). A maximum ductility of ∼525% was achieved at a strain rate of $10^{-2}\,s^{-1}$ and 430°C for the FSP 2024 Al. The parent 2024 Al alloy shows much lower ductility (∼70–100%) (Charit and Mishra, 2003).

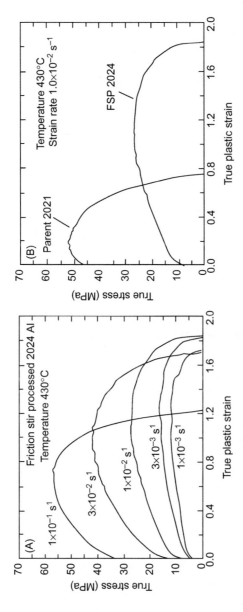

FIGURE 8.20 (A) Stress—strain behavior of FSP 2024 Al alloy at 430°C as a function of initial strain rates and (B) comparison between stress—strain behavior of parent and FSP 2024 Al alloys (Charit and Mishra, 2003).

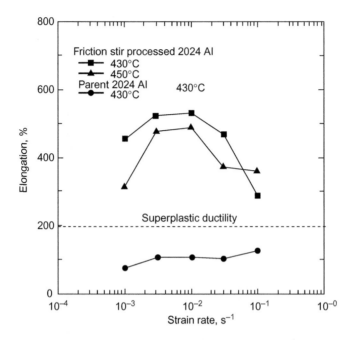

FIGURE 8.21   Ductility versus strain rate for FSP 2024 Al and the parent 2024-T4 Al (Charit and Mishra, 2003).

8.4.3 Cavity Density and Size Distribution

The finer the grain size, the lower is the amount of cavity formation. Figure 8.22 shows the variation of cavity density with true strain for two FSP 7075 Al alloys deformed at 480°C and an initial strain rate of $1 \times 10^{-2}\,s^{-1}$. The cavity density increases on increasing the superplastic strain for both FSP 7075 Al alloys. This clearly shows that new cavities were continuously nucleated with increasing strain. It can be assumed that both cavity growth and nucleation of new cavities contributed to increasing cavity volume fraction. However, at a certain value of strain rate, the cavity density decreases for both alloys. The reason for this drop is not clear. The 7075 Al alloy with a finer microstructure (3.8 μm) shows a lower tendency for cavity nucleation and growth than the 7.5-μm sized microstructure, which is consistent with what is already known in the superplasticity literature (Ma and Mishra, 2003).

Evidently, similar work should be attempted using Al–Li alloys of different compositions, with a view to commercial applications. A challenge that restricts the commercialization of Al–Li alloys is that it has lower and variable short-transverse fracture toughness in rolled, extruded, and forged products. Their cost is also three to five times greater than that of conventional Al alloys. The lower fracture toughness is due to one or more of the following reasons (Eswara

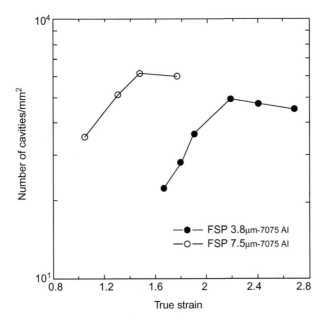

FIGURE 8.22   Cavity density versus true strain for 7.5 and 3.8 μm (starting grain sizes) 7075 Al alloys deformed at 480°C and an initial strain rate of $1 \times 10^{-2}\,\mathrm{s}^{-1}$ (Ma and Mishra, 2003).

Prasad et al., 2003; Hill et al., 1984; Lewandowski and Holroyd, 1990; Lynch, 1991a,b; Lynch et al., 1993; Sanders et al., 1980; Suresh et al., 1987; Vasudevan and Doherty, 1987; Webster, 1987):

1. Stress concentrations at grain boundaries due to coplanar slip
2. Weak area fractions—free zones at grain boundaries
3. Large area fraction of coarse grain boundary precipitates
4. The presence of hydrogen at grain boundaries
5. Lithium segregation at grain boundaries
6. The presence of discrete alkali metal impurity phases at the grain boundaries.

The fracture toughness problem(s) are discussed in detail in Chapter 13.

8.5 APPLICATIONS

Superplasticity in aluminum alloys was converted to a successful commercial venture by Superform Metals. The SUPRAL alloys promoted by them have a base composition of Al−6Cu−0.4Zr and have been assigned the commercial designation of Al-2004. These alloys exhibit elongations of 1000% or more under optimum (but commercially viable) conditions of strain rate and temperature. They can be heat treated after superplastic forming by solution heat treatment and aging to increase room temperature strength. Furthermore, complex shapes for a wide variety of aerospace and commercial applications have been successfully formed. An example of a component superplastically formed with SUPRAL is shown in Figure 8.23. Another component superplastically formed from a specially processed Al 2090 at Superform is shown in Figure 8.24 for comparison. The final dimensions of the part, shown in the figure, were 175 mm (maximum) in height, 170 mm in width, and 340 mm in length. Evidently, the superplastic properties of the Al—Li alloy were equally good.

The development of Al and Al—Li superplastically formed parts shows that economical and cost-effective production is possible. Cost and weight reductions are especially attractive for aerospace applications (Figure 8.24B). The post-superplastically formed mechanical properties can also meet the requirements of aircraft structures (Yuwei et al., 1997).

In recent years, there has been, and continues to be, considerable interest in SPF of the high strength AA 2195 alloy already much used for spacecraft components, including the Space Shuttle Super Lightweight External Tank. Figure 8.25 shows a 2195 SPF part produced at Boeing (Babel, 2005).

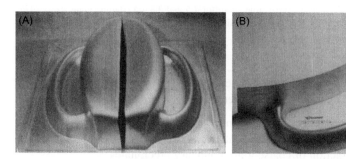

FIGURE 8.23 Superplastically formed component from a SUPRAL alloy in (A) the as-formed condition and (B) the final product (Henshall et al., 1987).

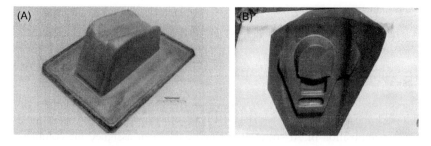

FIGURE 8.24 As superplastically formed (A) 2090 component and (B) aircraft structure made out of alloy 2091 (Yuwei et al., 1997).

FIGURE 8.25 SPF component made from AA 2195 sheet, dimensions 560 × 1450 mm (Babel 2005).

The use of Al-Li alloys for commercial transport aircraft is attracting increasing interest, see Chapters 2 and 15; and their use in helicopter airframes is also an attractive proposition. Thus the technological expansion in the use of Al-Li alloys is imminent, and they are likely to become one of the most important aerospace material classes of the twenty-first century (Cantor, 2001).

8.6 CONCLUDING REMARKS

This chapter discusses the phenomenon of superplasticity and the conditions under which superplastic flow and forming are possible in different Al−Li alloys. It also considers the processes and mechanisms that give rise to significant plasticity/superplasticity in these alloy systems, as well as those factors, including cost, that limit the use of Al−Li alloys.

REFERENCES

Adamczyk-Cieślak, B., Mizera, J., Kurzydłowski, K.J., 2010. Thermal stability of model Al−Li alloys after severe plastic deformation—effect of the solute Li atoms. Mater. Sci. Eng. A 527, 4716−4722.

Babel, H.W., 2005, *"Al-Li in Boeing products"*, 16th Annual Aeromat Advanced Aerospace Materials and Processes Conference and Exposition, Aeromat 2005, June 6−9, 2005, Orlando, FL, USA.

Bae, D.H., Ghosh, A.K., 2002a. Cavity growth in a superplastic Al−Mg alloy. II. An improved plasticity based model. Acta Metall. 50, 1011−1029.

Bae, D.H., Ghosh, A.K., 2002b. Cavity formation and early growth in a superplastic Al−Mg alloy. Acta Metall. 50, 511−523.

Bae, D.H., Ghosh, A.K., 2002c. Cavity growth during superplastic flow in an Al−Mg alloy. I. Experimental study. Acta Metall. 50, 993−1009.

Bairwa, M.L., Desai, S.G., Date, P.P., 2005. Identification of heat treatments for better formability in an aluminum−lithium alloy sheet. J. Mater. Eng. Perform. 14, 623−633.

Charit, I., Mishra, R.S., 2003. High strain rate superplasticity in a commercial 2024 Al alloy via friction stir processing. Mater. Eng. A 359, 290−296.

Chaudhuri, J., Gondhalekar, V., Inchekel, A., Talia, J.E., 1990. Study of precipitation and defor-mation characteristics of the aluminium—lithium alloy by X-ray double crystal diffractome-try. J. Mater. Sci. 25, 3938—3940.

Chockalingam, K.S.K., Neelakantan, M., Devaraj, S., Padmanabhan, K.A., 1985. On the pressure forming of two superplastic alloys. J. Mater. Sci. 20, 1310—1320.

Chokshi, A.H., 2000. The role of diffusion creep in the superplastic deformation of 3 mol% yttria stabilized tetragonal zirconia. Scr. Mater. 42, 241—248, http://www.sciencedirect.com/science/article/pii/S1359646299003401.

Chokshi, A.H., Mukherjee, A.K., 1988. A topological deformation in bimodal grain study of superplastic an Ai—Li alloy with a size distribution. Metall. Trans. A 19, 1621—1623.

Chokshi, A.H., Mukherjee, A.K., 1989. The cavitation and fracture characteristics of a superplas-tic Al—Cu—Li—Zr alloy. Mater. Sci. Eng. A 110, 49—60.

Chokshi, A.K., 2005. Cavity nucleation and growth in superplasticity. Mater. Sci. Eng. A 410—411, 95—99.

Chung, S.W., Higashi, K., Kim, W.J., 2004. Superplastic gas pressure forming of fine-grained AZ61 magnesium alloy sheet. Mater. Sci. Eng. A 372, 15—20.

Cocks, A.C.F., Ashby, M.F., 1982. On creep fracture by void growth. Prog. Met. Sci. 16, 465—474.

Commin, L., Dumont, M., Masse, J.-E., Barrallier, L., 2009. Friction stir welding of AZ31magnesium alloy rolled sheets: influence of processing parameters. Acta Mater. 57, 326—334.

Cui, Z., Zhong, W., Wei, Q., 1994. Superplastic behavior at high strain rate of rapidly solidified powder metallurgy Al—Li alloy. Scr. Metall. Mater. 30, 123.

Ehab, A., El-Danaf, 2012. Mechanical properties, microstructure and micro-texture evolution for 1050AA deformed by equal channel angular pressing (ECAP) and post ECAP plane strain compression using two loading schemes. Mater. Des. 34, 793—807.

Es-Said, O.S., Parrish, C.J., Bradberry, C.A., Hassoun, J.Y., Parish, R.A., Nash, A., et al., 2011. Effect of stretch orientation and rolling orientation on the mechanical properties of 2195 Al—Cu—Li alloy. J. Mater. Eng. Perform. 3, 292—299.

Eswara Prasad, N., Gokhale, A.A., Rama Rao, P., 2003. Mechanical behaviour of alumi-nium—lithium alloys. Sadhana 28, 209—246.

Garmestani, H., Kalu, P., Dingley, D., 1998. Characterization of Al-8090 superplastic materials using orientation imaging microscopy. Mater. Sci. Eng. A 242, 284—291.

Ghosh, A.K., Bae, D.H., 1997. Microstructure effects on cavitation, flow localization and frac-ture in superplastic metals. Mater. Sci. Forum 243—245, 89—98.

Gifkins, R.C., Snowden, K.U., 1966. Mechanism for viscous grain-boundary sliding. Nature 212, 916—917.

Giuliano, G., 2008. Constitutive equation for superplastic Ti—6Al—4V alloy. Mater. Des. 29, 1330—1333.

Gokhale, A.A., Singh, V., 2005. Effect of Zr content and mechanical working on the structure and tensile properties of AA8090 alloy plates. J. Mater. Process. Technol. 159, 369—376.

Gouthama, Padmanabhan, K.A., 2003. Transmission electron microscopic evidence for cavity nucleation during superplastic flow. Scr. Mater. 49, 761—766.

Gregson, P.J., Flower, H.M., 1985. Microstructural control of toughness in aluminium—lithium alloys. Acta Metall. 33, 527—537.

Henshall, C.A., Wadsworth, J., Reynolds, M.J., Barnes, A.D., 1987. Design and manufacture of a superplastic-formed aluminum—lithium component. Mater. Des. 8, 324—330.

Hill, D.P., Williams, D.N., Mobley, C.E., 1984. The effects of hydrogen on the ductility, tough-ness, and yield strength of an Al—Mg—Li alloy. In: Starke Jr., E.A., Sanders Jr., T.H. (Eds.), Aluminum—Lithium Alloys II. TMS-AIME, Warrendale, PA, pp. 201—208.

Hirai, K., Somekawa, H., Takigawa, Y., Higashi, K., 2007. Superplastic forging with dynamic recrystallization of Mg−Al−Zn alloys cast by thixo-molding. Scr. Mater. 56, 237−240.

Huang, J.C., Chuang, T.H., 1999. Progress on superplasticity and superplastic forming in Taiwan during 1987−1997. Mater. Chem. Phys. 57, 195−206.

Islamgaliev, R.K., Yunusova, N.F., Nurislamova, G.V., Krasil'nikov, N.A., Valiev, R.Z., Ovid'ko, I.A., 2009. Structure and mechanical properties of strips and shapes from ultrafine-grained aluminium alloy1421. Met. Sci. Heat Treat. 51, 82−86.

Ivanov, R., Boselli, J., Denzer, D., Larouche, D., Gauvin, R., Brochu, M., 2012. Hardening potential of an Al−Cu−Li friction stir weld. In: Weiland, H., Rollett, A. D., Cassada, W.A. (Eds.), 13th International Conference on Aluminum Alloys (ICAA13). TMS (John Wiley & Sons, Inc., Hoboken, NJ), pp. 659−664.

Jagan Reddy, G., Srinivasan, N., Gokhale, A.A., Kashyap, B.P., 2009. Processing map for hot working of spray formed and hot isostatically pressed Al−Li alloy (UL40). J. Mater. Process. Technol. 209, 5964−5972.

Jata, K.V., Semiatin, S.L., 2000. Continuous dynamic recrystallization during friction stir welding of high strength aluminum alloys. Scr. Mater. 43, 743.

Jiang, J.Q., Bate, P.S., 1996. Use of microstructural gradients in hot gas-pressure forming of Zn−Al sheet. Metall Mater. Trans. A 27A, 3250−3258.

Koh, H.J., Kim, N.J., Lee, S., Lee, E.W., 1998. Superplastic deformation behavior of a rapidly solidified Al−Li alloy. Mater. Sci. Eng. A 256, 208−213.

Lee, H.S., Mukherjee, A.K., 1991. Phenomenon of intergranular cavitation and failure in super-plasticity. Eng. Fract. Mech. 40, 843.

Lee, S., Horita, Z., Hirosawa, S., Matsuda, K., 2012. Age-hardening of an Al−Li−Cu−Mg alloy (2091) processed by high-pressure torsion. Mater. Sci. Eng. A 546, 82−89.

Lequeu, Ph., Smith, K.P., Danielou, A., 2010. Aluminum−copper−lithium alloy 2050 developed for medium to thick plate. J. Mater. Eng. Perform. 19, 841−848.

Lewandowski, J.J., Holroyd, N.J.H., 1990. Intergranular fracture of Al−Li alloys: effects of aging and impurities. Mater. Sci. Eng. A 123, 21−27.

Lieblich, M., Torralba, M., 1991. Cellular microstructure and heterogeneous coarsening of δ' in rapidly solidified Al−Li−Ti alloys. J. Mater. Sci. 26, 4361−4368.

Lynch, S.P., 1991a. Fracture of 8090 Al−Li plate I. Short transverse fracture toughness. Mater. Sci. Eng. A 136, 25−43.

Lynch, S.P., 1991b. Fracture of 8090 Al−Li plate II. Sustained-load crack growth in dry air at 50−200°C. Mater. Sci. Eng. A 136, 45−57.

Lynch, S.P., Wilson, A.R., Byrnes, R.T., 1993. Effects of ageing treatments on resistance to intergranular fracture of 8090 Al−Li alloy plate. Mater. Sci. Eng. A 172, 79−93.

Ma, Z.Y., Mishra, R.S., 2003. Cavitation in superplastic 7075Al alloys prepared via friction stir processing. Acta Mater. 51, 3551−3569.

Ma, Z.Y., Mishra, R.S., Mahoney, M.W., 2002. Superplastic deformation behaviour of friction stir processed 7075Al alloy. Acta Mater. 50, 4419−4430.

Maehara, Y., Ohmori, Y., 1987. Microstructural change during superplastic deformation of δ-ferrite/austenite duplex stainless steel. Metall. Mater. Trans. A 18, 663.

Mao, J., Kang, S.B., Park, J.O., 2005. Grain refinement, thermal stability and tensile properties of 2024 aluminum alloy after equal-channel angular pressing. J. Mater. Process. Technol. 159, 314−320.

Mishra, R.S., Ma, Z.Y., 2005. Friction stir welding and processing. Mater. Sci. Eng. R Rep. 50, 1.

Mishra, R.S., Mahoney, M.W., McFadden, S.X., Mara, N.A., Mukherjee, A.K., 2000. High strain rate superplasticity in a friction stir processed 7075 Al alloy. Scr. Mater. 42, 163−168, http://www.sciencedirect.com/science/article/pii/S1359646299003292.

Mogucheva, A.A., Kaibyshev, R.O., 2008. Ultrahigh superplastic elongations in an aluminum−lithium alloy. Physics 53, 431−433.

Moreira, P.M.G.P., De Jesus, A.M.P., De Figueiredo, M.A.V., Windisch, M., Sinnema, G., De Castro,, P.M.S.T., 2012. Fatigue and fracture behaviour of friction stir welded aluminium−lithium 2195. Theor. Appl. Fract. Mech. 60, 1−9.

Mukherjee, A.K., Bird, J.E., Dorn, J.E., 1969. Experimental correlation for high-temperature creep. Trans. Am. Soc. Met. 62, 155−179.

Needleman, A., Rice, J.R., 1980. Plastic creep flow effects in the diffusive cavitation of grain boundaries. Acta Metall. 28, 1315−1332.

Noda, M., Hirohashi, M., Funami, K., 2003. Low temperature superplasticity and its deformation mechanism in grain refinement of Al−Mg alloy by multi-axial alternative forging. Mater. Trans. 44, 2288−2297.

Ohishi, K., Boydon, J.F., McNelley, T.R., 2004. Deformation mechanisms and cavity formation in superplastic AA5083. In: Taleff, E.M., Krajewski, P.E., Friedman, P.A. (Eds.), Proceedings of 2004 TMS Annual Meeting on Advances in Superplasticity and Superplastic Forming. Wiley, Charlotte, NC, pp. 119−126.

Padmanabhan, K.A., 2009. Grain boundary sliding controlled flow and its relevance to superplasticity in metals, alloys, ceramics and intermetallics and strain-rate dependent flow in nanostructured materials. J. Mater. Sci. 44, 2226−2238.

Padmanabhan, K.A., Davies, G.J., 1980. Superplasticity. Springer Verlag, Berlin, Heidelberg, and New York.

Padmanabhan, K.A., Gleiter, H., 2004. Optimal structural superplasticity in metals and ceramics of microcrystalline- and nanocrystalline-grain sizes. Mater. Sci. Eng. A 381, 28−38.

Padmanabhan, K.A., Gleiter, H., 2012. A mechanism for the deformation of disordered states of matter, Current Opinion in Solid State and Mater. Sci. 16, 243−253.

Padmanabhan, K.A., Raviathul Basariya, M., 2009. Mesoscopic grain boundary sliding as the rate controlling process for high strain rate superplastic deformation. Mater. Sci. Eng. A 527, 225−234.

Padmanabhan, K.A., Schlipf, J., 1996. A model for grain boundary sliding and its relevance to optimal structural superplasticity: 1. Theory. Mater. Sci. Technol. 12, 391−399.

Padmanabhan, K.A., Vasin, R.A., Enikeev, F.U., 2001. Super Plastic Flow: Phenomenology and Mechanics. Springer Verlag, Berlin, Heidelberg, and New York, March.

Pancholi, V., Kashyap, B.P., 2007. Effect of layered microstructure on superplastic forming property of AA8090 Al−Li alloy. J Mater. Proc. Tech. 186, 214−220.

Pandey, M.C., Wadsworth, J., Mukherjee, A.K., 1986. Superplastic deformation behavior in ingot and powder metallurgically processed Al−Li-Based Alloys. Mater. Sci. Eng. A 80, 169−179.

Prangnell, P.B., Özkaya, D., Stobbs, W.M., 1994. Discontinuous precipitation in high Li content Al−Li−Zr alloys. Acta Metall. Mater. 42, 419−433.

Pu, H.P., Liu, F.C., Huang, J.C., 1995. Characterization and analysis of low-temperature superplasticity in 8090 Al−Li Alloys. Metall. Mater. Trans. A 26A, 1153.

Qing, L., Xiaoxu, H., Mei, Y., Jinfeng, Y., 1992. On deformation induced continuous recrystallization in a superplastic Al−Li−Cu−Mg−Zr alloy. Acta Metall. Mater. 40, 1753−1762.

Rabinovich, M.Kh., Trifonov, V.G., 1996. Dynamic grain growth during superplastic deformation. Acta Mater. 44, 2073–2078.

Rioja, R.J., Liu, J., 2012. The evolution of Al–Li base products for aerospace and space applications. Metall. Mater. Trans. A 43 (9), 3325–3337.

Salem, H.G., Lyons, J.S., 2002. Effect of equal channel angular extrusion on the microstructure and superplasticity of an Al–Li alloy. J. Mater. Eng. Perform. 11, 384–391.

Sanders, T.M., Ludwiczak, E.A., Sawtell, R.R., 1980. The fracture behavior of recrystallized Al–2.8% Li–0.3% Mn sheet. Mater. Sci. Eng. A 43, 247–260.

Shakesheff, A.J., Partridge, P.G., 1986. Superplastic deformation of Al–Li–Cu–Mg alloy sheet. J. Mater. Sci. 21, 1368–1376.

Sotoudeh, K., Bate, P.S., 2010. Diffusion creep and superplasticity in aluminium alloys. Acta Mater. 58, 1909–1920.

Starke, E.A., Staley, J.T., 1996. Application of modern aluminum alloys to aircraft. Prog. Aerosp. Sci. 32, 131.

Stowell, M.J., Liversy, D.W., Ridley, N., 1984. Cavity coalescence in superplastic deformation. Acta Metall. 32, 35–42.

Suresh, S., Vasudevan, A.K., Tosten, M., Howell, P.R., 1987. Microscopic and macroscopic aspects of fracture in lithium-containing aluminum alloys. Acta Metall. 35, 25–46.

Tang, S., Robbins, T.L., 1974. Bulging rupture of a superplastic sheet. J. Eng. Mater. Technol. (Trans. ASME) 96A, 77–79.

Thomsen, T.H., Holt, D.L., Backofen, W.A., 1970. Forming superplastic sheet metal in bulging dies. Met. Eng. Q. 10, 1–7.

Vasin, R.A., Enikeev, F.U., Mazurski, M.I., Munirova, O.S., 2000. Mechanical modeling of the universal superplastic curve. J. Mater. Sci. 35, 2455–2466.

Vasudevan, A.K., Doherty, R.D., 1987. Grain boundary ductile fracture in precipitation hardened aluminum alloys. Acta Metall. 35, 1193–1218.

Wadsworth, J., Palmer, I.G., Crooks, D.D., 1983. Superplasticity in Al–Li based alloys. Scr. Metall. 17, 347–352.

Wadsworth, J., Pelton, A.R., Lewis, R.E., 1985. Superplastic Al–Cu–Li–Mg–Zr alloys. Metall. Mater. Trans. A 16, 2319–2332.

Wang, Y.N., Huang, J.C., 2003. Comparison of grain boundary sliding in fine grained Mg and Al alloys during superplastic deformation. Scr. Mater. 48, 1117–1122.

Wanhill, R.J.H., 1994. Status and prospects for aluminium–lithium alloys in aircraft structures. Fatigue 16, 3–20.

Webster, D., 1987. The effect of low melting point impurities on the properties of aluminum–lithium alloys. Metall. Mater. Trans. A 18, 2181–2193.

Wu, H.Y., 2000. Cavitation characteristics of a superplastic 8090 Al alloy during equi-biaxial tensile deformation. Mater. Sci. Eng. A 291, 1–8.

Wu, H.Y., Hwang, J., Chiu, C., 2009. Deformation characteristics and cavitation during multiaxial blow forming in superplastic 8090 alloy. J. Mater. Process. Technol. 209, 1654–1661.

Xun, Y., Tan, M.J., Liew, K.M., 2005. EBSD characterization of cavitation during superplastic deformation of Al–Li alloy. J. Mater. Process. Technol. 162–163, 429–434.

Ye, L., Zhang, X., Zheng, D., Liu, S., Tang, J., 2009. Superplastic behavior of an Al–Mg–Li alloy. J. Alloys Compd. 487, 109–115.

Yuwei, X., Yiyuan, Z., Wenfeng, M., Jainzhong, C., 1997. Superplastic forming technology of aircraft structures for Al–Li alloy and high-strength Al alloy. J. Mater. Process. Technol. 72, 183–187.

Zelin, M.G., Bieler, T.R., Mukherjee, A.K., 1993. Cooperative grain-boundary sliding in mechanically alloyed in 90211 alloy during high strain rate superplasticity. Metall. Mater. Trans. A 24, 1208–1212.

Zhao, Z., Chen, Q., Chao, H., Hu, C., Huang, S., 2010. Influence of equal channel angular extrusion processing parameters on the microstructure and mechanical properties of Mg–Al–Y–Zn alloy. Mater. Des. 31, 1906–1916.

FURTHER READING

Barnes, A.J., 2007. Superplastic forming 40 years and still growing. J. Mater. Eng. Perform. 16, 440–454.

Bird, R.K., Dicus, D.L., Fridlyander, J.N., Sandler, V.S., 2000. Al–Li Alloy 1441 for fuselage application. Seventh International Conference on Aluminum Alloys. The Light Metals Centre at the University of Virginia, Charlottesville, VA, pp. 907–912.

Cantor, B., Assender, H., Grant, P. (Eds.), 2001. Series in Materials Science and Engineering: Aerospace Materials. Institute of Physics Publishing, Bristol and Philadelphia, PA.

Pacchione, M., Telgkamp, J., 2006. Challenges of the metallic fuselage. 25th International Congress of the Aeronautical Sciences, Hamburg, Germany, pp. 1–12.

Chapter 9

Welding Aspects of Aluminum–Lithium Alloys

G. Madhusudhan Reddy and Amol A. Gokhale

Defence Metallurgical Research Laboratory, Hyderabad, India

Contents

9.1 Introduction	260	
9.2 Weld Metal Porosity	260	
9.3 Solidification Cracking	264	
9.3.1 General Considerations	264	
9.3.2 Al–Li Alloy Solidification Cracking Guidelines	265	
9.3.3 Development of Weldable Al–Li Alloys	267	
9.4 Liquation Cracking	272	
9.5 EQZ Formation and Associated Fusion Boundary Cracking	273	
9.5.1 Experimental Observations	273	
9.5.2 Hypotheses of EQZ Formation and Their Evaluation	276	
9.5.3 Fusion Boundary Cracking	278	
9.6 Modification of Fusion Zone Microstructures	278	

9.6.1 Inoculation	279
9.6.2 Pulsed current	280
9.6.3 Magnetic Arc Oscillation	282
9.7 Mechanical Properties	284
9.7.1 Al–Li 1420 (First-Generation Al–Li Alloy)	286
9.7.2 Al–Li 1441, AA 8090, and AA 2090 (Second-Generation Al–Li Alloys)	286
9.7.3 Al–Li 2195 (Third-Generation Al–Li Alloy)	289
9.8 Corrosion	289
9.9 Solid-State Welding Processes	292
9.9.1 Friction Welding	292
9.9.2 Friction Stir Welding	295
9.10 Summary	296
Acknowledgments	296
References	297

9.1 INTRODUCTION

As discussed elsewhere in this book, aluminum–lithium (Al–Li) alloys are of great interest to aerospace designers because of their lower density and higher elastic modulus compared to conventional high-strength aluminum alloys. Both these benefits are due to lithium additions. Furthermore, lithium has another unique advantage by contributing to precipitation strengthening without forming low-melting eutectics that are deleterious to welding. Lithium may also reduce the solubility of copper in aluminum, thereby providing a beneficial increase in the quantity of eutectic generated in the Al–Cu–Li alloy system (Cross et al., 1990).

Conventionally, fusion-welded Al–Li alloy airframe structures have been used in Russian military aircraft such as the YAK-36 and MIG-29. The alloys in these structures, the 14XX series, belong to the first- and second-generation Al–Li alloys (Table 9.1). In the United States and Europe, the aircraft manufacturers and welding community have more recently been studying the potential of laser beam welding (LBW) and friction stir welding (FSW) for assembling airframe structures using third-generation Al–Li alloys.

The situation for spacecraft is somewhat different. Conventional fusion welding has been used for decades to manufacture cryogenic propellant tanks from established aluminum alloys (principally AA 2219) for launch vehicles and was also used for the Crew Compartment of the Space Shuttle. Over the last decade, fusion welding has been accompanied, or indeed supplanted, by FSW for fabricating tanks from both established and Al–Li alloys; and FSW has recently been used to join major Al–Li components of the Orion Crew Module. Some of the actual and potential aerospace applications using LBW and FSW are discussed in Chapter 15.

This chapter is concerned primarily with conventional fusion welding techniques and their application to Al–Li alloys, since these techniques are well established primarily for conventional alloys rather than Al–Li alloys, and there are many potential applications of Al–Li alloys besides those in aerospace (e.g., in marine hardware and lightweight armor). Research on conventional welding of Al–Li alloys has concentrated on the following areas: (i) porosity formation and control, (ii) weld cracking susceptibility, (iii) mechanical property characterization and optimization, and (iv) resistance to corrosion. These topics are essential parts of assessing the weldability of aluminum alloys and will be considered in the following sections of this chapter. A concise discussion of solid-state welding, including FSW, is also given.

9.2 WELD METAL POROSITY

Weld metal porosity is the most common defect for all fusion welding processes. Therefore, control of porosity and minimizing its adverse effects on weldment properties have been of great interest. Porosity may be gas-related

TABLE 9.1 Al–Li Alloys

Alloys	Li	Cu	Mg	Ag	Zr	Sc	Mn	Zn	Introduction
First Generation									
2020	1.2	4.5					0.5		Alcoa 1958
1420	2.1		5.2		0.11				Soviet 1965
1421	2.1		5.2		0.11	0.17			Soviet 1965
Second Generation (Li ≥ 2%)									
2090	2.1	2.7			0.11				Alcoa 1984
2091	2.0	2.0	1.3		0.11				Pechiney 1985
8090	2.4	1.2	0.8		0.11				EAA 1984
1430	1.7	1.6	2.7		0.11	0.17			Soviet 1980s
1440	2.4	1.5	0.8		0.11				Soviet 1980s
1441	1.95	1.65	0.9		0.11				Soviet 1980s
1450	2.1	2.9			0.11				Soviet 1980s
1460	2.25	2.9			0.11	0.09			Soviet 1980s
Third Generation (Li < 2%)									
2195	1.0	4.0	0.4	0.4	0.11				LM/Reynolds 1992
2196	1.75	2.9	0.5	0.4	0.11		0.35 max	0.35 max	LM/Reynolds/McCook Metals 2000
2297	1.4	2.8	0.25 max		0.11		0.3	0.5 max	LM/Reynolds 1997
2397	1.4	2.8	0.25 max		0.11		0.3	0.10	Alcoa 2002
2098	1.05	3.5	0.53	0.43	0.11		0.35 max	0.35	McCook Metals 2000
2198	1.0	3.2	0.5	0.4	0.11		0.5 max	0.35 max	Reynolds/McCook Metals/Alcan 2005
2099	1.8	2.7	0.3		0.09		0.3	0.7	Alcoa 2003
2199	1.6	2.6	0.2		0.09		0.3	0.6	Alcoa 2005
2050	1.0	3.6	0.4	0.4	0.11		0.35	0.25 max	Pechiney/Alcan 2004[a]
2296	1.6	2.45	0.6	0.43	0.11		0.28	0.25 max	Constellium Alcan 2010[a]

(Continued)

TABLE 9.1 (Continued)

Alloys	Li	Cu	Mg	Ag	Zr	Sc	Mn	Zn	Introduction
2060	0.75	3.95	0.85	0.25	0.11		0.3	0.4	Alcoa 2011
2055	1.15	3.7	0.4	0.4	0.11		0.3	0.5	Alcoa 2012
2065	1.2	4.2	0.50	0.30	0.11		0.40	0.2	Constellium 2012
2076	1.5	2.35	0.5	0.28	0.11		0.33	0.30 max	Constellium 2012

[a]*Pechiney acquired by Alcan 2003; Constellium formerly Alcan Aerospace.*
Source: After Rioja and Liu (2012) and other sources.

or result from solidification shrinkage. In terms of size, shape, and location, porosity can be described as interdendritic porosity or bulk porosity (D'Annessa, 1967; Devletian and Wood, 1983; Martukanitz and Michnuk, 1982). Interdendritic porosity occurs when gas bubbles are formed or entrapped between dendrite arms in the solidification substructure, whereas bulk porosity is the spherical pores that result from supersaturation of gases in the weld pool.

The porosity depends on both the amount of dissolved gases and the welding process variables. Hydrogen is the principal cause of gas porosity in aluminum welds (Erokhin and Obotorov, 1971; Ol'Shanskii and Dyachencho, 1977). Voids or porosity that are generally spherical in shape are caused by the sharp decrease in solubility of hydrogen during solidification: hydrogen solubility in molten aluminum is more than 10 times the solubility in the solid metal (Mondolfo, 1979; Ransley and Neufeld, 1948). Alloying additions to aluminum influence porosity formation by affecting the solubility of hydrogen in the matrix. Alloy additions of either copper or silicon to aluminum decrease the solid solubility of hydrogen and thus increase the propensity for pore formation (Devletian and Wood, 1983). Magnesium additions considerably increase the solid solubility of hydrogen in aluminum and therefore reduce the susceptibility to pore formation. Alloying additions also influence porosity formation in aluminum weld metal by affecting the solidification range and solidification mode (Kou, 1987).

The most experience of welding lithium-containing aluminum alloys comes from welding the first-generation Russian Al–Li–Mg alloy 1420. Reviews of a number of papers show that weld metal porosity is a greater problem for Li-bearing alloys than conventional aluminum alloys (Kostrivas and Lippold, 1999; Pickens, 1985, 1990). The porosity is mainly associated with a hygroscopic complex oxide skin on the components to be welded. It is believed that complex oxides of Li, Mg, and Al form at elevated temperatures during hot

TABLE 9.2 Effect of Surface Preparation and Shielding on Oxidation and Porosity in 1441 Al–Li–Cu–Mg Sheet Alloy Welds

S.No	Surface Condition and Shielding	Weld Quality
1	As-received	Heavy oxidation
2	Wire brushed	Gross porosity
3	0.1 mm machined	Gross porosity
4	0.2 mm machined	Dense fine porosity
5	0.2 mm machined + argon backing	Fine fusion line porosity + sparse fine porosity
6	0.2 mm machined + chemically cleaned + argon backing	Lowest porosity

Source: Madhusudhan Reddy and Gokhale (1993).

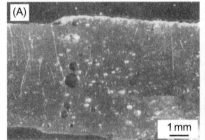 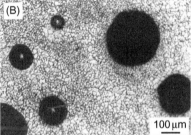

FIGURE 9.1 Effect of surface preparation on weld porosity in 1441 Al–Li–Cu–Mg alloy sheet, (A) as-received condition and (B) wire brushed condition (Madhusudhan Reddy and Gokhale, 1993).

working or solution treatment, and these are subsequently responsible for weld metal porosity (Fridlyander, 1970). If the oxide skin is removed by mechanical or chemical means, the porosity is greatly reduced (Fedoseev et al., 1978; Ramulu and Rubbert, 1990; Madhusudhan Reddy and Gokhale, 1993; Skillingberg, 1986). The combination of machining and chemical milling, with argon shielding during welding, gives the best results (Gittos, 1987; Madhusudhan Reddy and Gokhale, 1993) (Table 9.2), which is for the second-generation Russian Al–Li–Cu–Mg alloy 1441.

Sheets welded in the as-received condition or in the wire brushed condition resulted in gross porosity, as shown in Figure 9.1. The thickness of surface layer that needs to be removed for low-porosity welds depends on the nature and extent of prior heat treatments that result in surface oxidation.

Inert gas backing is not usually necessary when welding aluminum alloys, but the observation of reduced porosity listed in Table 9.2 is consistent with the results of Ischenko and Chayun (1977). The beneficial effect of inert gas backing is probably a further indication of the high reactivity of Li. The use of a vacuum heat treatment to minimize the weld zone porosity in alloy 1420 was studied by Mironenko et al. (1979a, 1979b). Vacuum heat treatment was performed for 12−24 h and in the temperature range of 450−500°C (Kostrivas and Lippold, 1999). This pretreatment appreciably reduced the weld zone porosity in both gas tungsten arc (GTA) and electron beam (EB) welds. The porosity reduction was attributed to the vacuum heat treatment driving off hydrogen entrapped in the surface layers. However, such thermal treatments will soften the base metal, and this means that ageing following welding will be required to recover the mechanical properties. In addition, this method cannot generally be used for large structural components and is expensive.

9.3 SOLIDIFICATION CRACKING

9.3.1 General Considerations

Aluminum alloys are more susceptible to solidification cracking compared to other alloy systems (e.g., iron- or titanium-base systems) owing to their large solidification temperature range, high coefficient of thermal expansion, high shrinkage stresses, and tendency to form low-melting constituents. These characteristics are also inherent in Al−Li alloys, therefore suggesting that these alloys are also susceptible to cracking during solidification.

Solidification cracking is known as hot cracking in the welding industry. The location for this cracking can be the weld centerline or anywhere in the weld and heat-affected zone (HAZ). Cracking can occur above the solidus temperature, in which case the type of cracking is called super-solidus cracking; or it can occur below the solidus temperature, and is then referred to as sub-solidus cracking. Matsuda (1990) has reviewed the various aspects of solidification cracking.

Weld solidification cracking occurs when there is a sufficient amount of restraint on contraction at a stage when the microstructure is susceptible to cracking. Solidification cracks usually occur towards the end of solidification, when the cumulative effect of restraint builds up to a maximum and the solid−solid contacts between neighboring dendrites are not strong enough to withstand the contraction stresses. Alloys with a narrow solidification range and with lower coherency temperature (at this temperature the alloy first acquires mechanical strength) are resistant to solidification cracking, since contraction restraint is less in both cases. The latter condition prevails when the volume fraction of the eutectic phase is large or when grain refinement takes place during weld solidification. The solidification crack resistance is also high in cases when the solid phase contiguity is large, since the contraction loads are divided over wider areas, leading to lower stresses.

Although weld design-induced restraint can be predicted to a reasonable extent, this may not always be possible for weld solidification structures because of the lack of material data, phase diagrams for multicomponent systems, etc. In addition, cracking criteria are not well established. As a result, the capability to predict solidification cracking is limited. Tests have therefore been devised to measure cracking susceptibility:

1. There are three types of tests for assessing solidification cracking tendency in alloys. They are the self-strained, externally strained, and externally stressed tests, respectively referred to as the Houldcroft test, Varestraint test, and SIGMA JIG test. The most commonly used test is the Varestraint test.

2. In the self-strained and externally strained tests, the extent of cracking is a measure of cracking tendency. In the externally stressed test, the critical stress for cracking determines the cracking tendency of a material. Since self-strained conditions represent actual weld situations, the results from these tests are considered to be more reliable.

Another kind of test, the Gleeble weld thermal simulation test, is used to measure hot ductility and thus indirectly assesses the hot cracking tendency of a material.

The susceptibility of some binary aluminum alloys to solidification cracking is shown in Figure 9.2. It can be seen that the cracking tendency depends on the composition. Certain compositions are highly susceptible: the susceptibility is least at low and high solute concentrations, which have short freezing ranges, and worst at intermediate compositions, which have wide freezing ranges.

The *hot tearing* susceptibility of a binary Al–Li alloy is at a maximum when Li is around 2.6 wt% (Figure 9.3). Unfortunately, some Al–Li alloys, notably the second-generation alloys 1440 and AA 8090, have been designed with the primary purpose of obtaining (very) low-density alloys with high mechanical properties, and the nominal compositions of these alloys are near this peak (Table 9.1). Hence the solidification cracking susceptibility of these alloys is inferred to be high. However, this indication of a potential weld cracking problem for AA 8090 is now more of general interest, since this alloy is no longer used.

9.3.2 Al–Li Alloy Solidification Cracking Guidelines

Solidification cracking is generally more pronounced in Al–Li alloys than in conventional aluminum alloys. Most cracks form along lithium- and copper-rich bands in the fusion zone. There are some general trends, the most important being (3) below.

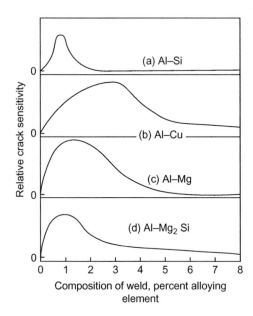

FIGURE 9.2 Effect of chemical composition of weld metal on solidification cracking susceptibility in various aluminum binary alloys (Matsuda, 1990).

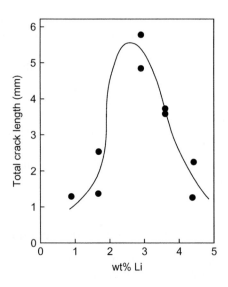

FIGURE 9.3 Hot tearing susceptibility dependence on Li content in binary Al–Li alloys (Pickens et al., 1989).

1. Since Al–Li alloys are multicomponent systems, their cracking tendency could be related to the extent of formation of eutectic phases and the available liquid fraction at any particular instance. Alloys which have a greater tendency to form eutectics, and in which the eutectic liquid is able to wet the grain boundaries, can be expected to be more prone to cracking. For example, 1420 alloy is least prone to solidification cracking, and this alloy does not contain copper (Table 9.1), whose presence results in a wider composition range for cracking (Cross et al., 1990).

2. Cracking can be minimized by addition of magnesium and zirconium. However, the susceptibility varies with specific alloy compositions and lithium content as well (Bittence, 1987). Compositions for which the contents of alloying elements match those of their binary counterparts of maximum cracking sensitivity are prone to cracking to a similar degree (Pickens et al., 1989).

3. If the compositions are so adjusted that the elements responsible for cracking are kept to their minimum cracking limits, then cracking problems could be solved. This is a most important consideration. The result of this approach is Weldalite™ 2195 (Kramer et al., 1989) and the Weldalite™ family of alloys. In Weldalite™ 2195 the levels of copper and lithium are kept in the minimum cracking range. Strength sacrifice by reduction in these elements was compensated by the addition of silver, and the alloy's strength is derived mainly from T_1 (Al_2CuLi) precipitation, since it does not form δ' (Al_3Li) on age hardening (Kramer et al., 1989). This alloy has proven to be weldable by all welding processes and is reported to be free from hot cracking.

9.3.3 Development of Weldable Al–Li Alloys

The first-generation Russian alloy 1420 and the third-generation Western alloy AA 2195 were developed specifically for weldability. Other Al–Li alloys with superior combinations of strength and toughness are considered non-weldable (Kostrivas and Lippold, 1999; Pickens, 1990). This being said, past efforts have concentrated on studying the weldability of 1420, 1441, AA 2090, AA 2195, and AA 8090 alloys. Of these, alloy 1420, which was derived from an Al–5Mg weldable alloy, was reported to be resistant to hot cracking (Bratukhin, 1995; Fridlyander et al., 1992).

Effects of Filler Metals

Alloy 1441 is a lower-Li content variant of alloy AA 8090 with improved in-plane mechanical property isotropy in thin sheets (Gokhale et al., 1994). The weldability of these alloys with and without filler has been extensively evaluated (Edwards and Stoneham, 1987; Gittos, 1987; Madhusudhan Reddy et al., 1998a,b, 1993; Skillingberg, 1986). Without filler, both alloys showed poor weldability, comparable to that of the hot cracking sensitive alloys AA 2014 and AA 6082. Studies using various filler metals showed that AA 1100 and AA 5556/5556A fillers gave crack-free welds in 1441 (Madhusudhan Reddy, 1988c) and AA 8090 (Gittos, 1987; Skillingberg, 1986), but parent filler and AA 2319 filler caused severe solidification cracking.

The weldability of the second-generation Al–Li alloy AA 2090 was studied with additions of different filler metals under GTA welding (Dvornak et al., 1989; Lippold and Lin, 1996; Martukanitz et al., 1987; Skillingberg et al., 1986). The dependence of cracking susceptibility on the filler metal increased in the order AA 4047, AA 4145, AA 2319, AA 4043, and AA 5356. With AA 4047 and AA 4145 fillers, the cracking was comparable to that of the well-known weldable alloy AA 6061. Hot cracking with AA 5356 filler gave unacceptable levels of cracking. Comparisons of AA 2090 with other AA aluminum alloys revealed that the hot cracking susceptibility decreased in the order: AA 2090, AA 2024/AA 2014, and AA 5083 (Dvornak et al., 1989; Lippold and Lin, 1996), i.e. AA 2090 is highly susceptible.

The very important third-generation Al–Li alloy AA 2195, which is used extensively in spacecraft cryogenic propellant tanks and other spacecraft structures (see Chapter 15) was studied by Pickens (1985), Kramer et al. (1989), and Srivatstan and Sudarshan (1991). With specially designed fillers, AA 2195 could be welded without any cracking. Comparison of AA 2195 alloy with lower Cu versions, and with other aluminum alloys, revealed that the alloys could be ranked with *decreasing* susceptibility to weld cracking in the order: AA 2014, 2090, and 2195 (low Cu), 2195 (medium Cu), 2195 (high Cu), and AA 2219 (Kramer et al., 1989). Improved crack resistance was related to greater amounts of eutectic liquid being present during the last stages of weld solidification.

Employing the Houldcroft test, Madhusudhan Reddy et al. (1998b) conducted weldability studies for the alloy 1441 using a range of filler alloys, namely, 1441 (parent metal), AA 4043, AA 2319, and AA 5356. It was found that weld cracking susceptibility was reduced in the order: autogenous (no filler), 1441, 4043, 2319, 5356, Figure 9.4. In more detail, some of the results were as follows:

1. Sound welds could be obtained with AA 5356 filler. However, the strength of the weld was low, and this cannot be improved by solution treatment and ageing, since AA 5356 is not a heat treatable alloy.
2. Using parent metal filler, a Li loss to an extent of 50% was observed, and there was also hot cracking. In order to obtain the maximum strength, it is necessary to compensate for the lithium loss and also make the weld respond to post-weld solution treatment and ageing. Keeping this in view, a special high-lithium 1441 filler metal (2.8 wt% Li) was employed to understand its effect on structure, weldability, and tensile properties. The hot cracking problem, observed from using the normal 1441 filler (1.9 wt% Li), was

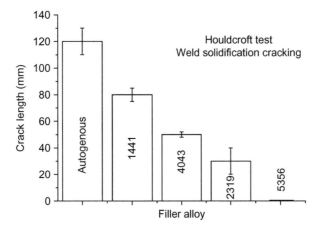

FIGURE 9.4   Cracking susceptibility of autogenous and filler metal welds in 1441-T8 aluminum alloy sheets (Madhusudhan Reddy et al., 1998b).

found to be absent with the high-lithium filler. Microstructural observations revealed that grain boundary T_2 was globular and less continuous than that from using the normal 1441 filler (Madhusudhan Reddy, 1998c), and this may indicate that the Li/Cu ratio is important in determining the T_2 morphology in the solidified weld (Madhusudhan Reddy, 1998c).

Varestraint Testing for Solidification Cracking Susceptibility

Solidification cracking susceptibility of Al–Li alloys 1420, 1441, AA 2090, AA 2195, and AA 8090 was were investigated with the aid of the Varestraint test. The results are showed in Figures 9.5 and 9.6 in terms of total crack length and maximum crack length, respectively, as functions of augmented strain (Madhusudhan Reddy and Gokhale, 2007). It is seen from Figures 9.5 and 9.6 that there was no threshold strain for cracking for any of the alloys, indicating that all were susceptible to weld solidification cracking. Also, the total and maximum crack lengths increased with increasing strain.

Assessment of cracking susceptibility using either total crack length or maximum crack length yielded similar rankings of the alloys. Alloy 1441 was found to be the most susceptible, while alloy 1420 was the least susceptible. The cracking tendency was found to be related to the solidification temperature range and the shape and distribution of the low-melting point eutectic phases present in the fusion zone (Figure 9.7). Specifically, these are the following:

1. The solidification cracking tendency increased with solidification temperature range, in confirmation of earlier theories (Matsuda, 1990). This explains the high resistance of alloys AA 2195 and 1420 to solidification cracking.

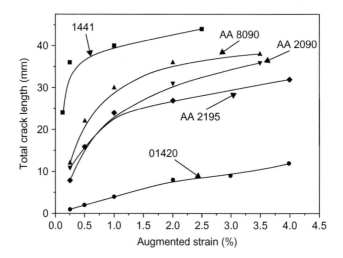

FIGURE 9.5 Total crack length in fusion zone versus augmented strain for several Al–Li alloys (Madhusudhan Reddy and Gokhale, 2007).

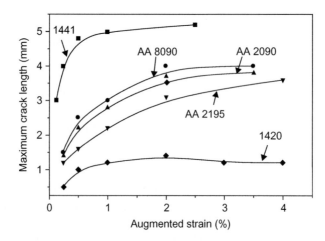

FIGURE 9.6 Maximum crack length in fusion zone versus augmented strain for several Al–Li alloys (Madhusudhan Reddy and Gokhale, 2007).

2. The weld solidification structure became finer in the order 1441, AA 8090, AA 2090, AA 2195, and 1420, leading to greater dispersion of the liquid phase and therefore greater matrix contiguity, as is clearly shown in Figure 9.7. This factor most probably improved the cracking resistance.

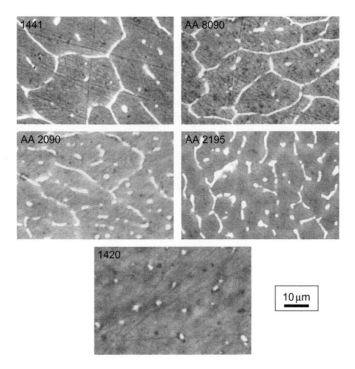

FIGURE 9.7 Backscattered scanning electron microscope (SEM) images of the weld micro-structures of several Al—Li alloys in the as-welded condition (Madhusudhan Reddy and Gokhale, 2007).

Effect of Pulsed Current

Al—Li alloy 1441 was found to be prone to hot cracking under continuous current (CC) autogenous GTA welding. A typical centerline crack is shown in Figure 9.8. To demonstrate the effect of welding conditions on solidification cracking, welding experiments were carried out by Madhusudhan Reddy et al. (1997 and 1999) on this alloy. They used basically CC GTA welding under conditions for which hot cracking took place: the hot cracks progressed in the welding direction along interdendritic boundaries. Subsequently, the current was changed to pulsed current (PC) mode. The change in arc conditions was found to result in a transition from a columnar grain structure to a band of equiaxed zones (EQZs) and subsequently to an equiaxed grain structure. This resulted in crack arrest, Figure 9.9.

When the CC mode was changed to arc oscillation mode or a combination of pulsing and arc oscillation mode, a similar crack arrest phenomenon was observed. Current pulsing and arc oscillation are discussed further in Section 9.6.

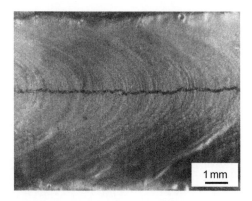

FIGURE 9.8 Typical centerline crack in an Al–Li 1441 alloy weld (Madhusudhan Reddy et al., 1997).

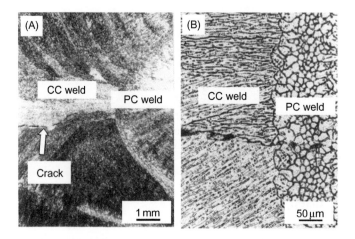

FIGURE 9.9 (A) Macrostructure of the area where switchover from continuous to PC was introduced, showing crack arrest and (B) columnar-to-equiaxed transition in grain structure near the crack arrest zone shown in (A) (Madhusudhan Reddy et al., 1997).

9.4 LIQUATION CRACKING

During aluminum alloy welding the base metal immediately adjacent to the fusion boundary is heated to a temperature between the liquidus and the eutectic temperatures. This results in melting of the base metal eutectic and formation of a partially melted zone (PMZ) within the HAZ. The presence of substantial amounts of eutectic at the grain boundaries in the PMZ can cause post-weld embrittlement. Hot cracking in the PMZ, known as *liquation cracking*, has been reported for many high-strength aluminum alloys (Dudas and Collins, 1966; Metzer, 1967; Schillinger et al., 1963). The severity of grain boundary melting and the thermal stresses causing cracking can be reduced by decreasing the power input, which is why lower power input processes are preferred for welding high-strength aluminum alloys.

The scarcity of data in the literature precludes a straightforward assessment of liquation cracking in Al–Li alloys. So far, related data have been generated using a variety of tests such as the Gleeble hot ductility test or the Varestraint test. However, results from different types of test cannot be directly compared, since each test uses different criteria in assessing weldability.

Using the Gleeble hot ductility test, Zacharia et al. (1989) studied liquation cracking of Al–Li alloys AA 2090 and AA 2091. They concluded that these alloys do not exhibit a strong tendency for liquation cracking, since the temperature range in which the material recovers its ductility on cooling from the peak (PMZ, HAZ) temperature is extremely small. The same conclusions were drawn by Yunjia et al. (1991), who studied the weldability of AA 2090 and the conventional alloy AA 2024 using the Varestraint test. AA 2090 was judged to be less susceptible to liquation cracking compared with 2024. In another study, Yunjia et al. (1991) studied the weldability of AA 8090 and AA 2024 alloys using the transverse Varestraint test. Again they concluded that the Al–Li alloy, AA 8090, was less prone to liquation cracking than AA 2024.

Studies were performed by Ilyushenko et al. (1991) to simulate the response of HAZs in Al–Li alloy 1420 welds under different thermal cycles and using isothermal soaking of specimens in a molten tin bath. It was deduced that heating in the range 550–580°C (below the liquidus temperature for complete melting: $T_L = 635°C$ for 1420) caused dissolution of the strengthening phases and extensive melting of primary intermetallic compounds. However, liquation cracking was not significant, and Ilyushenko et al. concluded that 1420 is not susceptible to liquation cracking.

Madhusudhan Reddy and Gokhale (2007) used Varestraint testing to evaluate the liquation cracking susceptibility of Al–Li alloys AA 2090, AA 2195, AA 8090, 1420, and 1441 and found that the weldments were free from liquation cracks: the solidification cracks generated during testing were confined to the fusion zone.

From the above discussion it may be concluded that Al–Li alloys are generally less prone to liquation cracking; and, by inference from the generally detrimental effect of substantial amounts of eutectic at grain boundaries, that the lesser susceptibility of Al–Li alloys is mainly due to the absence of a low-melting point eutectic.

9.5 EQZ FORMATION AND ASSOCIATED FUSION BOUNDARY CRACKING

9.5.1 Experimental Observations

Weld pool solidification generally begins by epitaxial growth from the PMZ (Kerr and Villafuerte, 1991; Kou, 1987; Savage and Aronson, 1965), since the base metal grains provide excellent growth sites and the growth rate far outweighs the nucleation rate in this zone. However, instead of epitaxial growth, a fine EQZ has been observed to form near the fusion line in Al–Li alloys welded with a variety of methods: LBW (Lee et al., 1996; Sriram et al., 1993), variable polarity plasma arc (VPPA) (Soni et al., 1996), electron beam (EB) (Bowden and Meschter, 1984), gas metal arc (GMA) (Padmanabham et al., 2005), and GTA (Gutierrez et al., 1996).

An EQZ typically consists of cells of size 3–6 μm and an overall width of 150–300 μm in the weld bead. The cell interiors are precipitate-free, while the cell boundaries are decorated with Li and Cu–Mg rich precipitates (Shah et al., 1992). Microsegregation of Cu and Mg occurs within the cells (Soni et al., 1996). A typical EQZ is shown in Figure 9.10 (Madhusudhan Reddy et al., 1998c). The structure suggests that this region has undergone complete melting. In weldments made using AA 4043 (Al–Si) filler, EDS analysis showed that there was no Si in the EQZ. Therefore, the EQZ is considered to be formed by re-solidified base material that is unaffected by the weld wire composition. Tensile tests of these weldments revealed that the fracture path is confined to this equiaxed grain region. Hence understanding the behavior of the EQZ is essential for the structural integrity of Al–Li alloy weldments. This is discussed in Section 9.5.3.

The formation of EQZs in Al–Li alloys has been attributed or linked to several mechanisms:

1. He et al. (1992) attributed EQZ formation in Al–Li alloys to a high nucleation rate due to the lower surface tension in these alloys.
2. Faster cooling adjacent to the fusion boundary was considered responsible for the high nucleation rate (Shah et al., 1992). The presence of "Li-aided greater nucleation sites" was also considered to be responsible for the formation of the EQZ (Shah et al., 1992).

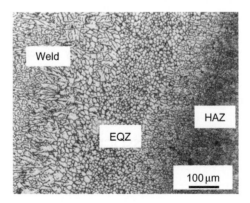

FIGURE 9.10 Microstructure near the fusion line showing an EQZ in Al–Li 1441 alloy (Madhusudhan Reddy et al., 1998c).

3. Experiments by Gutierrez et al. (1996) showed that EQZ formation was related to the initial microstructure of the base plate. If the base plate was as-cast or as-welded, no EQZ formed; and if the base plate was in the solution treated and aged or annealed condition, the EQZ did form.

Gutierrez et al. (1996) suggested that EQZ formation was closely linked to $Al_3(Li, Zr)$-aided nucleation. However, Bowden and Meschter (1984) did not observe Al_3Zr for a high ($\sim 0.5\%$) Zr-containing alloy in the as-welded condition, but it was present in the solution treated and aged condition.

Madhusudhan Reddy et al. (1998c) studied EQZ formation in Zr-containing as well as Zr-free aluminum alloys using conventional GTA welding and welding under PC and magnetic arc oscillation (MAO). Results for an Al–Li AA 8090-type alloy and AA 8090 using conventional GTA welding are illustrated in Figure 9.11:

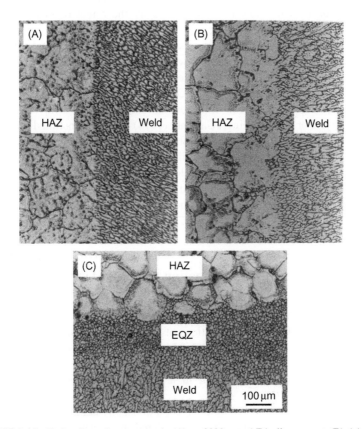

FIGURE 9.11 Fusion line microstructure in (A) an 8090-type (-Zr) alloy as-cast, (B) AA 8090 as-cast, and (C) AA 8090 cast + homogenized alloy. Note the epitaxial growth in (A) and (B), and the EQZ in (C) (Madhusudhan Reddy et al., 1998c).

a. Figure 9.11A shows the as-cast microstructure near the fusion line of a weld made with an Al−Li alloy of similar composition to AA 8090 but without Zr. No EQZ was seen, and epitaxial growth was clearly observed at the fusion line.

b. Similarly, for AA 8090 in the as-cast condition, no EQZ zone was observed and the growth was epitaxial (Figure 9.11B).

c. However, an EQZ did form when the AA 8090 base alloy was homogenized before welding (Figure 9.11C).

Madhusudhan Reddy et al. (1998c) conducted similar tests with Al−Li alloy 1441:

d. For CC welds, the EQZ was confined mainly to the region near the fusion line, where Al_3Zr particles are likely to survive. Dissolution of Al_3Zr particles in the weld interior, owing to the high temperatures reached during welding, was considered responsible for lack of an EQZ in the weld interior. This view is substantiated by the effect of heat input on EQZ width: as heat input decreases, the EQZ width increases.

e. Using PC and MAO, the EQZs appeared (i) as curved bands at the trailing edge of the weld pool for PC welds and (ii) scattered in many areas in MAO welds (Figure 9.12), i.e. the EQZs were present in the interior of the fusion zone. The presence of EQZs in the weld interiors could be the result of fluid flow-driven transport of chill crystals from the fusion boundary. For PC welding, Garland (1974) has demonstrated different types of fluid flow (e.g., radial, rotational, etc.), and only the equiaxed grains which remain near the trailing edge of the weld pool can be expected to survive, owing to the lower temperatures there, thus giving them the characteristic curved band appearance. For MAO welding, the fluid flow can be more complex due to the constantly changing orientation of the weld pool, leading to strong intermixing of equiaxed and dendritic grains.

9.5.2 Hypotheses of EQZ Formation and Their Evaluation

Two hypotheses have been proposed to explain the formation of EQZs. The first hypothesis (Shah et al., 1992) proposes that the EQZ is located in the PMZ and forms by recrystallization of preexisting grains. The second hypothesis (Lippold and Lin, 1996) proposes a solidification mechanism requiring both nucleation and growth of equiaxed grains. This suggestion is based on heterogeneous nucleation in a stagnant liquid region called the unmixed zone (UMZ).

Three possible factors: (i) welding conditions, (ii) composition, and (iii) base metal microstructure have been considered in an effort to better understand the EQZ formation (Cross et al., 1999; Gutierrez and Lippold, 1998; Reddy et al., 1998c). Based on the observations, a unified heterogeneous

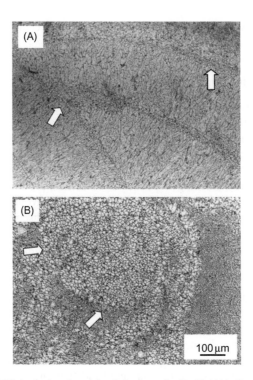

FIGURE 9.12 EQZs in the interior of the fusion zone in Al–Li 1441 alloy welded under (A) Pulsed Current PC and (B) Magnetic Arc Oscillation (MAO). The arrows indicate the EQZs (Madhusudhan Reddy et al., 1998c).

nucleation mechanism has been proposed that accounts for the effects of Li and Zr on the formation of EQZs:

1. Heterogeneous nuclei have been identified as Al_3Zr and $Al_3(Li_x,Zr_{1-x})$ precipitates, and the region where these nuclei may exist extends beyond the UMZ. According to this proposed mechanism, Li and Zr are primarily responsible for EQZ formation.
2. Zr forms metastable Al_3Zr precipitates that act as effective heterogeneous nucleation sites, while Li combines with Al_3Zr to produce a high volume fraction of $Al_3(Li_x,Zr_{1-x})$ precipitates, increasing the amount of nuclei in the UMZ.
3. Li also improves the wetting behavior at the precipitate/liquid interface by reducing the interfacial surface energy, increasing the effectiveness of the nucleation process.
4. Reduction in the Zr and Li contents of the base metal results in reduction or complete elimination of the EQZs.
5. The contributions of Cu, Mg, and Ag appear to be insignificant. This was inferred from similar patterns of EQZ formation in AA 2090 and AA

2195 alloys, which have different amounts of these elements (Mg and Ag absent in AA 2090) (Table 9.1).

9.5.3 Fusion Boundary Cracking

EQZs have been associated with severe fusion boundary cracking during the fabrication and repair of structures. Such cracks have been reproduced in Al—Li alloys using the Varestraint test (Lippold and Lin, 1996; Madhusudhan Reddy and Gokhale, 2007), showing clearly that the cracks occur preferentially through the EQZ.

Fusion boundary cracking is potentially a serious problem for the integrity of welds in Al—Li alloys, hence it is desirable to avoid EQZ formation. From the proposed mechanism of EQZ formation discussed above, one solution might be to eliminate Zr from the alloy composition. However, Zr is an important element both in strengthening and controlling recrystallization and texture in half-products (Rioja and Liu, 2012). Another possibility is to eliminate the narrow liquid layer which forms in the EQZ by enhanced weld pool stirring during welding. Aidun and Dean (1999) studied the effect of enhanced convection on the microstructure of welded AA 2195 by applying a high-gravity (centrifugal force) environment using a multi-gravity research welding system. The absence of an EQZ in weldments made at 10g qualitatively supports the mechanism proposed by Gutierrez and Lippold (1998), see Section 9.5.1. The altered thermal and fluid flow properties generated by the centrifuge effectively eliminated the nucleation sites for equiaxed grains by promoting rapid mixing throughout the weld pool, whereby dissolution of precipitates could also have occurred. Whether this method is feasible for manufacturing welded components remains an open question.

9.6 MODIFICATION OF FUSION ZONE MICROSTRUCTURES

One way of improving weldability in several alloy systems is modification of the weld microstructure. The columnar as-cast structure usually exhibited by weld fusion zones is associated with poor resistance to hot cracking and inferior weld mechanical properties. Refinement of fusion zone microstructures helps in reducing solidification cracking and may also improve the weld metal mechanical properties, such as ductility, fracture toughness, and fatigue strength. Thus the development of procedures for refining weld solidification structures is an important topic in welding research.

Although it is highly desirable to control the solidification structure in welds, such control is often very difficult because of the higher temperatures and higher thermal gradients in welds in relation to castings and also the epitaxial nature of the solidification process. However, despite these limitations several methods of grain refinement for welds have been used in the past: (i) inoculation with heterogeneous nucleants, (ii) surface nucleation induced by gas impingement, and (iii) introduction of physical disturbance through techniques such as PC and MAO (Section 9.5.1).

Inoculation is discussed in the next section. This is followed by discussions of the PC and MAO techniques, which have gained widespread popularity because of the relative ease of applying them to industrial situations with only minor modifications to the existing welding equipment.

9.6.1 Inoculation

The use of inoculants for refining the weld fusion zones is not as successful as in castings. This is because of the extremely high temperatures involved in welding, and also undesirable effects on weld mechanical properties at the inoculant levels required for refining the grains. Dvornak et al. (1989, 1991) have compared the effectiveness of titanium and zirconium in refining the solidification structure of Al—Li alloy welds and reducing their hot cracking tendency. The improved weldability due to Zr and Ti additions was attributed to grain refinement as well as alteration of the shapes and distributions of the eutectic phases in the weld microstructure. Their results showed that Zr improves weldability more than Ti for concentrations <0.3 wt% (Dvornak et al., 1989, 1991).

Janaki Ram et al. (2000) have achieved considerable refinement of the solidification structures of AA 2090 Al—Li alloy welds produced using AA 2319 (Al—6.3Cu) fillers containing three different inoculants: Ti, Ti + B, and Zr (Figure 9.13). Of the three inoculants, Zr gave the best results, and

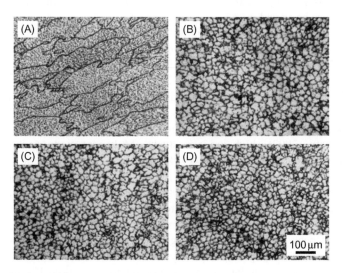

FIGURE 9.13 Microstructures of weld interiors for the Al—Li alloy AA 2090, using AA 2319 filler and three inoculants (A) untreated, (B) inoculated with Ti, (C) inoculated with Ti + B, and (D) inoculated with Zr (Janaki Ram et al., 2000).

the combined use of Ti + B was found to produce a slightly greater degree of grain refinement than Ti alone. The better performance of Ti + B was attributed to the role of TiB_2 as an additional nucleant and the effect of B in reducing the solubility of Ti in liquid aluminum.

Janaki Ram et al. (2000) also demonstrated that the grain refinement resulted in significant reductions in hot cracking susceptibility and improvements in weld tensile properties, especially ductility. In a subsequent study, Sundaresan et al. (2000) showed that the effectiveness of Ti inoculation can be improved by using MAO during welding. It was suggested that MAO-induced enhanced fluid flow and reduced thermal gradients in the weld pool-aided heterogeneous nucleation.

Other possibilities like surface nucleation and micro-cooler additions have been rejected because of the complicated welding setups and procedures needed for their use.

9.6.2 Pulsed current

PC techniques involve cycling the welding current from higher to lower levels with a regular frequency. PC influences the solidifying pool thermally and mechanically, causing periodic shaking of the liquid metal with a frequency equal to that of the pulses. The force of the mechanical action of the arc on the metal (the arc pressure) is proportional to the square of the current amplitude. The thermal energy depends on the shape and duration of the pulse, the frequency, and the current. The thermal effect on the liquid metal can therefore be varied over a relatively wide range.

PC has been used by several investigators to obtain grain refinement in weld fusion zones and improvement in weld mechanical properties. Significant refinement of the solidification structure has been reported in aluminum alloys (Davies and Garland, 1975; Garland, 1974; Madhusudhan Reddy and Gokhale, 1994; Madhusudhan Reddy et al., 1997; Yamamoto et al., 1992), austenitic stainless steels (Gokhale et al., 1982; Ravi Vishnu, 1995; Shinoda et al., 1990), nickel-base superalloys (Madhusudhan Reddy and Gokhale, 2007), high-strength steels (Mohandas and Madhusudhan Reddy, 1997), and tantalum (Grill, 1981).

The choice of pulse parameters is important. Several investigators have stressed the importance of using the optimum pulse parameters for maximizing the beneficial influence of PC. There are as many as five important parameters: peak current or pulse current level (I_p); background current level (I_b); the ratio of peak current time to background current time (or pulse on-time to off-time ratio), t_p/t_b; pulse frequency; and welding speed. It turns out that the pulse frequency is by far the most important consideration. The influence of pulse frequency on AA 8090 Al–Li alloy GTA weld microstructures is shown in Figure 9.14 (Madhusudhan Reddy et al., 1998d, 2002).

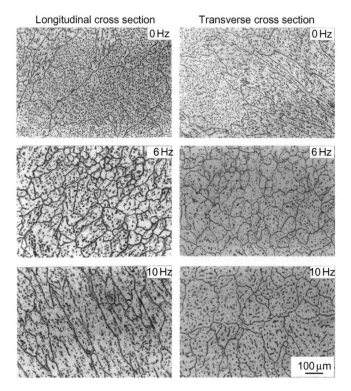

FIGURE 9.14   Influence of pulse frequency (Hz) on the fusion zone microstructure of GTA welds in AA 8090 Al–Li alloy (Madhusudhan Reddy et al., 1998d).

From Figure 9.14 it may be seen that conventional continuous welding (0 Hz) resulted in predominantly columnar grains, while PC resulted in grain refinement and a more equiaxed grain shape, with maximum refinement at the intermediate frequency of 6 Hz. Too high a frequency (in this example 10 Hz) resulted in larger grains. This effect has been attributed to diminished vibrations and temperature oscillations in the weld pool (Garland, 1974). It is given in more detail as follows:

1. Maximum physical disturbance can be induced in the weld pool by operating at an intermediate frequency close to the weld pool's resonant frequency.
2. The pulse frequency-welding speed combination should allow sufficient overlap of individual weld nuggets resulting from each peak current pulse. In other words, at a given pulse frequency the welding speed should be such that the torch does not travel beyond the diameters of weld nuggets resulting from peak current pulses. This ensures that the weld beads form a series of overlapping nuggets.

3. For a given pulse on-time to off-time ratio, t_p/t_b, the pulse frequency determines the durations of peak currents and background currents. The combination of pulse frequency and t_p/t_b allows through-thickness melting during the peak current time and at a given peak current level.

These considerations show that even though the pulse frequency is the most important PC parameter, a careful balance between the parameters is required to obtain the optimum welding conditions.

9.6.3 Magnetic Arc Oscillation

Convection always exists in the weld pool owing to various driving forces. Enhancement of the convection by artificial agitation/disturbance could result in grain refinement. One method of introducing a weld pool disturbance is to cause the welding arc to oscillate by using an alternating external magnetic field. It is well known that the interaction of an arc current with its own magnetic field leads to Lorentz forces that cause fluid flow and a self-induced stirring effect, particularly when the current density is large. Reinforcing the natural flow with an external magnetic field enhances this effect.

The forces on the arc by external magnetic fields depend on their active orientation:

1. If the field is coaxial with the arc, the induced forces will be perpendicular to both the magnetic field and the radial component of the diverging arc current through the arc. This will result in rotation of the arc and a circular flow of the molten metal in the weld pool. Using an alternating magnetic field reverses the flow periodically, resulting in electromagnetic stirring (EMS). This phenomenon is also called circular oscillation.
2. When the alternating external magnetic field is oriented parallel to the welding direction, the interaction of the magnetic field with the axial component of the diverging arc current results in oscillation of the welding arc transverse to the welding direction and in the plane of the sheet or plate being welded. This is commonly referred to as MAO. However, some investigators prefer to call this transverse arc oscillation in order to distinguish it clearly from longitudinal arc oscillation, which results when the alternating magnetic field is oriented perpendicular to the welding direction.

Several investigators have used external magnetic fields during welding. Brown et al. (1962) were among the first to study the effects of EMS and to obtain grain refinement in stainless steels, and aluminum and titanium alloys. Tseng and Savage (1971) reported that MAO refined the solidification substructure, decreased hot cracking, and produced a less undesirable pattern of microsegregation in high-tensile steel weld metal.

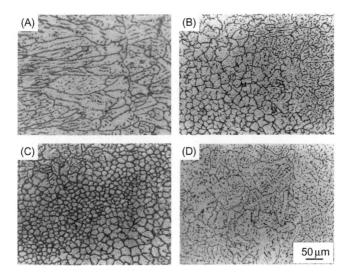

FIGURE 9.15 Effect of arc oscillation frequency on Al—Li 1441 alloy GTA weld zone microstructure (A) 0 Hz, (B) 4 Hz, (C) 6 Hz, and (D) 10 Hz (Madhusudhan Reddy, 1998a,b).

Madhusudhan Reddy et al. (1997) and Madhusudhan Reddy (1998) studied the influence of transverse arc oscillation frequency on the fusion zone grain size and morphology of Al—Li 1441 alloy autogenous welds. The results are presented in Figure 9.15. At "0Hz" frequency (non-oscillated condition), the fusion zone microstructure consisted of coarse columnar grains (Figure 9.15A). With the introduction of arc oscillation, a change in grain morphology from fully columnar to equiaxed, as well as refinement of the grain size was observed up to a frequency of 6 Hz (Figure 9.15C). Beyond 6 Hz the grain morphology was similar to that of non-oscillated welds (Figure 9.15D).

In all cases, the arc oscillation weld showed a mixed columnar + equiaxed structure (Figure 9.16). The proportion of equiaxed grains increased up to 60—70% as the frequency increased up to 6 Hz. At this optimum frequency, the equiaxed grain size was reduced to about 20 μm (Figure 9.15C). The variation of percentages of equiaxed grains with arc oscillation frequency is shown in Figure 9.17.

The effect of arc oscillation amplitude was also investigated by Madhusudhan Reddy (1998). The results are presented in Figure 9.18: increasing the arc amplitude up to 0.6 mm reduced the volume fraction of columnar grains, but further increases did not cause further refinement of the microstructure. It was also found that for arc amplitudes ≥ 0.6 mm, the columnar grains were mixed intimately with equiaxed grains, rather than forming distinct zones.

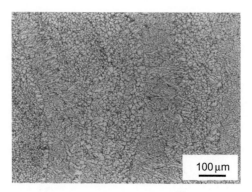

FIGURE 9.16 Typical Al–Li 1441 alloy arc oscillation GTA weld zone with alternating bands of equiaxed and columnar microstructure (Madhusudhan Reddy, 1998a,b).

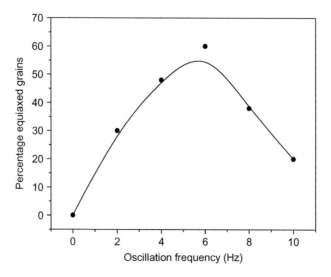

FIGURE 9.17 Effect of arc oscillation frequency on the percentage of equiaxed grains in the fusion zone for Al–Li 1441 alloy (Madhusudhan Reddy, 1998a,b).

9.7 MECHANICAL PROPERTIES

Welds are inherent sources of many defects (porosity, hot tears, undercuts, lack of fusion, etc.) that can serve as initiation sites for fatigue or fast fracture. However, even without these defects the weld metal and accompanying HAZ are typically lower in strength than the base metal. The low strength of the weld metal, associated with its as-cast structure, can be increased somewhat with solute content, owing to a corresponding increase in the amount of second-phase constituents. However, this results in loss of ductility.

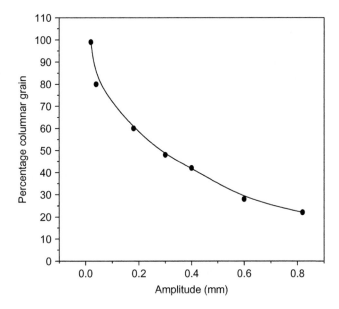

FIGURE 9.18 Effect of arc oscillation amplitude on the percentage of columnar grains in the fusion zone for Al–Li 1441 alloy (Madhusudhan Reddy, 1998a,b).

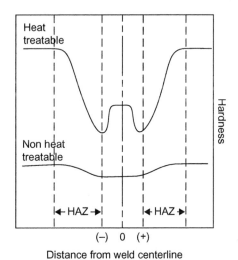

FIGURE 9.19 Schematic diagram of hardness traverses across aluminum weldments for heat treatable versus non-heat treatable aluminum alloys (Cross and Olson, 1988).

Loss of strength in the HAZ is the result of dissolution of precipitates and softening of cold-worked material, and is most noticeable when welding high-strength (precipitation hardenable) alloys (Figure 9.19): the strength and hardness in the as-welded condition remain usually in the range of 60–70% (termed as "joint efficiency"). This loss in strength and hardness is irreversible unless a full heat treatment comprising solutionizing and ageing is carried out, which is seldom practiced. The only exception is the Al–Zn–Mg alloy system, which undergoes natural ageing after welding. However, for all other heat treatable alloys the application of minimum heat input and faster cooling is required to minimize the loss in strength and hardness and thereby achieve better joint efficiency.

(Non-heat treatable alloys, such as alloy AA 5083, are popular for welded aluminum alloy structures because there is little difference in the properties of the base metal and HAZ, and the joints are nearly 100% efficient in the as-welded condition. However, these alloys have considerably lower strengths than the heat treatable alloys.)

9.7.1 Al–Li 1420 (First-Generation Al–Li Alloy)

Alloy 1420 (Al–2.1Li–5.2Mg–0.11Zr) conventional GTA welds can achieve joint efficiencies of up to 80% of the base metal strength without post-weld heat treatment and with the use of several filler metals (Pickens, 1985; Pickens et al., 1989). When specimens were re-solutionised and artificially aged, joint efficiencies of up to 99% were reported. Higher joint efficiencies could be achieved with the use of fillers containing 0.003% Be and other elements like Mn, Ti, Zr, and Cr (Pickens, 1985, 1990).

9.7.2 Al–Li 1441, AA 8090, and AA 2090 (Second-Generation Al–Li Alloys)

Al–Li 1441

Madhusudhan Reddy (1998) has shown that the weld properties for alloy Al–Li 1441 (Al–1.95Li–1.65Cu–0.9Mg–0.11Zr) autogenous welds can be improved by using PC and MAO during GTA welding (Table 9.4). In general, the PC, MAO, and PC + MAO techniques improved the weld tensile properties compared to CC welds, the beneficial effect being most pronounced for the PC + MAO technique:

1. The PC, MAO, and PC + MAO improvements in tensile properties in the as-welded condition can be related to refinement of the solidification structure. The order of increase was PC, MAO, and PC + MAO techniques.
2. The solution treated and aged weld properties showed considerable improvements over the as-welded properties. These improvements are due to the greater dissolution of nonequilibrium solidification products in the refined structures, allowing greater participation of the alloying elements in formation of strengthening phases during ageing.

TABLE 9.3 Tensile Properties of 1441 Alloy GTA Welds

Process	Condition	0.2% YS (MPa)	UTS (MPa)	Elongation (%)
CC-GTAW	AW	–	233	–
	STA	340	380	1.4
PC-GTAW	AW	240	259	2.0
	STA	354	433	7.0
MAO-GTAW	AW	248	290	1.5
	STA	357	432	8.0
PC + MAO-GTAW	AW	275	345	3.0
	STA	365	438	8.0
Base alloy	T81	398	450	9.0

AW, as welded; STA, solution treated and aged; UTS, ultimate tensile stress; YS, yield stress; CC, conventional continuous; PC, pulsed current; MAO, magnetic arc oscillation; GTAW, gas tungsten arc welding.
Source: Madhusudhan Reddy (1998a,b).

3. The tensile fracture surfaces in the as-welded condition showed a transition from EQZ grain boundary fractures in CC welds to ductile fracture for the PC, MAO, and PC + MAO welds. This change correlated well with the observed tensile elongation values in Table 9.3. Earlier work (Shah et al., 1992) showed that tensile failures of CC welds occurred preferentially along the fine EQZ present along the fusion boundary.

However, for the other welding conditions (PC, MAO, and PC + MAO), the EQZ was less concentrated at the fusion boundary and spread out into the bulk weld zone. In these cases, the tensile failure did not occur along equiaxed grain boundaries, implying that intergranular failure can be avoided if the EQZ is distributed into the bulk weld.

Table 9.4 compares the tensile properties of GTA welds deposited with high lithium (2.8 wt% Li) and 1441 fillers. The strengths and percentage elongations were higher from using the high-lithium filler. The response of the welds to solution treatment and ageing was also better.

The strength improvements in the welds using high-lithium filler were attributed to improved solid solution strengthening in the as-welded condition and enhanced δ' precipitation strengthening in the solution treated and aged condition (Namba and Sano, 1986). This was also reflected in the hardness trends and chemical compositions of the welds: high-lithium filler welds had higher hardness in the as-welded and solution treated and aged

TABLE 9.4 Comparison of Tensile Properties of Al–Li 1441 GTA Welds Using Special (High Li) and Parent Filler Metals

Filler	Test Condition	Weld Strength		Elongation %
		0.2% YS (MPa)	UTS (MPa)	
High-lithium filler	As-welded	240	336	3.5
	Solution treated and aged	290	430	8.0
1441 filler	As-welded	206	322	2.0
	Solution treated and aged	330	414	3.6

Source: Madhusudhan Reddy (1998a,b).

conditions compared to the parent filler welds; and the lithium contents of the high-lithium filler welds were 2–2.2 wt% (equal to the lithium content of the parent metal), while the lithium contents of the 1441 filler welds were much less, 1.1–1.25 wt%. (Madhusudhan Reddy, 1998) Thus, the mechanical property improvements of the high-lithium filler welds could be attributed to the filler compensating for the lithium loss that inevitably occurs during welding.

Al–Li AA 8090

Skillingberg (1986), Gittos (1987), and Edwards and Stoneham (1987) assessed the GTA weldability of AA 8090 (Al–2.4Li–1.2Cu–0.8Mg–0.11Zr) with AA 1100, 4043, 5356, 2319, and parent alloy AA 8090 fillers. In these studies, an as-welded strength of about 200–300 MPa was obtained. Re-solution treatment, quenching, and artificial ageing of the weldments made with AA 8090 parent filler produced the highest tensile strength (447 MPa), which was 85% of parent metal properties. However, hot cracking susceptibility would have effectively eliminated the viability of 8090 weldments made with parent alloy filler (Section 9.3.3).

Al–Li AA 2090

The first weldability study on alloy AA 2090 (Al–2.1Li–2.7Cu–0.11Zr), by Martukanitz et al. (1987), suggested that 2090 has undesirable as-welded and post-weld-aged mechanical properties and is unsuitable for welded applications. On the other hand, Kerr and Merino (1991) reported that the tensile and fracture properties of 2090 weldments at temperatures between 297 and

20 K were superior to those of AA 2219 weldments; and that on a density-compensated basis, the 2090 weldments were superior in structural efficiency. These are remarkable conclusions, given that AA 2219 is the mainstay aerospace alloy for high-strength welded structures (see Chapter 15, Section 15.3.3), and indeed AA 2090 has found some applications in launch vehicle cryogenic tankage components.

AA 2090 welds are mostly made with the peak-aged base metal, to maximize the post-weld properties, and then used as-welded. It is noteworthy that even with filler metal additions that provide some solid solution strengthening, the fusion zone is incapable of achieving the strength of the previously peak-aged base metal (Sunwoo and Morris, 1989).

9.7.3 Al–Li AA 2195 (Third-Generation Al–Li Alloy)

The third-generation Al–Li alloy 2195 (Al–1.0Li–4.0Cu–0.4Mg–0.4Ag–0.11Zr) was developed from the Weldalite™ alloy family (Pickens et al., 1989; Rioja et al., 1990), which in turn was developed using the conventional weldable alloy AA 2219 as a baseline, since it has good cryogenic properties. AA 2195 was commercially introduced in 1992 (Table 9.1) and subsequently used for the Space Shuttle Super Lightweight External Tank (SLW ET), first launched in June 1998.

The development history of the Weldalite™ alloy family and AA 2195 is concisely described in Chapter 1, Section 1.3.8. The composition of the first-patented Weldalite™ alloy was Al–1.3Li–(4.5–6.3) Cu–0.4Ag–0.4Mg–0.14Zr. Lithium was kept to no more than 1.3 wt%, facilitating the weldability and still giving high strength in combination with additions of Ag and Mg. Subsequently, Zn was added for corrosion resistance (Rioja et al., 1990), and the Li content was reduced to 1 wt% for AA 2195.

Although the Weldlite™ alloys and AA 2195 were specially developed for welded structures, the GTA post-weld solution treated and aged joint efficiency for Weldalite™ 049 (which has a similar composition to AA 2195) does not exceed 60% (Kostrivas and Lippold, 1999). This has led to much effort being invested in FSW (Section 9.9.2).

9.8 CORROSION

Aerospace Al–Li alloys and their welds can be susceptible to localized corrosion when exposed to aqueous chloride-containing environments. Although different aspects of corrosion of the base metal Al–Li alloys have been reported in the literature (Bavarani et al., 1989; Kumai et al., 1989; Sheppard and Parson, 1987; Yusheng et al., 1996), corresponding data for the weldments are lacking.

Welding can have a significant influence on the corrosion resistance owing to microstructural changes from the associated thermal cycles.

Besides other factors, the reliability of welded structures depends upon the presence of fabrication defects and service damage as a result of degradation processes such as fatigue cracking, corrosion, and stress corrosion. Corrosion pitting can be very destructive, since it can occur on any surface that is unprotected, or for which the surface protection has degraded or been damaged, and the corrosion pits can lead to fatigue cracking (Hoeppner, 2004). In particular, pits in the fusion zones of welds can have more deleterious effects, since the corrosion properties of the fusion zone are often inferior to those of the base metal (Srinivasa Rao and Prasad Rao, 2004).

Localized corrosion is strongly influenced by the presence of second-phase particles, whose size and distribution in the alloy depend upon the welding process, technique, and welding parameters. Madhusudhan Reddy et al. (2001) studied the influence of welding technique on the pitting corrosion resistance of Al–Li 1441 alloy autogenous GTA welds (Table 9.5). The potentials at which the current increased abruptly after the passive region was taken as pitting potential (E_{pit}). Specimens exhibiting higher pitting potentials were taken to have better pitting resistance. In all cases, the pitting potentials of the welds were more negative than that of the base metal, indicating that the welds would be more susceptible to corrosion. In fact this was the case (Madhusudhan Reddy et al., 2001).

The lesser pitting corrosion resistance of the welds can be attributed to the presence of segregation products in the as-solidified welds. This is discussed in more detail with the aid of Figures 9.20–9.24:

1. The relatively low corrosion resistance of CC welds can be attributed to the semi-continuous network of grain boundary precipitates, mainly the copper-rich T_2 (Al_6CuLi_3) phase (Figure 9.20A). Regions adjacent to the grain boundary would be expected to be depleted in copper owing to the presence of copper-rich T_2 phase at the grain boundaries. Furthermore, areas containing lower amounts of copper would be preferred locations for corrosion (Figure 9.21). The semi-continuous network

---

TABLE 9.5 Pitting Potentials for Al–Li 1441 Alloy GTA Weld Metal

Weld Technique	E_{pit} mV (SCE)
CC weld	−740
PC weld	−620
MAO weld	−570
Base metal (T81)	−540

Source: Madhusudhan Reddy et al. (2001)

of copper-rich grain boundary phase resulted in an analogous network-type of corrosion (Figure 9.22A).

2. The PC and MAO welds had greater resistance to pitting corrosion. This was due to discontinuous T_2 phase distributions (Figure 9.20B and C) and corresponding discrete areas of corrosion (Figure 9.22B and C).

3. Optical metallography revealed that corrosion was mainly confined to the fine EQZ in CC welds (Figure 9.23A) but was less regularly distributed in the PC and MAO welds (Figure 9.23B and C).

4. The extensiveness of pitting corrosion in the CC welds is demonstrated by the evidence of grain loss shown in Figure 9.24.

From the foregoing results, it is evident that microstructural changes during welding can drastically affect the corrosion behavior. The occurrence of copper-depleted regions, which are the result of non-equilibrium solidification, can be avoided by using solid-state welding techniques (e.g., FSW, which is discussed in the next section).

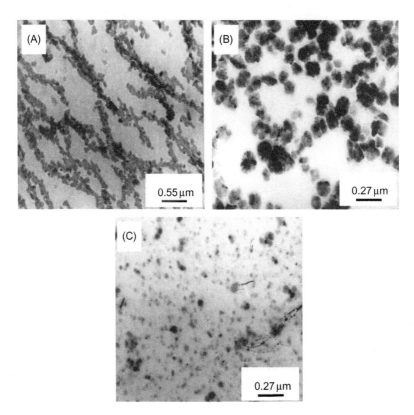

FIGURE 9.20 TEM microstructures of Al–Li alloy 1441 GTA welds (as-welded condition): (A) CC weld, (B) PC weld, and (C) MAO weld (Madhusudhan Reddy et al., 2001).

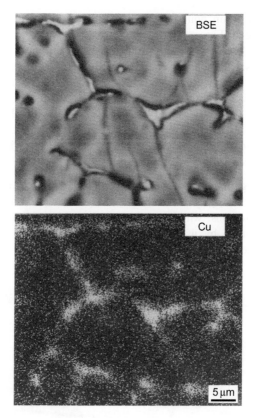

FIGURE 9.21 Electron probe microanalysis (EPMA) micrographs of the corroded surface of an Al—Li 1441 alloy GTA CC weld, showing attack adjacent to copper-rich grain boundaries (Madhusudhan Reddy et al., 2001).

9.9 SOLID-STATE WELDING PROCESSES

Solid-state welding is defined as a joining process (i) without any liquid or vapor phase, (ii) with the use of pressure, and (iii) with or without the aid of temperature. Solid-state welding is done over a wide range of pressure and temperature, with appreciable deformation and solid-state diffusion. In solid-state welding, the cohesive forces between metal atoms are utilized. Other forces, for example, van der Waals forces, are expected to be of lesser importance. Among the available solid-state welding processes, friction welding and FSW have rapidly gained wide importance and acceptability in industrial applications. FSW has become especially important for the aerospace industry in the last 20 years.

9.9.1 Friction Welding

Since friction welding is a solid-state process, it avoids detrimental weld features such as brittle eutectic and interdendritic phases that result from

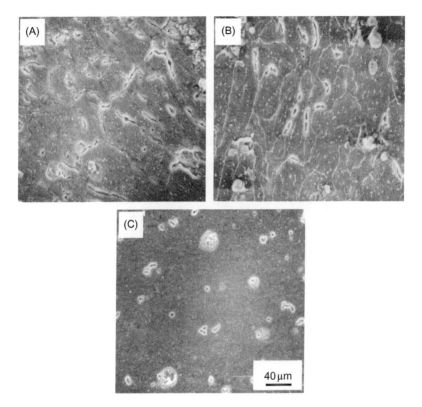

FIGURE 9.22   SEM micrographs of corroded surfaces of Al—Li 1441 alloy GTA welds: (A) CC weld, (B) PC weld, (C) MAO weld (Madhusudhan Reddy et al., 2001).

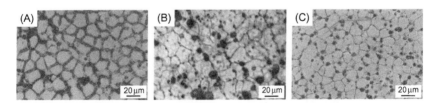

FIGURE 9.23   SEM micrographs of corroded surfaces of EQZs of Al—Li 1441 alloy GTA welds: (A) CC weld, (B) PC weld, and (C) MAO weld (Madhusudhan Reddy et al., 2001).

solidification of the fusion zones in conventional welds (Maalekian, 2007). Extensive work has been carried out by Madhusudhan Reddy et al. (2003) to evaluate the friction welding aspects of the second-generation Al—Li alloy AA 8090. The influences of joining parameters (friction pressure, forge pressure, burn-off) on microstructure, notch tensile strength, impact toughness, and hardness of friction joints were evaluated.

A typical friction weld bead configuration in AA 8090, together with details of the microstructure at the center and periphery, is shown in Figure 9.25. Macroscopically it is seen that deformation in the thermomechanically affected zone resulted in severe deformation (bending) of the grains. In contrast, the microstructure within the bond zone is fine and equiaxed owing to dynamic recrystallization. All the welds contained fine equiaxed

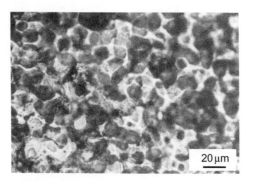

FIGURE 9.24 Optical micrographs of corroded surface of Al–Li 1441 alloy GTA CC weld showing grain fallout (Madhusudhan Reddy et al., 2001).

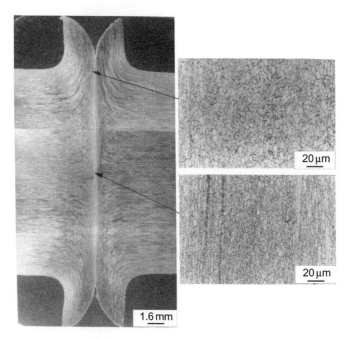

FIGURE 9.25 Macrostructure and microstructures of a friction weld in Al–Li alloy AA 8090 (Madhusudhan Reddy et al., 2003).

recrystallized zones (Madhusudhan Reddy et al., 2003). An optimum combination of strength and toughness was achieved by minimizing the heating time. This was done by using a high upsetting pressure in combination with low levels of friction pressure and burn-off.

9.9.2 Friction Stir Welding

FSW is an innovative and remarkable modification of the traditional friction welding process. In the past 20 years, FSW has experienced a significant growth in technological research and implementation. FSW is an environment-friendly and energy-efficient technique that can be used to weld high-strength aluminum alloys and other metallic materials that are difficult to join using conventional welding processes. As a solid-state joining method, FSW can avoid all the welding defects caused by melting and solidification in fusion welding and also increase the joint efficiency compared to conventional GTA welds, both by increasing the strength and improving the strength reliability (Threadgill et al., 2009).

An illustration of the FSW process is given in Figure 9.26. A rotating tool is forced down into the joint line under conditions where frictional heating is sufficient to locally increase the material temperature into the range where it readily plastically deforms. A base plate (backing bar, not shown) beneath the material to be welded provides constraint together with the material on either side of the rotating tool. Owing to this constraint, the plastically deforming material flows around the rotating tool in a complex pattern, back-filling the cavity formed when the tool is traversed along the weld line.

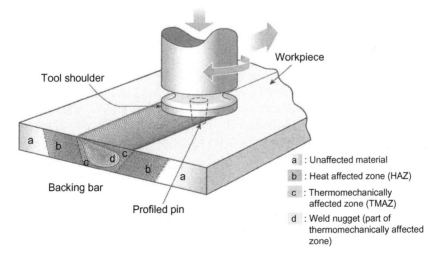

FIGURE 9.26   Schematic of the FSW process.

FSW is actually a combination of extrusion, forging, and stirring of the material and has been shown to result in strong and ductile joints. The maximum temperature reached is about 0.8 × the melting temperature. A detailed account of the FSW process has been provided by several researchers (Mishra and Ma, 2005; Nandan et al., 2008; Threadgill et al., 2009).

FSW is most suitable for simple joint geometries, notably butt joints, and needs special clamping systems. Originally, the components to be joined had to be flat, limiting FSW to plates and sheets. However, the attractiveness of the process has led to specialty jigs and frames to enable curved and annular welds to be made. Other possibilities are pipes and hollow sections.

FSW has been investigated for several Al—Li alloys, including 1420, AA 8090, and the third-generation alloys AA 2050, AA 2195, and AA 2198 (Arbegast and Hartley, 1999; Cavaliere et al., 2009; Hales et al., 2012; Lertora and Gambaro, 2010; Niedzinski and Thompson, 2010; Hatamlesh et al., 2008; Wei et al., 2007; Shukla and Baeslack, 2007; Windisch, 2009). Important aerospace applications of FSW are discussed in Chapter 15.

9.10 SUMMARY

Conventional fusion welding of Al—Li alloys can be done, but good welds with acceptable mechanical properties are difficult to achieve except for the specially developed Weldalite™ family of alloys and their successor, the third-generation Al—Li alloy AA 2195.

In this chapter we have concentrated on fusion welding of Al—Li alloys and the aspects and issues inherent to it. These aspects and issues are (i) obtaining sound, porosity-free welds, (ii) the occurrence and avoidance of several causes of weld cracking, and (iii) welding process modifications, including inoculation, PC, and MAO. The mechanical and corrosion properties of fusion-welded Al—Li alloys are also considered.

Lastly, solid-state friction welding and FSW are briefly discussed. These welding processes avoid several shortcomings of Al—Li alloy fusion welding, chiefly porosity, solidification cracking, and loss of lithium, which lowers the weld strengths. FSW, especially, has become an important technique to manufacture components and structures for aircraft and spacecraft.

ACKNOWLEDGMENTS

The authors profoundly thank Dr. RJH Wanhill for his numerous technical inputs and editorial corrections. They also thank DRDO, Ministry of Defence, Government of India, for the financial support to conduct most of the studies reported in this book.

REFERENCES

Aidun, K., Dean, J., 1999. Effect of enhanced convection on the microstructure of Al−Cu−Li welds. Weld. J. 78, 349s−354s.

Arbegast, W.J., Hartley, P., 1999. Friction stir weld technology development at Lockheed Martin, Michoud Space System, an overview. In: Vitek, J.M., David, S.A., Johnson, J.A., Smartt, H.B., DebRoy, T. (Eds.), Proceedings of the International Conference on Trends in Welding Research. ASM International, Materials Park, OH, pp. 541−546.

Bavarani, B., Becker, J., Parikh, S.N., Zamanazadeh, M., 1989. Localized corrosion of 2090 and 2091 Al−Li alloys. In: Sanders, T.H., Starke, E.A. (Eds.), Proceedings of the Fifth International Aluminium Lithium Conference, Williamsburg, Virginia, March 27−31, pp. 1227−1236.

Bittence, C., 1987. Welding the advanced alloys. Adv. Mater. Processes 12, 35−39.

Bowden, D.M., Meschter, P.J., 1984. Electron beam weld solidification structures and properties in Al−3Li−X alloys. Scr. Met. 18, 963−968.

Bratukhin, A.G., 1995. Weld. Res. Abroad 45, 29.

Brown, D.C., Crossley, F.A., Rudy, J.F., Schwartzbart, H., 1962. The effect of electromagnetic stirring and mechanical vibration on arc welds. Weld. J. 41, 241s−250s.

Cavaliere, P., Cabibbo, M., Panella, F., Squillace, A., 2009. 2198 Al−Li plates joined by friction stir welding: mechanical and microstructural behaviour. Mater. Des. 30, 3622−3631.

Cross, C.E., Olson, D.L., 1988. Compositional factors affecting weldability of extruded aluminium alloys. In: Proceedings of Fourth International Aluminium Extrusion Technology Seminar. April 11−14, Chicago, IL, pp. 391−393.

Cross, C.E., Capes, J.F., Olson, D.L., 1987. Characterization of binary aluminium alloy weld metal microstructures. Microstructural Sci. 14, 3−16.

Cross, C.E., Tack, T.K., Loechel, L.W., Kramer, L.S., 1990. Aluminum weldability and hot tearing theory. Proceedings of the Materials Weldability Symposium, Detroit, MI. ASM International, Materials Park, OH, pp. 275−282.

Cross, C.E., Grong, Ø., Mousavi, M., 1999. A model for equiaxed grain formation along the weld metal fusion line. Scr. Mater. 40, 1139−1144.

D'Annessa, A.T., 1967. Microstructural aspects of weld solidification. Weld. J. 46, 491s−499s.

Davies, G.J., Garland, J.G., 1975. Solidification structures and properties of fusion welds. Int. Metall. Rev. 20, 83−106.

Devletian, J.H., Wood, W.E., 1983. Factors affecting porosity in aluminium welds—a review. Welding Research Council Bulletin, The Welding Institute, UK, p. 290.

Dudas, J.H., Collins, F.R., 1966. Preventing weld cracks in high strength aluminium alloys. Weld. J. 45, 241s−249s.

Dvornak, M.J., Frost, R.H., Olson, D.L., 1989. The weldability and grain refinement of Al−2.2Li−2.7Cu. Weld. J. 68, 327s−335s.

Dvornak, M.J., Frost, R.H., Olson, D.L., 1991. Influence of solidification kinetics on aluminium weld grain refinement. Weld. J. 70, 271s−276s.

Edwards, M.R., Stoneham, V.E., 1987. The fusion welding of Al−Li−Cu−Mg (8090) alloy. In: Champier, G., Dubost, B., Miannay, D., Sabetay, L. (Eds.), Proceedings of the Fourth International Al−Li Conference. pp. C3 293−299.

Erokhin, A.A., Obotorov, V.O., 1971. Role of surface contaminant on formation of pores during welding. Weld. Prod. 18, 80−82.

Fedoseev, V.A., Ryazansev, V.I., Shiryaesa, N.V., Arbuzov, Yu.P., 1978. Weld. Prod. 6, 12−19.

Fridlyander, I.N., 1970. Properties of welded joints in 01420 alloy. Met. Oved Term Obrab Met. 4, 44−47.

Fridlyander, I.N., Bratukhin, A.G., Davydov, V.G., 1992. Russ. Metall. 3, 107.

Garland, J.G., 1974. Weld pool solidification control. British Weld. J. 22, 121−127.

Gittos, M.F., 1987. Gas shielded arc welding of the aluminium–lithium alloy 8090, Report No. 7944.01/87/556.2. Welding Institute, Abington, England.

Gokhale, A.A., Tzavaras, A.A., Brody, H.D., Ecer, G.M., 1982. Grain structure and hot cracking in pulsed current GTAW of AISI 321 stainless steel. Proceedings of the Conference Grain Refinement in Casting and Welds. The Metallurgical Society of AIME, St. Louis, MO, October 25−26, 1982, pp. 223−247.

Grill, A., 1981. Effect of arc oscillations on the temperature distribution and microstructure in GTAW tantalum welds. Metall. Trans. B 12B, 667−674.

Gutierrez, A., Lippold, J.C., 1998. A proposed mechanism for equiaxed grain formation along the fusion boundary in aluminium-copper-lithium alloys. Weld. J. 77, 123s−132s.

Hales, S.J., Tayon, W.A., Domack, M.S., 2012. Friction-stir-welded and spin formed end domes for cryogenic tanks, NASA Report NF1676L-13613. NASA Langley Research Center, Hampton, VA.

Hatamlesh, O., Rivero, I.V., Maredia, A., 2008. Residual stresses in friction stir-welded 2195 and 7075 aluminum alloys. Metall. Mater. Trans. A 39A, 2867−2874.

He, Y., Gao, D., Wu, L., Meng, L., 1992. Cracking behaviour during solidification of aluminium–lithium alloy 2091 welds. In: Proceedings of the Third International Conference on Aluminium alloys. Beijing, China, pp. 385−390.

Hoeppner, D.W., 2004. A review of corrosion fatigue and corrosion/fatigue considerations in aircraft structural design. In: Guillaume, M. (Ed.), ICAF 2003, Fatigue of Aeronautical Structures as an Engineering Challenge. Engineering Materials Advisory Services Publishing, Sheffield, UK, pp. 425−438.

Ilyushenko, R.V., Lozovskaya, A.V., Sklabinskaya, I.E., Paton, E.O., 1991. In: Peters, M., Winkler, P.J. (Eds.), Proceedings of the Sixth International Aluminium–Lithium Conference, Garmisch-Partenkirchen, Germany, vol. 2. DGM Informationsgesellschaft, Oberursel, pp. 381−386.

Ischenko, A.Y.A., Chayun, A.G., 1977. Special features of the fusion welding of the high-strength 01420 aluminium alloy. Autom. Weld. 30, 31−34.

Janaki Ram, G.D., Mitra, T.K., Raju, M.K., Sundaresan, S., 2000. Use of inoculants to refine weld solidification structure and improve weldability in type 2090 Al–Li alloy. Mater. Sci. Eng. A A 276, 48−57.

Kerr, H.W., Villafuerte, J.C., 1991. Columnar to equiaxed transitions in fusion welds. In: Proceeding of a Symposium, The Metal Science and Joining, Cincinnati, OH, October 20−24, pp. 11−20.

Kerr, J.R., Merino, R.F., 1991. Cryogenic properties of VPPA welded aluminium–lithium alloys. In: Sanders T.H., Jr., Starke E.A. Jr., (Eds.), Proceedings of the Fifth International Conference on Aluminium–Lithium Alloys, Williamsburg, Virginia, March 27−31, 1989, pp. 1491−1500.

Kostrivas, A., Lippold, J.C., 1999. Weldability of Li bearing aluminium alloys. Int. Mater. Rev. 44, 217−237.

Kou, S., 1987. Welding Metallurgy. John Wiley and Sons, New York.

Kramer, L.S., Heubaum, F.H., Pickens, J.R., 1989. The weldability of high strength Al–Cu–Li alloys. In: Sanders T.H., Jr., Starke E.A. Jr., (Eds.), Proceedings of the Fifth International Al–Li Conference, Williamsburg, Virginia, March 27−31, pp. 1415−1424.

Kumai, C., Kusinski, J., Thomas, G., Devine, T.M., 1989. Influence of aging at 200°C on the corrosion resistance of Al–Li and Al–Li–Cu alloys. Corrosion 45, 294−303.

Lee, M.F., Huang, J.C., Ho, N.J., 1996. Microstructural and mechanical characterization of laser beam welding of a 8090 Al—Li thin sheet. J. Mater. Sci. 31, 1455—1468.

Lertora, E., Gambaro, C., 2010. AA8090 Al—Li alloy FSW parameters to minimize defects and increase fatigue life. Int. J. Mater. Forum 3, 1003—1006.

Lippold, J.C., Lin, W., 1996. Weldability of commercial Al—Cu—Li alloys. Mater. Sci. Forum 217—222, 1685.

Maalekian, M., 2007. Friction welding—critical assessment of literature (review). Sci. Technol. Weld. Joining 12, 738—759.

Madhusudhan Reddy, G., 1998. Studies on the Application of Pulsed Current and Arc Oscillation Techniques on Aluminium—Lithium Alloy Welds, PhD Thesis. Department of Metallurgical Engineering, Indian Institute of Technology Madras, Chennai.

Madhusudhan Reddy, G., Gokhale, A.A., 1993. Gas tungsten arc welding of 8090 Al—Li alloy. Trans. Indian Inst. Met. 46, 21—30.

Madhusudhan Reddy, G., Gokhale, A.A., 1994. "Pulsed current gas tungsten arc welding of Al—Li alloy 1441", aluminium alloys: their physical and mechanical properties. In: Sanders, T.H., Starke, E.A. (Eds.), Proceedings of the Fourth International Conference, vol. 2. Georgia Institute of Technology, Atlanta, GA, pp. 496—503.

Madhusudhan Reddy, G., Gokhale, A.A., 2007. On weld solidification cracking in aluminum—lithium alloys. Met. Mater. Processes 19, 297—306.

Madhusudhan Reddy, G., Gokhale, A.A., Prasad Rao, K., 1997. Weld microstructure refinement in a 1441 grade Al—Li alloy. J. Mater. Sci. 32, 4117—4126.

Madhusudhan Reddy, G., Gokhale, A.A., Prasad Rao, K., 1998a. Porosity and hot cracking in aeronautical grade aluminum—lithium alloy 1441 welds. In: Sastry, D.H., Subramanian, S., Murthy, K.S.S., Abraham, K.P. (Eds.), Proceedings of the international Conference on Aluminum, INCAL 98; February 11—13, New Delhi, vol. 2. The Aluminum Association of India.

Madhusudhan Reddy, G., Gokhale, A.A., Prasad Rao, K., 1998b. Effect of filler metal composition on Weldability of Al—Li alloy 1441. Sci. Technol. Weld. Joining 3, 151.

Madhusudhan Reddy, G., Gokhale, A.A., Prasad, K.S., Prasad Rao, K., 1998c. Chill zone formation in Al—Li alloy welds. Sci. Technol. Weld. Joining 3, 208—212.

Madhusudhan Reddy, G., Gokhale, A.A., Prasad Rao, K., 1998d. Optimisation of pulse frequency in pulsed current gas tungsten arc welding of aluminium—lithium alloy weld. Mater. Sci. Technol. 14, 61—68.

Madhusudhan Reddy, G., Gokhale, A.A., Prasad Rao, K., 1999. Weldability aspects of an aluminium—lithium alloy. In: Proceedings of the International Welding Conference (IWC'99) on Welding & allied Technology "Challenges in 21st Century". pp. 659—671.

Madhusudhan Reddy, G., Gokhale, A.A., Narendra Janaki Ram, N., Prasad Rao, K., 2001. Influence of welding techniques on microstructure and pitting corrosion behaviour of 1441 grade Al—Li alloy gas tungsten arc welds. Br. Corr. J. 36, 304—309.

Madhusudhan Reddy, G., Gokhale, A.A., Prasad Rao, K., 2002. Effect of the ratio of peak and background current durations on the fusion zone microstructure of pulsed current gas tungsten arc welded Al—Li alloy. J. Mater. Sci. 21, 1623—1625.

Madhusudhan Reddy, G., Mohandas, T., Sobhana Chalam, P., 2003. Metallurgical and mechanical properties of AA 8090 Al—Li alloy friction welds. In: International Welding Symposium (IWS 2K3) on Emerging Trends in Welding Organized by Indian Welding Society. 22—23, February 2003, Hyderabad, India, pp. 147—158.

Madhusudhan Reddy, G., Srinivasa Murthy, C.V., Viswanathan, N., Prasad Rao, K., 2007. Effects of electron beam oscillation techniques on solidification behaviour and stress rupture properties of Inconel 718 welds. Sci. Technol. Weld. Joining 12, 106—114.

Martukanitz, R.P., Michnuk, P.R., 1982. Sources of porosity in gas metal arc welding of aluminium. Aluminium 5, 276—279.

Martukanitz, R.P., Natalie, C.A., Knoefel, J.O., 1987. The weldability of an Al—Li—Cu alloy. J. Met. 39, 38—42.

Matsuda, F., 1990. Hot cracking susceptibility of weld metal. In: Proceedings of the First United States—Japan Symposium on Advances in Welding Metallurgy. June 7—8, San Francisco, California, pp. 19—35.

Metzer, G.F., 1967. Some mechanical properties of welds in 6061 aluminium alloy sheet. Weld. J. 46, 457s—465s.

Mironenko, V.N., Evstifeev, V.S., Lubene, G.I., Karshukova, S.A., Zakhervov, V.V., Litvintsev, A.I., 1979a. The effect of vacuum heat treatment on the weldability of aluminium alloy 01420. Weld. Prod. 26, 30—31.

Mironenko, V.N., Kolgarova, I.F., 1979b. Conditions for vacuum heat treatment of the 01420 alloy before welding. Automat. Weld. 8, 44—48.

Mishra, R.S., Ma, Z.Y., 2005. Friction stir welding and processing. Mater. Sci. Eng. R 50, 78.

Mohandas, T., Madhusudhan Reddy, G., 1997. A comparison of continuous and pulsed current gas tungsten arc welds of an ultra high strength low alloy steel. J. Mater. Process. Technol. 9, 122—123.

Mondolfo, L.F., 1979. "Aluminium Alloys", Structure and Properties. Butterworth, Boston, Massachusetts.

Namba, K., Sano, H., 1986. Fusion Weldabilities of Al-4.7Mg-0.3 ~ 1.3Li alloys for fusion reactor. J. Light Met. Weld. Const. 24, 243—250.

Nandan, R., DebRoy, T., Bhadeshia, H.K.D.H., 2008. Recent advances in friction-stir welding—process, weldment structure and properties. Progress Mater. Sci. 53, 980—1023.

Niedzinski, M., Thompson, C., 2010. Airware 2198 backbone of the Falcon family of SpaceX launchers. Light Met. Age 68, 6—7, 55.

Ol'Shanskii, A.N., Dyachencho, V.V., 1977. Evaluation of the susceptibility of alloys to the formation of pores in welding. Weld. Prod. 24, 52—53.

Padmanabham, G., Pandey, S., Schaper, M., 2005. Pulsed gas metal arc welding of Al—Cu—Li alloy. Sci. Technol. Weld. Joining 10, 67—75.

Pickens, J.R., 1985. The weldability of lithium containing aluminium alloys. J. Mater. Sci. 20, 4247—4258.

Pickens, J.R., 1990. Recent developments in the weldability of lithium-containing aluminium alloys. J. Mater. Sci. 25, 3035—3047.

Pickens, J.R., Heubaum, F.H., Langan, T.J., Kramer, L.S., 1989. Al-(4.5—6.3) Cu-1.3 Li-0.4 Ag-0.4Mg-0.14 Zr alloy Weldalite 049. In: Sanders T.H., Jr., Starke E.A. Jr., (Eds.), "Aluminum—Lithium Alloys", Proceedings of the Fifth International Aluminum—Lithium Conference, vol. 3. Materials and Component Engineering Publications, Birmingham, UK, pp. 1397—1411.

Ramulu, M., Rubbert, M.P., 1990. Gas tungsten arc welding of Al—Li—Cu alloy 2090. Weld. J. 69, 100s—114s.

Ransley, C., Neufeld, J., 1948. The solubility of hydrogen in liquid and solid aluminium. J. Inst. Met. 74, 599—602.

Ravi Vishnu, P., 1995. Modelling of microstructural changes in the pulsed weldments. Weld. World 4, 214—222.

Rioja, R.J., Liu, J., 2012. The evolution of Al—Li base products for aerospace and space applications. Metall. Trans. A 43A, 3325—3337.

Rioja, R.J., Cho, A., Bretz, P.E., 1990. Al—Li alloys having improved corrosion resistance containing Mg and Zn. US Patent No 4,961,792.

Savage, W.F., Aronson, A.H., 1965. Preferred orientation in the weld fusion zone. Weld. J., 85s—87s.

Schillinger, D.E., Betz, I.G., Hussey, F.W., Markus, H., 1963. Improved weld strengths in 2000 series aluminium alloys. Weld. J. 42, 269s—275s.

Shah, S.R., Wittig, J.E., Hahn, G.T., 1992. Microstructural analysis of a high strength Al—Cu—Li (Weldalite049) alloy weld. In: Proceedings of the Third International Conference on Trends in Welding Research. June 1—5, Gatlinburg, TN, pp. 281—285.

Sheppard, T., Parson, N.C., 1987. Corrosion resistance of Al—Li alloys. Mater. Sci. Technol. 3, 345—352.

Shinoda, T., Ueno, Y., Masumoto, I., 1990. Effect of pulsed welding current on solidification cracking in austenitic stainless steel welds. Trans. JWRI 21, 18—23.

Shukla, A.K., Baeslack, W.A., 2007. Study of microstructural evolution in friction-stir welded thin-sheet Al-Cu-Li alloy using transmission-electron microscopy. Scr. Mater. 56, 513—516.

Skillingberg, M.H., 1986. Fusion welding of Al—Li—Cu—(Mg)—Zr plate. In: Proceedings of the Conference on Aluminium Technology, London, pp. 509—515.

Soni, K.K., Levi-Setti, R., Shah, S., Gentz, S., 1996. SIMS imaging of Al—Li alloy welds. Adv. Mater. Processes 4, 35—36.

Srinivasa Rao, K., Prasad Rao, K., 2004. Pitting corrosion of heat treatable aluminium alloys and welds—a review. Trans. Indian Inst. Met. 57, 593—610.

Sriram, S., Babu Vishwanathan, G., Prasad, K.S., Gokhale, A.A., Banerjee, D., Sivakumar, R., 1993. Solidification microstructures in an Al—Li—Cu—Mg—Zr alloy at different cooling rates. In: Banerjee, D., Jacobson, L.A. (Eds.), Metastable Microstructures. Oxford & IBH, New Delhi, pp. 103—100.

Srivatstan, T.S., Sudarshan, S., 1991. Welding of lightweight aluminum—lithium alloys. Weld. J. 70, 173s.

Sundaresan, S., Janaki Ram, G.D., Murugesan, R., Viswanathan, N., 2000. Combined effect of inoculation and magnetic arc oscillation on microstructure and tensile behaviour of type 2090 Al—Li alloy weld fusion zones. Sci. Technol. Weld. Joining 5, 257—264.

Sunwoo, A.J., Morris, J.W., 1989. Microstructure and properties of aluminium alloy 2090 weldments. Weld. J. 68, 262s—268s.

Threadgill, P.L., Leonard, A.J., Shercliff, H.R., Withers, P.J., 2009. Friction stir welding of aluminium alloys. Int. Mater. Rev. 54, 49—93.

Tseng, C., Savage, W.F., 1971. The effect of arc oscillation. Weld. J. 50, 777—786.

Wei, S., Hao, C., Chen, J., 2007. Study of friction stir welding of 01420 aluminum—lithium alloy. Mater. Sci. Eng. A 452—453, 170—177.

Windisch, M., 2009. Damage tolerance of cryogenic pressure vessels. European Space Agency ESA Technology and Research Programme Report ESA TRP DTA-TN-A250041-0004-MT, European Space Agency ESA/ESTEC, Noordwijk, The Netherlands.

Yamamoto, H., Kamiyama, Ogawo, 1992. Development of low frequency pulsed MIG welding process for aluminium and aluminium alloys. J. Light Weld. Constr. 10, 25—30.

Yunjia, H., Dalu, G., Zhixiong, 1991. In: Peters, M., Winkler, P.J. (Eds.), Proceedings of the Sixth International Aluminium—Lithium Conference Garmisch-Partenkirchen, Germany, vol. 2. DGM Informationsgesellschaft, Oberursel, pp. 1215—1220.

Yusheng, C., Ziyong, Z., Sue, L., Wei, K., 1996. The corrosion behaviour and mechanisms of 1420 Al—Li alloy. Scr. Mater. 34, 781—786.

Zacharia, T., David, S.A., Vitek, J.M., Martukanitz, R.P., 1989. Weldability and microstructural characterization of Al–Li alloys. In: Sanders, T.H., Starke, E.A. (Eds.), Proceedings of the Fifth International Conference on Aluminium–Lithium Alloys, vol. 3. Materials and Component Engineering Publications, Birmingham, UK, pp. 1387–139.

FURTHER READING

Gutierrez, A., Lippold, J.C., Lin, W., 1998. Nondendritic equiaxed zone formation in aluminium–lithium welds. Mater. Sci. Forum 217_212, 1891_1893.

Cross, C.E., Capes, J.F., Olson, D.L., 1987. Characterization of binary aluminium alloy weld metal microstructures. Microstructural Sci 14, 3–16.

Mechanical Behavior

Part IV

Mechanical Behavior

Quasi-Static Strength, Deformation, and Fracture Behavior of Aluminum—Lithium Alloys

T.S. Srivatsan*, Enrique J. Lavernia[†], N. Eswara Prasad**, and V.V. Kutumbarao[‡]

*Professor, Division of Materials Science and Engineering, Department of Mechanical Engineering, The University of Akron, Akron, Ohio 44325-3903, USA, [†]Distinguished Professor, Chemical Engineering and Materials Science, University of California Davis, Davis, CA 95616, USA, **Scientist 'G' & Regional Director, Regional Center for Military Airworthiness, CEMILAC, P.O. Box Kanchanbagh, Hyderabad 500 058, India, [‡]Visiting Scientist, Defence Metallurgical Research, Laboratory, Hyderabad 500 058, India

Contents

10.1 Introduction	306	Slip and Strain Localization 314
10.2 Mechanisms of Strengthening	307	10.3.3 Thermal Treatments for Improving Strength and Fracture Toughness 315
10.2.1 Strengthening by δ' Precipitates in Al–Li Alloys	307	
10.2.2 Strengthening by Other Phases	310	10.4 Anisotropy of Mechanical Properties 318
10.3 Ductility and Fracture Toughness	312	10.5 Tensile Properties of Selected Aluminum–Lithium Alloys 318
10.3.1 The Nature and Occurrence of Planar Slip Deformation	313	10.5.1 AA 2020: A First-Generation Al–Li Alloy 319
10.3.2 Methods for Reducing Planar		10.5.2 AA 8090: A Second-Generation Al–Li Alloy 322

Aluminum—Lithium Alloys.

10.5.3 AA 2198: A Third-   10.6 Summary and
       Generation Al–Li   Conclusions   331
       Alloy   330   Acknowledgments   334
                                References   334

10.1 INTRODUCTION

Aluminum–lithium (Al–Li) alloys represent a widely studied class of light-weight materials intended for aerospace structural applications. The history and progress in the development of Al–Li alloys since their emergence way back in 1924, when lithium was first added in small amounts to an Al–Zn–Cu alloy, have been described and reviewed several times (Balmuth and Chellman, 1987; Balmuth, 1994; Quist et al., 1984; Quist and Narayanan, 1989). Another authoritative review is given in Chapter 1.

Interest in Al–Li alloys stems from the fact that lithium not only reduces the alloy density but also contributes to increases in strength and elastic modulus. These synergistic property combinations lead to significant increases in specific strength (strength/density) and specific stiffness (E/density) (Peel et al., 1983; Sanders and Starke, 1989). These benefits offer the potential of significant weight reductions for aerospace structures, and this property improvement aspect is discussed in detail in Chapter 2 and also in Chapter 15.

The first alloy to take advantage of the synergistic effects of lithium additions was the Al–4.45 Cu–1.21 Li–0.51 Mn–0.2Cd alloy, designated AA 2020 by the Aluminium Association (LeBaron, 1945). Besides possessing high tensile strength (σ_{UTS}), high yield strength (σ_{YS}), low density, and improved elastic modulus, alloy 2020 also offered freedom from exfoliation corrosion and stress corrosion cracking (SCC). This made 2020 potentially superior to the other competing commercially available aluminum alloys for selection and use in high performance military aircraft structures.

Alloy 2020, which is classified as a first-generation Al–Li alloy, entered production in 1958 and was used for the wing skins and empennage of the Northrop RA-5C Vigilante aircraft (Quist et al., 1984). However, later concerns about the alloy's toughness led to its being withdrawn from production. In the mid-1960s, the Soviet Union developed the much lighter but relatively low strength 142X-type alloys, which contained no copper and were weldable. These too are classified as first-generation Al–Li alloys.

Beginning in the mid-1960s, it was projected that improved, newer, and emerging lithium-containing alloys, combined with innovative design approaches, could result in weight savings of the order of 10% (Divecha and Karmarkar, 1981; Grimes et al., 1985; Lewis et al., 1987; Peel et al., 1984; Sanders and Starke, 1984, 1989). This prospect led to the development of the second-generation alloys. These were developed primarily in the United

States and Europe, beginning in the 1970s and continuing through the 1980s, and also in the Soviet Union in the 1980s and 1990s. The aim was to obtain alloys 8−10% lighter (and stiffer) than equivalent conventional alloys by the addition of about 2 wt% lithium.

It was subsequently found that lithium contents of 2 wt% or more are linked to several disadvantages, including a tendency for strongly anisotropic mechanical properties, low short-transverse ductility and fracture toughness, and loss of toughness owing to thermal instability (Chen and Starke, Jr., E.A., 1984; Eswara Prasad et al., 2003; Lynch et al., 2003; Rioja and Liu, 2012). These adverse links between lithium content and properties are why the second-generation alloys have found only "niche" applications, and why third-generation alloys have been developed with reduced lithium contents (and higher densities).

The third-generation alloys have been developed mainly in the United States. Developments started in the late 1980s and early 1990s and have continued through the first decade of this century. These alloys have generally higher densities than the second-generation alloys, but they still offer up to 8% density reductions compared to non-lithium alloys. Also, they have much better combinations of engineering properties owing to improved compositions, thermomechanical processing, and heat treatments (Rioja and Liu, 2012). These latest Al−Li alloys offer replacements for many conventional aluminum alloys belonging to the AA 2XXX and AA 7XXX series, as indicated in Table 10.1.

In this chapter, we shall present and discuss some of the key aspects relevant to the tensile properties of aluminum−lithium alloys, spanning the domains of intrinsic microstructural features, an overview of the fundamental mechanisms contributing to strength, ductility, and fracture toughness, anisotropy in the mechanical properties, and tensile behavior of the first-, second-, and third-generations of Al−Li alloys.

10.2 MECHANISMS OF STRENGTHENING

10.2.1 Strengthening by δ' Precipitates in Al−Li Alloys

The solution of lithium atoms in aluminum produces only a small degree of solid solution strengthening. In comparison with the copper-containing aluminum alloys, the solid solution strengthening by lithium is caused principally by differences in size and/or differences in elastic modulus between atoms of the solute and the solvent. The general strength in Al−Li alloys is derived from the presence of a large volume fraction of the coherent δ' (Al$_3$Li) phase. The δ' (Al$_3$Li) phase has a high intrinsic modulus due to its ordered nature, and this contributes to the high value of elastic modulus observed in these alloys. When the element lithium is in solid solution, the elastic constants depend on both atomic interactions and interatomic potential. However, when lithium is present as a precipitated second phase, the elastic constants depend on both the volume fraction and the intrinsic modulus of the second phase (Sankaran and Grant,

TABLE 10.1 Proposed and Actual Uses of Third-Generation Al–Li Alloys to Replace Conventional Alloys

Product	Alloy/Temper	Substitute for	Applications
Sheet	2198-T8, 2199-T8E74, 2060-T8E30: damage tolerant/ medium strength	2024-T3, 2524-T3/351	Fuselage/pressure cabin skins
Plate	2199-T86, 2060-T8E86: damage tolerant	2024-T351, 2324-T39, 2624-T351, 2624-T39	Lower wing covers
	2098-T82P (sheet/ plate): medium strength	2024-T62	F-16 fuselage panels
	2297-T87, 2397-T87: medium strength	2124-T851	F-16 fuselage bulkheads
	2099-T86, medium strength	7050-T7451, 7×75-T7XXX	Internal fuselage structures
	2055-T8X, 2195-T82: high strength	7150-T7751, 7055-T7751, 7055-T7951, 7255-T7951	Upper wing covers
	2050-T84: medium strength	7050-T7451	Spars, ribs, other internal structures
	2195-T82: high strength	2219-T87	Launch vehicle cryogenic tanks
Forgings	2050-T852, 2060-T8E50: high strength	7175-T7351, 7050-T7452	Wing/fuselage attachments
Extrusions	2099-T81, 2076-T8511: damage tolerant	2024-T3511, 2026-T3511, 2024-T4312, 6110-T6511	Lower wing stringers, fuselage/pressure cabin stringers
	2099-T83, 2099-T81, 2196-T8511, 2055-T8E83: medium/high strength	7075-T73511, 7075-T79511 7150-T6511, 7175-T79511, 7055-T77511, 7055-T79511	Fuselage/pressure cabin stringers and frames, upper wing stringers, Airbus A380 floor beams and seat rails

See also Table 2.5.

1980). The effect of lithium additions to aluminum is unique because lithium substantially increases the elastic constants of the aluminum–lithium solid solution even though the values of its own constants are noticeably lower than those of aluminum (Webster, 1986, 1989; Huang and Ardell, 1988; Sankaran and Grant, 1980).

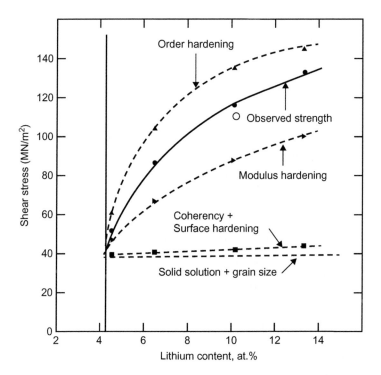

FIGURE 10.1   Contributions of various mechanisms to the strengthening by δ' precipitates in Al–Li alloys (Noble et al., 1982).

Several mechanisms contribute to the strengthening by δ' precipitates in Al–Li alloys. Figure 10.1 summarizes the contributions of these mechanisms to the overall strength in terms of the shear stress for slip to occur. Order hardening and modulus hardening contribute the most, while coherency and surface hardening contribute relatively less. Order hardening makes a major contribution to strength owing to the creation of antiphase boundaries (APBs) (Vasudevan and Doherty, 1987). In order to eliminate the extra energy required to create the APB, the dislocations in Al–Li alloys move in pairs connected by a region of APB such that passage of the second dislocation restores the disorder caused by the first (Noble et al., 1982a,b). The critical resolved shear stress (τ_{CRSS}) for such a process was found to be (Palmer et al., 1986)

$$\tau_{CRSS} \propto (\gamma_{APB})^{3/2} \cdot r^{1/2} \cdot f^{1/2} \tag{10.1}$$

In this expression, γ_{APB} is the APB energy of the δ' (Al$_3$Li) particles, r is the mean radius, and f is the volume fraction of the precipitate particles. Once sheared, the ordered precipitate particles would result in reduced contributions from order strengthening. This is essentially due to a reduction in cross-sectional area of the precipitate particles upon initial shearing. If n_d

dislocations, each having a Burger's vector b_v, shear a given particle, and if we assume shearing to take place across the diameter of the precipitate particle, then τ_{CRSS} for continued shearing becomes

$$\tau_{CRSS} \propto (\gamma_{APB})^{3/2} \cdot f^{1/2}[(r - n_d b_v)^{1/2}] \tag{10.2}$$

Thus, a reduction in the critical resolved shear stress (τ_{CRSS}) becomes significant, making further slip on that particular plane conducive. Hence, slip is favored to become planar, and the particular plane on which repeated slip occurs gradually becomes work-softened. Al–Li alloys artificially aged to the peak strength condition tend to exhibit such planar slip deformation behavior (Gregson and Flower, 1984; Sanders and Starke, 1982; Srivatsan et al., 1986a,b). Besides order/APB strengthening, the contributions to modulus hardening were also found to be significant for Al–Li alloys (Noble et al., 1982a,b) and can be estimated as (Dieter, 1986)

$$\Delta\sigma = \frac{\Delta G}{2\pi^2} \left| \frac{3I\Delta GI}{G_m b_v} \right| \cdot \left| 0.8 - 0.143 \ln \left| \frac{r}{b_v} \right| \right|^{3/2} r^{1/2} f^{1/2} \tag{10.3}$$

where $.\Delta G$ is the difference in the shear modulus values of the matrix (G_m) and the precipitate particles.

Overall, the relative degree of strengthening obtained from the different mechanisms was found to vary with both chemical composition and aging condition of the chosen alloy. During the early stages of age hardening, i.e., for alloys in the underaged (UA) condition, the key contributions to strengthening were found to be due to a synergism of the following: (i) coherency strain hardening, (ii) modulus hardening, and (iii) hardening from interfacial energy (Starke et al., 1981). Upon continued aging to peak strength or the peak-aged condition, the contributions from both order/APB strengthening and modulus hardening become dominant (Noble et al., 1982a,b). On the other hand, contribution to strengthening from both the solid solution and the grain size during the initial and subsequent aging stages was found to be only marginal (Figure 10.1).

10.2.2 Strengthening by Other Phases

The challenge of strengthening aluminum–lithium alloys with coherent lithium-rich phases, such as Al_3Li, which does not increase the density, has been met with only limited success. However, additional strengthening has been achieved by co-precipitation of other binary phases and ternary phases. The addition of various amounts of copper and magnesium to aluminum–lithium alloys has been shown to be effective in strengthening. These elements tend to modify the precipitation sequence either by altering the solubility of the principal alloying elements or by forming copper-rich and magnesium-rich phases and co-precipitating with the Al_3Li (δ'). Copper and magnesium also combine with lithium to precipitate as phases that exist in the ternary (Al$-$Li$-$Cu) and quaternary (Al$-$Li$-$Cu$-$Mg) alloy systems.

Ternary Al$-$Li$-$Cu Alloys

In ternary Al$-$Li$-$Cu alloys, additional strengthening is achieved by co-precipitation of copper-rich phases independently of δ' precipitation (Silcock, 1959$-$1960; Hardy and Silcock, 1955$-$1956). Six ternary compounds have been identified to be present in aluminum-rich alloys (Hardy and Silcock, 1955$-$1956; Silcock 1959$-$1960). Of these six compounds, the most important are (i) T_1 (Al_2CuLi), (ii) T_2 (Al_6CuLi_3), and (iii) T_B ($Al_{15}Cu_8Li_3$). In the leaner aluminum alloys, three other intermetallic compounds designated as P, Q, and R (Al_5CuLi_3) tend to be present. Studies of both the composition and structures of the equilibrium and nonequilibrium phases present in the ternary Al$-$Li$-$Cu system have been made by Noble and Thompson (1972) and Kang and Grant (1984). The phases present in several Al$-$Li alloys (including the ones that are mentioned here) and their details are given in Table 3.2, Section 3.5.

A number of interactions occur between the T_1 and the δ' (Al_3Li) precipitates, with T_1 either cutting or growing through the spherical δ' and the δ' on the θ' (Al_2Cu) (Tosten et al., 1986). In alloys that contain trace amounts of zirconium, the θ' and T_1 phases were found to nucleate on the Al_3Zr interface in addition to heterogeneous nucleation on the matrix dislocations. Although the nucleation of the T_1 precipitates occurred to a lesser degree than the θ', the T_1 phase was found to have a beneficial influence on the elastic modulus (Agyekum et al., 1986). The presence of T_1 precipitates is also beneficial to strength, primarily because they act as unshearable obstacles, which must be bypassed by dislocations during deformation (Tosten et al., 1986).

The types of strengthening phases that precipitate from the supersaturated solid solution strongly depend on the Cu:Li ratio (Noble and Thompson, 1972; Noble et al., 1970). The dependence of the precipitation sequence on the Cu:Li ratio has been discussed by Rinker et al. (1984). For high lithium and low copper-containing alloys (>2 wt% lithium, 2 wt% copper), the reaction sequence leading to the formation of the Al_2Cu precipitate is suppressed and precipitation of the T_1 (Al_2CuLi) phase occurs. The morphology of precipitates likely to be present in Al$-$Li$-$Cu alloys is shown in Figure 10.2.

Quaternary Al$-$Li$-$Cu$-$Mg Alloys

Both copper and magnesium tend to improve the overall strength of the binary alloy by coprecipitating with the δ' (Al_3Li) phase and/or incorporating lithium to form coherent and partially coherent ternary and more complex strengthening precipitates. Furthermore, magnesium additions result in precipitation of both S'' and S' phases near grain boundaries (GBs), thereby minimizing or even eliminating the formation of precipitate-free zones (PFZs) (Gregson et al., 1985, 1987). This is important because a study by Vasudevan and

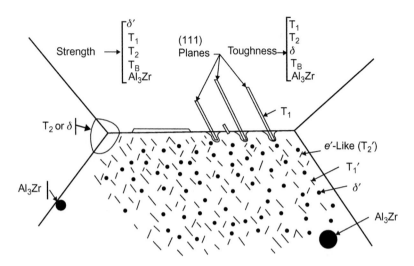

FIGURE 10.2 Schematic of the morphology of precipitates present in Al–Li–Cu alloys. The T_1' and T_2' precipitates are plate-like along the (111) and (100) matrix planes, respectively. *Source: Taken from Rioja and Liu (2012).*

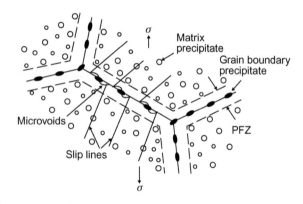

FIGURE 10.3 Schematic representation of void nucleation at grain boundary particles when PFZs are present (Vasudevan and Doherty, 1987).

Doherty (1987) showed that a combination of PFZs and coarse GB precipitates allows localized slip to produce stress concentrations that nucleate voids at the GB precipitates (Figure 10.3). These voids contribute to early failure and lower strength of the alloy.

10.3 DUCTILITY AND FRACTURE TOUGHNESS

Key microstructural features that tend to exert a significant *negative* influence on tensile ductility in Al–Li alloys have a similar influence on the overall fracture resistance and toughness of these alloys. These microstructural features and the associated deformation/fracture processes are described as follows:

1. Coplanar deformation and slip localization due to shearing of the coherent, ordered δ' (Al$_3$Li) precipitates, resulting in premature failure at the GBs (Gregson and Flower, 1985; Noble et al., 1982a,b; Starke et al., 1981).
2. Strain localization at the soft δ' PFZs that are formed during artificial aging (Lin et al., 1982).
3. Coarse equilibrium precipitates (T$_1$, T$_2$, and δ), and the coarse iron-rich and silicon-rich intermetallics along the GBs (Cassada et al., 1986; Suresh et al., 1987).
4. Segregation of alkali elements (Na and K) to GBs and formation of thin films of low melting eutectic phases along the GBs (Webster, 1987, 1988).
5. A higher hydrogen content that favors GB embrittlement (Hill et al., 1984).
6. Relatively easy crack propagation along both the grain and subgrain boundaries, particularly in unrecrystallized alloys (Starke and Lin, 1982; Srivatsan and Place, 1989).

We shall concentrate on items (1) and (2) in Sections 10.3.1−10.3.3. These items and also items (3) and (4) are also discussed in Chapter 13, Section 13.3.

10.3.1 The Nature and Occurrence of Planar Slip Deformation

Shearing of the strengthening precipitates results in pileups of dislocations at and along the GBs and grain boundary triple junctions (GBTJs). Increased grain size or an increase in the number of particles that are easily sheared increases the number of dislocations that pile up against a GB. This results in greater effective slip lengths and higher "local" stress concentrations at the GBs and GBTJs, as shown schematically in Figure 10.4A. Microvoid/crack nucleation is favored to occur at the slip band−GB intersections, and coalescence of these nucleation sites results in intergranular fracture (Lin et al., 1982; Srivatsan and Lavernia, 1991; Starke et al., 1981; Figure 10.4B).

Such planar slip deformation has also been observed in other precipitation-hardened high strength aluminum alloys (Jata and Starke, 1986). However, for Al−Li alloys, the effect is especially severe because strain localization at the GBs and GBTJs can be enhanced by the presence of δ'-PFZs. This enhanced strain localization favors substantial "localized" deformation that occurs before macroscopic deformation. When this highly localized deformation is combined with the "local" stress concentrations and associated microvoid nucleation at coarse and intermediate size intermetallic particles (which are dispersed along the GBs and also throughout the micro-structure), the result is inferior ductility and poor fracture toughness (Srivatsan et al., 1986a,b).

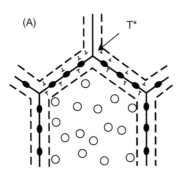
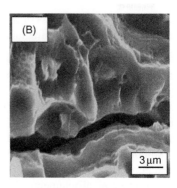

FIGURE 10.4 (A) Schematic showing PFZ at grain boundary and buildup of a stress concentration at a grain boundary triple junction (T*). (B) Scanning electron micrograph showing intergranular fracture and a population of microvoids adjacent to the grain boundary crack, obtained from AA 8090 alloy. *Source: From Srivatsan and Place (1989).*

10.3.2 Methods for Reducing Planar Slip and Strain Localization

Strain localization within the matrix can be minimized by changing the deformation mode from one of dislocation shearing of the strengthening precipitates to dislocation looping or bypassing the precipitates. This is readily accomplished (as in the case of other age-hardenable aluminum alloys) by overaging (OA). However, for lithium-containing aluminum alloys, there is a complication owing to the small coherency strains associated with the metastable δ' (Al$_3$Li) precipitates. These tend to grow in size prior to becoming noncoherent. This results in significant growth of the PFZs and a loss of both tensile ductility and fracture toughness. Hence, the use of OA as a means to induce and/or promote slip homogenization is not an attractive solution that can be easily adopted. There are three alternative possibilities: (i) decreasing the grain size, (ii) controlling the degree of recrystallization (Jata and Starke, 1986; Starke and Lin, 1982), and (iii) additions of copper and magnesium to result in nonshearable precipitates (Gregson and Flower, 1985).

Possibilities (i) and (ii) rely on decreasing the slip length and therefore the local stress concentrations caused by dislocation pileups. Both make use of the addition of grain-refining elements, which not only minimize grain growth while favoring a lower grain size but also reduce the degree of recrystallization and have an influence on slip dispersal. Using this approach, a significant improvement in both tensile ductility and resultant fracture resistance can be achieved through a change in fracture mode from predominantly low-energy intergranular fracture to high-energy-absorbing transgranular shear.

A major disadvantage of retaining a predominantly unrecrystallized microstructure is significant anisotropy in mechanical properties, especially in plate products (Peel et al., 1988). This disadvantage led to considering possibility (iii):

1. Sankaran and Grant (1980) tried using alloying additions, such as manganese and zinc, which resulted in a fine, nonshearable dispersion of the strengthening precipitates that favored the occurrence of cross-slip. Only marginal to no improvement in tensile ductility was observed, and there was a significant loss in strength. This undesirable behavior was rationalized to be due to a reduction in volume fraction of the δ' (Al_3Li) strengthening precipitates while concurrently minimizing cross-slip (Gregson and Flower, 1985).

2. The most useful method for homogenization of slip in Al–Li alloys was found to be the addition of copper (Cu) and magnesium (Mg), which resulted in precipitation of the nonshearable S′ (Al_2CuMg) phase (Gregson and Flower, 1985). According to these researchers, potential slip planes in the S′ phase are not parallel to slip planes of the matrix. Consequently, laths of the S′ phase are not likely to be easily sheared by dislocations moving through the alloy matrix. Bowing of dislocations around the S′ precipitates increases the local work hardening and reduces slip localization. However, a uniform and dense distribution of the S′ phase is essential to effectively induce and/or promote homogenization of slip at the fine microscopic level (Gregson and Flower, 1985; Gregson et al., 1986; Flower and Gregson, 1987; Dinsdale et al., 1981; Mukhopadhyay et al., 1990a). Such an S′ distribution results in a significant improvement in the ratio of notch tensile strength to yield strength as shown in Figure 10.5 (Gregson et al., 1987, 1988). This approach was also reported to result in isotropic properties in these highly textured alloys (Dinsdale et al., 1988): Figure 10.6 shows significant improvements in strength–ductility combinations owing to minor modifications in alloy chemistry.

We note here that Figure 10.6 presents a key finding. The large improvements in strength–ductility combinations achieved by minor additions of zirconium have led to all modern Al–Li alloys being of the Al–Li–Cu–Mg–Zr type (see Table 2.1 in Chapter 2).

10.3.3 Thermal Treatments for Improving Strength and Fracture Toughness

Various thermal as well as thermomechanical treatments have been sequentially developed and tried in order to obtain improved strength–toughness combinations in Al–Li alloys. Artificial aging at an elevated temperature leading to peak strength was observed to result in a gradual decrease in plane strain fracture toughness K_{Ic} (Vasudevan and Suresh, 1985). On the other hand, the influence of OA was found to depend on the chemistry of the alloy.

An attempt has been made to systematically study and document the variation of fracture toughness (K_{Ic}) with aging of Al–Li–Cu ternary alloys

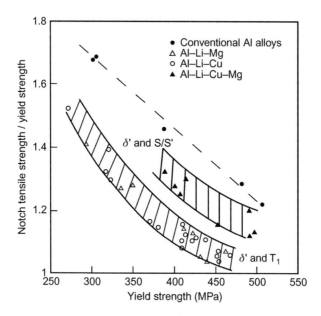

FIGURE 10.5 Variation of the ratio of notch tensile strength to yield strength as a function of yield strength for ternary and quaternary Al–Li alloys (Gregson and Flower, 1985).

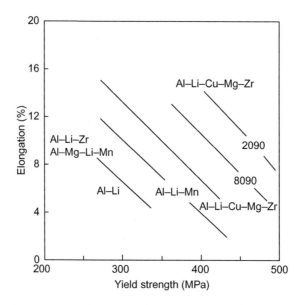

FIGURE 10.6 Variation of elongation (ductility) with yield strength as function of Al–Li alloys with minor variations in alloy composition (Dinsdale et al., 1988).

with varied lithium and copper content (Suresh et al., 1987). In this study, the lithium to copper (Li:Cu) atomic ratio varied from 2.2 to 25.2. For the alloys with lower Li:Cu ratios, the fracture toughness increased rapidly with increased OA. On the other hand, for alloys with high Li:Cu ratios, the continuation of aging beyond peak strength resulted in a marginal decrease in fracture resistance. This study also revealed a similar trend in the effect of OA on *stable crack growth* toughness measured in terms of the tearing modulus (T_R). However, the Li:Cu ratio giving the highest K_{Ic} did not give the highest T_R value. This disparity was attributed to a change in the mode of crack extension through the microstructure (Suresh et al., 1987). Crack bifurcation was observed in alloys with high Li:Cu ratios, while nominally straight crack paths occurred in alloys with low Li:Cu ratios.

In a similar contemporary study by Jata and Starke (1986), a consistent reduction in K_{Ic} occurred when going from the UA to peak-aged (PA) condition. They attributed this K_{Ic} decrease to a gradual increase in slip band spacing and a concurrent decrease in slip band width, both as consequences of an aging-induced increase in planar slip deformation (Figure 10.7).

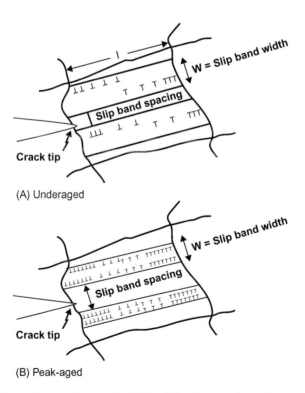

FIGURE 10.7 A schematic showing the slip band developments ahead of a crack tip for (A) underaged and (B) peak-aged conditions of an Al–Li alloy (Jata and Starke, 1986).

10.4 ANISOTROPY OF MECHANICAL PROPERTIES

Al–Li alloys, especially those having a predominantly unrecrystallized grain structure, can show considerable anisotropy in different products (Peters et al., 1986; Sadananda and Jata, 1988; Vasudevan et al., 1988). The anisotropy in mechanical properties can be attributed to one or a combination of the following (Eswara Prasad, 1993; Askeland et al., 2012):

1. Crystallographic texture, i.e., the nature and degree of alignment of individual grains (crystals) in a polycrystalline material.
2. The nature of the major strengthening precipitates.
3. Features collectively referred to as "mechanical fibering." These include the grain shapes (widths and aspect ratios); banding of small grains; and alignments of the coarse and intermediate size intermetallic particles, equilibrium precipitates, and other second-phase particles in the microstructure.

Crystallographic texture effects are not usually strong for aluminum alloys, which have a fairly isotropic face centered cubic (FCC) crystal structure. However, commercial Al–Li alloys, notably the second-generation alloys AA 2090, AA 2091, and AA 8090, were often found to develop a strong deformation texture (Rioja and Liu, 2012). There was also a high variability in through-thickness strength.

Contemporary studies of the seond-generation alloys showed that the tensile anisotropy could be reduced by prestretching before artificial aging (Peters et al., 1986), aging beyond peak strength (Gregson and Flower, 1985), and increasing the degree of recrystallization (Peel et al., 1988). However, the problems were not eliminated. These problems, which extended to other properties, including manufacturing difficulties, made second-generation Al–Li alloy products less competitive as replacements for conventional aluminum alloys. In fact, the manifestation of a number of problems owing to anisotropy is the main reason why second-generation alloys have found only limited application, as mentioned already in Section 10.1.

Subsequently, the anisotropy problems plaguing the second-generation Al–Li alloys have been extensively studied during development of the third-generation Al–Li alloys. Besides lowering the lithium content to below 2 wt%, other modifications of the alloy compositions (use of zirconium and manganese to control recrystallization and texture) and use of innovative thermomechanical processing have much reduced the tensile anisotropy (Rioja and Liu, 2012) and, indeed, the anisotropy of other mechanical properties. These developments, beginning with the first-generation Al–Li alloys, are discussed next.

10.5 TENSILE PROPERTIES OF SELECTED ALUMINUM–LITHIUM ALLOYS

In this section, key aspects governing the tensile response of three different aluminum–lithium alloys, categorized as first-generation (ternary

Al–Li–Cu alloys), second-generation (quaternary Al–Li–Cu–Mg–Zr alloys with lithium contents at and above 2 wt%), and third-generation (emerging alloys) with lower lithium contents, are presented and discussed.

10.5.1 AA 2020: A First-Generation Al–Li Alloy

As mentioned in Section 10.1, the first alloy to take advantage of the synergistic effects of lithium additions was the Al–4.45 Cu–1.21 Li–0.51 Mn–0.2Cd alloy, designated as AA 2020. This first-generation Al–Li alloy possessed high strength in the peak-aged (T651) condition and was corrosion resistant. AA 2020 performed satisfactorily for some 20 years in service as the wing and empennage skins on the Northrop RA-5C Vigilante aircraft, but the alloy was withdrawn from production already in the mid-1960s owing to concern about its lack of ductility and toughness.

Despite its withdrawal from production, AA 2020 remained the subject of several investigations because of its superiority in specific strength (strength/density) and specific stiffness (E/density) over the then widely used aircraft structural alloys AA 2024 and AA 7075. In particular, it was thought that if the ductility could be improved, then 2020 would become a very attractive alloy for a spectrum of aerospace structural applications.

Influence of Different Aging Treatments

Sanders (1979) found that both tensile ductility and fracture toughness could be substantially improved, with only a small sacrifice in strength, by giving commercial AA 2020 an underaging (UA) treatment. In a subsequent study, improved ductility for the underaged temper of AA 2020 was established (Starke and Lin, 1982). In both these studies, a single underaging treatment was used. Therefore, optimization of the heat treatment was not relevant. A detailed study of the influence of aging variations on the mechanical properties was made by Rinker et al. (1984). Their results are summarized in Table 10.2, where the mechanical properties and short-transverse fracture toughnesses of 2020 alloy variants are compared with those of the industry standard AA 2024 and AA 7075 alloys, each in two tempers. Isothermal aging (UA) of AA 2020 at 422 K strongly increased the yield strength but at the expense of ductility and fracture toughness. Overaging (OA) reduced the strength, but the ductility and fracture toughness were still lower than those of AA 2024 and AA 7075 at similar strength levels.

Some of the underaged strength and toughness data are shown in Figure 10.8 together with data for several conventional 2XXX aluminum alloys. AA 2020 in the UA12, UA15, and UA18 conditions had better combinations of strength and short-transverse toughness; and the AA 2020-UA/15 condition was superior overall, with approximately twice the tensile elongation of higher strength conditions. Table 10.2 also gives that AA 2020 in the commercial peak-aged T651 temper had both low ductility and toughness. Based on fractographic observations, Rinker et al. (1984) attributed this inferior combination of

TABLE 10.2 Influence of Aging on Mechanical Properties of Al–Li Alloy
AA 2020 (Rinker et al., 1984)

Alloy/Temper/ Aging Time	Yield Strength (MPa)	Elongation (%)	Plane Strain Short-Transverse Fracture Toughness K_Q (MPa$\sqrt{m}$)
2020-UA/5	333	14−15	−
2020-UA/10	398	15−16	−
2020-UA/12	413	9−10	26.9
2020-UA/15	466	9−13	25.4
2020-UA/18	480	7	22.6
2020-UA/21	510	−	19.3
2020-UA/26	534	5−6	17.3
2020-T651	530	3−6	14.1
2020-OA/3	499	5−8	16.8
2020-OA/10	474	6−7	n.d.
2020-OA/14	462	−	16.6
2020-OA/24	444	6−9	−
2020-OA/38	421	−	18.9
2024-T351	310	17−19	22−29
2024-T86	450	5−6	18−22
7075-T651	500	7−9	17−22
7075-T7351	430	6−7	20−23

UA, underaged at 422 K for the indicated number of hours; OA, 2020-T651 overaged at 463 K for the indicated number of hours.

properties to an increasing tendency toward strain localization and intergranular fracture, apparently caused by localized shear. In fact, AA 2020-T651 has been found to be partially recrystallized (26%), and the recrystallized grains were as large as 1.5 mm (Srivatsan and Coyne, 1987; Srivatsan et al., 1986a,b). These features in combination with planar slip associated with shearable strengthening precipitates and PFZs along GBs have been suggested to lead to low ductility problems (Starke and Lin, 1982). The planar slip and the tendency toward strain localization lead to slip bands impinging upon GBs and can cause stress concentrations, the magnitude of which depends on the slip length and therefore the grain size (Starke et al., 1981). Thus, the planar slip and large recrystallized grains had a negative effect on ductility and fracture toughness.

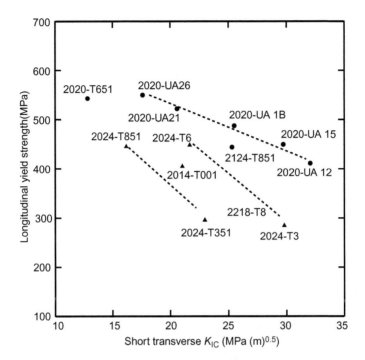

FIGURE 10.8 Variation of short-transverse K_{Ic} and longitudinal yield strength for AA 2020 and other AA 2XXX alloys (Rinker et al., 1984).

Unrecrystallized and Recrystallized Microstructures — Thermomechanical Processing Studies

A thermomechanical processing treatment was used to obtain AA 2020 with a completely unrecrystallized microstructure (Starke and Lin, 1982; Srivatsan et al., 1986a,b). Tensile tests were performed on partially recrystallized (PR) commercial 2020-T651 and the unrecrystallized (UR) thermomechanically processed alloy at room (298 K) and elevated (433 K) temperatures. The results are summarized in Table 10.3.

At 298 K (25°C), the UR material showed a 140% improvement in ductility over the commercial PR alloy. At 433 K (150°C), the UR material retained most of its strength and ductility, while the PR alloy gained ductility but with loss of strength. The differences in tensile behavior were attributed to differences in the fracture mechanisms:

1. The T651 PR fracture surfaces were macroscopically flat, normal to the tensile axis, and predominantly intergranular. Shallow near-equiaxed dimples, reminiscent of "locally" ductile failure mechanisms, were found covering the intergranular fracture regions. The dimples were associated with incoherent dispersoids ($Al_{20}Cu_2Mn_3$) and equilibrium precipitates along the GB. At the

TABLE 10.3 Influence of Test Temperature on Tensile Properties of AA 2020 (Srivatsan et al., 1986a,b)

2020 Condition	Temperature (K)	Yield Strength (MPa)	Ultimate Tensile Strength (MPa)	Elongation (%)	Reduction in Area (%)
As-received (T651): PR	298	526	567	5.0	5.0
	433	483	513	6.0	8.0
Thermomechanically processed: UR	298	462	509	12.0	20.0
	433	468	505	11.0	23.0

elevated temperature, there was more secondary cracking, which occurred along GBs separating regions of intergranular and transgranular fracture.

2. The UR fracture surfaces were macroscopically at about 45° to the tensile axis and predominantly ductile and transgranular, especially at the elevated temperature (Srivatsan and Coyne, 1986; Srivatsan et al., 1986a,b). Shallow dimples were observed in the regions separating the transgranular and intergranular fracture areas, indicating microvoid formation along the GBs.

The predominance of higher-energy-absorbing ductile transgranular fracture for the UR material, as compared to the low energy intergranular rupture for the commercial PR alloy, provided convincing evidence of macroscopically homogeneous deformation in the UR material (Srivatsan et al., 1986a,b).

10.5.2 AA 8090: A Second-Generation Al–Li Alloy

Anisotropy

An aspect that has received much attention for second-generation Al–Li alloys is the anisotropy of mechanical properties, especially the yield strength (Vasudevan et al., 1988) and low short-transverse ductility and fracture toughness (Eswara Prasad et al., 2003; Lynch et al., 2003; Rioja and Liu, 2012).

The tensile properties of second-generation plate and sheet products were found to exhibit a significant degree of in-plane anisotropy as well as through-thickness anisotropy. In most cases, the alloy plates possessed significantly higher yield strength (10–20% higher) at the midsection when compared to the surface. The observed variations in yield strength were attributed to the synergistic and mutually interactive influences of (i) changes in the nature and degree of crystallographic texture, (ii) preferential surface recrystallization, and (iii) intrinsic differences in volume fractions of the major strengthening precipitates.

Anisotropy of tensile ductility and fracture toughness also depend on the "banding" (fibering) of coarse and intermediate-size constituent particles and dispersoids in the short-transverse plane. Much research effort leading to the development of third-generation Al−Li alloys has been directed to reducing crystallographic texturing and the severity of fibering of the grains. In this section, we present and discuss the results of a study undertaken to evaluate the in-plane anisotropy in tensile properties of the quaternary Al−Li−Cu−Mg alloy 8090 in both plate and sheet product forms. The observed variations in tensile anisotropy are correlated with the various metallurgical parameters.

The orientation dependences of the tensile properties of AA 8090 plate and sheet alloys are summarized in Table 10.4. Table 10.5 gives the degree of anisotropy as a percentage of the longitudinal tensile properties (Eswara Prasad et al., 1992, 1993). The properties include the following: 0.2% offset yield strength (YS), ultimate tensile strength (UTS), total elongation, uniform strain, and strain hardening exponent (n). The data show considerable anisotropy:

1. A change in specimen orientation from 0° to 60° with respect to the rolling direction (RD) resulted in decreasing yield and ultimate tensile strengths with a fairly consistent increase in ductility.
2. For specimen orientations beyond 60°, the strength values of the higher strength tempers (8090-T8E51 and 8090-T6) increased while the ductility decreased. The strengths and ductilities of the lower strength tempers (8090-T81 and 8090-T3) remained essentially the same.
3. The orientation dependences of the strain hardening exponent (n) and uniform strain (ϵ_U) are similar. Both "n" and ϵ_U attain a maximum value at an angle of 45−60° to the RD.

The degree of tensile anisotropy in AA 8090 (Table 10.5) is significantly higher than that reported for other high strength aerospace aluminum alloys (Smith and Scully, 2000). The observed anisotropy trends for AA 8090 (in plate and sheet forms) accord reasonably well with those reported in the published literature for forgings (Doorbar et al., 1986). However, the amount of yield stress anisotropy can be product dependent: 8090 extrusions showed the highest degree of anisotropy (27%) in an underaged condition (Tempus et al., 1991).

Anisotropy in Yield Strength

Yield strength anisotropy has attracted much interest. Two major influences are the crystallographic texture and the final heat-treatment condition. The

TABLE 10.4 Orientation Dependence of Tensile Properties of 8090 Alloy Plates and Sheet (Eswara Prasad, 1993; Eswara Prasad and Malakondaiah, 1992)

Alloy Designation and Product	Property	Specimen Orientation with Respect to the Rolling Direction, RD ($°$)				
		$0°$ (L)	$30°$	$45°$	$60°$	$90°$ (T)
8090-T8E51 plate (12.5 mm thick)	0.2%Y.S, MPa	485	447	393	382	467
	UTS, MPa	555	509	478	513	534
	Total elongation, % (25-mm gauge length)	5.4	6.1	11.5	9.6	7.0
	Uniform elongation, %	0.054	0.058	0.075	0.07	0.06
	Strain hardening exponent, n	0.053	0.063	0.077	0.11	0.05
	UTS/σ_y	1.14	1.14	1.22	1.34	1.14
8090-T8E51 plate (8 mm thick)	0.2%Y.S, MPa	471	–	370	–	457
	UTS, MPa	561	–	472	–	547
	Total elongation, % (10-mm gauge length)	5.0	–	5.7	–	7.5
	Uniform elongation, %	0.05	–	0.057	–	0.07
	Strain hardening exponent, n	0.065	–	0.09	–	0.088
	UTS/σ_y	1.19	–	1.28	–	1.20
8090-T6 sheet (3 mm thick)	0.2%Y.S, MPa	456	424	352	350	432
	UTS, MPa	528	487	476	464	538
	Total elongation, % (25-mm gauge length)	>2.3	4.4	14.0	14.1	6.2
	Uniform elongation, %	0.023	0.044	0.102	0.096	0.062
	Strain hardening exponent, n	0.068	0.058	0.111	0.101	0.086
	UTS/σ_y	1.16	1.15	1.35	1.33	1.25
8090-T81 sheet (1.6 mm thick)	0.2%Y.S, MPa	355	326	324	315	310
	UTS, MPa	425	461	448	450	449
	Total elongation, % (25-mm gauge length)	6.0	10.4	11.6	10.0	10.0

(Continued)

TABLE 10.4 (Continued)

Alloy Designation and Product	Property	Specimen Orientation with Respect to the Rolling Direction, RD (°)				
		0° (L)	30°	45°	60°	90° (T)
	Uniform elongation, %	0.06	0.08	0.097	0.088	0.092
	Strain hardening exponent, n	0.108	0.111	0.107	0.123	0.131
	UTS/σ_y	1.27	1.41	1.38	1.43	1.45
8090-T3 sheet (1.6 mm thick)	0.2%Y.S, MPa	211	204	206	198	203
	UTS, MPa	328	328	330	315	331
	Total elongation, % (25-mm gauge length)	17.0	17.4	17.8	19.0	16.2
	Uniform elongation, %	0.145	0.14	0.16	0.15	0.14
	Strain hardening exponent, n	0.16	0.161	0.155	0.162	0.163
	UTS/σ_y	1.55	1.61	1.60	1.59	1.63

The data for each condition are averages of a minimum of three tests.

use of high levels of zirconium (Zr) as a grain refiner in quaternary Al–Li alloys such as 8090 generally suppresses the occurrence of recrystallization, and this is the principal reason for this alloy's high degree of anisotropy. Efforts directed toward reducing anisotropy have considered cross-rolling as a possible means. However, Table 10.5 shows that cross-rolling and unidirectional (straight) rolling resulted in similar percentages of yield strength anisotropy. In other words, cross-rolling was ineffective in reducing AA 8090 strength anisotropy. A similar conclusion was reached by Gregson and Flower (1985).

With respect to heat treatment, Tempus et al. (1991) observed that 8090 extrusions had a higher degree of yield strength anisotropy when in the solution-treated condition compared to the artificially aged condition. Also, Engler and Lücke (1991) found that artificial aging, and hence matrix precipitation, altered the crystallographic texture. Hence, the effect of aging on yield strength anisotropy can be attributed to the nature of the precipitates and their concomitant influence on crystallographic texture.

TABLE 10.5 Degrees of Tensile Anisotropy (($\pm$ %) of L-Value from the Maximum Observed Variation) of 8090 Alloy Plates and Sheet (Eswara Prasad, 1993; Eswara Prasad and Malakondaiah, 1992)

Description	8090-T8E51 (12.5 mm thick)	8090-T8E51 (8 mm thick)	8090-T6 (3 mm thick)	8090-T81 (1.6 mm thick)	8090-T3 (1.6 mm thick)
Aging condition	Peak aged	Peak aged	Peak aged	Overaged	Underaged
Recrystallization (%)	PR⁺(20–30%)	PR⁺(20–25%)	PR⁺(10–20%)	R*(80–90%)	R*(80–90%)
Crystallographic texture	Strong, cross-rolled	Strong, straight rolled	Strong, straight rolled	–	–
Anisotropy ($\pm$ % of L-value from the maximum observed variation)					
Yield strength	21	21	23	13	6
UTS	14	16	12	– 8	– 1
Ductility (elongation)	113	50	513	93	12
Strain hardening exponent, n	45	38	63	21	2

+ PR, partially recrystallized (10–30% recrystallization); *R, fully recrystallized (80–90% recrystallization).

Anisotropy in UTS

Table 10.5 gives that the yield strength anisotropy was much larger than the anisotropy in UTS for all the alloys, being highest for the peak-aged conditions (T8E51 and T6). The lesser UTS anisotropy correlates with the dependences of plastic flow behavior for the different test directions. For example, the strain hardening exponent (n) is almost constant for AA 8090-T3 sheet and so is the UTS. On the other hand, there is considerable variation in n for the AA 8090-T8E51 plates and 8090-T6 sheet. The decrease in anisotropy in going from yield to the UTS has been found to depend on both the nature of the matrix precipitates and the resultant microscopic deformation behavior (Fox et al., 1986).

Anisotropy in Tensile Ductility

The 8090 data in Tables 10.4 and 10.5 show large differences in tensile ductility, with minimum values for the longitudinal (L) test direction and maximum values at L + (45°−60°). Apart from this trend, there was no *consistent* difference between the type of product (plate or sheet) and aging condition. However, the sheet alloys differed greatly in tensile ductility anisotropy depending on the degree of aging. Thus, the artificially aged 8090-T81 sheet had a much higher degree of ductility anisotropy (93%) than the naturally aged 8090-T3 sheet (12%). This was attributed to the 8090-T81 longitudinal specimens failing before the onset of (theoretical) plastic instability, while for all other conditions, failure occurred due to plastic instability (Eswara Prasad, 1993; Eswara Prasad et al., 1992).

More generally, the anisotropy in tensile ductility of Al−Li alloys has been attributed to the independent or mutually interactive influences of the following factors:

- Extent of shearing of the δ' strengthening precipitates and orientation of the resultant flow localization with respect to the applied far-field stress axis (Fox et al., 1986).
- Density and distribution of the coarse and intermediate-size intermetallic particles, which can favor orientation-dependent stress concentrations (Starke, 1977).
- Nature, morphology, and distribution of the major strengthening precipitates, which are controlled by both alloy chemistry and thermomechanical processing treatments (Tempus et al., 1991).
- Processing history, degree of recrystallization, and the nature and amount of cold-work imparted to the alloy before artificial aging (Takahashi et al., 1987).
- Mode of fracture (Srivatsan and Place, 1989).

Besides these factors, the tensile ductility and its anisotropy may be influenced by (i) the strength of the GB, (ii) PFZ widths, and (iii) the density of the equilibrium precipitates along the GBs. Clearly, the orientation dependence of ductility in AA 8090 plate and sheet is not amenable to any straightforward analysis since it is influenced by many parameters and test conditions (Flower and Gregson, 1987).

Strength Differential

Most commercial and emerging Al–Li plate and sheet alloys are often subjected to a tensile prestretch as part of thermomechanical processing. Prestretching takes advantage of the beneficial effects of heterogeneous S' (Al$_2$CuMg) precipitation on overall mechanical properties. Prestretching not only improves the strength–fracture resistance relationship but also results in a strength differential effect that is related to the residual stresses induced.

The influence of prestretching on the tensile and compression properties of AA 8090 plate has been studied by Grimes et al. (1985). This study clearly showed that imposition of a prestretch resulted in an increase in the strength differential (Table 10.6).

The data in Table 10.6 also show (i) a 7% prestretch increased both the tensile and compressive yield strengths and (ii) the damage tolerant temper had a much higher longitudinal (L) strength differential, $\sigma_C - \sigma_T$, than the near peak-aged temper.

However, no convincing explanation for the observed strength differential effect was provided by this study. Hence, a systematic *ad hoc* investigation was conducted with the primary objective of understanding the observed strength differential effect (Eswara Prasad et al., 1992, 1994).

TABLE 10.6 Effect of Tensile Prestretch on "Strength Differential" for Al–Li Alloy 8090 Plates (Grimes et al., 1985)

Temper	Test Direction	0.2 % Offset Yield Strength					
		Without Prestrain/ Stretch			With 7% Stretch		
		σ_T	σ_C	$\sigma_C - \sigma_T$	σ_T	σ_C	$\sigma_C - \sigma_T$
Damage tolerant	Longitudinal (L)	390	338	−52	425	356	−69
	Long transverse (LT)	358	372	+14	368	486	+18
	Short transverse (ST)	280	337	+57	292	357	+65
Near peak aged	Longitudinal (L)	442	423	−19	495	463	−32
	Long transverse (LT)	429	441	+12	457	490	+33
	Short transverse (ST)	353	390	+37	360	455	+95

Table 10.7 presents flow stress anisotropy data and the strain hardening exponents for one of the AA 8090-T8E51 plates discussed in Section 10.5.2. The compressive yield stress in the longitudinal (L) direction is 8% lower than the tensile yield strength, showing a reverse "strength differential" effect. In the long-transverse (LT) direction, the compressive yield strength is higher (6−7%) than the tensile yield strength, as would be expected for a normal "strength differential" effect. However, the orientation dependence of the work hardening exponent under compressive loading (n_C) is different from that for tensile loading (n_T): while both n_C and n_T were minimum in the LT direction, the maximum in n_C was obtained in the L direction while that in n_T was found in the L + 60° direction.

Several mechanisms have been proposed to explain the "strength differential" effect. These are based on (i) residual stresses, (ii) microscopic cracks, (iii) interactions between dislocations and interstitial solute atoms, and (iv) stress concentration at particle−matrix interfaces. Mechanisms (iii) and (iv) may be relevant to AA 8090, owing to the type of strengthening precipitate, i.e., δ', which is coherent and spherical. For a spherical particle, the stress concentration at the particle−matrix interface is approximately 9 times greater in tension than in compression. When the level of normal stress at the particle−matrix interface is significantly greater than the interface bond strength, then the stress required to cause dislocation shearing of the particle or bowing around it will not be reached, and only "localized" yielding of the matrix adjacent to the particle is likely to occur. This, in addition to the differential stress concentration, is principally responsible for the observed strength differential in AA 8090.

TABLE 10.7 Flow Stress (0.2% Plastic Strain) Anisotropy for 12.5-mm Thick 8090-T8E51 Plate (Eswara Prasad et al., 1992).

Property	Orientation with Respect to Rolling Direction (°)				
	0° (L)	30°	45°	60°	90° (LT)
Compressive yield strength, σ_C (MPa)	445	–	420	–	498
Tensile yield strength, σ_T (MPa)	485	447	393	382	467
Strength differential, $\sigma_C - \sigma_T$ (MPa) at $\varepsilon_P = 0.002$	−40	–	+27	–	+31
Compressive work hardening exponent, n_C	0.104	–	0.083	–	0.056
Tensile work hardening exponent, n_T	0.053	0.063	0.077	0.11	0.05

10.5.3 AA 2198: A Third-Generation Al–Li Alloy

The third-generation Al–Li alloy AA 2198, introduced in 2005, has a nominal composition (wt%) of Al–1.0Li–3.2Cu–0.5Mg–0.4Ag–0.11Zr–0.5maxMn–0.35maxZn and a density of 2.69 g/cm³. It has potential application as a damage tolerant/medium strength alloy for transport aircraft fuselage skins (Table 10.1).

The tensile behavior of alloy AA 2198 in 6-mm thick sheet stock was first studied and reported in the open literature by Chen et al. (2011). The as-supplied material was in the naturally aged T3 temper, which included 3% stretch. Subsequent artificial aging resulted in the T8 temper. Flat tensile test specimens of 2 mm thickness in the T3 and T8 tempers were prepared from the sheet stock. The ambient temperature test results are given in Figure 10.9 and Table 10.8.

The alloy AA 2198 was found to exhibit complex anisotropic behavior, with the longitudinal direction (L) exhibiting the highest strength (YS and UTS) and the L + 45° direction (designated by the authors as D) the lowest strength, both in the underaged (T3) and peak-aged (T8) conditions. Ductility showed the opposite trend in the underaged (T3) condition, but little anisotropy in the peak-aged (T8) condition. The alloy in the T8 condition exhibited higher yield and tensile strengths than in the T3 condition but a

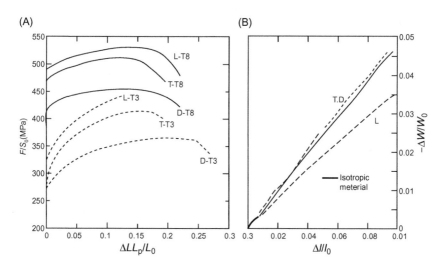

FIGURE 10.9 (A) Variation in nominal tensile stress (ratio of applied force (F) to the initial cross-sectional area (S_0))—tensile strain (plastic load line displacement (ΔLL_p) normalized with initial gauge length (L_0)) data with specimen orientation (longitudinal, L; long transverse, LT (designated in the figure as T) and L + 45° (designated in the figure as D) in AA 2198 alloy; and (B) width reduction ($-\Delta W/W_0$, the ratio of change in width to the initial width) as a function of axial strain ($\Delta l/l_0$) in three different in-plane directions of L, LT (T), and L + 45° (D). *Source: Data from Chen et al. (2011).*

lower work hardening ability. Reduction of specimen width as a function of tensile strain (Figure 10.9B) showed a large deviation from the isotropic behavior in the L direction but not in the LT and L + 45° directions. Further, the data in Table 10.8 show that the alloy AA 2198 exhibits a lower degree of anisotropy in tensile properties in the peak-aged (T8) condition as compared to the underaged (T3) condition: clearly, the higher amounts of T_1 precipitates present in the T8 condition reduced the tensile anisotropy significantly.

Values of the Lankford coefficient (an index of plastic anisotropy) given in Table 10.8 clearly indicate that for testing along the L direction the thickness reduction was higher than the width reduction. However, this is not the case for testing along the LT and L + 45° directions. Since such complex anisotropic plastic deformation behavior could not be satisfactorily described by a simple quadratic yield surface (Hill, 1950), Chen et al. (2011) used the yield criterion proposed by Bron and Besson (2004) to simulate the observed yield anisotropy and explain the variation in the Lankford coefficient. By doing so, Chen and coworkers (2011) obtained excellent agreement between the experimental behavior and the simulation results (Figure 10.10).

Chen et al. (2011) also investigated the fracture behavior of AA 2198 in the T3 and T8 heat treatment conditions. They found that the fracture path mainly depended on the loading direction (L, LT, or L + 45) and less on the heat treatment condition or sheet thickness. The behavior was analyzed using a strain localization indicator based on Rice's bifurcation analysis (Rice, 1976). The analysis achieved partial success by explaining the T8 specimen fractures in terms of anisotropic plasticity. However, the T3 fractures were not explained by the analysis.

10.6 SUMMARY AND CONCLUSIONS

The increasing need for materials for both safe and efficient use in weight-critical and stiffness-critical applications has generated much interest in studies aimed at understanding the physical metallurgy, metallurgical characteristics, mechanical properties, and failure or fracture behavior of aluminum—lithium (Al—Li) alloys. The potential use of the third-generation alloys in a spectrum of aircraft structural applications looks promising. Not only do these alloys offer the benefit of substantial savings in weight, but they should also help maintain the preeminence of aluminum alloys as the primary structural materials for civilian aircraft.

In this chapter, an overview is provided of the strengthening contributions to these alloys, and studies on composition—processing—microstructure—mechanical property relationships are discussed. Rationalization for the observed enhancement in strength of the solid solution is made with specific reference to the presence of lithium in solid solution and to the presence of coherent and ordered precipitates in the aluminum alloy matrix.

TABLE 10.8 Tensile Properties for 6 mm Thick Al–Li Alloy 2198 Sheet in the Longitudinal (L), Long-Transverse (LT; authors designate the direction as T) and L + 45° Direction (authors designate the direction as D)

Test Direction	T3				T8			
	Yield Strength (MPa)	UTS (MPa)	Uniform Elongation (%)	Lankford Coefficient for 5%	Yield Strength (MPa)	UTS (MPa)	Uniform Elongation (%)	Lankford Coefficient For 5%
L	324	442	13.0	0.52	490	530	14.0	0.64
LT	300	416	15.4	1.63	470	512	12.3	1.25
L + 45°	266	363	21.1	1.00	404	453	13.0	1.06

Lankford coefficient (l_k) is defined as the ratio of the true deformation along the width of the specimen to the same along the thickness of the specimen (Chen et al., 2011).

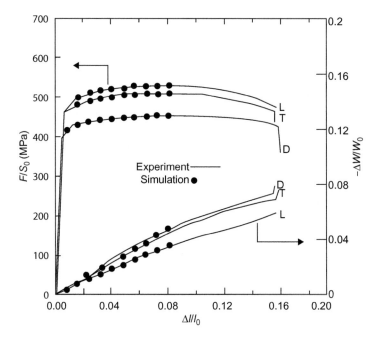

FIGURE 10.10   Variation in the experimentally obtained and theoretically derived nominal tensile stresses and strains for AA 2198 alloy in the three in-plane directions L, LT, and L + 45°. *Source: Data from Chen et al. (2011).*

Three specific examples drawn from the published literature, and relating specifically to the first-, second-, and third-generation Al–Li alloys, suggest that additional strengthening is achieved by the co-precipitation and presence of other binary and ternary phases in the microstructure. The addition of various amounts of copper and magnesium to these alloys modifies the precipitation sequence by altering the solubility of the alloying elements and by co-precipitating with the lithium-rich ordered matrix precipitates. Co-precipitation of the ternary and more complex matrix strengthening phases is found to be beneficial because in addition to contributing to higher strength, it facilitates homogeneous deformation at the fine microscopic level.

The overall strength and deformation characteristics of these alloys are governed by metallurgical variables. Such variables include (i) intrinsic microstructural features such as the nature and type of matrix strengthening precipitates and PFZs adjacent to the GBs and (ii) interactions between dislocations generated during deformation and the intrinsic microstructural features.

The broad observations made in this chapter provide convincing evidence that there are several mechanisms that contribute to controlling the strength, deformation characteristics, and overall mechanical response of Al—Li alloys. For a particular aging condition, the intrinsic mechanisms governing deformation characteristics at the fine microscopic level appear to be influenced by alloy chemistry, the processing, and intrinsic microstructural effects with minimum influence of stress or load history. The development of novel processing techniques has made possible the ability to tailor the properties of an alloy for a particular application. It is hoped that continued exchange of results and ideas among researchers, combined with sustained and continuing research efforts to resolve any of the anomalies that exist, will aid in better understanding of the mechanical response of these alloys.

ACKNOWLEDGMENTS

The authors would like to thank profoundly Dr. R.J.H. Wanhill of NLR, Amsterdam, The Netherlands, for his extensive modifications of the manuscript and also for suggesting many valuable changes. One of the authors (EJL) would like to acknowledge financial support by the Office of Naval Research (Grant Number N00014-12-1-0237) with Dr. Lawrence Kabacoff as program officer. Dr. T.S. Srivatsan extends gracious thanks and appreciation to Mr. M. Kannan (Graduate Student, University of Akron) for able and skilful assistance with the graphics.

REFERENCES

Agyekum, E., Ruch, W., Starke Jr., E.A., Jha, S.C., Sanders Jr., T.H., 1986. Effect of precipitate type on elastic properties of Al—Li—Cu and Al—Li—Cu—Mg alloys. In: Baker, C., Gregson, P.J., Harris, S.J., Peel, C.J. (Eds.), Aluminium—Lithium Alloys III: Proceedings of the Third International Conference on Aluminium—Lithium Alloys, 1986. The Institute of Metals, London, pp. 448—454.

Ashton, R.F., Thompson, D.S., Starke Jr., E.A., Lin, F.S., 1986. Processing Al—Li—Cu—(Mg) alloys. In: Baker, C., Gregson, P.J., Harris, S.J., Peel, C.J. (Eds.), Proceedings of the Third International Conference on Aluminium—Lithium Alloys. The Institute of Metals, London, pp. 66—77.

Askeland, D.R., Phule, P.P., Wright, W.J., 2012. The Science and Engineering of Materials, sixth ed. Thompson Publishers, Canada.

Balmuth, E.S., 1994. The status of Al—Li alloys. In: Sanders, T.H., Starke, E.A. (Eds.), Aluminium Alloys, Their Physical and Mechanical Properties:, Proceedings of the Fourth International Conference on Aluminium Alloys. Georgia Institute of Technology, Atlanta, GA, pp. 82—89.

Bron, F., Besson, J., 2004. A yield function for anisotropic materials. Application to aluminium alloys. Int. J. Plast. 20, 937—963.

Bull, M.J., Lloyd, D.J., 1986. *Textures developed in Al—Li—Cu—Mg alloy*. In: Baker, C., Gregson, P.J., Harris, S.J., Peel, C.J. (Eds.), Proceedings of the Third International Conference on Aluminium—Lithium Alloys III. The Institute of Metals, London, pp. 402—410.

Cassada, W.A., Shiflet, G.J., Starke, E.A., 1986. Grain boundary precipitates with five-fold diffraction symmetry in an Al—Li—Cu alloy. Scr. Metall. 20, 751—756.

Chen, J., Mady, Y., Morgeneyer, F., Besson, J., 2011. Plastic flow and ductile rupture of a 2198 Al–Cu–Li aluminium alloy. Comput. Mater. Sci. 50, 1365–1371.

Chen, R.T., Starke, E.A., 1984. Microstructure and mechanical properties of mechanically alloyed, ingot metallurgy and powder metallurgy Al–L–Cu–Mg alloys. Mater. Sci. Eng. 67, 229–245.

Crooks, R., Wang, Z., Levit, A.V.I., Shenoy, R.N., 1998. Microtexture, microstructure and plastic anisotropy of AA 2195. Mater. Sci. Eng. A257, 145–152.

Dieter, G.E., 1986. Mechanical Metallugry. McGraw-Hill Inc., London.

Dinsdale, K., Harris, S.J., Noble, B., 1981. Relationship between microstructure and mechanical properties of aluminium–lithium–magnesium alloys. In: Sanders, T.H., Starke Jr., E.A. (Eds.), Proceedings of the First International Conference on Aluminium–Lithium Alloys. The Metallurgical Society of AIME, Warrendale, PA, pp. 101–118.

Dinsdale, K., Noble, B., Harris, S.J., 1988. Development of mechanical properties in Al–Li–Zn–Mg–Cu alloys. Mater. Sci. Eng. A 104, 75–84.

Divecha, A.P., Karmarkar, S.D., 1981. Casting problems specific to aluminium–lithium alloys. In: Sanders, T.H., Starke, E.A. (Eds.), Proceedings of the First International Conference on 'Aluminium–Lithium Alloys'. The Metallurgical Society of AIME, New York, NY, pp. 49–62.

Doorbar, P.J., Borradaile, J.B., Driver, D., 1986. Evaluation of aluminium–lithium–copper–magnesium–zirconium alloy as forging material. In: Baker, C., Gregson, P.J., Harris, S.J., Peel, C.J. (Eds.), Proceedings of the Third International Conference on Aluminium–Lithium Alloys. The Institute of Metals, London, pp. 496–508.

Engler, O., Lücke, K., 1991. Influence of the precipitation state on the cold rolling texture in 8090 Al–Li material. Mater. Sci. Eng. A A148, 15–23.

Eswara Prasad, N., 1993. In-Plane Anisotropy in the Fatigue and Fracture Properties of Quaternary Al–Li–Cu–Mg Alloys. Doctoral Thesis, Banaras Hindu University, Varanasi, India.

Eswara Prasad, N., Malakondaiah, G., 1992. Anisotropy in the mechanical properties of quaternary Al–Li–Cu–Mg alloys. Bull. Mater. Sci. 15, 297–310.

Eswara Prasad, N., Malakondaiah, G., Rama Rao, P., 1992. Strength differential in Al–Li alloy 8090. Mater. Sci. Eng. A 150, 221–229.

Eswara Prasad, N., Kamat, S.N., Malakondaiah, G., 1993. Effect of crack deflection and branching on the R-curve behaviour of an Al–Li alloy 2090 sheet. Inter. J. Fract. 61, 55–69.

Eswara Prasad, N., Kamat, S.V., Malakondaiah, G., Kutumbarao, V.V., 1994. Static and dynamic fracture toughness of an Al–Li 8090 alloy plate. Fatigue Fract. Eng. Mater. Struct. 17, 441–450.

Eswara Prasad, N., Gokhale, A.A., Rama Rao, P., 2003. Mechanical behaviour of aluminium–lithium alloys. Sādhanā 28 (1 & 2), 209–246.

Flower, H.M., Gregson, P.J., 1987. Solid state phase transformations in aluminium alloys containing lithium. Mater. Sci. Technol. 3, 81–90.

Fox, S., Flower, H.M., McDarmaid, D.S., 1986. Formation of solute-depleted surfaces in Al–Li–Cu–Mg–Zr alloys and their influence on mechanical properties. In: Baker, C., Gregson, P.J., Harris, S.J., Peel, C.J. (Eds.), Proceedings of the Third International Conference on Aluminium–Lithium Alloys, vol. 3. The Institute of Metals, London, pp. 263–272.

Gregson, P.J., Flower, H.M., 1984. δ' precipitation in Al–Li–Mg–Cu–Zr alloys. J. Mater. Sci. Lett. 3, 829–834.

Gregson, P.J., Flower, H.M., 1985. Microstructural control of toughness in aluminium–lithium alloys. Acta Metall. 33, 527–537.

Gregson, P.J., Flower, H.M., Tete, C.N.J., Mukhopadhyay, A.K., 1986. Role of vacancies in precipitation of δ'- and S-phases in Al—Li—Cu—Mg alloys. Mater. Sci. Technol. 2, 349—353.

Gregson, P.J., Dinsdale, K., Harris, S.J., Noble, B., 1987. Evolution of microstructure in Al—Li—Zn—Mg—Cu alloys. Mater. Sci. Technol. 3, 7—13.

Gregson, P.J., McDarmaid, D.S., Hunt, E., 1988. Post-yield deformation characteristics in Al—Li alloys. Mater. Sci. Technol. 4, 713—718.

Grimes, R., Cornish, A.J., Miller, W.S., Reynolds, M.A., 1985. Aluminium—lithium based alloys for aerospace applications. Met. Mater. 1, 357—363.

Hill, D.P., Williams, D.N., Mobley, C.E., 1984. The effect of hydrogen on the ductility, toughness and yield strength of an Al—Mg—Li alloy. In: Sanders, T.H., Starke Jr., E.A. (Eds.), Proceedings of the Second International Conference on Aluminum—Lithium Alloys. The Metallurgical Society of AIME, Warrendale, PA, pp. 201—218.

Hill, R., 1950. The Mathematical Theory of Plasticity. Clarendon Press, Oxford.

Huang, J.G., Ardell, A.J., 1988. Precipitation strengthening of binary Al—Li alloys by δ' precipitates. Mater. Sci. Eng. 104A, 149—156.

Jata, K.V., Starke, E.A., 1986. Fatigue crack growth and fracture toughness behaviour of an Al—Li—Cu alloy. Metall. Trans. A 17A, 1011—1026.

Kang, S., Grant, N.J., 1984. Mechanical properties of rapidly solidified X2020 aluminium alloys. In: Sanders, T.H., Starke, E.A. (Eds.), Proceedings of the Second International Conference on Aluminum—Lithium Alloys II. The Metallurgical Society of AIME, Warrendale, PA, pp. 469—484.

Lavernia, E.J., Grant, N.J., 1987. Review aluminium—lithium alloys. J. Mater. Sci. 22, 1521—1529.

Lavernia, E.J., Srivatsan, T.S., Mohamed, F.A., 1990. Review-strength, deformation, fracture behaviour and ductility of aluminium—lithium alloys. J. Mater. Sci. 25, 1137—1158.

LeBaron, I.M., 1945. U.S. Patent Number 2,381,219, Granted.

Lewis, R.F., Starke, E.A., Coons, W.C., Shiflet, G.J., Willner, E., Bjeletich, J.G., et al., 1987. Microstructure and properties of Al—Li—Cu—Mg—Zr(8090) heavy section forgings. In: Champier, G., Dubost, B., Miannay, D., Sabatay, L. (Eds.), Proceedings of the Fourth International Aluminium—Lithium Conference. J. Phys. Colloque 48, C3.643—C3.652.

Lin, F.S., Chakraborty, S.B., Starke, E.A., 1982. Microstructure-property relationships of two Al—3Li—2Cu—0.2Zr—X Cd alloys. Metall. Trans. 13A, 401—410.

Lynch, S.P., Shekhter, A., Moutsos, S., Winkelman, G.B., 2003. Challenges in developing high performance Al—Li alloys. LiMAT 2003: Third International Conference in Light Materials for Transportation Systems. Center for Advanced Aerospace Materials, Pohang University of Science and Technology, Pohang, Korea.

Mukhopadhyay, A.K., Flower, H.M., Sheppard, T., 1990a. Development of microstructure in AA 8090 alloy produced by extrusion processing. Mater. Sci. Technol. 6, 461—468.

Mukhopadhyay, A.K., Flower, H.M., Sheppard, T., 1990b. Development of mechanical properties in AA 8090 alloy produced by extrusion processing. Mater. Sci. Technol. 6, 611—620.

Noble, B., Thompson, G.E., 1971. Precipitation characteristics of aluminium—lithium alloys. Met. Sci. J. 5, pp. 114—120.

Noble, B., Thompson, G.E., 1972. $T_1(Al_2CuLi)$ precipitation in aluminium—copper—lithium alloys. Met. Sci. J. 6, 167—174.

Noble, B., McClaughlin, I.R., Thompson, G.E., 1970. Solute atom clustering processes in aluminium—copper—lithium alloys. Acta Metall. 18, 339—345.

Noble, B., Harris, S.J., Dinsdale, K., 1982a. The elastic modulus of aluminium—lithium alloys. J. Mater. Sci. 17, 461—468.

Noble, B., Harris, S.J., Dinsdale, K., 1982b. Yield characteristics of aluminium–lithium alloys. Met. Sci. J. 16, 425–430.

Peel, C.J., Evans, B., Baker, C.A., Bannet, D.A., Gregson, P.J., Flower, H.M., 1983. The development and application of improved aluminium–lithium alloys. Proceedings of the Second International Conference on Aluminum–Lithium Alloys. The Metallurgical Society of AIME, Warrendale, PA, pp. 363–392.

Peel, C.J., Evans, B., Baker, C.A., Bennett, B.A., Gregson, P.J., Flower, H.M., 1984. In: Sanders, T.H., Starke Jr., E.A. (Eds.), Aluminum Lithium Alloys II. Metallurgical Socierty of AIME, Warrendale, PA, USA, pp. 363–383.

Peel, C.J., McDarmaid, D., Evans, B., 1988. Considerations of critical factors for the design of aerospace structures using current and future Al–Li alloys. In: Kar, R.J., Agarwal, S.P., Quist, W.E. (Eds.), Aluminium–Lithium Alloys: Design, Development and Applications Update. ASM International, Metals Park, OH, pp. 315–337.

Peters, M., Eschweiler, J., Welpmann, K., 1986. Strength profile in Al–Li plate material. Scr. Metall. 20, 259–264.

Quist, W.E., Narayanan, G.H., 1989. Aluminium–lithium alloys, Treatise on Material Science and Technology, vol. **31**. Academic Press Inc., New York, NY, pp. 219–254.

Quist, W.E., Narayanan, G.H., Wingart, A.L., 1984. Aluminium–lithium alloys for aircraft structures—an overview. In: Sanders, T.H., Starke Jr., E.A. (Eds.), Proceedings of the Second International Conference on Aluminum–Lithium Alloys II. The Metallurgical Society of AIME, Warrendale, PA, pp. 313–334.

Rice, J.R., 1976. The localisation of plastic deformation. In: Koiter, W. (Ed.), The Proceedings of the 14th International Conference on Theoretical and Applied Mechanics. North-Holland, Delft, Amsterdam, pp. 207–220.

Rinker, J.G., Marek, M., Sanders Jr., T.H., 1983. Microstructure, toughness and SCC behaviour of 2020. In: Sanders, T.H. and Starke E.A. Jr., (Eds.), Proceedings of the Second International Conference on Aluminum–Lithium alloys II. The Metallurgical Society of AIME, Warrendale, PA. pp. 597–626.

Rinker, J.G., Marek, M., Sanders Jr., T.H., 1984. Microstructure toughness and stress corrosion cracking behavior of aluminum alloy 2020. Mater. Sci. Eng. 64, 203–221.

Rioja, R.J., Liu, J., 2012. The evolution of Al–Li base products for aerospace and space applications. Metall. Trans. A 43A, 3325–3337.

Sadananda, K., Jata, K.V., 1988. Creep crack growth behaviour of two Al–Li alloys. Metall. Trans. 19A, 847–854.

Sanders Jr., T.H., 1979. Factors Influencing Fracture Toughness and Other Properties of Aluminum-Lithium Alloys, Final Report, Contract Number N62269-76-C-0271, Naval Air Development Center, Warminster, PA.

Sanders Jr., T.H., 1981 Al–Li–X alloys: an over view. In: Sanders, T.H., Starke E.A. Jr., (Eds.), Proceedings of the First International Conference on Aluminium–Lithium Alloys. The Metallurgical Society of AIME, Warrendale, PA., Pp. 63–68.

Sanders Jr., T.H., Starke Jr., E.A., 1982. The effect of slip distribution on the monotonic and cyclic ductility of Al–Li binary alloys. Acta Metall. 30, 927–939.

Sanders, T.H., Starke, E.A., 1984. Overview of the physical metallurgy in the Al–Li–X systems. In: Sanders, T.H., Starke, E.A. (Eds.), Proceedings of the Second International Conference on Aluminium–Lithium Alloys. The Metallurgical Society of AIME, Warrendale, PA, pp. 1–16.

Sanders, T.H., Starke, E.A., 1989. The physical metallurgy of aluminium–lithium alloys—a review. In: Sanders, T.H., Starke, E.A. (Eds.), Aluminium–Lithium Alloys, vol. **1**. Materials and Component Engineering Publications, Birmingham, pp. 1–37.

Sankaran, K.K., Grant, N.J., 1980. The structure and properties of splat-quenched aluminium alloy 2024 containing lithium additions. Mater. Sci. Eng. 44, 213−227.

Silcock, J.M., 1959−1960. J. Inst. Met. 88, 357−364.

Smith, S.W., Scully, J.R., 2000. The identification of hydrogen trapping states in an Al−Li−Cu−Zr alloy using thermal desorption spectroscopy. Metall. Mater. Trans. A 31a, 179−193.

Srivatsan, T.S., Coyne, E.J., 1986. Cyclic stress response and deformation behaviour of precipitation-hardened aluminium−lithium alloys. Inter. J. Fatigue 8, 201−208.

Srivatsan, T.S., Coyne, E.J., 1987. Mechanisms governing cyclic fracture in an Al−Cu−Li alloy. Mater. Sci. Technol. 3, 130−138.

Srivatsan, T.S., Lavernia, J., 1991. The presence and consequences of precipitate free zones in an aluminium−copper−lithium alloy. J. Mater. Sci. 26, 940−950.

Srivatsan, T.S., Place, T.A., 1989. Microstructure, tensile properties and fracture behaviour of an Al−Cu−Li−Mg−Zr alloy 8090. J. Mater. Sci. 24, 1543−1551.

Srivatsan, T.S., Coyne, E.J., Strake, E.A., 1986a. Microstructural characterization of two lithium containing aluminium alloys. J. Mater. Sci. 21, 1553−1560.

Srivatsan, T.S., Yamaguchi, Y., Starke Jr., E.A., 1986b. The elevated temperature low cycle fatigue behavior of aluminum alloy 2020. Mater. Sci. Eng. 83, 87−107.

Starke, E.A., 1977. Aluminum alloys of the 70's: scientific solutions to engineering problems. Mater. Sci. Eng. 29, 99−114.

Starke, E.A., Lin, F.S., 1982. The influence of grain structure on the ductility of the Al−Cu−Li−Mn−Cd alloy 2020. Metall. Trans. 13A, 2259−2269.

Starke, E.A., Sanders, T.H., Palmer, I.G., 1981. New approaches to alloy development in the Al−Li system. J. Met. 33, 24−36.

Suresh, S., Vasudevan, A.K., Tosten, M., Howell, P.R., 1987. Microscopic and macroscopic aspects of fracture in lithium containing aluminium alloys. Acta Metall. 35, 25−46.

Takahashi, K., Minakawa, K., Ouchi, C., 1987. The effect of thermomechanical processing variables on anisotropy in mechanical properties of Al−Li alloys. In: Champier, G., Dubost, B., Miannay, D., Sabetay, L. (Eds.), Proceedings of the Fourth International Conference on Aluminium−Lithium Alloys. J. Phys. Colloque 48, C3.163−C3.169.

Tempus, G., Calles, W., Scharf, G., 1991. Influence of extrusion process parameters and texture on mechanical properties of Al−Li extrusions. Mater. Sci. Technol. 7, 937−945.

Tosten, M.H., Vasudevan, A.K., Howell, P.R., 1986. Microstructural development in Al−2% Li−3%Cu alloy. In: Baker, C., Gregson, P.J., Harris, S.J., Peel, C.J. (Eds.), Proceedings of the Third International Conference on Aluminium−Lithium Alloys. The Institute of Metals, London, pp. 483−489.

Vasudevan, A.K., Doherty, R.D., 1987. Grain boundary ductile fracture in precipitation hardened aluminum alloys. Acta Metall. 35 (6), 1193−1203.

Vasudevan, A.K., Suresh, S., 1985. Microstructural effects on quasi-static fracture mechanisms in Al−Li alloys: the role of crack geometry. Mater. Sci. Eng. 72, 37−49.

Vasudevan, A.K., Fricke Jr., W.G., Malcolm, R.C., Bucci, R.J., Przystupa, M.A., Barlat, E., 1988. On through thickness crystallographic texture gradient in Al−Li−Cu−Zr alloy. Metall. Trans. A 19a, 731−732.

Webster, D., 1986. Temperature dependence of toughness in various aluminium−lithium alloys. In: Baker, C., Gregson, P.J., Harris, S.J., Peel, C.J. (Eds.), Proceedings of the Third International Conference on Aluminium−Lithium Alloys. The Institute of Metals, London, pp. 602−609.

Webster, D., 1987. The effect of low melting point impurities on the properties of aluminum–lithium alloys. Metall. Trans. 18A, 2187–2193.

Webster, D., 1988. Aluminum–lithium powder metallurgy alloys with improved toughness. Metall. Trans. A 19a, 603–615.

Webster, D., 1989. Effect of alkali metal impurities on the toughness of aluminium–lithium alloys. In: Sanders Jr., T.H., Starke Jr., E.A. (Eds.), Proceedings of the Fifth International Conference on Aluminum–Lithium Alloys, vol. **1**. Materials and Component Engineering Publications, Birmingham, pp. 497–518.

Chapter 11

Fatigue Behavior of Aluminum–Lithium Alloys

N. Eswara Prasad*, T.S. Srivatsan**, R.J.H. Wanhill[†], G. Malakondaiah[‡], and V.V. Kutumbarao[§]

*Regional Centre for Military Airworthiness (Materials), CEMILAC, Hyderabad, India, **Department of Mechanical Engineering, The University of Akron, Akron, USA, [†]NLR, Emmeloord, the Netherlands, [‡]Office of the Chief Controller R&D (Aero), DRDO HQs., New Delhi, India, [§]Defence Metallurgical Research Laboratory, Hyderabad, India

Contents

11.1 Introduction	**342**	
11.2 The Phenomenon of Fatigue	**342**	
Part A: LOW CYCLE FATIGUE (LCF)	**344**	
11.3 LCF Behavior	**344**	
11.4 Test Methods and Analyses	**345**	
11.4.1 Characterization of Cyclic Stress Response Behavior	347	
11.4.2 Characterization of CSS Behavior	347	
11.5 LCF Behavior of Aluminum–Lithium Alloys	**348**	
11.5.1 General Survey/ Microstructural and Environmental Effects	348	

11.5.2 Fatigue Life Power Law Relationships 349
11.5.3 CSR Behavior 355
11.5.4 CSS Behavior 357
11.5.5 Fatigue Toughness 359
11.5.6 The LCF Resistance of Aluminum– Lithium Alloys 362
Part B: HIGH CYCLE FATIGUE (HCF) **363**
11.6 Introduction to the HCF Behavior of Aluminum Alloys **363**
11.7 Background on Test Methods **366**
11.8 HCF Behavior of Aluminum–Lithium Alloys **366**
11.8.1 General Survey 366

11.8.2 Effects of Lithium
       Content, Aging,
       and Cold Work   366
11.8.3 HCF Behavior of
       AA 2020-T651—
       A First-Generation
       Al–Li Alloy   367
11.8.4 HCF Behavior of
       AA 8090-T651—
       A Second-
       Generation
       Al–Li Alloy   369
11.8.5 HCF Behavior of
       AA 2098—A Third-

       Generation Al–Li
       Alloy   370
11.8.6 The HCF Resistance
       of Al–Li Alloys:
       Smooth and
       Notched
       Properties, and
       a Note of
       Caution   371
11.9 Summary and
      Conclusions   371
11.10 Final Remarks   374
List of Symbols   375
References   375

11.1 INTRODUCTION

In recent years, the technologies related to the emerging alloys of aluminum, including Al–Li alloys, have grown stronger with competition, with a strong emphasis on both physical and mechanical properties of the alloys and the dominant role they play in structural applications. A majority of these structural parts are subjected to fatigue. New Al–Li based alloys in different tempers have evolved to offer enhanced fatigue resistance with or without simultaneous improvements in other mechanical properties and corrosion resistance. The objective of this chapter is to present and discuss the fatigue behavior of the family of aluminum–lithium alloys to aid in their selection and appropriate use for a variety of performance-critical applications. The fatigue behavior of the emerging Al–Li alloys is compared one-on-one with their earlier versions and also with the most widely used AA 2XXX and AA 7XXX series aluminum alloys.

11.2 THE PHENOMENON OF FATIGUE

Fatigue often involves microstructural damage and the resultant failure of materials and structures when subjected to cyclically varying loads. Structural materials are carefully designed with compositions and microstructures optimized for achieving enhanced fatigue resistance. While metallic alloys are generally designed for strength, their structural integrity is often limited by their mechanical performance under the influence of cyclic loads. In fact, it is believed that over 80% of all service-related failures can be traced to mechanical fatigue, often in association with (a) cyclic plasticity, (b) sliding or physical contact (fretting fatigue and rolling contact fatigue), (c) environment-induced damage (corrosion and corrosion fatigue), or (d) elevated temperatures (creep-fatigue).

Despite abundant information on aspects related to fatigue of metals and their composite counterparts, there exists a need to carefully study the fatigue behavior of the emerging alloys for both selection and use in structural applications where cyclic loading is critical. It is important to appreciate that the fatigue process is conditioned by cyclic plastic deformation. Without repeated plastic deformation, at least on a microstrain level, there would be no fatigue at all. When applied only once, this plastic strain does not cause any significant changes in the substructure of the material. However, multiple repetitions of even small amounts of plastic deformation lead to cumulative damage culminating in failure. When the plastic strain is very small, not measurable macroscopically, fatigue failure occurs after a large number of load cycles, typically 10^8 or more. This is termed *high cycle fatigue (HCF)*. On the other hand, if the plastic strain per cycle is a measurable quantity (typically $0.05-2\%$), fatigue failure occurs after the application of only a small number of load cycles, typically from 10 to 10^4. This is termed *low cycle fatigue (LCF)*.

On the basis of the types of irreversible changes caused by cyclic plastic deformation, it is possible to divide the fatigue process into three consecutive and partly overlapping stages:

a. *Cyclic hardening and/or cyclic softening*: This takes place in the cyclically loaded volume and depends both on the initial state of the material and the applied stress amplitude (or strain amplitude, as the case may be). Such hardening or softening behavior is reflected in increasing or decreasing trends, respectively, in the observed values of the stress amplitude (or strain amplitude, as the case may be) with increasing number of load cycles.

b. *Microscopic crack initiation*: This takes place in a small part of the cyclically loaded volume, mostly at slip bands grain boundaries, inclusions, or pores in the surface layers. In ductile solids microscopic fatigue crack initiation is almost always preceded by the localization of slip.

c. *Crack propagation culminating in final failure*: The controlling factor of crack propagation is highly concentrated cyclic plastic deformation within the plastic zone at the crack tip.

The relative preponderance of the three stages depends on the test conditions, such as the temperature, the strain rate, and the applied stress or strain amplitude.

In the present chapter the strain-controlled and stress-controlled approaches to fatigue that essentially govern the kinetics behind crack initiation, stages (a) and (b) referred to above, are the basis for presentation and discussion. The fatigue crack propagation characteristics, stage (c), are discussed in **Chapter 12**.

PART A: LOW CYCLE FATIGUE (LCF)

11.3 LCF BEHAVIOR

The majority of components and structures known or expected to undergo fatigue in service are designed to withstand a finite number of load or stress excursions, referred to in the engineering context as fatigue cycles, turning points, or reversals. Well-known and extensively used examples of service components and structures subjected to fatigue include: (i) rotating axles and shafts of automobiles, (ii) aircraft wings and fuselages, and (iii) aeroengine turbine vanes, blades, and disks. Some of these components, notably those used in aeroengines, are designed based on resistance to damage by LCF since they experience a significant amount of plastic strain during each reversal of loading. The LCF resistance of structural materials is analyzed in terms of the variation of fatigue life (number of cycles to failure (N_f) or the number of (stress) reversals to failure ($2N_f$)) with respect to the following parameters:

i. Average stress amplitude ($\Delta\sigma/2$)
ii. Plastic strain amplitude ($\Delta\varepsilon_p/2$)
iii. Plastic strain energy per cycle (ΔW_p).

Several empirical power law relationships have been proposed to describe the above. The most important ones are:

$$\frac{\Delta\sigma}{2} = \sigma_f'(2N_f)^b \quad \text{Basquin (1910)} \tag{11.1}$$

$$\frac{\Delta\varepsilon_p}{2} = \varepsilon_f'(2N_f)^c \quad \text{Coffin (1954) and Manson (1953)} \tag{11.2}$$

$$\Delta W_p = W_f'(2N_f)^\beta \quad \text{Morrow (1965) and Halford (1966)} \tag{11.3}$$

where σ_f', ε_f' are fatigue strength and fatigue ductility coefficients; b and c are fatigue strength and fatigue ductility exponents; W_f' is the plastic strain energy coefficient; and β is the plastic strain energy exponent ($\beta = b + c$).
ΔW_p is computed as

$$\Delta W_p = \Delta\varepsilon_p\Delta\sigma\left[\frac{1 - n'}{1 + n'}\right] \tag{11.4}$$

where n' is the cyclic work hardening exponent, numerically equal to b/c.

These power law relationships are extremely useful in designing components that have finite lives ($N_f < 10^4$ cycles) owing to damage induced by cyclic plastic strain. The basic design diagram is shown in Figure 11.1, which is an idealized plot showing the variation of alternating strain (ε_a) with the number of reversals to failure ($2N_f$). The alternating elastic strain is given by the expression $\Delta\sigma/2E$. The alternating plastic strain is derived by subtracting the elastic strain from the total strain. Further,

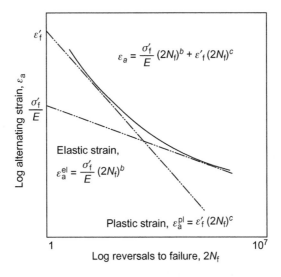

FIGURE 11.1 Basic LCF strain-life diagram (applied strain (ε_a) indicated in the figure corresponds to elastic (ε_a^{el}), plastic (ε_a^{pl}), and total (ε_a^{t}), respectively, for the elastic, plastic, and total strain-life equations).

1. The elastic strain equation is derived from Eq. (11.1), where σ_f' is the fatigue strength coefficient and b is the Basquin fatigue strength exponent (Basquin equation).
2. The plastic strain equation is derived from Eq. (11.2), where ε_f' is the fatigue ductility coefficient and c is the fatigue ductility exponent (Coffin–Manson equation).
3. Equations (11.1) and (11.2) may be combined to give the overall LCF strain-life equation:

$$\varepsilon_a = \frac{\sigma_f'}{E}(2N_f)^b + \varepsilon_f'(2N_f)^c \tag{11.5}$$

11.4 TEST METHODS AND ANALYSES

The parameters in Eqs. (11.1), (11.2), and (11.5) are determined from fully reversed strain-controlled tests performed on smooth laboratory-scale specimens. Equation (11.5) is used in LCF design analysis in conjunction with the cyclic stress–strain (CSS) curve, which is also determined from fully reversed strain-controlled tests performed on smooth laboratory-scale test specimens. Test results are plotted in the controlled strain amplitude (ε_a) versus dependent stress amplitude (σ_a) space. Because of the irreversibility of the plastic strain component of the total applied strain amplitude, the variation of σ_a with ε_a exhibits "hysteresis" and the resultant graph is called a "hysteresis loop" (Figure 11.2).

Determination of the parameters mentioned above (discussed in detail in Sections 11.4.1 and 11.4.2) requires the formation of stable hysteresis loops during cyclic strain-controlled testing. Stable hysteresis loops are used to obtain the stress amplitude ($\Delta\sigma/2 = \sigma_a$). The plastic strain amplitude is obtained from the expression

$$\Delta\varepsilon_p/2 = 1/2(\Delta\varepsilon_t - \Delta\sigma/E) \tag{11.6}$$

where $\Delta\varepsilon_t$ is the total strain range and E is the elastic modulus of the chosen material. Then from the plots showing (i) the variation of stress amplitude (σ_a) with reversals to failure ($2N_f$) and (ii) the variation of plastic strain amplitude ($\Delta\varepsilon_p/2$) with reversals to failure ($2N_f$), the coefficients σ_f' and ε_f' and the exponents b and c are determined in conformance with Eqs. (11.1) and (11.2).

In following the above procedure, complications arise due to the occurrence of cyclic softening or hardening. Furthermore, the idealized power law relationship represented by Eq. (11.2) may be invalid, particularly in the case of Al–Li alloys. These aspects are discussed in Section 11.5.2 and further elaborated in Section 11.5.5.

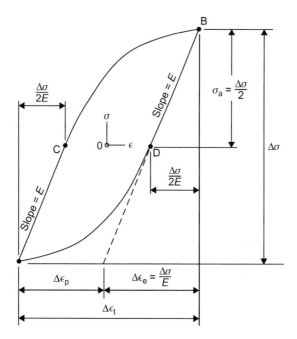

FIGURE 11.2 Typical LCF hysteresis loop showing the experimentally determined parameters.

11.4.1 Characterization of Cyclic Stress Response Behavior

From fully reversed strain-controlled tests performed on smooth specimens we obtain the number of reversals (one reversal = ½ cycle) to failure ($2N_f$) for different total strain amplitudes ($\Delta\varepsilon_t/2 = \varepsilon_a$). The data recorded for each test as a function of the number of elapsed cycles include the following:

a. Tensile peak stress (σ_T)
b. Compressive peak stress (σ_c)
c. Stress range ($\Delta\sigma = \sigma_T + |\sigma_c|$)
d. Plastic strain amplitude ($\Delta\varepsilon_p/2$).

These parameters are defined in Figure 11.2.

A variation of average stress amplitude ($\Delta\sigma/2$) with elapsed fatigue cycles (N or a fraction of fatigue life, N/N_f) indicates whether the chosen material undergoes cyclic softening, defined by a gradual decrease in stress amplitude ($\Delta\sigma/2$) with cycles (N or N/N_f), or cyclic hardening, defined as a gradual increase in stress amplitude ($\Delta\sigma/2$) with cycles (N or N/N_f). If the stress amplitude during fully reversed strain cycling remains constant, that is, there is no observable change in stress amplitude ($\Delta\sigma/2$) with cycles (N or N/N_f), then the material shows cyclic stability. These three possibilities represent the basic cyclic stress response (CSR) behavior. However, most engineering alloys tend to show a mixed fourth type of stress response behavior during fully reversed strain cycling, with an initial increase in stress amplitude ($\Delta\sigma/2$), attaining a peak in $\Delta\sigma/2$, followed by a gradual decrease in stress amplitude ($\Delta\sigma/2$) culminating in fracture. This mixed type of behavior may or may not have a region of constant $\Delta\sigma/2$.

11.4.2 Characterization of CSS Behavior

The CSS curve is defined as the locus of the tips of stable hysteresis loops, obtained from fully reversed strain-controlled tests on smooth laboratory scale test specimens for different applied strain amplitudes (ε_a), as shown in Figure 11.3. The CSS data are analyzed in conformance with the power law relationship

$$\Delta\sigma/2 = K'(\Delta\varepsilon_p/2)^{n'} \qquad (11.7)$$

where K' is the cyclic work hardening coefficient and n' is the cyclic strain-hardening exponent.

Higher or lower values of K' and n' when compared to the monotonic tensile or compressive work hardening coefficient (K) and exponent (n), respectively, provide an indication as to whether a material is prone to exhibit cyclic hardening ($K' > K$ and $n' > n$) or cyclic softening ($K' < K$ and $n' < n$). Also, variation of stress amplitude ($\Delta\sigma/2$) with plastic strain amplitude ($\Delta\varepsilon_p/2$) reveals whether or not the cyclic strain-hardening exponent (n')

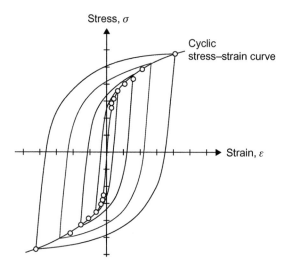

FIGURE 11.3 Derivation of the CSS curve from stable hysteresis loops obtained at different applied strain amplitudes.

changes as a function of plastic strain amplitude ($\Delta\varepsilon_p/2$). Changes in the cyclic strain-hardening exponent (n') are clear indications of changes in deformation mode or mechanism.

11.5 LCF BEHAVIOR OF ALUMINUM–LITHIUM ALLOYS

11.5.1 General Survey/Microstructural and Environmental Effects

The LCF behavior of lithium-containing aluminum alloys has been the subject of several studies in recent years. These studies have attempted to examine the intrinsic influence of microstructural features and the extrinsic influence of both environment and test temperature. The microstructural features that have been investigated include the following:

i. *The precipitate state that emerges with artificial aging*: (Coyne et al., 1981; Dhers et al., 1986; Eswara Prasad and Rama Rao, 2000; Eswara Prasad et al., 1996a,b, 1997, 2004; Srivatsan, 1984; Srivatsan and Coyne, 1986a, 1987; Srivatsan et al., 1986a,b, 1991a,b).

We note here that the microstructural effects arising from artificial aging of Al–Li alloys are complex. Intrinsic variations in the extent and/or degree of aging were found to result in: (a) different amounts and size of the strengthening precipitates Al_3Li (δ'); (b) varying amounts of Al_2CuMg (S'), Al_2CuLi (T_1); and (c) coarse equilibrium precipitates, such as Al_6CuLi_3 (T_2), AlLi (δ), and Al_2Cu (θ). Aging for prolonged periods of time or at higher temperatures was found to result in δ' precipitate-free zones (δ'-PFZs).

The important findings here are that the lithium additions without Cu and/or Mg as well as high degree of artificial aging drastically lower the LCF resistance, leading to significant reductions in all the LCF properties (particularly, ε'_f as well as b and c). This is attributed to the coarsening of δ' precipitates and the resultant higher degree of slip localization through precipitate shearing and strain localization along PFZs at high angle grain boundaries. The PFZs result in lower energy ductile intergranular fracture.

ii. *Influence of co-precipitation*: (Khireddine et al., 1989; Sanders and Starke, 1982)

Noteworthy here is the beneficial effect of S' phase (formed due to simultaneous additions of Cu and Mg), which results in a considerable degree of slip dispersal.

iii. *Degree of recrystallization*: (Srivatsan et al., 1986a)

It has been shown that extensive recrystallization lowers the LCF resistance.

iv. *Anisotropy in LCF properties*: Studies aimed at investigating and understanding anisotropy in LCF properties are limited. Khireddine et al. (1989) showed that the LCF resistance of extruded flat bars of Al–Li alloy AA 8090 was significantly anisotropic. This alloy exhibited longer LCF fatigue lives (up to a factor of 2) at the higher cyclic strain amplitudes in the longitudinal (L) direction of the bar cores when compared to the long-transverse (LT) direction of the bar. The observed anisotropy was considered to be due to variations in both grain size and crystallographic texture as a function of test specimen orientation. Similar anisotropy studies were conducted and reported for in-plane directions by Eswara Prasad and coworkers (Eswara Prasad, 1993; Eswara Prasad et al., 1996a,b). These studies showed that the degree of in-plane anisotropy in strain amplitude-controlled LCF properties was as pronounced as that for the tensile properties.

11.5.2 Fatigue Life Power Law Relationships

An observation that could be significant from an engineering design perspective is that, during strain amplitude-controlled LCF tests, most Al–Li alloys show bilinear (also known as "dual slope") behavior instead of a linear variation in log–log plots suggested by Eqs. (11.1)–(11.3) (see Section 11.3). This is illustrated with the help of LCF life data of an AA 8090 alloy (Figure 11.4). Clearly, the hypotransition and hypertransition regions shown in Figure 11.4 have different power law constants. Values of the constants reported for pure aluminum and a variety of Al–Li alloys are given in Table 11.1. Most of the alloys exhibit a bilinear behavior; the exceptions

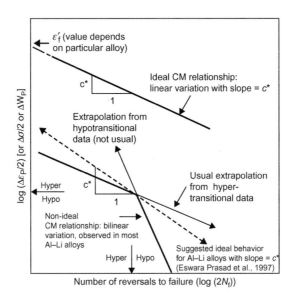

FIGURE 11.4 Schematic diagram showing ideal (single slope) and nonideal (bilinear) LCF power law relationships. Common extrapolations employed in the literature are also shown. Arrows in the hypo- and hyper-transitional regions of the bilinear plot show possible explanations for the degradation of LCF resistance in various structural alloys, especially in Al–Li alloys (both axes are on log–log scales.)

being pure aluminum, Al–0.7 wt% Li, Al–2.5 wt% Li (PA), Al–3 wt% Li + Mn (UA), and AA 8090-T61, which exhibit a simple linear behavior.

It follows that extrapolations to obtain the fatigue life values at untested strain amplitudes for the purpose of engineering design must be done with caution. It is possible to overestimate the LCF life in *both* the hypotransitional and hypertransitional regions, if the extrapolations are based solely on test data from one of the two regions.

Eswara Prasad et al. (1997) have suggested an extrapolation procedure that may be more appropriate in such cases: let c^* denotes the fatigue ductility exponent when the ideal Coffin–Manson power law relationship is obeyed (Figure 11.4). Now draw a straight line with slope equal to c^*, passing through the transition point in the bilinear log–log plot (shown as a dashed line in Figure 11.4). This dashed line is suggested to be the appropriate ideal extrapolation line, which theoretically predicts the maximum possible LCF resistance. The deviation of this ideal extrapolation from the two extrapolations based on the actual data shows the extent of degradation that occurs in the LCF resistance of any particular Al–Li alloy.

Bilinear behavior may also be observed in the Basquin and Halford–Morrow equations (Eqs. (11.1) and (11.3)) when they follow a similar fatigue life dependence on $\Delta\sigma/2$ and ΔW_p, respectively.

TABLE 11.1 Fatigue Life Power Law Constants (σ_f', ϵ_f', W_f', b, c, and β) for Al–Li Alloys Exhibiting Bilinear Behavior.

Alloy Designation and Heat-Treatment Condition	Test Condition and Specimen Location and Orientation	Lower Strains (Hypotransition Regions)						Higher Strains (Hypertransition Regions)					
		σ_f' MPa	ϵ_f'	W_f' MJ m^{-3}	$-b$	$-c$	$-\beta$	σ_f' MPa	ϵ_f'	W_f' MJ m^{-3}	$-b$	$-c$	$-\beta$
Al (99.98%)[a]	Standard	—	0.87	25.2[b]	—	0.62	0.59[b]	—	0.87	25.2[b]	—	0.62	0.59[b]
Al–0.7Li[a]	Standard	—	0.42	36.6[b]	—	0.60	0.56[b]	—	0.42	36.6[b]	—	0.60	0.56[b]
Al–2.5Li, UA	Standard	—	>10^3	(>10^5)[b]	—	2.38	2.31[b]	—	2.1	811	—	0.798	0.85[b]
Al–2.5Li, PA[a]	Standard	—	0.032	36.1[b]	—	0.71	0.77[b]	—	0.032	36.1[b]	—	0.71	0.77[b]
Al–3Li + Mn, UA[a]	LA or distilled H$_2$O	—	0.059	85.1[b]	—	0.76	0.74[b]	—	0.059	85.1[b]	—	0.76	0.74[b]
Al–3Li + Mn, PA	LA or distilled H$_2$O	—	0.146	220[b]	—	0.96	0.96[b]	—	0.021	37.0[b]	—	0.48	0.52[b]
AA 2020-T651, PR	298 K, LA, V	—	97	71,660[b]	—	1.60	1.47[b]	—	0.038	83.6[b]	—	0.451	0.45[b]
	433 K, LA	—	245	(>10^5)[b]	—	2.0	1.94[b]	—	0.052	99.3[b]	—	0.62	0.62[b]
	433 K, V	—	2.5	—	—	1.3	—	—	0.052	—	—	0.62	—
AA 2020-T6X51, UR	298 K, LA, V	—	250	(>10^4)[b]	—	1.56	1.57[b]	—	0.058	106[b]	—	0.448	0.45[b]
	433 K, LA	—	379	(>10^5)[b]	—	3.0	3.57[b]	—	0.23	414[b]	—	0.64	0.64[b]
	433 K, V	—	>10^3	(>10^{10})[b]	—	1.75	1.72[b]	—	0.23	414[b]	—	0.64	0.64[b]
AA 2090-T8E41	Standard, L	—	>10^3	(>10^5)[b]	—	3.8	3.4[b]	—	0.08	227[b]	—	0.60	0.63[b]

(Continued)

TABLE 11.1 (Continued)

Alloy Designation and Heat-Treatment Condition	Test Condition and Specimen Location and Orientation	Lower Strains (Hypotransition Regions)						Higher Strains (Hypertransition Regions)					
		σ'_f MPa	ε'_f	W'_f MJ m^{-3}	$-b$	$-c$	$-\beta$	σ'_f MPa	ε'_f	W'_f MJ m^{-3}	$-b$	$-c$	$-\beta$
AA 8090-T3	Standard, L	—	4.07	32,328	—	1.17	1.41	—	0.051	99.2	—	0.48	0.51
AA 8090-T61,	L, core	—	0.35	981[b]	—	0.97	1.0[b]	—	0.35	981	—	0.97	1.0
CP271[a,c]	L, surface	—	0.076	192[b]	—	0.74	0.8[b]	—	0.076	192[b]	—	0.74	0.8[b]
	LT, core	—	0.062	157[b]	—	0.74	0.8[b]	—	0.062	157[b]	—	0.74	0.8[b]
AA 8090-T8E51	L	887	5.5	30,043	0.093	1.15	1.35	575	0.06	110.1	0.035	0.46	0.49
									0.054[d]				
	L + 45°	662	2.7	80,168	0.67	1.02	1.44	501	0.12	473.0	0.037	0.56	0.66
									0.105[d]				
	LT	832	1.77	2835	0.8	0.95	0.94	562	0.096	258.8	0.042	0.52	0.57
									0.07[d]				
AA 8090-T3	L	631	4.1	3x10^4	0.06	1.17	1.40	521	0.051	100	0.033	0.48	0.514
							1.23[e]		0.063[d]				0.513[e]

[a] Alloys exhibit ideal (single slope) power law relationships.
[b] Computed values ($\Delta\sigma/2$ is approximated by the yield strength values).
[c] Limited strain amplitude range investigated.
[d] Monotonic tensile ductility values.
[e] Values of β obtained from $\beta = b + c$.
UA, underaged; PA, peak aged; PR, partially recrystallized; UR, unrecrystallized; LA, laboratory air; V, vacuum; L, longitudinal direction; LT, long-transverse direction. Data from Coyne et al. (1981), Dhers et al. (1986), Eswara Prasad (1993), Eswara Prasad and Rama Rao (2000), Eswara Prasad et al. (1996a, 1997, 2004), Khireddine et al. (1989), and Srivatsan et al. (1986a, 1991a,b).

The observed bilinear behavior can be due to any one or a combination of the following mechanisms (Chen and Starke, 1984; Eswara Prasad et al., 1989, 1994, 1996a,b,c, 2004; Feng et al., 1984; Khireddine et al., 1988, 1989; Oh et al., 1999; Sanders et al., 1980; Srivatsan and Coyne, 1986a,b; Srivatsan et al., 1986a,b, 1991a,b; Xu et al., 1990):

a. change in the nature of deformation, as reflected by a change in n', the cyclic work hardening exponent or directly observed differences in slip homogeneity;
b. changes in the deformation and/or environment-assisted fracture mechanisms;
c. varying extent and/or severity of degradation due to the environment.

Additional details pertaining to this specific aspect are given by Eswara Prasad et al. (1996a, 1997) and Eswara Prasad and Rama Rao (2000). The mechanisms responsible for changes in the deformation behavior occurring at the fine microscopic level and the resultant bilinear behavior are summarized in Table 11.2. The observed difference in deformation mechanisms is largely influenced by the nature, size, and volume fraction of the major strengthening precipitates. Some of these mechanisms are mentioned in Section 11.5.3 and micrographic illustrations are provided in Figure 11.5.

Use of Eq. (11.3) (Halford–Morrow) is appropriate for analyzing the fatigue data in terms of plastic strain energy, since this cannot be done by Eq. (11.1) (Basquin) and Eq. (11.2) (Coffin–Manson). Use of this approach has shown that for AA 8090 the peak-aged (PA) temper has better resistance to LCF damage than the underaged (UA) temper (Eswara Prasad et al., 2004)—a result not normally expected. The inferior LCF resistance experienced by the UA condition was attributed to the mutually interactive influences of the following: (a) a more recrystallized grain structure, (b) lower content of S' precipitates, and (c) finer δ' precipitate size. In a comprehensive study based on ΔW_p, Eswara Prasad and Rama Rao (2000) have analyzed the effects of alloy composition, degree of aging, degree of recrystallization, and crystallographic texture on LCF resistance of the lithium-containing aluminum alloys, and have concluded that most of the Al–Li alloys exhibit degradation in LCF resistance due to the combined effects of: (i) mechanical fatigue, (ii) strain localization through dislocation–precipitate interactions (δ' and T_1), (iii) environmental effects, and (iv) strain localization through high angle grain boundaries.

It is interesting to note that the slope transition in log–log plots based on the above power laws often occurs at plastic strain amplitudes of $(1 - 3) \times 10^{-3}$. This observation has prompted fatigue researchers to relate the transition to a change from either microscopic to macroscopic deformation mode, or a change from elastic-dominant deformation to plastic-dominant deformation behavior (Radhakrishnan, 1992; Thielen et al., 1976).

TABLE 11.2 Changes in Deformation Behavior Causing LCF Power Law Bilinearity in Al–Li Alloys

Al–Li Alloy Series	Major Strengthening Precipitates	CSR Behavior and n′ Values		Deformation Mode	
		Low Strain Amplitudes (Hypotransition)	High Strain Amplitudes (Hypertransition)	Low Strain Amplitudes (Hypotransition)	High Strain Amplitudes (Hypertransition)
Al–2.5 wt% Li (UA)	Al_3Li (δ')	Cyclic hardening followed by near stable behavior; slightly higher n′ than at high strain amplitudes	Cyclic hardening followed by near stable behavior; low n′	Shearing of δ' precipitates suggested, but no experimental evidence	Strain localization along soft δ' PFZs suggested, but no experimental evidence
Al–3Li + Mn (PA)	Al_3Li (δ'), Al_6Mn	Predominantly cyclic hardening	Predominantly cyclic hardening	Strain localization at precipitate-free zones; decreasing strain localization with increase in strain amplitude; no shear bands	Strain localization via precipitate-free zones; decreasing strain localization with increase in strain amplitude; no shear bands
AA 2020-T651, partially recrystallized	Al_2Cu (θ'), Al_2CuLi (T_1), Al_3Li (δ')	High degree of cyclic softening; n′ independent of strain amplitude	Low degree of cyclic softening; n′ independent of strain amplitude	Inhomogeneous deformation with intense shear bands; single slip	Homogeneous deformation; multiple slip
AA 2090-T8E41	Al_3Li (δ'), Al_2CuLi (T_1), Al_2Cu (θ')	Near stable behavior; small degree of cyclic softening	Cyclic softening; increased softening with increase in strain amplitude	—	—
AA 8090-T61, CP271	Al_3Li (δ'), Al_2CuMg (S)	Cyclic hardening followed by stability; high n′	Cyclic hardening followed by softening; low n′	Uniform dislocation structure with sinuous deformation bands; no δ' shearing	Intense and multiple shearing of δ'; densely distributed coarse shear bands
AA 8090-T8E51	Al_3Li (δ'), Al_2CuMg (S)	Cyclic hardening followed by stability; high n′	Cyclic hardening followed by softening; low n′	Uniform dislocation cell-like structure; no δ' shearing (Figure 11.5A–C)	Intense δ' shearing; densely distributed coarse shear bands; single slip (Figure 11.5D–F)

Data from Coyne et al. (1981), Dhers et al. (1986), Eswara Prasad (1993), Eswara Prasad and Rama Rao (2000), Eswara Prasad et al. (1996a, 1997, 2004), Khireddine et al. (1989), and Srivatsan et al. (1986a, 1991a,b).

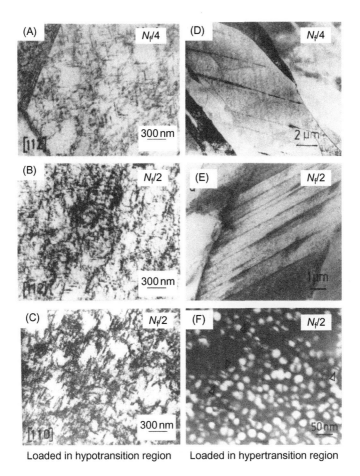

Loaded in hypotransition region   Loaded in hypertransition region

FIGURE 11.5 Transmission electron micrographs obtained from LCF specimens of Al–Li alloy AA 8090-T8E51 loaded to different fractions of fatigue life (N/N_f) at low (A–C, hypotransition region) and high (D–F, hyper-transition region) strain amplitudes. Note the change in deformation mode from (i) a cell-like uniform dislocation structure at both $N_f/4$ (A) and $N_f/2$ (B and C) for a low strain amplitude ($\Delta\varepsilon_p/2 = 4.5 \times 10^{-5}$) to (ii) a highly heterogeneous deformation structure of intense single slip that resulted in widely spaced shear bands at $N_f/4$ (D) and closely spaced shear bands at $N_f/2$ (E), together with (iii) intense shearing of δ' precipitate (F) at a high strain amplitude of ($\Delta\varepsilon_p/2 = 9 \times 10^{-3}$).

11.5.3 CSR Behavior

As mentioned in Section 11.4.1, the variation of average stress amplitude ($\Delta\sigma/2$) with elapsed fatigue cycles (N or a fraction of fatigue life, N/N_f) indicates whether the material has the propensity to undergo cyclic softening or cyclic hardening, or a combination of both. Results of fully reversed cyclic loading tests on δ'-strengthened Al–Li alloys show a general trend of:

i. an initial hardening with the occurrence of cyclic stability at the lower strain amplitude or **ii.** an initial hardening followed by cyclic softening at the higher strain amplitudes (Dhers et al., 1986; Eswara Prasad et al., 1989, 1992, 1994, 1996a,b; Khireddine et al. 1988; Srivatsan et al., 1991a,b).

This general trend is illustrated in Figure 11.6, which shows the CSR results for the quaternary Al–Li–Cu–Mg alloy AA 8090 in the T8E51 temper. CSR plots at the lower strain amplitudes, i.e., curves A–C, reveal cyclic hardening followed by stability for most of the fatigue life. However, CSR plots at the higher strain amplitudes, i.e. curves D and E, reveal a considerable degree of hardening followed by softening.

The microscopic mechanisms responsible for the observed cyclic hardening or softening during fully reversed loading have been extensively studied for many materials, including Al–Li alloys. Cyclic hardening is due to an increase in dislocation density, coupled with an increase in dislocation–dislocation and dislocation–precipitate interactions, with small yet observable contributions from the creation and presence of antiphase

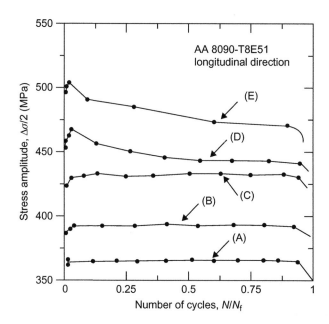

FIGURE 11.6 CSR of Al–Li alloy AA 8090-T8E51 longitudinal specimens tested at several $\Delta\varepsilon_t/2$ or ε_a values. The corresponding $\Delta\varepsilon_p/2$ values are: (A) 0.009%; (B) 0.03%; (C) 0.075%; (D) 0.2%; (E) 0.53%. *Source: Data from Eswara Prasad et al. (1992). Similar behavior has been reported for other alloys by Dhers et al. (1986), Khireddine et al. (1988), and Srivatsan et al. (1991a,b).*

boundaries (Sanders and Starke, 1977; Sanders et al., 1980). The cyclic softening in Al–Li alloys has been attributed to the shearing of the fully ordered
and coherent δ' precipitates, with or without a significant amount of local disordering of the precipitates (Calabrese and Laird, 1974a,b; Eswara Prasad
et al., 1996a; Sanders and Starke, 1982; Srivatsan and Coyne, 1986a,b). There
is an insignificant contribution, if any, from the other mechanisms responsible
for facilitating softening during fully reversed cyclic deformation, namely:

a. overaging (proposed by Lynch and Ryder (1973));
b. aging inhomogeneities (proposed by Laird and Thomas (1967));
c. precipitate reversion or resolutionizing (proposed by McEvily et al.
    (1963) and Abel and Ham (1966)).

Although the trend shown in Figure 11.6 seems to be generally applicable, the ternary Al–Li–Cu alloy AA 2020 exhibits softening when cyclically
deformed at all strain amplitudes (Srivatsan et al., 1986a). The reason lies in
the lower Li/Cu ratio of this first-generation alloy in comparison with the
2nd generation Al–Li alloys. The 2020 alloy is strengthened by the presence
of fully ordered, but semi-coherent Al_2Cu (θ') precipitates with additional
contributions from the fully coherent δ' precipitates. For this microstructure,
softening during fully reversed cyclic loading occurs from the very first
cycle and can be attributed to the shearing of θ' precipitates with significant
contributions from the disordering of θ' and δ'. Contribution from shearing
of the δ' precipitates is small and/or marginal. The results reveal that the differences in CSR behavior can be attributed to the nature and extent of dislocation interactions with the major strengthening precipitates, which for AA
2020 alloy are θ' and to a small extent δ'.

11.5.4 CSS Behavior

The cyclic stress versus strain test data are analyzed in conformance with the
power law relationship:

$$\Delta\sigma/2 = K'(\Delta e_p/2)^{n'} \tag{11.7}$$

Changes in the cyclic work hardening exponent (n') as a function of
strain amplitude are an indication of changes in deformation mode or mechanism occurring at a microscopic level. The three basic possibilities of cyclic
work hardening behavior are shown in Figure 11.7.

i. The first possibility (A), in which n' is higher at higher strain amplitudes,
    is observed in pure aluminum and Al–0.7 wt% Li alloy, where lithium
    is in solid solution (Dhers et al., 1986).
ii. The opposite trend (C), in which n' is lower at higher strain amplitudes,
    is observed in the δ' (Al_3Li) precipitate strengthened Al–Li alloys,

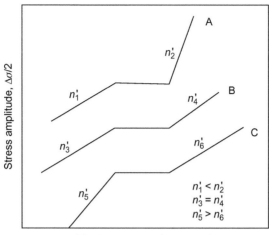

FIGURE 11.7. Schematic diagram showing the general trends in CSS behavior: (A) $n_2' > n_1'$, (B) equal n', and (C) $n_6' < n_5'$.

namely the PA Al−2.5 wt% Li alloy and AA 8090-T8E51 (Dhers et al., 1986; Eswara Prasad et al., 1996a).

iii. The intermediate behavior (B), where n' is the same at all strain amplitudes, is characteristic of the low Li/Cu ratio alloy AA 2020 (Srivatsan, 1988a,b).

Note the central horizontal region in these plots. This is representative of ideal behavior, where n' is near zero because cyclic hardening due to dislocation−dislocation and dislocation−precipitate interactions is more or less nullified by the softening due to the shearing of the strengthening precipitates, with or without local disordering of the precipitates.

The experimentally derived cyclic stress versus strain response of a δ' strengthened aluminum−lithium alloy, AA 8090-T8E51, is shown in Figure 11.8. The bilinearity is a sharp transition rather than a three-stage trend usually exhibited by many engineering alloys.

The different values of the cyclic strain-hardening exponent (n') at different strain amplitudes can be attributed to intrinsic differences in the amount of cyclic hardening and cyclic softening experienced by 8090-T8E51 during fully reversed cyclic loading. In turn, these are controlled by the different deformation mechanisms occurring at the fine microscopic level that prevail at the different strain amplitudes. The deformation mechanisms include the following:

i. Dislocation−dislocation interactions and dislocation interaction with the strengthening precipitates (δ', T_1 with or without S')

ii. Shearing and eventual re-solution of the strengthening precipitates

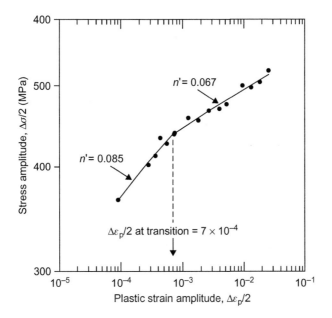

FIGURE 11.8.   CSS data for Al–Li alloy AA 8090-T8E51 (Eswara Prasad et al., 1996a).

iii. Planar slip

iv. Localization of slip at the δ' PFZs associated with high angle grain boundaries.

In addition, the CSS behavior of extruded and textured rods of AA 8090 was found to be controlled primarily by grain size rather than the texture (Khireddine et al., 1989).

11.5.5 Fatigue Toughness

A significant part of the irreversible plastic strain energy (ΔW_p) generated during LCF is absorbed and contributes to exacerbating the damage due to fully reversed cyclic loading. Failure occurs when the accumulated strain energy (and the resultant damage) exceeds a critical value, known as the *fatigue toughness* or the total cyclic plastic strain energy to fracture (W_f). Chung and Lee (1994) have shown that the following simple relationship based on Eq. (11.3) (Halford–Morrow relationship) predicts the value of W_f accurately (within 5% variation):

$$W_f = \Delta W_p \cdot N_f \qquad (11.8)$$

Eswara Prasad et al. (1994, 1996a) have shown that Eq. (11.8) may be rewritten in the form of a power law relationship as

$$W_f = 2\varepsilon_f'\sigma_f'\left[\frac{c-b}{c+b}\right](2N_f)^{1+b+c} \tag{11.9}$$

or

$$W_f = W_f^*(2N_f)^{1+b+c} \tag{11.10}$$

where W_f^* represents the toughness at $N_f = 1/2$, that is, in monotonic loading. Values of W_f may be computed from the test data given in Table 11.1. It is possible to determine the true value of W_f using present-day testing capabilities by numerically adding up the cyclic hysteresis energy values obtained from each fatigue cycle. The data so computed have been used to show the variation of fatigue toughness (W_f) with LCF life, plotted on a bi-logarithmic scale, shown in Figure 11.9. This figure clearly reveals a bilinear relationship

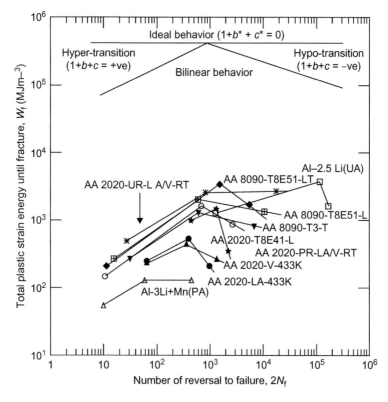

FIGURE 11.9 Variation of fatigue toughness (total plastic strain energy to fracture, W_f) with number of reversals to fracture ($2N_f$) for different Al–Li alloys. Some W_f values have been computed with yield stress as an index of stress amplitude. Note that the alloys Al–3 wt% Li + Mn and AA 8090-T8E51-LT are exceptions, since they do not show a degradation of fatigue toughness in the hypotransition region. Explanations of the abbreviations UA, PA, PR, UR, LA, and V are given in the footnote of Table 11.1.

for Eqs. (11.9) and (11.10), with a distinct peak in W_f at an intermediate value of $2N_f$. It is interesting to note that this peak corresponds to the bilinear transition in the other three power law relationships (Eqs. (11.1)–(11.3)).

Figure 11.10 is a schematic of experimentally observed fatigue toughness power law relationships based on W_f and also the proposed ideal behavior for Al–Li alloys:

1. Halford (1966) analyzed nearly 200 sets of LCF data obtained from different grades of steels and suggested that engineering materials should exhibit a continuously increasing fatigue toughness with increase in cyclic fatigue life or decrease in the LCF parameters, such as (i) $\Delta\sigma/2$, (ii) $\Delta\varepsilon_p/2$, and (iii) ΔW_p. This suggestion was based on the observation that for longer lives the rate of accumulation of damage progressively decreases, and the contribution of the LCF parameters to the overall damage process decreases. This behavior is indicated in Figure 11.10 by the rising dashed

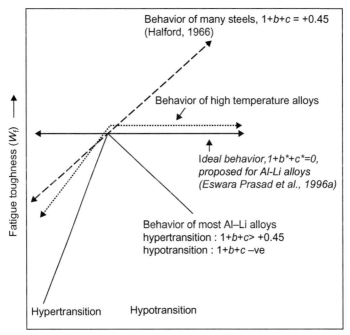

FIGURE 11.10 Schematic diagram showing the variation of fatigue toughness (total cyclic plastic strain energy till fracture, W_f) with number of reversals to failure ($2N_f$) for most of the aluminum–lithium alloys that were studied for LCF resistance to date. Also included are the trends in fatigue toughness variation with life for several structural steels and high temperature structural alloys. The ideal behavior proposed by Eswara Prasad and coworkers is also shown in this figure. *Source: After Halford (1966) and Eswara Prasad et al. (1996a).*

line and is well described by the experimentally observed and statistically derived value of the exponent $1 + b + c = +0.45$.

2. Tests on high temperature alloys in different environments revealed a bilinear behavior with values of the exponent $(1 + b + c)$ significantly different from $+0.45$. In the hypertransition region the exponent $(1 + b + c)$ is greater than $+0.45$, whereas in the hypotransition region it is close to zero (Chung and Lee, 1994; Malakondaiah and Nicholas, 1995; Mediratta et al., 1988; Raman and Padmanabhan, 1995).

3. Most of the Al–Li alloys that have been studied to date show a bilinear behavior with the hypertransitional exponent $(1 + b + c)$ greater than 0.45, and widely varying negative values of the exponent in the hypotransition region, see Figure 11.9. The negative exponents cannot be easily explained, primarily because the alloys exhibit cyclic stability, homogeneous deformation, and a high energy absorbing transgranular fracture mode in the hypotransitional region.

4. In order to explain this paradoxical behavior of Al–Li alloys empirically, Eswara Prasad et al (1996a) and Eswara Prasad and Rama Rao (2000) proposed that the observed behavior is a deviation from ideal behavior, characterized by $1 + b + c$ being independent of LCF life and the value of $1 + b + c = 0$. This independence implies that the experimentally observed behavior of Al–Li alloys can be characterized by a distinct degradation in fatigue toughness in both the hypotransition and hypertransition regions. The degradation can be attributed to the mutually interactive influences of the following:

   i. Mechanical fatigue (especially in the hypotransition region)
   ii. Strain localization due to dislocation–precipitate interactions
   iii. Environment and test temperature effects
   iv. Different degrees of strain localization along high angle grain boundaries, especially when they are associated with large PFZs.

The relative contribution of these four degradation mechanisms in the transition region depends on both the chemistry of the alloy and intrinsic microstructural effects (Eswara Prasad and Rama Rao, 2000).

11.5.6 The LCF Resistance of Aluminum–Lithium Alloys

Figure 11.11 depicts the LCF life of various Al–Li alloys in terms of the plastic strain amplitude ($\Delta\varepsilon_p/2$; Coffin–Manson plots) and compares them with the widely used conventional aluminum alloys AA 2024 and AA 7075. The data trend in this figure reveals the Al–Li alloys to have an inferior resistance to strain amplitude-controlled fatigue deformation. The binary Al–Li alloys have the lowest LCF resistance, while the first-generation (AA 2020) and second-generation (AA 2090 and AA 8090) semicommercial and fully commercial Al–Li alloys show improved LCF resistance, though still

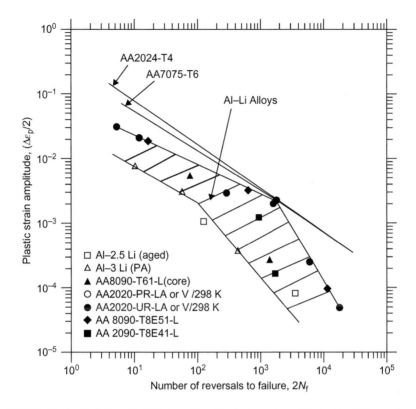

FIGURE 11.11 Variation of LCF fatigue life with plastic strain amplitude ($\Delta\varepsilon_p/2$) for various Al—Li alloys, compared with data for the widely used conventional AA 2024 and AA 7075 alloys. Explanations of the abbreviations PA, PR, UR, LA, and V are given in the footnote of Table 11.1. *Source: The data are from Endo and Morrow (1969) and Eswara Prasad and Rama Rao (2000).*

less than that of the conventional alloys. This is obviously a cause for concern when considering the possibility of both selection and use of Al—Li alloys as a viable alternative to the conventional alloys for components whose LCF properties are important.

PART B: HIGH CYCLE FATIGUE (HCF)

11.6 INTRODUCTION TO THE HCF BEHAVIOR OF ALUMINUM ALLOYS

Wöhler's seminal work on the stress-controlled cyclic loading effects on the life of railway carriage axles led to the archetypal basic HCF plot showing the variation of stress (σ) with number of cycles to failure (N_f), commonly known as an S—N curve or Wöhler curve (Wöhler, 1860). A schematic of basic stress

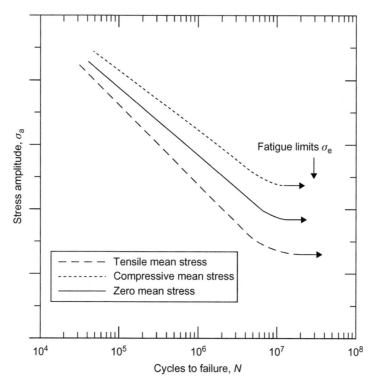

FIGURE 11.12　Schematic S–N diagram (Wöhler curves) showing plot of stress amplitude (alternating stress), σ_a, versus cycles to failure, N.

(S)–fatigue life (N) curves is shown in Figure 11.12. Such plots show the variation of cyclic (alternating) stress amplitude (σ_a) with fatigue life (N_f) using either linear–log or log–log coordinates, and generally run to about 10^7 cycles. The value of stress amplitude (σ_a) at $5 \times 10^6 – 10^7$ cycles is known or referred to as the fatigue limit or endurance limit (σ_e). Figure 11.12 also shows the variation in σ_a values when the materials or components are subjected to HCF with tensile or compressive mean stresses. Beyond this limit, the specimen or component is deemed to have infinite fatigue life. It is often assumed that the fatigue limit (also called the fatigue strength) corresponds to the case where the specimen or component is under the influence of only elastic stresses, that is, there is no plasticity at the global level. Both these assumptions are incorrect, but for a spectrum of current engineering applications the HCF fatigue limit is sufficient. Several fundamental, applied, and mechanistic approaches to both analyze and apply the HCF properties of structural materials have become part of the basic design philosophy for purposes of material selection and component design (Hertzberg, 1989; Nicholas, 2006; Suresh, 2001; Wöhler, 1870).

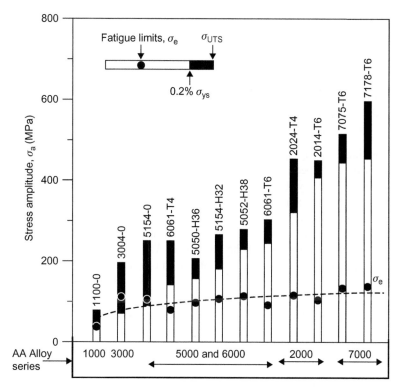

FIGURE 11.13   Smooth specimen (stress concentration $K_t = 1$) zero mean stress fatigue limits of conventional wrought aluminum alloys (Wanhill, 1994a).

Figure 11.13 shows the smooth specimen (stress concentration, $K_t = 1$) fatigue limit for conventional wrought aluminum alloys belonging to the AA 1XXX series to AA 7XXX series. Unlike steels, the fatigue limits/fatigue strengths of aluminum alloys are not proportional to the tensile strength. Instead, there is a gradual increase and eventually a plateau over a wide range of alloy strengths. This is important primarily because it shows that an increase in the yield/tensile strength of an aluminum alloy does not necessarily mean an increase in fatigue strength.

Furthermore, it is also important to note that for aerospace applications the notched fatigue strength is more appropriate, since fatigue cracks in aircraft structures generally initiate at locations of high "local" stress concentration, i.e. at rivet and bolt holes, cutouts, corners, and other sharp changes in cross-section. Thus fatigue data for K_t values ranging from about 1.5 to 4, typically 3 for fastener holes, are generally used for purposes of design. This issue is also mentioned in Chapter 2 and in Section 11.8.6, which includes a practical example of the potential problems that may arise.

11.7 BACKGROUND ON TEST METHODS

HCF fatigue studies are usually done under load (stress) control. Two common test methods are as follows:

i. Fully reversed ($R = -1$, the ratio of minimum to maximum stress) rotating bending
ii. Axial push–pull loading that may be fully reversed, partially reversed ($-1 < R < 0$), or completely in tension ($R \geq 0$).

Rotating bending specimens typically have an hourglass shape, whereas axial push–pull specimens can have either a cylindrical or a parallel-sided rectangular geometry. Basic HCF tests are done on smooth, unnotched specimens with a stress concentration factor $K_t = 1$. Notched specimens having a higher value of K_t are sometimes used, in recognition of the practical significance mentioned in Section 11.8.6. Finite fatigue lives are usually determined by complete fracture of the specimens. The fatigue or endurance limit (σ_e) is defined by the attainment of fatigue life in the range of $5 \times 10^6 - 10^8$ cycles.

11.8 HCF BEHAVIOR OF ALUMINUM–LITHIUM ALLOYS

11.8.1 General Survey

The beneficial effects of lithium additions on HCF resistance (fatigue strength) of high strength aluminum alloys have long been recognized. However, few studies have attempted to conduct a detailed investigation and to provide a convincing explanation of the observed behavior. Some of the studies that have attempted to investigate the HCF behavior of Al–Li alloys include an evaluation of (i) effect of lithium content, (ii) effect of prestrain before artificial aging, (iii) degree of aging, (iv) crystallographic texture, and finally (v) the role of impurity elements resulting in the presence of coarse intermetallic particles in the alloy microstructure (Bischler and Martin, 1987; Bretz et al., 1984; De et al., 2011; Di et al., 1987; Farcy et al., 1987; Peters et al., 1986).

11.8.2 Effects of Lithium Content, Aging, and Cold Work

Di et al. (1987) investigated the influence of lithium additions (1.82 wt%) on the HCF resistance of aluminum, including the effects of artificial aging with and without cold rolling (CR) subsequent to solution treatment (ST) but prior to artificial aging. The results obtained from rotating bending tests on smooth hourglass specimens under conditions of fully reversed loading ($R = -1$) are summarized in Figure 11.14. The tests were done in a room temperature, laboratory air environment using a cyclic frequency of 30 Hz. From Figure 11.14 it is clear that lithium addition greatly increases the fatigue strength of both the

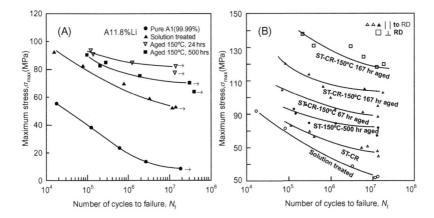

FIGURE 11.14 HCF smooth specimen $R = -1$ data for an Al–1.82 wt% Li binary alloy (A) compared to pure aluminum and (B) showing the effect of CR with or without subsequent aging (Di et al., 1987). The data in (B) also show the directionality in HCF resistance: the fatigue strength is significantly higher when the loading is normal to the rolling direction ($\perp$ RD).

solution-treated and aged specimens. This increase has been attributed to the influence of δ' (Al$_3$Li) precipitates. However, straightforward δ' precipitation promotes localized planar slip deformation resulting in the formation of coarse slip bands. This is not desirable since fatigue cracks tend to nucleate preferentially at slip band intersections and grow along these bands. This disadvantage was alleviated by CR prior to artificial aging.

Figure 11.14B shows that CR noticeably increased the HCF strength, depending on the final heat-treatment condition of the chosen aluminum–lithium alloy. The increase in fatigue strength was attributed to the introduction and distribution of dislocation substructures that (i) provided marginal strengthening to the alloy in the solution-treated condition, and (ii) during artificial aging strengthened the alloy further by promoting a high density of uniformly distributed δ' precipitates that prevented the occurrence of localized planar slip, thereby (iii) inhibiting and retarding the nucleation of fatigue cracks.

11.8.3 HCF Behavior of AA 2020-T651—A First-Generation Al–Li Alloy

One of the earliest investigations of HCF in Al–Li alloys was made by Bretz et al. (1984). They studied the first-generation Al–Li–Cu–Mn alloy AA 2020, which has a lower Li/Cu ratio (1.2/4.45, both in wt%) than the more recent alloys. AA 2020 derives its strength principally from Al$_2$Cu (θ') and Al$_2$CuLi (T$_1$) precipitates, with minor contributions from Al$_3$Li (δ') and Al$_{15}$Cu$_8$Li$_2$ (T$_B$). In the PA (T651) and nearly recrystallized condition, this Al–Li alloy possessed a smooth specimen HCF resistance equivalent to that

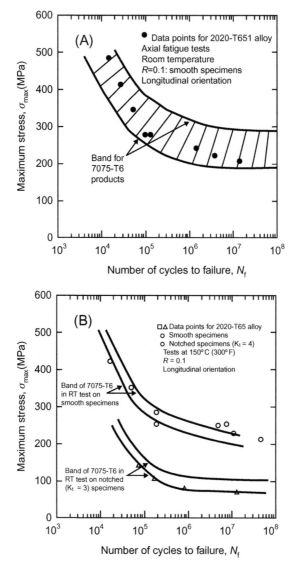

FIGURE 11.15 (A) HCF smooth specimen ($K_t = 1$) and (B) notch specimen ($K_t = 3$ and 4) data for the first-generation PA Al–Li alloy AA 2020-T651 compared with a data band for the conventional PA alloy AA 7075-T6 (Bretz et al., 1984). Note that the data are for a positive stress ratio, $R = 0.1$, and that the y-axis plots maximum stress rather than alternating stress.

of the widely used high strength aluminum alloy AA 7075 in the PA maximum strength condition (T6), see Figure 11.15A. This is remarkable because 2020 has much lower tensile ductility, which might be thought to be detrimental to fatigue strength. Bretz et al. (1984) extended their study to

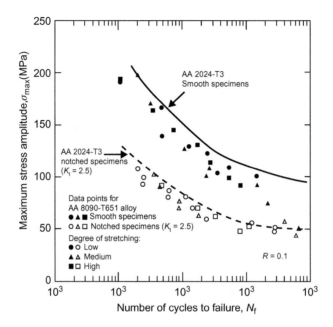

FIGURE 11.16 HCF smooth ($K_t = 1.0$) and notched ($K_t = 2.5$) specimen data for the second-generation Al–Li alloy AA 8090-T651 compared with data for the conventional naturally aged alloy AA 2024-T3 (Peters et al., 1986). Note that the data are for a positive stress ratio, $R = 0.1$. Data points for AA 8090-T651 alloy are superimposed on trend lines for AA 2024-T3 alloy.

compare the notched fatigue and fatigue crack growth behavior of aluminum alloy AA 2020 with aluminum alloy 7075. Again, they found similar or even slightly superior fatigue properties for aluminum–lithium alloy 2020 in the T651 temper condition (see the data in Figure 11.15B).

11.8.4 HCF Behavior of AA 8090-T651—A Second-Generation Al–Li Alloy

Peters et al. (1986) studied the HCF behavior of a high Li/Cu ratio (2.4/1.24, both in wt%) Al–Li–Cu–Mg alloy AA 8090. The alloy was studied in both the PA(T6) and stretched plus PA (T651) conditions. This alloy contains copious amounts of δ' precipitates that are the major strengthening phase in the PA condition. The results are summarized in Figure 11.16, which reveals slightly lower smooth and notched specimen fatigue strength when compared to the widely used naturally aged and lower strength alloy AA 2024-T3. Peters et al. (1986) ascribed the lower fatigue strength of the AA 8090 alloy to the relatively early nucleation of fine microscopic cracks during fatigue loading. Peters et al. (1986) extended their study to combinations of different amounts of stretch (low, medium, and high) and degree of aging, with the primary purpose of obtaining enhanced precipitation of the S' phase, which is conducive to the homogenization of slip. They found both the thermal and thermomechanical treatments to have negligible influence on fatigue resistance, although there was a pronounced effect on tensile properties and a favorable reduction in tensile anisotropy.

11.8.5 HCF Behavior of AA 2098—A Third-Generation Al–Li Alloy

Since the middle of the 1990s, the development of third-generation Al–Li alloys has been the focus of global research and development efforts to enable their selection and use for a variety of applications in industries related to aerospace. These alloys contain significantly reduced lithium (<1.8 wt%) in comparison with the previous generation of Al–Li alloys. The primary objective is to get the maximum possible advantage of higher specific strength (σ/ρ) and specific modulus (E/ρ) without significantly compromising the mechanical properties with respect to the conventional aluminum alloys.

In a recent study, De et al. (2011) reported the smooth specimen HCF behavior of one such Al–Li–Cu–Mg–Ag–Zr alloy, AA 2098, in both the UA and PA conditions. The UA alloy, designated UA-T3, contained Al_3Zr (β') and unresolved Al_3Li (δ') precipitates. The PA alloy, designated PA-T8 and aged at 160°C for 19 h, contained Al_2CuMg (S') and Al_2CuLi (T_1), as well as Al_3Li (δ'). Figure 11.17 summarizes the LCF–HCF data obtained by De et al. (2011) for the two chosen aging conditions. The data include loading both in the longitudinal (L) and transverse (LT) directions, and also results for near-surface specimens (designated PAS). It is evident that the PA specimen data indicate a generally higher fatigue strength than their UA counterparts. This is

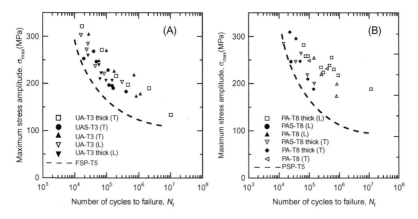

FIGURE 11.17 LCF–HCF smooth specimen $R = -1$ data for the third-generation Al–Li alloy AA 2098 in two heat-treatment conditions: (A) UA (UA-T3) and (B) PA (PA-T8) (De et al., 2011).

consistent with the beneficial influence of aging on a binary Al–Li alloy discussed in Section 11.8.2.

Further analysis of the test data in terms of the specific influence of precipitates, texture, and grain size showed that:

1. In the UA condition the effects of slip planarity and orientation were minimal.
2. In the PA condition the effect of texture was found to be minimal. This was attributed to homogenization of slip caused by the presence of S' and T_1 precipitates.
3. In both the UA and PA conditions the larger near-surface grain size resulted in noticeably shorter fatigue lives.

11.8.6 The HCF Resistance of Al–Li Alloys: Smooth and Notched Properties, and a Note of Caution

The data presented and discussed in Sections 11.8.2–11.8.5 clearly show that lithium additions to aluminum alloys coupled with an aging heat treatment can improve the smooth specimen fatigue strength (see the data in Figures 11.14. and 11.17). However, such improvements do not necessarily result in better HCF strengths than those of the conventional aluminum alloys, see Figures 11.15 and 11.16. Furthermore, as mentioned in Section 11.6, it is important to consider notched fatigue strength for applications in the aerospace industries. From Figure 11.16 it is seen that the notched fatigue strength of PA 8090-T651 alloy is similar to that of the naturally aged 2024-T3 alloy. Some additional results comparing the second-generation and third-generation Al–Li alloys with the conventional alloys are shown in Figures 11.18–11.20. The results and comparisons reveal the notched fatigue strengths of Al–Li alloys to be similar to and essentially no better than those of the conventional alloys.

Finally, we note that the essential equivalence of the fatigue strengths of Al–Li alloys and conventional alloys could give rise to problems in actual applications. Owing to their higher elastic moduli and specific stiffnesses, see **Chapter 15**, components made from Al–Li alloys could attract more load than those made from conventional alloys. Without careful design this could lead to premature fatigue cracking in Al–Li components.

11.9 SUMMARY AND CONCLUSIONS

A. Low cycle fatigue

The LCF behavior of Al–Li alloys is influenced primarily by the microstructural characteristics and to a lesser extent by crystallographic texture. To an appreciable extent, a significant influence is exerted by (i) composition of the alloy, especially the lithium content; (ii) the volume

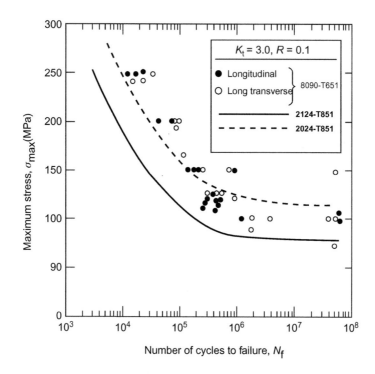

FIGURE 11.18 HCF-notched specimen data for the second-generation PA Al–Li alloy AA 8090-T651 compared with data for conventional artificially aged (T851) AA 2024 and 2124 alloys (McDarmaid, 1985). Note that the data are for a positive stress ratio, $R = 0.1$.

fraction, size, and distribution of the major strengthening precipitates that result in improvements in both strength and slip homogeneity; (iii) the nature and degree of aging; and (iv) the incorporation of tensile stretching, with or without natural aging, after solution heat treatment but prior to artificial aging. In this context, the presence of S' precipitates and retention of an unrecrystallized grain structure can be beneficial.

Alloy development programs specifically attempting to investigate the strain amplitude-controlled fatigue behavior are limited. More work is definitely needed to study the following:

a. The effects of major and minor alloying additions (especially rare earths, beryllium, silver, and TiB).

b. The effects of various thermomechanical treatments.

c. Favorable crystallographic textures.

d. The allowable amounts of the coarse equilibrium precipitates δ and T_2.

e. The influence of both high and low angle grain boundaries and the widths of precipitate-free zones.

The available data for Al–Li alloys indicate that their LCF properties are generally inferior to those of the conventional alloys established and

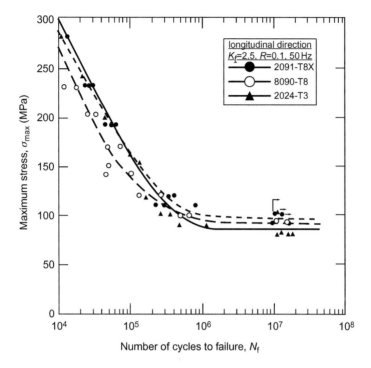

FIGURE 11.19   HCF-notched specimen data for the second-generation artificially aged Al–Li alloys AA 2091-T8X and 8090-T8 compared with data for the conventional naturally aged alloy AA 2024-T3 (Wanhill, 1994b). Note that the data are for a positive stress ratio, $R = 0.1$.

used for applications in the aerospace industry. This is a disadvantage in the design and use of components where LCF properties and resistance to damage induced by strain amplitude-controlled deformation are important. Studies have pointed out that factors improving the tensile strength and ductility, and hence the tensile toughness, have similar beneficial effects on LCF resistance. Where there exist pronounced variations in strength, ductility, work hardening, and LCF resistance, it may be useful to develop an alloy design philosophy based on either plastic strain energy per cycle (ΔW_p) or fatigue toughness (i.e., total plastic strain energy to fracture (W_f)). This approach should be given priority in continuing studies on alloy development for the purpose of optimizing the LCF resistance, especially for the third-generation Al–Li alloys.

B. High cycle fatigue

The HCF behavior of Al–Li alloys is enhanced by solid solution strengthening and coarsening of the δ' precipitates. Additional contributions to enhancing HCF resistance are obtained from thermal and thermo-mechanical treatments involving artificial aging and tensile prestraining or cold work prior to artificial aging. A recent study (De et al., 2011) on

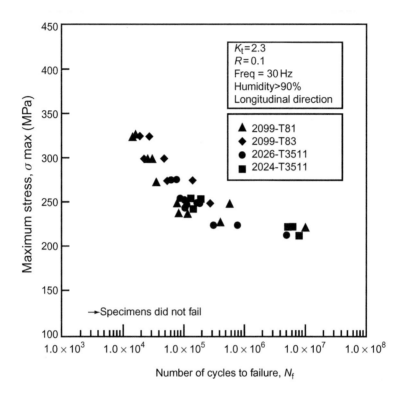

FIGURE 11.20 HCF-notched specimen data for the third-generation artificially aged Al–Li alloys AA 2099-T81 and 2099-T83 compared with data for conventional naturally aged AA 2024-T3511 and 2026-T351 alloys (Alcoa Aerospace Technical Fact Sheet, 2008). Note that the data are for a positive stress ratio, $R = 0.1$.

the third-generation alloy AA 2098 showed that (i) grain size has a significant effect and (ii) crystallographic texture also has an effect for the UA (T3) condition, but not for the PA (T8) temper.

The available data for Al–Li alloys reveal that their HCF properties are generally equivalent to but not significantly better than those of conventional aluminum alloys. This is notably the case for notched fatigue behavior, which is of major importance for aerospace structures and components.

11.10 FINAL REMARKS

At the present time, there are no openly reported studies aimed at obtaining design data/fatigue life diagrams for the third-generation Al–Li alloys that include the effects of the following:

  i. Stress ratio (R)
 ii. Tensile and compressive mean stresses

iii. LCF–HCF interactions

iv. Different stress concentration factors (notch fatigue)

v. Surface treatments, coatings, and fretting

vi. Synergism of fatigue and corrosion

vii. Occurrence of fatigue under realistic variable amplitude (VA) loading.

At least some, if not all, of these aspects need to be investigated, preferably before instigating widespread selection and use of the newer and improved Al–Li alloys.

LIST OF SYMBOLS

β	plastic strain energy exponent
$\Delta\varepsilon_p$	plastic strain range
$\Delta\varepsilon_p/2$	plastic strain amplitude
$\Delta\sigma$	stress range
$\Delta\sigma/2$	stress amplitude
ΔW_p	plastic strain energy per cycle
ε_a	applied strain or alternate strain
ε_f'	fatigue ductility coefficient (Coffin–Manson)
σ_a	applied stress or alternate stress
σ_e	fatigue or endurance limit
σ_f'	fatigue strength coefficient (Basquin's)
σ_c	compressive peak stress
σ_{max}	maximum stress
σ_{min}	minimum stress
σ_T	tensile peak stress
$2N_f$	number of reversals to failure
E	Young's modulus
K'	cyclic work hardening coefficient
K_t	stress concentration factor
N	number of cycles or number of turning points in fatigue
R	stress ratio or ratio of minimum to maximum stress
W_f	total cyclic plastic strain energy to fracture or fatigue toughness
W_f'	plastic strain energy coefficient
W_f^*	total plastic strain energy coeffiecient
b	fatigue strength exponent (Basquin's)
c	fatigue ductility exponent (Coffin–Manson)
n'	cyclic work hardening exponent

REFERENCES

Abel, A., Ham, R.K., 1966. Cyclic strain behaviour of crystals of aluminium—4 wt% copper. Acta Metall. 14, 1495–1503.

Alcoa Aerospace Technical Fact Sheet, 2008. Alloy 2099-T83 and 2099-T8E67 Extrusions. AEAP-Alcoa Engineered Aerospace Products, Lafayette, IN.

Basquin, O.H., 1910. The exponential law of endurance tests. Proceedings of ASTM 10, 625−630.

Bischler, P.J.E., Martin, J.W., 1987. The temperature dependence of the high cycle fatigue properties of an Al−Li−Cu−Mg alloy. In: Champier, G., Dubost, B., Miannay, D., Sabatay, L. (Eds.), Proceedings of the Fourth International Aluminium−Lithium Conference. J. Phys. (Paris) 48, C3.761−C3.767.

Bretz, P.E., Mueller, L.N., Vasudevan, A.K., 1984. Fatigue properties of 2020-T651 aluminium alloy. In: Sanders, T.H., Starke, E.A. (Eds.), Proceedings of the Second International Conference on Aluminum−Lithium Alloys II. The Metallurgical Society of AIME, Warrandale, PA, pp. 543−559.

Calabrese, C., Laird, C., 1974a. Cyclic stress−strain response of two-phase alloys. Part I. Microstructures containing particles penetrable by dislocations. Mater. Sci. Eng. 13, 141−157.

Calabrese, C., Laird, C., 1974b. Cyclic stress−strain response of two-phase alloys. Part II. Particles not penetrated by dislocations. Mater. Sci. Eng. 13, 159−174.

Chen, R.T., Starke, E.A., 1984. Microstructure and mechanical properties of mechanically alloyed, ingot metallurgy and powder metallurgy Al−Li−Cu−Mg alloys. Mater. Sci. Eng. 67, 229−245.

Chung, Y.W., Lee, W.J., 1994. Cyclic plastic strain energy as a damage criterion and environmental effect in Nb-bearing high-strength low-alloy steel. Mater. Sci. Eng. A 186, 121−128.

Coffin, L.F., 1954. A study of the effects of cyclic thermal stresses on a ductile metal. Trans. ASME 76, 931−950.

Coyne, E.J., Sanders, T.H., Starke, E.A., 1981. The effect of microstructure and moisture on the LCF and FCP resistance of two Al−Li−X alloys. In: Sanders, T.H., Starke, E.A. (Eds.), Aluminum−Lithium Alloys I: Proceedings of the First International Conference. The Metallurgical Society of AIME, Warrendale, PA, pp. 293−306.

De, P.S., Mishra, R.S., Baumann, J.A., 2011. Characterisation of high cycle fatigue behaviour of a new generation aluminum lithium alloy. Acta Mater. 59, 5946−5960.

Dhers, J., Driver, J., Fourdeux, A., 1986. Cyclic deformation of binary Al−Li alloys. In: Baker, C., Gregson, P.J., Harris, S.J., Peel, C.J. (Eds.), Aluminum −Lithium Alloys III: Proceedings of the Third International Conference. The Institute of Metals, London, pp. 233−238.

Di, Z., Sajji, S., Hori, S., 1987. Effect of microstructure on high cycle fatigue behaviour of Al−Li binary alloy. In: Champier, G., Dubost, B., Miannay, D., Sabatay, L. (Eds.), Proceedings of the Fourth International Aluminium−Lithium Conference. J. Phys. (Paris) 48, C3.753−C3.759.

Endo, T., Morrow, J.D., 1969. Cyclic stress−strain and fatigue behaviour of representative aircraft metals. J. Met. 4, 159−175.

Eswara Prasad, N., 1993. In-plane anisotropy in the fatigue and fracture properties of quaternary Al−Li−Cu−Mg alloys. Doctoral Thesis, Indian Institute of Technology (BHU) [formerly Institute of Technology, Banaras Hindu University], Varanasi, India.

Eswara Prasad, N., 1997. The metallurgy of Al−Li alloys—an overview. Banaras Metall. 14&15, 69−87.

Eswara Prasad, N., Malakondaiah, G., 1992. Anisotropy of mechanical properties in quaternary Al−Li−Cu−Mg alloys. Bull. Mater. Sci. 15, 297−310.

Eswara Prasad, N., Rama Rao, P., 2000. Low cycle fatigue resistance of Al−Li alloys. Mater. Sci. Technol. 16, 408−426.

Eswara Prasad, N., Malakondaiah, G., Raju, K.N., Rama Rao, P., 1989. Low cycle fatigue behavior of an Al–Li alloy. In: Salama, K., Ravi-Chander, K., Taplin, D.M.R., Rama Rao, P. (Eds.), Advances in Fracture Research: Proceedings of International Conference on Fracture. Pergamon, New York, NY, pp. 1103–1112.

Eswara Prasad, N., Malakondaiah, G., Rama Rao, P., 1992. Strength differential in Al–Li alloy 8090. Mater. Sci. Eng. A 150, 221–229.

Eswara Prasad, N., Paradkar, A.G., Malakondaiah, G., Kutumbarao, V.V., 1994. An analysis based on plastic strain energy for bilinearity in Coffin–Manson plots in an Al–Li alloy. Scr. Metall. Mater. 30, 1497–1502.

Eswara Prasad, N., Malakondaiah, G., Kutumbarao, V.V., Rama Rao, P., 1996a. In-plane anisotropy in low cycle fatigue properties of and bilinearity in Coffin–Manson plots for quaternary Al–Li–Cu–Mg 8090 alloy plate. Mater. Sci. Technol. 12, 563–577.

Eswara Prasad, N., Malakondaiah, G., Kutumbarao, V.V., 1996b. On the bilinearity in fatigue power-law relationships in Al–Li alloys. Trans. Indian Inst. Met. 49, 465–469.

Eswara Prasad, N., Malakondaiah, G., Kutumbarao, V.V., 1996c. Fracture mode transition corresponding to dual slope Coffin–Manson relationship in Al–Li alloys. In: Somashekar, B.R., Parida, B.K., Dattaguru, B., Rajaiah, K. (Eds.), Fatigue and Fracture of Materials and Structures: Proceedings of Sixth National Seminar on Aerospace Structures (6th NASAS), vol. 1. Allied Publishers Ltd., New Delhi, India, pp. 195–202.

Eswara Prasad, N., Malakondaiah, G., Kutumbarao, V.V., 1997. On the micromechanisms responsible for bilinearity in fatigue power-law relationships in aluminium–lithium alloys. Scr. Mater. 37, 581–587.

Eswara Prasad, N., Gokhale, A.A., Rama Rao, P., 2003. Mechanical behavior of Al–Li alloys. Sadhana 28, 209–246.

Eswara Prasad, N., Malakondaiah, G., Rama Rao, P., 2004. Low cycle behaviour of an underaged Al–Li–Cu–Mg alloy. Trans. Indian Inst. Met. 57, 181–194.

Eswara Prasad, N., Vogt, D., Bidlingmaier, T., Wanner, A., Arzt, E., 2010. Low cycle fatigue and creep-fatigue interaction in a short fibre-reinforced aluminium alloy composite. Mater. Sci. Technol. 26, 1363–1372.

Farcy, L., Carre, C., Clavel, M., Barbaux, Y., Aliaga, D., 1987. Factors of crack initiation and microcrack propagation in aluminum–lithium 2091 and in aluminium 2024. In: Champier, G., Dubost, B., Miannay, D., Sabatay, L. (Eds.), Proceedings of the Fourth International Aluminium–Lithium Conference. J. Phys. (Paris) 48, C3.769–C3.775.

Feng, W.X., Lin, F.S., Starke, E.A., 1984. The effect of minor alloying elements on the mechanical properties of Al–Cu–Li alloy. Metall. Trans. A 15, 1209–1220.

Gregson, P.J., Flower, H.M., Tete, C.N.J., Mukhopadhyay, A.K., 1986. Role of vacancies in precipitation of δ' and S phases in Al–Li–Cu–Mg alloys. Mater. Sci. Technol. 2, 349–353.

Halford, G.R., 1966. The energy required for fatigue. J. Mater. 1, 3–18.

Hertzberg, R.W., 1989. Deformation and Fracture Mechanics of Engineering Materials, third ed. John Wiley & Sons, New York, NY.

Khireddine, D., Rahouadj, R., Clavel, M., 1988. Evidence of S' phase shearing in an aluminum–lithium alloy. Scr. Metall. 22, 167–172.

Khireddine, D., Rahouadj, R., Clavel, M., 1989. The influence of δ' and S' precipitation on low cycle fatigue behaviour of an aluminium alloy. Acta Metall. 37, 191–201.

Laird, C., Thomas, G., 1967. On fatigue-induced reversion and overaging in dispersion strengthened alloy systems. Int. J. Fract. Mech. 3, 81–97.

Lynch, S.P., Ryder, D.A., 1973. The fatigue behaviour of a super pure and of a commercial aluminum–zinc–magnesium alloy. Aluminium 49, 748–755.

Malakondaiah, G., Nicholas, T., 1995. High-temperature low-cycle fatigue and lifetime prediction of Ti–24Al–11Nb alloy. Metall. Mater. Trans. A 26, 1113–1121.

Manson, S.S., 1953. Behaviour of materials under the conditions of thermal stress. NASA Technical Note No. 2933, National Advisory Committee for Aeronautics. In: Heat Transfer Symposium. Engineering Research Centre, University of Michigan, Ann Arbour, MI, pp. 9 – 75.

McDarmaid, D.S., 1985. Fatigue and fatigue crack growth behaviour of medium and high strength Al–Li–Cu–Mg–Zr alloy plate. Royal Aircraft Establishment Technical Report 85016, Farnborough, Hampshire.

McEvily, A.J., Clark, J.B., Utley, E.C., Hernstein, W.H., 1963. Evidence of reversion during cyclic loading of an aluminum alloy. Trans. Metall. Soc. AIME 221, 1093–1110.

Mediratta, S.R., Ramaswamy, V., Rama Rao, P., 1988. On the estimation of the cyclic plastic strain energy of dual-phase steels. Int. J. Fatigue 10, 13–19.

Morrow, J.D., 1965. Cyclic plastic strain energy and fatigue of metals. In: Internal Friction, Damping and Cyclic Plasticity, ASTM STP 378. ASTM, Philadelphia, PA, pp. 45–87.

Nicholas, T., 2006. High Cycle Fatigue, first ed. Elsevier Limited, New York, NY.

Oh, Y.J., Lee, B.S., Kwon, S.C., Hong, J.H., Nam, S.W., 1999. Low-cycle fatigue crack initiation and break in strain-life curve of Al–Li 8090 alloy. Metall. Mater. Trans. A 30, 887–890.

Peters, M., Welpmann, K., Zink, W., Sanders, T.H., 1986. Fatigue behaviour of Al–Li–Cu–Mg alloy. In: Baker, C., Gregson, P.J., Harris, S.J., Peel, C.J. (Eds.), Aluminium–Lithium Alloys III: Proceedings of the Third International Conference on Aluminium–Lithium Alloys. The Institute of Metals, London, pp. 239–246.

Radhakrishnan, V.M., 1992. On the bilinearity of the Coffin–Manson low-cycle fatigue relationship. Int. J. Fatigue 14, 305–311.

Raman, S.G.S., Padmanabhan, K.A., 1995. A comparison of the room-temperature behaviour of AISI 304LN stainless steel and nimonic 90 under strain cycling. Int. J. Fatigue 17, 271–277.

Sanders, R.E., Starke, E.A., 1977. The effect of grain refinement on the low cycle fatigue behavior of an aluminum–zinc–magnesium–(zirconium) alloy. Mater. Sci. Eng. 28, 53–68.

Sanders, T.H., Starke, E.A., 1982. The effect of slip distribution on the monotonic and cyclic ductility of Al–Li binary alloys. Acta Metall. 30, 927–939.

Sanders, T.H., Ludwiczak, E.A., Sawtell, R.R., 1980. The fracture behavior of recrystallized Al–0.8%Li–0.3%Mn sheet. Mater. Sci. Eng. 43, 247–260.

Singh, V., Sundararaman, M., Chen, W., Wahi, R.P., 1991. Low cycle fatigue behaviour of NIMONIC PE16 at room temperature. Metall. Trans. A 22, 499–506.

Srivatsan, T.S., 1984. The influence of grain structure, testing temperature and environment on the monotonic and low cycle fatigue behavior of high strength aluminum-copper-lithium alloy. Doctoral thesis. Georgia Institute of Technology, Atlanta, GA.

Srivatsan, T.S., 1988a. The effect of grain-refining additions to lithium-containing aluminum alloys. J. Mater. Sci. (Lett.) 1, 940–943.

Srivatsan, T.S., 1988b. Mechanisms of damage in high-temperature, low cycle fatigue of an aluminum alloy. Int. J. Fatigue 2, 91–99.

Srivatsan, T.S., Coyne, E.J., 1986a. Cyclic stress response and deformation behaviour of precipitation-hardened aluminum–lithium alloys. Int. J. Fatigue 8, 201–208.

Srivatsan, T.S., Coyne, E.J., 1986b. Mechanisms governing the high strain fatigue behavior of Al–Li–X alloys. In: Proceedings of the International Symposium for Testing and Failure Analysis, American Society for Materials, OH, vol. 12, pp. 281–293.

Srivatsan, T.S., Coyne, E.J., 1987. Mechanisms governing cyclic fracture in an Al—Cu—Li alloy. Mater. Sci. Technol. 3, 130—138.

Srivatsan, T.S., Coyne, E.J., Starke, E.A., 1986a. Microstructural characterization of two lithium containing aluminum alloys. J. Mater. Sci. 21, 1553—1560.

Srivatsan, T.S., Yamaguchi, K., Starke, E.A., 1986b. The effect of environment and temperature on the low cycle fatigue behavior of aluminum alloy 2020. Mater. Sci. Eng. 83, 87—107.

Srivatsan, T.S., Hoff, T., Prakash, A., 1991a. Cyclic stress response characteristics and fracture behavior of aluminum alloy 2090. Mater. Sci. Technol. 7, 991—997.

Srivatsan, T.S., Hoff, T., Prakash, A., 1991b. The high strain cyclic fatigue behaviour of 2090 aluminum alloy. Eng. Fract. Mech. 40, 297—309.

Sundararaman, M., Chen, W., Singh, V., Wahi, R.P., 1990. TEM investigation of γ' free bands in Nimonic PE16 under LCF loading at room temperature. Acta Metall. Mater. 38, 1813—1822.

Suresh, S., 2001. Fatigue of Materials, second ed. Cambridge University Press, New York, NY.

Thielen, P.N., Fine, M.E., Fournelle, R.A., 1976. Cyclic stress strain relations and strain-controlled fatigue of 4140 steel. Acta Metall. 24, 1—10.

Wöhler, A., 1860. Versuche uber die festigkeit der eisenbahnwagenachsen. Zeitschrift für Bauwesen, 10; English Summary (1967), Engineering, 4, pp. 160 — 161.

Wöhler, A., 1870. Uber die festigkeits versuche mit eisen and stahl (On strength tests of iron and steel). Zeitschrift fur Bauwesen 20, 73—106.

Wanhill, R.J.H., 1994a. Fatigue and fracture of aerospace aluminum alloys: a short course. National Aerospace Laboratory NLR Technical Publication 94034, Amsterdam, The Netherlands.

Wanhill, R.J.H., 1994b. Status and prospects for aluminium—lithium alloys in aircraft structures. Int. J. Fatigue 16, 3—20.

Xu, Y.B., Wang, X., Wang, Z.G., Luo, L.M., Bai, Y.L., 1990. Localised shear deformation of an aluminum—lithium 8090 alloy during low cycle fatigue. Scr. Metall. 25, 1149—1154.

Chapter 12

Fatigue Crack Growth Behavior of Aluminum−Lithium Alloys

R.J.H. Wanhill* and G.H. Bray**

*NLR, Emmeloord, the Netherlands, **Alcoa Technical Center, Alcoa Center, USA*

Contents

12.1 Introduction	382	
12.2 Background on Test Methods and Analysis	383	
12.2.1 Testing	383	
12.2.2 Analysis	384	
12.3 Survey of FCG of Al−Li Alloys	384	
12.3.1 Long/Large Cracks: CA/CR Loading	385	
12.3.2 Long/Large Cracks: Flight Simulation Loading	386	
12.3.3 Short/Small Cracks	387	
12.4 FCG Comparisons of Al−Li and Conventional Alloys I: Long/Large Cracks, CA/CR Loading	387	
12.4.1 First-Generation Al−Li Alloys	387	
12.4.2 Second-Generation Al−Li Alloys	388	
12.4.3 Third-Generation Al−Li Alloys	392	
12.5 FCG Comparisons of Al−Li and Conventional Alloys II: Long/Large Cracks, Flight Simulation Loading	395	

12.5.1 Second-Generation Al−Li Alloys: Gust and Maneuver Spectrum Loading 395
12.5.2 Third-Generation Al−Li Alloys: Gust Spectrum Loading 401
12.6 FCG Comparisons of Al−Li and Conventional Alloys III: Short/Small Cracks 402
12.6.1 CA Loading 402
12.6.2 Flight Simulation Loading 402
12.7 Differing FCG Behaviors and Advantages for Second- and Third-Generation Al−Li Alloys 405
12.8 Summary and Conclusions 406
12.8.1 Al−Li Alloy FCG 406
12.8.2 Practice-Related Crack Growth Regimes 408
References 409

Aluminum−Lithium Alloys.

12.1 INTRODUCTION

Fatigue crack growth (FCG) in aluminum–lithium (Al–Li) and other high strength aerospace aluminum alloys may be discussed in the context of several crack growth regimes that relate to the practical aspects of cracks in components and structures. Figure 12.1 is a schematic of these FCG regimes. In the first regime, short/small (i.e., short and small) fatigue cracks nucleate at a variety of initial discontinuities usually in the 10–50 μm size range (Barter et al., 2012). Importantly, these cracks often exhibit approximately exponential FCG (Molent et al., 2011) that continues into the next regime of noninspectable slow FCG. The transition to this second regime represents a change from short/small crack growth to long/large (i.e., long and large) crack growth, occurring at crack sizes of about 0.25–0.5 mm in aluminum alloys (Anstee, 1983; Anstee and Edwards, 1983).

The subsequent two regimes represent in-service inspectable FCG and crack growth dominated by the monotonic crack growth properties of fracture toughness and crack resistance. The boundary between the non-inspectable and inspectable FCG regimes is defined by the in-service *reliably* detectable crack sizes, which—except in special cases—correspond to cracks larger than about 5 mm surface length. The crack size boundary between FCG and monotonic fracture can vary widely, since it depends on the structural configuration as well as the material properties (fracture toughness and crack growth).

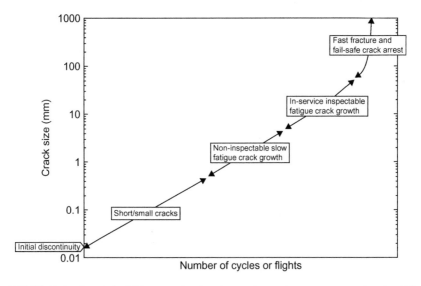

FIGURE 12.1 Schematic FCG curve showing four regimes of crack growth. After Wanhill (1994a) and modified to account for actual initial discontinuity sizes (Barter et al., 2012) and the approximately exponential behavior of short/small cracks growing from these discontinuities (Molent et al., 2011).

The first two regimes of short/small and non-inspectable crack growth are of most importance for assessing the FCG *durability* of components and structures: more than 70% of the FCG life can be spent in these regimes, a fact known for nearly 50 years (Schijve, 1967). The third and fourth regimes are important for determining the *damage tolerance* properties, including the provision of fail-safe crack arrest in stiffened sheet and plate structures. These terms, *durability* and *damage tolerance*, refer to fatigue and fracture design methodologies for aircraft. The interested reader is recommended to consult Swift (1983) and Gallagher et al. (1984) for discussions and explication of the methodologies for civil and military aircraft, respectively.

Within the context of these practice-related crack growth regimes, it can be stated that most of the FCG data for aerospace aluminum alloys refer to the second and third regimes. Short/small crack growth studies are in limited supply, owing partly to technical difficulties and also—with the benefit of experience and hindsight—to insufficient recognition of the importance of short/small crack growth for the analyses and predictions of FCG behavior (Molent et al., 2011; Wanhill et al., 2013).

12.2 BACKGROUND ON TEST METHODS AND ANALYSIS

12.2.1 Testing

Most FCG studies on aluminum alloys are of long/large crack growth obtained from notched or precracked coupons and flat panel specimens tested under constant amplitude (CA) or constant stress ratio (CR) loading. Use of the terms "constant stress ratio" and "constant R" instead of "constant amplitude" is strictly necessary for including test data obtained from load-shedding techniques whereby the stress ratio, $R = \sigma_{min}/\sigma_{max}$, is kept constant. Load-shedding is used to determine very low crack growth rates and FCG thresholds, and when R is kept constant (CR loading) the data may be considered equivalent to those obtained under CA loading. (Use of the term "constant R" is also necessary when a K-gradient approach is used to control either increasing-K or decreasing-K tests (see the American Society for Testing and Materials ASTM Standard E647-13e1, Standard Test Method for Measurement of Fatigue Crack Growth Rates (2013)).

Limited data are available to compare CA short/small crack growth in Al–Li and conventional alloys, and the same is generally true for FCG under realistic variable amplitude (VA) loading, namely gust and maneuver flight simulation load histories:

1. Short/small FCG tests require special techniques, including microscopical measurements of crack sizes on specimen surfaces and quantitative fractography (QF) of the post-test fatigue fracture surfaces (Edwards and Newman 1990a,b). There are some CA and flight simulation data for Al–Li alloys (e.g., Venkateswara Rao and Ritchie (1989) and Blom (1990)).

2. Flight simulation FCG data can be obtained from coupon/specimen, component, and full-scale tests. The data are usually for long/large cracks, but there is a need for short/small crack growth QF data (Molent et al., 2011; Wanhill and Hattenberg, 2006). At present, only specimen FCG data appear to be available for comparing Al–Li and conventional alloys (Bretz and Sawtell, 1986; Bucci et al., 1989; Shaw, 1985; Wanhill, 1994b; Wanhill and Schra, 1989; Wanhill et al., 1990, 1991).

It is worth noting here that flight simulation FCG tests on coupons and specimens have to take account of a number of guidelines in order to give representative results (Wanhill, 1994c).

12.2.2 Analysis

CA and CR FCG data are generally analyzed by correlating the crack growth rates with ΔK, the stress intensity factor range (Paris et al., 1961). It is sometimes useful to compare the FCG lives over specified ranges and regimes of crack growth.

Flight simulation (VA) FCG rates and lives are usually compared over similar ranges of crack length, since correlations of crack growth rates using the stress intensity factor approach are subject to several restrictions (Wanhill, 1994b). For long/large crack growth tests using the same VA load history and the same (or similar) stress levels, the main restrictions are:

1. Crack growth should be a regular, quasi-stationary process. In general, this means that the peak loads are either frequent and of similar magnitude, as in maneuver load histories, or else infrequent and having only minor effects on crack growth.
2. The crack tip constraint and any changes in constraint, especially during peak loads, should be the same. In practice, this means that thin-section specimens must be of similar thickness and geometry.

For short/small cracks, these restrictions do not necessarily apply, owing to characteristically large data "scatter" and little or no change in constraint.

12.3 SURVEY OF FCG OF Al–Li ALLOYS

This survey has two main aspects: firstly, a summary of FCG properties with respect to the crack growth regimes considered in Section 12.1; and secondly, a distinction, where appropriate, among first-, second-, and third-generation Al–Li alloys. These three alloy generations and the reasons for their development are discussed in Chapter 1 and also in Chapter 2.

12.3.1 Long/Large Cracks: CA/CR Loading

Since the early 1980s many results have shown that CA and CR FCG rates for long/large cracks in first- and second-generation Al—Li alloys tend to be lower than those of conventional alloys at similar strength levels (Coyne et al., 1981; Vasudévan et al., 1984; Venkateswara Rao and Ritchie, 1989). At first, this trend was at least partly attributed to the generally beneficial effect of a higher elastic modulus on FCG rates (Coyne et al., 1981; Pearson, 1966). However, the main reason is now known to be the development of rough fracture surfaces that cause high levels of crack closure in the wakes of long/large fatigue cracks and greatly reduce the crack tip driving force. This has been called "crack tip shielding" (Venkateswara Rao et al., 1988a).

The rough fatigue fracture surfaces in the second-generation Al—Li alloys are due to inhomogeneous plastic deformation owing to slip planarity, the development of intense slip bands, and strong crystallographic textures (Jata and Starke, 1986; Peters et al., 1989; Rioja and Liu, 2012; Starke and Quist, 1989; Vasudévan et al., 1984; Venkateswara Rao and Ritchie, 1989). In turn, slip planarity and intense slip bands result from dislocation shear of coherent Al_3Li (δ') matrix precipitates (Starke and Quist, 1989). Also, these alloys were typically underaged in order to improve the fracture toughness, and this underaged condition probably increased the tendency for planar slip (Venkateswara Rao et al., 1988a).

The FCG behavior of third-generation Al—Li alloys is much more similar to that of conventional alloys. There are several reasons for this. Firstly, the Li contents have been reduced significantly: all third-generation alloys contain from 0.75 to 1.8 wt% Li, compared to the more than 2 wt% Li typical of second-generation alloys. The lower Li content has reduced and in some alloys essentially eliminated the coherent Al_3Li (δ') matrix precipitates responsible for high slip planarity in the second-generation alloys. Instead, the main strengthening phases are θ'-like, such as T1 (Al_2CuLi) and Ω phases. Secondly, slip planarity in third-generation alloys has also been reduced because they are closer to peak-aged compared to second-generation alloys; and thirdly, the crystallographic texture is less as well. As a result of these changes, the FCG fracture surfaces of third-generation Al—Li alloys are much less rough than those of second-generation Al—Li alloys and similar in appearance to those of conventional alloys. Nevertheless, the CA and CR FCG rates for long/large cracks in third-generation Al—Li alloys are at least equivalent, and mostly superior, to those of the conventional alloys that they are intended to replace.

It is important to note that the FCG behavior and fracture surface appearances of third-generation Al—Li alloys can be divided into two types: alloys with Li content from about 1.4 to 1.8 wt% Li, and alloys with less than about 1.3 wt% Li. (These ranges should be taken as approximate and will be influenced not only by Li content but also by the amounts and ratios of other alloying elements, the precipitate type, aging practice, and possibly other factors.) The distinction between these two types is as follows:

1. Alloys with Li contents from about 1.4 to 1.8 wt% (e.g., AA 2099 and AA 2199) behave similarly to conventional naturally aged AA 2XXX-T3 alloys strengthened by coherent, shearable clusters like Guinier–Preston (GP) zones. This suggests a similar degree of slip planarity, and this is reflected in similar fracture surface roughnesses for the Al–Li and 2XXX-T3 alloys.

   The FCG behavior and fracture surface appearance of all these alloys (both Al–Li and conventional) can be significantly influenced by grain structure and texture. For example, an unrecrystallized extrusion having a low aspect ratio (similar height and width) and strong fiber texture will have lower FCG rates and rougher fracture surfaces than an extrusion with a high aspect ratio (wide and thin) and more plate-like texture (Garratt et al., 2001; Tchitembo Goma et al., 2012).

   Similarly, a plate product processed to have an unrecrystallized grain structure and strong texture will have significantly rougher fracture surfaces and better FCG properties than a plate subjected to special thermomechanical processing to reduce texture (Rioja and Liu, 2012). However, the strongly textured plate will also have greater in-plane and through-thickness anisotropy in tensile and other mechanical properties. This anisotropy may be undesirable, depending on the envisaged application.

2. Alloys with Li contents less than about 1.3 wt% Li (e.g. AA 2060) have fatigue fracture surfaces very similar to those of artificially aged AA 2XXX-T8 or 7XXX-T6/T7X alloys. This is because slip is more homogeneous in all these alloys.

   This second type of third-generation Al–Li alloys typically has flatter fatigue fracture surfaces and higher FCG rates than the first type, and texture and grain structure have much less influence. Even so, and as mentioned above, these alloys have at least equivalent and mostly superior FCG properties compared to those of the conventional alloys they are intended to replace. These trends are discussed further in Section 12.4.3.

12.3.2 Long/Large Cracks: Flight Simulation Loading

Flight simulation FCG data are available for second-generation Al–Li alloys. Most results have been obtained for gust spectrum (transport aircraft) load histories, mainly at the National Aerospace Laboratory NLR in the Netherlands. There are also limited data for maneuver spectrum (tactical aircraft) load histories (Bretz and Sawtell, 1986; Bucci et al., 1989; Shaw, 1985).

Comparisons of the flight simulation FCG results for the second-generation Al–Li alloys and conventional alloys depend on the spectrum

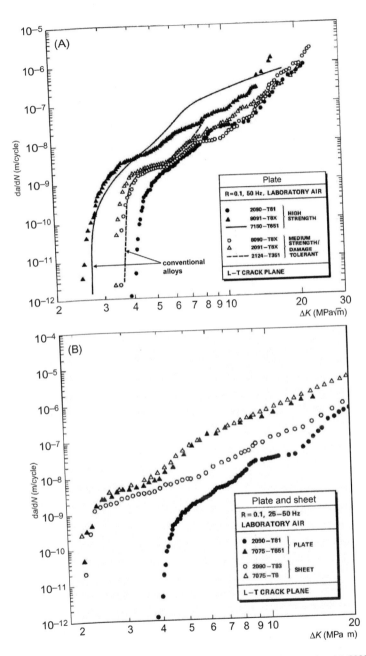

FIGURE 12.4   CA and CR long/large crack FCG rates for the second-generation AA 2090 and AA 2091 Al–Li alloys compared with equivalent strength conventional alloys (Wanhill, 1994d). *Source: Data compiled from Venkateswara Rao and Ritchie (1989) and Venkateswara Rao et al. (1990, 1991).*

type. Under gust spectrum loading, the overall FCG performance of the Al–Li alloys was usually inferior to that of equivalent conventional alloys. Under maneuver spectrum loading, the Al–Li alloy AA 2090 in T8 tempers was generally better. These trends are discussed in Section 12.5.

Only limited flight simulation (gust spectrum) FCG data are currently available for third-generation Al–Li alloys. These data indicate superior performance compared to the equivalent conventional alloys. Reasons for the differences in the gust spectrum flight simulation behavior between second- and third-generation Al–Li alloys are discussed in Section 12.5.

12.3.3 Short/Small Cracks

Some short/small FCG data are available for comparing second-generation Al–Li alloys and conventional alloys tested under CA loading (Farcy and Clavel, 1990; Farcy et al., 1987; Venkateswara Rao and Ritchie, 1989; Venkateswara Rao et al., 1988a, 1989) and flight simulation loading (Edwards and Newman, 1990a). The results have been correlated by characteristic stress intensity factors or plotted against crack size. In all cases, the FCG rates for Al–Li and conventional alloys are similar, irrespective of alloy composition, microstructure and strength level, and the type of fatigue load history. For CA loading this similarity is attributed primarily to a limited influence of roughness-induced crack closure on the crack tip driving force, since short/small cracks have only small wakes (Venkateswara Rao et al., 1988a). Note that this contrasts with the large influence of roughness-induced crack closure on the crack tip driving force for long/large cracks in Al–Li alloys tested under CA and CR loading (see Sections 12.3.1, 12.4.1, and especially Section 12.4.2).

12.4 FCG COMPARISONS OF Al–Li and Conventional Alloys I: Long/Large Cracks, CA/CR Loading

12.4.1 First-Generation Al–Li Alloys

Some information on long/large FCG under CA loading is available for the first-generation Al–Cu–Li–Mn–Cd alloy AA 2020, which was developed in the 1950s and used as plate in the wings of the US Navy's RA-5C North American Vigilante aircraft. In the peak-aged condition, 2020-T651, this alloy has yield and tensile strengths comparable to those of AA 7075-T651 (Vasudévan et al., 1984).

Figure 12.2 compares 2020-T651 FCG rate data with data bands for several conventional 2XXX and 7XXX alloys. The 2020 data are generally below or close to the lower bounds of the conventional alloy data bands. This relatively good behavior of 2020 is mainly due to rough fracture

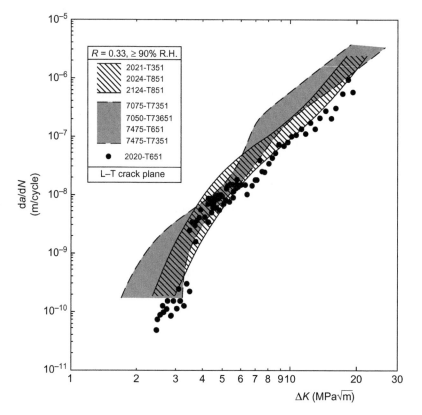

FIGURE 12.2 CA and CR long/large crack FCG rate comparisons for the first-generation Al–Li alloy AA 2020-T651 (Vasudévan et al., 1984) and several conventional alloys (Chanani et al., 1982) tested in humid air with relative humidity ≥ 90%.

surfaces causing high levels of crack closure in the wakes of the fatigue cracks and significantly reducing the crack tip driving force (see Section 12.3.1).

12.4.2 Second-Generation Al–Li Alloys

From the mid-1980s to mid-1990s, there were many publications on the properties of second-generation Al–Li–Cu–Mg–Zr alloys. In particular, the long/large crack CA/CR FCG properties were shown to be often better than those of conventional alloys (Wanhill, 1994d). As stated in Section 12.3.1, the main reason is that these Al–Li alloys develop rough fracture surfaces that cause high levels of crack closure in the wakes of long/large fatigue cracks and greatly reduce the crack tip driving force.

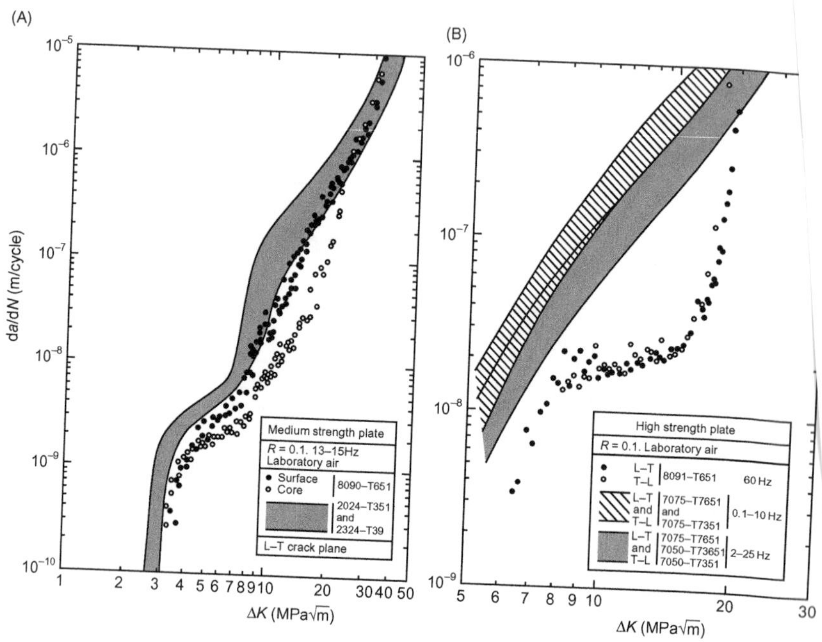

FIGURE 12.3 CA and CR long/large crack FCG rates for the second-generation AA 8090 and AA 8091 Al–Li alloys compared with equivalent strength conventional alloys (Wanhill, 1994d). *Source: Data compiled from NLR results and McDarmaid (1985).*

In Figures 12.3–12.5, the FCG rates (and FCG lives in Figure 12.5) for a variety of second-generation Al–Li alloys are compared with those of equivalent strength conventional alloys in order to show several trends:

1. Figure 12.3 gives two examples of Al–Li alloys (AA 8090 and AA 8091) having lower FCG rates especially at intermediate ΔK values and a higher strength level (8091). However, crystallographic texture and hence the orientations of fatigue fracture planes (Venkateswara Rao et al., 1988b) are complicating factors. This is why the 8090 surface and core data shown in Figure 12.3A differ at intermediate ΔK values: the core specimens had a stronger texture ('t Hart et al., 1989) that led to rougher fatigue fracture surfaces in this ΔK range (Wanhill et al., 1991).

2. Figure 12.4 compares FCG rates for several Al–Li and conventional plate and sheet alloys over a wide range of ΔK values. Figure 12.4A shows trends similar to those shown in Figure 12.3A, namely lower FCG rates for Al–Li alloys at intermediate ΔK values—with a notable exception: the high strength AA 2090 alloy is superior to all the other alloys over almost the entire ΔK range. This is because the alloy had a very strong texture that resulted in very rough fatigue fracture surfaces and high levels of crack closure (Yoder et al., 1989). Figure 12.4B shows the differing effect of product form (plate or sheet) for Al–Li and conventional alloys: the FCG rates for

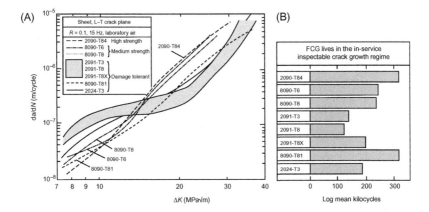

FIGURE 12.5 CA long/large crack FCG rates and lives for the second-generation AA 2090, AA 2091, and AA 8090 Al—Li alloys compared with the conventional aircraft standard AA 2024-T3 alloy (Wanhill, 1994d; Wanhill et al., 1991).

AA 7075-T651 plate and 7075-T6 sheet are the same, but the FCG rates for 2090-T81 plate are much lower than for 2090-T83 sheet, resulting in a relatively high FCG threshold ($da/dN \sim 10^{-11}$ m/cycle) of about 3.8 MPa$\sqrt{m}$. Again, this difference is due to a very strong texture in the plate alloy, resulting in very rough fatigue fracture surfaces and high levels of crack closure (Venkateswara Rao et al., 1990, 1991).

3. Figure 12.5 compares FCG rates and lives for several Al—Li sheet alloys and the industry standard *damage tolerant alloy* AA 2024-T3. Figure 12.5A shows that (i) the AA 2090 and AA 8090 alloys had the lowest FCG rates at lower ΔK values, (ii) the AA 2091 alloys became superior at higher ΔK, and (iii) the 2024-T3 FCG rates were intermediate to this trend. The change in Al—Li alloy FCG rankings was probably caused mainly by changes in fatigue fracture surface roughness.

Figure 12.5B is a basic illustration of how FCG rate data like those in Figure 12.5A may be interpreted for practical applications. The FCG rate curves have been used to derive CA FCG lives for the in-service inspectable crack growth regime, e.g., for transport aircraft pressure cabins. Figure 12.5B shows that the 2090 and 8090 alloys had FCG lives longer than 2024-T3, even in *medium* and *high strength* tempers. This is because these alloys had the lowest FCG rates at lower ΔK values, most probably owing to greater fracture surface roughness and enhanced crack closure (Wanhill, 1994b; Wanhill et al., 1991). However, actual use of *any* alloy would also have to consider the fourth crack growth regime of fast fracture and fail-safe crack arrest (see Figure 12.1). In this regime, 2024-T3 (and its derivative alloys) has established an excellent reputation. The reader is referred to Chapter 13 for further consideration of the fracture properties of Al—Li and conventional alloys.

12.4.3 Third-Generation Al—Li Alloys

Although the long/large crack CA/CR FCG properties of third-generation Al—Li alloys have been studied much less extensively than those of second-generation alloys, the available results are very encouraging. Figure 12.6 compares the CA FCG rates of third-generation Al—Li and equivalent conventional alloys in several product forms. The individual product comparisons are discussed here, making a distinction between the two types of third-generation Al—Li alloys (see Section 12.3.1).

1. Figure 12.6A compares the Al—Li *damage tolerant* plate products AA 2199-T86 and AA 2060-T8E86 (an E-temper is an Alcoa internal designation and indicates the temper has not yet been registered with the Aluminum Association) with the conventional damage tolerant plate products AA 2024-T351, AA 2324-T39, and the more recent AA 2624 in both the T351 and T39 tempers. These products are typically for use in lower wing and other tension-dominated structures requiring high damage tolerance.

   The FCG rates of 2199 plate (first type, 1.4—1.8 wt% Li) are slightly lower than those of 2060 plate (second type, 0.6—0.9 wt% Li). The FCG behavior of both alloys is very similar to that of alloy AA 2624, the most damage tolerant 2X24 alloy, and superior to that of alloys AA 2024 and AA 2324. The improvement over 2024 and 2324 is even greater at high ΔK, reflecting the better fracture toughness of 2199 and 2060 relative to the older 2X24 alloys.

2. Figure 12.6B compares the FCG rates for several *damage tolerant* extrusion alloys, typically used for fuselage and lower wing stringers. The Al—Li AA 2099 extrusion (first type, 1.6—2.0 wt% Li) FCG rates are similar to those of the most damage tolerant conventional alloy, AA 2026, and superior to those of the older AA 2024 and AA 2224 alloys.

3. Figure 12.6C compares the Al—Li *damage tolerant* sheet products AA 2199-T8E74 and AA 2060-T8E30 with the conventional damage tolerant sheet products AA 2024 and AA 2524, both in the T3 naturally aged temper. This type of product is used typically for fuselage skins. The FCG performance of 2199 (first type, 1.4—1.8 wt% Li) is superior to those of the 2X24 alloys over the entire ΔK range. The FCG rates for 2060 (second type, 0.6—0.9 wt% Li) are similar to those of the 2X24 alloys at low-to-medium ΔK, but lower at high ΔK owing to better fracture toughness.

4. Figures 12.6D and E compare the Al—Li *high strength* plate and extrusion products AA 2055-T8X, AA 2055-T8E3 and AA 2099-T83 with the equivalent conventional high strength alloys AA 7075-T6511, AA 7150-T7751, and the more recent very high strength alloy AA 7055 in two tempers. These products are typically for use in applications requiring high tensile or compressive strength, such as upper wing skins, fuselage stringers, floor beams, and keel beams.

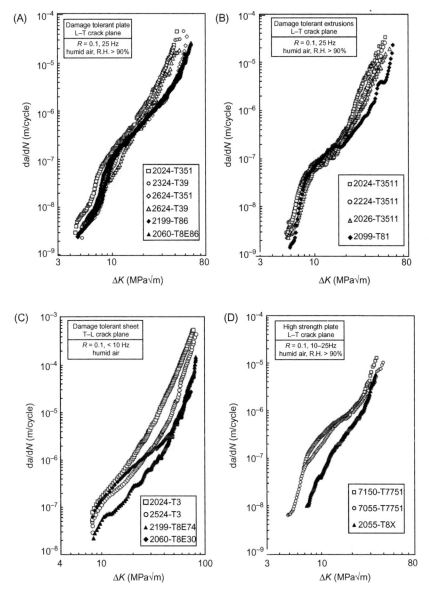

FIGURE 12.6 CA long/large crack FCG rates for third-generation Al−Li alloys compared to conventional alloys: (A−C) damage tolerant plate, extrusion, and sheet products; (D and E) high strength plate and extrusion products; and (F) medium-to-high strength plate and forging products. *Source: After Denzer et al. (2012), Magnusen et al. (2012), and from Alcoa in-house data.*

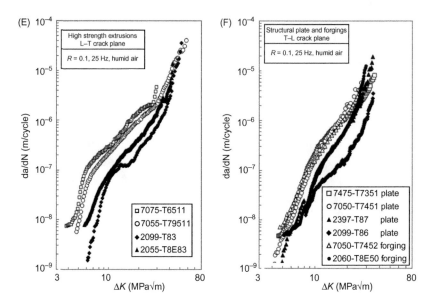

FIGURE 12.6 (Continued)

The FCG performances of the 2055 and 2099 alloys are much superior to those of the conventional alloys over most of the ΔK range. In Figure 12.6E it is seen that 2099 (first type, 1.6−2.0 wt% Li) has FCG rates generally lower than those of 2055 (second type, 1.0−1.3 wt% Li).

5. Figure 12.6F compares the Al–Li *medium/high strength* AA 2397-T87 and AA 2099-T86 plate and 2060-T8E50 forging products with the equivalent conventional high strength alloys 7475-T7351 and 7050-T7451 plate and 7050-T7452 forgings. These products are typically for use in internal aircraft structure, such as ribs, bulkheads, thick frames, attach fittings, and landing gear components.

Similar to the case of high strength plate and extrusions, the FCG performances of the 2397, 2099, and 2060 alloys are significantly superior to those of the conventional alloys over most of the ΔK range. Furthermore, 2099 (first type, 1.6−2.0 wt% Li) has FCG rates in the intermediate-to-high ΔK range much lower than those of the second-type lower-Li content alloys 2397 (1.1−1.7 wt% Li) and 2060 (0.6−0.9 wt% Li), which have similar FCG rates.

These results can be summarized as follows. The long/large crack CA FCG properties of third-generation Al–Li alloys *in all product forms* are at least equivalent and mostly superior to those of the best conventional alloys (2XXX and 7XXX) they are intended to replace. However, a distinction has to be made between the two types of third-generation Al–Li alloys. The first type has higher Li contents, typically about 1.4−1.8 wt%, and the second type has lower Li contents, less than about 1.3 wt%. The FCG performance of the first type is generally better than that of the second type: see Section 12.3.1 for other distinctions and characteristics of the two types.

Comparison of Figure 12.6 with Figures 12.3–12.5 shows that at low-to-medium ΔK, the improvements in CA FCG properties offered by the third-generation Al–Li alloys relative to conventional alloys are not as great as those provided by some of the second-generation Al–Li alloys. However, the third-generation alloys are generally better at higher ΔK, owing to significantly improved fracture toughness.

It is important to note that the FCG improvements in third-generation products do not result from the development of the very rough fracture surfaces characteristic of most second-generation alloys. This means that the third-generation alloys are expected to retain their superior FCG performances in a greater variety of product forms and thicknesses, and under a greater variety of loading conditions than the second-generation alloys (see Section 12.5). Also, the third-generation alloys do not suffer from undesirable anisotropic mechanical properties, unlike the second-generation alloys.

12.5 FCG COMPARISONS OF Al–Li AND CONVENTIONAL ALLOYS II: LONG/LARGE CRACKS, FLIGHT SIMULATION LOADING

As mentioned in Section 12.3.2, flight simulation FCG data are available for second-generation Al–Li alloys. Most results have been obtained for gust spectrum (transport aircraft lower wing skin) load histories. There are also a few maneuver spectrum (tactical aircraft) data, limited to AA 2090 in T8 tempers (Bretz and Sawtell, 1986; Bucci et al., 1989; Shaw, 1985).

The results of comparing the second-generation Al–Li alloys with conventional alloys differ according to the spectrum type. Under gust spectrum loading the FCG behavior of Al–Li plate and sheet alloys was usually inferior, but under maneuver spectrum loading the 2090 alloy data show generally better performance.

Some flight simulation (gust spectrum) FCG data are also available for third-generation Al–Li alloys. These data indicate superior performance compared to the equivalent conventional alloys.

12.5.1 Second-Generation Al–Li Alloys: Gust and Maneuver Spectrum Loading

Gust Spectrum Loading

Figure 12.7 compares MINITWIST (Lowak et al., 1979) gust spectrum flight simulation FCG rates and lives, in the in-service inspectable crack growth regime, for several second-generation Al–Li alloys and the conventional

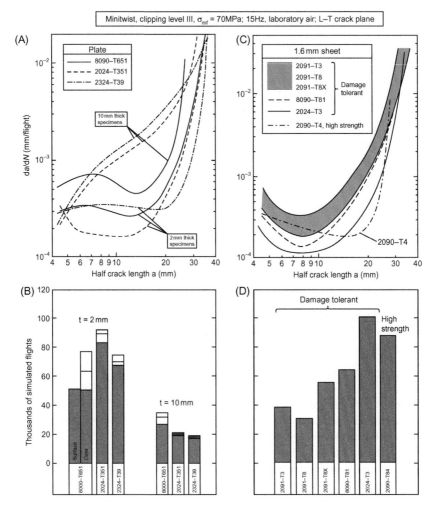

FIGURE 12.7 Gust spectrum (MINITWIST) FCG rates and lives in the in-service inspectable crack growth regime: (A and B) for second-generation damage tolerant AA 8090 Al–Li medium strength plate compared with conventional damage tolerant AA 2024 and AA 2324 plates; and (C and D) for second-generation damage tolerant AA 2091 and AA 8090, and high strength AA 2090 Al–Li sheet, compared with conventional damage tolerant AA 2024 sheet. Note: σ_{mf} = mean stress in the simulated flight load history. *Source: Data compiled from Wanhill (1994d).*

damage tolerant alloys AA 2024-T3 and AA 2324-T39. There are several points to be made. The first two points refer mainly to particular aspects of gust spectrum flight simulation FCG testing:

1. The shapes of the FCG curves for the 2- and 1.6-mm-thick specimens shown in Figures 12.7A and C are due to (i) reduced crack tip constraint

during peak loads, resulting in larger plastic zones, and (ii) overlapping of the peak load plastic zones to cause persistent crack growth retardation and decreasing growth rates (Wanhill, 1979).

2. The constraint changes due to peak loads depend on specimen thickness: thicker specimens are normally less affected (Wanhill, 1994b,c). This is why the 10-mm-thick 2024 and 2324 specimen data shown in Figure 12.7A do not show decreasing crack growth rates. However, the behavior of 10-mm-thick specimens of AA 8090-T651 is unusual, since the FCG curve resembles that of the 2-mm-thick 8090-T651 specimens. The reason is constraint reduction owing to several through-thickness delaminations in the 10-mm-thick specimens during crack growth (Wanhill, 1994b; Wanhill et al., 1991).

Bearing in mind these particular aspects, the following additional points are made:

3. Figures 12.7A and C shows that the performance of the 2-mm-thick specimens of medium strength 8090-T651 plate was inferior to that of the conventional alloys. However, owing to through-thickness delaminations, the 10-mm-thick 8090 specimens developed lower FCG rates, resulting in longer lives than those of the 10-mm-thick 2024 and 2324 specimens.

4. Figures 12.7B and D show that the Al–Li sheet alloys were inferior to 2024-T3. Nevertheless, the relatively good performance of the high strength AA 2090-T84 alloy is remarkable, since high strength increases the crack tip constraint and would therefore be expected to reduce crack growth retardation due to peak loads (Wanhill, 1994b,c).

Taken overall, these results contrast with the trend for CA/CR loading discussed in Section 12.4.2, namely that second-generation Al–Li alloys often show better CA/CR FCG performance than conventional alloys.

This effect of load history on the Al–Li alloy FCG rankings with respect to conventional alloys is largely reflected in changes in fracture surface roughness (Ranganathan et al., 1990; Wanhill, 1994b,d; Wanhill and Schra, 1989; Wanhill et al., 1990, 1991) and hence the amount of crack closure in the wakes of the cracks. Figure 12.8 illustrates that MINITWIST loading reduced R_ℓ, the fracture surface linear roughness, of three of the Al–Li sheet alloys to similar values, which were also within the range of R_ℓ for 2024-T3 sheet tested under CA and gust spectrum loading (Wanhill et al., 1991). A similar effect occurred for the 8090-T651 plate alloy. Figure 12.9 shows the FCG topographies corresponding approximately to the profiles shown in Figure 12.8.

Based on the foregoing results, it was originally proposed that peak loads and peak load interactions during gust spectrum loading activated more slip

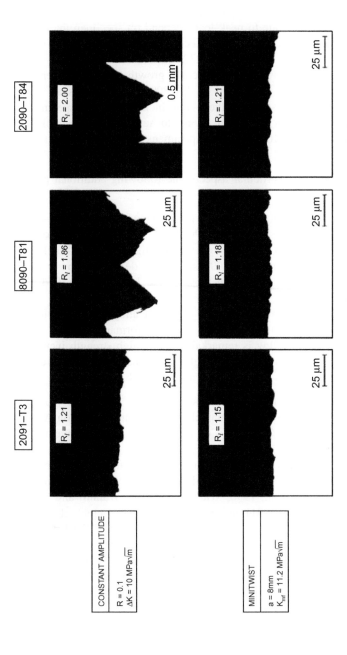

FIGURE 12.8 Comparisons of FCG fracture surface roughness, R_ℓ, for three Al–Li sheet alloys tested under CA and gust spectrum (MINITWIST) loading in the in-service inspectable crack growth regime. The MINITWIST roughness values are similar and also within the range of R_ℓ for 2024-T3 sheet tested under CA and MINITWIST loading (Wanhill et al., 1991). Note: R_ℓ is the fracture surface linear roughness, given by the ratio of the true length of fracture surface profile to the projected length.

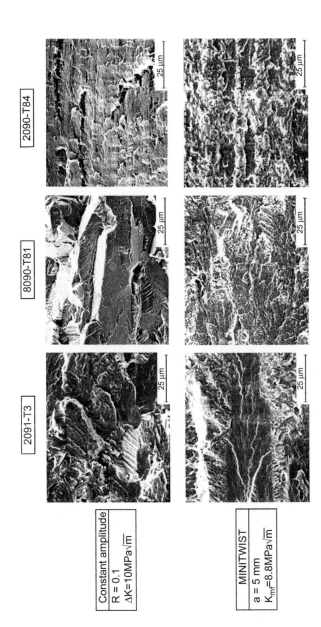

FIGURE 12.9 Comparisons of FCG fracture surface topographies (SEM fractographs) for three Al–Li sheet alloys tested under CA and gust spectrum (MINITWIST) loading in the in-service inspectable crack growth regime (Wanhill et al., 1991). The fractograph for 2090-T84 under CA loading corresponds to the midsection of the profile shown in Figure 12.8.

planes in the monotonic plastic zones ahead of the crack tips, leading to more homogeneous cyclic plastic deformation and less rough fatigue fractures (Wanhill and Schra 1989). However, recent CA, CA + peak load, and CA + underload tests on strongly textured 8090-T81 sheet suggest that underloads are primarily responsible for flattening the fatigue fracture surfaces (Barter and Wanhill, 2013).

Maneuver Spectrum Loading

Figure 12.10 compares maneuver spectrum flight simulation FCG life data, in the in-service inspectable crack growth regime, for second-generation high strength AA 2090 Al–Li plate alloys, conventional high strength and damage tolerant plate alloys, and the first-generation high strength AA 2020 Al–Li plate alloy (Bretz and Sawtell, 1986; Shaw, 1985). These data are of a preliminary nature, and the testing conditions were not fully specified in the source publications.

The results show that the 2090 FCG lives were similar to—or longer than—the lives of the conventional and 2020 alloys. The result for 2090-T8E41 plate shown in Figure 12.10B is extraordinary. It is possibly due to a very strong texture, resulting in very rough fatigue fracture surfaces and high levels of crack closure, as observed for 2090-T81

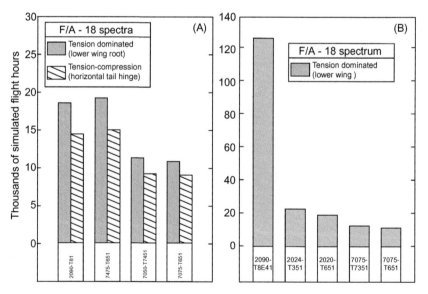

FIGURE 12.10 F/A-18 maneuver spectrum FCG lives in the in-service inspectable crack growth regime for second-generation high strength AA 2090 Al–Li plate alloys compared with conventional high strength and damage tolerant plate alloys, and also the first-generation high strength AA 2020 Al–Li plate alloy. *Data compiled from (A) Shaw (1985) and (B) Bretz and Sawtell (1986).*

plate tested under CA/CR loading (see the discussion of Figure 12.4 in Section 12.4.2).

12.5.2 Third-Generation Al—Li Alloys: Gust Spectrum Loading

Figure 12.11 compares gust spectrum flight simulation FCG curves, in the in-service inspectable crack growth regime, for the third-generation damage tolerant Al—Li alloys AA 2199-T86 and AA 2060-T8E86 and the conventional damage tolerant alloys AA 2024, AA 2324, and AA 2624 in T3XX tempers. Both Al—Li alloys gave longer FCG lives than the conventional alloys, but the higher-Li content 2199 alloy (first type, 1.4—1.8 wt% Li) was significantly better than the lower-Li content 2060 alloy (second type, 0.6—0.9 wt% Li). The best conventional alloy was 2624-T351 followed by 2624-T39. The older 2024 and 2324 alloys gave the shortest FCG lives.

The differences between the T351 and T39 tempers in 2624 and also between 2024-T351 and 2324-T39 are due to the lower yield strengths of the T351 tempers. The T351 lower yield strengths resulted in more crack growth retardation after peak loads and hence longer FCG lives. This "yield strength effect" makes it even more remarkable that the 2199-T86 and 2060-T8E86 alloys had longer FCG lives than 2024-T351 and 2624-T351, since their respective yield strengths were more than 20% and 40% higher. However, it

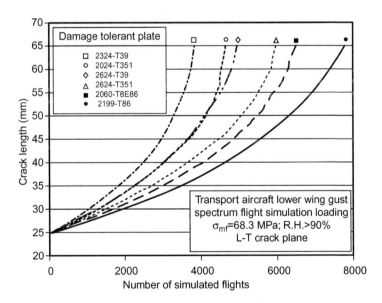

FIGURE 12.11   Gust spectrum FCG curves in the in-service inspectable crack growth regime for third-generation damage tolerant AA 2199 and 2060 Al—Li alloys compared with the conventional older damage tolerant alloys AA 2024 and AA 2324, and the newer AA 2624 alloy (Karabin et al., 2012). The specimen thickness for all alloys was 12mm.

must be noted that the relative performances of different alloys can depend on several other factors besides yield strength (Wanhill, 1994b,c).

12.6 FCG COMPARISONS OF Al–Li AND CONVENTIONAL ALLOYS III: SHORT/SMALL CRACKS

As mentioned in Section 12.3.3, some short/small FCG data are available for comparing second-generation Al–Li alloys and conventional alloys tested under CA loading and flight simulation loading. In all cases, the FCG rates for Al–Li and conventional alloys are similar, irrespective of alloy composition, microstructure, and strength level. For CA loading this similarity is attributed primarily to a limited influence of roughness-induced crack closure on the crack tip driving force, since short/small cracks have only small wakes (Venkateswara Rao et al., 1988a). This explanation probably applies to the flight simulation data also.

12.6.1 CA Loading

Figure 12.12 compares the CA FCG rates for short/small cracks in damage tolerant and high strength Al–Li and conventional plate alloys over a fairly wide range of ΔK values. There is considerable data scatter, which is characteristic of short/small crack growth, but the Al–Li and conventional alloy FCG rates are essentially similar. Also, both sets of data are well fitted by lines with slope $m = 2$, where m is the exponent in the well-known Paris equation (Paris and Erdogan, 1963; Paris et al., 1961). This equation was originally proposed to account for long/large crack FCG rates at intermediate ΔK values. However, when $m = 2$, which represents exponential crack growth, this equation—or a more general exponential relationship (Molent et al., 2011)—has widespread application to short/small crack FCG in actual components and structures. This is an important aspect of short/small crack growth and has already been mentioned in Section 12.1.

12.6.2 Flight Simulation Loading

The load histories FALSTAFF (Van Dijk and de Jonge, 1975), TWIST (De Jonge et al., 1973), and FELIX (Edwards and Darts, 1984) were used to obtain short/small crack flight simulation FCG data for high strength AA 2090-T8E41 Al–Li alloy sheet and conventional damage tolerant AA 2024-T3 sheet (Blom, 1990; Cook, 1990).

Figure 12.13 compares the short/small FCG rates for both alloys, using characteristic stress intensity factors to correlate the data obtained at different spectrum stress levels. As in the case of CA loading, Section 12.6.1, the Al–Li and conventional alloy FCG rates are essentially similar for each load history. This is remarkable in view of the different flight simulation load

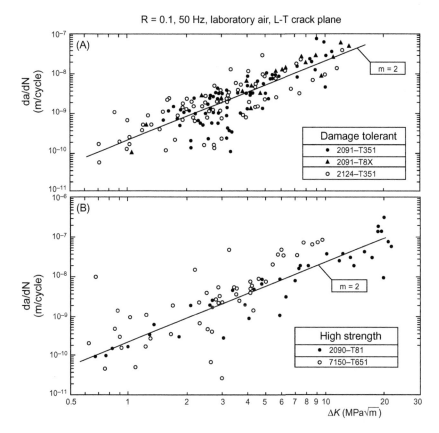

FIGURE 12.12 CA short/small crack FCG rates for (A) second-generation damage tolerant AA 2091 Al—Li alloy plate compared with conventional damage tolerant AA 2124 plate and (B) second-generation high strength AA 2090 plate compared with conventional high strength AA 7150 plate. Note: *m* is the exponent in the Paris equation (Paris and Erdogan, 1963; Paris et al., 1961). *Source: Data compiled from Venkateswara Rao and Ritchie (1989).*

histories, and also because the 2090 alloy had a much higher strength and was strongly textured (Bowen, 1990).

An additional feature was that the 2090 FALSTAFF and FELIX fatigue fracture surfaces were much rougher (or rather, more slanted, like the 2090-T84 CA profile in Figure 12.8) than those obtained with TWIST, where the roughness was similar to that for the 2024 fracture surfaces (Edwards and Newman, 1990b). This latter result parallels that for long/large cracks in Al—Li alloys subjected to MINITWIST (see Section 12.5.1).

In view of all the results, it appears that the conclusion that short/small crack FCG is not significantly influenced by roughness-induced crack closure can be extended from CA loading (Venkateswara Rao et al., 1988a) to flight simulation loading, irrespective of the spectrum type.

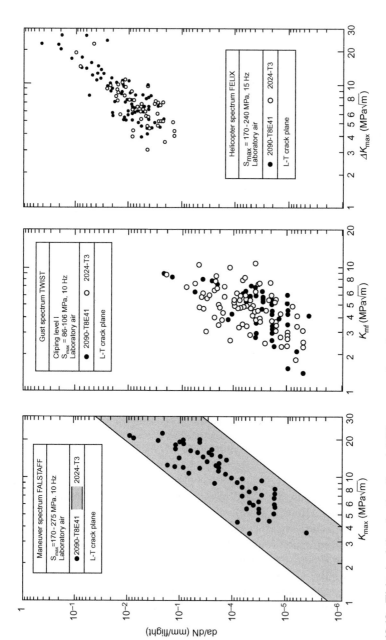

FIGURE 12.13 Flight simulation short/small crack FCG rate comparisons for second-generation high strength AA 2090 Al–Li alloy sheet and conventional aircraft standard AA 2024-T3 sheet. Note: S_{max} is the maximum stress describing the overall severity of the applied flight simulation load histories; K_{max}, K_{mf}, and ΔK_{max} are characteristic stress intensity factors enabling correlation of data obtained with different S_{max} levels. *Source: Data compiled from Blom (1990) and Cook (1990).*

A final point is that the data slopes in Figure 12.13 appear to be about $m = 3$, that is, crack growth would appear to be "hyper-exponential." However, such overall slope values must be treated with caution because the data were collated for significantly different stress levels, and it is known that this can lead to less-than-perfect correlations by characteristic stress intensity factors (Schijve et al., 1972).

12.7 DIFFERING FCG BEHAVIORS AND ADVANTAGES FOR SECOND- AND THIRD-GENERATION Al—Li ALLOYS

The information presented in Sections 12.4.2, 12.4.3, 12.5.1, and 12.5.2 shows that there are major differences in the long/large crack FCG behavior of second- and third-generation Al—Li alloys under CA/CR and gust spectrum flight simulation loading:

1. Under CA/CR loading second-generation alloys can have much lower FCG rates compared to equivalent conventional alloys. These much lower FCG rates reflect the development of rough fatigue fracture surfaces that cause high levels of crack closure in the crack wakes and reduce the crack driving force. The rough fracture surfaces are due to inhomogeneous plastic deformation owing to slip planarity, the development of intense slip bands, and strong crystallographic textures; and the slip planarity is itself a consequence of dislocation shear of coherent Al_3Li (δ') matrix precipitates, probably exacerbated by the typically underaged tempers used to improve fracture toughness. However, there is no consistency in the effect on crack growth rates. In some cases, the FCG rates are lowered over the entire range of ΔK; in others, the lower FCG rates occur only at lower or intermediate ΔK.

   The third-generation alloys have similar or lower CA FCG rates compared to equivalent conventional alloys. As mentioned in Section 12.3.1, the improvements are not due to greater fracture surface roughness, and they are much more consistent. This consistency is maintained in the distinction between the two types of third-generation alloys: type 1 alloys have higher Li contents and generally lower FCG rates than type 2 alloys.

2. Under gust spectrum flight simulation loading, the overall FCG performance of second-generation alloys is usually inferior to that of equivalent conventional alloys. This contrasts with the (inconsistent) advantage under CA/CR loading and is largely due to the elimination of the fracture surface roughness effect responsible for better CA/CR performance.

   Although gust spectrum flight simulation FCG data for third-generation alloys are limited, the data are consistent in showing that these alloys retain or even improve their FCG advantage over equivalent conventional alloys. Again, these improvements are not due to differences in fracture surface roughness.

On the whole, differences in fracture surface roughness are not an issue for the FCG ranking of third-generation Al–Li and conventional alloys. In other words, crack closure and the crack driving force are likely to be similar. The question then remains: why do third-generation alloys have generally better FCG performance than conventional alloys under CA and gust spectrum loading? It is always difficult to assess FCG advantages and disadvantages, since many factors influence the crack driving force and FCG resistance (see Figure 12.14). However, three possible reasons are suggested here. Firstly, the Al–Li alloys may benefit from their higher elastic modulus, as originally suggested by Pearson (1966) and Coyne et al. (1981). Secondly, the particular advantage over AA 7XXX alloys, see Figures 12.6C–E, may also be due to less homogeneous slip. Thirdly, improved FCG behavior at higher ΔK, and at larger crack lengths under gust simulation loading, is probably due to improved fracture toughness. For example, the AA 2199-T86 and AA 2199-T8E86 specimens represented in Figure 12.11 exhibited far less tensile crack jumping during peak loads than AA 2024-T351 and AA 2324-T39, and tensile crack jumping is known to shorten the FCG life (Wanhill, 1984; Wanhill et al., 1976).

12.8 SUMMARY AND CONCLUSIONS

12.8.1 Al–Li Alloy FCG

Most of the available data for Al–Li alloy FCG have been obtained for second-generation alloys, specifically under CA and CR loading, and for long/large cracks. These data show that the Al–Li alloy FCG rates are often lower than those of equivalent conventional alloys over intermediate ΔK ranges of 3–5 to 15–20 MPa$\sqrt{m}$. The main reason is "crack tip shielding," that is, the development of rough fracture surfaces that cause high levels of crack closure in the wakes of the fatigue cracks and greatly reduce the crack tip driving force.

This FCG "advantage" tends to be much reduced or nullified by flight simulation loading, and is also absent for short/small cracks. Furthermore, it is due to inhomogeneous plastic deformation owing to slip planarity, the development of intense slip bands, and strong crystallographic textures. These features have undesirable consequences for other important mechanical properties, including variable and anisotropic tensile strengths and fracture toughness, and have greatly restricted the practical applications of these alloys.

FCG data for third-generation Al–Li alloys are becoming more available. Many of the issues associated with second-generation alloys have been eliminated or greatly alleviated in third-generation alloys as a result of several changes:

Crack driving force

Factors	Effects
Crack deflection	Reduced crack tip ΔK
Crack branching	Reduced crack tip ΔK
Homogeneous plastic deformation	Reduced amounts of crack deflection and branching and hence increased crack tip ΔK
Inhomogeneous plastic deformation	Increased amounts of crack deflection and branching and hence decreased crack tip ΔK
Plasticity - induced crack closure	Reduced effective ΔK
Crack closure induced by roughness and crack deflection and branching	Reduced effective ΔK
Irregular crack front (Often related to roughness)	Reduced effective ΔK
Crack closure induced by oxide or corrosion debris	Reduced effective ΔK in the near-threshold regime
precipitates, dispersoids, dislocation substructures, grain boundaries } microstructural barriers; crystallographic texture (especially Al-Li alloys); environment	Changes in crack tip plasticity, crack deflection, branching, roughness and irregularity, and hence changes in crack tip ΔK and effective ΔK
higher E (Al-Li alloys)	Decrease in G (crack driving force per unit thickness)

Fatigue crack growth resistance

Factors	Effects
Lattice friction stress (τ)	Increased flow stress and hence resistance to fatigue crack growth
Homogeneous plastic deformation	(1) Multiple slip and crack blunting with less slip available for crack extension, i.e. increased crack growth resistance (2) Less slip reversibility, i.e. decreased crack growth resistance
Inhomogeneous plastic deformation	(1) Concentrated slip bands available for crack extension, i.e. decreased crack growth resistance (2) Slip reversibility, i.e. increased crack growth resistance
Precipitates, Dispersoids, Dislocation Substructures, Grain boundaries } Micro-structural barriers	(1) Increased crack growth resistance per se (2) Changes in crack tip plasticity and hence crack growth resistance
Crystallographic texture (especially Al-Li alloys)	Changes in crack tip plasticity and hence crack growth resistance
Environment	Aggressive environments reduce crack growth resistance (e.g. by hydrogen embrittlement or adsorption-induced embrittlement); also slip reversibility is decreased

FIGURE 12.14 Survey of factors contributing to and influencing the crack driving force and FCG resistance of aluminum alloys under stationary or quasi-stationary load histories (Wanhill, 1994b).

1. The Li content in third-generation alloys has been reduced to less than about 1.8 wt% and in some alloys to about 1% or less, compared to Li contents of 2 wt% or more and as high as 2.4 wt% in second-generation alloys.
2. Lower Li contents mean that third-generation alloys have inherently higher fracture toughness. This enables them to be artificially aged closer to the peak-aged condition compared to second-generation alloys, which were typically underaged to improve the fracture toughness.
3. Compositional changes and special thermomechanical processes in some products have reduced crystallographic texture or its influence.

The results of these changes are that slip planarity and slip band intensity are less in third-generation alloys and deformation is more homogeneous. Consequently, their FCG behavior in terms of both growth rates and fracture characteristics is more similar to that of conventional alloys than the behavior of second-generation alloys.

Under CA/CR loading, the FCG rates of third-generation alloys are typically higher than those of second-generation alloys at low-to-intermediate ΔK. At higher ΔK, the third-generation alloy FCG rates are lower owing to improved fracture toughness. Nevertheless, the third-generation alloys have at least equivalent and mostly superior CA/CR FCG properties compared to those of the conventional alloys they are intended to replace. Furthermore, the limited available flight simulation test data for third-generation alloys indicate that the FCG advantage over conventional alloys is retained, unlike the situation for second-generation alloys.

The improvements in FCG and other properties in third-generation Al—Li alloys relative to both second-generation and conventional alloys have already led to many aircraft applications, including floor beams, seat tracks, frames, fuselage and wing stringers, ribs, and bulkheads. These latest Al—Li alloys are also being used for cryotankage in spacecraft. A survey of these applications is given in Chapter 2 and also in Chapter 15.

12.8.2 Practice-Related Crack Growth Regimes

Most FCG test data are collected and compared for what would be the non-inspectable and in-service inspectable crack growth regimes, see Figure 12.1. This includes load-shedding CR tests down to the long/large crack FCG thresholds. The long/large crack near-threshold FCG and threshold data are customarily used for fracture mechanics modeling of early crack growth. However, this approach does not describe the actual behavior of short/small cracks growing into the non-inspectable crack growth regime (Molent et al., 2011; Wanhill et al., 2013), and so it cannot provide reliable and accurate assessments of the FCG *durability* of components and structures.

Because FCG *durability* represents the major part of the FCG life (Schijve, 1967), there is a general need of data for cracks growing from natural or realistically-sized initial discontinuities and progressing through the short/small and non-inspectable crack growth regimes. It would be especially valuable to obtain these types of data for both CA and flight simulation loading; and in the context of this chapter and Book it would be appropriate to acquire these data for third-generation Al—Li alloys and equivalent conventional alloys.

REFERENCES

Alcoa Aerospace Technical Fact Sheet, 2008. Alloy 2099-T83 and 2099-T8E67 Extrusions. AEAP-Alcoa Engineered Aerospace Products, Lafayette, IN.

Anstee, R.F.W., 1983. An assessment of the importance of small crack growth to aircraft design. In: Behaviour of Short Cracks in Aircraft Components. AGARD Conference Proceedings No. 328, Advisory Group for Aerospace Research and Development, Neuilly-sur-Seine, France, pp. 3-1—3-9.

Anstee, R.F.W., Edwards, P.R., 1983. A review of crack growth threshold and crack propagation rates at short crack lengths. In: Some Considerations on Short Crack Growth Behaviour in Aircraft Structures. AGARD Report No. 696, Advisory Group for Aerospace Research and Development, Neuilly-sur-Seine, France, pp. 2-1—2-12.

Barter, S.A., Wanhill, R.J.H., 2013. Crack growth path studies: aluminium—lithium 8090-T81 preliminary tests. Air Vehicles Division, Defence Science and Technology Organisation DSTO, Melbourne, Australia.

Barter, S.A., Molent, L., Wanhill, R.J.H., 2012. Typical fatigue-initiating discontinuities in metallic aircraft structures. Int. J. Fatigue 41, 11—22.

Blom, A.F., 1990. Short crack growth under realistic flight loading: model predictions and experimental results for Al 2024 and Al—Li 2090. In: Short-Crack Growth Behaviour in Various Aircraft Materials. AGARD Report No. 767, Advisory Group for Aerospace Research and Development, Neuilly-sur-Seine, France, pp. 6-1—6-15.

Bowen, A.W., 1990. Annex: texture analysis of 2090-T8E41aluminium—lithium alloy sheet. In: Short-Crack Growth Behaviour in Various Aircraft Materials. AGARD Report No. 767, Advisory Group for Aerospace Research and Development, Neuilly-sur-Seine, France, pp. 11-1—11-5.

Bretz, P.E., Sawtell, R.R., 1986. Alithalite alloys: progress, products and properties. In: Baker, C., Gregson, P.J., Harris, S.J., Peel, C.J. (Eds.), Aluminium—Lithium Alloys III: Proceedings of the Third International Conference on Aluminium—Lithium Alloys. The Institute of Metals, London, pp. 47—56.

Bucci, R.J., Malcolm, R.C., Colvin, E.L., Murtha, S.J., James, R.S., 1989. Cooperative test programme for the evaluation of engineering properties of Al—Li alloy 2090-T8X sheet, plate and extrusion products. Naval Surface Warfare Center Technical Report 89-106, Silver Spring, MD.

Chanani, G.R., Telesman, J., Bretz, P.E., Scarich, G.V., 1982. Methodology for evaluation of fatigue crack growth resistance of aluminum alloys under spectrum loading. Northrop Corporation Technical Report NOR 82-54, Northrop Corporation Aircraft Division, Hawthorne, CA.

Cook, R., 1990. The growth of short fatigue cracks in 2024 and 2090 aluminium alloys under variable amplitude loading. In: Short-Crack Growth Behaviour in Various Aircraft Materials. AGARD Report No. 767, Advisory Group for Aerospace Research and Development, Neuilly-sur-Seine, France, pp. 5-1—5-11.

Coyne Jr., E.J., Sanders Jr., T.H., Starke Jr., E.A., 1981. The effect of microstructure and moisture on the low cycle fatigue and fatigue crack propagation of two Al—Li—X alloys.

In: Sanders Jr., T.H., Starke Jr., E.A. (Eds.), Aluminium–Lithium Alloys I: Proceedings of the First International Aluminium–Lithium Conference. The Metallurgical Society of AIME, Warrendale, PA, pp. 293–305.

De Jonge, J.B., Schütz, D., Lowak, H., Schijve, J., 1973. A standardized load sequence for flight simulation tests on transport aircraft wing structures. National Aerospace Laboratory NLR Technical Report NLR-TR-73029, Amsterdam, The Netherlands.

Denzer, D.K., Rioja, R.J., Bray, G.H., Venema, G.B., Colvin, E.L., 2012. The evolution of plate and extruded products with high strength and toughness. In: Weiland, H., Rollett, A.D., Cassada, W.A. (Eds.), Proceedings of the 13th International Conference on Aluminum Alloys (ICAA13). The Minerals, Metals and Materials Society (TMS) and John Wiley & Sons, Hoboken, NJ, pp. 587–592.

Edwards, P.R., Darts, J., 1984. Standardised fatigue loading sequences for helicopter rotors (Helix and Felix), part 2: final definition of Helix and Felix. Royal Aircraft Establishment Technical Report 84085, Farnborough, UK.

Edwards, P.R., Newman Jr., J.C. Eds., 1990a. Short-Crack Growth Behaviour in Various Aircraft Materials. AGARD Report No. 767, Advisory Group for Aerospace Research and Development, Neuilly-sur-Seine, France.

Edwards, P.R., Newman, Jr., J.C., 1990b. An AGARD supplemental test programme on the behaviour of short cracks under constant amplitude and aircraft spectrum loading. In: Short-Crack Growth Behaviour in Various Aircraft Materials. AGARD Report No. 767, Advisory Group for Aerospace Research and Development, Neuilly-sur-Seine, France, pp. 1-1–1-43.

Farcy, L., Clavel, M., 1990. Small fatigue crack growth behaviour in a 2091 aluminium lithium alloy in comparison with 2024 alloy. In: Khan, T., Effenberg, G. (Eds.), Advanced Aluminium and Magnesium Alloys: Proceedings of the International Conference on Light Metals. ASM International European Council, Brussels, pp. 173–180.

Farcy, L., Carre, C., Clavel, M., Barbaux, Y., Aliaga, A., 1987. Factors of crack initiation and microcrack propagation in aluminum lithium 2091 and in aluminum 2024. In: Champier, G., Dubost, B., Miannay, D., Sabatay, L. (Eds.), Fourth International Aluminium Lithium Conference. J. Phys. Colloque C3 48, C3-769–C3-775.

Gallagher, J.P., Giessler, F.J., Berens, A.P., Engle Jr., R.M., 1984. USAF damage tolerant design handbook: guidelines for the analysis and design of damage tolerant aircraft structures. Air Force Wright Aeronautical Laboratories Technical Report AFWAL-TR-82-3073, Dayton, OH.

Garratt, M.D., Bray, G.H., Koss, D.A., 2001. Influence of texture on fatigue crack growth behavior. In: Tiryakioğlu, M. (Ed.), Advances in the Metallurgy of Aluminum Alloys: Proceedings from Materials Solutions Conference 2001, The James T. Staley Honorary Symposium on Aluminum Alloys. ASM International, Materials Park, pp. 151–159.

Jata, K.V., Starke Jr., E.A., 1986. Fatigue crack growth and fracture toughness behavior of an Al–Li–Cu alloy. Metall. Trans. A 17A, 1011–1026.

Karabin, L.M., Bray, G.H., Rioja, R.J., Venema, G.B., 2012. Al–Li–Cu–Mg–(Ag) products for lower wing skin applications. In: Weiland, H., Rollett, A.D., Cassada, W.A. (Eds.), Proceedings of the 13th International Conference on Aluminum Alloys (ICAA13). The Minerals, Metals and Materials Society (TMS) and John Wiley & Sons, Hoboken, NY, pp. 529–534.

Lowak, H., De Jonge, J.B., Franz, J., Schütz, D., 1979. MINITWIST: a shortened version of TWIST. National Aerospace Laboratory NLR Miscellaneous Publication NLR-MP-79018U, Amsterdam, The Netherlands.

Magnusen, P.E., Mooy, D.C., Yocum, L.A., Rioja, R.J., 2012. Development of high toughness sheet and extruded products for airplane fuselage structures. In: Weiland, H., Rollett, A.D., Cassada, W.A. (Eds.), Proceedings of the 13th International Conference on Aluminum Alloys (ICAA13). The Minerals, Metals and Materials Society (TMS) and John Wiley & Sons, Hoboken, NJ, pp. 535–540.

McDarmaid, D.S., 1985. Fatigue and fatigue crack growth behaviour of medium and high strength Al–Li–Cu–Mg–Zr alloy plate. Royal Aircraft Establishment Technical Report 85016, Farnborough, UK.

Molent, L., Barter, S.A., Wanhill, R.J.H., 2011. The lead crack fatigue lifing framework. Int. J. Fatigue 33, 323–331.

Paris, P.C., Erdogan, F., 1963. A critical analysis of crack propagation laws. J. Basic Eng. Trans. ASME 85D, 528–534.

Paris, P.C., Gomez, M.P., Anderson, W.E., 1961. A rational analytic theory of fatigue. Trend Eng. 13, 9–14.

Pearson, S., 1966. Fatigue crack propagation in metals. Nature 211, 1077–1078.

Peters, M., Welpmann, K., McDarmaid, D.S.,'t Hart, W.G.J., 1989. Fatigue properties of Al–Li alloys. In: New Light Alloys. AGARD Conference Proceedings No. 444, Advisory Group for Aerospace Research and Development, Neuilly-sur-Seine, France, pp. 6-1–6-18.

Ranganathan, N., Ait Abdedaim, M., Petit, J., 1990. Microscopic load interaction effects observed in an Al–Li alloy as compared to classical damage tolerant alloys. In: Khan, T., Effenberg, G. (Eds.), Advanced Aluminium and Magnesium Alloys: Proceedings of the International Conference on Light Metals. ASM International European Council, Brussels, pp. 165–172.

Rioja, R.J., Liu, J., 2012. The evolution of Al–Li base products for aerospace and space applications. Metall. Mater. Trans. A 43A, 3325–3337.

Schijve, J., 1967. Significance of fatigue cracks in micro-range and macro-range. Fatigue Crack Propagation, ASTM STP 415. American Society for Testing and Materials, Philadelphia, PA, pp. 415–457.

Schijve, J., Jacobs, F.A., Tromp, P.J., 1972. Fatigue crack growth in aluminium alloy sheet material under flight-simulation loading. Effects of design stress level and loading frequency. National Aerospace Laboratory NLR Technical Report NLR-TR-72018, Amsterdam, The Netherlands.

Shaw, P., 1985. WESTEC'85—aluminum–lithium alloy sessions. GAC Memorandum M&ME-TS441A-85K-01, Grumman Aerospace Corporation, Bethpage, NY.

Starke E.A., Jr., Quist, W.E., 1989. The microstructure and properties of aluminum–lithium alloys. In: New Light Alloys. AGARD Conference Proceedings No. 444, Advisory Group for Aerospace Research and Development, Neuilly-sur-Seine, France, pp. 4-1–4-23.

Swift, T., 1983. Verification of methods for damage tolerance evaluation of aircraft structures to FAA requirements. 12th ICAF Symposium. Centre d'Essais Aéronautique de Toulouse (CEAT), Toulouse, pp. 1.1/1–1.1/87.

't Hart, W.G.J., Schra, L., McDarmaid, D.S., Peters, M., 1989. Mechanical properties and fracture toughness of 8090-T651 plate and 2091 and 8090 sheet. In: New Light Alloys. AGARD Conference Proceedings No. 444, Advisory Group for Aerospace Research and Development, Neuilly-sur-Seine, France, pp. 5-1–5-17.

Tchitembo Goma, F.A., Larouche, D., Bois-Brochu, A., Blais, C., Boselli, J., Brochu, M., 2012. Fatigue crack growth behavior of 2099-T83 extrusions in two different environments. In: Weiland, H., Rollett, A.D., Cassada, W.A. (Eds.), Proceedings of the 13th International

Conference on Aluminum Alloys (ICAA13). The Minerals, Metals and Materials Society (TMS) and John Wiley & Sons, Hoboken, NJ, pp. 517–522.

Van Dijk, G.M., de Jonge, J.B., 1975. Introduction to a fighter aircraft loading standard for fatigue evaluation FALSTAFF. National Aerospace Laboratory NLR Miscellaneous Publication NLR-MP-75017, Amsterdam, The Netherlands.

Vasudévan, A.K., Bretz, P.E., Miller, A.C., Suresh, S., 1984. Fatigue crack growth behavior of aluminum alloy 2020 (Al–Cu–Li–Mn–Cd). Mater. Sci. Eng. 64, 113–122.

Venkateswara Rao, K.T., Ritchie, R.O., 1989. Mechanical properties of Al–Li alloys: part 2. Fatigue crack propagation. Mater. Sci. Technol. 5, 896–907.

Venkateswara Rao, K.T., Yu, W., Ritchie, R.O., 1988a. On the behavior of small cracks in commercial aluminum–lithium alloys. Eng. Fract. Mech. 31, 623–635.

Venkateswara Rao, K.T., Yu, W., Ritchie, R.O., 1988b. Fatigue crack propagation in aluminum–lithium alloy 2090. Part I: Long crack behavior. Metall. Trans. A 19A, 549–561.

Venkateswara Rao, K.T., Piascik, R.S., Gangloff, R.P., Ritchie, R.O., 1989. Fatigue crack propagation in aluminum–lithium alloys. In: Sanders Jr., T.H., Starke Jr., E.A. (Eds.), Aluminium–Lithium Alloys: Proceedings of the Fifth International Aluminium–Lithium Conference, vol. 2. Materials and Component Engineering Publications, Birmingham, pp. 955–971.

Venkateswara Rao, K.T., Bucci, R.J., Ritchie, R.O., 1990. On the micromechanisms of fatigue-crack propagation in aluminium–lithium alloys: sheet versus plate material. In: Kitagawa, H., Tanaka, T. (Eds.), Fatigue 90: Proceedings of the Fourth International Conference on Fatigue and Fatigue Thresholds, vol. 2. Engineering Materials Advisory Services, Warley, UK, pp. 963–970.

Venkateswara Rao, K.T., Bucci, R.J., Jata, K.V., Ritchie, R.O., 1991. A comparison of fatigue-crack propagation behavior in sheet and plate aluminum–lithium alloys. Mater. Sci. Eng. A A141, 39–48.

Wanhill, R.J.H., 1979. Gust spectrum fatigue crack propagation in candidate skin materials. Fatigue Eng. Mater. Struct. 1, 5–19.

Wanhill, R.J.H., 1984. Fatigue and fracture resistance of wrought PM 7091 and IM 7050, PM Aerospace Materials, vol. 1. MPR Publishing Services, Shrewsbury, UK, pp. 36-1–36-19.

Wanhill, R.J.H., 1994a. Fatigue and fracture properties of aerospace aluminium alloys. In: Carpinteri, A. (Ed.), Handbook of Fatigue Crack Propagation in Metallic Structures. Elsevier Science Publishers, Amsterdam, pp. 247–279.

Wanhill, R.J.H., 1994b. Damage tolerance engineering property evaluations of aerospace aluminium alloys with emphasis on fatigue crack growth. National Aerospace Laboratory NLR Technical Publication NLR-TP-94177, Amsterdam, The Netherlands.

Wanhill, R.J.H., 1994c. Flight simulation fatigue crack growth testing of aluminium alloys. Specific issues and guidelines. Int. J. Fatigue 16, 99–110.

Wanhill, R.J.H., 1994d. Status and prospects for aluminium–lithium alloys in aircraft structures. Int. J. Fatigue 16, 3–20.

Wanhill, R.J.H., Hattenberg, T., 2006. Fractography-based estimation of fatigue crack "initiation" and growth lives in aircraft components. National Aerospace Laboratory NLR Technical Publication NLR-TP-2006-184, Amsterdam, The Netherlands.

Wanhill, R.J.H., Schra, L., 1989. Fracture and fatigue of damage tolerant Al–Li sheet alloys. National Aerospace Laboratory NLR Contractor Report 89349C, Amsterdam, The Netherlands.

Wanhill, R.J.H., Jacobs, F.A., Schijve, J., 1976. Environmental fatigue under gust spectrum loading for sheet and forging aircraft materials. In: Bathgate, R.G. (Ed.), Fatigue Testing and Design, vol. 1. The Society of Environmental Engineers, Buntingford, UK, pp. 8.1–8.33.

Wanhill, R.J.H., Schra, L., 't Hart, W.G.J., 1990. Fracture and fatigue evaluation of damage tolerant Al—Li alloys for aerospace applications. In: Firrao, D. (Ed.), Fracture Behaviour and Design of Materials and Structures: Proceedings of the Eighth European Conference on Fracture, ECF 8, vol. 1. Engineering Materials Advisory Services, Warley, UK, pp. 257—271.

Wanhill, R.J.H., 't Hart, W.G.J., Schra, L., 1991. Flight simulation and constant amplitude fatigue crack growth in aluminium—lithium sheet and plate. In: Kobayashi, A. (Ed.), Aeronautical Fatigue: Key to Safety and Structural Integrity. Ryoin Co., Tokyo, and Engineering Materials Advisory Services, Warley, UK, pp. 393—430.

Wanhill, R.J.H., Molent, L., Barter, S.A., 2013. Fracture mechanics in aircraft failure analysis: uses and limitations. Eng. Fail. Anal. Available online, 15 November 2012 (Corrected Proof).

Yoder, G.R., Pao, P.S., Imam, M.A., Cooley, L.A., 1989. Micromechanisms of fatigue fracture in Al—Li 2090. In: Sanders Jr., T.H., Starke Jr., E.A. (Eds.), Aluminum—Lithium Alloys: Proceedings of the Fifth International Aluminum—Lithium Conference, vol. 2. Materials and Component Engineering Publications, Birmingham, UK, pp. 1031—1041.

Chapter 13

Fracture Toughness and Fracture Modes of Aerospace Aluminum–Lithium Alloys*

S.P. Lynch*, R.J.H. Wanhill**, R.T. Byrnes*, and G.H. Bray†
*DSTO, Melbourne, Australia, **NLR, Emmeloord, the Netherlands, †Alcoa Technical Center, Alcoa Center, USA

Contents

13.1 Introduction	**416**	
13.2 Test Methods for Determining Fracture Toughness (and Terminology)	**418**	
13.2.1 Plane-Stress/Plane-Strain Considerations	419	
13.2.2 Testing Thick Products	420	
13.2.3 Testing Sheet and Thin Plate	422	
13.3 Effects of Microstructural Features on Fracture Toughness and Fracture Modes	**423**	
13.3.1 Extrinsic Inclusions and Porosity	424	
13.3.2 Constituent Particles	424	
13.3.3 Alkali-Metal Impurity Phases	426	
13.3.4 Dispersoids	427	

13.3.5 Matrix Precipitates 427
13.3.6 Grain-Boundary Precipitates and Precipitate-Free Zones 430
13.3.7 Grain-Boundary Segregation 430
13.3.8 Overview and Microstructural Design of Third-Generation Al–Li Alloys 433
13.4 Fracture Toughness of Second-Generation Al–Li Alloys Versus Conventional Al Alloys 435
13.4.1 Room-Temperature Data 435
13.4.2 Effects of Testing Temperature and Strain Rate 439

Aluminum–Lithium Alloys.

13.5 Fracture Toughness of Third-   **13.6 Uses and Potential Uses of**
      Generation Al–Li Alloys 442   **Third-Generation Al–Li**
      13.5.1 Thick Plate and Other   **Alloys   448**
             Thick Products   442   **13.7 Conclusions   449**
      13.5.2 Sheet and Thin   **References   451**
             Plate   444   **Specific references   451**
                                            Other references   455

13.1 INTRODUCTION

Specifying and obtaining adequate fracture toughness (combined with high strength) are essential parts of modern damage-tolerant aircraft design processes. Other essential damage-tolerant properties include fatigue and stress-corrosion cracking (SCC) resistance, topics that are covered in other chapters of this book. Sub-critical cracking by fatigue and SCC is, of course, usually a precursor to fast fracture, with fracture toughness determining the critical crack lengths at which fast fracture ensues under the applied operating loads.

For aircraft fuselage structures, for example, it is a fail-safe certification requirement that unstable crack growth, should it occur, be arrested within two bays (Figure 13.1). Crack-arrest is achieved by using materials with adequate toughness and by careful design of the structure, including the addition of doublers acting as tear straps between frames and skin. If crack-arrest

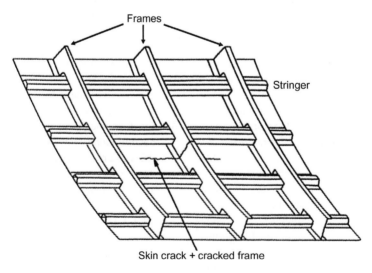

FIGURE 13.1 Schematic diagram showing cracking in the skin (across two bays) and a frame in a typical stiffened fuselage. The extent of fast fracture depends on the panel design, applied stresses (mainly from pressurization), and fracture toughness.

does not occur, the fuselage can fail catastrophically, as happened for two de Havilland Comet aircraft in 1954 causing 56 fatalities (Swift, 1987; Wanhill, 2003).

Using materials with high fracture toughness is also beneficial in minimizing impact damage from foreign objects or during crashes (when preexisting cracks may not be present). If materials with low fracture toughness are used, as in early model AgustaWestland helicopters where AA 8090 Al−Li was used extensively to reduce weight, "hard landings" can result in extensive damage or complete disintegration of the airframe (Figures 13.2 and 13.3) (Pasang et al., 2012; Wanhill et al., 2013). The low toughness of the AA 8090 Al−Li alloy was associated with "brittle" intergranular fracture, and the reasons why this fracture mode occurs and how it has largely been eliminated in the latest Al−Li alloys are discussed in this chapter.

The catastrophic fire (55 fatalities out of 137 passengers and crew) after an aborted take-off of a Boeing 737 British Airtours flight from Manchester airport in 1985 is another example where inadequate fracture toughness (impact resistance) of an Al alloy was a contributing factor. In this case, the failure of an engine combustor-can (owing to fatigue associated with a poor weld repair) led to part of the combustor-can impacting/fracturing/holing a brittle Al alloy wing-tank access panel, leading to the escape of fuel and a fire that consumed the aircraft. The Air Accidents Investigation Branch report (No. 8/88) suggested, *inter alia*, that airframe manufacturers and airworthiness authorities should specify that materials with adequate impact resistance be used for structures vulnerable to impact by engine or wheel/tyre debris.

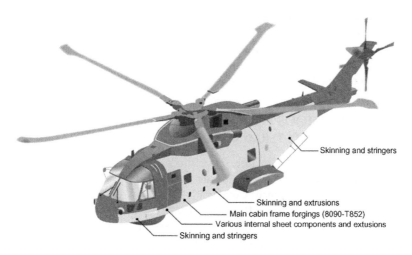

Skinning and stringers

Skinning and extrusions
Main cabin frame forgings (8090-T852)
Various internal sheet components and extusions
Skinning and stringers

FIGURE 13.2   Schematic diagram showing usage of second-generation Al−Li−Cu−Mg (AA 8090) alloy in the airframe of early-model EH101 AgustaWestland helicopters.

FIGURE 13.3 AgustaWestland EH101 helicopter after a "hard landing" from 4 to 5 m after a loss of control due to a tail-rotor malfunction. Some crew were injured, but there were no fatalities (although fatalities have occurred in other cases). The low fracture toughness of the AA 8090 Al–Li–Cu–Mg alloy used in the airframe was a major factor contributing to its disintegration.

Fracture toughness must not only be sufficient to prevent catastrophic failure of structures and components during service, but must also be high enough to enable them to be manufactured without cracking (e.g., during hole drilling and installation of interference-fit fasteners (Rioja and Liu, 2012) and "joggling" of extrusions (Zink and Weilke, 1992)). The limited use of first- and second-generation Al–Li alloys (developed in the 1940s–1960s and 1980s–1990s, respectively) for aircraft structures was partly due to such cracking problems, which were the result of low fracture toughness in the short-transverse crack-plane orientation. Fracture toughness needs to be adequate for all crack-plane orientations (Figure 13.4), and mechanical properties in general should be as isotropic as possible.

Microstructural factors that control fracture toughness and the recent developments of third-generation Al–Li alloys with less anisotropic properties than first- and second-generation Al–Li alloys are discussed after briefly describing test methods for determining fracture toughness.

13.2 TEST METHODS FOR DETERMINING FRACTURE TOUGHNESS (AND TERMINOLOGY)

For aerospace structures, fracture under mode-I loading (where the applied load is perpendicular to the plane of the notch or fatigue pre-crack) has the most engineering significance. Combinations of crack-opening (mode I) and crack-sliding/shearing (modes II and III) displacements do occur in some circumstances, but mixed-mode fracture toughness of Al alloys is only occasionally measured (Eswara Prasad and Kamat, 1995; Eswara Prasad et al., 1994; Shashidhar et al., 1995) and will not be considered further in this chapter.

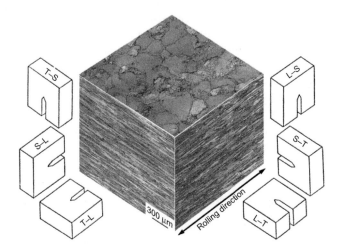

FIGURE 13.4   Optical micrographs of grain structure in an AA 8090 alloy rolled plate, plus crack-plane orientation codes for rectangular sections such as plate, extrusions, and forgings (ASTM Standard E399-09). The first letter of the code designates the direction normal to the crack plane and the second letter gives the expected direction of crack growth. For the fuselage sheet and frame shown in Figure 13.1, it is the T−L fracture toughness that is relevant.

13.2.1 Plane-Stress/Plane-Strain Considerations

Fracture-toughness values, and the test methods used to obtain them, usually depend on the thickness of specimens. For thin specimens (e.g., 1−2 mm thick sheet) where material ahead of cracks is subjected to plane-stress conditions, fracture toughness is much higher than for thick specimens subjected to plane-strain conditions. Fracture toughness therefore decreases with increasing thickness toward a minimum (plateau) value under full plane-strain. This value is defined as the (opening-mode) plane-strain fracture toughness, K_{Ic}, although other subscripts are also used depending on the type of specimen.

Decreasing fracture toughness with increasing thickness is associated with increasing constraint on plasticity for material further away from the side surfaces. The increasing constraint leads to a change from plane-stress (biaxial stress state) to plane-strain (triaxial stress state) conditions. The higher constraint/stress-triaxiality in plane-strain reduces the size of the crack-tip plastic zone and the amount of plastic work required for microvoid nucleation and coalescence processes, thereby reducing fracture toughness. Furthermore, the high level of constraint under plane-strain results in macroscopically flat fracture at 90° to the specimen sides. This contrasts with plane-stress fracture, which occurs on planes at 45° to the specimen sides, resulting in single or double slant fractures. More detailed considerations of fracture under plane-stress and plane-strain conditions can be found in fracture mechanics books (Ewalds and Wanhill, 1984; Thomason, 1990).

For some materials, including second-generation Al–Li alloys, fracture toughness does *not* decrease with increasing thickness for L–T and T–L crack-plane orientations when the S–L/S–T fracture toughness is low. In these circumstances, S–L delaminations occur ahead of (and normal to) L–T and T–L cracks (owing to stresses arising from lateral constraint) so that fracture occurs by shear of thin ligaments essentially under plane-stress conditions regardless of specimen thickness (McKeighan et al., 1992). The effects of specimen thickness (or a lack thereof) on fracture toughness, along with schematic diagrams of fracture-surface profiles, are summarized in Figure 13.5.

13.2.2 Testing Thick Products

Since the 1960s, numerous specimen designs and test methods have been considered for K_{Ic} determinations by the American Society for Testing and Materials ASTM E08 Fatigue and Fracture Committee (Figure 13.6). ASTM Standard E399-09 describes several specimen types, of which the fatigue pre-cracked Compact Tension (CT) specimen is the most commonly used. This standard also describes the analysis for deriving K_{Ic} and for checking whether requirements for obtaining valid K_{Ic} values have been met. Codes that indicate the specimen crack-plane orientations are also included, as indicated in Section 13.1 (Figure 13.4).

Another standard, ASTM E1304-97, describes the determination of plane-strain fracture toughness K_{Iv} or K_{Ivj} using "short bar" and "short rod"

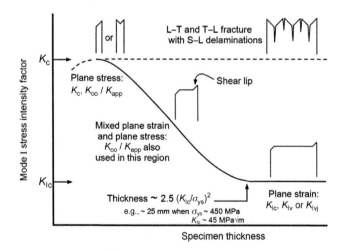

FIGURE 13.5 Schematic plots of the effects of specimen thickness on fracture toughness and associated fracture-surface profiles. The lack of an effect of specimen thickness on fracture toughness (dotted line) (owing to delaminations) has been observed for some second-generation Al–Li alloys.

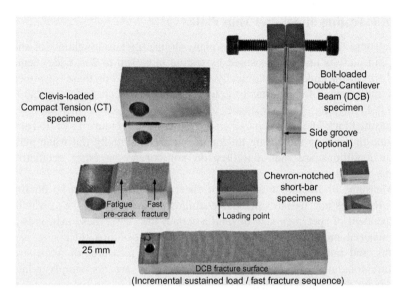

FIGURE 13.6 Some of the specimen types used for determining plane-strain fracture toughness.

chevron-notched specimens. These fracture-toughness values are not regarded as valid K_{Ic}, but comparative tests have shown the results to be within about $\pm 10\%$ of values obtained using CT specimens (Brown, 1984). The chevron-notched specimens have the advantage of not requiring fatigue pre-cracking, and their small size enables them to be taken from material with limited thickness, such that S−L fracture toughness can be determined for material with only 12.5 mm thickness. These advantages make this test method especially suitable for establishing the effects of metallurgical and testing variables on the fracture toughness of small volumes of newly developed or existing materials.

Bolt-loaded, double-cantilever-bend (DCB) specimens (which do not require testing machines) have also been used to determine S−L fracture toughness, providing that S−L toughness values are less than about 25 MPa$\sqrt{m}$ and toughness values for other crack-plane orientations are significantly higher. Otherwise, cracks deviate from the mid-plane position and the DCB arms break off. Side grooves can sometimes inhibit such behavior but are generally not required. For DCB specimens, the stress-intensity factors decrease with increasing crack length so that small increments of crack-opening displacement result in small increments (several millimeters) of crack growth and, hence, multiple measurements can be made on the same specimen. Moreover, measurements can be made for a particular ageing condition, then specimens re-aged, and further measurements made on the same specimen to determine the effects of re-ageing.

13.2.3 Testing Sheet and Thin Plate

Plane-stress (and mixed plane-stress/plane-strain) fracture toughness of sheet and thin plate is usually measured by testing large (up to 2 m wide) center-cracked panels under monotonically increasing loads. For these specimens, large increments of slow (stable) crack growth often occur before instability and fast fracture. The resistance to stable crack growth increases with increasing crack length, as characterized by R-curves (Figure 13.7). R-curves become flatter with increasing thickness and are essentially flat when plane-strain conditions are met, but they do not depend on other geometrical factors.

Measures of fracture toughness of sheet and thin plate can be obtained from R-curves but, unlike valid plane-strain fracture toughness, K_{Ic}, which is independent of specimen size beyond a certain thickness, these measures are not material properties because they depend on specimen thickness, panel widths, and initial crack length/width ratios. Two measures of plane-stress fracture toughness are commonly used, and they are not simply related. Moreover, different shorthand terminologies are used in the literature for the same measurement, which can also cause confusion. The fracture toughness designated as K_c (or K_{Rc}) is calculated from the maximum load and the effective crack length (physical crack length + plastic zone size) and is the crack-tip stress-intensity factor corresponding to rapid unstable fracture. The toughness designated as K_{app} (or K_{co}) is the apparent stress-intensity factor at maximum load using the pre-crack length, a_0, and takes no account of stable crack growth.

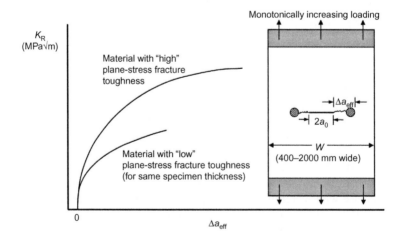

FIGURE 13.7 Schematic R-curves for materials with high and low plane-stress fracture toughness. The inset shows a center-cracked panel (designated middle-cracked tension M(T) configuration) that is often used for testing, see ASTM E561-05.

K_c values can be much higher than K_{app} values, and both may increase markedly with increasing panel width, especially in highly damage-tolerant materials (Fournier et al., 1992). Thus, one must be careful when comparing the plane-stress (and mixed plane-stress/plane-strain) fracture toughness of different alloys and heat treatments (and different crack-plane orientations) to ensure that "like is being compared with like". For the data presented in Sections 13.4 and 13.5, this is the case, although in some cases, K_c values are compared, while in other cases, K_{app} (K_{co}) values are compared.

Further information on plane-stress (and mixed plane-stress/plane-strain) fracture-toughness testing can be found in ASTM standard E561-05 and in the book by Ewalds and Wanhill (1984).

13.3 EFFECTS OF MICROSTRUCTURAL FEATURES ON FRACTURE TOUGHNESS AND FRACTURE MODES

The fracture toughness of high-strength precipitation-hardened Al alloys depends on numerous microstructural features (Staley, 1976; Hahn and Rosenfield, 1975; Starke and Quist, 1989) (Figures 13.8 and 13.9). These features occur in both conventional Al alloys and Al−Li alloys, and are obviously

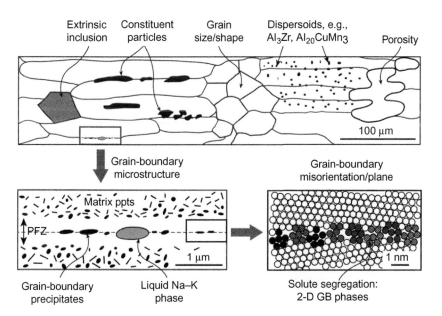

FIGURE 13.8 Schematic diagram showing the principal microstructural features that can affect fracture toughness (and other properties) in Al alloys. The atomic structure of grain boundaries (lower right-hand side) depends on the grain-boundary (GB) misorientation, GB plane, GB segregation, and temperature. Weak bonding across grain boundaries due to specific types of atomic arrangements/compositions (2-D GB phases) can facilitate brittle intergranular fracture.

dependent on the alloy composition, processing, and heat-treatment procedures. Some features, however, may be subtly different or more significant in Al–Li alloys than in conventional alloys. The relative importance of these microstructural features on fracture toughness and fracture modes is discussed here, with an emphasis on Al–Li alloys and the evolution of third-generation alloys.

13.3.1 Extrinsic Inclusions and Porosity

Extrinsic microstructural features such as entrapped refractory inclusions, aluminum oxide films, and porosity are not common in modern, commercially processed Al alloy products (plate, extrusions, sheet, etc.). If they are present, they are likely to be few and far between and, hence, have little effect on fracture toughness (although they could facilitate fatigue crack initiation if they were at, or near, surfaces). Remnants of casting porosity can sometimes be present in worked products if the degree of working is small (e.g., as can occur in thick plate (Figure 13.9)). For aerospace products, ultrasonic inspection is carried out to reduce the risk of using products containing significant defects.

13.3.2 Constituent Particles

Constituent particles often occur as clusters or stringers up to $100\,\mu m$ long and are detrimental to fracture toughness because the particles are brittle and

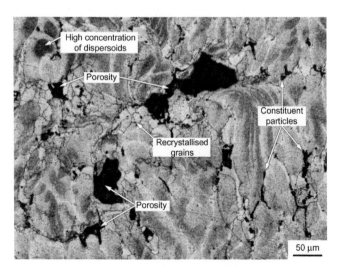

FIGURE 13.9 Optical micrograph of polished and etched section of a rolled 76-mm thick AA 7079-T651 Al–Zn–Mg–Cu plate showing various microstructural features, namely, remnants of interdendritic casting porosity, constituent particles, elongated grains, equiaxed recrystallized grains, and remnants of dendrites (compositional variations) formed during solidification, with dispersoid particles concentrated in the center of dendrites (darker etching regions) (Knight, 2008).

cleave easily, or could have been broken during processing. Some of these particles are rich in iron and silicon, and minimizing Fe and Si impurity levels has been the main approach in reducing the size and number of these insoluble particles, and thereby improving toughness, for both conventional Al alloys and Al–Li alloys. For soluble (or partially soluble) constituent particles with high concentrations of alloying elements, solution-treatment for longer times or at higher temperatures can decrease the number of particles and thereby increase toughness. A solution-treatment temperature just below the solidus must be approached in a series of steps (times at progressively increasing temperatures) to prevent the occurrence of incipient melting.

For the Fe and Si levels (typically <0.1 wt%) present in modern Al alloys that are solution-treated using commercial practices, the number and size of constituent particles are still significant (Figure 13.10), but further reduction in their volume fraction may not be economically viable. Constituent particles are usually present as stringers along the direction of hot/cold working and, hence, might be expected to have greater detrimental effects on the S–L and S–T toughness than the L–T and T–L toughness, especially when they are present along the boundaries of pancake-shaped grains and intergranular fracture occurs. However, if *low-energy* intergranular fracture occurs, then the presence of constituent particles would probably decrease short-transverse toughness to only a small extent. Addressing the problem of brittle intergranular fracture, which is more prevalent in Al–Li alloys than in conventional alloys, is therefore more important in developing tough Al–Li alloys than reducing the number of constituent particles further.

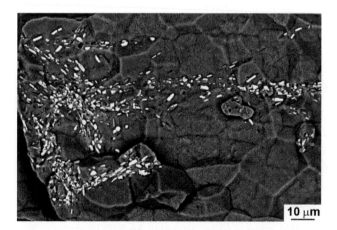

FIGURE 13.10 SEM (back-scattered electron mode) image of an S–L fracture surface of AA 8090-T8771 (0.04 wt%Fe + 0.02 wt%Si) showing distribution of constituent particles (lighter areas) and intergranular facets, after testing at 20°C (Byrnes and Lynch, 1990–2012).

13.3.3 Alkali-Metal Impurity Phases

Alkali-metal impurities (AMIs) such as sodium and potassium can be intro-duced into Al alloys through feedstock and pickup from refractories. In most Al alloys, these AMIs are tied up in innocuous solid compounds by bismuth or silicon, but this appears not to be the case in Al–Li alloys because lithium reacts preferentially with these elements. In Al–Li alloys with (deliberately introduced) high levels of AMIs (>60 wt ppm), discrete, *liquid* Na–K phases (at 20°C) have been observed directly by transmission electron microscopy (TEM) (Webster, 1987). These phases can result in liquid-metal embrittlement (LME), involving weakening of interatomic bonds by adsorp-tion of Na and K atoms at the tips of cracks nucleated at liquid-phase sites (Sweet et al., 1996).

The embrittling liquid is rapidly drawn along with the advancing crack tips by capillary action so that extensive "brittle" cracking can occur even when the volume of AMI phases is small. A thin layer of liquid is left behind on fracture surfaces so that cracks can run out of liquid when the volume of phases is small and they are widely spaced. When this occurs, a transition to a "more ductile" fracture mode occurs, and brittle intergranular or cleavage-like islands centered on sites of the liquid phase, surrounded by dimpled areas, are observed on fracture surfaces (Figure 13.11). The number and size of brittle islands decrease with decreasing temperature as the Na–K-rich phases progressively solidify. The number and size of brittle islands also decrease with decreasing AMI content (for testing at 20°C) and are few and far between when AMI levels are less than about 5 wt ppm (Sweet et al., 1996). For commercially produced AA 8090 plates, Na + K levels were ≤5 ppm (Lynch, 1991; Lynch et al., 1993). Thus, the occurrence of brittle intergranular fracture in second-generation alloys with high Li contents (such as AA 8090) is not associated with AMIs.

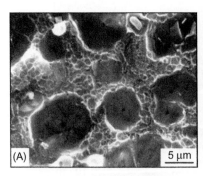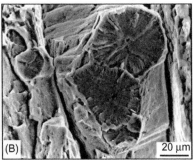

FIGURE 13.11   SEM images of fracture surfaces of AA 8090-T8, containing high (>10 ppm) Na + K levels, tested at 20°C, showing (A) brittle intergranular islands and (B) cleavage-like islands, surrounded by dimpled areas. Brittle islands are centered on sites where liquid Na–K phases had been present (Sweet et al., 1996).

13.3.4 Dispersoids

Dispersoid particles such as Al_3Zr, $Al_{20}CuMn_3$, or $Al_{12}Mg_2Cr$ are formed during homogenization (providing, of course, that the alloys contain Zr, Mn, or Cr, respectively). The sizes of dispersoids range from 40 to 100 nm for Al_3Zr to several hundred nanometers (and occasionally several micrometers) for Mn- and Cr-rich particles. These particles inhibit recrystallization and grain growth by pinning grain boundaries during subsequent heat treatments. This is generally beneficial because material with unrecrystallized grain structures has higher strength than material with recrystallized grains, although mechanical properties are more anisotropic. Unrecrystallized grain structures also inhibit intergranular cracking and increase toughness for L–T and T–L crack-plane orientations.

If there are only Al_3Zr dispersoids (which are fully coherent with the matrix), as in some second-generation Al–Li alloys such as AA 8090, strong crystallographic textures tend to be present. Moreover, if most age-hardening precipitates are coherent with the matrix, and there are only coherent Al_3Zr dispersoids present (as in the 8090 alloy), then deformation can be localized into narrow slip bands, resulting in slip-band fractures (which can extend across numerous grains if strong textures are present). The fracture energy associated with localized microvoid-coalescence in slip bands, which results in shallow shear dimples on fracture surfaces, may be less than desirable.

Incoherent dispersoids, such as $Al_{20}CuMn_3$, which are present in some third-generation Al–Li alloys in addition to Al_3Zr, help disperse slip and result in more isotropic mechanical properties (Rioja et al., 1991). The Mn-rich dispersoids do, however, initiate voids more easily than Al_3Zr dispersoids during crack growth but, providing the Mn-rich dispersoids are not too coarse or too closely spaced, they probably have overall beneficial effects on the fracture toughness of Al–Li alloys. Fractographic observations of deep, equiaxed dimples containing Mn-rich dispersoids, rather than shallow shear dimples when only Al_3Zr dispersoids are present (as in second-generation Al–Li alloys), support this view (Figure 13.12). For conventional 7XXX Al–Zn–Mg–Cu alloys, data in the literature (e.g. Staley, 1976) indicate that Mn-based dispersoids have adverse effects on toughness compared with Zr-based ones. However, differences in dispersoid volume-fraction, grain structure/crystallographic texture, strength, and fracture mode in different alloys complicate comparison of the influence of various types of dispersoids on fracture toughness, and differences in their effects for different alloys/ heat-treatments would not be surprising.

13.3.5 Matrix Precipitates

The main effect of age-hardening precipitates on fracture toughness is via their effect on strength, with higher strengths generally resulting in lower

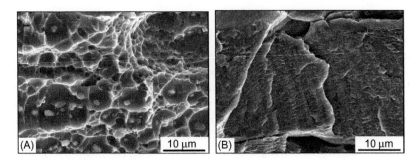

FIGURE 13.12 SEM images of transgranular T−L fracture surfaces after testing at 20°C for (A) AA 2050-T8X showing deep dimples associated with void formation around (larger than average) incoherent $Al_{20}CuMn_3$ dispersoids and (B) AA 8090-T8771 showing shallow shear dimples (on a facet inclined to the stress axis) associated with localized slip due to dislocations cutting through coherent δ' (Al_3Li) precipitates and coherent Al_3Zr dispersoids (Byrnes and Lynch, 1990−2012).

plane-stress and plane-strain fracture toughness, as documented in a later section. This relationship between toughness and strength applies to most materials, and is the result of greater amounts of strain localization (smaller plastic zones) around crack tips and the reduced ability of the material to accommodate strain without fracture, for example, owing to more numerous void initiation sites, for higher strength materials. The exact relationship, however, will depend on the fracture modes (or change in fracture modes with strength) which depend on the microstructural features discussed in this section. With respect to matrix precipitation, important characteristics affecting strength and toughness include coherency, volume fraction, size, spacing, and aspect ratio.

As mentioned in Section 13.3.4, when matrix precipitates are coherent with the matrix (such as Al_3Li δ' precipitates in AA 8090), they are easily sheared by dislocations, thereby reducing their effectiveness as obstacles to further slip, so that slip is concentrated in narrow bands leading to low-energy fracture by localized shear. The resistance to fracture by localized shear decreases with increasing strength (as for most fracture modes), and a concern with second-generation Al−Li alloys with high Li contents has been that secondary ageing (and strength increases) due to additional matrix precipitation could occur in service, thereby decreasing toughness below specified (as manufactured) values. Tests on near peak-aged AA 8090-T81 and AA 2091-T84 sheets subsequently aged for up to 6 months at 70°C (to simulate service exposure) decreased toughness substantially (Figure 13.13A) (Fournier et al., 1992; Pitcher et al., 1992; Noble et al., 1994; Starink et al., 2000).

It has often been suggested that the prevalence of brittle intergranular fracture in second-generation Al−Li alloys such as AA 8090 is associated with easily sheared coherent δ' matrix precipitates since they promote coarse,

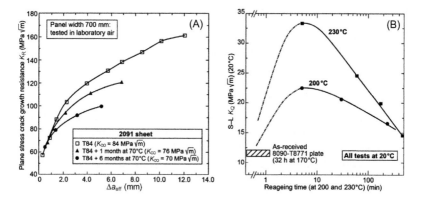

FIGURE 13.13 (A) R-curves (and plane-stress fracture toughness) for AA 2091-T84 sheet showing that long times at 70°C (simulating service exposure) decrease toughness (Fournier et al., 1992). Similar behavior has been observed for AA 8090-T81 sheet (Pitcher et al., 1992). (B) Plot of plane-strain S–L fracture toughness (from DCB tests) of AA 8090-T8771 plate (aged 32 h at 170°C) versus re-ageing times at 200°C and 230°C, showing that short re-ageing times dramatically increased fracture toughness, whereas longer times were less effective (and eventually ineffective) (Lynch, 1991).

planar slip. Specifically, it has been proposed that dislocations in narrow, planar slip bands pile-up at grain boundaries so that locally high stress concentrations are produced at grain boundaries, resulting in brittle intergranular fracture (Blankenship and Starke, 1993; Csontos and Starke, 2005; Eswara Prasad et al., 1993; Gregson and Flower, 1985). However, this hypothesis is questionable (Lynch, 1991; Lynch et al., 1993; Lynch et al., 2001, 2002; Moutsos et al., 2010; Pasang et al., 2012) for a number of reasons:

1. For AA 8090 plate (S–L orientation), the occurrence of brittle intergranular fracture depends on ageing treatments (and testing temperature, Section 13.4), which do not significantly affect slip characteristics. Thus, 8090 plate aged to near peak strength (T8771 condition/32 h at 170°C) exhibits predominantly brittle intergranular fracture and low S–L fracture toughness, whereas material re-aged for 5 min at 200°C exhibits high toughness and predominantly dimpled intergranular fracture—with matrix precipitation, strength, and slip mode not being greatly affected by the second ageing treatment at 200°C (Figure 13.13B).

2. Some other precipitation-hardened materials, for example Waspaloy, Ni−Cr−Mo−Ti, which is strengthened by γ' (Ni$_3$−(Al,Ti)) (L1$_2$) coherent precipitates (analogous to δ' precipitates in Al−Li alloys), can exhibit marked slip planarity without undergoing brittle intergranular fracture.

Brittle intergranular fracture is therefore more likely to be associated with grain-boundary microstructures rather than with slip-mode/matrix precipitation, as discussed in the following sections.

13.3.6 Grain-Boundary Precipitates and Precipitate-Free Zones

Precipitate-free zones (PFZs) are usually present adjacent to grain boundaries in precipitation-hardened Al alloys (including Al–Li alloys) because nucleation and growth of precipitates occur preferentially at grain boundaries during quenching (for slow cooling rates) and ageing, resulting in solute depletion adjacent to the boundaries so that matrix precipitates are not formed. Strain localization in soft PFZs results in nucleation and growth of voids around grain-boundary precipitates (GBPs) so that dimpled intergranular fracture surfaces are produced (e.g. Vasudevan and Doherty, 1987). The size and depth of dimples decrease with decreasing spacing of GBPs and decreasing PFZ width, as would be expected. For an intergranular fracture mode, toughness values would, correspondingly, be expected to decrease with decreasing spacing of GBPs and decreasing PFZ width, although PFZ width would have little effect if large area fractions of GBPs were present (Lynch, 1988).

For near peak-aged AA 8090-T8771 plate material, PFZs are reasonably wide and GBPs are quite well spaced (Figure 13.14A). Thus, well-defined dimples would be expected on intergranular fracture surfaces, and are indeed observed on some facets. Many facets, however, exhibit relatively featureless areas except for GBPs and some very shallow dimples (Figure 13.14B). Since PFZ widths and GBP spacings do not appear to vary that much from one grain boundary to another, some other factor must be responsible for the brittle nature of many facets.

The effects of re-ageing mentioned previously (Figure 13.13B) also indicate that PFZs and GBPs are not primarily responsible for *brittle* intergranular facets because PFZ width and GBP spacings were not significantly affected by the short re-ageing treatment that largely eliminated such facets. Rather, the evidence suggests that segregation of lithium (and possibly also magnesium) at some grain boundaries is primarily responsible for *brittle* intergranular fracture, although strain localization in PFZs and around GBPs (when they are present) is undoubtedly a contributing factor (Lynch, 1991; Lynch et al., 1993, 2001, 2002; Miller et al., 1986; Moutsos et al., 2010; Pasang et al., 2012).

13.3.7 Grain-Boundary Segregation

Segregation of certain elements at grain boundaries is known to induce brittle intergranular fracture in some materials (e.g., metalloid impurities in tempered martensitic steels). However, there is no evidence (from high-resolution TEM or from surface analysis of fracture surfaces by Auger Electron Spectroscopy) that impurities (besides hydrogen) segregate at grain boundaries in Al alloys (Lewandowski and Holroyd, 1990; Miller et al., 1986). Hydrogen at grain boundaries is known to embrittle Al alloys, but comparison of AA 2090 Al–Li alloys with low and high hydrogen contents

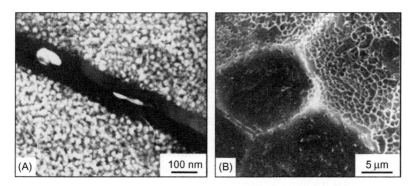

FIGURE 13.14 (A) Dark-field TEM image of near peak-aged AA 8090-T8771 alloy showing δ' (Al$_3$Li) matrix precipitates, grain-boundary precipitates, and a precipitate-free zone. (B) SEM of S−L fracture surface of AA 8090-T8771 plate tested at 20°C, showing well-defined dimples on some grain-boundary facets and relatively featureless facets in other areas (Lynch, 1991; Lynch et al., 2002).

suggested that hydrogen embrittlement is probably not a significant factor affecting fracture toughness in Al−Li alloys—perhaps because hydrogen is tied up as innocuous lithium hydride particles (Bennett et al., 1998).

Segregation of alloying elements, on the other hand, is known to occur in Al alloys, and theoretical calculations (Lewandowski and Holroyd, 1990; Seah, 1980) suggest that some elements (e.g., Li and Mg) could be embrittling if they segregated at grain boundaries in aluminum. For some Al alloys, segregation of Mg has been observed directly using TEM and atom-probe techniques (Liddicoat et al., 2009; Valiev et al., 2010). Al−Mg alloys with ≥9 wt% Mg (which are essentially single-phase solid solutions except for isolated GBPs) exhibit brittle intergranular fracture at low testing temperatures, and since there are no other reasonable explanations, Mg segregation is probably responsible (Pasang et al., 2012).

There is some evidence for Li segregation at grain boundaries in high-Li second-generation AA 8090 alloys, but it is limited (Lewandowski and Holroyd, 1990; Lynch et al., 1993; Wilson, 1990), and more recent studies have not been able to establish whether or not it is present (Moutsos, 2012). However, it must be remembered that monolayer segregation of lithium (atomic number 3) is difficult or impossible to *unambiguously* detect using standard techniques for showing segregation at grain boundaries, especially when Li-containing GBPs are present. Indirect evidence that Li segregation, rather than slip planarity, is primarily responsible for *brittle* intergranular fracture in high-Li second-generation alloys such as AA 8090, on the other hand, is compelling.

Firstly, brittle intergranular fracture occurs (at low testing temperatures) in "binary" Al−2.7 wt% Li (0.1 wt% Zr) alloys in the *as-quenched* condition where there are no GBPs or PFZs, and explanations other than solute segregation at grain boundaries can be discounted (as for Al–Mg alloys) (Lynch, 1991; Lynch et al., 1993, 2001; Miller et al., 1986). Secondly, explanations for brittle intergranular fracture (in AA 8090 plate) based on Li segregation are consistent with:

1. the effects of ageing treatments on toughness *when fracture is intergranular* (Figure 13.13B);[1]
2. the kinetics (activation energies) of embrittlement during ageing or re-ageing;
3. the effects of grain-boundary misorientation (texture) on toughness;
4. the effects of testing temperature on toughness and the extent of brittle intergranular fracture (Section 13.4.2).

The proposed explanation for brittle intergranular fracture based on Li segregation at grain boundaries also invokes the formation of 2D Li-rich grain-boundary phase(s) with specific structural units that have weak bonding (Lynch et al., 2001, 2002). The formation of 2D grain-boundary phases has a sound thermodynamic and theoretical basis (Cahn, 1982; Guttmann, 1977; Rottman, 1988), but their importance in regard to fracture resistance has often been overlooked. The formation of a Li-rich 2D phase would probably depend not only on Li segregation levels, but also on the presence of other solute segregants, the grain-boundary misorientation, and the temperature.

The toughening effects of short re-ageing treatments for AA 8090 plate (Figure 13.13B) can be explained in terms of the above as follows. Embrittling 2D Li-rich grain-boundary phases formed during the first ageing treatment undergo reversion (like small, 3D matrix precipitates) during the short ageing times at the secondary, somewhat higher temperatures, so that embrittlement is largely eliminated. For longer re-ageing times at the higher secondary ageing temperature, or during re-ageing of toughened material at lower ageing temperatures, the Li-rich 2D grain-boundary phases become re-established so that re-embrittlement occurs. For the AA 8090 plate, the kinetics of re-embrittlement during low temperature ageing are consistent with Li diffusion from the adjacent matrix to grain boundaries since the activation energy for re-embrittlement is similar to the activation energy for Li diffusion (Lynch et al., 1993).

The dimpled and brittle fracture modes for different grain-boundary facets for AA 8090 plate (Figure 13.14B) can best be explained on the basis that the levels of Li segregation (and atomic structure at grain boundaries) vary from one grain boundary to another owing to their different misorientations. It is well established that segregation (of impurities) in other materials is often less for lower angle grain boundaries, as might be expected due to

---

1. The effects of ageing/re-ageing treatments on intergranular and transgranular fracture resistance have different explanations: Transgranular shear is associated with matrix precipitation, as indicated previously.

their "more compact" structures. For AA 8090-T8771 plate, the toughness was higher, with fewer brittle facets, in the mid-thickness position compared with the near-surface positions, probably because the former had a more marked crystallographic texture and, hence, had a greater proportion of low-angle boundaries (with less Li segregation) compared with the near-surface positions (Lynch et al., 1992).

This explanation for embrittlement, toughening, and re-embrittlement of AA 8090 plate is supported by the remarkably similar behavior observed in a temper-embrittled martensitic steel. Thus, re-tempering the steel for short times (5 min) at 525°C, following an initial tempering at 425°C for 24 h, produced significant toughening that was associated with a decrease in the extent of brittle intergranular fracture (Shinoda et al., 1986). For temper-embrittled martensitic steels, it is well established that brittle intergranular fracture is associated with segregation of metalloid impurities to prior-austenite grain boundaries (as already mentioned), and it was proposed that re-tempering for short times resulted in reversion of a 2D phosphorus-rich grain-boundary phase, with this phase re-establishing itself after longer times (Shinoda et al., 1986).

13.3.8 Overview and Microstructural Design of Third-Generation Al−Li Alloys

Clearly, the fracture toughness and fracture modes of Al alloys depend on numerous microstructural features, and Al−Li alloys can differ significantly from conventional Al alloys in some respects. In summary, the most common fracture modes, and the microstructural features associated with them, in Al−Li alloys are as follows:

1. Cleavage of large constituent particles.
2. Brittle intergranular fractures, involving either decohesion or a very localized microvoid-coalescence process along grain boundaries, possibly weakened by the segregation of alloying elements (and the formation of Li-rich 2-D grain-boundary phases), in combination with strain localization in PFZs and around GBPs.
3. Localized, transgranular shear fractures (exhibiting shallow shear dimples) along slip bands, due to shearing of coherent age-hardening precipitates and dispersoids, which can extend over quite large length scales when crystallographic textures are strong.
4. "Ductile" intergranular fractures involving strain localization in soft PFZs and nucleation of voids at GBPs, such that well-defined dimples are present on intergranular facets when PFZs are reasonably wide and GBP spacings are not too small.
5. Dimpled transgranular fractures involving void nucleation and growth around incoherent Mn-rich dispersoids, such that large, deep, equiaxed

dimples are produced when dispersoids are not too large or too closely spaced.

The plastic work required for these different fracture modes, illustrated schematically in Figure 13.15, ranges from very small for cleavage of constituent particles and for brittle intergranular fracture to quite large for transgranular fractures with deep dimples. Intermediate levels of plastic work are required for the other fracture modes—other things such as strength being equal (which is rarely the case in practice, of course). The development of third-generation Al–Li alloys (and some conventional Al alloys) with improved fracture toughness has been based on microstructural (and compositional) modifications designed to minimize the low-energy fracture modes associated with second-generation Al–Li alloys.

Producing third-generation Al–Li alloys with adequate toughness (in all crack-plane orientations) has involved the following (Rioja and Liu, 2012):

1. Controlling Fe and Si levels to less than 0.1 wt% (to minimize the size and number of coarse constituent particles) and keeping AMI levels low to prevent LME (as was done for second-generation Al–Li alloys).
2. Introducing incoherent Mn-rich dispersoids (but keeping coherent Zr-based dispersoids which more effectively prevent recrystallization) to homogenize slip and control crystallographic texture/grain size so that properties are more isotropic than those of second-generation Al–Li alloys containing only Zr-based dispersoids. Mn- rather than Cr-based dispersoids are probably preferred for Al–Li alloys because the latter result in greater "quench-sensitivity."

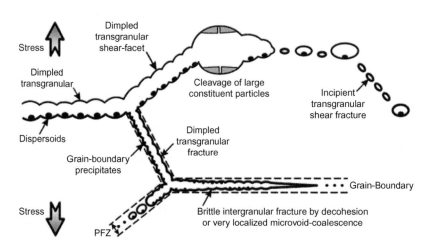

FIGURE 13.15 Schematic diagram of common fracture modes in age-hardenable aluminum alloys (Lynch et al., 2008).

3. Reducing the Li content, in some cases quite markedly (e.g., from ~ 2.4 wt% (~ 8.78 at%) in AA 8090 to ~ 0.75 wt% (~ 2.94 at%) in AA 2060) for a number of reasons. Decreasing Li levels will decrease coherent δ' (Al_3Li) precipitation (or avoid it completely for Li ≤ 1.3 wt%) so that slip planarity is reduced and, hence, low-energy slip-band fractures are less likely. Decreasing Li contents should, of course, also decrease Li segregation at grain boundaries so that brittle intergranular fracture is not a problem (assuming that Li segregation occurs and is responsible).

4. Increasing Cu contents, and having sufficient Li to produce a high number density of small, high-aspect-ratio T_1 (Al_2CuLi) precipitates, along with θ'-type ($\sim Al_2Cu$) precipitates, to provide high strength. Both T_1 and θ' have plate-like morphologies (but different crystallographic relationships with the matrix) and are more resistant to shear than δ' (Al_3Li) so that strain localization in slip bands is further reduced.

5. Adding Ag to some alloys, e.g., ~ 0.25 wt% in AA 2060, to stimulate T_1 precipitation and provide some solid-solution strengthening. Cold work prior to ageing also strongly promotes T_1 matrix precipitation (since T_1 nucleates preferentially on dislocations).

6. Adding Zn to some alloys, e.g., ~ 0.4 wt% in AA 2060, for solid-solution strengthening and corrosion resistance, while retaining some Mg (up to 0.85 wt% in AA 2060) to help offset density increases associated with increases in Cu and Zn and decreases in Li, as well as provide some additional solid-solution strengthening.

These points are, of course, general guidelines, and the relative importance of each factor will depend on the specific alloy composition, processing, and heat-treatment/ageing practices, which are generally proprietary and not divulged by manufacturers. As regards ageing treatments, they should be sufficiently long to produce a high number density of matrix hardening precipitates without excessive grain-boundary precipitation. The concentration of solute participating in matrix precipitation should also be low after ageing in order to minimize additional ageing during low-temperature service exposure (i.e., the alloys should be "thermally stable").

13.4 FRACTURE TOUGHNESS OF SECOND-GENERATION AL−LI ALLOYS VERSUS CONVENTIONAL AL ALLOYS

13.4.1 Room-Temperature Data

The extensive research carried out on second-generation Al−Li alloys in the 1980s and 1990s was aimed at developing alloys with ≥ 2 wt% Li that were 8−10% lighter (and also stiffer) than conventional Al alloys and had at least equivalent strength and other properties (see Peel, 1989, and general references). In terms of mechanical properties, the requirements were to develop (i) damage-tolerant Al−Li alloys with tensile yield strengths, σ_y, of

330−370 MPa to replace the (then) widely used AA 2024-T3 Al−Cu−Mg alloy, (ii) medium strength ($\sigma_y \sim 450$ MPa) Al−Li alloys to replace conventional Al−Cu−Mg alloys such as AA 2014-T651, and (iii) high strength alloys ($\sigma_y \approx 500-550$ MPa) to replace 7XXX Al−Zn−Mg−Cu alloys. In addition to achieving strength targets, equivalent or better fracture toughness (or at least adequate values) was required, along with equivalent or better fatigue and corrosion/SCC resistance.

Numerous alloys (and manufacturing processes) were investigated during the 1980s, and three of the most promising alloys (Table 13.1) were subjected to more extensive testing to optimize compositions, processing (e.g., casting, hot and cold rolling), and heat treatments (e.g., solution treatment, quenching, stretching, and ageing). Despite such extensive development of these high-Li second-generation alloys, none of them succeeded in achieving mechanical properties equivalent to those of conventional alloys, and fracture toughness was especially problematic. Thus, the second-generation Al−Li alloys have not been widely used, and the use of AA 8090 for early models of the EH101 helicopter was discontinued owing to the low fracture toughness of the material—which manifested itself during "hard landings"/crashes (Figure 13.3). The minimum specified S−L, T−L, and L−T plane-strain (20°C) fracture toughness of AA 8090-T852 forgings used for the helicopter substructures were ≥ 11, ≥ 16, and ≥ 18 MPa$\sqrt{\text{m}}$, respectively, and "actual" values might have been lower due to secondary ageing in service, especially at low in-service temperatures (Pasang et al., 2012). An S−L toughness value of 11 MPa$\sqrt{\text{m}}$ is *less than half* that of comparable conventional Al alloys, and the L−T and T−L values of the forgings were also much inferior.

The plane-strain (20°C) S−L fracture toughness of AA 8090-T8771 plate was also sometimes as low as 11 MPa$\sqrt{\text{m}}$, although higher values of ~ 18 MPa$\sqrt{\text{m}}$ were often quoted or plotted on graphs and bar charts ('t Hart et al., 1989). The higher values were probably average ones for (i) alloys

---

TABLE 13.1 Composition Ranges (wt%) of the Major Alloying Elements, Plus Typical Values of Minor Elements and Impurities, for the Three Most Widely Studied (and Then Most Promising) Second-Generation Al−Li Alloys (with Li Contents $\geq$ about 2.0 wt%)

Alloy (Density, g/cm^3)	Li	Cu	Mg	Zr	Ti	Fe	Si
AA 8090 (2.54)	2.2−2.7	1.0−1.6	0.6−1.3	∼0.1	0.10	<0.10	<0.10
AA 2090 (2.60)	1.9−2.6	2.4−3.0	0.25	∼0.1	0.15	<0.10	<0.10
AA 2091 (2.59)	1.7−2.3	1.8−2.5	1.1−1.9	∼0.1	0.10	<0.10	<0.10

Density values are for a mid-range composition.

with a Li content toward the lower end of the range, (ii) mid-thickness positions of plate, and (iii) material not subjected to low-temperature ageing for long times. In general, short-transverse fracture toughness tended to be lower for material with higher Li (and Mg) contents, positions nearer the surface, and after long times at ambient temperatures (Lynch et al., 1992; Peel, 1989). For example, the (mid-thickness) S−L fracture toughness of an AA 8090-T8771 plate (of unspecified Li content) tested in 1987 was ~17.5 MPa√m, which decreased to 13.7 MPa√m in 1991 after storage at ambient temperature (Pitcher et al., 1992).

For L−T and T−L orientations of medium-to-high strength AA 8090 and AA 2090 plate (with more elongated grain structures than forgings), "plane-strain" fracture-toughness values were similar to those of conventional alloys (about 30−40 MPa√m). However, such comparisons are (arguably) spurious because, unlike that of conventional Al alloys, T−L and L−T fracture of Al−Li plates was associated with deep S−L delaminations ahead of L−T and T−L cracks, resulting in plane-stress conditions, as described previously (Figure 13.5). These S−L delaminations are, of course, associated with low S−L fracture toughness and, as indicated in Section 13.1, fracture toughness needs to be adequate for all crack-plane orientations.

For sheet (typically ~1.6 mm thick), plots of the plane-stress fracture toughness, K_c, versus yield strength, and R-curves for high-strength and damage-tolerant material, show (i) significantly lower toughness (~30%) for the high-strength AA 2090-T83 compared with AA 7075-T6, and (ii) decreasing differences in toughness with decreasing strength such that damage-tolerant, recrystallized AA 2091 and AA 8090 sheet had about the same toughness as AA 2024-T3 (~150 MPa√m for 400−500 mm wide specimens)—albeit with somewhat lower (~5%) yield strength for the Al−Li alloys (Figures 13.16 and 13.17) (Wanhill, 1994). However, these reasonably favorable comparisons (taking density into account) for damage-tolerant alloys need to take other factors into consideration. For example, the data are for material not subjected to secondary low-temperature ageing in service, which can substantially decrease the toughness of AA 8090 and AA 2091 alloys (Figure 13.13A) (Fournier et al., 1992; Noble et al., 1994; Pitcher et al., 1992) but not affect the toughness of AA 2024.

In summary, the lower fracture toughness of second-generation Al−Li alloys (for both plane-stress and plane-strain conditions) compared with conventional Al alloys was a major factor in preventing their widespread use. There were also other issues with some second-generation Al−Li alloys such as (i) abnormal, macroscopic deviations of fatigue cracks at large angles (up to 60°) to the applied tensile-stress direction in sheet material, which were not acceptable to aircraft designers, (ii) higher spectrum (flight-simulation) fatigue crack growth rates compared with conventional alloys (even though constant-amplitude fatigue crack growth rates were lower in some cases, and (iii) lower resistance to SCC for some product forms and heat treatments, as

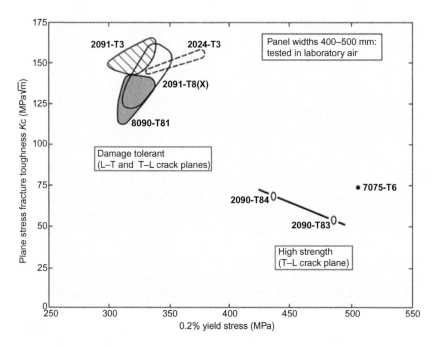

FIGURE 13.16 Plot of plane-stress fracture toughness versus yield strength for some second-generation Al–Li alloys compared with conventional alloys, showing that the properties of the former became increasingly inferior with increasing strength (Wanhill, 1994). Differences in panel width between 400 and 500 mm make little difference to K_c values.

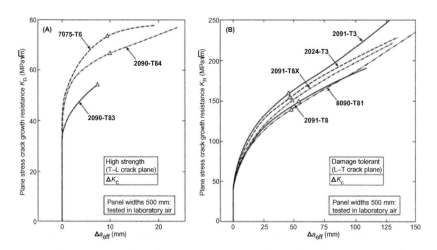

FIGURE 13.17 Comparisons of R-curves and plane-stress fracture toughness, K_c, of second-generation Al–Li alloy sheet versus conventional alloys (Wanhill, 1994) for (A) high strength alloys (T–L crack plane) showing inferior properties for Al–Li alloys, and (B) damage-tolerant alloys (L–T crack plane) showing fairly similar properties for Al–Li alloys compared with AA 2024-T3—but note the caveats mentioned in the text (Section 13.4.1).

discussed in other chapters. These issues and the experience with developing second-generation Al−Li alloys did, however, provide a basis for developing the third-generation Al−Li alloys that have a balance of properties that exceeds those of conventional Al alloys, as quantified in Section 13.5.

13.4.2 Effects of Testing Temperature and Strain Rate

Testing Temperature

The fracture behavior of some second-generation high-Li alloys such as AA 8090 can be particularly sensitive to the testing temperature. Perhaps the most dramatic manifestation of this sensitivity is the sharp transitions from 100% ductile transgranular fracture to 100% brittle intergranular fracture with decreasing temperature observed in very under-aged AA 8090 plate for the S−L crack-plane orientation. The ductile-to-brittle transition temperatures increased with increasing ageing times and were higher for material with weaker textures (more high-angle grain boundaries) (Lynch et al., 1993). These observations are interesting, even though very under-aged alloys are not used, since they support the view that Li segregation/2D phases are responsible for brittle intergranular fracture, as discussed in Section 13.3.7 for near peak-aged AA 8090 alloys.

For under-aged alloys, explanations for brittle intergranular fracture involving GBPs and PFZs can be discounted since these features are not present, and changes in slip planarity (and strength) are not significant over the observed 10−20°C transition temperature range. Rather, the sharp transitions are best explained by a 2D grain-boundary phase change, which occurs at higher temperatures for grain boundaries with higher levels of Li segregation (resulting from longer ageing times or higher angle grain boundaries for a given ageing time/temperature). Similar ductile-to-brittle intergranular fracture transitions with decreasing temperature occur in other systems (e.g., Cu−Sb, where segregation at grain boundaries is known to be involved (McLean, 1952−53)).

For second-generation (low AMI) Al−Li alloys such as near peak-aged AA 8090-T8771 plate, the S−L fracture toughness decreases with decreasing temperature (Lynch, 1991). The proportion of brittle intergranular fracture (vis-à-vis "ductile" intergranular fracture) correspondingly increases with decreasing temperature. For L−T and T−L crack-plane orientations, on the other hand, decreasing temperature results in increased fracture toughness. This increase is mainly associated with increasing extents (number and depth) of short-transverse intergranular delaminations ahead of (and normal to) growing L−T and T−L cracks with decreasing temperature owing to the decrease in S−L toughness with decreasing temperature (Venkateswara Rao and Ritchie, 1989). The increased splitting (thinner and longer ligaments) reduces the through-thickness constraint, leading to greater amounts of plasticity (shear) associated with failure of the ligaments. More homogenous slip or greater work-hardening rates with decreasing temperature may also result in more plasticity associated with ligament shear (Glazer et al., 1987).

For near peak-aged AA 8090-T81 recrystallized sheet, fast fracture at 20°C occurred almost entirely by transgranular shear, but increasing proportions of brittle intergranular fracture and cleavage-like fracture occurred with decreasing temperature—unlike the behavior of AA 2024-T3 sheet (Figure 13.18) (Byrnes and Lynch, 1990–2012). The fracture toughness of the AA 8090 sheet would, therefore, be expected to decrease in correspondence with increasing proportions of brittle fracture modes. Given that the external temperature of aircraft structures can be as low as −55°C in (very) cold climates or at high altitudes, fracture toughness should be measured at low temperatures, not just in laboratory air. In other words, it should not be assumed, just because the toughness of some "old" alloys is not that sensitive to temperature, that "new" alloys will behave similarly. The possible influence of secondary ageing due to thermal exposure during service on the temperature dependence of fracture toughness should also have been considered.

The increase in the extent of brittle intergranular fracture with decreasing temperature in the near peak-aged 8090-T8771 plate and 8090-T81 sheet can

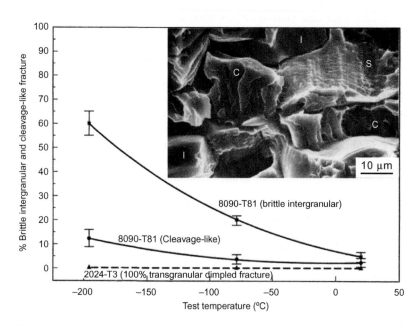

FIGURE 13.18 Plot showing increasing extents of brittle intergranular and cleavage-like cracking with decreasing testing temperature for an AA 8090-T81 recrystallized sheet (in comparison with AA 2024-T3 sheet). The inset shows the fracture surface (after testing at −196°C) exhibiting areas of brittle intergranular fracture (I), cleavage-like fracture (C), and transgranular shear fracture (S) (Byrnes and Lynch, 1990–2012).

be explained in terms of the 2D grain-boundary phase transitions related to Li segregation, as discussed previously. However, the reason why cleavage-like fractures occur at low temperatures is not clear. They are not associated with liquid AMI phases (as shown in Figure 13.11B) since AMI contents were low and any such phases would be solid at $-196°C$. Intrinsic cleavage-like fractures in face centered cubic (FCC) metals (in inert environments) are certainly unusual, but have been observed in several FCC metals (e.g., Ir) where it has been suggested that interatomic bonding at crack tips is different from normal (Lynch, 2007). Maybe the high Li contents in AA 8090 affects surface bonding at low temperatures in a similar way.

Strain Rate

The effect of strain rate on fracture toughness (at 20°C) has not been widely studied for Al alloys, but observations for some second-generation Al−Li alloys (e.g., AA 2091-T81) suggest that dynamic toughness values (for unstable, fast crack growth) could be significantly lower than the quasi-static ones for stable crack growth. Moreover, fracture occurred by transgranular shear during stable crack growth and by shear and brittle intergranular fracture during unstable cracking (Figure 13.19). The second-generation Al−Li sheet alloys also exhibited a tendency for increased amounts of short-transverse intergranular delamination when slow stable monotonic crack growth changed to unstable fast crack growth (Wanhill et al., 1992). However, it is not clear why high strain rates associated with fast fracture facilitated intergranular fracture in Al−Li alloys.

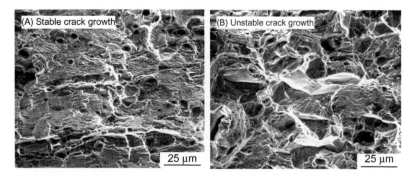

FIGURE 13.19   Fractographic characteristics of stable and unstable monotonic crack growth in 1.6 mm thick AA 2091-T8X sheet tested at 20°C: main crack-plane orientation L−T (Wanhill and Schra, 1991). Stable crack growth (A) was 100% transgranular, while unstable crack growth (B) was about 50% intergranular and 50% transgranular.

13.5 FRACTURE TOUGHNESS OF THIRD-GENERATION AL–LI ALLOYS

The compositions and densities of third-generation Al–Li alloys discussed in the following sections are given in Table 13.2. A more comprehensive table of compositions for third- (and first- and second-) generation Al–Li alloys can be found in Chapter 2. The rationale behind the various alloying additions and amounts in third-generation Al–Li alloys was outlined in Section 13.3.8.

13.5.1 Thick Plate and Other Thick Products

For medium-to-high strength plates of AA 2050-T84 and AA 2060-T8E33, the plane-strain fracture-toughness anisotropy is similar to that of the widely used conventional plate alloy AA 7050-T7451 (Figure 13.20) and much less than that of second-generation Al–Li alloy plates (Rioja and Liu, 2012). For the S–L (weakest) orientation of these third-generation Al–Li alloys, fracture toughness values were $23-28$ MPa$\sqrt{m}$—similar to or higher than the value for AA 7050-T7451, and at somewhat higher strength levels (Figure 13.20). The potential weight savings from the use of these third-generation Al–Li alloys in place of conventional 7XXX alloys is also significant ($\sim$4% lower density), although not as high as the targeted $8-10\%$ for second-generation alloys. The plane-strain fracture-toughness values of other thick product forms (forgings and extrusions) for third-generation Al–Li alloys also compare favorably against various conventional Al alloys, and have similar or higher strength along with $5-6\%$ density reductions and a higher elastic modulus (Figure 13.21).

TABLE 13.2 Composition Ranges (wt%) of Alloying Elements for Some Third-Generation Al–Li Alloys

Alloy (Density, g/cm³)	Li	Cu	Mg	Zn	Mn	Ag
AA 2195 (2.71)	0.7–1.5	3.9–4.6	0.25–0.8	–	–	0.25–0.6
AA 2297 (2.65)	1.1–1.7	2.5–3.1	0.25 max	0.5 max	0.1–0.5	–
AA 2050 (2.70)	0.7–1.3	3.2–3.9	0.2–0.6	0.25 max	0.2–0.5	0.2–0.7
AA 2060 (2.72)	0.6–0.9	3.4–4.5	0.6–1.1	0.3–0.5	0.1–0.5	0.05–0.5
AA 2099 (2.63)	1.6–2.0	2.4–3.0	0.1–0.5	0.4–1.0	0.1–0.5	–
AA 2055 (2.70)	1.0–1.3	3.2–4.2	0.2–0.6	0.3–0.7	0.1–0.5	0.2–0.7

Zr, Ti, Fe, and Si levels are similar to those in second-generation Al–Li alloys. Densities are for mid-range compositions (AA Teal sheets 2009, 2012).

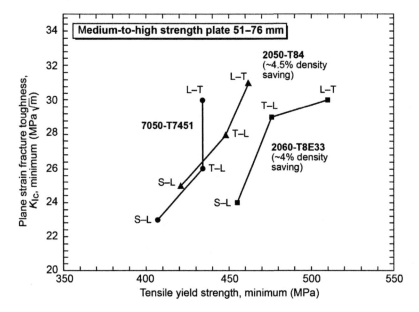

FIGURE 13.20 Plot of plane-strain fracture toughness versus tensile yield strengths for medium-to-high strength plates in S−L, L−T, and T−L crack plane orientations, comparing the third-generation Al−Li alloys (AA 2050-T84 and AA 2060-T8E33) with the conventional AA 7050-T7451 alloy, showing generally superior properties for the former. *Source: Data from AMS 4050H (2003) and AMS 4413 (2007) and also Alcoa in-house data.*

The much higher short-transverse fracture toughness (associated with a much higher resistance to brittle intergranular fracture) of third-generation Al−Li plate compared with second-generation material is reflected in the appearance of fracture surfaces. For example, the S−L fracture surfaces of AA 2297-T87 (1.2 wt% Li) and AA 2050-T8X (0.75 wt% Li) plate exhibited mostly deep dimples, with only small areas of brittle intergranular fracture compared with second-generation alloys such as AA 8090-T8771 (Figure 13.22). The S−L fracture surface of AA 2060-T8E33 also exhibited mostly deep dimples (Figure 13.23).

The L−T fracture surfaces (and fracture-surface profiles) of AA 2297-T87 and AA 2050-T8X plate were also quite different from the AA 8090-T8771 L−T fractures. The third-generation alloys exhibited mostly well-defined, equiaxed dimples with only isolated patches of intergranular fracture, and there were only shallow S−L delaminations. The AA 8090 plate, on the other hand, exhibited considerable amounts of low-energy intergranular and transgranular shear fractures, and there were numerous deep S−L delaminations (which resulted in plane-stress conditions such that the fracture toughness was quite high) (Figures 13.24 and 13.25).

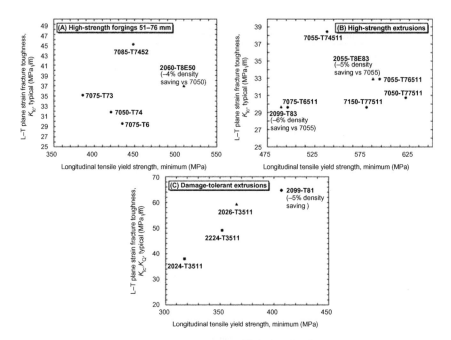

FIGURE 13.21 Plots of L−T plane strain fracture toughness versus longitudinal yield strength for (A) high strength forgings, (B) high-strength extrusions, and (C) damage-tolerant extrusions, comparing some third-generation Al−Li alloys and conventional Al alloys, showing that the Al−Li alloys have equivalent or better strength−toughness combinations, along with significant density reductions (Schmidt et al., 2012 and Alcoa in-house data). Note: in (C) K_Q is the nominal plane-strain fracture toughness.

13.5.2 Sheet and Thin Plate

Measurements of plane-stress fracture toughness versus yield strength for some third-generation Al−Li alloy sheet and thin plate in damage-tolerant conditions show that these alloys also have toughness−strength−density advantages over conventional Al alloys (and second-generation Al−Li alloys) (Figures 13.26 and 13.27). In terms of toughness−strength relationships, the third-generation Al−Li alloys often do not follow the general trend of decreasing toughness with increasing strength. The reasons for the general trend were touched on previously in Section 13.3, as were the principles underlying the development of third-generation alloys. However, a more detailed explanation for the outstanding strength/toughness of some third-generation alloys would require more knowledge of heat-treatment practices, and more detailed metallographic and fractographic observations, than are publically available.

Unlike some second-generation Al−Li alloys, the third-generation alloys (given proprietary multi-step ageing treatments) are thermally stable, i.e.,

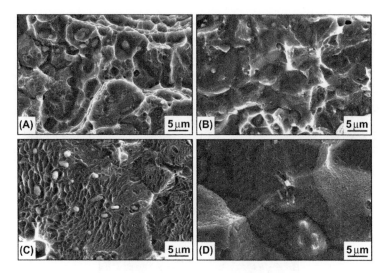

FIGURE 13.22   SEM images of S−L fracture surfaces after tests at 20°C for (A) AA 2050-T8X (0.75−0.9 wt% Li) showing mixture of deep and relatively shallow dimples, (B) AA 2050-T8X showing brittle intergranular facets that were present in a few areas, (C) AA 2297-T87 (1.2 wt% Li) ($K_{Ic} \sim 29.5$ MPa$\sqrt{m}$) showing mostly deep dimples with small areas of very shallow dimples, and (D) AA 8090-T8771 (2.4 wt% Li) ($K_{Ic} = 11-13$ MPa$\sqrt{m}$) that exhibited predominantly brittle intergranular facets with very shallow dimples resolvable on some facets (Byrnes and Lynch, 1990−2012).

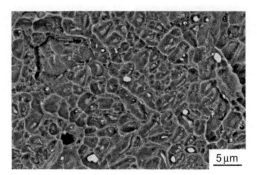

FIGURE 13.23   SEM (back-scattered electron) image of S−L fracture surface after tests at 20°C for thick plate (63.5 mm) AA 2060-T8E33 (0.75 wt% Li) ($K_{Ic} = 28.8$ MPa$\sqrt{m}$) showing deep dimples associated with Mn-rich dispersoids and grain-boundary precipitates (Boselli, 2012).

they do not experience decreases in toughness after long-term ageing (at up to 85°C) simulating exposures that could occur in service in hot climates (Figure 13.28). The good thermal stability of third-generation alloys in regards to fracture toughness is probably, at least partly, due to their lower Li content than second-generation Al−Li alloys, such that additional

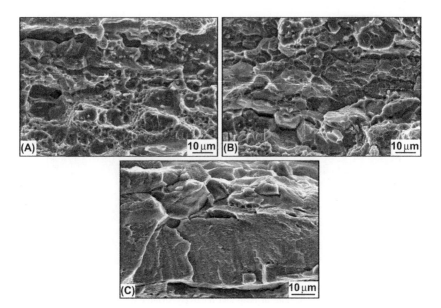

FIGURE 13.24 SEM image of L–T fracture surfaces after tests at 20°C for (A) AA 2050-T8X, (B) AA 2297-T83, and (C) AA 8090-T8771 thick plate, showing mostly equiaxed dimples for the third-generation alloys in comparison with the low-energy intergranular and transgranular shear fracture modes for the second-generation alloy (Byrnes and Lynch, 1990−2012).

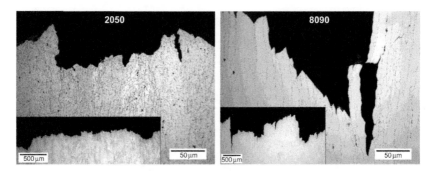

FIGURE 13.25 Metallographic sections through L–T fracture surfaces after tests at 20°C for AA 2050-T8X and AA 8090-T8771 alloys, showing only minor S–L delaminations for the third-generation alloy compared with the second-generation alloy (Byrnes and Lynch, 1990−2012).

fine-scale Al₃Li precipitation (which could promote slip-planarity/low-energy shear fracture) or increases in Li segregation at grain boundaries (which could promote brittle intergranular fracture) are not significant during low-temperature ageing. The detrimental effects of low testing temperatures and high strain rate on fracture toughness of some second-generation Al−Li alloys such as AA 8090, which were potential issues in using them in

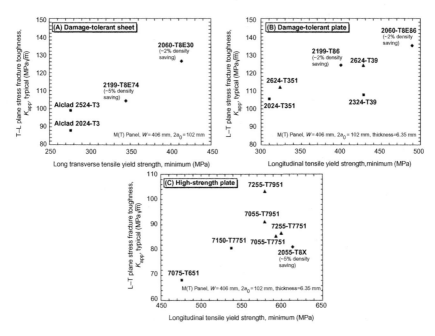

FIGURE 13.26 Plane-stress fracture toughness (K_{app}) versus yield-strength data for third-generation Al−Li alloys: (A,B) damage-tolerant Al−Li sheet and plate (AA 2060-T8X and AA 2199-T8X), and (C) high-strength plate (AA 2055) (L−T orientation) in comparison with relevant conventional Al alloys, showing the superior strength/toughness/density combinations of the third-generation Al−Li alloys. Note that the comparisons in (A) have been made between Alclad 2024 and bare 2199 and 2060, since the corrosion resistance of the bare third-generation alloys is acceptable to users currently considering these products, whereas 2024 needs to be clad for adequate corrosion resistance (Alcoa in-house data).

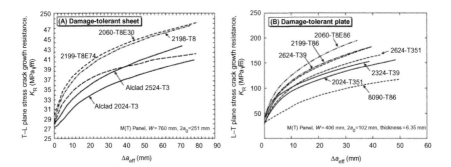

FIGURE 13.27 R-curves for damage-tolerant sheet (A) and plate (B) of some third-generation Al−Li alloys (AA 2060 and AA 2199) compared with relevant conventional Al alloys (and the second-generation AA 8090-T86 alloy in (B)), showing the better plane-stress crack growth resistance of the third-generation Al−Li alloys, especially compared with the second-generation AA 8090-T86 material (Newman et al., 1992; MMPDS-05, 2010; Magnusen et al., 2012; Alcoa in-house data).

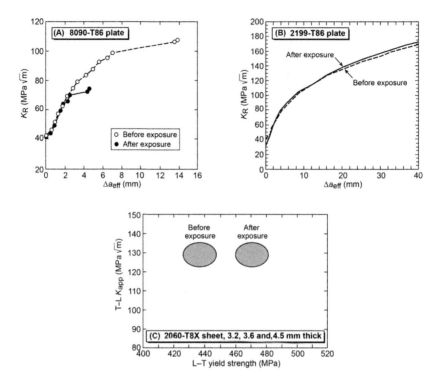

FIGURE 13.28   (A,B) L–T R-curves for damage-tolerant AA 2199 and AA 8090 plates before and after thermal exposure for 1000 h at 80°C, and (c) T–L fracture toughness versus yield strength for AA 2060 sheet, before and after thermal exposure for 1000 h at 85°C, showing that the third-generation Al–Li alloys are "thermally stable"–unlike the second-generation AA 8090 (Rioja and Liu, 2012).

aircraft, do not occur for third-generation Al–Li alloys, but not much data are presently available in the open literature (Rioja and Bray, personal communication, 2013).

13.6 USES AND POTENTIAL USES OF THIRD-GENERATION AL–LI ALLOYS

Uses and potential uses of third-generation Al–Li alloys are described in detail in Chapter 2. Noteworthy examples of the use of early third-generation alloys were AA 2195 for the external tank of the space shuttle (which experiences cryogenic temperatures), and AA 2297 for bulkheads and other parts of military aircraft. The more recently developed third-generation alloys have the potential for more widespread use in both military and commercial aircraft owing to their better properties, especially their ability to match or exceed the strength–toughness combinations of

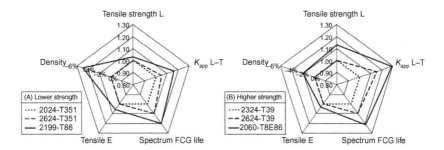

FIGURE 13.29 Examples of "spider" charts showing that third-generation Al–Li alloys have a good balance of properties—not just good fracture toughness. These examples are property comparisons for the Al–Li plate alloys AA 2199-T86 and AA 2060-T8E86 and the baseline conventional 2X24-T3XX alloys (Karabin et al., 2012): L, longitudinal loading direction; L–T, longitudinal loading direction and transverse crack growth; K_{app}, plane-stress fracture toughness, FCG, fatigue crack growth. Other examples (and this one) are shown in Chapter 2.

the latest conventional Al alloys, as documented in the previous section. The modulus, density, and spectrum-fatigue crack-growth properties are also better (Figure 13.29).

A significant contribution to the potential of third-generation Al–Li alloys is the development of a range of ageing treatments that allow optimizations and trade-offs of properties so that there is considerable flexibility in matching the alloys to particular applications. The recently developed AA 2055 alloy in the T8E83 condition, for example, is used for high-strength (~600 MPa) extrusions, and has a combination of properties (including corrosion resistance) that could make it more competitive than conventional Al alloys for applications such as upper-wing and fuselage stringers (Denzer et al., 2012; Magnusen et al., 2012). High-strength AA 2050-T852 and AA 2060-T8E50 forgings have a better combination of L–T toughness and strength than any of the conventional alloys, and also have lower densities (4–5%) and higher elastic moduli (7–9%) than the conventional alloys, making them candidates for wing-to-fuselage attachments (Schmidt et al., 2012). Other potential uses of third-generation Al–Li alloys, and the conventional Al alloys that they could replace, are summarized in Table 13.3.

13.7 CONCLUSIONS

1. Fracture toughness is a critical property that needs to be considered when selecting materials for use in aerospace applications, not only to minimize the possibility of unstable fracture in service but also to increase "crashworthiness". Preventing cracking during manufacturing processes—a significant problem with second-generation Al–Li alloys—is also obviously important.

TABLE 13.3 Examples of Proposed Uses of Third-Generation Al–Li Alloys That Could Replace Conventional Alloys in Fracture-Resistant Applications

Product	Alloy/Temper	Substitute for	Applications
Sheet	2198-T8, 2199-T8E74, 2060-T8E30: damage tolerant/medium strength	2024-T3, 2524-T3/351	Fuselage/pressure cabin skins
Plate	2199-T86, 2060-T8E86: damage tolerant	2024-T351, 2324-T39, 2624-T351, 2624-T39	Lower wing covers
	2055-T8X, 2195-T82: high strength	7150-T7751, 7055-T7751, 7055-T7951, 7255-T7951	Upper wing covers
	2050-T84, 2060-T8E33: medium/high strength	7050-T7451	Spars, ribs, other internal structures made from thick plate
	2195-T82: high strength	2219-T87	Launch vehicle cryogenic tanks
Forgings	2050-T852, 2060-T8E50: high strength	7175-T7351, 7050-T7452	Wing-to-fuselage attachments
Extrusions	2099-T81, 2076-T8511: damage tolerant	2024-T3511, 2026-T3511, 2024-T4312, 6110-T6511	Lower wing stringers, fuselage/pressure cabin stringers, and frames
	2099-T83, 2196-T8511, 2055-T8E83: medium/high strength	7075-T79511, 7150-T6511, 7175-T79511, 7055-T77511	Fuselage/pressure cabin stringers and frames, floor beams, upper wing stringers

E tempers are Alcoa in-house designations not yet registered with the Aluminium Association.

2. The microstructural features that control fracture toughness of Al alloys in general, and Al–Li alloys specifically, are reasonably well understood, although the evidence (mostly indirect, but nevertheless persuasive) that Li segregation at grain boundaries is mainly responsible for brittle intergranular fracture and low toughness for some second-generation Al–Li alloy products is still a matter of debate.

3. Second-generation Al–Li alloys did not achieve significant success in replacing conventional Al alloys partly due to their inability to achieve

equivalent strength-toughness requirements, and their susceptibility to low-energy intergranular fracture was a major factor in this regard. Decreases in fracture toughness due to (i) secondary ageing in service, (ii) dynamic effects, and (iii) decreasing temperature, were also potential problems.

4. Third-generation Al−Li alloys, especially those developed over the last 10 years, have toughness−strength combinations that are superior to those of the best available conventional Al alloys. The third-generation alloys are also "thermally stable" and have an excellent overall balance of properties that should see their widespread use for many aerospace applications.

REFERENCES

GENERAL REFERENCES FOR SECOND-GENERATION AL−LI ALLOYS

[1] Sanders, T.H. Jr. and Starke E.A. Jr. (Eds.), (1981) Aluminum−Lithium Alloys I: Proceedings of the First International Conference on Aluminum−Lithium Alloys. Met. Soc. AIME, Warrendale, PA.

[2] Starke, E.A. Jr. and Sanders, T.H. Jr. (Eds.), (1984) Aluminum−Lithium Alloys II: Proceedings of the Second International Conference on Aluminum−Lithium Alloys. Met. Soc. AIME, Warrendale, PA.

[3] Baker, C., Gregson, P.J., Harris, S.J., Peel, C.J. (Eds.), (1986) Aluminium−Lithium Alloys III: Proceedings of the Third International Conference on Aluminium−Lithium Alloys. Inst. of Metals, London.

[4] Champier, G., Dubost, B., Miannay, D., Sabetay, E. (Eds.), (1987) Proceedings of the Fourth International Conference on Aluminium−Lithium Alloys. J. de Phys. (Paris), 48, Supp. C3.

[5] Starke, E.A. Jr. and Sanders, T.H. Jr. (Eds.), (1984) Aluminum−Lithium Alloys 5: Proceedings of the Fifth International Conference on Aluminum−Lithium Alloys. Met. Soc. AIME, Warrendale, PA.

[6] Peters, M. and Winkler, P.J. (Eds.), (1992) Aluminium−Lithium: Proceedings of the Sixth International Aluminium−Lithium Conference, DGM Germany.

[7] New Light Alloys, AGARD Conference Proceedings No. 444, Advisory Group for Aeronautical Research and Development, Neuilly-sur-Seine, France, 1989.

SPECIFIC REFERENCES

Bennett, C.G., Lynch, S.P., Nethercott, R.B., Kerr, M., Sweet, E.D., 1998. Fracture toughness of 2090 Al−Li−Cu extrusions with high and low hydrogen contents. Mater. Sci. Eng. A247, 32−39.

Blankenship Jr., C.P., Starke Jr., E.A., 1993. Mechanical behavior of double-aged AA8090. Metall. Trans. A 24A, 833−841.

Boselli, J., 2012. Alcoa in-house data.

Brown, K.R., 1984. The use of the chevron-notched short-bar specimen for plane-strain toughness determination in aluminum alloys. In: Underwood, J.H., Freiman, S.W., Baratta, F.I.

(Eds.), Chevron-Notched Specimens: Testing and Stress Analysis, ASTM STP 855. American Society for Testing and Materials, Philadelphia, PA, pp. 237–254.

Byrnes, R.T., Lynch, S.P., 1990–2012. Defence Science and Technology Organisation (DSTO), unpublished work.

Cahn, J.W., 1982. Transitions and phase equilibria among grain boundary structures. J. de Phys. 43 (12), C6-199–C6-213.

Csontos, A.A., Starke Jr., E.A., 2005. The effect of inhomogeneous plastic deformation on the ductility and fracture behavior of age hardenable aluminum alloys. Int. J. Plast. 21, 1097–1118.

Denzer, D.K., Rioja, R.J., Bray, G.H., Venema, G.B., Colvin, E.L., 2012. The evolution of plate and extruded products with high strength and toughness. In: Weiland, H., Rollett, A.D., Cassada, W.A. (Eds.), Proceedings of the 13th International Conference on Aluminum Alloys (ICAA13). The Minerals, Metals and Materials Society (TMS) and John Wiley & Sons, Hoboken, NJ, pp. 587–592.

Eswara Prasad, N., Kamat, S.V., 1995. Fracture behaviour of quaternary Al–Li–Cu–Mg alloys under mixed-mode I/III loading. Metall. Mater. Trans. A 26A, 1823–1833.

Eswara Prasad, N., Kamat, S.V., Prasad, K.S., Malakondaiah, G., Kutumbarao, V.V., 1993. In-plane anisotropy in fracture toughness of an Al–Li 8090 plate. Eng. Fract. Mech. 46, 209–223.

Eswara Prasad, N., Kamat, S.V., Prasad, K.S., Malakondaiah, G., Kutumbarao, V.V., 1994. Fracture toughness of quaternary Al–Li–Cu–Mg alloy under mode I, mode II and mode III loading conditions. Metall. Mater. Trans. A 25A, 2439–2452.

Ewalds, H.L., Wanhill, R.J.H., 1984. Fracture Mechanics, first ed. Edward Arnold, Delftse Uitgevers Maatschappij, London, Delft.

Fournier, P., Barbaux, Y., Rendigs, K.H., 1992. Status of qualification work on aluminium–lithium alloys for application on Airbus aircraft. In: Gen. Ref. [6], pp. 1271–1288.

Glazer, J., Verzasconi, S.L., Sawtell, R.R., Morris Jr., J.W., 1987. Mechanical behavior of aluminum–lithium alloys at cryogenic temperatures. Metall. Trans. A 18A, 1695–1701.

Gregson, P.J., Flower, H.M., 1985. Microstructural control of toughness in aluminium–lithium alloys. Acta Metall. 33, 527–537.

Guttmann, M., 1977. Grain boundary segregation, two dimensional compound formation, and precipitation. Metall. Trans. A 8A, 1383–1401.

Hahn, G.T., Rosenfield, A.R., 1975. Metallurgical factors affecting fracture toughness of aluminum alloys. Metall. Trans. A 6A, 653–668.

Karabin, L.M., Bray, G.H., Rioja, R.J., Venema, G.B., 2012. Al–Li–Cu–Mg–(Ag) products for lower wing skin applications. In: Weiland, H., Rollett, A.D., Cassada, W.A. (Eds.), Proceedings of the 13th International Conference on Aluminum Alloys (ICAA13). The Minerals, Metals and Materials Society (TMS) and John Wiley & Sons, Hoboken, NJ, pp. 529–534.

Knight, S., 2008. Stress Corrosion Cracking of Al–Zn–Mg–Cu Alloys. PhD Thesis, Monash University, Melbourne, Australia.

Lewandowski, J.J., Holroyd, N.J.H., 1990. Intergranular fracture of Al–Li alloys: Effects of aging and impurities. Mater. Sci. Eng. A123, 219–227.

Liddicoat, P.V., Liao, X.-Z., Ringer, S.P., 2009. Novel grain boundary solute architecture in a nanostructured ultra-high strength 7075 aluminium alloy. Mater. Sci. Forum 618-619, 543–546.

Lynch, S.P., 1988. Ductile and brittle crack growth: fractography, mechanisms and criteria. Mater. Forum 11, 268–283.

Lynch, S.P., 1991. Fracture of 8090 Al–Li plate I. Short transverse fracture toughness. Mater. Sci. Eng. A136, 25–43.

Lynch, S.P., 2007. Concerning the anomalous brittle fracture behaviour of iridium. Scr. Mater. 57, 85–88.

Lynch, S.P., Byrnes, R., Nethercott, R.B., Bragianos, A., Crosky, A., 1992. Effects of double ageing treatments on fracture toughness of an 8090 Al–Li alloy plate. In: Gen. Ref. [6], pp. 391–396.

Lynch, S.P., Wilson, A.R., Byrnes, R.T., 1993. Effects of ageing treatments on resistance to intergranular fracture of 8090 Al–Li alloy plate. Mater. Sci. Eng. A172, 79–93.

Lynch, S.P., Muddle, B.C., Pasang, T., 2001. Ductile-to-brittle fracture transitions in 8090 Al–Li alloys. Acta Mater. 49, 2863–2874.

Lynch, S.P., Muddle, B.C., Pasang, T., 2002. Mechanisms of brittle intergranular fracture in Al–Li alloys and comparison with other alloys. Philos. Mag. A 82 (17/18), 3361–3373.

Lynch, S.P., Knight, S.P., Birbilis, N., Muddle, B.C., 2008. Heat-treatment, grain-boundary characteristics and fracture resistance of some aluminium alloys. In: Hirsch, J., Skrotski, B., Gottstein, G. (Eds.), Aluminium Alloys—Their Physical and Mechanical Properties: Proceedings of the 11th International Conference on Aluminum Alloys (ICAA11). Wiley-VCH Verlag GmbH & Co. KGaA, Weinheim, pp. 1409–1415.

Magnusen, P.E., Mooy, D.C., Yocum, L.A., Rioja, R.J., 2012. Development of high toughness sheet and extruded products for airplane fuselage structures. In: Weiland, H., Rollett, A.D., Cassada, W.A. (Eds.), Proceedings of the 13th International Conference on Aluminum Alloys (ICAA13). The Minerals, Metals and Materials Society (TMS) and John Wiley & Sons, Hoboken, NJ, pp. 535–540.

McKeighan, P.C., Hillberry, B.M., Sanders Jr., T.H., 1992. The role of grain boundary delamination on the fracture behaviour of an extruded Al–Li–Zr alloy. In: Gen. Ref. [6], pp. 247–252.

McLean, D., 1952. The embrittlement of copper–antimony alloys at low temperatures. J. Inst. Met. 81, 121–123.

Miller, W.S., Thomas, M.P., Lloyd, D.J., Creber, D., 1986. Deformation and fracture in Al–Li base alloys. Mater. Sci. Technol. 2, pp. 1210–1216, and Gen. Ref. [3], pp. 584–594.

Moutsos, S., 2012. Some Aspects of Age-Hardening, Grain Refinement, Deformation and Fracture in Al–Li Based Alloys. PhD Thesis, Monash University, Melbourne, Australia.

Moutsos, S., Pasang, T., Muddle, B.C., Lynch, S.P., 2010. The effect of slip mode on brittle intergranular fracture. In: Kumai, S., Umezawa, O., Takayama, Y., Tsuchida, T., Sato, T. (Eds.), Proceedings of the 12th International Conference on Aluminium Alloys (ICAA12). The Japan Institute of Light Metals, Tokyo, Japan, pp. 314–321.

Newman, J.M., Goodyear, M.D., Witters, J.J., Veciana, J, Platts, G.K., 1992. Age forming 8090 damage tolerant plate for lower wing skin applications. In: Gen. Ref. [6], pp. 1371–1376.

Noble, B., Harris, S.J., Dinsdale, K., 1994. Low temperature embrittlement of 8090 in the damage tolerance condition. In: Sanders Jr., T.H., Starke Jr., E.A. (Eds.), Fourth International Conference on Aluminium Alloys: Their Physical and Mechanical Properties. Georgia Inst. Tech., Atlanta, GA, pp. 460–466.

Pasang, T., Symonds, N., Moutsos, S., Wanhill, R.J.H., Lynch, S.P., 2012. Low-energy intergranular fracture in Al–Li alloys. Eng. Fail. Anal. 22, 166–178.

Peel, C.J., 1989. Current status of the application of conventional aluminium–lithium alloys and the potential for future developments. In: Gen. Ref. [7], pp. 21-1–21-9.

Pitcher, P.D., McDarmaid, D.S., Peel, C.J., 1992. Effect of thermal processing on the fracture toughness of 8090 alloy sheet and plate. In: Gen. Ref. [6], pp. 235–240.

Rioja, R.J., Liu, J., 2012. The evolution of Al—Li base products for aerospace and space applications. Metall. Mater. Trans. A 43A, 3325–3337.

Rioja, R.J., Bowers, J.A., James, R.S., 1991. Aluminum—Lithium Alloys and Method of Making the Same. U.S. Patent No. 5,066,342.

Rottman, C., 1988. Theory of phase transitions at internal interfaces. J. de Phys. 49 (10), C5.313–C5.326.

Schmidt, T., Yocum, L., Bush, D., Caratelli, J., 2012. Advanced aluminum alloys enabling performance improvements. Presentation at the Joint Armaments Conference. May 2012, Seattle.

Seah, M.P., 1980. Adsorption-induced interfacial decohesion. Acta Metall. 28, 955–962.

Shashidhar, S.R., Kumar, A.M., Hirth, J.P., 1995. Fracture toughness of an Al—Li alloy at ambient and cryogenic temperatures. Metall. Mater. Trans. A 26A, 2269–2274.

Shinoda, T., Mashima, Y., Kobayashi, A., Suzuki, T., 1986. Reversion of temper embrittlement. Z. für. Metallkunde 77, 433–441.

Staley, J.T., 1976. Microstructure and toughness of high-strength aluminium alloys: properties related to fracture toughness. Properties Related to Fracture Toughness, ASTM STP 605. American Society for Testing and Materials, Philadelphia, PA, pp. 71–103.

Starink, M.J., Hobson, A.J., Sinclair, I., Gregson, P.J., 2000. Embrittlement of Al—Li—Cu—Mg alloys at slightly elevated temperatures: microstructural mechanisms of hardening. Mater. Sci. Eng. A289, 130–142.

Starke Jr., E.A., Quist, W.E., 1989. The microstructure and properties of aluminium—lithium alloys. In: Gen. Ref. [7], pp. 4.1–4.23.

Sweet, E.D., Lynch, S.P., Bennett, C.G., Nethercott, R.B., Musulin, I., 1996. Effects of alkali-metal impurities on fracture toughness of 2090 Al—Li—Cu extrusions. Metall. Mater. Trans. A 27A, 3530–3541.

Swift, T., 1987. Damage tolerance in pressurized fuselages. In: Simpson, D.L. (Ed.), 11th Plantema Memorial Lecture on New Materials and Fatigue Resistant Aircraft Design. Engineering Materials Advisory Services Ltd., Warley, pp. 1–77.

't Hart, W.G.J., Schra, L., McDarmaid, D.S., Peters, M., 1989. Mechanical properties and fracture toughness of 8090-T651 plate and 2091 and 8090 sheet. In: Gen. Ref. [7], pp. 5-1–5-17.

Thomason, P.F., 1990. Ductile Fracture of Metals, first ed. Pergamon Press, Oxford.

Valiev, R.Z., Enikeev, N.A., Murashkin, M.Y., Kazykhanov, V.U., Sauvage, X., 2010. On the origin of the extremely high strength of ultrafine-grained Al alloys produced by severe plastic deformation. Scr. Mater. 63, 949–952.

Vasudévan, A.K., Doherty, R.D., 1987. Grain boundary ductile fracture in precipitation hardened aluminum alloys. Acta Metall. 35, 1193–1219.

Venkateswara Rao, K.T., Ritchie, R.O., 1989. Mechanical properties of Al—Li alloys: part 1. Fracture toughness and microstructure. Mater. Sci. Technol. 5, 882–895.

Wanhill, R.J.H., 1994. Status and prospects for aluminium—lithium alloys in aircraft structures. Int. J. Fatigue 16, 3–20.

Wanhill, R.J.H., 2003. Milestone case histories in aircraft structural integrity. In: Milne, I., Ritchie, R.O., Karihaloo, B.L. (Eds.), Comprehensive Structural Integrity Volume 1: Structural Integrity Assessment—Examples and Case Studies. Elsevier Science Publishers, London, pp. 61–72.

Wanhill, R.J.H., Schra, L., 1991. Fracture toughness and crack resistance of damage tolerant and high strength Al−Li alloy sheet. National Aerospace Laboratory NLR Technical Publication NLR-TP-91245, Amsterdam, The Netherlands.

Wanhill, R.J.H., Schra, L., 't Hart, W.G.J., 1992. Crack resistance, fracture toughness and instability in damage tolerant aluminum−lithium alloys. In: Mitchell, M.R., Buck, O. (Eds.), Cyclic Deformation, Fracture and Nondestructive Evaluation of Advanced Materials, ASTM STP 1157. American Society for Testing and Materials, Philadelphia, PA, pp. 224−240.

Wanhill, R.J.H., Symonds, N., Merati, A., Pasang, T., Lynch, S.P., 2013. Five Helicopter Accidents with evidence of material and/or design deficiencies. Eng. Fail. Anal. (In Press: Corrected proofs available on-line 3 Jan 2013).

Webster, D., 1987. The effect of low melting point impurities on the properties of aluminum−lithium alloys. Metall. Trans. A 18, 2128−2193.

Wilson, A.R., 1990. Lithium detection in aluminium−lithium alloys using PEELS. In: Michael, J.R., Ingram, P. (Eds.), Microbeam Analysis. San Francisco Press, USA, pp. 39−42.

Zink, W., Weilke, J., 1992. Influence of further processing at final user on mechanical−technological properties of aluminium−lithium 8090 extrusions. In: Gen. Ref. [6], pp. 987−992.

OTHER REFERENCES

Aerospace Material Specification AMS 4050H, 2003. SAE International, Warrendale, PA.

Aerospace Material Specification AMS 4413, 2007. SAE International, Warrendale, PA.

Alcoa Aerospace Technical Fact Sheet, 2008. Alloy 2099-T83 and 2099-T8E67 extrusions. AEAP-Alcoa Engineered Aerospace Products, Lafayette, IN.

Aluminum Association Registration Record Series Teal Sheets, 2009 and Addendum, 2012. International Alloy Designations and Chemical Composition Limits for Wrought Aluminum and Wrought Aluminum Alloys (The Aluminum Association, Inc., Arlington).

Metallic Materials Properties Development and Standardization MMPDS, 2010. MMPDS-05, Federal Aviation Administration, pp. 3−252.

Chapter 14

Corrosion and Stress Corrosion of Aluminum−Lithium Alloys

N.J.H. Holroyd*, G.M. Scamans, R.C. Newman[†] and A.K. Vasudevan[‡]**
*Consultant, Riverside, California, CA 92506, **Innoval Technology Limited, Banbury, Oxon, OX16 1TQ, UK, [†]Chemical Engineering & Applied Chemistry, University of Toronto, Ontario, M5S 3E5, Canada, [‡]Office of Naval Research, Arlington, VA 22203, USA*

Contents

14.1 Introduction and Historical Background   457
14.2 Localized Corrosion of Al−Li Based Alloys   461
   14.2.1 Al−Li Binary Alloys   461
   14.2.2 Al−Li−Cu and Al−Li−Cu−Mg Alloys   462
   14.2.3 Al−Mg−Li Alloys   478
14.3 Stress Corrosion Cracking   479
14.3.1 Al−Li Binary Alloys   482
14.3.2 Al−Li−Cu and Al−Li−Cu−Mg Alloys   483
14.3.3 Mechanistic Implications   490
14.4 Summary and Conclusions   492
Acknowledgment   493
References   493

14.1 INTRODUCTION AND HISTORICAL BACKGROUND

Interest in wrought heat-treatable aluminum−lithium (Al−Li) based alloys dates back to around 1919 in Germany, with lithium additions made to Al−Cu and Al−Cu−Zn based alloys to increase alloy strength (Assmann, 1926; Czochralski, 1927; Czochralski and Welter, 1927). Commercialization was confounded, however, by inappropriate choices of alloy composition and heat treatment, and the alloys almost immediately fell from favor. The first successful aircraft application of lithium-containing aluminum alloys, in 1958, was AA 2020 plate (Al−4.5Cu−1.1Li−0.5Mn−0.2Cd) used in the wing skins and tail of the US Navy RA-5C Vigilante aircraft (Balmuth and Schmidt, 1981) based on work conducted by LeBaron at Alcoa during the early 1940s (LeBaron, 1945). This was followed in the Soviet Union by the introduction of a slightly lower-strength alloy, VAD-23 (Al—5.3Cu—1.1Li—0.6Mn—0.17Cd), into several aerospace applications (Rioja, 1998). The poor fracture toughness characteristics of these "first-generation" Al—Li alloys led to their commercial demise, with the AA 2020 alloy being withdrawn from registration in 1974 and subsequent Russian alloy development focussing on the Al—Mg—Li alloys, with the emergence of 1420 (Al—5.3Mg—2.0Li—0.5Mn) in 1965 (Fridlyander, 2003; Fridlyander et al., 1969, 1983).

Further alloy development in the West required an improved understanding of strengthening and deformation mechanisms, coupled with commercially viable solutions to ingot quality issues encountered during large-scale casting (Grimes et al., 1985, 1987). This stimulated the development of the second-generation Al—Li based alloys during the 1980s, AA 2090 (Al—Li—Cu—Zr), AA 2091 (Al—Li—Cu—Mg—Zr), and AA 8090 (Al—Li—Cu—Mg—Zr) (Table 14.1) with mechanical properties to compete against incumbent aerospace medium and high-strength aluminum alloys such as AA 2024-T3, AA 2014-T6, AA 7075-T6, and AA 7075-T73.

The second-generation alloys offered density, modulus, and fatigue performance advantages over conventional aluminum alloys (Rioja and Liu, 2012; Zakharov, 2003) and were extensively used in certain aircraft applications, for example, AA 8090 in the AgustaWestland EH101 helicopter (Merati, 2011; Smith, 1987). However, general adoption by the industry was curtailed by anisotropy in in-plane and through-thickness mechanical properties, low short-transverse fracture toughness, potentially poor localized corrosion resistance (intergranular, exfoliation, and stress corrosion), and temperature stability issues (Pasang et al., 2006; Zakharov, 2003).

During the late 1980s and early 1990s, a family of weldable Al—Li alloys with an Ag addition ($\sim$0.2—0.4 wt%), known as Weldalite™ alloys, (Langan and Pickens, 1991; Tack et al., 1990) were developed with chemical compositions ranging from copper-free alloys with $\sim$2 wt% lithium and $\sim$5 wt% magnesium (Weldalite™ 050) through to alloys with high copper contents, up to $\sim$6 wt%, combined with a low lithium content, $\sim$1.25 wt% (Weldalite™ 049), see AA 2094, AA 2095, and AA 2195 in Table 14.1. Weldalite™ 050 targeted potential marine applications (Ahmad and Aleem, 1996; Kramer et al., 1994) and offered an improved corrosion performance with respect to the copper-containing second-generation Al—Li alloys although inferior to the incumbent AA5XXX marine grade aluminum alloys (Carmody and Shaw, 1997). Weldalite™ 049 (AA 2195) with its exceptional strength level was successfully used commercially as the main structural alloy for the external tank of the Space Shuttle, providing an improved toughness in welded structures at cryogenic temperatures (Pickens and Tack, 1995).

TABLE 14.1 Nominal Chemical Compositions of Commercial Al–Li Based Alloys

Alloys		Li	Cu	Mg	Zr	Zn	Ag	Other	Approximate Date
First generation	2020	1.2	4.5	–	–	–	–	0.5 Mn + 0.2Cd	1958
	VAD-23	1.2	5	–	–	–	–	0.6Mn + 0.17Cd	1961
	1420	2.1	–	5.2	0.11	–	–		1965
Second generation (Li > 2 wt%)	2090	2.1	2.7	–	0.11	–	–		1984
	2091	2.0	2.0	1.3	0.11	–	–		1985
	8090	2.4	1.2	0.8	0.11	–	–		1984
	1421	2.1	–	5.1	0.11	–	–	0.17 Sc	1984
	1423	1.9	–	3.5	0.10	–	–	0.08 Sc	1985
	1430	1.7	1.6	2.7	0.11	–	–	0.17 Sc	1980s
	1440	2.4	1.5	0.8	0.11	–	–		1980s
	1450	2.1	2.9	–	0.11	–	–		1980s
	1460	2.25	2.9	–	0.11	–	–	0.09 Sc	1980s
Third generation (Li < 2 wt%)	2094	1.0	4.8	0.5	0.11	0.25*	0.4		1990
	2095	1.1	4.2	0.5	0.11	0.25*	0.4		1990
	2195	1.0	4.0	0.4	0.11	–	0.4		1992
	2197	1.5	2.8	0.25	0.11	0.05*	–	0.3 Mn	1993

(Continued)

TABLE 14.1 (Continued)

Alloys	Mid-Range Composition (wt%)							Approximate Date
	Li	Cu	Mg	Zr	Zn	Ag	Other	
2196	1.75	2.9	0.5	0.11	0.35*	0.4	0.35* Mn	2000
1424	1.7	–	5.35	0.	0.1	0.08	0.08 Sc	1998
2297	1.4	2.8	0.25*	0.11	0.5*	–	0.3 Mn	1997
2397	1.4	2.8	0.25*	0.11	0.1	–	0.3 Mn	2002
2198	1.0	3.2	0.5	0.11	0.35*	0.4	0.5* Mn	2005
2099	1.8	2.7	0.3	0.09	0.7	–	0.3 Mn	2003
2199	1.6	2.6	0.2	0.09	0.6	–	0.3 Mn	2005
1461	1.7	2.8	0.4	0.08	0.5	–	0.06 Sc	~2000
2050	1.0	3.6	0.4	0.11	0.25*	0.4	0.35 Mn	2004
2060	0.75	3.95	0.85	0.11	0.4	0.25	0.3 Mn	2011
2055	1.15	3.7	0.4	0.11	0.5	0.4	0.3 Mn	2012
2296	1.6	2.45	0.6	0.11	0.25*	0.43	0.25*	2010
2076	1.5	2.35	0.5	0.11	0.30*	0.28	0.33	2012

* Max. wt%

The Weldalite™ alloys anticipated the compositional modifications required for the subsequent third-generation Al—Li alloys, but did not surmount the in-plane tensile strength/ductility (Ruschau and Jata, 1993) and stress corrosion cracking (SCC) (Lee et al., 2002; Moshier et al., 1992) issues that restricted the extensive commercialization of second-generation Al—Li alloys.

In the past 10 years, our understanding of how alloy composition, microstructure, and thermomechanical processing influence the mechanical and corrosion properties has improved. This knowledge has been used to make appropriate choices of alloy chemistry, thermomechanical processing, and multistep ageing practices (Kolobnev et al., 2004) that have resulted in third-generation Al—Li alloys suitable for more extensive usage in the aerospace and space industries (Giummarra et al., 2008; Noble et al., 2008; Rioja and Liu, 2012). These alloys and their potential applications are discussed in detail in Chapters 2 and 15.

The nominal chemical compositions of some of the first-, second-, and third-generation commercial Al—Li based alloys are listed in Table 14.1.

We have reviewed the corrosion characteristics of these Al—Li based alloys with an emphasis on localized corrosion (intergranular and exfoliation) and SCC as presented in the following sections.

14.2 LOCALIZED CORROSION OF AL—LI BASED ALLOYS

General corrosion of Al—Li based alloys usually only occurs on exposure to environments with solution pHs outside the range of pH ~ 3 to pH ~ 10, and use in these environments is inappropriate other than in specialized applications. However, localized corrosion may occur in relatively benign environments owing to reactive phases precipitated within matrix and grain boundary (g.b.) regions (Hu et al., 1993; Moran et al., 2012; Zhu et al., 2000) (Table 14.2). These phases and their distributions depend on the alloy system, thermomechanical processing, and heat treatment (temper) conditions.

14.2.1 Al—Li Binary Alloys

Al—Li and Al—Li—Zr alloys with sufficiently low lithium content to avoid δ' (Al_3Li) precipitation have a high resistance to localized corrosion, irrespective of alloy temper, that is equivalent to commercially pure aluminum, AA 1100 (Niskanen et al., 1982). Higher lithium content promotes δ (AlLi) precipitation in g.b. regions, which increases susceptibility to local pitting and intergranular attack that increase with further ageing, paralleled with a decreasing pitting potential (Niskanen et al., 1982; Ohsaki et al., 1988).

These binary alloys are highly resistant to exfoliation corrosion but are potentially susceptible to stress corrosion based on the available test results.

TABLE 14.2 Phases Precipitated in Al–Li Based Alloys

Alloy System	Main Phases	
	Grain Boundary	Matrix
Al–Li	δ (AlLi)	δ' (Al$_3$Li)
Al–Li–Mg (1420, 1421, 1423, 1424, 1460)	S$_1$ (Al$_2$MgLi)	δ' (Al$_3$Li)
Al–Li–Cu	T$_1$ (Al$_2$CuLi)	δ' (Al$_3$Li)
(2020, 2090)	T$_2$ (Al$_6$CuLi$_3$)	T$_1$ (Al$_2$CuLi)
(1451,1461)	θ (Al$_2$Cu)	θ' (Al$_2$Cu)
Al–Li–Cu–Mg	S (Al$_2$Cu(LiMg))	δ' (Al$_3$Li)
(8090, 2091)	T$_2$ (Al$_6$CuLi$_3$) (Al$_6$Cu(LiMg)$_3$)	S' (Al$_2$Cu(LiMg)
(1430)	T$_2$ (Al$_6$CuLi$_3$)	T$_1$ (Al$_2$CuLi)
(2099, 2199)	T$_1$ (Al$_2$CuLi)	δ' (Al$_3$Li) θ' (modified)

Source: Giummarra et al. (2007); Kolobnev (2002); Quist and Narayanan (1989).

14.2.2 Al–Li–Cu and Al–Li–Cu–Mg Alloys

14.2.2.1 First-Generation Alloys

The first generation of commercial Al–Li–Cu alloys with nominal lithium and copper contents of 1.2 and 5 wt%, respectively, performed adequately, with no reported corrosion-related service life issues (Balmuth and Schmidt, 1981; Rioja, 1998), although their inherent corrosion resistance was relatively poor, with peak-aged AA 2020 plate material having an ASTM G34 EXCO 48 h test exfoliation rating of "EC" (Colvin, 1988; Vasudevan et al., 1985) and "ED" (Thompson, 1992). However, as will be discussed the EXCO test probably has little value for the assessment of corrosion susceptibility of these alloys.

14.2.2.2 Second- and Third-Generation Alloys

Unlike the limited quantitative localized corrosion data available for first-generation Al–Li alloys, a wealth of data has been generated for second-generation alloys. Initial thoughts on the pitting and intergranular corrosion (IGC) susceptibility of Al–Li–Cu type alloys (e.g., AA 2090) were based on a proposal that intergranular attack in re-solution heat-treated and re-aged commercial AA 2090-T8E51 (Al–2Li–2.9Cu–0.12Zr) is strongly associated with dissolution of copper-depleted zones that form adjacent to grain and sub-grain boundaries owing to local precipitation of copper-rich phases (Kumai et al., 1989). This proposal was countered by the suggestion that continuous sub-boundary dissolution occurs by localized galvanic attack of active

TABLE 14.3 Electrochemical Characteristics in Aqueous NaCl of Typical Phases in Al–Li–Cu Alloys

	V_{SCE} FCP (Ep)	V_{SCE} FCP (Ep)
	Aerated 0.6M NaCl	De-Aerated 0.6M NaCl
SHT AA2090 (α-Al)	−0.729	−0.731
Pure Al (Cu-depleted zone)	(− 0.749)[a]	−1.008 (− 0.762)
Bulk T₁	−1.096 (− 0.723)	−1.094 (− 0.756)
AA2090-UA	−0.725 (− 0.710)	−0.916 (− 0.690)
AA2090-PA	−0.735 (− 0.720)	−0.892 (− 0.725)

[a]Wide range of values quoted in the literature, highly sensitive to surface or test environment contamination.
Source: Buchheit et al. (1994)

precipitates, thereby generating a locally acidified crevice environment that stimulates sufficient further local attack to expose additional T_1 sub-grain boundary precipitates, and the process then repeats itself (Buchheit et al., 1990, 1994).

The latter hypothesis is supported by Kumai et al.'s observation that localized corrosion for re-solution heat-treated and re-aged commercial AA 2090-T8E41 material was significantly greater than that for as-received *commercially* heat-treated T8E41 material. The rationale for this difference is that solution heat-treated commercial material is given a 6% stretch before ageing, which promotes the precipitation of T_1 within grains, retarding its precipitation at grain and sub-grain boundaries and reducing corrosion susceptibility.

Further experimental work established the dissolution characteristics and electrochemical behavior of bulk forms of the various alloy phases (Buchheit et al., 1994; Li et al., 2007, 2008). These investigations revealed complexities not related to the bulk alloy's free-corrosion and breakdown potentials, which are relatively insensitive to ageing (Table 14.3). The T_1 intermetallic compounds have rapid anodic and cathodic reaction kinetics and inevitably corrode by de-alloying of the more active Li, Mg, and Al components, generating Cu-rich remnants that coarsen to establish 10−100 nm copper-rich clusters. These clusters are often unstable and become detached and facilitate copper liberation and its local surface redistribution in corrosion products or as surface films formed over particles (Buchheit, 2000; Li et al., 2007). The T_1 particles liberate Cu ions by a similar mechanism as proposed for S and θ-phases (Buchheit et al., 2000).

Studies to characterize the de-alloying process for T_1 and T_2 and the composition-modified variants that precipitate in the g.b.'s of third-generation

Al–Li alloys have not been published to date. The composition modifications of T_1 and T_2 precipitates in third-generation alloys are due to minor alloying additions such as zinc.

The second-generation Al–Li alloy development in the United Kingdom added magnesium to the Al–Li–Cu system to stimulate co-precipitation of S-phase (Al_2CuMg) with δ' and T_1 (Al_2CuLi). Magnesium additions combined with stretching after solution heat treatment enabled S-phase to dominate over T_1 precipitation during subsequent ageing and offered the potential to exploit the S-phase's superior ability to disperse slip (Grimes et al., 1987) with a consequent improvement in fracture toughness-related properties. An example of this behavior is AA 8090, which contains ~ 0.8 wt% Mg (Table 14.1). However, the addition of Mg was found to slightly decrease the IGC resistance (Semenov, 2001; Semenov and Sinyavskii, 2001).

The influence of copper and magnesium additions on the IGC resistance of Al–Li alloys containing 2 wt% lithium is shown in Table 14.4 and Figure 14.1. IGC susceptibility increases with the additions of copper and magnesium and/or ageing, and the level of copper is the dominant factor. Similar trends have been observed for a wider range of lithium contents (1.5–2.5 Li) (Table 14.4), with IGC susceptibility consistently increasing with ageing (UA < PA < OA) (Figure 14.2).

Recently published data for the third-generation Al–Li alloys AA 2197 (Jiang et al., 2005) and AA 2195 (Xu et al., 2011) and a laboratory Al–1.5Li–3.5Cu–0.22(Zr + Sc) (Liang et al., 2008) alloy indicate that their IGC susceptibility can be significantly lower than for the higher lithium content second-generation alloys, particularly when aged at lower temperatures (Figure 14.2). This is an interesting finding because none of the alloys cited above contain a deliberate zinc addition, which potentially could provide a further opportunity to improve localized corrosion resistance by reducing electrochemical potential differences between the matrix and the potentially active g.b. precipitates such as T_1. Zinc incorporation into the aluminum alloy matrix will increase its passive current, whereas Zn incorporation in T_1 or T_2 g.b. precipitates (Gable et al., 2002; Kertz et al., 2001) will probably reduce the isolated phases electrochemical potential.

Conventional EXCO tests (ASTM G34) on Al–Li–Cu (Colvin, 1988; Colvin and Lifka, 1990; Colvin and Murtha, 1989; Lee and Lifka, 1992; Vasudevan et al., 1985) and Al–Li–Cu–Mg (Braun, 1995a; Gray, 1987; Habashi et al., 1993; Lane et al., 1985; Reboul and Mayer, 1987) second-generation alloys invariably incorrectly predicted the alloy-temper's exfoliation corrosion performance established in real-life marine environments and/or less aggressive exfoliation test media after longer exposure times (see the data provided in Table 14.4 for the MASTMAASIS (Modified ASTM Acetic Acid Salt Intermittent Spray, ASTM G85 Annex 2)). Studies for both AA 2090 and AA 2091 in various tempers revealed that the visual ratings obtained from EXCO failed to correlate with those for natural environments

TABLE 14.4 Localized Corrosion of Al–Li–Cu and Al–Li–Cu–Mg Alloys in Various Standard Test Environments

Alloy	Temper	Li wt%	Cu wt%	Mg wt%	Zr wt%	Product Thickness (mm)	Depth of Attack (µm)			
							IGC[a]	Exfoliation[b]	Pitting	Marine[c]
Semenov and Sinyavsavskii (2001)										
Lab	UA	2	0.7	1.4	0.1	Extrusion (10)	26	50	–	30
Lab	PA						45	–	–	–
Lab	OA						51	–	–	–
Lab	UA	2	1.4	0.7	0.1		40	95	–	110
Lab	PA						132	–	–	–
Lab	OA						137	–	–	–
Lab	UA	2	2.2	1.4	0.1		182	123	–	135
Lab	PA						192	–	–	–
Lab	OA						220	–	–	–
Lab	UA	2	2.9	0.25	0.1	Extrusion (10)	140	120	–	130
Lab	UA	2	2.9	0.7	0.1		260	–	–	–
Lab	PA						272	–	–	–
Lab	OA						280	–	–	–

(Continued)

TABLE 14.4 (Continued)

Alloy	Temper	Li wt%	Cu wt%	Mg wt%	Zr wt%	Product Thickness (mm)	Depth of Attack (μm)			Marine[c]
							IGC[a]	Exfoliation[b]	Pitting	
Semenov (2001)										
Lab	UA	2	2.8	0	0.1	Extrusion (10)	90	–	–	–
				0.15	0.1		150			
				0.25	0.1		160			
				0.35	0.1		180			
				0.7	0.1		200			
				1	0.1		270			
Liang et al. (2008)							GB 7998-87[d]	48 h EXCO[e]		
Lab		2.34	1.15	0.61	0.13	Plate (2.3)	89	EC	–	–
							127	ED	–	–
							138	ED	–	–
Gray (1987)							2 weeks MASTAASIS[f] (μm)	48 h EXCO[e] (μm)		Marine[g]
8090	UA	2.34	1.15	0.61	0.13	Plate (30)	EA (40)	P (30)	–	P (20)[e]
	PA						EA (50)	P/EA (530)	–	P (130)[e]
	T651						EA (200)	P/EA (320)	–	P (130)[e]

Reference / Alloy	Temper						P/EA	EA/EB	–
Habashi et al. (1993)									
8090	PA	2.4	1.15	0.67	0.11	Plate (50)	–	–	–
Hu et al. (1993); Zhu et al. (2000)									
8090	NA	2.5	1.3	0.7	0.1	Extrusion (4)	(0)	P (P)	4.7
8090	UA						(41.3)	EA (ED)	19.5
8090	PA						(153)	EB (ED)	136
8090	OA						(72)	EA (ED)	98
2090	NA	2.2	2.7	–	0.12		(0)	P (P)	38
2090	UA						(51)	EB (ED)	49
2090	PA						(165)	EC (ED)	165
2090	OA						(234)	EC (ED)	192
Vasudevan et al. (1985)									
2020	T651					Plate (36) T/10	–	EC	–
						T/2	–	EC	–
Colvin (1988)									
Al–Li–Cu	T3	2.2	3.3	–	–	Extrusion (25)	P	EA	N
	PA						EA	EA	N
	T3	2.0	3.0	–	0.12	Plate (25)	–	N	N
	UA						–	EA	N
	OA						–	EB	N

(Continued)

TABLE 14.4 (Continued)

| Alloy | Temper | Li wt% | Cu wt% | Mg wt% | Zr wt% | Product Thickness (mm) | Depth of Attack (μm) | | Pitting | Marine[c] |
							IGC[a]	Exfoliation[b]		
2020	T3					Plate (35)	P	EA		N
	T651						EA	EC		N
2090	T8E41	2.2	2.5	–	0.12	Plate (6)	P	EB		N
			3.0				P	EC		N
		2.5	2.5				P	EC		N
			3.0				P	EC		N
		2.2	2.5			Plate (25)	P	EA		N
			3.0				P	EA		N
		2.5	2.5				P	EA		N
			3.0				P	EA		N
LAB	UA	2.5	1.5	1.0	0.12	Plate	P	EA		EA
	PA						EA	EA		EA
	PA	2.8	0.5	1.0	0.12		EA	P		N
	T3	3	1.2	1.5	0.08		EA	EA		EA
	PA						EA	EC		EA

(Continued)

Thompson (1992)

2020	T651	—	—	—	Plate (21) T/10	P	ED	—	P/EA
					T/2	P	ED	—	P/EA
2090	T8E41	—	—	—	Plate (13) T/10	P	ED	—	P
					T/2	P	ED	—	P
8090	T851	—	—	—	Plate (46) T/10	P	ED	—	P/EA
					T/2	P	ED	—	P/EA
8090	T851	—	—	—	Plate (16.5) T/10	P/EA	ED	—	EA/EB
					T/2	P/EA	ED	—	EA/EB

Jiang et al. (2005) — GB 7998-87[d]

2197	UA[h]	1.48	2.87	0.24	0.11	Plate (2)	186	EB	—	—
	PA[h]						200	EC	—	—
	OA[h]						212	EC	—	—
	vUA[i]						55	P	—	—
	UA[i]						64	EA	—	—
	PA[i]						103	EC	—	—

Xu et al. (2011) — IGC[a]

2195	PA[i]	1.0	4.0	0.4	0.11	Plate (2)	79	—	—	—

TABLE 14.4 (Continued)

Alloy	Temper	Li wt%	Cu wt%	Mg wt%	Zr wt%	Product Thickness (mm)	Depth of Attack (µm)			
							IGC[a]	Exfoliation[b]	Pitting	Marine[c]
Moran et al. (2012)							2 weeks MASTAASIS[f]			
2060	T3	0.8[j]	4[j]	0.2[j]	0.1[j]	Plate	ED	EA	–	EB
	UA						ED	EA	–	EB/EC
	T86						P	EB	–	P
2099	T3	1.8[j]	2.7[j]	0.3[j]	0.1[j]		P	EA	–	P
	UA						EC	EA	–	EB/EC
	T86						P	EB	–	P
Henon and Rouault (2012)							4 weeks MASTAASIS[f]			
2050	vUA	1[j]	3.5[j]	0.4[j]	0.1[j]	Plate (50)	EA/EB	–	–	–
	UA						P	–	–	–
	PA						P	–	–	–
	OA						EA/EB	–	–	–

Rioja et al. (1990,1992)	Li	Cu	Mg	Zn	Plate (25−38 mm)		48 h EXCO/ 96 h EXCO		
Al–Li–Cu–Mg–Zn T8 (6% Stretch + 20 h 163°C)	2.2	2.5	0	0	–		EA/EB	–	–
	2.2	2.5	0.3	0	–		EC/EC	–	–
	2.1	2.5	0.6	0	–		EC/ED	–	–
	2.2	2.6	0.6	0.6	–		EC/EC	–	–
	2.2	2.5	0.5	1	–		EB/EB	–	–
	2.1	2.6	0.3	0.5	–		EB/EB	–	–
	2.2	2.6	0.3	0.9	–		EB/EB	–	–

Test environments:
[a] 6 h 1M aqueous NaCl + 1% H_2O_2.
[b] ISO 11881.
[c] 312 days on ship.
[d] 24 h 0.53M aqueous NaCl + 1% HCl.
[e] 48 h ASTM G34.
[f] ASTM G85 Annex 2.
[g] 32 months Point Judith.
[h] 175°C.
[i] 160°C.
[j] Nominal composition.
EXCO, MASTAASIS, and Marine Exposure Corrosion Ratings (representative micrographs of relative corrosion attack (Davis, 1999)); N, no appreciable attack; P, pitting; EA, slight; EB, moderate; EC, severe; ED, very severe.

such as the seacoast (Colvin and Lifka, 1990; Colvin and Murtha, 1989), whereas those obtained by Boeing and Alcoa using dry bottom MASTMAASIS provided very good correlation with seacoast exposure (Colvin and Murtha, 1989).

For peak-aged tempers, which perform well in seacoast exposure, the visual EXCO ratings often indicated a high susceptibility not experienced during use in applications such as AA 2090-T8 in the McDonnell Douglas C17 cargo door. Conversely, T3 and under-aged tempers showed less exfoliation after EXCO exposure than they showed at the seacoast (Colvin and Lifka, 1990; Colvin and Murtha, 1989).

These findings promoted responses, ranging from proposing a standard EXCO test with a 48 h rather than a 96 h exposure time (Reboul and Mayer, 1987) to considering use of a modified EXCO test environment (Lee and Lifka, 1992), but it was inevitably concluded that the EXCO test is inappropriate for Al–Li alloys (Lane et al., 1985; Moran et al., 2010, 2012; Thompson, 1992). EXCO tests suggest that increasing alloy copper content is detrimental (Figure 14.3A), whereas increasing magnesium content for a given copper level is beneficial, and susceptibility is high when the copper/magnesium ratio exceeds ~ 1.5 (Figure 14.3B), and that susceptibility for first-, second-, and third-generation Al–Li alloys increases with ageing (UA < PA < OA).

The inability of EXCO test exfoliation ratings to correlate with natural outdoor and service exposures extends also to third-generation Al–Li alloys. Again, conventional EXCO testing shows that there is a reversed temper dependency (Figure 14.4) in contrast to the rankings from testing in dry bottom MASTMAASIS and seacoast environments (Figure 14.5). In general EXCO testing predicts good exfoliation performance when outdoor exposure shows severe susceptibility, and can have extremely poor exfoliation resistance when outdoor exposure shows only pitting (Figures 14.6 and 14.7) (Moran et al., 2012). Conversely, dry bottom MASTMAASIS testing accurately predicts outdoor seacoast corrosion results for AA 2099, 2199, and 2060 (Figures 14.6 and 14.7). It is therefore strongly recommended that MASTMAASIS is used to assess the resistance of an Al–Li alloy to exfoliation corrosion, and that EXCO should not be used, as originally proposed by Lane et al. (1985).

During the late 1980s, it was realized that combined minor alloying additions of zinc and magnesium to an AA2090 base alloy composition (Al–2.2Li–2.5Cu) could provide improved exfoliation corrosion and SCC resistance (Figure 14.8 and Table 14.5) along with higher strength levels for a given temper (Table 14.5; Rioja et al., 1992). This provided one of the key elements for the development of third-generation Al–Li alloys, where Mg remains to provide strength while maintaining low alloy density; and Zn is added to offset the loss in corrosion resistance when adding Mg. Zinc additions provide strength while substantially improving the corrosion resistance (Rioja et al., 1990, 1992). However, the Mg/Zn wt% ratio must be kept at less than 1, and optimally near 0.5, to maintain the improved corrosion resistance

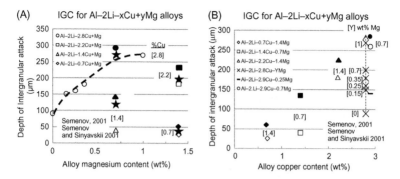

FIGURE 14.1 Depth of IGC for Al−2Li−Cu−Mg alloys after 6 h exposure to 1M aqueous NaCl + 1% H_2O_2 at 30°C with respect to (A) magnesium content and (B) copper content. (Symbols: open, under-aged; solid, over-aged; and star, peak-aged.). *Source: Data provided by Semenov (2001) and Semenov and Sinyavsavskii (2001).*

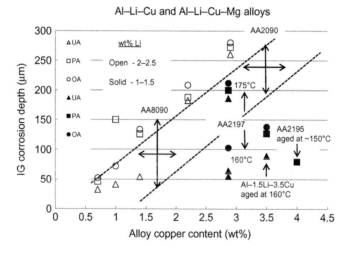

FIGURE 14.2 Depth of IGC for various Al−Li based alloys presented as a function of alloy copper content. *Source: Data from references provided in Table 14.4.*

and SCC properties (Figure 14.8). The effect of zinc and magnesium on exfoliation resistance can be seen in Tables 14.4 and 14.5. Results from Rioja et al. (1990, 1992) show that while 0.3 and 0.6 wt% Mg additions degrade exfoliation performance, the addition of Zn, either in equal or greater amount, improves the resistance to exfoliation. Additionally, results in Table 14.4 show that alloys with Mg/Zn wt% ratios of less than 1 perform better in a 96 h EXCO (ASTM G34) test than alloys with no Zn and a Mg/Zn wt% of 1. However, these EXCO derived conclusions must be treated with caution.

Exfoliation corrosion data generated from MASTMAASIS testing and long-term outdoor seacoast exposures suggest that the exfoliation resistance

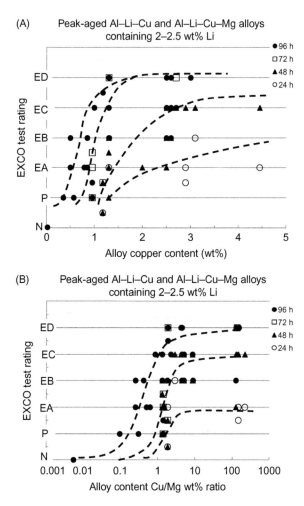

FIGURE 14.3 EXCO exfoliation corrosion ratings (24–96 h) for various peak-aged Al–Li alloys containing 2–2.5 wt% lithium: (A) influence of alloy copper content and (B) influence of alloy Cu/Mg ratio. *Source: Data provided in Table 14.4.*

of third-generation Al–Li alloys is superior to that of the second-generation alloys (Giummarra et al., 2008; Henon and Rouault, 2012; Jiang et al., 2005; Moran et al., 2012). This contrasts with EXCO test ratings up to 48 h immersion, which suggest a similar performance for second- and third-generation alloys. From the results in Figures 14.6 and 14.7 (Moran et al., 2012), those from Colvin and Murtha (1989) and Colvin and Lifka (1990) on AA 2090, and Henon and Rouault (2012) on AA 2050, the conclusions drawn from EXCO testing are unlikely to be representative for service performance.

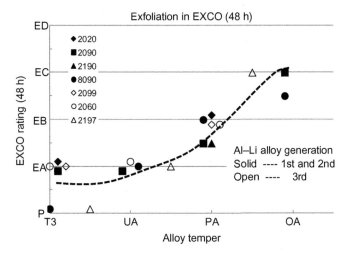

FIGURE 14.4 EXCO exfoliation corrosion ratings (48 h) for first-, second-, and third-generation Al—Li alloys. *Source: Data from references provided in Table 14.4.*

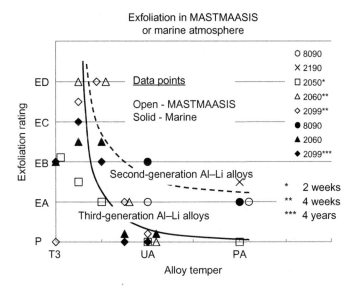

FIGURE 14.5 Exfoliation corrosion for second- and third-generation Al—Li alloys after 14 days exposure to MASTMAASIS testing or an outdoor marine environment. *Source: Data from references provided in Table 14.4.*

14.2.2.3 Further Discussion on Alloy Susceptibilities

The localized corrosion susceptibilities of Al—Li—Cu and Al—Li—Cu—Mg alloys are significantly influenced by alloy composition, temper and micro-structure, grain shape, and aspect ratio: IGC is favored by recrystallized

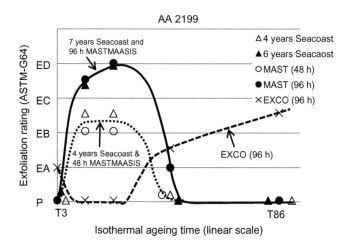

FIGURE 14.6 Comparison of exfoliation corrosion rankings as a function of temper for AA 2199 in various accelerated laboratory tests and seacoast exposure (Moran et al., 2012). (Test environments: seacoast exposure at Atlantic Ocean test station at Point Judith, USA; dry bottom MASTMAASIS (ASTM G85); EXCO (ASTM G34).)

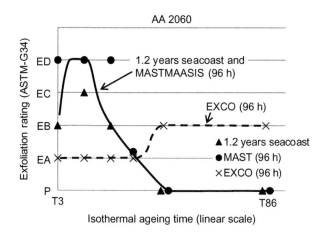

FIGURE 14.7 Exfoliation corrosion rankings as a function of temper for AA2060 in various accelerated laboratory tests and seacoast exposure (Moran et al., 2012). Test methods same as used for results shown in Figure 14.6.

equiaxed grains, and exfoliation corrosion is favored by elongated grain structures.

Factors influencing the exfoliation corrosion of aluminum alloys have been studied by Robinson and co-workers for Al–Zn–Mg–Cu (McNaught et al., 2003; Robinson, 1993), Al–Cu (Robinson and Jackson, 1999a,b), and Al–Li–Cu–Mg (Kelly and Robinson, 1993), who concluded that exfoliation

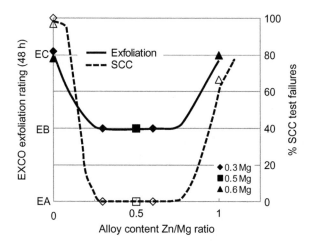

FIGURE 14.8   Exfoliation corrosion and SCC (ASTM-G44) performance for an AA2090 base alloy (Al–2.2Li–2.5Cu) with minor Zn and Mg additions (Rioja et al., 1992) (further details given in Table 14.5).

TABLE 14.5 Exfoliation Corrosion Rating (EXCO 48 h) and SCC Performance (30 Days Alternate Immersion in 0.6M Aqueous NaCl at 276 MPa) for 25–38 mm Thick Plate AA 2090 (Al–2.2Li–2.5Cu) Alloy with Minor Zn and Mg Additions, Aged for 20 h at 163°C

Zn wt%	0.3Mg wt%	0.5Mg wt%	0.6Mg wt%
0	EC [3/3] (537)		EC [1/1] (562)
0.5	EB [0/3] (584)		
0.6			EC [2/3] (528)
0.9	EB [0/3] (617)		
1		EB [0/3] (579)	

Key: (X/X) SCC performance (no. failures/no. tests); (yyy) longitudinal tensile yield stress (MPa).
Source: Data from Rioja et al. (1992).

susceptibility is dictated by the magnitude of the forces generated by voluminous grain boundary corrosion products (corrosion product wedging, CPW). CPW increases with increasing grain aspect ratio for a given alloy/temper. For example, for EXCO tested 67 mm thick unrecrystallized AA 8090 plate and 1.6 mm thick unrecrystallized and recrystallized AA 8090 sheet, the CPW forces (and observed exfoliation susceptibility) were higher for (i) the T/2 and T/4 plate sections with respect to the plate surface, which had less elongated grains, and (ii) unrecrystallized compared to recrystallized sheet material (Kelly and Robinson, 1993).

Habashi et al. (1993), inspired by Robinson's work, conducted similar studies on a range of aluminum alloys exposed to an EXCO test environment, including Al−2.3Li−1.2Cu−0.09Zr, AA 8090, and AA 2091 in various tempers. They noted that significant incubation times were consistently required before the CPW forces were detected.

In addition to evaluating forces generated in AA 8090 exposed to an EXCO test environment, Kelly and Robinson (1993) also measured the CPW forces developed during stress corrosion crack propagation, under decreasing load conditions, in plate and sheet material continuously immersed in 0.6M NaCl. CPW forces that were still generated well after crack growth rates had decreased to sufficiently low levels to indicate that the applied stress intensity factor had fallen to its threshold level, K_{Iscc}. Kelly and Robinson concluded that these CPW forces were consistent with exfoliation corrosion propagation, being promoted by a stress-assisted corrosion mechanism. Habashi et al. (1993) also reached a similar conclusion.

Economic and efficiency requirements are now driving an increased usage of mixed material assemblies (hybrid structures) in aircraft designs. This trend necessitates an enhanced awareness of the potential need for galvanic protection. Recent corrosion studies on third-generation Al−Li alloys galvanically coupled with carbon fiber composites display a significantly improved resistance compared to those found for lithium-free AA 2XXX and 7XXX series aerospace alloys (Zhang et al., 2012).

14.2.3 Al−Mg−Li Alloys

The relatively poor corrosion characteristics (particularly SCC resistance) of ternary Al−Mg−Li alloys in the $(\alpha + S_1)$ phase field were significantly improved by additions of manganese (~ 0.5 wt%) or zirconium (~ 0.1 wt%) when lithium contents exceeded 1.8 wt% (Fridlyander et al., 1983).

Typical mechanical and corrosion resistance properties quoted for peak-aged wrought 1420 are UTS 450−460 MPa, YS 290 MPa, elongation 6−10% in the longitudinal direction ($\sim 4\%$ in the short-transverse direction), combined with an "EB" EXCO exfoliation corrosion rating, an IGC performance similar to peak-aged AA 2090 (Yusheng et al., 1996), and a stress corrosion susceptibility similar to 2024-T3.

The corrosion characteristics of 1420 are highly dependent on the quantity and nature of the S_1 phase present on g.b.'s, and IGC resistance is consistent with a dissolution model based on preferential dissolution of the g.b. S_1 (Yusheng et al., 1996). Thus, low-strength cladding layers or protective coatings will be required during the use of 1420 in any commercial applications involving the potential exposure to marine environments.

In the mid-1980s, scandium additions to 1420-type alloy compositions led to the alloy 1421 and a lower-magnesium version, 1423 (Table 14.1). The 1423 alloy provided 20–25 MPa higher yield strengths, improved weldability, and a slight corrosion resistance benefit (EXCO rating EB to EA) compared to 1420, albeit with a slight loss of elongation. More recently, alloy development has led to a third-generation alloy, 1424, which is a weldable alloy from the Al–Mg–Li–Zn–Zr–Sc system. The 1424 has an alloy chemistry based on 1420 and 1421 and with a zinc addition and lower lithium content (Table 14.1).

An unexpected observation for peak-aged 1424 (aged using a three-stage ageing practice) is that its fatigue crack growth rates in 0.6M aqueous NaCl under constant amplitude loading at 8 and 0.03 Hz are significantly accelerated in comparison to those for AA 2024. Fractographic observations revealed significant regions of intergranular crack growth for 1424, and these were not present after similar testing in laboratory air. As concluded by Fridlyander et al. (2002), further work is needed to explain this phenomenon, which is most likely to be unacceptable for application of 1424 in fatigue-loaded aerospace structures.

Al–Mg–Li alloy development work in the United States during the late 1980s and early 1990s conducted at Martin Marietta Laboratories evaluated ~0.2–0.4 wt% Ag additions to Al–Mg alloys with 3–5 wt% Mg and led to the alloy Weldalite™ 050. This alloy, although offering higher tensile properties for a given magnesium concentration than the Russian non-Ag containing Al–Mg–Li alloys,1420 or 1421 (Kramer et al., 1994) provided no significant corrosion performance improvement in marine environments and arguably promoted a slight degradation (Table 14.6).

14.3 STRESS CORROSION CRACKING

Environment-enhanced crack initiation and growth in Al–Li based alloys under static, monotonic, or cyclic loading (depending on the circumstances) may result from anodic dissolution and/or hydrogen-induced embrittlement via cathodic charging or pre-exposure.

Processes controlling crack initiation and propagation during the SCC of Al–Li alloys differ. Data from early stress corrosion studies clearly indicate that crack initiation is significantly more difficult under continuous rather than alternate immersion conditions (Pizzo et al., 1984a,b); and care is required to ensure that any assessment of an alloy's relative susceptibility is not biased by the test method, loading orientation, and/or the environmental test conditions. For example, crack initiation in smooth test specimens is rarely detected in peak-aged Al–Li binary, Al–Li–Cu, or Al–Li–Cu–Mg alloys tested in 0.6M aqueous NaCl under continuous immersion conditions, using constant strain (Holroyd et al., 1986; Ohsaki and Takahashi, 1991), constant load (Buis and Schijve, 1992; Ohsaki and Takahashi, 1991), or slow

TABLE 14.6 Pitting Corrosion for Weldalite™ 050
(Al−4.2Mg−2.2Li−0.4Ag−0.1Li−0.02Cu) and Various Other Aluminum
Alloys After 12 Months Exposed to Filtered Natural Seawater (ASTM-G52)
Under Alternate Immersion (ASTM-G44) Conditions

Alloy	Product Thickness (mm)	Maximum Pit Depth (μm)	Maximum Pit Size Ranking[a]	Pitting Factor[a]
AA 1100-H14	Sheet (1.6)	12	1	0.10
AA 5053-H116	Pate (6.25)	0	0	0
1420	−	62	3	0.55
1421	−	73	3[b]	0.76
AA 2090-T84	Plate	−	5[b]	−
AA 8090	Sheet	128	>5	1.51
Weldalite™ 050	Extrusion (12.7)	98	3	0.89

[a]Pit ranking is based on the ASTM standard pit ranking values and reflects the single largest pit found on each specimen evaluated.
[b]Measurements involve areas displaying uniform-localized corrosion.
Source: Carmody and Shaw, 1997.

strain rate loading (Braun, 1994, 1995b, 2003; Buis and Schijve, 1992); whereas crack initiation and propagation usually occur in copper-containing Al−Li alloys (i) under alternation immersion test conditions or (ii) when a saline test environment used for continuous immersion testing is modified with additions of hydrogen peroxide (Ohsaki and Takahashi, 1991), sulfate and carbonate ions (Braun, 1994, 1995a,b, 2003; Craig et al., 1987a,b; Marsac et al., 1992), or chromate ions with acidification (Holroyd et al., 1986) or by applying anodic polarization (Balasubramaniam and Duquette, 1992; Ghosh et al., 2007).

A simplified schematic summary of events occurring on smooth Al−Li alloy SCC samples during continuous (total) or alternate immersion in 0.6M NaCl is provided in Figure 14.9. SCC initiation rarely results during continuous (total) immersion conditions, even with extended immersion times, despite localized corrosion promoting acidified local environments within partially occluded pits/fissures that are slightly deeper for stressed samples. However, under alternate immersion the electrochemical conditions within the localized corrosion sites differ during the wet and drying stages of the test cycle (Figure 14.10): crack initiation occurs when the appropriate local environmental conditions and stress state are established, with enhanced

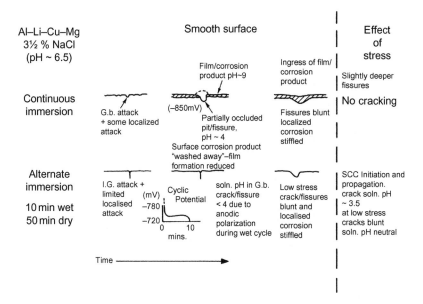

FIGURE 14.9   Typical behavior for Al−Li−Cu−Mg alloys on smooth surfaces exposed to 0.6M saline environments: continuous (total) immersion versus alternate immersion (Holroyd et al., 2013).

crack growth rates consistently being detected immediately after the removal of the bulk environment (Newton and Holroyd, 1992).

An unexpected observation, initially reported by Holroyd and co-workers (Holroyd et al., 1986), was that removal of stressed peak-aged AA 8090 from continuous immersion in a bulk aqueous NaCl environment, followed by exposure to laboratory air, often led to crack initiation and rapid crack growth in smooth test specimens. However, crack initiation was prevented if pre-exposed stressed samples were exposed to CO_2-free laboratory air (Craig et al., 1987a,b), or to exposed dry air or vacuum, for around 28 days prior to exposure to laboratory air (Holroyd et al., 1986). According to Craig et al., the chemistry of the pre-exposure effect, and its mitigation by removing CO_2, can be explained as follows: after removal from the chloride solution, thin-layer atmospheric corrosion of the 8090 alloy initially produces an alkaline solution based on LiOH, which causes a fairly uniform dissolution of the material. In the absence of CO_2, this chemistry persists and there is no SCC. When CO_2 is present, it partially neutralizes the alkalinity, producing a carbonate-chloride solution that creates a borderline condition of passivity and initiates SCC very rapidly. The chemistry in the crack becomes acidic later on, as in other aluminum alloys, but the pre-exposure effect overcomes the barrier to crack initiation. If full-immersion tests are done in a simulation of the carbonate-chloride thin-layer solution, SCC again occurs very easily

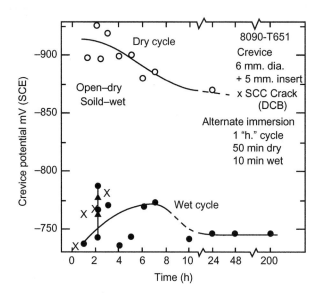

FIGURE 14.10 Electrochemical potential for AA 8090-T651, measured within the notch region of a DCB test specimen or at the base of an artificial crevice during alternate immersion exposure to 0.6M aqueous NaCl (Holroyd et al., 2013).

(Braun, 1994; Craig et al., 1987b). These observations were reproduced by Moran et al. for AA 2090 (Moran, 1990; Moran and Stoner, 1989; Moran et al., 1992), and they focused attention on processes influencing the development of the local environmental conditions associated with crack initiation and growth.

14.3.1 Al–Li Binary Alloys

SCC initiation from smooth surfaces for binary Al–Li alloys exposed to 0.6M aqueous NaCl rarely occurs (Holroyd et al., 1986), irrespective of alloy temper or the test method employed, unless some form of an initiation site is provided, such as a pre-existing notch or a pre-crack, or a suitable site is generated *in situ* by the local environmental test conditions (e.g., by the application of anodic polarization to promote localized corrosion).

Anodic polarization provides an explanation why Balasubramaniam et al. (1991a) observed crack initiation in smooth Al–2.1Li–0.05Zr tensile specimens subjected to slow strain rate testing (SSRT) while continuously immersed in 0.5M aqueous NaCl. During SSRT testing, the electrochemical potential of the tensile samples when held potentiostatically at -0.750 V_{SCE} (Balasubramaniam and Duquette, 1992). Maintaining this potential (reportedly the free-corrosion potential) during straining provides sufficient anodic polarization for localized corrosion and the development of local environmental conditions capable of triggering SCC. The results presented by Balasubramaniam et al. (1991a) are important because they provide convincing arguments for the possible involvement of a brittle hydride (LiAlH$_4$) during the SCC of Al−Li based alloys.

Christodoulou et al. (1984) observed intergranular SCC growth in fatigue pre-cracked Al−2.8Li−0.12Zr double cantilever beam (DCB) specimens loaded initially to about 80% K_{Ic} and subjected to alternate immersion in an inhibited acidified saline environment at room temperature. Crack growth was observed for under- and peak-aged materials, but not for over-aged material. Crack growth is also known to occur in a peak-aged Al−2.5Li−0.12Zr alloy in less aggressive test environments, including water vapor saturated air (WVSA), laboratory air, 35% relative humidity, and vacuum (Behnood et al., 1989; Figure 14.11). This observation, while of considerable mechanistic interest for understanding sustained load cracking (SLC) and creep crack growth in commercial Al−Li alloys, is unlikely to be a major influence on SCC at room temperature. This is because the environmental temperature would need to exceed about 50°C for significant SLC crack growth to occur in commercial Al−Li alloys at applied stress intensities (K_I values) well below K_{Ic} (Holroyd et al., 2014). Nevertheless, it should be noted that SLC crack growth rates for peak-aged AA 8090 at 80°C are in excess of 10^{-9} m/s for K_I values as low as 10 MNm$^{-3/2}$ (Behnood et al., 1989).

14.3.2 Al−Li−Cu and Al−Li−Cu−Mg Alloys

SCC data for smooth and pre-cracked AA 2090 and AA 8090 tested in aerated saline environments (Ohsaki and Takahashi, 1991) provide a good summary of the SCC characteristics of second-generation commercial Al−Li alloys.

14.3.2.1 Crack Initiation

Constant-load continuous immersion SCC testing of AA 2090 or AA 8090 in 0.9M aqueous NaCl fails to initiate and promote any failures within 3 months, irrespective of alloy temper or the applied load up to at least 1.1 times the alloy's yield stress. Similar testing with a 0.3% H$_2$O$_2$ addition resulted in cracks eventually emerging from localized corrosion sites, with crack initiation occurring earlier in under-aged tempers, after a significant incubation time (Ohsaki and Takahashi, 1991).

It is an extremely important observation that SCC initiation in Al−Li based alloys exposed to 0.6M aqueous NaCl requires generation of a local corrosion site, within which the local environmental and stress states will differ from those experienced by the alloy surface exposed to the bulk solution.

Based on a detailed study of crack initiation from a 2 mm radius machined notch in AA 2096 sheet material exposed to 0.6M aqueous NaCl, using controlled alternate immersion conditions, it was concluded that crack

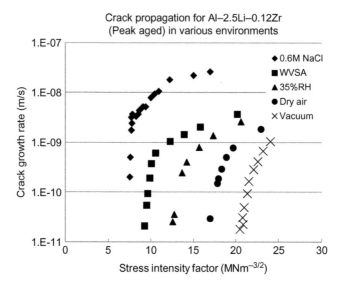

FIGURE 14.11 Crack propagation as a function of the applied stress intensity factor for peak-aged Al–2.5Li–0.12Zr exposed to various bulk environments at room temperature (Behnood et al., 1989; Holroyd et al., 2014).

initiation is a multistep process (Connolly and Scully, 2000, 2005). The initial phase is electrochemically driven and essentially independent of the stress state (pitting at constituent particles). This is followed by pit coalescence to form elongated corrosion fissures, which under stress may slowly grow in concert with local environmental changes within these restricted geometries that, when in contact with an SCC-susceptible alloy microstructure under an appropriate loading condition, can lead to a transition to intergranular SCC (Connolly and Scully, 2000).

Although the influence of alloy temper on SCC susceptibility of Al–Li–Cu–Mg alloys is generally regarded as similar to that of Al–Zn–Mg–Cu alloys (Holroyd et al., 1986) (i.e., with under-aged susceptible, peak-aged less so, and resistance increasing with further ageing, some controversy exists for Al–Li–Cu alloys (Connolly and Scully, 2000)). For AA 2090 and AA 2096, Connolly and Scully (2000) observed two regimes of SCC susceptibility in 0.6M NaCl under alternate immersion conditions, one for severely under-aged tempers and another for over-aged tempers, separated by a regime including peak-aged tempers that had a high SCC resistance (Figure 14.12).

Dorward and Hasse (1988) reported a similar temper dependency for SCC initiation in smooth tensile and C-ring specimens of under- and peak-aged AA 2090 plate when tested under alternate immersion or marine atmosphere conditions. (Over-aged tempers were not tested owing to their poor fracture toughness.) Interestingly, they noted that while the peak-aged

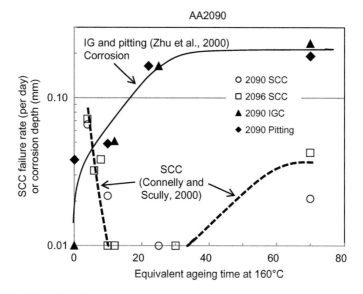

FIGURE 14.12   The influence of AA2090 temper on the SCC in 0.6M aqueous NaCl (ASTM G-47) (Connolly and Scully, 2000) and intergranular and pitting corrosion in 0.6M aqueous NaCl + 1% H_2O_2 solution (Zhu et al., 2000).

material did not show any significant IGC, it did contain deep fissures with some inter-subgranular attack, which visually could be mistaken as SCC. These authors suggested that this may explain why SSRT test data for AA 2090, interpreted solely on the basis of loss of tensile elongation, can incorrectly suggest that peak-aged material has a significant SCC susceptibility. Braun (1995a) provided support for this suggestion by demonstrating that the elongation loss during SSRT of AA 2190-T8 in 0.6M aqueous NaCl is similar to the loss caused by localized corrosion of unstressed test samples exposed to the test environment for the same length of time.

Although AA 2090 has two regimes of significant SCC susceptibility on either side of peak-aged material, as in Figure 14.12, the susceptibility of AA 2090 to (i) pitting and IGC in 0.6M NaCl + 1% H_2O_2 and (ii) exfoliation corrosion during EXCO testing increases progressively with ageing (UA < PA < OA) (Figure 14.4; Table 14.4). On the other hand, Figure 14.5 shows that exfoliation corrosion in less aggressive environments, such as MASTMAASIS and outdoor marine environments, is more pronounced for the under-aged temper, agreeing with the ranking found for SCC (Figure 14.12).

Recent data presented by Henon and Rouault (2012) for a third-generation Al–Li alloy, AA 2050, containing significantly less lithium (~1 wt%) but increased copper and magnesium (Table 14.1), suggest that it has a similar SCC temper dependency to AA 2090 (~2.1 wt% Li) (i.e., a significant SCC susceptibility for ageing times up to the equivalent of 1 h at

175°C, falling significantly with further ageing but beyond 20 h once again becoming susceptible). This SCC behavior was paralleled by severe IGC for the under- and over-aged materials in a MASTMAASIS test environment, but not for intermediate tempers, including peak-aged material. Supporting data (Moran et al., 2012) for two other third-generation alloys, AA 2099 (∼1.8% Li—see Table 14.1) and AA 2060 (∼0.75% Li—see Table 14.1), continue to suggest a SCC temper dependency similar to AA 2090, as shown in Figure 14.13. It should be noted, however, for peak-aged material that deep/wide fissures of significant depth were generated in both "unfailed" stress corrosion tests and corrosion tests (ASTM G110) samples (Henon and Rouault, 2012; Moran et al., 2010; Warner-Locke et al., 2013).

The deep/wide corrosion fissures generated during SCC and IGC ASTM G110 tests result from an intragranular or inter-subgranular corrosion process (Figure 14.14), although their effect on fatigue performance for AA 2099-T86 shows that they are no more harmful than conventional pits for AA 7050-T76, i. e., no additional loss of fatigue life (Warner-Locke et al., 2013). Further evaluation is required to establish whether long-term incubation periods can lead to SCC initiation or whether minor modifications to the bulk environmental conditions, such as sulfate additions, can transform the mode of attack to IGC or SCC, as was found for second-generation Al−Li alloys (Craig et al., 1987b).

Ohsaki and Takahashi (1991) reported that under total immersion conditions in 0.6M aqueous NaCl there were significant incubation times (e.g., >3 months) for SCC initiation from fatigue pre-cracks in under-aged AA 2090 and AA 8090 DCB test specimens loaded to 50−80% of K_{Ic}. However, if the saline environment was added dropwise on a daily basis (a kind of alternate immersion), then cracks initiated within ∼24 h. Similar observations for peak-aged fatigue pre-cracked AA 8090 DCB samples initially loaded to a $K_I \sim 13$ MNm$^{-3/2}$ and exposed to various test environments are given in Figure 14.15. Closer examination of the events occurring during SCC initiation under alternate immersion conditions revealed that crack growth initiated within a few hours and propagated at a low growth rate for 20−40 h, during which the solution pH of local environments in the crack tip region acidified, and crack growth rates subsequently increased by at least an order of magnitude, to those typical for SCC in saline environments (Figure 14.16). Similar observations for Al−Zn−Mg−Cu (AA 7XXX series) alloys have attributed the initial low crack growth rates to cracking in distilled water, following chloride adsorption on an oxide layer covering the fatigue pre-crack surfaces and a consequent depletion of chloride ions in the crack tip environment (Holroyd and Scamans, 2013).

14.3.2.2 Crack Propagation

Reported SCC growth rates for AA 2090 (Dorward and Hasse, 1988; Ohsaki and Takahashi, 1991) and AA 8090 (Ahmad, 1990; Dorward and Hasse,

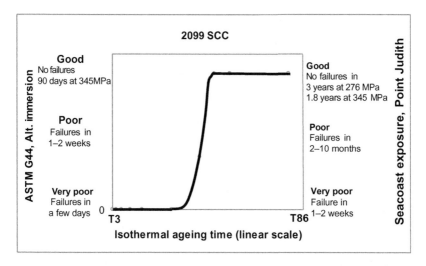

FIGURE 14.13   Schematic representation of short-transverse SCC data for AA 2099 plate in various tempers tested under alternate immersion (ASTM-G44) and under seacoast exposure at stress levels in the range 172–345 MPa (Moran et al., 2012).

FIGURE 14.14   "Elongated corrosion fissure" generated in an AA 2199-T86 C-ring test specimen subjected to ASTM-G44 alternate immersion SCC testing at 345 MPa in 0.6M NaCl for 5 days (Warner-Locke et al., 2013). *Source: Photograph courtesy of Dr. J.S. Warner-Locke.*

1987; Holroyd et al., 2014; Wang et al., 1992) are remarkably similar (Figures 14.17 and 14.18). Maximum crack growth rates tend to decrease with ageing (UA > PA > OA), whereas the SCC threshold stress intensity factor K_{Iscc} displays a reverse trend (Figure 14.17). Similar findings reported for AA 2090 (Dorward and Hasse, 1988) and AA 8090 (Dorward and Hasse, 1987) led to the suggestion that the SCC threshold may better be represented by a normalized factor, such as K_{Iscc}/K_{Ic}, thereby taking into account the alloy's fracture toughness, which usually decreases significantly with ageing.

Following the realization in the late 1980s that zinc additions could significantly improve the SCC characteristics of Al–Li–Cu–Mg alloys (Gray

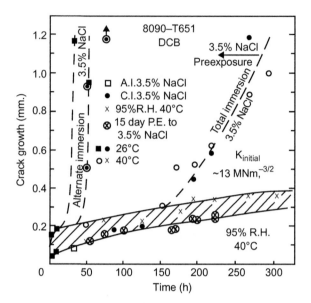

FIGURE 14.15 Influence of various bulk environmental conditions on the initial stages of SCC growth for AA8090-T651 DCB specimens loaded to an initial stress intensity factor of ~ 13 MNm$^{-3/2}$ (Holroyd et al., 2014).

et al., 1989), microstructural studies indicated that zinc additions below ~ 1 wt% promoted zinc incorporation into δ and T_2 g.b. precipitates, rather than changing the type of precipitates or their precipitate-free zone widths (Gregson et al., 1987; Jiang et al., 1993; Kilmer and Stoner, 1991; Singh et al., 1997). Subsequent SCC studies have confirmed the potential beneficial effects of minor zinc additions to AA 8090 (Kilmer et al., 1991), AA 2090 (Rioja et al., 1990; Ohsaki et al., 1996; Osaki et al., 1999), and AA 2091 (Ohsaki et al., 1996).

Despite the limited quantitative data available to characterize the SCC characteristics of third-generation Al–Li alloys (Colvin et al., 2012; Connolly and Scully, 2000, 2005; Henon and Rouault, 2012; Li and Zheng, 2012; Moran et al., 2012; Rioja and Liu, 2012), it is reasonable to conclude, for product thicknesses up to ~ 30 mm, that alloys with reduced lithium contents (typically ~ 1–1.5 wt% compared to 2–2.5 wt% for second-generation alloys) combined with zinc and/or silver additions are capable of providing SCC resistances in peak-aged tempers that will outperform those of Al–Cu–Mg (2XXX series) and Al–Zn–Mg–Cu (7XXX series) high-strength aerospace aluminum alloys. Further evidence supporting this proposal is provided by comparison of the SCC propagation rates as function of the applied stress intensity factor for various aluminum alloys subjected to

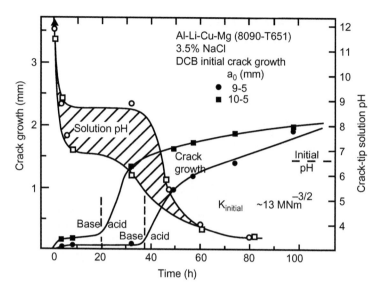

FIGURE 14.16   Initial crack growth and the solution pH developed at the base of the notch for fatigue pre-cracked AA 8090-T651 DCB specimens loaded to an initial stress intensity factor of ~ 13 MNm$^{-3/2}$ and exposed to 0.6M aqueous NaCl under alternate immersion conditions (Holroyd et al., 2014).

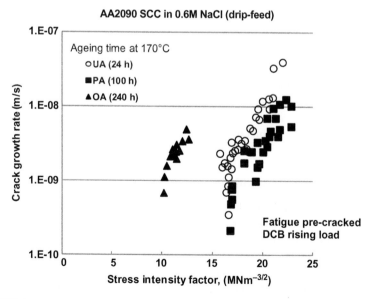

FIGURE 14.17   SCC crack growth rate as a function of the applied stress intensity factor for peak-aged AA 2090 exposed to a saline environment (Ohsaki and Takahashi, 1991).

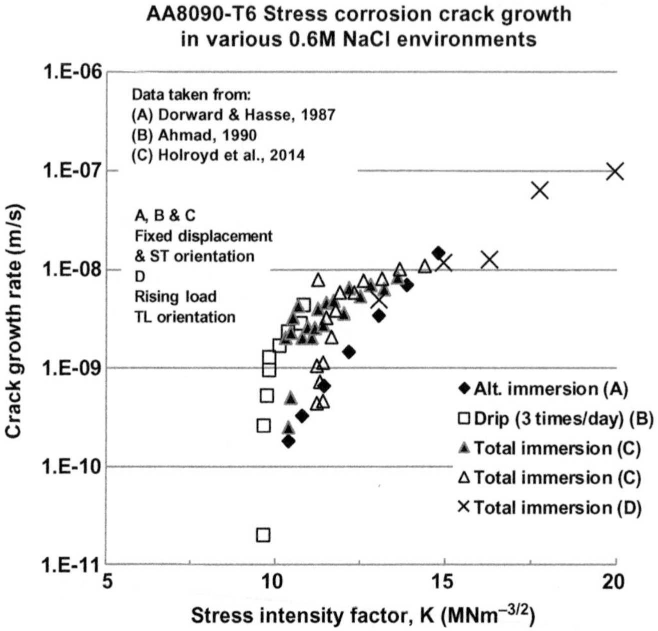

FIGURE 14.18 SCC crack growth rate as a function of the applied stress intensity factor for peak-aged AA 8090 exposed to 0.6M aqueous NaCl under various conditions. *Source: Data taken from (A) Dorward and Hasse, 1987; (B) Ahmad, 1990; (C) Holroyd et al., 2014.*

testing using DCB specimens exposed to 0.6M NaCl or artificial seawater (Figure 14.19).

Whether equivalent SCC resistances can be guaranteed in thicker third-generation Al—Li products (e.g., $\sim >30$ mm) remains an open question. The answer may depend upon the success of the thermomechanical processing and multistage ageing practices used to manipulate and control an alloy's g.b. microstructure. The ability to do this will become increasingly difficult as the product thickness increases.

14.3.3 Mechanistic Implications

Controversy remains over which processes control the SCC of Al—Li based alloys. There are advocates for (i) anodic dissolution driven processes, involving dissolution of precipitates such as T_1, T_2, and δ on g.b.'s (Buchheit, 2000; Buchheit et al., 1990; Meletis, 1986; Rinker et al., 1984) or copper-depleted zones along g.b.'s (Kumai et al., 1989; Wall and Stoner, 1997) or (ii) hydrogen-induced fracture processes involving hydride generation promoted by lithium or trace impurity segregation at g.b.'s (Lewandowski and Holroyd, 1990; Lynch et al., 2002; Roth and Kaesche, 1989) or the decohesion of weak

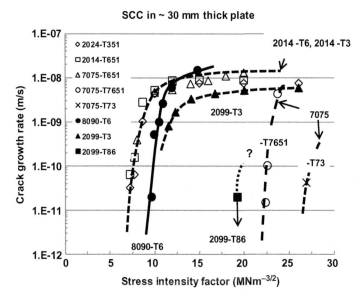

FIGURE 14.19   Stress corrosion crack growth rates as a function of the stress intensity factor for various high-strength aluminum aerospace alloys exposed to aqueous 0.6M NaCl or artificial seawater. *Source: Data from: AA 2024-T351 and AA 2014-T651 (Speidel, 1975); AA 7075-T651 means result taken from several sources (Holroyd and Scamans, 2013); AA 7075-T7651 and AA 7075-T73 (Dorward and Hasse, 1976); AA 8090 average taken from Figure 14.18; and AA 2099 (Moran et al., 2012).*

interfaces or hydrides precipitated on phases such as T_1, T_2, and δ (Balasubramaniam and Duquette, 1992; Balasubramaniam et al., 1991b; Meletis and Huang, 1991; Smith and Scully, 2000; Speidel, 1974).

Many researchers have demonstrated that cathodic pre-charging of Al–Li binary (Lee and Pyun, 1990), Al–Li–Cu (Balasubramaniam et al., 1991b; Kim et al., 1988; Shin et al., 1990), or Al–Li–Cu–Mg (Chen et al., 1993; Jang et al., 2002) alloys with hydrogen can promote a hydrogen-induced loss of mechanical properties, and in many cases, the induced detrimental effects parallel those observed for SCC. For instance, Jang et al. (2002) report that the maximum loss of properties for cathodically charged AA 8090 occurs for (i) loading in the short-transverse orientation, (ii) decreasing strain rates, and (iii) under-aged tempers; and that susceptibility decreases with further ageing.

Experimental evidence indicates that different processes control crack initiation and propagation during the SCC of Al–Li alloys. It is likely that crack initiation is controlled by an anodic dissolution dominated process, akin to those described by Connolly and Scully (2005), whereas crack propagation may be controlled by anodic dissolution and/or hydrogen-related processes, depending on the specific conditions.

14.4 SUMMARY AND CONCLUSIONS

IGC susceptibility for second- and third-generation Al–Li–Cu and Al–Li–Cu–Mg based alloys in saline environments increases with alloy copper content, and the depth of attack generally progressively increases with thermal ageing (i.e., UA < PA < OA, albeit the mode of attack, particularly for the zinc-containing third-generation alloys, becomes less truly intergranular and more predominantly intragranular or inter-subgranular in nature). Susceptibility to exfoliation corrosion shows a similar temper dependence when assessed by EXCO testing, but a differing one following exposure to either a "real-life" outdoor marine environment or less aggressive test media such as a MASTMAASIS environment. In these less aggressive environments, there are two regimes of significant potential exfoliation susceptibility corresponding to under- and over-aged tempers, and an intermediate regime, including peak-aged tempers, providing significantly improved exfoliation corrosion resistance (i.e., UA > PA < OA).

SCC of AA 8090 (Al–Li–Cu–Mg) in moist air or saline environments typically displays the temper dependence found for Al–Zn–Mg–Cu alloys, i.e., crack initiation and propagation susceptibility decrease with ageing (i.e., UA > PA > OA). Although the temper dependence for SCC of AA 2090 (Al–Li–Cu) is less well established, a study of SCC initiation in AA 2090 and limited data for third-generation Al–Li alloys provide grounds for suggesting that the temper dependency corresponds to that for exfoliation corrosion in "real-life" outdoor marine environments (i.e., UA > PA < OA). It is possible that previously published data, particularly for peak- and over-aged tempers, have been compromised by considering increased localized corrosion to be SCC.

SCC initiation from smooth surfaces is rare for copper-free Al–Li alloys exposed to saline environments, irrespective of alloy temper or whether exposure is under continuous or alternate immersion conditions. For copper-containing Al–Li alloys, SCC initiation remains difficult under continuous immersion conditions, unless a suitable local initiation site is generated within which an appropriate local environment develops that comes in contact with a susceptible alloy microstructure subjected to a sufficient stress. This sequence of events is triggered far more readily under alternate immersion conditions, and hence SCC initiation is a more common event during alternate immersion testing.

SCC initiation in notched or pre-cracked test specimens exposed to aqueous saline environments follows the same sequence of events. Crack initiation under continuous immersion conditions without accelerant additions or anodic polarization usually requires extended periods (days to months); whereas with the accelerants, or under alternate immersion conditions, crack initiation occurs typically within 24 h.

On the basis of available information, the lower lithium content third-generation Al–Li alloys, with zinc and/or silver additions, will be capable of providing commercial products in near peak-aged tempers with acceptable resistances to localized corrosion (IGC and exfoliation) and SCC. For thicker products ($\sim > 30$ mm), this conclusion assumes that the plate manufacturers will be able to apply the required tailored thermomechanical processing and multistage ageing practices needed to manipulate grain boundary microstructures.

ACKNOWLEDGMENT

The authors profoundly thank Dr. RJH Wanhill for many of his critical comments and numerous inputs, which have become an integral part of this book.

REFERENCES

Ahmad, M., 1990. Effects of environment and microstructure on stress corrosion crack propagation in an Al–Li–Cu–Mg alloy. Mater. Sci. Eng. A125, 1–14.

Ahmad, M., Aleem, A., 1996. Evaluation of corrosion behavior of Al–Mg–Li alloys in seawater. J. Mater. Eng. Perform. 5, 235–240.

Assmann, P., 1926. Age hardening aluminum–lithium alloys. Zeitschrift für Metallkunde 18, 51–54.

Balasubramaniam, R., Duquette, D.J., 1992. Stress corrosion of an aluminum–lithium copper alloy in chloride solution. In: Bruemmer, S.M., Meletis, E.I., Jones, R.H., Gerberich, W.W., Ford, F.P., Staehle, R.W. (Eds.), Parkins Symposium on Fundamental Aspects of Stress Corrosion Cracking. TMS, Warrendale, PA, pp. 131–139.

Balasubramaniam, R., Duquette, D.J., Rajan, K., 1991a. On stress corrosion cracking in aluminum–lithium alloys. Acta Metall. Mater. 39, 2597–2605.

Balasubramaniam, R.D., Duquette, D.J., Rajan, K., 1991b. Hydride formation in an Al–Li–Cu alloy. Acta Metall. Mater. 39, 2607–2613.

Balmuth, E.S., Schmidt, R., 1981. A perspective of the development of aluminum–lithium alloys. In: Sanders Jr., T.H., Starke Jr., E.A. (Eds.), Aluminum–Lithium Alloys. Metallurgical Society AIME, Warrendale, PA, pp. 69–88.

Behnood, N., Cai, H., Evans, J.T., Holroyd, N.J.H., 1989. Microcrack nucleation and sustained-loading crack growth in Al–Li alloys. Mater. Sci. Eng. A A119, 23–32.

Braun, R., 1994. Exfoliation corrosion and stress corrosion cracking behaviour of 8090-T81 sheet. In: Sanders Jr., T.H., Starke Jr., E.A. (Eds.), Aluminum Alloys—Their Physical and Mechanical Properties (ICAA4), vol. 2. The Georgia Institute of Technology, Atlanta, GA, pp. 511–518.

Braun, R., 1995a. Exfoliation corrosion and stress corrosion cracking behaviour of Al–Li Alloys. In: Corrosion Detection and Management of Advanced Airframe Materials, AGARD Conference Proceedings No.565, Advisory Group for Aeronautical Research and Development (AGARD), Neuilly-sur-Seine, France, pp. 2.1–2.12.

Braun, R., 1995b. Evaluation of stress corrosion cracking behaviour of damage-tolerant Al–Li sheet using the slow strain rate technique. Mater. Sci. Eng. A A190, 143–154.

Braun, R., 2003. Anion effects on the stress corrosion cracking behaviour of aluminium alloys. Mater. Corros. 54, 157–162.

Buchheit, R.G., 2000. The electrochemistry of θ (Al₂Cu), S (Al₂CuMg) and T₁ (Al₂CuLi) and localized corrosion and environment assisted cracking in high strength Al alloys. Mater. Sci. Forum 331–337, 1641–1646.

Buchheit, R.G., Moran, J.P., Stoner, G.E., 1990. Localized corrosion behavior of alloy 2090—the role of microstructural heterogeneity. Corrosion 46, 610–617.

Buchheit, R.G., Moran, J.P., Stoner, G.E., 1994. Electrochemical behavior of the T₁ (Al₂CuLi) intermetallic compound and its role in localized corrosion of Al–2%Li–3%Cu alloys. Corrosion 50, 120–130.

Buchheit, R.G., Martinez, M.A., Montes, L.P., 2000. Evidence of Cu ion formation by dissolution and dealloying the Al₂CuMg intermetallic compound in rotating ring-disk collection experiments. J. Electrochem. Soc. 147, 119–124.

Buis, A., Schijve, J., 1992. Stress corrosion cracking behaviour of Al Li 2090-T83 in artificial seawater. Corrosion 48, 898–909.

Carmody, T.A., Shaw, B.A., 1997. The determination of the corrosion behavior of a developmental Al–Mg–Li–Ag alloy. In: 1997 Tri-Service Corrosion Conference, AMPTIAC Document No. AM022865. National Materials Information System.

Chen, L., Chen, W., Liu, Z., Shao, Y., Hu, Z., 1993. Effect of hydrogen on mechanical properties and fracture mechanism of 8090 Al–Li Alloy. Metall. Trans. A 24A, 1355–1361.

Christodoulou, L., Struble, L., Pickens, J.R., 1984. Stress corrosion cracking in Al–Li binary alloys. In: Sanders Jr., T.H., Starke Jr., E.A. (Eds.), Aluminum–Lithium Alloys II. Metallurgical Society AIME, Warrendale, PA, pp. 561–579.

Colvin, E.L., 1988. Exfoliation and stress corrosion performance of Al–Li alloys. In: Kar, R.J., Agrawal, S.P., Quist, W.E. (Eds.), "Aluminum–Lithium Alloys" Design, Development and Application Update. ASM International, Metals Park, OH, pp. 273–289.

Colvin, E.L., Lifka, B.W., 1990. Accelerated and atmospheric exfoliation corrosion performance of AlLi alloys. In: Isaacs, H.S., Bertocci, U., Kruger, J., Smialowski, S. (Eds.), Advances in Localized Corrosion (NACE-9). NACE, Houston, TX, pp. 215–220.

Colvin, E.L., Murtha, S.J., 1989. Exfoliation corrosion testing of Al–Li alloys 2090 and 2091. In: Sanders Jr., T.H., Starke Jr., E.A. (Eds.), Aluminum–Lithium V, vol. III. Materials and Component Engineering Publications, Birmingham, UK, pp. 1251–1260.

Colvin, E.L., Rioja, R.J., Yocum, L.A., Denzer, D.K., Cogswell, T.G., Bray, G.H., et al., 2012. Aluminum–copper–lithium alloys. US Patent No. 8,118,950, February 21, 2012.

Connolly, B.J., Scully, J.R., 2000. Corrosion cracking susceptibility of Al–Li–Cu alloys 2090 and 2096 as a function of isothermal ageing time. Scr. Mater. 42, 1039–1045.

Connolly, B.J., Scully, J.R., 2005. Transition from localized corrosion to stress corrosion cracking in an Al–Li–Cu–Ag alloy. Corrosion 61, 1145–1166.

Craig, J.D., Newman, R.C., Jarrett, M.R., Holroyd, N.J.H., 1987a. Stress corrosion and pre-exposure effects in Al–Li–Cu–Mg alloy 8090. In: Louthan Jr., M.R., McNitt, R.P., Sissons Jr., R.D. (Eds.), Environmental Degradation of Engineering Materials III. Pennsylvania State University, University Park, pp. 313–320.

Craig, J.D., Newman, R.C., Jarrett, M.R., Holroyd, N.J.H., 1987b. Local chemistry of stress corrosion cracking in Al–Li–Cu–Mg alloys. J. Phys. 48, C3.825–C3.833.

Czochralski, J., 1927. Aluminum alloy containing lithium. US Patent No. 1,620,082, March 8, 1927.

Czochralski, J., Welter, G., 1927. Alloy of lithium and aluminum. US Patent No. 1,620,081, March 8, 1927.

Davis, J.R., 1999. Corrosion testing. In: Davis, J.R. (Ed.), Corrosion of Aluminum and Aluminum Alloys. ASM International, Materials Park, OH, pp. 227–233.

Dorward, R.C., Hasse, K.R., 1976. Flaw growth of 7075, 7475, 7050, and 7049 aluminum plate in stress corrosion environments. In: Final NASA Report Contract No. NAS8-30890, October 1976.

Dorward, R.C., Hasse, K.R., 1987. Stress corrosion cracking behavior of an AlLiCuMg alloy. Corrosion 43, 408—413.

Dorward, R.C., Hasse, K.R., 1988. Stress corrosion characteristics of AlLiCu AA2090 alloy. Corrosion 44, 932—941.

Fridlyander, I.N., 2003. Aluminum alloys with lithium and magnesium. Met. Sci. Heat Treat. 45 (9—10), 344—347.

Fridlyander, I.N., Cherkizovskaya, B.N., Shiryaeva, V., Ambartasumyan, S.M., Gorokhova, T. A., Gabidullin, R.M., et al., 1969. Aluminium base alloy. GB Patent No. 1,172,736, December 3, 1969.

Fridlyander, I.N., Sandler, V.S., Nikolskaya, T.I., 1983. Specific features of structure and properties of 1420 aluminum alloy. Met. Sci. Heat Treat. 25 (7), 495—498.

Fridlyander, I.N., Khokhlatova, L.B., Kolobnev, N.I., Rendiks, K., 2002. Thermally stable aluminum—lithium1424 for application in welded fuselage. Met. Sci. Heat Treat. 44 (1—2), 3—8.

Gable, B.M., Pana, M.A., Shiflet, G.J., Starke Jr., E.A., 2002. The role of trace additions on the T_1 coarsening in Al—Li—Cu—X alloys. Mater. Sci. Forum 396—402, 699—704.

Ghosh, K.S., Das, K., Chatterjee, U.K., 2007. Environmental induced cracking of Al—Li—Cu—Mg—Zr alloys of peak aged and retrogressed and reaged tempers under applied potential. In: Shifler, D., Christodoulou, J.A., Moran, J.P., Perez, A., Zhen, W. (Eds.), Degradation of Light Weight Alloys. TMS, Warrendale, PA, pp. 11—20.

Giummarra, C., Thomas, B., Rioja, R.J., 2007. New aluminum lithium alloys for aerospace applications. In: Proceedings of the Light Metals Technology Conference. Bombardier Aerospace and Alcoa Trade Study.

Giummarra, C., Rioja, R.J., Bray, G.H., Magnusen, P.E., Moran, J.P., 2008. Al—Li alloys: development of corrosion resistant, high-toughness aluminum—lithium aerospace alloys. In: Hirsch, J., Skrotzki, B., Gottstein, G. (Eds.), Aluminium Alloys—Their Physical and Mechanical Properties, vol. 1. Wiley-VCH GmbH & Co. KGaA, Weinheim, Germany, pp. 176—188.

Gray, A., 1987. Factors influencing the environmental behaviour of aluminium—lithium alloys. J. Phys. 48 (C3), 891—904.

Gray, A., Holroyd, N.J.H., White, J., 1989. The influence of microstructure on the environmental cracking behaviour of Al—Li—Cu—Mg—Zr alloys. In: Sanders Jr., T.H., Starke Jr., E.A. (Eds.), Aluminum—Lithium V, vol. III. Materials and Component Engineering Publications, Birmingham, UK, pp. 1175—1186.

Gregson, P.J., Dinsdale, K., Harris, S.J., Noble, B., 1987. Evolution of microstructure in Al—Li—Zn—Mg—Cu alloys. Mater. Sci. Technol. 3, 7—13.

Grimes, R., Cornish, A.J., Miller, W.S., Reynolds, M.A., 1985. Aluminium—lithium based alloys for aerospace applications. Met. Mater. 1 (6), 357—563.

Grimes, R., Davis, T., Saxty, H.J., Feoron, J.F., 1987. Progress to aluminium—lithium semi-fabricated products. J. Phys. 48 (C3), 11—24.

Habashi, M., Bonte, E., Galland, J., Bodu, J.J., 1993. Quantitative measurements on the degree of exfoliation on aluminium alloys. Corros. Sci. 35, 169—183.

Henon, C., Rouault, S., 2012. Comparison of corrosion performance and mechanisms of Al—Cu alloys with and without lithium additions. In: Weiland, H., Rollett, A.D., Cassada, A. (Eds.),

13th International Conference on Aluminum Alloys (ICAA13). The Minerals, Metals and Materials Society (TMS) and John Wiley & Sons, Hoboken, NJ, pp. 431–436.

Holroyd, N.J.H., Scamans, G.M., 2013. Stress corrosion of Al–Zn–Mg–Cu aluminum alloys in saline environments. Metall. Mater. Trans. A 44A, 1230–1253.

Holroyd, N.J.H., Gray, A., Scamans, G.M., Hermann, R., 1986. Environmental-sensitive fracture of Al–Li–Cu–Mg alloys. In: Baker, C., Gregson, P.J., Harris, S.J., Peel, C.J. (Eds.), Aluminium–Lithium Alloys III. The Institute of Metals, London, UK, pp. 310–320.

Holroyd, N.J.H., Newman, R.C., Vasudevan, A.K., Scamans, G.M., 2014. Stress corrosion cracking of aluminumlithium based alloys. To be Presented at the International Symposium on Environmental Damage in Structural Materials Under Static/Cyclic Loads at Ambient Temperature. June 2014, Tuscany, Italy.

Hu, Z.Q., Zang, Y., Liu, L., Zhu, Z.Y., 1993. Corrosion behavior of 8090 Al–Li alloy. Corrosion 49, 491–498.

Jang, W., Kim, S., Shin, K., 2002. Hydrogen-assisted deformation and fracture behaviors of Al 8090. Metall. Mater. Trans. A 33A, 1755–1763.

Jiang, N., Li, J.F., Zheng, Z., Wei, X.Y., Li, Y.F., 2005. Effect of ageing on mechanical properties and localized corrosion behaviors of Al–Li–Cu alloy. Trans. Non-Ferrous Met. Soc. China 15, 23–29.

Jiang, X.J., Li, Y.Y., Deng, W., Gui, Q.H., Xiong, L.Y., Shi, C.X., 1993. Effects of zinc on the microstructure and tensile properties of Al–Li alloys. J. Mater. Sci. Lett. 12, 1375–1377.

Kelly, D.J., Robinson, M.J., 1993. Influence of heat treatment and grain size on exfoliation corrosion of Al–Li alloy 8090. Corrosion 49, 787–795.

Kertz, J.E., Gouma, P.I., Buchheit, R.G., 2001. Localized corrosion susceptibility of Al–Li–Cu–Mg–Zn alloy AF/C458 due to interrupted quenching from solutionizing temperatures. Metall. Mater. Trans. A 32A, 2561–2573.

Kilmer, R.J., Stoner, G.E., 1991. Effect of Zn additions on precipitation during ageing of alloy 8090. Scr. Metall. Mater. 25, 243–248.

Kilmer, R.J., Witters, J.J., Stoner, G.E., 1991. Effect of Zinc additions on the precipitation events and implications to stress corrosion cracking behaviour in Al–Li–Cu–Mg–Zr alloys. In: Peters, M., Winkler, P.J. (Eds.), Proceeding of Sixth International Aluminium–Lithium Conference. Deutsche Gesellschaft für Metallkunde, Frankfurt, Germany, pp. 755–760.

Kim, S.S., Lee, E.W., Shin, K.S., 1988. Effect of cathodic hydrogen charging on tensile properties of 2090 Al–Li alloy. Scr. Metall. 22, 1831–1834.

Kolobnev, N.I., 2002. Aluminum–lithium alloys with scandium. Met. Sci. Heat Treat. 44 (7–8), 297–299.

Kolobnev, N.I., Khokhlatova, Fridlyander, I.N., 2004. Ageing of Al–Li alloys having composite particles as hardening phases. Mater. Forum 28, 207–211.

Kramer, L.S., Langan, T.J., Pickens, J.R., Last, H., 1994. Development of Al–Mg–Li alloys for marine applications. J. Mater. Sci. 29, 5826–5832.

Kumai, C., Kusinski, J., Thomas, G., Devine, T.M., 1989. Influence of aging at 200°C on the corrosion of Al–Li and Al–Li–Cu alloys. Corrosion 45, 294–302.

Lane, P., Gray, J.A., Smith, C.J.E., 1985. Comparison of corrosion behaviour of lithium-containing aluminium alloys and conventional aerospace alloys. In: Baker, C., Gregson, P.J., Harris, S.J., Peel, C.J. (Eds.), Aluminium–Lithium Alloys III. The Institute of Metals, London, UK, pp. 273–281.

Langan, T.J., Pickens, J.R., 1991. Microstructure–property relationships in Al–Cu–Li–Ag–Mg Weldalite® Alloys. In: NASA Langley Research Center Contract Report 4364, NASA.

LeBaron, I.M., 1945. Aluminum alloy. US Patent No. 2,381,219, August 7, 1945.

Lee, C.S., Choi, Y., Park, I.G., 2002. Stress corrosion cracking behaviour of Al–Cu–Li–Mg–Zr–(Ag) alloys. Met. Mater. Int. 8, 191–196.

Lee, S., Lifka, B.W., 1992. Modification of EXCO test method for exfoliation corrosion suscep-tibility in 7xxx, 2xxx and aluminum–lithium alloys. In: Agarwala, V.S., Ugiansky, G.M. (Eds.), "New Methods for Corrosion Testing of Aluminum Alloys" ASTM STP 1134. American Society for Testing and Materials, Philadelphia, PA, pp. 1–19.

Lee, S., Pyun, S., 1990. Effects of hydrogen redistribution on hydrogen-assisted cracking in Al-1.9Li and Al–4.5Zn–2.3Mg alloys. Scr. Metall. Mat. 24, 1629–1634.

Lewandowski, J.J., Holroyd, N.J.H., 1990. Intergranular fracture of Al–Li alloys: effects of age-ing and impurities. Mater. Sci. Eng. A A123, 219–227.

Li, J.F., Zheng, Z.Q., 2012. Corrosion and potentiodynamic polarization of an Al–Cu–Li alloy under tensile stress. In: Weiland, H., Rollett, A.D., Cassada, A. (Eds.), 13th International Conference on Aluminum Alloys (ICAA13). The Minerals, Metals and Materials Society (TMS) and John Wiley & Sons, Hoboken, NJ, pp. 443–450.

Li, J.F., Zheng, Z.Q., Li, S.C., Chen, W.J., Ren, W.D., Zhao, X.S., 2007. Simulation study on function mechanism of some precipitates in localized corrosion of Al alloys. Corros. Sci. 49, 2436–2449.

Li, J.F., Li, C.X., Peng, Z.W., Chen, W.J., Zheng, Z.Q., 2008. Corrosion mechanism associated with T_1 and T_2 precipitates of Al–Cu–Li alloys in NaCl solutions. J. Alloys Compd. 460, 688–693.

Liang, W.J., Pan, Q., He, Y., Zhou, Y., Lu, C., 2008. Effect of ageing on mechanical properties and corrosion susceptibility of Al–Cu–Li–Zr alloy containing scandium. Rare Met. 27, 146–152.

Lynch, S.P., Muddle, B.C., Pasang, T., 2002. Mechanisms of brittle intergranular fracture in Al–Li alloys and comparison with other alloys. Philos. Mag. A 82, 3361–3373.

Marsac, S., Mankowoski, G., Darbosi, F., 1992. Stress corrosion cracking behaviour of 2091 Al–Li alloy in sulphate and chloride containing solutions. Br. Corros. J. 27, 50–58.

McNaught, D., Worsfold, M., Robinson, M.J., 2003. Corrosion product force measurements in the study of exfoliation and stress corrosion in high strength aluminium alloys. Corros. Sci. 45, 2377–2389.

Meletis, E.I., 1986. Stress corrosion properties of 2090 Al–Li alloy. In: Geol, V.S. (Ed.), Corrosion Cracking. ASM, Metals Park, OH, pp. 315–326.

Meletis, E.L., Huang, W., 1991. The role of the T_1 phase in the pre-exposure and hydrogen embrittlement of Al–Li–Cu alloys. Mater. Sci. Eng. A A148, 197–209.

Merati, A., 2011. Materials replacement for aging aircraft. "Corrosion Fatigue and Environmentally Assisted Cracking in Aging Military Vehicles", RTO AGARDograph AG-AVT-140. Research and Technology Organisation (NATO), Neuilly-sur-Seine, France, pp. 24-1–24-22.

Moran, J.P., 1990. Mechanisms of Localized Corrosion and Stress Corrosion Cracking of Al–Li–Cu Alloy 2090, Ph.D Dissertation. University of Virginia.

Moran, J.P., Stoner, G.E., 1989. Solution chemistry effects on the SCC behavior of 2090 and 2024. In: Starke Jr., E.A., Sanders Jr., T.H. (Eds.), Aluminum–Lithium V. Materials and Component Engineering Publications, Birmingham, UK, pp. 1187–1196.

Moran, J.P., Buchheit, R.G., Stoner, G.E., 1992. Mechanisms of SCC of alloy 2090 (Al–Li–Cu)—A comparison of interpretations from static and slow-strain-rate techniques. In: Brummer, S.M., Melitis, E.L., Jones, R.H., Gerberich, W.W., Ford, F.P., Staehle, R.W. (Eds.), Fundamental Aspects of Stress Corrosion Cracking. The Minerals, Metals and Materials Society (TMS), Warrendale, PA, pp. 141–158.

Moran, J.P., Bovard, F.S., Chrzan, J.D., Rioja, R.J., Colvin, E.L., 2010. Improvements in corrosion resistance offered by newer generation 2×99 aluminum–lithium alloys for aerospace applications. Proceedings of the 12th International Conference on Aluminium Alloys (ICAA12). The Institute of Japanese Light Metals, pp. 1492–1497.

Moran, J.P., Bovard, F.S., Chrzan, J.D., Vandenburgh, P., 2012. Corrosion performance of new generation aluminum–lithium alloys for aerospace applications. In: Weiland, H., Rollett, A. D., Cassada, A. (Eds.), 13th International Conference on Aluminum Alloys (ICAA13). The Minerals, Metals and Materials Society (TMS) and John Wiley & Sons, Hoboken, NJ, pp. 425–430.

Moshier, W.C., Shaw, B.A., Tack, W.T., Phull, B., 1992. Stress corrosion cracking behavior of two high-strength Al–xCu–Li–Ag–Mg–Zr alloys. Corrosion 48, 306–308.

Newton, C.J., Holroyd, N.J.H., 1992. Time-lapse video techniques in the corrosion testing of aluminum alloys. In: Agarwala, V.S., Ugiansky, G.M. (Eds.), "New Methods for Corrosion Testing of Aluminum Alloys", ASTM STP 1134. American Society for Testing and Materials, Philadelphia, PA, pp. 153–179.

Niskanen, P., Sanders, T.H., Rinker, J.G., Marek, M., 1982. Corrosion of aluminium alloys containing lithium. Corros. Sci. 22, 283–304.

Noble, B., Harris, S.J., Katsikis, S., Dinsdale, K., 2008. Aerospace applications: analysis of the corrosion resistance of the potential of lithium as an alloying addition to aluminum aerospace alloys. In: Hirsch, J., Skrotzki, B., Gottstein, G. (Eds.), Aluminium Alloys—Their Physical and Mechanical Properties, vol. 1. Wiley-VCH GmbH & Co. KGaA, Weinheim, Germany, pp. 215–221.

Ohsaki, S., Takahashi, T., 1991. Stress corrosion cracking behavior of Al–Li alloys. Key Eng. Mater. 51/52, 107–112.

Ohsaki, S., Sato, T., Takahashi, T., 1988. Effect of ageing on pitting corrosion behavior of Al–Li alloys. J. Jpn. Inst. Light Met. 38 (5), 264–269.

Ohsaki, S., Kobayashi, K., Iino, M., Sakamoto, T., 1996. Fracture toughness and stress corrosion cracking of aluminium–lithium alloys 2090 and 2091. Corros. Sci. 38, 793–802.

Osaki, S., Iino, M., Kobayashi, K., Sakamoto, T., 1999. Effect of zinc addition on static- and dynamic-SCC properties of Al–Li–Cu 2090 alloy. Jpn. Soc. Mech. Eng., Ser. A 42 (2), 288–293.

Pasang, T., Lynch, S.P., Moutsos, S., 2006. Challenges in developing high performance Al–Li alloys. Int. J. Soc. Mater. Eng. Resour. 14 (1/2), 7–11.

Pickens, J.R., Tack, W.T., 1995. Al–Cu–Li alloys with improved cryogenic toughness. US Patent No. 5,455,003, October 3, 1995.

Pizzo, P.P., Galvin, R.P., Nelson, H.G., 1984a. Utilizing various test methods to study the stress corrosion behavior of Al–Li–Cu alloys. In: Dean, S.W., Pugh, E.N., Ugiansky, G.M. (Eds.), "Environment-Sensitive Fracture: Evaluation and Comparison of Test Methods", ASTM STP 821. American Society for Testing and Materials, Philadelphia, PA, pp. 173–201.

Pizzo, P.P., Galvin, R.P., Nelson, H.G., 1984b. Stress corrosion behavior of aluminum–lithium alloys in aqueous salt solutions. In: Sanders Jr., T.H., Starke Jr., E.A. (Eds.), Aluminum–Lithium Alloys II. AIME, NewYork, NY, pp. 627–656.

Quist, W.E., Narayanan, G.H., 1989. Aluminum–lithium alloys. In: Vasudevan, A.K., Doherty, R.D. (Eds.), Aluminum Alloys—Contemporary Research and Applications. Academic Press, San Diego, CA, pp. 219–254.

Reboul, M., Mayer, P., 1987. Intergranular and exfoliation corrosion of Al–Li–Cu–Mg alloys. J. Phys. 48 (C3), 881–889.

Rinker, J.G., Marek, M., Sanders Jr., T.H., 1984. Microstructure, toughness and stress corrosion cracking behavior of aluminum alloy 2020. Mater. Sci. Eng. A 64, 203–221.

Rioja, R.J., 1998. Fabrication methods to manufacture isotropic Al–Li alloys and products for space and aerospace applications. Mater. Sci. Eng. A A257, 100–107.

Rioja, R.J., Liu, J., 2012. The evolution of Al–Li based products for aerospace and space applications. Metall. Mater. Trans. A 43A, 3325–3337.

Rioja, R.J., Cho, A., Bretz, P.E., 1990. Aluminum–lithium alloys having improved corrosion resistance containing Mg and Zn. US Patent No. 4,961,792, January 30, 1990.

Rioja, R.J., Cho, A., Colvin, E.J., Vasudevan, A.K., 1992. Aluminum–lithium alloys. US Patent No. 5,137,686, August 11, 1992.

Robinson, M.J., 1993. The role of wedging stresses in the exfoliation of high strength aluminium alloys. Corros. Sci. 23, 887–899.

Robinson, M.J., Jackson, N.C., 1999a. Exfoliation corrosion in high strength Al–Cu–Mg alloys: effect of grain structure. Br. Corros. J. 34, 45–49.

Robinson, M.J., Jackson, N.C., 1999b. The influence of grain structure and intergranular corrosion rate on exfoliation and stress corrosion cracking of high strength Al–Cu–Mg alloys. Corros. Sci. 41, 1013–1028.

Roth, A., Kaesche, H., 1989. Electrochemical investigation of technical aluminum–lithium alloys-Part II: The lithium depleted zone. In: Sanders Jr., T.H., Starke Jr., E.A. (Eds.), Aluminum–Lithium V, vol. 3. Materials Component Engineering Publications, Birmingham, UK, pp. 1207–1216.

Ruschau, J.J., Jata, K.V., 1993. Weldalite 049 (Al 2095-T8): anisotropy effects at cryogenic conditions. Interim Report for August 1991–June 1993—WL-TR-4096. Materials Directive, Wright Laboratory. Wright-Patterson Air Force Base, Dayton, Ohio, USA.

Semenov, A.M., 2001. Effect of Mg additions and thermal treatment on corrosion properties of Al–Li–Cu base alloys. Prot. Met. 37, 126–131.

Semenov, A.M., Sinyavskii, V.S., 2001. Effect of the copper to magnesium ratio and their summary content on the corrosion properties of Al–Li alloys. Prot. Met. 37, 132–137.

Shin, K.S., Austin, N.J., Kim, S.S., 1990. Effect of hydrogen on mechanical behavior of a 2090 Alloy. In: Moody, N.R., Thompson, A.W. (Eds.), Hydrogen Effects on Material Behavior. TMS, Warrendale, PA, pp. 1023–1031.

Singh, V., Mukhopadhyay, A.K., Prasad, K.S., 1997. Influence of small additions of zinc on the nature of the grain boundary precipitates in an AA 8090 alloy. Scr. Mater. 37, 1519–1523.

Smith, A.F., 1987. Aluminium–lithium alloys for helicopter structures. Met. Mater. 3 (7), 438–444.

Smith, S.W., Scully, J.R., 2000. The identification of hydrogen trapping states in an Al–Li–Cu–Zr alloy using thermal desorption spectroscopy. Metall. Mater. Trans. A 31A, 179–193.

Speidel, M.O., 1974. Hydrogen embrittlement of aluminum alloys. In: Thompson, A.W. (Ed.), Hydrogen in Metals. ASM, Metals Park, OH, pp. 249–276.

Speidel, M.O., 1975. Stress corrosion cracking of aluminum alloys. Metall. Trans. A 6A, 631–651.

Tack, W.T., Heubaum, F.H., Pickens, 1990. Mechanical property evaluations of a new, ultra-high strength Al–Cu–Li–Ag–Mg alloy. Scr. Metall. Mater. 24, 1685–1690.

Thompson, J.J., 1992. Exfoliation corrosion testing of aluminum–lithium alloys. In: Agarwala, V.S., Ugiansky, G.M. (Eds.), 'New Methods for Corrosion Testing of Aluminum Alloys', ASTM STP 1134. American Society for Testing and Materials, Philadelphia, PA, pp. 70–81.

Vasudevan, A.K., Malcolm, R.C., Frinkle, W.G., Rioja, R.J., 1985. In: Resistance to fracture, fatigue and stress-corrosion cracking of Al–Cu–Li–Zr Alloys. Naval Air Systems Command Final Report Contract N00019-80-C-0569, June 30, 1985.

Wall, F.D., Stoner, G.E., 1997. The evolution of critical electrochemical potentials influencing environmentally assisted cracking of Al–Li–Cu alloys in selected environments. Corros. Sci. 39, 835–853.

Wang, Z.F., Zhu, Z.Y., Zhang, Y., Ke, W., 1992. Stress corrosion cracking of an Al–Li Alloy. Metall. Trans. A 23A, 3337–3341.

Warner-Locke, J.S., Moran, J.P., Hull, B., Reilly, L., 2013. The effect of corrosion pit morphology on SCC and fatigue of 2×99 alloys compared to 7xxx alloys. Presented at National Association of Corrosion Engineers (NACE) Corrosion 2013 Conference. March 2013, Orlando, FL.

Xu, Y., Wang, X., Yan, Z., Li, J., 2011. Corrosion properties of light-weight and high-strength 2195 Al–Li alloy. Chin. J. Aeronaut. 24, 681–686.

Yusheng, C., Ziyong, Z., Sue, L., Wei, K., Yun, Z., Wanming, Z., 1996. The corrosion behaviors andmechanism of 1420 Al–Li alloy. Scr. Mater. 34, 781–786.

Zakharov, V.V., 2003. Some problems of the use of aluminum–lithium alloys. Met. Sci. Heat Treat. 45 (1/2), 49–54.

Zhang, W., Moran, J.P., Morales, R.S., Vandenburgh, P.K., June 2012. Galvanic corrosion of Al–Li alloys coupled to CFRP composite. Presented at 13th International Conference on Aluminum Alloys (ICAA13). Pittsburgh, PA.

Zhu, Z., Zhang, Y., Zhang, W., 2000. Intergranular corrosion and exfoliation corrosion of Al–Li–alloys. Mater. Sci. Forum 331–337, 1671–1676.

Applications

Part V

Applications

Chapter 15

Aerospace Applications of Aluminum−Lithium Alloys

R.J.H. Wanhill

NLR, Emmeloord, The Netherlands

Contents

15.1 Introduction	**503**	
15.2 Weight Savings	**504**	
15.2.1 Density	506	
15.2.2 Density and Elastic Moduli: Specific Stiffnesses	506	
15.2.3 Modern/Innovative Structural Concepts and a Case Study	509	
15.3 Materials Selection	**514**	
15.3.1 Aluminum Alloys, CFRPs, and FMLs: Advantages and Disadvantages	514	
15.3.2 Materials for Aircraft Structures	517	

15.3.3 Aluminum Alloys for Spacecraft Structures 519
15.3.4 Materials Qualification: an Example 521
15.4 Applications of Al−Li Alloys (Third Generation) 521
15.4.1 Aircraft 521
15.4.2 Spacecraft 525
15.5 Summary and Conclusions 530
Acknowledgments 531
References 531

15.1 INTRODUCTION

Previous chapters of this book have shown that aluminum−lithium (Al−Li) alloy development has a long history going back to the 1920s. However, it is only since the 1990s that the fundamental understanding of these alloys has matured and enabled the development of a family of alloys with excellent combinations of engineering properties, the so-called third-generation Al−Li alloys. The developments that make these alloys serious contenders for many aerospace applications have been concisely summarized by Professor Rama Rao in the Foreword to this book. These developments are (i) the production technologies for large-scale melting and casting of numerous Al—Li alloys with optimized chemistry, (ii) advanced processing based on process modeling, (iii) thermal and thermomechanical treatments for widely ranged microstructure—mechanical property combinations (strength, fracture toughness, fatigue and crack growth resistances), and (iv) manufacturing and fabrication technologies, including new welding techniques and superplastic forming (SPF).

The development of third-generation Al—Li alloys is very much in progress. Most have been introduced since 2000 (Table 15.1). The densities are 2—8% less than the densities of conventional AA 2XXX and 7XXX aerospace aluminum alloys and offer potential weight savings. Furthermore, greater weight savings are possible by taking advantage of higher specific stiffnesses and strengths and using more efficient structural concepts, including the use of laser beam welding (LBW), friction stir welding (FSW), and SPF. These aspects are considered in Section 15.2, which also includes results from an extensive case study.

The third-generation Al—Li alloys have to compete with conventional aluminum alloys, carbon fiber composites, and fiber metal laminates (FMLs) for aerospace applications, particularly in transport aircraft structures. This competition involves many other factors besides weight savings. These factors, classified as relative advantages and disadvantages, are reviewed in the first part of Section 15.3. This review is followed by (i) a general description of materials selection for aircraft structures and aluminum alloy selections for spacecraft, and (ii) an example program to qualify a material for component manufacture and service.

Some of the earlier third-generation Al—Li alloys have already been used in aerospace structures, including civil and military aircraft and spacecraft (launch vehicles). Examples are AA 2297 and AA 2397, which have been used for Lockheed Martin F-16 bulkheads and other parts in military aircraft (Acosta et al., 2002; Balmuth, 2001), and AA 2195 plate, which was used for the Space Shuttle Super Lightweight External Tank (SLW ET) (Williams and Starke, 2003). AA 2297 was also used for the intertank thrust panels of the SLW ET owing to its high elastic modulus and low density. More recently, several alloys have been used, qualified, or proposed for applications in transport aircraft and spacecraft. These actual and potential applications will be discussed in Section 15.4, which is followed by the summary in Section 15.5.

15.2 WEIGHT SAVINGS

Besides engineering properties, the material density (ρ) is very important for the efficiency of aerospace structures. This is seen from quantification of the weight savings potentially achievable from property improvements (Ekvall et al., 1982; Peel et al., 1984). The results obtained by Ekvall et al. (1982) are summarized in Figure 15.1. Reducing density is the most effective way of saving weight. Next are the strength and stiffness increases, which combine with reduced density to give improvements in specific strength and stiffness. Finally, improvements in engineering damage tolerance (DT) properties have the least potential for saving weight.

TABLE 15.1 Third-Generation Al–Li Alloys

Alloys	Li	Cu	Mg	Ag	Zr	Mn	Zn	Density ρ (g/cm^3)	Introduction
2195	1.0	4.0	0.4	0.4	0.11			2.71	LM/Reynolds 1992
2196	1.75	2.9	0.5	0.4	0.11	0.35 max	0.35 max	2.63	LM/Reynolds/McCook Metals 2000
2297	1.4	2.8	0.25 max		0.11	0.3	0.5 max	2.65	LM/Reynolds 1997
2397	1.4	2.8	0.25 max		0.11	0.3	0.10	2.65	Alcoa 2002
2098	1.05	3.5	0.53	0.43	0.11	0.35 max	0.35	2.70	McCook Metals 2000
2198	1.0	3.2	0.5	0.4	0.11	0.5 max	0.35 max	2.69	Reynolds/McCookMetals/Alcan 2005
2099	1.8	2.7	0.3		0.09	0.3	0.7	2.63	Alcoa 2003
2199	1.6	2.6	0.2		0.09	0.3	0.6	2.64	Alcoa 2005
2050	1.0	3.6	0.4	0.4	0.11	0.35	0.25 max	2.70	Pechiney/Alcan 2004[a]
2296	1.6	2.45	0.6	0.43	0.11	0.28	0.25 max	2.63	Constellium Alcan 2010[a]
2060	0.75	3.95	0.85	0.25	0.11	0.3	0.4	2.72	Alcoa 2011
2055	1.15	3.7	0.4	0.4	0.11	0.3	0.5	2.72	Alcoa 2011
2065	1.2	4.2	0.50	0.30	0.11	0.40	0.2	2.70	Constellium 2012
2076	1.5	2.35	0.5	0.28	0.11	0.33	0.30 max	2.64	Constellium 2012

[a]Pechiney acquired by Alcan 2003; Constellium formerly Alcan Aerospace.
Source: After Rioja and Liu (2012) and information from G.H. Bray and M. Niedzinski.

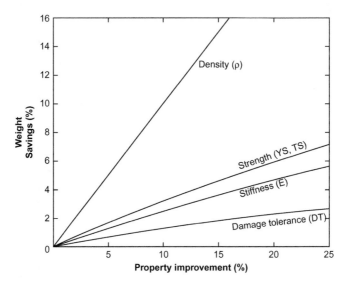

FIGURE 15.1 Effects of property improvements on potential weight savings for aircraft structures (Ekvall et al., 1982). Note: This diagram is also given as Figure 2.2 in Chapter 2.

15.2.1 Density

Lower density was one of the main reasons for developing the second-generation Al–Li alloys. The goal was to obtain alloys 8–10% lighter (and stiffer—see below) than equivalent conventional alloys by the addition of about 2 wt% lithium. This goal was achieved, but it was subsequently found that lithium contents of 2 wt% or more are linked to several disadvantages. The main disadvantages are a tendency for strongly anisotropic mechanical properties, low short-transverse ductility and fracture toughness, and loss of toughness owing to thermal instability (Eswara Prasad et al., 2003; Lynch et al., 2003; Rioja and Liu, 2012). These disadvantages are discussed elsewhere in this book.

The adverse links between lithium content and properties are why the third-generation alloys have been developed with the reduced lithium contents (and higher densities) shown in Table 15.1. These densities are 2–8% less than those typical of conventional AA 2XXX alloys ($2.77–2.80$ g/cm^3) and AA 7XXX alloys ($2.80–2.85$ g/cm^3). However, even a 2% density advantage, which translates into a direct 2% weight savings (Figure 15.1), can be worthwhile for aerospace structures, especially launch vehicles, satellites, and helicopters.

15.2.2 Density and Elastic Moduli: Specific Stiffnesses

Specific stiffnesses are basic design parameters. The specific stiffness E/ρ is important for lower wing surfaces, spars, ribs, and frames, and the specific buckling resistance, $E^{1/3}/\rho$, is important for upper wing surfaces and the

TABLE 15.2 Specific Stiffness Improvements Obtainable from Third-Generation Al—Li Alloys with Respect to Equivalent Conventional Aluminum Alloys

Alloy Families	Specific Stiffness GPa/ (g/cm³)	Specific Buckling Resistance (GPa)$^{1/3}$/(g/cm³)	Average Improvements Due to Third-Generation Al—Li Alloys	
			Specific Stiffness	Specific Buckling Resistance
2XXX	26.1−27.1	1.48−1.52	+13%	+8.0%
7XXX	25.9−26.4	1.46−1.50	+15%	+9.5%
Third-generation Al—Li alloys	28.9−31.2	1.58−1.65		

Source: Data from MMPDS-07 (2012).

fuselage. The specific stiffnesses of aluminum alloys are synergistically improved by additions of lithium, which decrease the density and increase the elastic modulus. This is illustrated in Table 15.2 for the third-generation Al—Li alloys. The values in the table were derived from the densities and compression elastic moduli for a variety of alloys and products, including the Al—Li alloys AA 2050, 2098, 2099, 2195, 2196, 2198, 2297, and 2397. The improvements translate into generally higher weight savings, 8−15%, than the densities alone (2−8%, see table 15.2).

Figure 15.2 compares the specific stiffnesses of the third-generation Al—Li alloys with those for conventional AA 2XXX and 7XXX alloys, and also some straightforward examples of high-fiber-density carbon fiber composites (carbon fiber reinforced plastics, CFRPs). The aluminum alloys match only the 25% aligned-fiber composites in specific stiffness (Figure 15.2A). However, there is a strong dependence of CFRP stiffnesses on the amount of fibers aligned in the test direction. This is because individual layers of unidirectional fibers are extremely orthotropic.

In practice, CFRP components are assembled from layers with different fiber orientations. Also, most aircraft structures are subjected to multidirectional loads, and this means that mechanical property isotropy may often be important. (As mentioned in Section 15.2.1, this is one of the main reasons why second-generation Al—Li alloys were unsuccessful, and why the third-generation Al—Li alloys were developed.) For CFRP components a requirement for mechanical isotropy means that the amount of fibers

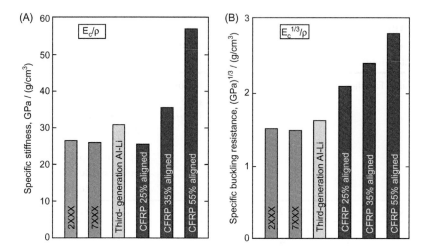

FIGURE 15.2 Specific stiffnesses (average values) for third-generation Al–Li alloys, equivalent conventional aluminum alloys, and examples of high-fiber-density (60% volume) carbon fiber composites tested in the aligned-fiber direction: CFRP, carbon fiber reinforced plastics. E_c = elastic modulus in compression; ρ= density. *Source: Data from Table 15.2 and Peel (1990).*

aligned in the principal loading direction will be about 25% (Mouritz, 2012), which is the *lowest* value in Figure 15.2. On the other hand, if mechanical property isotropy is not necessary, the CFRP layers of a component can be "tailored" to preferentially align some or most of the fibers in the principal load direction, thereby increasing the structural efficiency. This is an advantageous flexibility of long fiber composites, and also FMLs (Schijve, 2009), although similar effects should be possible for metallic wing structures by orientating stiffening elements to tailor the elastic response (Harmin et al., 2011).

The foregoing discussion is most important. It shows that although aluminum alloys cannot match the very high specific stiffnesses of CFRPs when they have high percentages of aligned fibers, a direct translation to high weight savings in CFRP components is not possible. Mouritz (2012) suggests 10–20% weight savings from using CFRP instead of aluminum, by which he presumably means conventional aluminum alloy aerospace structures assembled mainly by mechanical fastening. This 10–20% weight savings range is only slightly higher than the 8–15% potentially achievable by substituting the third-generation Al–Li alloys for conventional aluminum alloys. Furthermore, the weight savings potential of Al–Li alloy structures can be increased by employing modern and innovative structural concepts, notably LBW, FSW, and SPF, discussed in subsequent sections in this chapter.

15.2.3 Modern/Innovative Structural Concepts and a Case Study

Survey of the Concepts

There are some modern structural concepts suitable for Al–Li alloys that can reduce the potential advantages of CFRPs. These are LBW, FSW, and SPF:

1. *LBW*: This has the most potential for fabricating aircraft fuselage panels, since it has low heat input and results in narrow weld seams; and it has a high process speed and can be highly automated (Lenczowski, 2002). Fuselage panels using AA 6XXX alloys have been fabricated for the A318, A340, and A380 aircraft since 2001. More recently, the development work has turned to the third-generation Al–Li AA 2196 and 2198 alloys (Enz et al., 2012; Kashaev et al., 2013) and is continuing as part of the European Aeronautical Research Programme "Clean Sky." Alcoa is also investigating the possibility of employing LBW for Al–Li fuselage panels.
2. *FSW*: This is suitable for simple joint geometries, notably butt joints, and needs special clamping systems (Threadgill et al., 2009). The pioneering FSW work was done on large cryogenic tankage panels of Al–Li alloy AA 2195 by Lockheed Martin Michoud (Arbegast and Hartley, 1999). Also, FSW has been used for (i) integrally stiffened panels in an extensive case study of innovative structural concepts (Heinimann et al., 2007), (ii) longitudinal joints in AA 2024 fuselage panels of the Eclipse 500 business jet (Dubois, 2007), (iii) Airbus and Embraer components, (iv) joining Al–Li components of the Orion Crew Module, including a final 11.3 m circumferential weld joining the forward cone assembly and crew tunnel to the aft assembly (NASA, 2010), and (v) manufacturing Al–Li AA 2198 first- and second-stage fuselages (barrels) of the SpaceX Falcon 9 (Niedzinski and Thompson, 2010).

   FSW has been further developed for manufacturing Al–Li AA 2195 propellant tank domes (Windisch, 2009; Hales et al., 2012) and will be introduced for joining parts of the conventional aluminum alloy main tank of the Ariane 5 launcher. These development activities and the manufacturing innovations (iv) and (v) are examples of current major efforts to use FSW for spacecraft assemblies instead of conventional welding.
3. *SPF*: Al–Li and other aluminum alloys can be processed to obtain superplastic properties, see Chapters 7 and 8. The SPF process does, however, require low forming rates (Davis and Hryn, 2007) and is best suited to production of complex sheet metal components. Hence SPF has been used only in "niche" applications, and this may well remain so.

All three of these processes reduce the weight and decrease the part count of "traditional" aluminum alloy airframe structures, which are built up by mechanical fastening with rivets and bolts, sometimes in combination with adhesive bonding. Figure 15.3 gives a schematic example of how LBW and

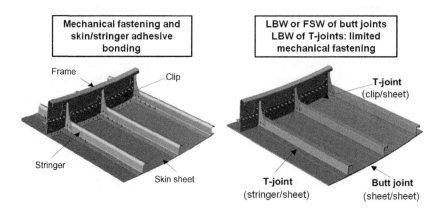

FIGURE 15.3 Illustration of reduced part count by using LBW and FSW instead of mechanical fastening and adhesive bonding for a stiffened panel structure.

FSW can reduce the part count even when the stringers have been adhesively bonded (as well as mechanically fastened) to the skin sheet.

Advantages proposed for welded airframe structures are decreased weight, cost reductions owing to a high degree of automation and fewer manufacturing steps, and better corrosion resistance owing to fewer fastener holes, gaps, and crevices. All these benefits may well be possible and to a large extent achievable. However, fully-welded airframe structures are unacceptable in commercial transport aircraft, which must have a high degree of DT and fail-safety, whereby partial structural failure can be sustained without immediately leading to complete failure. This is achieved by designing structures to have (i) multiple load paths consisting of discrete structural elements, (ii) residual strengths meeting established requirements if failure of an element occurs, and (iii) good inspectability to allow detection of element damage or failure before safety is compromised. These safety principles are discussed, with case histories that helped define them, by Wanhill (2003).

Case Study: DT of Stiffened Panel Concepts for Fuselage and Lower Wing Applications

Alcoa has undertaken several case studies of innovative concepts for airframe structures, e.g. Heinimann et al. (2007). Some results that include third-generation Al–Li alloys are presented here to illustrate design options for achieving high DT as well as weight savings. The results are for fatigue crack growth tests on skin–stringer panel configurations, representing fuselage and lower wing locations in large transport aircraft.

1. *Fuselage panels*: Mechanically fastened panels representing fuselage crown locations were tested under uniaxial constant amplitude fatigue loading with a baseline maximum stress of 117 MPa and a stress ratio of

TABLE 15.3 Fuselage Skin–Stringer Panel Concepts

Concept	Skin Material	Stringer Material	Selective Reinforcement	Maximum Fatigue Stress (MPa)
F1	2524-T3	7150-T77511	No	117
F3	2199-T8E74	2099-T83	No	117
F5	2199-T8E74	2099-T83	Bonded FML straps under stringers	147 (+ 25%)

Source: Heinimann et al. (2007).

$R = 0.1$. Three of the panel configurations are listed in Table 15.3. The baseline F1 panel used conventional aluminum alloys. The F3 and F5 panels used the AA 2099 and AA 2199 third-generation Al–Li alloys and were about 5% lighter than the F1 panel. Also, the F5 panel was reinforced by unidirectional FML straps under the stringers. The FML was a type of GLARE (GLAss REinforced aluminum laminates), which is discussed in Section 15.3 also.

Each panel had five Z-stringers and was tested with a completely severed central stringer and a visually detectable fatigue precrack in the skin. The test set-up is shown in Figure 15.4A, which gives an impression of the panel size (2030 × 762 mm).

Figure 15.5 presents the fatigue crack growth results. Both of the Al–Li panels gave significant improvements in crack growth life, with a dramatic improvement for the selectively reinforced F5 panel, even though the stress level for F5 was 25% higher. The selective reinforcement concept therefore provides an opportunity for significant weight savings, though there are, of course, extra production steps that would increase the commercial manufacturing costs.

2. *Lower wing panels*: Mechanically fastened and integral panels representing inboard lower wing skins were tested under uniaxial gust spectrum flight simulation loading with a baseline mean stress of 83 MPa. Five of the panel configurations are listed in Table 15.4. The two baseline panels W1 and W2 used conventional advanced aluminum plate and extrusion alloys. The W4 panel used the third-generation Al–Li alloy AA 2099 extrusions for stringers and a conventional advanced plate alloy for the skin, and the integral W7 and W8 panels were made entirely of 2099 extrusions. The fully Al–Li panels were about 5% lighter than the baseline W1 and W2 panels. In addition, the W4 and W8 panels were reinforced by unidirectional FML (GLARE) straps under (W4) or between (W8) the stringers.

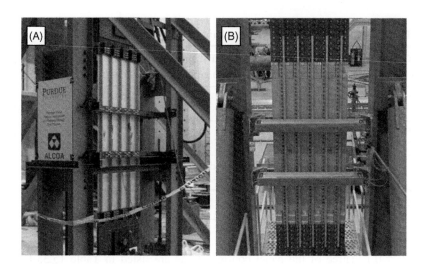

FIGURE 15.4 (A) Fuselage crown panel test set-up and (B) lower wing panel test set-up (Heinimann et al., 2007).

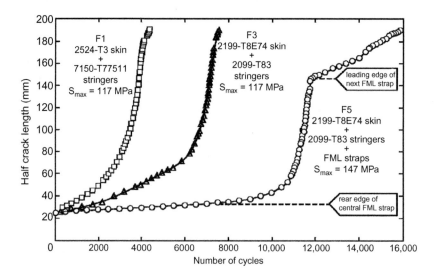

FIGURE 15.5 Constant amplitude fatigue crack growth results for fuselage crown panels. *Source: After Heinimann et al. (2007).*

As before, each panel had five Z-stringers and was tested with a completely severed central stringer and a visually detectable fatigue precrack in the skin. The test set-up is shown in Figure 15.4B: the panel size was 2280 × 762 mm.

TABLE 15.4 Lower Wing Skin–Stringer Panel Concepts

Concept	Skin Material	Stringer Material	Panel Configuration	Selective Reinforcement
W1	2624-T351	2026-T3511	Mechanically fastened	No
W2	2624-T39	2224-T3511		No
W4	2624-T39	2099-T81		Bonded FML straps under stringers
W7	2099-T81	2099-T81	Integral skin–stringer extrusions joined by FSW	No
W8	2099-T81	2099-T81		Bonded FML straps between stringers

Source: Heinimann et al. (2007).

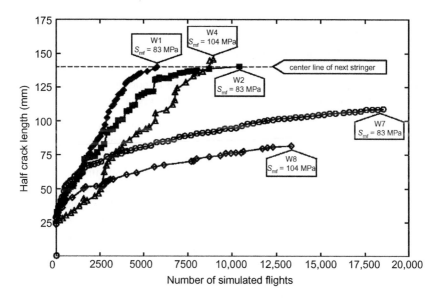

FIGURE 15.6 Gust spectrum flight simulation fatigue crack growth results for lower wing panels, after Heinimann et al. (2007). The flight-by-flight gust load history is after MINITWIST (Lowak et al., 1979): S_{mf}, mean stress in flight.

Figure 15.6 gives the fatigue crack growth results. At the same stress levels the W7 Al–Li panel was a significant improvement on the baseline W1 and W2 panels. At 25% higher stress levels the W4 selectively reinforced panel results were equivalent to those of the baseline panels, which is a good result; however, the crack growth resistance of the W8 panel was outstanding. Again it is seen that the selective reinforcement concept provides opportunities for significant weight savings.

15.3 MATERIALS SELECTION

As mentioned in the introduction to this chapter, third-generation Al–Li alloys have to compete with conventional aluminum alloys, carbon fiber composites (i.e., CFRPs), and FMLs for aerospace applications, particularly in transport aircraft structures. The FMLs currently to be considered, because they are the most developed and used, have the generic name GLARE. The "competition" between all the materials involves many other factors besides weight savings. These factors, classified and considered as relative advantages and disadvantages, are discussed in the following section.

15.3.1 Aluminum Alloys, CFRPs, and FMLs: Advantages and Disadvantages

Table 15.5 lists the factors considered as relative advantages and disadvantages of conventional and third-generation Al–Li alloys, CFRPs, and GLARE for application in aircraft structures. The importance of these factors differs greatly, but some general indications can be given:

1. *Engineering properties*: These must obviously meet the aircraft load requirements, which vary for different structural areas. Examples are shown in Figure 15.7.
2. *Improved material and engineering properties*: The importance of improved properties, or combinations of properties, depends on both the structural load requirements and the potential for weight savings. As discussed in Section 15.2, the greatest weight savings potentials come from low-density materials. This is basically why CFRPs are so attractive, and also why Al–Li alloys are candidates for replacing conventional aluminum alloys.

    Other properties, especially strength and stiffness (elastic modulus), also contribute to weight savings. For aluminum alloy sheet and plate, the corrosion resistance is particularly important: the use of Al–Li alloys may obviate the need for low-strength cladding layers that prevent corrosion of conventional AA 2XXX and 7XXX alloys, but add extra weight. Another example is the use of GLARE instead of aluminum alloy sheet. GLARE has excellent DT properties that enable weight savings in fatigue critical areas like the upper fuselage.

TABLE 15.5 Relative Advantages and Disadvantages of Conventional Aerospace Aluminum Alloys, Third-Generation Al–Li Alloys, CFRPs, and GLARE FMLs for Aircraft Structures

Conventional Al Alloys	Third-Generation Al–Li Alloys	CFRPs	GLARE
Advantages • moderate material, labor, and manufacturing costs • moderate specific stiffness • isotropic mechanical properties • good DT and property control by thermomechanical processing • generally recyclable **Disadvantages** • poor corrosion resistance • stress corrosion susceptibilities • most alloys difficult or unsuitable to weld except with non-fusion techniques (e.g., FSW and linear friction welding (LFW))	**Advantages** • moderate labor and manufacturing costs • 8–15% higher specific stiffness than conventional aluminum alloys and weldable: weight savings and reduced part count • isotropic and improved mechanical properties • good DT and property control by thermomechanical processing • good/excellent corrosion and stress corrosion resistance: sheet cladding may be unnecessary • recyclable, *but see below* **Disadvantages** • higher material costs compared to conventional aluminum alloys • property dependences on multistage thermomechanical treatments • separate recycling	**Advantages** • higher and much higher specific stiffness, depending on percentages of aligned fibers in actual components (greatly) reduced part count • 10–20% weight savings in actual components • high fatigue and corrosion resistance: reduced maintenance costs • greater flexibility in designing structurally efficient components **Disadvantages** • high material, labor and manufacturing costs • possible delaminations and other flaws during fabrication • intrinsically anisotropic: complex components difficult to analyze, sometimes giving poor failure predictions • higher notch sensitivity (e.g., fastener holes) • non-destructive inspection difficult • conservative design and safety factors owing to: (i) high susceptibility to damage from impacts (bird strikes and ground operation vehicles), (ii) damage growth difficult to control and predict, and (iii) difficult validation of repairs • low electrical conductivity requires metal (Cu) mesh in external surface layers to protect from lightning strikes • flammable and non-recyclable	**Advantages** • fabricability generally similar to aluminum alloys, *but see below* • high strength: weight savings in tension-loaded structures • excellent DT (fatigue crack growth, impact) • burn-through (firewall) resistance **Disadvantages** • high material and manufacturing costs • difficult to form • only sheet products (skins, stringers, butt straps) at present • lower buckling resistance: unsuitable for compression-loaded structures

Source: Information from various sources, including Mouritz (2012).

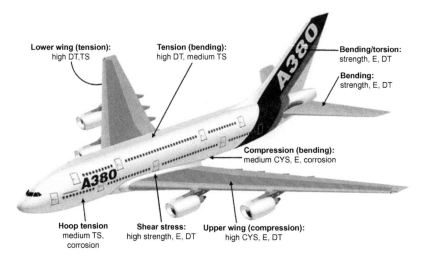

FIGURE 15.7 Typical load conditions and engineering property requirements for main structural areas in a transport aircraft (Airbus A380): CYS, compressive yield strength; E, elastic modulus; TS, tensile strength; DT, damage tolerance properties (fatigue, fatigue crack growth, fracture toughness).

3. *Costs and weight savings*: Material and manufacturing costs are relatively high for CFRPs and GLARE, but can be offset by their weight savings potentials. In fact, costs and weight savings are inextricably linked through the effects that weight savings have on fuel savings and other operating costs. Thus all material choices are strongly driven by the combination of costs and weight savings.

4. *Design principles and safety*: Not only must aircraft have lightweight and highly efficient structures, but they must also be safe, as is discussed in Chapter 16. The design principles developed for the aircraft industry are based on mature engineering experience and well-established safety factors for metallic aircraft structures. This is an important advantage for aluminum alloys and also GLARE, which can be treated basically like a metallic material (Beumler, 2004).

CFRPs, however, cannot use the design principles for metallic aircraft structures. In addition to problems in analyzing complex components and predicting the onset of failure, even in nominally undamaged components, CFRPs are susceptible to impact damage and subsequent fatigue cracking and delamination. The growth of this damage is difficult to predict, and this also makes it difficult to validate repairs. The overall result of these problems is that safety must be ensured by over-designing CFRPs according to the "no growth" DT principle (Mouritz, 2012). This increases their weight and manufacturing costs.

15.3.2 Materials for Aircraft Structures

The selection of different classes of aircraft structural materials depends on (i) the types of aircraft, (ii) which components and parts of the overall structure are under consideration, and (iii) the material properties, including their advantages and disadvantages. Examples for points (ii) and (iii) are given in the previous section, which is particularly relevant to transport aircraft.

The choices of material classes and amounts for different types of aircraft vary greatly. Aluminum alloys still account for about 60% of the structural weight of transport aircraft, with the notable exception of the Boeing 787 *Dreamliner*, which uses 50% composites and only 20% aluminum. For high-performance military aircraft, the emphasis is more on composites and lesser percentages of titanium and aluminum alloys and steels. For example, the British Aerospace/DASA/CASA/Alenia Eurofighter and Lockheed Martin JSF use 35–40% composites, most of which are used for the external skin and substructure.

Even higher percentages of composites may be used for helicopters, for which weight savings are most important. The Bell-Boeing V-22 Osprey airframe consists of nearly 80% composites (Deo et al., 2003), and the NH Industries NH90 airframe is over 90% composites. However, hybrid structures using metallic (aluminum alloy) frames and composite skins could be good choices for future helicopters. An example from the present generation of helicopters is shown in Figure 15.8: highly loaded structural components, particularly if they have complex geometries like the main cabin frames, are prime candidates for the continuing use of aluminum alloys.

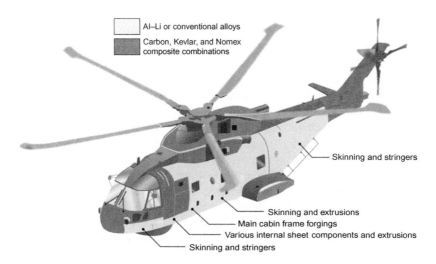

FIGURE 15.8   AgustaWestland EH101 hybrid metal/composite airframe.

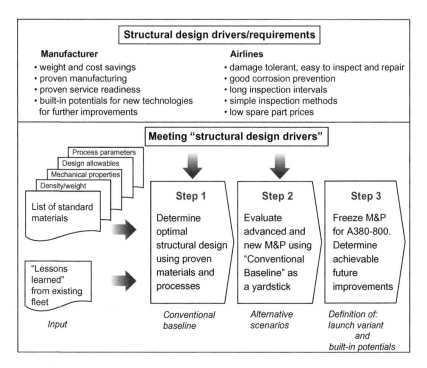

FIGURE 15.9 Materials and processes technology selection for the Airbus A380. *Source: After Hinrichsen (2003).*

Actual selection of the materials, and their associated process technologies, is a complex task. A schematic of how Airbus made these selections for the A380 is given in Figure 15.9. This diagram is more or less self-explanatory—at least as an overview. However, as Giummarra et al. (2007) have pointed out, a direct comparison of material properties between aluminum alloys and composites is not possible. The properties and design principles are very different, and detailed analyses and (iterative) trade studies must be made. These studies are used in conjunction with manufacturing evaluations, supply chain strategy, and life cycle cost models to make a final decision on material choice (Giummarra et al., 2007).

The results of some of the A380 selection processes are shown in Figure 15.10. The most notable and important general aspect of these selections is the result of a hybrid airframe structure using conventional AA 2XXX, 6XXX, and 7XXX aluminum alloys; the third-generation Al–Li alloys 2050, 2099, and 2196; large CFRP components and empennage; and GLARE for parts of the upper fuselage. Hybrid airframe structures may well become the norm for future transport aircraft as well as helicopters, though the relative amounts of metals and composites are likely to vary, depending

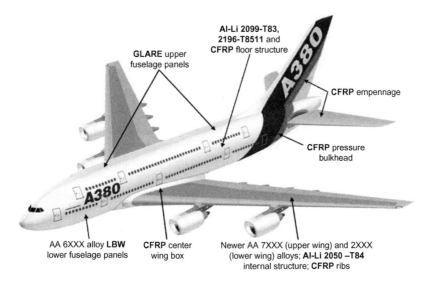

GLARE upper
fuselage panels

**Al-Li 2099-T83,
2196-T8511 and
CFRP** floor structure

CFRP empennage

CFRP pressure
bulkhead

AA 6XXX alloy **LBW**
lower fuselage panels

CFRP center
wing box

Newer AA 7XXX (upper wing) and 2XXX
(lower wing) alloys; **Al-Li 2050 −T84**
internal structure; **CFRP** ribs

FIGURE 15.10 Advanced materials/processes for major structural areas of the Airbus A380.
Source: Data from various sources, mainly Airbus SAS.

on aircraft sizes. Smaller transport aircraft in, say, the Airbus A320 and
Boeing 737 category, may continue to have basically all-metal airframes,
since down-scaling composite structures, which are relatively costly to manufacture, could be uneconomic.

15.3.3 Aluminum Alloys for Spacecraft Structures

Aluminum alloys have been used for launch and space vehicles and satellite
structures since the beginning of the Space Age in the late 1950s. The selection of aluminum alloys has been based on their relatively high structural
efficiency, the ability to withstand the loads and stresses during launch and
space operation, excellent fabricability, and moderate costs. The main applications have been for cryogenic propellant tanks, and this is still the case.

Until the late 1990s, all the aluminum alloys were conventional. For example, AA 2219 and 7075 were used in much of the Apollo Lunar Module structure (Weiss, 1973); AA 2024, 2124 and 2219 were used for parts of the Space
Shuttle Orbiter fuselage, wings and vertical tail (Liu, 1974); and conventional
AA 2XXX and 7XXX alloys are used in satellites (Köble, 2003).

The AA 2219 alloy is particularly important. It has been the mainstay for
tankage (Srinivasa Rao et al., 1996), since it is weldable and has good
strength and fracture toughness at liquid hydrogen and oxygen temperatures.
Because of these properties, 2219 became the baseline for developing Al—Li
alloy propellant tanks, resulting firstly in the Weldalite™ alloy family

(Pickens et al., 1989; Rioja et al., 1990). Continued development led to the weldable third-generation Al–Li alloy AA 2195, which was introduced in 1992 (Table 15.1). This alloy was subsequently used for the Super Lightweight External Tank (SLW ET), first launched in June 1998. Direct substitution of 2195 for 2219 reduced the tank weight by 3175 kg, a weight savings of over 10%. This excellent result was achieved by exploiting 2195's advantageous combination of reduced density, increased stiffness, and higher strength.

The design load drivers for spacecraft structures include static, dynamic, and thermal loads and stresses. The cyclic load histories can be very complicated, consisting of handling and ground loads; random and acoustic vibrations during launch; in-flight dynamic loads, including shocks due to ignition cutoff and stage separation; and more dynamic loads and vibrations for re-entry vehicles.

The material and engineering property requirements include (i) high structural efficiency based on density, strength, stiffness (elastic modulus), and careful design, (ii) good DT properties, also at cryogenic temperatures for tankage and pressure vessel materials, (iii) reliability, i.e. proven service readiness, and (iv) moderate material and manufacturing costs. Apart from the cryogenic property requirements, this list of properties is similar to what is expected of aircraft structures.

There are, however, two important differences in the design requirements for spacecraft and aircraft structures:

1. *Non-inspectability*: Spacecraft structures are essentially non-inspectable in service. There are generally no opportunities for maintenance, repair, or replacement. This, and the need for high structural efficiency, has an impact on the *fracture control methodology*. Most space structures are non-redundant, i.e., there are no alternative load paths. In terms of fracture control this means that structural safety must be ensured by demonstrating that (assumed) initial discontinuities/defects at critical locations will not grow by fatigue, or some other cracking mode, to cause fast fracture and failure during the service life. This is called the "safe life" approach.

   On the other hand, many aircraft structures, especially in commercial transport aircraft, are designed to be "fail-safe," whereby partial structural failure can be sustained without immediately leading to complete failure. As mentioned in Section 15.2.3, this is achieved by designing structures to have (i) multiple load paths consisting of discrete structural elements, (ii) residual strengths meeting established requirements if failure of an element occurs, and (iii) *good inspectability* to allow detection of element damage or failure before safety is compromised.

2. *Stress corrosion cracking (SCC)*: This is considered a major hazard for space structures, owing to numerous problems encountered in the early 1960s during preservice testing and verification, and also later on (Johnson, 1973; Korb, 1987; Weiss, 1973; Williamson, 1969). The space

agencies ESA (2008, 2009) and NASA (2005) have stringent design requirements that can cause problems in selecting suitable structural materials. This is discussed in Section 15.4.2.

15.3.4 Materials Qualification: an Example

Qualification of an alloy, or alloys, for aerospace structures involves many considerations relating to the engineering properties and also the manufacturing of components and complete structures. As an example, Table 15.6 outlines a service qualification program for an Al–Li sheet/plate alloy (Wanhill, 1994) with some updates (Babel and Parrish, 2004; Lequeu et al., 2010; Pacchione and Telgkamp, 2006; Rioja and Liu, 2012). Another example of a qualification program is included in Chapter 16, which is a review of the procedures for certifying the metallic materials to be used in aerospace structures. A very important certification document for material allowables is MMPDS, formerly known as MIL-HDBK-5, e.g. Niedzinski et al. (2013a).

One consideration in Table 15.6 is particularly relevant to spacecraft structures, namely the SCC resistance in thicker sections. Although third-generation Al–Li alloys have generally good-to-excellent SCC resistance (Denzer et al., 2012; Henon and Rouault, 2012; Karabin et al., 2012; Rioja and Liu, 2012) that considerably exceeds that of conventional AA 2XXX plate alloys, it remains an open question, despite the best efforts of the alloy producers, whether high SCC resistance can be guaranteed in sections thicker than about 30 mm. This is briefly considered in Chapter 14, section 14.3.2; and also in Chapter 7, sections 7.6.1 and 7.7.2.

15.4 APPLICATIONS OF Al–Li ALLOYS (THIRD GENERATION)

15.4.1 Aircraft

The history of Al–Li alloys and the problems encountered with the first- and second-generation alloys have been discussed in several chapters of this book. As mentioned at the beginning of this chapter, it is only since the 1990s that the fundamental understanding of Al–Li alloys has matured and enabled the development of a family of alloys with excellent combinations of engineering properties, the third-generation Al–Li alloys. Most of these alloys have been introduced since the year 2000 (Table 15.1), and their development continues. However, some of the earlier ones have already been used in military aircraft. These include AA 2297 and AA 2397, which have been used for Lockheed Martin F-16 bulkheads and other parts (Acosta et al., 2002; Balmuth, 2001).

Over the last decade, several third-generation alloys have been qualified for applications in commercial transport aircraft, notably the Airbus A380, and there are ongoing developments (Figure 15.11; Table 15.7) which have

TABLE 15.6 Example of a Qualification Program for an Al–Li Sheet/Plate Alloy

Properties	Property Details	Special Considerations
Mechanical	• TYS, CYS, TS, E • Shear and bearing strengths	• Multiangle (texture and anisotropy effects) • Temperature: 225–424 K; cryogenic (space)
Fatigue strength	• HCF of notched coupons and structural joints • CA and VA load histories	• Texture and anisotropy effects • Environmental effects • Corrosion protection (coatings and primers)
Fatigue crack growth	• Long and short cracks • CA and VA load histories	• Texture and anisotropy effects • Environmental effects • Modeling and prediction of crack growth
Fracture	• Plane strain/plane stress fracture toughness • K_R and R-curves • Residual strength of stiffened panels	• Texture and anisotropy effects • Temperature: 225–295 K; cryogenic (space) • Thermal stability during service • Possible dynamic effects on toughness
Corrosion	• Pitting and exfoliation • Stress corrosion (SCC) • Accelerated and natural environments	• Intergranular attack • Microbiological attack in fuel tanks • Structural joints • SCC resistance in thicker sections
Manufacturing and process technologies	• Forming: stretch and spin forming • Machining, cutting, hole drilling • Mechanical fastening: rivets and bolts • Welding: FSW and LBW • Surface treatments: chemical milling, etching, anodizing, priming, adhesive bonding	• Texture and anisotropy • SPF • Combinations of forming and heat treatment • Automated 3D welding; post-weld heat treatments and stress relief

HCF, high cycle fatigue; CA, constant amplitude; VA, variable amplitude; FSW, friction stir welding; LBW, laser beam welding.

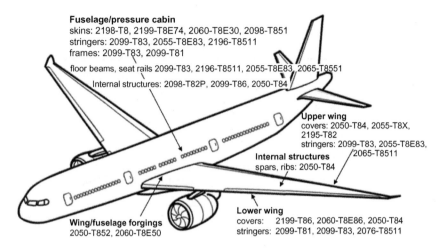

Fuselage/pressure cabin
skins: 2198-T8, 2199-T8E74, 2060-T8E30, 2098-T851
stringers: 2099-T83, 2055-T8E83, 2196-T8511
frames: 2099-T83, 2099-T81
floor beams, seat rails 2099-T83, 2196-T8511, 2055-T8E83, 2065-T8551
Internal structures: 2098-T82P, 2099-T86, 2050-T84

Upper wing
covers: 2050-T84, 2055-T8X, 2195-T82
stringers: 2099-T83, 2055-T8E83, 2065-T8511
Internal structures
spars, ribs: 2050-T84

Wing/fuselage forgings
2050-T852, 2060-T8E50

Lower wing
covers: 2199-T86, 2060-T8E86, 2050-T84
stringers: 2099-T81, 2099-T83, 2076-T8511

FIGURE 15.11   Proposed use of third-generation Al—Li alloys for main structural areas in a transport aircraft.

been compiled from several sources (Daniélou et al., 2012; Denzer et al., 2012; Giummarra et al., 2007, 2008; Karabin et al., 2012; Lequeu, 2008; Lequeu et al., 2007, 2010; Magnusen et al., 2012; Pacchione and Telgkamp, 2006; Rioja and Liu, 2012; Tchitembo Goma et al., 2012) and Alcoa in-house knowledge.

Alloy producers develop basically similar alloys for different product forms and applications. Data given in Table 15.7 shows that the potential exists for replacing virtually all the current conventional alloys by Al—Li alloys, and this is emphasized by Figure 15.11. The most important contribution to this potential is the development of a range of alloy tempers that allow optimizations and trade-offs of properties, and hence affords considerable flexibility in matching the alloys to particular applications. Examples of property developments and trade-offs for fuselage and wing structures are discussed in detail in Chapter 2, Section 2.5.

Another important aspect is the generally good-to-excellent corrosion resistance (Denzer et al., 2012; Henon and Rouault, 2012; Karabin et al., 2012; Rioja and Liu, 2012). This is especially of interest for sheet and (thin) plate applications since, as mentioned in Section 15.3.1 and Table 15.5, the use of Al—Li alloys may obviate the need for low-strength cladding layers that prevent corrosion of conventional AA 2XXX and 7XXX alloys, but add extra weight. An additional disadvantage of cladding is that it decreases the fatigue durability of mechanically fastened joints (e.g. riveted lap splices (Wanhill, 1986)).

The Airbus A380 has led the use of third-generation Al—Li alloys in transport aircraft, with AA 2099-T83 and 2196-T8511 floor beams, 2196-T8511 for fuselage stringers, and AA 2050-T84 for wing internal structure. However, as mentioned in Section 15.3.2 and illustrated in Figures 15.10 and

TABLE 15.7 Actual and Proposed Uses of Third-Generation Al—Li Alloys to Replace Conventional Alloys in Aircraft

Product	Alloy/Temper	Substitute for:	Applications
Sheet	2098-T851, 2198-T8, 2199-T8E74, 2060-T8E30: damage tolerant/medium strength	2024-T3, 2524-T3/351	Fuselage/pressure cabin skins
Plate	2199-T86, 2050-T84, 2060-T8E86: damage tolerant	2024-T351, 2324-T39, 2624-T351. 2624-T39	Lower wing covers
	2098-T82P (sheet/plate): medium strength	2024-T62	F-16 fuselage panels
	2297-T87, 2397-T87: medium strength	2124-T851	F-16 fuselage bulkheads
	2099-T86: medium strength	7050-T7451, 7X75-T7XXX	Internal fuselage structures
	2050-T84, 2055-T8X, 2195-T82: high strength	7150-T7751, 7055-T7751, 7055-T7951, 7255-T7951	Upper wing covers
	2050-T84: medium strength	7050-T7451	Spars, ribs, other internal structures
Forgings	2050-T852, 2060-T8E50: high strength	7175-T7351, 7050-T7452	Wing/fuselage attachments, window and crown frames
Extrusions	2099-T81, 2076-T8511: damage tolerant	2024-T3511, 2026-T3511 2024-T4312, 6110-T6511	Lower wing stringers, fuselage/pressure cabin stringers
	2099-T83, 2099-T81, 2196-T8511, 2055-T8E83, 2065-T8511: medium/high strength	7075-T73511, 7075-T79511 7150-T6511, 7175-T79511, 7055-T77511, 7055-T79511	Fuselage/pressure cabin stringers and frames, upper wing stringers, Airbus A380 floor beams and seat rails

15.12, the A380 has a hybrid airframe structure using conventional aluminum alloys, Al—Li alloys, CFRP, and GLARE.

The A380 example in Figure 15.12 is especially interesting, as it illustrates different material choices for nominally identical applications. However, there are important differences. To maximise usable space, the upper deck floor must span the entire fuselage without the intermediate supports commonly used for

FIGURE 15.12 Al–Li alloy and CFRP floor beams in the Airbus A380 (see Figure 15.10 also). *Source: Image source is from Airbus (public domain).*

lower/main deck flooring. This puts great demands on the floor beam stiffness, and the higher specific stiffness attainable with CFRPs, e.g. Figure 15.2, favours their use. On the other hand, the lower/main deck beams *can* be supported, as may be seen in Figure 15.12. This has enabled Al-Li AA 2196-T8511 extrusions (Lequeu et al., 2007) to be competitive for the floor beams, in a trade-off between the engineering and material properties, the resulting structural weight, and the manufacturing costs.

The use of third-generation Al–Li alloys for smaller transport aircraft, helicopters, and high-performance military aircraft is not yet established. The aluminum industry is making great efforts to develop and offer competitive materials and prices, and so it is likely that they will be used. Nevertheless, the competition between different material classes has reached a development stage where hybrid airframe structures may become the rule rather than the exception.

15.4.2 Spacecraft

To date, there is limited information, mostly concerning the launchers, about the use of Al–Li alloys in contemporary space programs. The programs for which Al–Li alloys have been proposed are (i) the *canceled* NASA Constellation program, consisting of the Ares I Crew Launch Vehicle, including the Orion crew and service modules, and the Ares V Heavy Cargo Launch Vehicle; and (ii) the NASA heavy-lift Space Launch System (SLS) which, as the successor to Constellation, will be able to take the Orion. The application of Al–Li alloys for future cryogenic tanks has also been studied by ESA (Windisch, 2009, 2013). The programs actually using Al—Li alloys are the Orion (NASA, 2010; Niedzinski et al., 2013b) and the SpaceX Falcon 9 launcher (Niedzinski and Thompson, 2010).

Before discussing individual programs, it is worth emphasizing that FSW is a key technology for manufacturing Al—Li alloy components and structures for spacecraft, especially the tanks and tank domes. As mentioned in Section 15.2.3, FSW has been used for (i) joining Al—Li components of the Orion Crew Module, including a final 11.3 m circumferential weld joining the forward cone assembly and crew tunnel to the aft assembly (NASA, 2010), and (ii) manufacturing Al—Li AA 2198 first- and second-stage fuselages of the SpaceX Falcon 9 (Niedzinski and Thompson, 2010). FSW has also been developed for manufacturing spin-formed Al—Li AA 2195 propellant tank domes (Hales et al., 2012; Windisch, 2009) and is currently (2013) planned to be introduced for joining parts of the conventional AA 2219 alloy main tank of the Ariane 5 launcher.

Constellation

1. *Ares I and Ares V*: Expanded views of the Ares I and Ares V concepts are shown in Figure 15.13. The Ares I upper stage was to have had an Al—Li tank and internal structures. The Ares V core and earth departure stages would also have had Al—Li tanks. The intention was to use AA 2195 plate for the tanks (Rioja and Liu, 2012; Rioja et al., 2012). However, there was already an initiative to evaluate Al—Li AA 2050 plate as a replacement for 2195 in the Ares V core stage (NASA, 2009).
2. *Orion*: This has a Crew Module and a Service Module, to be seen in the Ares I illustration in Figure 15.13. Both use Al—Li alloys: the CM structure uses AA 2195-T8511 extrusions for the main load-bearing longerons, and AA 2050-T84 plate for other components, including frames, ribs, and window sections. FSW is used extensively throughout the structure.

Space Launch System (SLS)

This heavy-lift launch vehicle system is intended to carry the Orion as well as cargo, equipment and science experiments beyond Earth orbit for long-duration deep space missions, and also to provide backup transportation, if needed, to the International Space Station (NASA, 2012a). The SLS will have multiple launch configurations, illustrated in Figure 15.14.

The core stage, which is common to all configurations, is a modification of the Space Shuttle External Tank and was initially intended to be fabricated using Al—Li AA 2195 plate. However, the core stage has to carry

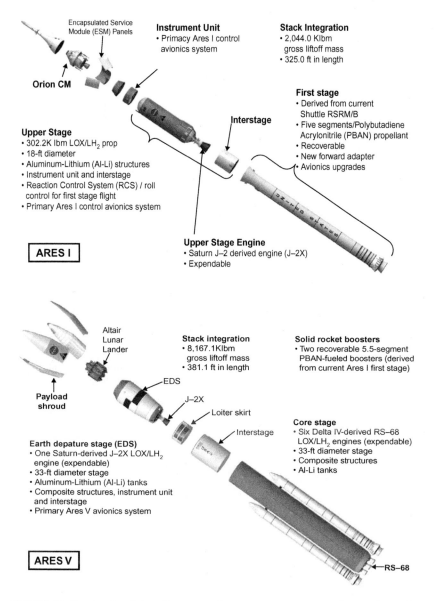

Encapsulated Service Module (ESM) Panels

Instrument Unit
• Primacy Ares I control avionics system

Stack Integration
• 2,044.0 Klbm gross liftoff mass
• 325.0 ft in length

Orion CM

First stage
• Derived from current Shuttle RSRM/B
• Five segments/Polybutadiene Acrylonitrile (PBAN) propellant
• Recoverable
• New forward adapter
• Avionics upgrades

Interstage

Upper Stage
• 302.2K lbm LOX/LH$_2$ prop
• 18-ft diameter
• Aluminum-Lithium (Al-Li) structures
• Instrument unit and interstage
• Reaction Control System (RCS) / roll control for first stage flight
• Primary Ares I control avionics system

Upper Stage Engine
• Saturn J–2 derived engine (J–2X)
• Expendable

ARES I

Altair Lunar Lander

Stack integration
• 8,167.1Klbm gross liftoff mass
• 381.1 ft in length

Solid rocket boosters
• Two recoverable 5.5-segment PBAN-fueled boosters (derived from current Ares I first stage)

Payload shroud

EDS

J–2X

Loiter skirt

Interstage

Core stage
• Six Delta IV-derived RS–68 LOX/LH$_2$ engines (expendable)
• 33-ft diameter stage
• Composite structures
• Al-Li tanks

Earth depature stage (EDS)
• One Saturn-derived J–2X LOX/LH$_2$ engine (expendable)
• 33-ft diameter stage
• Aluminum-Lithium (Al-Li) tanks
• Composite structures, instrument unit and interstage
• Primary Ares V avionics system

ARES V

RS–68

FIGURE 15.13   Ares I and Ares V concepts. *Source: Image sources are from NASA (public domain).*

much higher loads than the Space Shuttle External Tank and needs a thicker and more ductile plate material for tank fabrication. In the first instance, this has resulted in a change back to the mainstay tankage alloy AA 2219: later it may be possible to use Al–Li AA 2050 plate in an upgrade (Payne, 2013).

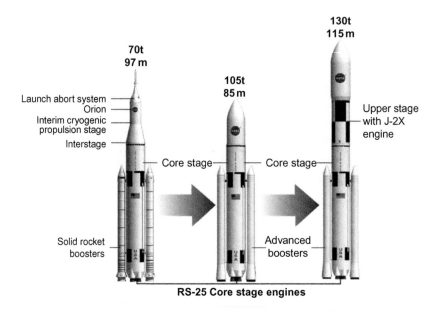

FIGURE 15.14 SLS concepts. *Source: Image source is from NASA (public domain).*

In this respect, it is notable that NASA has been evaluating AA 2050-T84 thick plate for tankage use since 2009 (Hafley et al., 2011).

The upper stage is less highly loaded, such that thinner gauge materials can be used. The selected tankage plate material is again the Al–Li AA 2195 alloy (Niedzinski et al., 2013b).

SpaceX Falcon 9 Launcher

The SpaceX Falcon 9 launcher is part of a major commercial initiative, under the aegis of NASA, for cost-efficient transport of satellites to low Earth orbit and geosynchronous transfer orbit, as well as sending the SpaceX Dragon cargo and crew spacecraft to the International Space Station (NASA, 2012b). The complete assembly of the Falcon 9 and Dragon is approximately 48 m high.

The Falcon 9 is a two-stage liquid propellant rocket 3.66 m in diameter. The tanks of both stages are fabricated from Al–Li AA 2198 sheet and plate using FSW (Niedzinski and Thompson, 2010). An as-manufactured first-stage tank barrel and FSW + spin-formed dome are shown in Figure 15.15. The choice of 2198 was based on its advantages in specific stiffness and compressive strength with respect to the "legacy" conventional alloys AA 2219 and AA 2014, and there are also advantages in toughness and fatigue and corrosion resistance (Niedzinski and Thompson, 2010).

FIGURE 15.15   Al—Li alloy AA 2198 FSW first-stage barrel and spin-formed dome for the Falcon 9. *Source: Image source from C. Thompson, SpaceX, and M. Niedzinski, Constellium.*

Some Additional Remarks on Alloy Property Requirements for Spacecraft

Two property requirements for spacecraft structural alloys that stand out as very important are high structural efficiency and SCC resistance:

1. *Structural efficiency*: As mentioned in Section 15.3.3, the structural efficiency is based on density, strength, stiffness (elastic modulus), and careful design. The most general combined property is the specific stiffness (modulus/density). Specific strength (strength/density) is also important, but more variable, and also not determining for the buckling resistance under compression.

   Table 15.8 compares the specific stiffnesses of several third-generation Al—Li alloys with that of the conventional mainstay spacecraft tankage alloy AA 2219. All the third-generation alloys have significantly higher specific stiffnesses than AA 2219, and hence all are candidates to replace AA 2219. The actual selections will depend very much on other properties, including the strength, cryogenic fracture toughness, fatigue crack growth, corrosion and SCC resistances of *both base and FSW materials* (Windisch, 2009). The determination of all these properties is a lengthy and costly process.

2. *SCC resistance*: ESA (2008, 2009) and NASA (2005) attach great importance to the SCC resistance of metallic materials; and rightly so, since there have been many SCC failures in the aerospace industry (Johnson, 1973; Korb, 1987; Weiss, 1973). However, the ESA and NASA system of classifying materials into high, moderate, and low SCC resistance has deficiencies that can work to the disadvantage of new materials, including Al—Li alloys. This classification system is under review to eliminate these deficiencies (Niedzinski et al., 2013b).

TABLE 15.8 Actual and Proposed Uses of Third-Generation Al–Li Alloys to Replace Conventional Alloys in Spacecraft

Alloys	Current or Proposed Uses	Density ρ (g/cm^3)	Compression Modulus (GPa)	Specific Stiffness GPa/(g/cm^3)
2219-T851	Conventional alloy: Ariane V, SLS	2.85	74.5	26.1
2050-T84	Thick plate: Orion, SLS	2.70	77.9	28.85
2098-T82	Sheet and thin plate: Falcon 9	2.70	79.0	29.3
2195-T82	Thin-to medium gauge plate: SLS	2.71	78.6	29.0
2297-T87	Thick plate: SLS	2.65	77.2	29.1
2055-T8X	—	2.70	78.5*	29.1
2099-T86	—	2.63	79.3	30.15

*Provisional value.
Source: After Rioja et al. (2012) and Niedzinski et al. (2013b). Data sourced from MMPDS-07 (2012).

15.5 SUMMARY AND CONCLUSIONS

This chapter has focussed on the material and manufacturing property requirements for selection and application of third-generation Al–Li alloys in aircraft and spacecraft. Modern structural concepts using LBW, FSW, SPF, and selective reinforcement by FMLs are mentioned; and it is evident that FSW is now a major fabrication process, especially for spacecraft.

Al–Li alloys have to compete with conventional aluminum alloys, CFRPs, and GLARE, particularly for transport aircraft structures. The third-generation Al–Li alloys are therefore compared with conventional aluminum alloys, CFRPs, and GLARE before discussing the materials selection processes for aircraft. This is followed by a review of the aluminum alloy selection process for spacecraft.

Actual and potential applications of third-generation Al–Li alloys are presented. For aircraft it is concluded that the competition between different material classes (aluminum alloys, CFRPs, and FMLs) has reached a development stage where hybrid structures may become the rule rather than the exception. However, aluminum alloys are still the main contenders for spacecraft liquid propellant launchers.

This chapter is complementary to Chapter 2, which discusses the engineering property requirements for using Al–Li alloys in aircraft, especially the proposed uses of third-generation alloys in commercial transport aircraft.

ACKNOWLEDGMENTS

It is a particular pleasure to acknowledge the advice and support of Dr. Gary Bray (Alcoa Technical Center, Alcoa Center, PA), who has made many expert contributions to several chapters of this book, including this one. The assistance of Michael Niedzinski (Constellium, Rosemont, IL) is also much appreciated.

REFERENCES

Acosta, E., Garcia, O., Dakessian, A., Aung Ra, K., Torroledo, J., Tsang, A., et al., 2002. On the effect of thermomechanical processing on the mechanical properties of 2297 plates. Mater. Sci. Forum 396–402, 1157–1162.

Arbegast, W.J., Hartley, P., 1999. Friction stir weld technology development at Lockheed Martin Michoud space system—an overview. In: Vitek, J.M., David, S.A., Johnson, J.A., Smartt, H. B., DebRoy, T. (Eds.), Proceedings of the International Conference on Trends in Welding Research. ASM International, Materials Park, OH, pp. 541–546.

Babel, H., Parrish, C., 2004. Manufacturing considerations for Al–Li alloys. Technical Paper 03H0108 in WESTEC 2004 Conference Proceedings. Society of Manufacturing Engineers, Dearborn, MI, Los Angeles, March 2004.

Balmuth, E.S., 2001. Application of aluminum alloy 2297 in fighter aircraft structures. In: Kim, N., Lee, C.S., Eylon, D. (Eds.), Proceedings of the International Conference on Light Materials for Transportation Systems (LiMAT–2001), vol. 2. Pohang Institute of Science and Technology, Pohang, Korea, pp. 589–596.

Beumler, T., 2004. Flying GLARE: A Contribution to Aircraft Certification Issues on Strength Properties in Non-Damaged and Fatigue Damaged GLARE Structures. Delft University Press, Delft, The Netherlands.

Daniélou, A., Ronxin, J.P., Nardin, C., Ehrstrom, J.C., 2012. Fatigue resistance of Al–Cu–Li and comparison with 7XXX aerospace alloys. In: Weiland, H., Rollett, A.D., Cassada, W.A. (Eds.), Proceedings of the 13th International Conference on Aluminum Alloys (ICAA13). The Minerals, Metals and Materials Society (TMS) and John Wiley & Sons, Hoboken, NJ, pp. 511–516.

Davis, B., Hryn, J., 2007. Innovative forming and fabrication technologies: new opportunities. Final Report, Energy Systems Division, Argonne National Laboratory, Chicago, IL.

Denzer, D.K., Rioja, R.J., Bray, G.H., Venema, G.B., Colvin, E.L., 2012. The evolution of plate and extruded products with high strength and toughness. In: Weiland, H., Rollett, A.D., Cassada, W.A. (Eds.), Proceedings of the 13th International Conference on Aluminum Alloys (ICAA13). The Minerals, Metals and Materials Society (TMS) and John Wiley & Sons, Hoboken, NJ, pp. 587–592.

Deo, R.B., Starnes Jr., J.H., Holzwarth, R.C., 2003. Low-cost composite materials and structures for aircraft applications. Low Cost Composite Structures and Cost effective Application of Titanium Alloys in Military Platforms, RTO Meeting Proceedings 69(II). NATO Research and Technology Organisation, Neuilly-sur-Seine, France, pp. (SMI) 1-1–1-11.

Dubois, T., 2007. New materials in business aircraft: composites, metals vie for supremacy. Aviat. Int. News 50–54, March 2007.

Enz, J., Riekehr, S., Ventzke, V., Kashaev, N., 2012. Influence of the local chemical composition on the mechanical properties of laser beam welded Al–Li alloys. Phys. Procedia 39, 51–58.

Ekvall, J.C., Rhodes, J.E., Wald, G.G., 1982. Methodology for evaluating weight savings from basic material properties. Design of Fatigue and Fracture Resistant Structures, ASTM STP 761. American Society for Testing and Materials, Philadelphia, PA, pp. 328–341.

Eswara Prasad, N., Gokhale, A.A., Rama Rao, P., 2003. Mechanical behaviour of aluminium–lithium alloys. Sādhanā 28 (1 & 2), 209–246.

European Space Agency (ESA), 2008. Space product assurance. Determination of the susceptibility of metals to stress-corrosion cracking. European Cooperation for Space Standardization ECSS-Q-ST-70-37C, ECSS Secretariat, ESA-ESTEC Requirements & Standards Division, Noordwijk, The Netherlands.

European Space Agency (ESA), 2009. Space product assurance. Material selection for controlling stress-corrosion cracking. European Cooperation for Space Standardization ECSS-Q-ST-70-36C, ECSS Secretariat, ESA-ESTEC Requirements & Standards Division, Noordwijk, The Netherlands.

Giummarra, C., Thomas, B., Rioja, R.J., 2007. New aluminum–lithium alloys for aerospace applications. In: Sadayappan, K., Sahoo, M. (Eds.), Proceedings of the Third International Conference on Light Metals Technology. CANMET, Ottawa, ON, Canada, pp. 41–46.

Giummarra, C., Rioja, R.J., Bray, G.H., Magnusen, P.E., Moran, J.P., 2008. Development of corrosion resistant, high toughness aluminum–lithium aerospace alloys. In: Hirsch, J., Skrotski, B., Gottstein, G. (Eds.), Proceedings of the 11th International Conference on Aluminum Alloys (ICAA11), vol. 1. Wiley-VCH Verlag GmbH & Co. KGaA, Weinheim, Germany, pp. 176–188.

Hafley, R.A., Domack, M.S., Hales, S.J., Shenoy, R.N., 2011. Evaluation of aluminum alloy 2050-T84 microstructure and mechanical properties at ambient and cryogenic temperatures. NASA Technical Memorandum NASA/TM-2011-217163. NASA Langley Research Center, Hampton, VA.

Hales, S.J., Tayon, W.A., Domack, M.S., 2012. Friction-stir-welded and spin-formed end domes for cryogenic tanks. NASA Report NF1676L-13613. NASA Langley Research Center, Hampton, VA.

Harmin, M., Abdelkader, A., Cooper, J., Bron, F., 2011. Aeroelastic tailoring of metallic wing structures, 52nd AIAA/ASME/ASCE/AHS/ASC Structures, Structural Dynamics and Materials Conference 2011, vol. 1. Curran Associates Inc., Red Hook, NY, pp. 183–201.

Heinimann, M., Kulak, M., Bucci, R., James, M., Wilson, G., Brockenbrough, J., et al., 2007. Validation of advanced metallic hybrid concept with improved damage tolerance capabilities for next generation lower wing and fuselage applications. In: Lazzeri, L., Salvetti, A. (Eds.), ICAF 2007—Durability and Damage Tolerance of Aircraft Structures: Metals vs Composites, vol. 1. Pacini, Naples, Italy, pp. 206–223.

Henon, C., Rouault, R., 2012. Comparison of corrosion performance and mechanisms of Al–Cu alloys with and without Li addition. In: Weiland, H., Rollett, A.D., Cassada, W.A. (Eds.), Proceedings of the 13th International Conference on Aluminum Alloys (ICAA13). The Minerals, Metals and Materials Society (TMS) and John Wiley & Sons, Hoboken, NJ, pp. 431–436.

Hinrichsen, J., 2003. Praxis-Seminar Luftfahrt, Hochschule für Angewandte Wissenschaften Hamburg, DGLR, VDI. Airbus A380: Vertical Tailplane, Hamburg, Germany, April 10, 2003.

Johnson, R.E., 1973. Apollo experience report—the problem of stress-corrosion cracking. National Aeronautics and Space Administration, NASA Technical Note TN D-7111, NASA Lyndon B. Johnson Space Center, Houston, TX.

Karabin, L.M., Bray, G.H., Rioja, R.J., Venema, G.B., 2012. Al–Li–Cu–Mg–(Ag) products for lower wing skin applications. In: Weiland, H., Rollett, A.D., Cassada, W.A. (Eds.), Proceedings of the 13th International Conference on Aluminum Alloys (ICAA13). The Minerals, Metals and Materials Society (TMS) and John Wiley & Sons, Hoboken, NJ, pp. 529–534.

Kashaev, N., Enz, J., Horstmann, M., Groth, A., Ventzke, V., Riekehr, S., et al., 2013. Quality assessment of laser beam welded AA2198 integral structures. Paper to be Presented at the 27th ICAF Symposium. June 5–7, 2013, Jerusalem, Israel.

Köble, K., 2003. Small satellite design: satellite structures. EADS Astrium Presentation, Institut für Raumfahrtsysteme, University of Stuttgart, Stuttgart, Germany, October 2003.

Korb, L.J., 1987. Corrosion in the aerospace industry. Metals Handbook Ninth Edition, Vol. 13 Corrosion. ASM International, Metals Park, OH, pp. 1058 – 1100.

Lenczowski, B., 2002. New lightweight alloys for welded aircraft structures, Paper 4101. In: 23rd International Congress of the Aeronautical Sciences, ICAS 2002 CD-ROM Proceedings. September 2002, Toronto, Canada.

Lequeu, P., 2008. Advances in aerospace aluminum. Adv. Mater. Processes 166 (2), 47–49.

Lequeu, P., Lassince, P., Warner, T., 2007. Aluminum alloy development for the Airbus A380 – Part 2. Adv. Mater. Processes 165 (7), 41–44.

Lequeu, P., Smith, K.P., Daniélou, A., 2010. Aluminum–copper–lithium alloy 2050 developed for medium to thick plate. J. Mater. Eng. Perform. 19 (6), 841–847.

Liu, A.F., 1974. Fracture control methods for space vehicles volume I, fracture control design methods. National Aeronautics and Space Administration NASA Contractor Report NASA CR-134596, NASA Lewis Research Center, Cleveland, OH.

Lowak, H., De Jonge, J.B., Franz, J., Schütz, D., 1979. MINITWIST: a shortened version of TWIST. National Aerospace Laboratory NLR Miscellaneous Publication NLR-MP-79018U, Amsterdam, the Netherlands.

Lynch, S.P., Shekhter, A., Moutsos, S., Winkelman, G.B., 2003. Challenges in developing high performance Al–Li alloys. LiMAT 2003, Third International Conference in Light Materials for Transportation Systems. Published on a CD by the Center for Advanced Aerospace Materials, Pohang University of Science and Technology, Pohang, Korea.

Magnusen, P.E., Mooy, D.C., Yocum, L.A., Rioja, R.J., 2012. Development of high toughness sheet and extruded products for airplane fuselage structures. In: Weiland, H., Rollett, A.D., Cassada, W.A. (Eds.), Proceedings of the 13th International Conference on Aluminum Alloys (ICAA13). The Minerals, Metals and Materials Society (TMS) and John Wiley & Sons, Hoboken, NJ, pp. 535–540.

MMPDS-07, 2012. Metallic Materials Properties Development and Standardization (MMPDS). Battelle Memorial Institute, Columbus, OH.

Mouritz, A.P., 2012. Introduction to Aerospace Materials. Woodhead Publishing Limited, Cambridge, UK.

National Aeronautics and Space Administration, 2005. Guidelines for the selection of metallic materials for stress corrosion cracking resistance in sodium chloride environments. NASA Standard MSFC-STD-3029 Revision A, George C. Marshall Space Flight Center, Huntsville, AL.

National Aeronautics and Space Administration, 2009. NASA Engineering & Safety Center NESC 2009 Technical Update, NASA Publication NP-2009-11-225-LaRC, NASA Langley Research Center, Hampton, VA.

National Aeronautics and Space Administration, 2010. Orion: America's next generation space-craft. NASA Publication NP-2010-10-025-JSC, NASA Lyndon B. Johnson Space Center, Houston, TX.

National Aeronautics and Space Administration, 2012a. Space Launch System: building America's next heavy-lift launch vehicle. NASA Facts FS-2012-06-59-MSFC, NASA George C. Marshall Space Flight Center, Huntsville, AL.

National Aeronautics and Space Administration, 2012b. SpaceX CRS-1 mission: first cargo resupply services mission. NASA Press Kit/October 2012, <www.nasa.gov>.

Niedzinski, M., Thompson, C., 2010. Airware 2198 backbone of the Falcon family of SpaceX launchers. Light Met. Age 68, 6–7, December 2010, 55.

Niedzinski, M., Kabarra, J., Rubadue, J., Rice, R., 2013a. MMPDS handbook: a key element in aircraft and aerospace design. Light Met. Age 71, 38–40, February 2013.

Niedzinski, M., Ebersolt, D., Schulz, P., 2013b. Review of Airware alloys currently used for space launchers. Constellium Presentation, Technical Interchange Meeting, European Space Agency ESA/ESTEC, Noordwijk, The Netherlands, March 2013.

Pacchione, M., Telgkamp, J., 2006. Challenges of the metallic fuselage. Paper ICAS 2006-4.5.1, 25th International Congress of the Aeronautical Sciences (ICAS 2006), Published on a CD by the German Society for Aeronautics and Astronautics (DGLR), Bonn, Germany.

Payne, M., 2013. SLS takes on new buckling standards, drops Super Light alloy. NASA <Spaceflight.com>, February 18, 2013.

Peel, C.J., 1990. The development of aluminium–lithium alloys: an overview. In: "New Light Alloys", AGARD Lecture Series No. 174, Advisory Group for Aerospace Research and Development, Neuilly-sur-Seine, France, pp. 1-1–1-55.

Peel, C.J., Evans, B., Baker, C.A., Bennet, D.A., Gregson, P.J., Flower, H.M., 1984. The development and application of improved aluminium–lithium alloys. In: Sanders, T.H., Starke E.A. Jr., (Eds.), "Aluminium–Lithium Alloys II", Proceedings of the Second International Conference. Metallurgical Society of AIME, Warrendale, PA, pp. 363–392.

Pickens, J.R., Heubaum, F.H., Langan, T.J., Kramer, L.S., 1989. Al-(4.5-6.3) Cu-1.3 Li-0.4 Ag-0.4Mg-0.14 Zr alloy Weldalite 049. In: Sanders T.H., Jr., Starke E.A. Jr., (Eds.), "Aluminum–Lithium Alloys", Proceedings of the Fifth International Aluminum–Lithium Conference, vol. 3. Materials and Component Engineering Publications, Birmingham, UK, pp. 1397–1411.

Rioja, R.J., Liu, J., 2012. The evolution of Al–Li base products for aerospace and space applications. Metall. Mater. Trans. A 43A, 3325–3337.

Rioja, R.J., Cho, A., Bretz, P.E., 1990. Al–Li alloys having improved corrosion resistance containing Mg and Zn. U.S. Patent No. 4,961,792, October 9, 1990.

Rioja, R.J., Denzer, D.K., Mooy, D., Venema, G., 2012. Lighter and stiffer materials for use in space vehicles. In: Weiland, H., Rollett, A.D., Cassada, W.A. (Eds.), Proceedings of the 13th International Conference on Aluminum Alloys (ICAA13). The Minerals, Metals and Materials Society (TMS) and John Wiley & Sons, Hoboken, NJ, pp. 593–598.

Schijve, J., 2009. Fatigue of Structures and Materials, second ed. Springer Science + Business Media, B.V., Dordrecht, The Netherlands.

Srinivasa Rao, P., Sivadasan, K.G., Balasubramanian, P.K., 1996. Structure-property correlation on AA 2219 aluminium alloy weldments. Bull. Mater. Sci. 19, 549–557.

Tchitembo Goma, F.A., Larouche, D., Bois-Brochu, A., Blais, C., Boselli, J., Brochu, M., 2012. Fatigue crack growth behavior of 2099-T83 extrusions in two different environments. In: Weiland, H., Rollett, A.D., Cassada, W.A. (Eds.), Proceedings of the 13th International

Conference on Aluminum Alloys (ICAA13). The Minerals, Metals and Materials Society (TMS) and John Wiley & Sons, Hoboken, NJ, pp. 517–522.

Threadgill, P.L., Leonard, A.J., Shercliff, H.R., Withers, P.J., 2009. Friction stir welding of aluminium alloys. Int. Mater. Rev. 54 (2), 49–93.

Wanhill, R.J.H., 1986. Effects of cladding and anodising on flight simulation fatigue of 2024-T3 and 7475-T761 aluminium alloys. In: Barnby, J.T. (Ed.), Fatigue Prevention and Design. Engineering Materials Advisory Services Ltd., Warley, UK, pp. 323–332.

Wanhill, R.J.H., 1994. Status and prospects for aluminium–lithium alloys in aircraft structures. Int. J. Fatigue 16, 3–20.

Wanhill, R.J.H., 2003. Milestone case histories in aircraft structural integrity. In: Milne, I., Ritchie, R.O., Karihaloo, B.L. (Eds.), Comprehensive Structural Integrity Volume 1: Structural Integrity Assessment—Examples and Case Studies. Elsevier Science Publishers, London, UK, pp. 61–72.

Wanhill, R.J.H., Byrnes, R.T., Smith, C.L., 2011. Stress corrosion cracking (SCC) in aerospace vehicles. In: Raja, V.S., Shoji, T. (Eds.), Stress Corrosion Cracking: Theory and Practice. Woodhead Publishing Limited, Cambridge, UK, pp. 608–650.

Weiss, S.P., 1973. Apollo experience report—lunar module structural subsystem. National Aeronautics and Space Administration NASA Technical Note NASA TN D-7084. Manned Spacecraft Center, Houston, TX.

Williams, J.C., Starke E.A., Jr., 2003. Progress in structural materials for aerospace systems. Acta Mater. 51, 5775–5799.

Williamson, J.G., 1969. Stress corrosion cracking of Ti-6Al-4V titanium alloy in various fluids. National Aeronautics and Space Administration NASA Technical Memorandum NASA TM X-53971. George C. Marshall Space Flight Center, Huntsville, AL.

Windisch, M., 2009. Damage tolerance of cryogenic pressure vessels. European Space Agency ESA Technology and Research Programme Report ESA TRP DTA-TN-A250041-0004-MT, European Space Agency ESA/ESTEC, Noordwijk, The Netherlands.

Windisch, M., 2013. Damage tolerance characterization of 2195 base material and friction stir welds. MT Aerospace Presentation, Technical Interchange Meeting, European Space Agency ESA/ESTEC, Noordwijk, The Netherlands, March 2013.

Chapter 16

Airworthiness Certification of Metallic Materials

B. Saha*, R.J.H. Wanhill†, N. Eswara Prasad*, G. Gouda** and K. Tamilmani‡

*Regional Centre for Military Airworthiness (Materials), CEMILAC, Hyderabad, India, †NLR, Emmeloord, the Netherlands, **Centre for Military Airworthiness and Certification (CEMILAC), Bangalore, India, ‡Office of the Chief Controller R&D (Aero), DRDO HQs., New Delhi, India

Contents

16.1 Introduction	537	16.4 Example of Certification of an Al–Li Alloy	549
16.2 Aviation and Airworthiness Regulatory Bodies	538	16.4.1 Certification Methodology	549
16.2.1 Civil Aviation	538	16.4.2 Certification of Al–Li Alloy 1441M Sheet	550
16.2.2 Military Aviation	539		
16.3 Airworthiness of Metallic Materials	540	16.5 Summary	553
16.3.1 Fatigue Design Philosophies	541	References	554
		General References	553
16.3.2 Materials and Structures Certification Methodology	544	Specific References	554

16.1 INTRODUCTION

Aviation is a "hazardous" mode of transportation that requires safety regulations and the guarantee of "airworthiness," i.e. an aircraft and its crew are "fit to fly." Safety is a fundamental requirement related to all human activities. It is an obligation and at the same time a practical demand, since accidents causing injuries to people and damage to properties have social and economic impacts.

Aluminum–Lithium Alloys.

Aircraft airworthiness implies, in the first instance, that the aircraft as a whole, and also all of its structural components, electrical, hydraulic and mechanical systems, and other equipment, are capable of fulfilling the mission requirements with an acceptable level of safety and reliability. However, airworthiness is much more than this, since it also covers safety and reliability during aircraft design, manufacture, and maintenance.

There is a basic difference between military and civil (commercial) aircraft airworthiness. Military aircraft design and manufacture often precedes the development of well-matured technologies acceptable for use in civil aircraft and there is also a difference in the emphasis on safety. For military aircraft the priority is given to mission capability, while the basic aim of civil aircraft airworthiness is to ensure the safety and comfort of the passengers (and crew). In practice this means that the required safety levels are high for military aircraft and even higher for civil aircraft.

For both military and civil aircraft the airworthiness safety level generally depends on: (i) airworthiness of the aircraft, (ii) operating crew and their skills, (iii) inspectability and maintainability of the aircraft, (iv) maintenance of crew skills, (v) air traffic control systems, (vi) effectiveness of the navigational aids, and (vii) effectiveness of weather forecasting. Clearly, the task of achieving airworthiness is a complex one. This has resulted in setting up a number of international and national regulatory bodies, both military and civil. A survey of some of these regulatory bodies and their tasks is given in Section 16.2.

There are many Regulation and Specification Handbooks, Airworthiness Directives, and Technical Orders containing essential and valuable information on aircraft airworthiness (e.g. the general references at the end of this chapter). However, in the present context of a book on aluminum−lithium (Al−Li) alloys, we shall concentrate on (i) the airworthiness of metallic materials in Section 16.3 and (ii) an example of the certification of Al−Li alloys in Section 16.4.

16.2 AVIATION AND AIRWORTHINESS REGULATORY BODIES

16.2.1 Civil Aviation

The beginnings of international and national aviation organizations and regulatory bodies go back to WWII. In 1944 the United States convened a meeting of allied and neutral states, resulting in the establishment of the International Civil Aviation Organization (ICAO) in 1947. Two more international organizations were set up in Europe. The Joint Aviation Authorities (JAA) started in 1970 as the Joint Airworthiness Authorities; and the European Aviation Safety Agency (EASA) was created in 2002.

There are also national agencies, of which the best known is the Federal Aviation Administration (FAA) in the United States. The FAA was created by the Federal Aviation Act of 1958 and was called the "Federal Aviation

Agency" until 1966. Many other countries have civil airworthiness authorities, including India with its Directorate General of Civil Aviation (DGCA), which was established pursuant to the Indian Aircraft Act of 1934.

All these regulatory bodies fulfill a number of tasks associated with airworthiness, though not necessarily with the same emphasis and scope:

1. ICAO: The ICAO was created with the main objective of achieving standardization during the operation of safe, regular, and efficient air services. The outcome of this standardization is a high level of reliability in many areas that collectively shape international civil aviation, particularly with respect to aircraft, their crew, and the ground-based facilities and services.
2. JAA: The JAA mainly looks into operations, maintenance, and licensing and certification design standards for all categories of aircraft. The JAA is also an associated body of the European Civil Aviation Conference and represents the civil aviation regulatory authorities of several European countries.
3. EASA: The EASA is an independent European community body created to put in place a system of air safety and environmental regulations with a legal identity and autonomy. EASA develops its own known-how in the field of aviation safety and environmental protection to assist legislations for issuance of rules and directives.
4. FAA: The FAA organization is very complex owing to its many tasks and the size of the United States, and its interaction and relationships with the rest of the world. The main responsibility of the FAA is to promote the safety of civil aviation. However, the FAA has a broader authority to combat aviation hazards, including the sole responsibility for developing and maintaining a common civil–military system of air navigation and air traffic control. The FAA also regulates the US commercial space transportation industry and licenses commercial space launch facilities and private launches of space payloads on expendable launch vehicles.
5. DGCA: The DGCA primarily deals with safety issues in India. It is responsible for the regulation of air transport services to, from, and within India and for enforcement of civil air regulations, air safety, and airworthiness standards. The DGCA also coordinates with the ICAO on all regulatory functions. The DGCA organization is large, with 10 Directorates and 14 regional or subregional Airworthiness Offices.

16.2.2 Military Aviation

There are, of course, many regulatory bodies for the airworthiness of military aircraft, but information about them is often less accessible than that for civil organizations. The Indian authors of this chapter are familiar with the Indian military aviation regulatory system described in the subsequent sections. This system began with the formation of a Resident Technical Office (RTO) in Bangalore in 1958. More RTOs were established in many regions of India to meet the regulatory demands of various certification disciplines.

In 1995, the Center for Military Airworthiness and Certification (CEMILAC) was founded in Bangalore to coordinate and consolidate the activities of the RTOs (Tamilmani, 2012). CEMILAC currently has 14 Regional Centers for Military Airworthiness (RCMAs, previously the RTOs), each with unique core competence. These competences include the following (Tamilmani, 2012):

- Capability to undertake certification of fully-fledged combat aircraft, helicopters, engines, systems, equipment, materials, and software programs.
- Knowledge bases to assess performance and evaluate designs after identifying the implications of technology advances in aerodynamics, structures, systems, and equipment.
- Ability to devise qualification requirements for airworthiness certification at the component, systems, subsystems, equipment, and whole aircraft or engine level.
- Security and impartiality to safeguard the intellectual property of customers by ensuring the confidentiality of information.

CEMILAC executes the airworthiness requirements through three of its functional groups. These are the Aircraft Group, Propulsion Group, and System Group, consisting of various RCMAs. Together these have vast experience and expertise concerning the airworthiness requirements for certification of Indian military aircraft and airborne systems.

CEMILAC also undertakes:

- Life evaluations of aircraft, helicopters, engines, and accessories to fully exploit their operational potential. These evaluations have enabled the extension of the lives of many aircraft and engines, resulting in very large savings in terms of foreign exchange.
- Certification of mid-life upgrades of aircraft, weapon systems, early warning (EW) systems, and avionics systems. These upgrades are necessary for fully realizing the aircraft and systems potentials.
- Design and test house approvals. More than 140 firms are approved by CEMILAC to carry out certification activities till date.
- Type approvals, covering a wide spectrum of aeronautical stores and fuel, oil, and lubricants.

16.3 AIRWORTHINESS OF METALLIC MATERIALS

The design philosophies developed for the aircraft industry are based on well-established safety factors that, compared with those of other industries, are relatively small. This is essential because aircraft must have lightweight and highly efficient structures. The combination of relatively small safety factors and highly efficient structures demands that high-strength materials with verified consistent performance must be available. Furthermore, the relatively high costs of manufacturing, the large actual and potential markets, and the thrust for military aerial supremacy have stimulated a competitiveness that has brought airworthiness requirements into ever sharper focus. The results are unprecedented improvements in materials and aerospace structural technologies, the use of sophisticated manufacturing techniques, and incorporation of mandatory systematic characterization procedures (Gupta et al., 1991).

16.3.1 Fatigue Design Philosophies

The topic of fatigue design requires special mention, since practically any material used for aircraft components and structures will experience fatigue loading conditions. For example, Figure 16.1 shows the major design criteria for the fuselage panels of a commercial transport aircraft; and Figure 16.2 shows the engineering property requirements for the main structural areas, most of which have to take account of damage tolerance (DT) properties that include fatigue, fatigue crack growth, and fracture toughness. These topics are discussed in Chapters 11–13. It is also worth noting that helicopters are "flying fatigue machines," since they experience very large numbers of fatigue loading cycles owing to the power and lift system (rotor head and blades) and vibration.

The importance of fatigue is reflected in the extensive literature and specification handbooks (e.g., MIL-A-83444) on fatigue design philosophies and methods for assessing aircraft fatigue lives. There are several possible design philosophies and life assessment methods, and a survey of actual and proposed methods is given in Table 16.1. Some comments on these methods are made in the following text:

1. Stress-life (S–N): This method has been used extensively in the past. It relies on simple cumulative damage rules to obtain "safe" fatigue lives under variable amplitude (VA) load histories. These safe lives are then factored down using scatter factors based on engineering judgement.

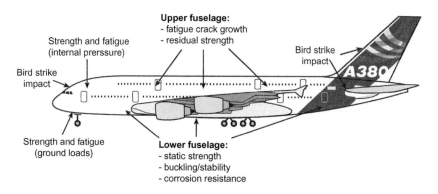

FIGURE 16.1   Major structural design criteria for a transport aircraft.

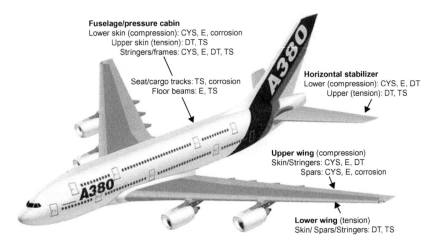

Fuselage/pressure cabin
Lower skin (compression): CYS, E, corrosion
Upper skin (tension): DT, TS
Stringers/frames: CYS, E, DT, TS

Seat/cargo tracks: TS, corrosion
Floor beams: E, TS

Horizontal stabilizer
Lower (compression): CYS, E, DT
Upper (tension): DT, TS

Upper wing (compression)
Skin/Stringers: CYS, E, DT
Spars: CYS, E, corrosion

Lower wing (tension)
Skin/ Spars/Stringers: DT, TS

FIGURE 16.2 Engineering property requirements for main structural areas in a transport aircraft: CYS, compressive yield strength; E, elastic modulus; TS, tensile strength; DT, damage tolerance properties (fatigue, fatigue crack growth, fracture toughness). This illustration is also given in Chapter 2.

TABLE 16.1 Current and Proposed Methods for Aircraft Fatigue Life Assessment (Barter et al., 2010)

Stress-life (S–N)

Fatigue limits, S_e ; unnotched and notched (K_t); constant amplitude (CA) data
Modifications to S_e
Mean stress effects (described in terms of the stress ratio, $R = S_{min}/S_{max}$)
Linear damage rule, also for variable R
Scatter factors

Strain-life (ε–N)

Strain-life equation, unnotched data, $R = -1$
Cyclic stress–strain curve analysis
Rainflow cycle counting (closed hysteresis loops)
Stress–strain at critical location (notch analysis)
Mean stress effects (R) via equivalent strain equations, leading to equivalent strain amplitudes
Damage accumulation rule

Damage Tolerance (DT)

Specified equivalent initial flaw sizes based on non-destructive inspection (NDI)
Back-extrapolation of long crack growth data to derive short crack growth
LEFM long crack growth models (non-interaction, yield zone, crack opening, strip yield) to derive variable amplitude (VA) crack growth from CA data
Possible use of crack opening model for short cracks (FASTRAN); differences in long and short crack thresholds need to be included
Mainly deterministic: stochastic approach becoming accepted

(*Continued*)

TABLE 16.1 (Continued)

Holistic approach: proposed

Fatigue initiation mechanisms (also as functions of notch stress concentrations, K_t)
Fatigue initiation lives (S−N and/or ε−N assessments)
Marker load strategies for quantitative fractography (QF) of short-to-long crack growth
Actual short-to-long long crack growth using marker loads and QF
Establishment, validation, and choice of appropriate crack growth models and "laws"
Deterministic ("upper bound") and stochastic approaches
Environmental effects, notably corrosion

DSTO approach: proposed and implemented for the RAAF

Actual initial discontinuity/flaw sizes and their equivalent pre-crack sizes
Actual short-to-long crack growth data using QF
Data compilations to establish empirical relationships describing crack growth behavior
Deterministic ("upper bound") estimates of *lead crack* growth
Scatter factors

2. Strain-life (ε−N): A common method for tracking fatigue damage in aircraft fleets. There are several manufacturers' and publicly available tools for doing this. There is always a need to know the effectiveness of these tools and what their weaknesses are.

3. DT: Originally developed by the United States Air Force (USAF) and currently used for most military aircraft. However, there are problems. The equivalent initial flaw size (EIFS) concept is dubious, especially with respect to the assumed minimum crack dimensions. The EIFS requirements can lead to predicted fatigue crack growth lives that are (much) too conservative. This has led to setting up the NATO Research and Technology Organization working group AVT − 125 "Future Airframe Lifing Methodologies."

   Another serious problem is that the predicted early crack growth behavior is highly questionable, since it is derived from back-extrapolation of (i) VA long crack growth data or (ii) VA growth curves derived from long crack CA data, with both methods using analytical models "tuned" to long crack growth behavior.

4. Holistic approach: This is a very ambitious program (Figure 16.3) and is intended for all types of aircraft. The inclusion of corrosion effects is especially noteworthy. The project leader is the Canadian National Research Council.

5. Defence Science and Technology Organisation (DSTO) approach: Much less ambitious than the HOLSIP program and especially applicable to high-performance aircraft. This approach has been developed by the Australian DSTO for the Royal Australian Air Force (RAAF).

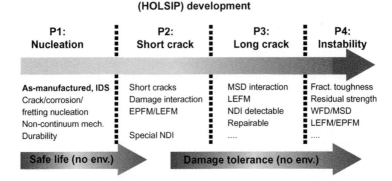

FIGURE 16.3 The HOListic Structural Integrity Process (HOLSIP) program. IDS, initial defect state; EPFM, elastic-plastic fracture mechanics; LEFM, linear elastic fracture mechanics; MSD, multisite damage; NDI, nondestructive inspection; WFD, widespread fatigue damage.

16.3.2 Materials and Structures Certification Methodology

The most well-known and widely used standard for aviation materials has been Military Handbook MIL-HDBK-5, which was canceled in 2004 and superseded for metallic materials by MMPDS-01 (see the general references at the end of this chapter: the MMPDS is regularly updated and the 8th edition, MMPDS-8, was due in 2013). These standards provide data and guidelines for the required properties of aerospace metals and alloys, and it is up to the certificating authorities as to how they qualify these materials. In the following paragraphs we present a generic approach based on the authors' experience.

Approval of Suppliers

There are two distinct stages involved in the approval of a new supplier/manufacturer. Firstly, there is a quality audit of the supplier's organization. This begins with the supplier submitting the detailed documents about the organization, emphasizing the technical aspects. After scrutiny of these documents, a team of quality and engineering experts visits the organization and investigate its facilities in detail, in particular the following:

- Personnel employed and their technical background
- Production facilities
- Performance of equipment and system of monitoring
- Quality control and its check system
- Discipline of check system and checks on equipment and control systems
- Handling of nonconformity in production
- Systems of studying specifications and filing
- Laboratory and R&D facilities (if any).

The team prepares a report after being satisfied about the supplier's capability of manufacturing aerospace materials and/or components. This report is submitted to the regulatory body's quality assurance agency for approval of the supplier's organization as a potential manufacturer. Only after this approval has been obtained is the second stage undertaken, namely to approve the materials and components themselves.

Approval of Materials

In the first instance, the supplier provides general data to indicate competence in making a particular material or component, together with a declared procedure from start to finish (despatch to the customer). This is followed by several evaluation stages. For any Class A (i.e., primary and premium quality) metallic component, there are three distinct stages of evaluation for ensuring consistent behavior in service (Gupta et al., 1991):

1. Material evaluation
   a. Semi-product (mill form) evaluation: bars, sheets, and plates
   b. raw-part cutup testing: forgings, extrusions, and castings.
2. Product evaluation
   a. Component testing.
3. Performance evaluation
   a. Full-scale structural unit and complete airframe tests
   b. Ground and flight testing of the aircraft.

Material Evaluation

There are two distinct stages of material evaluation for each Class A component, namely evaluations of mill products and raw parts:

1. Mill products: Evaluation of mill products could be either for *ab initio* development or second source approval. In the case of *ab initio* development the material is tested for all possible properties. From the results the use of the material for specific applications is assessed. If a second source for an already-known material is to be approved, then only the properties that depend on processing are checked. The list of properties includes tensile and compression strengths and elastic moduli, mechanical anisotropy, fatigue, crack growth, and fracture. A major aspect of the evaluation is correlation of the properties with processing variables, microstructures, and chemical composition.

   Repeat tests are done by the customer/certification authority to check the properties. If satisfactory, then the supplier is requested to provide raw parts from 3–5 batches of material for evaluation. For Class A components the usual number of batches is 5. For Class B (secondary) components two or more batches may be evaluated. The supplier also has to provide details of processing factors and procedures that, if changed, would significantly affect the final product.

2. Raw parts: Evaluation of raw parts (also termed cutup testing) is different in emphasis from mill product evaluation. The cutup specimen tests are designed to determine the material performance under load and environmental conditions representative of service.

Product and Performance Evaluations

These evaluation stages represent successive scale-ups in the size, complexity, and costs of testing. For aircraft structures the material evaluations concentrate on the DT properties of fatigue, fatigue crack growth, and fracture toughness.

An outstanding example of a full-scale material evaluation test is shown in Figure 16.4. This was the Airbus A380 MegaLiner Barrel (MLB) test, which evaluated fuselage panels made from several aluminum alloys and GLARE (GLAss REinforced aluminum laminates). This test ran for 45,402 simulated flights with conservatively high fatigue loads, and cost the equivalent of about €50,000,000 in the mid-to-late 1990s. Teardown, inspection, and quantitative fractography (QF) of the F4, F6, and F7 locations confirmed the exceptional DT capability of GLARE, which had already been selected for large parts of the A380 fuselage.

Additional Requirements and Considerations

a. Sealing the production route: The supplier has to control the production according to a fully documented procedure determined and specified during the approval stages. The documentation covers all significant aspects from the raw material stage to the finished product. Any in-process changes resulting in improvements are introduced by re-evaluation followed by documented modifications and amendments to the procedure.

These controls, and mandatory compliance to them, ensure the following:

- The basic mechanical properties not only exceed material and process specification minima, but do so in a consistent manner.
- All components of a given type will respond to service-imposed stresses and environments in similar manner.
- Minimization of the likelihood of undetected residual stresses and limits on defect sizes in the finished components.
- In the unlikely event of component malfunction, the entire material history can be rapidly traced, thereby facilitating the identification of the cause(s) of failure.

b. Aerospace and general engineering specifications: Aerospace alloys, even though they have similar general engineering counterparts, are covered by tighter specifications reflecting their better quality:

- Narrower ranges of alloying elements.
- Tighter controls on impurity and trace elements; very low limits for embrittling elements.

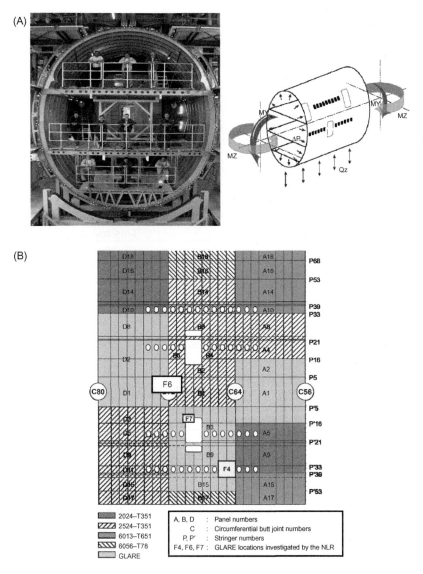

FIGURE 16.4 Airbus A380 MLB full-scale fatigue test: (A) MLB cross-sectional view and general loading conditions; (B) "opened-out" view of the fuselage panels, passenger doors, windows, and GLARE teardown locations (Wanhill et al., 2009).

Ground and flight tests

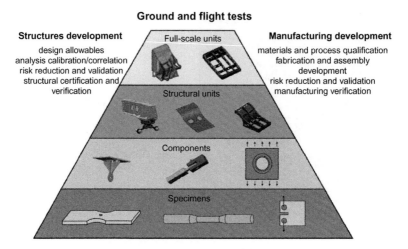

Structures development	Full-scale units	Manufacturing development

Structures development: design allowables, analysis calibration/correlation, risk reduction and validation, structural certification and verification

Full-scale units

Manufacturing development: materials and process qualification, fabrication and assembly development, risk reduction and validation, manufacturing verification

Structural units

Components

Specimens

FIGURE 16.5 Schematic "building block" test approach for materials and components certification. *Source: Adapted from Ball et al. (2006).*

- Stringent microstructural acceptance norms.
- Higher mechanical property limits.
- Checks on fatigue and DT properties.
- Closer dimensional tolerances.

c. Use of general engineering materials: It may be permissible to use well-documented general engineering materials for Class B components if aerospace quality materials are not readily available. This provision is made, for example, in the UK Defence Standard 00-970. The case for such substitutions is benefited by material characterization to aerospace material standards. Nevertheless, design calculations must allow for greater scatter in properties. Also, clearance from airworthiness agencies is mandatory.

In summary, the successive stages of aerospace materials and structures certification processes are increasingly costly, as demonstrated dramatically by the MLB test illustrated in Figure 16.4. The basis for certification lies in the initial material evaluation. By the time the product and performance stages are reached, there should ideally be little or no doubt about a material's suitability for the intended application. However, large-scale testing and flight evaluation are mandatory parts of the certification process.

A schematic view of the materials and structures certification process is given in Figure 16.5. This is a "building block" (BB) approach adapted from one for the Lockheed Martin F-35 Joint Strike Fighter (JSF), but this approach is generically valid. The BB approach may be viewed as a pyramid whose base is the initial material evaluation. Each level of the pyramid is the foundation for the next, and the structural complexity and costs increase with each level. The final phase of certification is ground and flight testing of the aircraft.

16.4 EXAMPLE OF CERTIFICATION OF AN Al—Li ALLOY

The continuing pressures for increased fuel efficiency and higher payloads and performance are major issues for the aerospace industry and have boosted the demands for more advanced high-strength materials with high-specific properties. There is a trend toward using composites extensively in the airframes of transport (passenger and cargo) aircraft, e.g. 50% in the Boeing 787 *Dreamliner*, but for most transport aircraft, including the Airbus A380, aluminum alloys still account for about 60% of the structural weight. For high-performance military aircraft the emphasis is on composites and titanium alloys, and the latest aircraft (Eurofighter and JSF) use about 40% composites. Nevertheless, there are many opportunities for Al—Li alloys to replace conventional aluminum alloys. This is particularly the case for transport aircraft, as discussed in Chapters 2 and 15, but also for military aircraft, e.g. in Lockheed Martin F-16 bulkheads and other parts (Acosta et al., 2002; Balmuth, 2001).

In this main section we shall discuss certification of the second-generation Al—Li alloy 1441M, which has been recommended for use in a light combat aircraft (LCA) for the Indian Air Force (IAF). This alloy is a modification of 1441, which belongs to the low-density Al—Li—Cu—Mg—Zr group of second-generation alloys introduced in the Soviet Union in the 1980s. The low density makes 1441M very attractive for use in aluminum alloy airframe components.

A major emphasis of the certification process has been to determine whether a good balance of properties could be achieved through optimization of alloy processing. This emphasis is due to the problems experienced with some of the second-generation Al—Li alloys, including a tendency for strong anisotropic mechanical properties, low short-transverse ductility and fracture toughness, and loss of toughness owing to thermal instability (Eswara Prasad et al., 2003; Lynch et al., 2003; Rioja and Liu, 2012): these problems are discussed in several chapters of this book.

16.4.1 Certification Methodology

The certification of a new or an existing material for aerospace applications is a necessarily complex and thorough procedure, consisting of many steps and stages. Whenever there is (or appears to be) a new material requirement for the IAF, the general procedure is as follows:

1. The aircraft operator approaches an RCMA. Based on the RCMA's advice, the operator selects a potential supplier/manufacturer. After assessing the supplier's capabilities, the RCMA circulates a Draft Development/Type Test Schedule (DTS/TTS) for a Local Type Certification Committee (LTCC) meeting chaired by the Regional Director of the RCMA. Members of the LTCC include the relevant agencies and the supplier and operator. After the LTCC meeting the RCMA finalizes the DTS/TTS, which is the final qualification document for the material.

2. As described in Section 16.3.2, there are two stages involved in the approval of a supplier/manufacturer: approval of the supplier and approval of the material. Both tasks are carried out by teams from the RCMA and the Directorate General of Aeronautical Quality Assurance (DGAQA). Approval of the material requires satisfactory tests' typically from three batches, and verification of compliance with the DTS/TTS. The RCMA issues provisional clearances for each batch.

3. After three provisional clearances the supplier/manufacturer and the RCMA prepare the "type record." The RCMA then forwards the type record, along with a recommendation, to CEMILAC in order to obtain the type approval for the supplier. Once the type approval is issued, the DGAQA and RCMA prepare the release specification and issue it to the supplier. The production route is also sealed by the RCMA.

4. All material and component batches processed according to the sealed production route are released by the DGAQA, which issues release specifications and forwards copies of the test certificates to the RCMA.

5. Any deviations or problems encountered during series production are to be referred by the supplier/manufacturer to the DGAQA and via the DGAQA to the RCMA. Based on the RCMA's advice about production modifications, the supplier prepares revised process sheets for approval by the RCMA.

6. The type approval issued to the supplier is renewed periodically by CEMILAC, contingent upon a mandatory reapplication from the supplier and a subsequent recommendation from the RCMA.

Figure 16.6 is a detailed illustration of just how complex and thorough the certification methodology has to be.

16.4.2 Certification of Al–Li Alloy 1441M Sheet

As stated earlier, 1441M alloy is a modification of 1441, which belongs to the high-lithium Al–Li–Cu–Mg–Zr group of second-generation alloys introduced in the Soviet Union in the 1980s. 1441M sheets are heat treated in an air atmosphere which is a much easier option than the salt bath treatment normally practiced for Al–Li alloys, and was jointly developed by VIAM, Russia and the Defence Metallurgical Research Laboratory, India, for long-term application in an LCA program. Owing to its high lithium content ∼2 wt%, 1441M has a density 7–9% less than those of equivalent conventional AA 2XXX and 7XXX alloys, respectively, and also a higher elastic modulus (78 vs 71–73 GPa), making it very attractive for aluminum alloy airframe use.

Chemical Composition and Density

The chemical composition of 1441M is given in Table 16.2. The density was determined to be 2.57 g/cm^3. This is lower than the standard (MMPDS) value of 2.59 g/cm^3 for alloy 1441 (see Table 2.1).

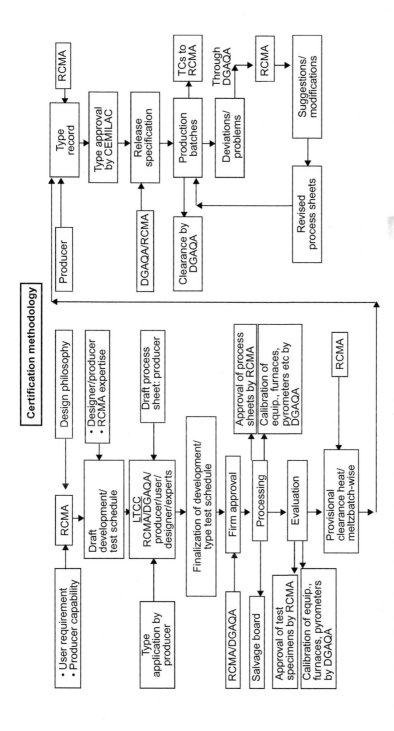

FIGURE 16.6 Certification methodology and procedure.

TABLE 16.2 Composition of Al–Li Alloy 1441M (wt%)

Li	Cu	Mg	Zr	Si	Fe
1.85–1.90	1.6–2.0	0.90–1.10	0.09 max	0.05 max	0.09–0.11

TABLE 16.3 Tensile Properties of Al–Li Alloy 1441M-T6 and 1441-T6 Sheet

Thickness (mm)	Orientation	0.2% YS (MPa)		UTS (MPa)		% Elongation	
		Alloy 1441M	Alloy 1441	Alloy 1441M	Alloy 1441	Alloy 1441M	Alloy 1441
1.2 (Unclad)	L	399–403	387–421	452–458	446–470	9.7–9.9	6.7–10
	LT	360–370	360–421	440–445	451–470	10.9–12.2	7–10
	L + 45°	370–380	387–407	445–446	431–456	12.2–12.4	6–9
1.2 (Clad)	L	390–398	377–426	436–446	446–456	8.8–9.3	8–9
	LT	401–407	360–421	451–453	451–471	8.3–8.4	7–9
	L + 45°	376–381	362–392	425–435	421–456	9.6–11.2	6.7–13.5
2.0 (Unclad)	L	408–414	397–461	462–469	441–490	7.8–9.6	8.3–10
	LT	388–390	392–456	468–475	475–500	9.2–10.3	6.7–8.3
	L + 45°	394–399	426–446	468–477	465–480	9.1–9.3	5.4–9.2

TABLE 16.4 Proposed Certification Test Program for Al–Li alloy 1441M Sheet

Material Properties	Quality Control	Mechanical Properties and Formability	Engineering Properties
Composition Density Elastic modulus	Dimensional checks U-NDI Hardness Macrostructure Microstructure	Tensile tests: −150°C to +150°C Notch tensile tests Shear and bend tests Erichsen cupping test	LCF and HCF: RT & −50°C FCG K_c and R-Curves Creep and stress rupture (120°C) up to 1000 hours Corrosion and SCC

Notes: U-NDI, ultrasonic non-destructive inspection; LCF, low-cycle fatigue; HCF, high-cycle fatigue; FCG, fatigue crack growth; K_c and R-Curves, fracture toughness parameters; SCC, stress corrosion cracking.

Room Temperature Tensile Properties

The room temperature tensile properties of peak-aged (T6 temper) 1441M and 1441 sheet are given in Table 16.3. The 1441 sheet was obtained from the All-Russian Institute of Aviation Materials (VIAM), formerly of the Soviet Union.

The strength properties of 1441M and 1441 are equivalent (Table 16.3). This is an encouraging result for scaling-up production of 1441M on an industrial scale, specifically to support the LCA program, since heat treatment in air is an industrially more acceptable option. A development-type test schedule has been prepared for 1441M sheet applications, with the broad test requirements as given in Table 16.4.

16.5 SUMMARY

Airworthiness regulatory bodies are authorized and responsible for verifying and ensuring the safety and reliability of aircraft. Global airworthiness agencies are more or less affiliated to the ICAO, which was created with the main objective of achieving standardization during the operation of safe, regular, and efficient air services.

There are many civil and military aviation organizations and regulatory bodies, including EASA, FAA, JAA, and DGCA (civil) and the USAF and CEMILAC (military), whose main objectives are to achieve and ensure a high level of safety and reliability. The functions and responsibilities of these organizations are briefly summarized in this chapter.

Next is a survey of fatigue design philosophies, owing to the importance and ubiquity of fatigue in aircraft structures. This is followed by a discussion of the airworthiness certification methodology for materials and structures, starting with the initial mill products and proceeding via incremental levels to the finished aircraft.

An example of material certification has been given for a second-generation Al—Li alloy. The development of this alloy, 1441M, is promising for its application as sheet materials in the airframes of light combat aircraft.

The materials certification methodology presented in this chapter, and the specific certification program for 1441M sheet, provide illustrations of the procedures and tests required to ensure the airworthiness of high-quality metallic materials in the next generation of aircraft.

REFERENCES

GENERAL REFERENCES

Defence Standard 00-970, Design and Airworthiness Requirements for Service Aircraft, Ministry of Defence, UK (undated).

European Aviation Safety Agency (EASA) Regulations and Certifications (undated).

Federal Aviation Administration (FAA) Federal Aviation Regulations, FAR (undated).

International Civil Aviation Organisation (ICAO) Standards, Recommended Practices and Procedures, Safety Management Manual (SMM), DOC 9859-AN 460, 2007.

Joint Aviation Authorities Joint Aviation Requirements, JAR (undated).

Military Specification Airplane Damage Tolerance Requirements, MIL-A-83444 (USAF), 1974.

Military Handbook MIL-HDBK-5H: Metallic Materials and Elements for Aerospace Vehicle Structures, US Department of Defense, Cancelled May 2004.

MMPDS-01, Metallic Materials Properties Development and Standardization (MMPDS), DOT-FAA-AR-MMPDS-01, US Department of Transport, January 2003.

SPECIFIC REFERENCES

Acosta, E., Garcia, O., Dakessian, A., Aung Ra, K., Torroledo, J., Tsang, A., et al., 2002. On the effect of thermomechanical processing on the mechanical properties of 2297 plates. Mater. Sci. Forum 396—402, pp. 1157—1162.

Ball, D.L., Norwood, D.S. TerMaath, S.C., 2006. Joint Strike Fighter airframe durability and damage tolerance certification. American Institute of Aeronautics and Astronautics Paper AIAA 2006-1867, 47th AIAA/ASME/ASCE/AHS/ASC Structures, Structural Dynamics, and Materials Conference, Newport, RI, May 1—4, 2006.

Balmuth, E.S., 2001. Application of aluminum alloy 2297 in fighter aircraft structures. In: Kim, N., Lee, C.S., Eylon, D. (Eds.), Proceedings of the International Conference on Light Materials for Transportation Systems (LiMAT-2001), vol. 2. Pohang Institute of Science and Technology, Pohang, Korea, pp. 589—596.

Barter, S.A., Molent, L., Wanhill, R.J.H., 2010. Fatigue life assessment for high performance metallic airframe structures—an innovative practical approach. In: Ho, S.-Y. (Ed.), Structural Failure Analysis and Prediction Methods for Aerospace Vehicles and Structures. Benthem Science Publishers Ltd., Sharjah, UAE, pp. 1—17.

Eswara Prasad, N., Gokhale, A.A., Rama Rao, P., 2003. Mechanical behaviour of aluminium—lithium alloys. Sādhanā 28 (1 & 2), 209—246.

Gupta, B., Yadav, J.S., Krishna Rao, M.S., 1991. Metallurgical evaluation: the first step in type certification. J. Inst. Eng. (India) 72, 7—13.

Lynch, S.P., Shekhter, A., Moutsos, S., Winkelman, G.B., 2003. Challenges in developing high performance Al—Li alloys. In: LiMAT 2003, Third International Conference in Light Materials for Transportation Systems. Published on a CD by the Center for Advanced Aerospace Materials, Pohang University of Science and Technology, Pohang, Korea.

Rioja, R.J., Liu, J., 2012. The evolution of Al—Li base products for aerospace and space applications. Metall. Trans. A 43A, 3325—3337.

Tamilmani, K., 2012. Indian military airworthiness approval process by CEMILAC. Aeromag Asia VI, 24—26.

Wanhill, R.J.H., Platenkamp, D.J., Hattenberg, T., Bosch, A.F., De Haan, P.H., 2009. GLARE teardowns from the MegaLiner Barrel (MLB) fatigue test. In: Bos, M.J. (Ed.), ICAF 2009: Bridging the Gap Between Theory and Operational Practice. Springer, Netherlands, Dordrecht, the Netherlands, pp. 145—167.

Interconversion of Weight and Atomic Percentages of Lithium and Aluminum in Aluminum— Lithium Alloys

The weight fraction or percentage, w_I, of any alloying element I in an alloy system $A-B - \ldots I - \ldots N$, can be converted to atomic fraction or percentage, a_I, by considering its weighted average in terms of atomic weight (AW), using the basic correlation:

$$a_I = \frac{w_I/(AW)_I}{(w_A/(AW)_A) + (w_B/(AW)_B) + \cdots + (w_I/(AW)_I) + \cdots + (w_N/(AW)_N)}$$

(A1.1)

Equation (A1.1) has been used to calculate the atomic percentages of major alloying elements in commercial Al–Li alloys from the weight percentages given in Table 2.1. The complete results are listed in Table A1.1. The results for the lithium contents have been used to plot Figure A1.1, which shows the weight and atomic percentages of lithium in the alloys. Note (1) the approximately linear relationship between the weight and atomic percentages and (2) that the third generation alloys have significantly lower lithium contents than most of the first and second generation alloys. As discussed in this book, the lower lithium range (below 2 wt%) for third generation alloys has been introduced to improve several properties, including yield strength isotropy, short-transverse ductility, short-transverse fracture toughness, and thermal stability.

TABLE A1.1 Compositions of Commercial Al–Li Alloys: wt% Data from Table 2.1 and Calculations of at.% as per Eq. (A1.1).

Alloy	Li		Cu		Mg		Ag		Mn		Zn		Al (Balance)	
	wt%	at.%	wt%	at.%	wt%	at.%	wt%	at.%	wt%	at.%	wt%	at.%	wt%	at.%
First generation														
2020	1.2	4.65	4.5	1.91					0.5	0.245			93.8	93.3
1420	2.1	7.61			5.2	5.43							92.7	87.3
1421	2.1	7.61			5.2	5.43							92.7	87.3
Second generation (Li ≥ 2%)														
2090	2.1	7.8	2.7	1.09									95.2	91.1
2091	2.0	7.4	2.0	0.80	1.3	1.37							94.7	90.4
8090	2.4	8.8	1.2	0.48	0.8	0.84							95.6	89.8
1430	1.7	6.3	1.6	0.65	2.7	2.87							94	90.2
1440	2.4	8.8	1.5	0.60	0.8	0.84							95.4	89.8
1441	1.95	7.2	1.65	0.67	0.9	0.95							95.5	91.1
1450	2.1	7.74	2.9	1.18									95	90.8
1460	2.25	8.35	2.9	1.18									94.85	90.5
Third generation (Li < 2%)														
2195	1.0	3.84	4.0	1.68	0.4	0.44	0.4	0.1					94.2	93.04

ID														
2196	1.75	6.62	2.9	1.20	0.5	0.544	0.4	0.1	0.35 max.	0.17 max.	0.35 max.	0.15 max.	93.75	91.22
2297	1.4	5.20	2.8	1.14	0.25 max.	0.26 max.			0.3	0.14	0.5	0.20	94.75	91.26
2397	1.4	5.28	2.8	1.16	0.25 max.	0.264 max.			0.3	0.144	0.10	0.040	95.15	94.22
2098	1.05	4.06	3.5	1.48	0.53	0.59	0.43	0.11	0.35 max.	0.17 max.	0.35	0.145	93.79	93.44
2198	1.0	3.88	3.2	1.29	0.5	0.56	0.4	01	0.50 max.	0.245 max.	0.35 max.	0.145 max.	94.05	93.78
2099	1.8	6.69	2.7	1.08	0.3	0.535			0.3	0.141	0.7	0.28	94.2	90.18
2199	1.6	6.05	2.6	1.03	0.2	0.215			0.3	0.143	0.6	0.24	94.7	92.4
2050	1.0	3.9	3.6	1.53	0.4	0.45	0.4	0.105	0.35	0.173	0.25 max.	0.1 max.	94.0	94.0
2296	1.6	6.01	2.45	0.96	0.6	0.64	0.43	0.104	0.28	0.13	0.25 max.	0.1 max.	94.4	91.25
2060	0.75	2.99	3.95	1.69	0.85	0.95	0.25	0.06	0.3	0.15	0.4	0.17	93.5	94.0
2055	1.15	4.46	3.70	1.56	0.4	0.44	0.4	0.1	0.3	0.147	0.5	0.21	93.55	93.1
2065	1.2	4.64	4.20	1.77	0.50	0.56	0.30	0.075	0.40	0.196	0.2	0.082	93.2	92.7
2076	1.5	5.7	2.35	0.975	0.5	0.544	0.28	0.068	0.33	0.16	0.30 max.	0.12 max.	94.74	92.5

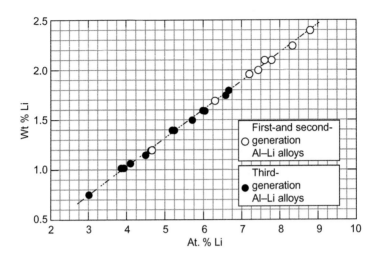

FIGURE A1.1   Lithium contents of the commercial Al–Li alloys in Table A1.1.

Selected Conversion Factors For SI Units

Convert From	Convert To	Multiplication Factor
Angstrom	Meter (m)	1×10^{-10}
Atmosphere (760 Torr)	Pascal (Pa)	1.013×10^5
Bar	Pascal (Pa)	1×10^5
Calorie	Joule (J)	4.184
Degree Celsius	Kelvin (K)	$t_k = t_c + 273.15$
Degree Fahrenheit	Kelvin (K)	$t_k = (t_f - 32)/1.8 + 273.15$
Degree Rankine	Kelvin (K)	$t_k = t_r/1.8$
Dyne	Newton (N)	1×10^{-5}
Dyne-centimeter	Newton-meter (N-m)	1×10^{-7}
Dyne/centimeter^2	Pascal (Pa)	0.1
Electron volt (eV)	Joule (J)	1.602×10^{-19}
Erg	Joule (J)	1×10^{-7}
Ergs/centimeter^2	Joule/meter^2 (J/m^2)	1×10^{-3}
Foot	Meter (m)	0.3048
Foot-pound	Joule (J)	1.356
Gram-force/centimeter^2	Pascal (Pa)	9.807
Inch	Meter (m)	0.024
Inch of mercury at 0°C	Pascal (Pa)	3386
Kilocalorie	Joule (J)	4184
Kilogram-force	Newton (N)	9.807
Kilogram-force-meter	Newton-meter (N-m)	9.807
Kilogram-force/centimeter^2	Pascal (Pa)	9.807×10^4
Kilogram-force/meter^2	Pascal (Pa)	9.807
Kilogram-force/millimeter^2	Pascal (Pa)	9.807×10^6
kip (1000 Pounds)	Newton (N)	4.448×10^3
kip/inch^2 (ksi)	Pascal (Pa)	6.895×10^6
kip/inch$^2 \cdot \sqrt{inch}$ (ksi $\sqrt{inch}$)	Pascal. $\sqrt{m}$ (Pa $\sqrt{m}$)	1.099×10^6
mil	Meter (m)	2.54×10^{-5}
Poise	Pascal-second (Pa-s)	0.1
Pound-force	Newton (N)	4.448
Pound-force/inch^2 (psi)	Pascal (Pa)	6.895×10^3
Pound-force/inch^2-$\sqrt{inch}$ (psi$\sqrt{in}$)	Pascal-$\sqrt{m}$ (Pa$\sqrt{m}$)	1.099×10^3

Index

Note: Page numbers followed by "*f*" and "*t*" refer to figures and tables, respectively.

β′ phase, 121−124
  crystal structure, 122−123
  precipitation behavior, 123−124
Δ phase, 125−126
  crystal structure, 125
  formation of, 125−126
Δ′ phase, 107−111
Δ′ precipitate-free zones (PFZs), 63−65, 71,
  88−89
Δ′ precipitates, 307−310
Ω phase, 77−78
θ′ phase, 111

A

Aerospace applications of Al−Li alloys, 501
  aircraft, 521−525
    materials for, 517−519
  aluminum alloys, CFRPs and FMLs,
    514−516
  GLARE, 510−511
  materials qualification, 521
  spacecraft, 525−529
    alloy property requirements for, 529
    aluminum alloys for, 519−521
    constellation, 526
    Orion, 526
    space launch system, 526−528
    SpaceX Falcon 9 Launcher, 528
  weight savings, 504−514
    density, 506
    density and elastic moduli, 506−508
    modern/innovative structural concepts,
      509−514
Aerostructural design and its application to
  Al−Li alloys, 27
  aircraft structural property requirements,
    28−30
    empennage, 30
    fuselage/pressure cabins, 29
    wings, 29−30
  engineering property requirements for,
    30−39

damage tolerance, 35−39
density and stiffness, 31−33
strength and ductility, 34−35
families of Al−Li alloys, 39−43
  second-generation alloys, 39−40
  third-generation alloys, 40−43
service qualification programs, 53−55, 54*t*
third-generation alloy property
    developments and trade-offs, 44−53
  fuselage/pressure cabins, 44−47
  lower wings, 49−52
  spars, ribs, and other internal structures,
    52−53
  upper wings, 47−49
Age hardening behavior, 4, 102−106
  age hardening curve, 103
  microstructures
    after high-temperature exposure,
      105−106
    at different stages of aging, 104−105
Aged Al−Li−X alloys, deformation behavior
  in, 15−17
AgustaWestland EH101, 42*f*, 418*f*, 458, 517*f*
Airbus A380, 28, 547*f*
  advanced materials/processes for, 519*f*
  materials and processes technology
    selection for, 518*f*
  MegaLiner Barrel (MLB) test, 546
  third-generation Al−Li alloys use in,
    523−524
Aircraft, 521−525
  design requirements for, 520−521
  noninspectability, 520
  stress corrosion cracking (SCC), 520−521
Aircraft fatigue life assessment methods, 542*t*
  current and proposed methods for, 542*t*
  Defence Science and Technology
    Organisation (DSTO) approach, 543
  DT method, 543
  Strain−life method, 543
  Stress−life (S−N), 541−542
Aircraft structural alloys, 28
Aircraft wings, 29−30

Airworthiness certification, of metallic
  materials, 537
  aviation and airworthiness regulatory
    bodies, 538–540
    civil aviation, 538–539
    military aviation, 539–540
  certification of Al–Li alloy, 549–553
    1441M sheet, 550–553
    certification methodology, 549–550, 551f
  fatigue design philosophies, 541–543
  materials and structures certification
    methodology, 544–548
Al-2004, 251–252
Al$_2$Cu (θ') phase, 111
Al$_2$Cu precipitates, 151
Al$_2$CuLi (T$_1$) phase
  in quaternary Al–Li–Cu–Mg alloys,
    115–116
  in ternary Al–Li–Cu alloys, 112–115
    crystal structure, 113–114
    nucleation, 114–115
    precursors, 112–113
Al$_2$CuMg (S$'$) phase and equilibrium S phase,
    116–121
Al$_6$CuLi$_3$ (T$_2$) phase, 127–130
Al$_3$Li (δ') phase, 107–111
Al$_3$Zr (β') phase, 121–124
Alcan Global ATI, 181
Alclad 2524-T3, 44, 45f, 46
Alcoa, 10, 21–22, 188
Alcoa/AFRL process flow chart, for reducing
    texture, 151f
Alcoa's 2020, 6
  ductility problem of, 7
Al–Cu–Li, 63
Al–Cu–Li–X alloys, precipitate structure of,
    13f, 14–15
Al–Cu–Mg alloys, 76–78, 116–117
Aleris Aluminum Koblenz GmbH,
    Germany, 181
Alkali metal impurity (AMI) phases, 426
Al–Li (δ') phase, 125–126
Al–Li–Cu and Al–Li–Cu–Mg alloys,
    462–478
  first generation alloys, 462
  localized corrosion of, 462–478
  second- and third-generation alloys,
    462–474
  stress corrosion cracking, 483–490
Al–Li–Cu system, 73–76
Al–Li–Cu–Mg alloys, 117–118
Al–Li–Cu–Mg–Zr, 63

Al–Li–Mg alloys, 63
  strengthening in, 72–73
Al–Li–Mg system, 71–73
Al–Li–X alloys
  background information that led to
    improvements in, 20–22
  precipitate structure in, 11–12
  predicting strain localization in, 17–19
Al–Li–Zr system, 78–79
Alloying elements and composition, influence
    of, 63–65
Al–Mg–Li alloys, 478–479
  localized corrosion of, 478–479
Alpha fiber components, 149–150
Aluminum alloy fatigue and fracture data
    requirements for aerospace structures,
    36t
Aluminum–lithium alloys, 28, 514–516
  1420, 8, 32t, 62–63, 64t, 100–101, 143t,
    223t, 238, 261t, 262–263, 267,
    269–271, 286, 459t, 556t
    liquation cracking of, 273
  1421, 32t, 64t, 143t, 223t, 261t, 459t, 556t
  1430, 32t, 64t, 143t, 223t, 261t, 459t, 556t
  1440, 32t, 64t, 143t, 223t, 261t, 459t, 556t
  1441, 32t, 64t, 143t, 261t, 263t, 267–271,
    276, 286–289
    hot cracking, 271
  1441M alloy certification, 550–553
    chemical composition and density,
      550–552, 552t
    proposed certification test program
      for, 552t
    room-temperature tensile properties,
      552t, 553
  1450, 261t
  1460, 73, 261t
  2020, 6–7, 62, 73, 100–101, 261t, 306
    influence of different aging treatments,
      319–320
    tensile properties of, 319–322
    thermomechanical processing studies,
      321–322
  2020-T651, 367–369
  2050, 261t
  2050-T84, 49–50
  2050-T84, 52
  2055, 49, 261t
  2055-T8E83, 49
  2055-T8X, 48–49, 48f
  2060, 44, 49–50, 261t
  2060-T8E30, 45f

2060-T8E86, 49−50
2076, 261*t*
2090, 12−14, 14*f*, 39−40, 62, 100−101, 112−113, 140−141, 209−210, 261*t*
  liquation cracking of, 273
  microstructures of weld interiors for, 279*f*
  weldability of, 268, 288−289
2091, 62, 100−101, 153, 261*t*
2096, 89
2098, 261*t*, 370−371
2098 billets, 233
2099, 261*t*, 392
2099-T81 extrusion alloy, 52
2195, 34−35, 88−89, 261*t*, 267−268, 289
2195 plate, 40−42
2196, 261*t*
2198, 44, 88, 261*t*
  tensile properties of, 330−331
2199, 44, 49−52, 156−157, 261*t*
2199-T86, 49−50
2199-T8E74, 45*f*
2219, 19−20, 519−520
2296, 261*t*
2297, 261*t*
2397, 261*t*
2965, 261*t*
5083, 286
5356, 268
7050-T7451, 52
7055-T7751, 48*f*
7055-T77511 alloy, 49
7075-T6511 alloy, 49
7075-T6651, 48*f*
7255-T7751, 48*f*
7XXX plate alloys, 47, 47*f*
8090, 39−40, 62, 100−101, 105−106, 128, 140−141, 151−153, 177−178, 209−210, 261*t*, 288
  anisotropy, 322−323
  strength differential, 328−329
  tensile properties of, 322−329
8090 forgings, 211
8090-T651, 369−370, 372*f*, 373*f*
8091, 39, 100−101, 209−210
Al−Zn−Mg alloy system, 285−286
Al−Zn−Mg−Cu system, 4−5
Anisotropy
  in AA 8090, 322−323
  ductility, 327−328
  of mechanical properties, 318
  reducing, 154−155
  thick plate, 156−157
  UTS, 327
  yield strength, 152−154, 323−326
Apollo Lunar Module structure, 519

B

Bell-Boeing V-22 Osprey airframe, 517
Beta fiber components, in binary and ternary Al−Li alloys, 146−149
Binary Al−Li alloys, 66−71
  beta fiber components in, 146−149
  hot workability of, 192−193
  localized corrosion of, 461
  stress corrosion cracking, 482−483
Boeing 247, 28
Boeing 787 *Dreamliner*, 28, 517
British Alcan, 10

C

CA loading, 402
Cadmium, 7, 86−87
Carbon fiber reinforced plastics (CFRPs), 514−516
  advantages and disadvantages of, 515*t*
Casting practices, 178−181
  Direct Chill (DC) casting, 180−181
  metal/mold reactions, castability and shape casting, 178−180
Center for Military Airworthiness and Certification (CEMILAC), 540
Certification of Al−Li alloy, 549−553
  1441M sheet, 550−553
    chemical composition and density, 550−552
    room-temperature tensile properties, 553
  certification methodology, 549−550, 551*f*
Circular oscillation, 282
Civil aviation, 538−539
Closed-die forgings, 188, 210−211
Coherent Al$_3$Li (δ'), 11−12
Commercial Al−Li alloys, 193−195
  compositions of, 556*t*
  nominal chemical compositions of, 459*t*
  precipitate phases in, 80−85
Composite materials, 173
Composite structures, 28
Constant amplitude (CA), 383
Constant stress ratio (CR), 383
Constellium, 188
Constellium consortium, 181
Constituent particles, 424−425

Conventional alloys I and Al−Li alloys, 387−395
  first-generation Al−Li alloys, 387−388
  second-generation Al−Li alloys, 388−391
  third-generation Al−Li alloys, 392−395
Conventional alloys II and Al−Li alloys, 395−402
  second-generation Al−Li alloys, 395−401
    gust spectrum loading, 395−400
    maneuver spectrum loading, 400−401
  third-generation Al−Li alloys, 401−402
Conventional alloys III and Al−Li alloys, 402−405
  CA loading, 402
  flight simulation loading, 402−405
Copper, addition of, 100
Corrosion, 289−291
Crack arrest, 416−417
Crack deviations, 157−159
Crack propagation, 343
"Crack tip shielding," 385
Cracking during solidification.
  See Solidification cracking
Creation of antiphase boundaries (APBs), 309
Crucible materials, 172−174
Crystallographic texture, 140
Cu:Mg ratio, 116−117
Cyclic hardening, 343
Cyclic softening, 343
Cyclic stress response (CSR) behavior, 355−357
  characterization of, 347
Cyclic stress−strain (CSS) behavior, 357−359
  characterization of, 347−348

D

Damage tolerance (DT), 35−39, 383
  fatigue, 35−37
  fatigue crack growth, 37−38
  fracture toughness, 38−39
  of stiffened panel concepts for fuselage and lower wing applications, 510−514
Damage tolerant alloy, 391
Damage tolerant extrusion alloys, 392
Damage tolerant sheet products, 392
Deformation mechanism
  superplastic behavior and, 246−249
Density and stiffness of Al−Li alloys, 31−33
DIFFRACT, 113−114
Diffusion creep, 227
Direct Chill casting, 180−181

Directorate General of Civil Aviation (DGCA), 538−539
Discontinuous dynamic recrystallization (DDRX), 196−201
Dispersoids, 65, 427
Double cantilever beam, 483
Double-cantilever-bend, 421
Douglas DC-2, 28
Draft Development/Type Test Schedule (DTTS), 549
"Dual slope" behavior, Al−Li alloys, 349−350
Ductility and fracture toughness, 312−317
  planar slip and strain localization, methods for reducing, 314−315
  planar slip deformation, nature and occurrence of, 313
  thermal treatments, 315−317
Ductility anisotropy, 327−328
Durability, 383
Dynamic materials modeling (DMM), 192

E

Electromagnetic stirring (EMS), 282
Electron backscatter diffraction orientation imaging microscopy (EBSD−OIM), 196−202
Electron diffraction patterns (EDPs), 113
Empennage, 30
Energy dispersive X-ray analysis (EDX), 87
Equiaxed zone (EQZ) formation, in Al−Li alloys, 273−278
  and associated fusion boundary cracking, 278
  experimental observations, 273−276
  hypotheses of, 276−278
Equilibrium S phase, 116−121
Equivalent initial flaw size (EIFS) concept, 543
European Aviation Safety Agency (EASA), 538−539
Exfoliation Corrosion Rating (EXCO) test, 462, 475*f*, 477*t*, 478
Extrinsic inclusions and porosity, 424
"Extrinsic workability," 203
Extrusion characteristics, 209−210

F

Face centered cubic (FCC) metals, 440−441
FALSTAFF, 402
Families of Al−Li alloys, 39−43

second-generation alloys, 39–40
third-generation alloys, 40–43
Fatigue behavior, of Al–Li alloys, 35–37, 341
  crack propagation, 343
  cyclic hardening, 343
  cyclic softening, 343
  cyclic stress response (CSR) behavior,
    355–357
  cyclic stress–strain (CSS) behavior,
    357–359
  fatigue toughness, 359–362
  high cycle fatigue (HCF), 343, 363–365
    aging, 366–367
    behavior of, 366–371
    cold work, 366–367
    general survey, 366
    lithium content effects, 366–367
    test methods, 366
  life power law relationships, 349–354
  low cycle fatigue (LCF), 343–345
    anisotropy in, 349
    behavior of Al–Li alloys, 344–345,
      348–363
    resistance of Al–Li alloys, 362–363
  microscopic crack initiation, 343
  phenomenon of fatigue, 342–343
  test methods and analyses, 345–348
    cyclic stress response (CSR) behavior,
      characterization of, 347
    cyclic stress–strain (CSS) behavior,
      characterization of, 347–348
Fatigue crack growth (FCG) behavior of
    Al–Li alloys, 37–38, 381, 406–408
  Al–Li and conventional alloys I, 387–395
    first-generation Al–Li alloys, 387–388
    second-generation Al–Li alloys,
      388–391
    third-generation Al–Li alloys, 392–395
  Al–Li and conventional alloys II, 395–402
    second-generation Al–Li alloys,
      395–401
    third-generation Al–Li alloys, 401–402
  Al–Li and conventional alloys III,
    402–405
    CA loading, 402
    flight simulation loading, 402–405
  analysis, 384
  practice-related crack growth regimes,
    408–409
  survey of, 384–387
    long/large cracks, 385–387
    short/small cracks, 387

testing, 383–384
Fatigue design philosophies, 541–543
Fatigue limit, 363–364
Fatigue toughness, 359–362
  variation of, 360f, 361f
Federal Aviation Act of 1958, 538–539
Federal Aviation Administration (FAA),
    538–539
Federal Aviation Agency, 538–539
FELIX, 402
Fiber metal laminates (FMLs), 514–516
  advantages and disadvantages of, 515t
First-generation Al–Li alloys, 55, 62, 306
  and conventional alloys I, FCG
    comparisons with, 387–388
Flight simulation loading, 402–405
Fluxes, in melting practices, 170–171
Forgings, 188, 210–211
Fracture toughness, 38–39, 315–317, 319
Fracture toughness and fracture modes of
    aerospace Al–Li alloys, 415
  microstructural features' effects on, 423–435
    alkali metal impurity (AMI) phases, 426
    constituent particles, 424–425
    dispersoids, 427
    extrinsic inclusions and porosity, 424
    grain boundary precipitates and
      precipitate-free zones, 430
    grain boundary segregation, 430–433
    matrix precipitates, 427–429
    microstructural design of third-generation
      Al–Li alloys, 433–435
  second-generation Al–Li alloys versus
    conventional AL alloys, 435–441
    room-temperature data, 435–439
    strain rate, 441
    testing temperature, 439–441
  test methods for determination of, 418–423
    plane-stress/plane-strain considerations,
      419–420
    testing sheet and thin plate, 422–423
    testing thick products, 420–421
  third-generation Al–Li alloys, 442–448
    sheet and thin plate, 444–448
    thick plate and other thick products,
      442–443
    uses and potential uses of, 448–449
Friction stir processing (FSP), 245
  superplasticity in, 245–246
Friction stir welding (FSW), 260, 295–296,
    504, 509
Friction welding, 292–295

Fridlyander, I.N., 8
Fuselage panel with 1441 Al−Li alloy skin, 253*f*
Fuselage skins, 44−46
Fuselage stringers and frames, 46−47
Fuselage/pressure cabins, 29
Fusion zone microstructures, 278−283
Fusion-welded Al−Li alloy airframe
        structures, 260

G

Gas tungsten arc (GTA) welds, 271
    Al−Li 1441, 288*t*
GLARE (GLAss REinforced aluminum
        laminates), 510−511, 546
Gleeble hot ductility test, 273
Gleeble weld thermal simulation test, 265
Grain boundaries (GBs), 99−100, 313
Grain boundary precipitates (GBPs),
        118−119, 430
Grain boundary precipitation, 88−89
Grain boundary segregation, 430−433
Grain boundary sliding (GBS), 227−228, 237
Grain boundary triple junctions (GBTJs), 313
Grain refinement, 174−178
    minor alloying additions for, 85−86
Graphite as crucible material, 173
Guinier−Preston (GP) zones, 386
Gust spectrum loading, 395−400

H

Hand forgings, 188
Helium atmosphere for melting, 171
High cycle fatigue (HCF) behavior, 343,
        363−371
    AA 2020-T651, 367−369
    AA 2098, 370−371
    AA 8090-T651, 369−370
    general survey, 366
    lithium content, aging, and cold work,
        effects of, 366−367
    resistance of Al−Li alloys, 371
    test methods, 366
High-angle grain boundaries (HAGBs), 115
High-cycle fatigue (HCF), 35−36
High-resolution electron microscope
        (HREM), 87
High-strain rate superplasticity (HSRSP), 225,
        246
High strength plate and extrusion products,
        392−394
High-temperature exposure, 105

Historical development, of aluminum−lithium
        alloys, 4−8
    Alcoa's 2020, 6
        ductility problem of, 7
    first modern Al−Li Alloys, 5
    in the Soviet Union, 7−8
HOListic Structural Integrity Process
        (HOLSIP) program, 544*f*
Hot cracking, in welding industry.
        See Solidification cracking
Hot deformation characteristics of Al−Li
        alloys, 192−196
    binary alloy, 192−193
    multicomponent alloys, 193−195
    superplastic behavior, 195−196
Houldcroft test, 265, 268−269
Hydrogen pickup and melt degassing, 174
Hysteresis loops, 345−346

I

Impurities, 87−88
In, minor addition of, 86−87
Indian Aircraft Act of 1934, 538−539
Industrial-scale processing, 211−215
    available information for third-generation
        alloys, review and discussion of,
        212−215
Inert gas backing, 264
Ingot metallurgy (I/M) approach, 10
Innovative structural concepts, of Al−Li
        alloys aerospace application, 509−514
    case study, 510−514
    survey of concepts, 509−510
Insoluble particles in Al−Li alloys, 102*t*
Instability Map, 192
Intergranular corrosion (IGC), 462−463
Intermetallic constituent particles, 65
International Civil Aviation Organization
        (ICAO), 538−539
Intrinsic workability, 191−192

J

Joint Aviation Authorities (JAA), 538−539
Joint Strike Fighter (JSF), 548

L

Lankford coefficient, 332*t*
Laser beam welding (LBW), 260, 504, 509
LeBaron, I.M., 4−5
Lifshitz−Wagner kinetic model, 70

Liquation cracking, 272—273
Local Type Certification Committee (LTCC), 549
Localized corrosion of Al—Li based alloys, 461—479
  Al—Li—Cu and Al—Li—Cu—Mg Alloys, 462—478
    first generation alloys, 462
    second- and third-generation alloys, 462—474
  Al—Mg—Li alloys, 478—479
  binary alloys, 461
  historical background, 457—461
Long/large cracks
  CA/CR loading, 385—386
  flight simulation loading, 386—387
Low cycle fatigue (LCF) behavior, 35—36, 343—345
  environmental effects, 348—349
  general survey, 348—349
  LCF resistance, 362—363
  microstructural features, 348—349
Low-temperature superplasticity (LTSP), 234—236

M

M value, 237
Magnetic arc oscillation (MAO), 282—283
Maneuver spectrum loading, 400—401
Manufacturing issues, 11
Master alloys, 86
  lithium addition via, 170
MASTMAASIS, 472
Matrix precipitates, 427—429
Mechanical working of Al—Li alloys, 187—191
  overview, 189—191
  processing, 203—211
    extrusions and forgings, 209—211
    industrial-scale processing, 211—215
    rolled products, 205—208
  workability, 191—192
    binary Al—Li alloy, hot workability of, 192—193
    hot deformation investigations, 203
    hot deformation regions favoring superplasticity, 195—196
    multicomponent Al—Li alloys, hot workability of, 193—195
    UL40, hot deformation and process mapping of, 196—202
Medium/high strength Al—Li products, 394
Melting and casting, 167—168

casting practices, 178—181
  Direct Chill (DC) casting, 180—181
  metal/mold reactions, castability and shape casting, 178—180
crucible materials, 172—174
grain refinement, 174—178
hydrogen pickup and melt degassing, 174
melt protection from atmosphere, 168—172
  addition of lithium via an Al—Li master alloy, 170
  inert atmosphere, 171—172
  Li addition and further melt processing under flux cover, 170—171
  lithium reactivity, 168—169
Metal/mold reactions, castability and shape casting, 178—180
Microscopic crack initiation, 343
Microstructural design of third-generation Al—Li alloys, 433—435
Microstructures, 99—101
  after high-temperature exposure, 105—106
  age hardening behavior, 102—106
  age hardening curve, 103
  at different stages of aging, 104—105
  example of, for third-generation Al—Li alloys, 159
  in solution-treated condition, 101—102
Military aviation, 539—540
MINITWIST, 395—397, 396f, 398f, 399f
Minor alloying additions, 85—87
  for grain refinement, 85—86
  for strengthening, 86—87
Modern aluminum—lithium alloy development, 8—22
  aged Al—Li—X alloys, deformation behavior in, 15—17
  Al—Li—X alloys
    background information that led to improvements in, 20—22
    precipitate structure in, 11—12
    predicting strain localization in, 17—19
  manufacturing issues, 11
  prior deformation effect on precipitation during ageing, 12—15
  second generation, 10
    applications and problems of, 19
  third generation, 19—20
Multicomponent Al—Li alloys, hot workability of, 193—195
  experimental alloys, 193
  near-commercial and commercial alloys, 193—195

N

NASA heavy-lift Space Launch System
(SLS), 525–526
Near-commercial alloys, 193–195
"Niche" applications, 39–40
Nominal compositions of Al–Li alloys, 223*t*
Nominally plane stress fracture toughness, 39
Nonoctahedral slip, 152

O

Orientation distribution function (ODF)
plot, 142
Overageing, 154–155

P

Partially melted zone (PMZ), 272–273
hot cracking in. *See* Liquation cracking
Pechiney (France), 10
Phase diagrams and phase reactions, 59
Al–Li binary system, 66–71
alloy development history, 62–63
impurities and grain boundary precipitation,
87–89
influence of alloying elements and
composition, 63–65
minor alloying additions to Al–Li alloys,
85–87
for grain refinement, 85–86
for strengthening, 86–87
nature of phases, 65–66
quaternary Al–Li–Cu–Mg system, 79–85
commercial Al–Li alloys, overview of
precipitate phases in, 80–85
phase equilibria, 79–80
ternary systems, 71–79
Al–Cu–Mg system, 76–78
Al–Li–Cu system, 73–76
Al–Li–Mg system, 71–73
Al–Li–Zr system, 78–79
Planar slip deformation, nature and occurrence
of, 313
Plane strain fracture toughness, 39
Plane-stress fracture toughness, 447*f*
Plane-stress/plane-strain considerations,
419–420
Powder metallurgy (P/M) approcah, 10
Power Dissipation Map, 192
Power law relationships, 344, 349–354
Precipitate particles, 66, 309–310
Precipitate phases in commercial Al–Li
alloys, 80–85

Precipitate-free zones (PFZs), 71, 104, 430
Precipitates
characteristics of, 106–130
Al_2Cu (θ') phase, 111
Al_2CuLi (T_1) phase in quaternary
Al–Li–Cu–Mg alloys, 115–116
Al_2CuLi (T_1) phase in ternary
Al–Li–Cu alloys, 112–115
Al_6CuLi_3 (T_2) phase, 127–130
Al_2CuMg (S') phase and the equilibrium
S phase, 116–121
Al–Li (δ') phase, 125–126
Al_3Li (δ') phase, 107–111
Al_3Zr (β') phase, 121–124
grain boundary, 88–89
in hot and cold forming, 151–152
Precipitation hardening, 99–100
Primary processing, 140
texture evolution during, 145–152
Process Maps, 192, 196–203
limitations to, 203
uses of, 203
Processing of Al–Li alloys, 203–211
extrusions and forgings, 209–211
extrusion characteristics, 209–210
forging characteristics, 210–211
industrial-scale processing, 211–215
rolled products, 205–208
grain structure and texture, 205–206
quench sensitivity, 206–208
Pulsed current (PC) technique, 280–282

Q

Quaternary Al–Li–Cu–Mg alloys, 79–85,
311–312
Al_2CuLi (T_1) phase in, 115–116
commercial Al–Li alloys, overview of
precipitate phases in, 80–85
phase equilibria, 79–80

R

Regional Centers for Military Airworthiness
(RCMAs), 540, 549–550
Rolled products, 188, 205–208
Room-temperature data, 435–439

S

S' phase, 116–121, 315
Al–Cu–Mg alloys, 116–117
Al–Li–Cu–Mg alloys, 117–118

heterogeneous and homogeneous
precipitation, 121
S′ and S crystal structure, 118
S′ and S nucleation and growth, 118−121
Samray, F.I., 8
Sandler, V.S., 8
Scleron, 4
Secondary processing, 140
Second-generation Al−Li alloys, 10, 62, 222,
306−307
in aerostructural design, 39−40
applications and problems of, 19
versus conventional Al alloys
fracture toughness of, 435−441
room-temperature data, 435−439
strain rate, 441
testing temperature, 439−441
and conventional alloys I, FCG
comparisons with, 388−391
development of, 33
differing FCG behaviors and advantages
for, 405−406
gust spectrum loading, 395−400
maneuver spectrum loading, 400−401
microstructural features in, 130f
Selected area electron diffraction (SAED), 87
Service qualification programs, 53−55, 54t
Severe plastic deformation (SPD) processes,
231−232, 235
Shear stress concentration, 18−19
Sheet and thin plate, 444−448
testing, 422−423
Sheet product fatigue crack deviations, 157−159
Shiryaeva, M.V., 8
Short/small cracks, 387, 402−405
CA loading, 402
flight simulation loading, 402−405
SI Units, conversion factors for, 559
SIGMA JIG test, 265
Silcock model, 111, 111f
Slip character, 151−152
Slip intensity, calculating, 17−18
Small-angle X-ray scattering (SAXS), 121
Sn, minor addition of, 86−87
S−N curve, 363−364, 364f
Solidification cracking
development of, 267−271
effect of pulsed current, 271
effects of filler metals, 267−269
varestraint testing, 269−270
general considerations, 264−265
guidelines, 265−267

Solid-state welding processes, 292−296
friction stir welding, 295−296
friction welding, 292−295
Soviet Union, development of Al−Li Alloys
in, 7−8
Space Launch System (SLS), 526−528, 528f
Space Shuttle Super Lightweight External
Tank (SLW ET), 289
Spacecraft
alloy property requirements for, 529
aluminum alloys for, 519−521
applications of Al−Li alloys, 525−529
constellation, 526
design requirements for, 520−521
Orion, 526
space launch system, 526−528
SpaceX Falcon 9 Launcher, 528
State of stress (SOS) workability, 191
STEM, 87
Strain localization, 314−315
Strain rate, effect on fracture toughness, 441
"Strength differential" effect, 328−329
Strength−ductility relationships for Al−Li
alloys, 34−35
Strengthening of Al−Li alloys, 307−312
by δ′ precipitates, 307−310
by binary/ternary phases, 310−311
by lithium-rich phase, 310−311
by magnesium-rich phases, 310−311
minor alloying additions for, 86−87
quaternary Al−Li−Cu−Mg alloys,
311−312
ternary Al−Li−Cu alloys, 311
Stress corrosion cracking (SCC), 416, 461,
478−491
Al−Li−Cu and Al−Li−Cu−Mg alloys,
483−490
crack Initiation, 483−486
crack propagation, 486−490
binary alloys, 482−483
mechanistic Implications, 490−491
Structural superplasticity, 222
Super Lightweight External Tank (SLW ET),
504
Super Lightweight Tank of the Space
Shuttle, 20
Superplastic behavior, 195−196
Superplastic forming (SPF), 239−244,
503−504, 509
applications, 251−253
cavity nucleation and growth, 240−244
role of friction stir processing on, 245−251

Superplastic forming (SPF) (*Continued*)
  cavity density and size distribution,
      250−251
  FSP materials, superplasticity in,
      245−246
  superplastic behavior and deformation
      mechanism, 246−249
Superplasticity, 221−224
  applications, 251−253
  effects of strain rate and strain rate
      sensitivity on, 237−239
  experimental investigations, 230−234
  low-temperature superplasticity, 234−236
  phenomenon of, 225−230
  stress−strain rate relationship in, 225
Super-solidus cracking, 264
SUPRAL alloys, 251−252

T

T_1 phase. *See* Al_2CuLi (T_1) phase
T_2 phase. *See* Al_6CuLi_3 (T_2) phase
Tantalum as crucible material, 172−173
Taylor−Bishop−Hill (TBH) polycrystal
      model theory, 153−154
Temperature, effect on fracture toughness,
      439−441
Tensile properties of Al−Li alloys, 318−331
  AA 2020, 319−322
      influence of different aging treatments,
          319−320
      thermomechanical processing studies,
          321−322
  AA 2198, 330−331
  AA 8090, 322−329
      anisotropy, 322−323
      strength differential, 328−329
Ternary Al−Li alloys
  beta fiber components in, 146−149
Ternary Al−Li−Cu alloys, 311
  Al_2CuLi (T_1) phase in, 112−115
Ternary systems, 71−79
  Al−Cu−Mg system, 76−78
  Al−Li−Cu system, 73−76
  Al−Li−Mg system, 71−73
  Al−Li−Zr system, 78−79
Texture in Al-Li alloys, 140−145
  evolution of, during primary processing,
      145−152
      alpha fiber components, 149−150
      beta fiber components, 146−149
      nonoctahedral slip, influence of, 152

precipitates and slip character, role of,
          151−152
      macroscopic anisotropy of yield strength,
          152−154
      practical methods to reduce texture,
          154−159
      processing methods, 154−155
      sheet product fatigue crack deviations,
          157−159
      thick plate anisotropy, 156−157
      third-generation Al−Li alloys, example
          microstructures for, 159
Thermal treatments, for improving strength
      and fracture toughness, 315−317
Thermomechanical processing (TMP), 195,
      205−206
Thermomechanical treatment (TMT), 227
Thick plate, 442−443
  anisotropy of, 156−157
Thick products, 442−443
  testing, 420−421
Third-generation Al−Li alloys, 19−20, 63,
      100−101, 307
  in aerostructural design, 40−43
  and conventional alloys I, 392−395
  and conventional alloys II, 401−402
  development of, 33
  differing FCG behaviors and advantages
      for, 405−406
  example microstructures for, 159
  fracture toughness of, 442−448
  microstructural features in, 130*f*
  processing information for, 212−215, 213*t*
  property developments and trade-offs,
      44−53
      fuselage/pressure cabins, 44−47
      lower wings, 49−52
      spars, ribs, and other internal structures,
          52−53
      upper wings, 47−49
  sheet and thin plate, 444−448
  "spider" charts for, 449*f*
  thick plate and other thick products,
      442−443
  uses and potential uses of, 448−449
Three-dimensional atom probe (3DAP), 87
TiBAl grain-refined 8090 alloy,
      microstructures of, 178*f*
Titanium alloys, 28
Transition elements, 86
Transitional alloys, processing information
      for, 213*t*

Transmission electron microscope (TEM), 121
Transport aircraft
　design criteria for, 541*f*
　fuselage of, 29, 44
　lower wing covers, 49−52
　main structural areas in, 542*f*
　third-generation Al−Li alloys in, 44*f*,
　　523−525
　upper wing covers, 47−49
　wing construction, 189*f*
TWIST, 402
Tyurin, B.V., 8

U

UL40, hot deformation and process mapping
　　of, 196−202
　EBSD−OIM, 196−202
Ultrafine-grained (UFG) materials, 235
United States Air Force (USAF), 543
Unrecrystallized/recrystallized
　　microstructures, 321−322
US Air Force Wright Laboratory, Materials
　　Directorate, 21−22
UTS anisotropy, 327

V

VAD 23, 62, 73
Varestraint test, 265, 269−270
Variable amplitude (VA) loading, 383−384
Very high cycle fatigue (VHCF), 363−364
Viscoplastic self-consistent (VPSC)
　　polycrystal model, 153−154

W

Weight and atomic percentages of lithium and
　　aluminum, interconversion of, 555
Weldalite alloys, 289, 458, 519−520
　049 alloys, 194, 211
Weldalite-type alloys, 88−89
Welding aspects of Al−Li alloys, 260
　corrosion, 289−291
　equiaxed zone (EQZ) formation, 273−278
　　and associated fusion boundary cracking,
　　278
　　experimental observations, 273−276
　　hypotheses of, 276−278
　fusion zone microstructures, modification
　　of, 278−283

　　inoculation, 279−280
　　magnetic arc oscillation, 282−283
　　pulsed current, 280−282
　liquation cracking, 272−273
　mechanical properties, 284−289
　　Al−Li 1420, 286
　　Al−Li 1441, 286−288
　　Al−Li 2090, 288−289
　　Al−Li 2195, 289
　　Al−Li 8090, 288
　solidification cracking, 264−271
　　development of, 267−271
　　general considerations, 264−265
　　guidelines, 265−267
　solid-state welding processes, 292−296
　　friction stir welding, 295−296
　　friction welding, 292−295
　weld metal porosity, 260−264
Wohler curve. *See* S−N curve
Workability, 191−192
　binary Al−Li alloy, hot workability of,
　　192−193
　hot deformation investigations, 203
　hot deformation regions favoring
　　superplasticity, 195−196
　multicomponent Al−Li alloys, hot
　　workability of, 193−195
　　experimental alloys, 193
　　near-commercial and commercial alloys,
　　193−195
　UL40, hot deformation and process
　　mapping of, 196−202
　EBSD−OIM, 196−202
Wrought product forms developed for Al−Li
　　alloys, 188, 190*t*

X

X8092 alloy, 39
X8192 alloy, 39
2XXX extrusion alloys, 52
7XXX high-strength plate alloys, 48−49

Y

Yb addition to Al−Li alloy, 87
Yield strength anisotropy, 152−154, 323−326

Z

Zirconium, 100, 102, 142, 175−176

Bibliography

Abis, S., Evangelista, E., Mengucci, P., Riontino, G., 1987. Calorimetric evaluation of the microstructure of Al−Cu−Li−Mg 2091 alloy. In: Champier, G., Dubost, B., Miannay, D., Sabetay, L. (Eds.), Proceedings of the Fourth International Conference on Aluminium−Lithium Alloys, J. Phys., 48, C3, 447−C3, 456.

Abis, S., Evangelista, E., Mengucci, P., Riontino, G., 1989. Microstructural evolution of Al−Li−Cu−Mg AA 8090 alloy during ageing at 160°C. In: Sanders Jr., T.H., Starke Jr., E.A. (Eds.), Proceedings of the Fifth International Conference on Aluminium−Lithium Alloys, vol. 2. Materials and Component Engineering Publications Ltd, Birmingham, UK, pp. 681−690.

Adam, C.M., 1981. Overview of DC casting. In: Sanders Jr., T.H., Starke Jr., E.A. (Eds.), Proceedings of the First International Conference on Aluminium−Lithium Alloys. The Metallurgical Society of AIME, Warrendale, PA, USA, pp. 37−48.

Adamczyk-Cieślak, B., Mizera, J., Kurzydłowski, K.J., 2010. Thermal stability of model Al−Li alloys after severe plastic deformation—effect of the solute Li atoms. Mater. Sci. Eng. A, A527, 4716−4722.

Agarwal, S.P., Kar, R.J. (Eds.), 1986. Aluminium-lithium development, application and superplastic forming. Proceedings of the WESTEC'86. American Society of Metals, Materials Park, OH, USA.

Agyekum, E., Ruch, W., Starke Jr., E.A., Jha, S.C., Sanders Jr., T.H., 1986. Effect of precipitate type on elastic properties of Al−Li−Cu and Al−Li−Cu−Mg alloys. In: Baker, C., Gregson, P.J., Harris, S.J., Peel, C.J. (Eds.), Proceedings of the Third International Conference on Aluminium−Lithium Alloys. The Institute of Metals, London, UK, pp. 448−454.

Ahmad, M., 1987. Thermal oxidation behaviour of an Al−Li−Cu−Mg−Zr alloy. Metall. Trans. A, 18A, 681−689.

Ahmad, M., Ericsson, T., 1986. Coarsening of δ', T₁, S' phases and mechanical properties of two Al−Li−Cu−Mg alloys. In: Baker, C., Gregson, P.J., Harris, S.J., Peel, C.J. (Eds.), Proceedings of the Third International Conference on Aluminium−Lithium Alloys. The Institute of Metals, London, UK, pp. 509−515.

Ahmad, Z., Aleem, A.J.W., 1996. Evaluation of corrosion behaviour of Al−Mg−Li alloys in sea water. J. Mater. Eng. Perform. 5, 235−240.

Ahmadi, S., Arabi, H., Shokuhfar, A., 2010. Effects of multiple strengthening treatments on mechanical properties and stability of nanoscale precipitated phases in an aluminum−copper−lithium alloy. J. Mater. Sci. Technol. 26, 1078−1082.

Ahmed, M., 1987, Correlation between aging heat treatments, microstructure and stress corrosion properties of Al−Li−Cu−Mg alloys. In: Champier, G., Dubost, B., Miannay, D., Sabetay, L. (Eds.), Proceedings of the Fourth International Conference on Aluminium−Lithium Alloys, J. Phys., 48, C3, 871−C3, 879.

Ahrens, T., Starke Jr., E.A., 1989. Effect of prior cold rolling on the mechanical properties and ageing behaviour of the Al−Li alloy 8090. In: Sanders Jr., T.H., Starke Jr., E.A. (Eds.), Proceedings of the Fifth International Conference on Aluminum−Lithium Alloys, vol. 1. Materials and Component Engineering Publications Ltd., Birmingham, UK, pp. 385−396.

Ainsdale, K.D., Noble, B., Harris, S.J., 1989. Al−Li−Zn−Mg−Zr alloys with high magnesium contents. In: Sanders Jr., T.H., Starke Jr., E.A. (Eds.), Proceedings of the Fifth International

Conference on Aluminum−Lithium Alloys, vol. 3. Materials and Component Engineering Publications Ltd., Birmingham, UK, p. 1707.

Albrecht, J., Lütjering, G., 1994. Elevated temperature properties of aluminium−lithium alloys. In: Sanders Jr., T.H., Starke Jr., E.A. (Eds.), Proceedings of the Fourth International Conference on Aluminum Alloys: Their Physical and Mechanical Properties, vol. 2. Georgia Institute of Technology, Atlanta, GA, USA, pp. 452−459.

Alcoa Aerospace Technical Fact Sheet, 2008. Alloy 2099-T83 and 2099-T8E67 extrusions. In: AEAP-Alcoa Engineered Aerospace Products. Lafayette, IN, USA.

Altman, M.B., 1974. High-strength, heat-resistant and structural alloys of aluminium with lithium. In: AD/A-005 977. Army Foreign Science and Technology Center, Charlottesville, VA, USA, Translation from Alyuminiyevyyesplavy, 1972, Moscow, Chapter 7, pp. 204−230.

Ambat, R., Prasad, R.K., Dwarakadasa, E.S., 1995. The influence of aging at 180°C on the corrosion behaviour of a ternary Al−Li−Zr alloy. Corros. Sci. 37, 1253−1265.

Amichi, R., Ridley, N., 1989. Superplastic behaviour and microstuctural evolution in Al−Li−8090 (Lital A). In: Sanders Jr., T.H., Starke Jr., E.A. (Eds.), Proceedings of the Fifth International Conference on Aluminum−Lithium Alloys, vol. 1. Materials and Component Engineering Publications Ltd., Birmingham, UK, pp. 159−168.

Aminul Islam, M.D., Haseeb, A.S.M.A., 1999. A comparative wear study on heat-treated aluminium−lithium alloy and pure aluminium. Mater. Sci. Eng. A, A268, 104−108.

Anand, K.G., Ramnarayan, V., Chaudhury, P.K., 1994. Formability of aluminium 2024-0 sheet. In: Sanders Jr., T.H., Starke Jr., E.A. (Eds.), Proceedings of the Fourth International Conference on Aluminum Alloys: Their Physical and Mechanical Properties, vol. 2. Georgia Institute of Technology, Atlanta, GA, USA, pp. 64−71.

Anderson, W.E., Quist, W.E., 1966. Aluminum Alloy. US Patent Number 3,284,193.

Angerman, C.L., 1964. The aging of aluminum−lithium alloys before and after neutron irradiation. J. Nucl. Mater. 11, 41−52.

Annessa, A.T.D., 1967. Microstructural aspects of weld solidification. Weld. J. 46, 491−499.

Anyalebechi, P.N., Tabolt, D.E.J., Granger, D.A., 1988. The solubility of hydrogen in liquid binary Al−Li alloys. Metall. Trans. B 19B, 227−232.

Anyalebechi, P.N., Talbot, D.E.J., Granger, D.A., 1990. Hydrogen diffusion in Al−Li alloys. Metall. Trans. B 21B, 649−655.

Archakova, Z.N., Romanova, O.A., Archakova, Z.I., 1960. Bulletins of AN SSSR OTN (Akademiyanauk.SSSR, Academy of Sciences of the USSR, Otdaleniyetechnicheskikhnauk, Department of Technical Sciences), No. 4, p. 106.

Armstrong, R.W., Javadpour, S., 1994. Dislocation mechanics description of plastic anisotropy and fracturing observations in textured Al−Li 2090-T8E41 alloy. In: Sanders Jr., T.H., Starke Jr., E.A. (Eds.), Proceedings of the Fourth International Conference on Aluminum Alloys: Their Physical and Mechanical Properties, vol. 2. Georgia Institute of Technology, Atlanta, GA, USA, pp. 405−412.

Arrowsmith, D.J., Clifford, A.W., Moth, D.A., Davies, R.J., 1986. Adhesive bonding of aluminium−lithium alloys. In: Baker, C., Gregson, P.J., Harris, S.J., Peel, C.J. (Eds.), Proceedings of the Third International Conference on Aluminium−Lithium Alloys. The Institute of Metals, London, UK, pp. 148−151.

Ashton, R.F., Thompson, D.S., Starke Jr., E.A., Lin, F.S., 1986a. Processing Al−Li−Cu−(Mg) alloys. In: Baker, C., Gregson, P.J., Harris, S.J., Peel, C.J. (Eds.), Proceedings of the Third International Conference on Aluminium−Lithium Alloys. The Institute of Metals, London, UK, pp. 66−77.

Ashton, R.F., Thompson, D.S., Gayle, F.W., 1986b. The effect of processing on the properties of Al–Li alloys. In: Starke Jr., E.A., Sanders Jr., T.H. (Eds.), Aluminum Alloys—Their Physical and Mechanical Properties. Engineering Materials Advisory Services, Warley, UK, pp. 403–417.

Assmann, P., 1926. Age hardened aluminium–lithium alloys. Z. Metallk. 18, 51.

Audier, M., Janot, M., de Boissieu, M., Dubost, B., 1989a. Structural relationships in intermetallic compounds of the Al–Li–(Cu, Mg, Zn) system. Philos. Mag. B 60, 437–466.

Audier, M., Janot, M., de Boissieu, M., Dubost, B., 1989b. Structural correlations between periodic and quasiperiodic Al–Li–(Cu, Mg, Zn) intermetallic compounds. In: Sanders Jr., T.H., Starke Jr., E.A. (Eds.), Proceedings of the Fifth International Conference on Aluminium–Lithium Alloys, vol. 2. Materials and Component Engineering Publications Ltd., Birmingham, UK, pp. 565–574.

Avalos-Borja, M., Pizzo, P.P., Larson, L.A., 1984. Transmission electron microscopy characterisation of microstructural features in aluminium–lithium–copper alloys. In: Sanders Jr., T.H., Starke Jr., E.A. (Eds.), Proceedings of the Second International Conference on Aluminum–Lithium Alloys. The Metallurgical Society of AIME, Warrendale, PA, USA, pp. 287–302.

Averill, W.A., Olsen, D.L., Matlock, D.K., Edwards, G.R., 1981. Lithium reactivity and containment. In: Sanders Jr., T.H., Starke Jr., E.A. (Eds.), Proceedings of the First International Conference on Aluminium–Lithium Alloys. The Metallurgical Society of AIME, Warrendale, PA, USA, pp. 9–28.

Avramovic-Cingara, G., McQueen, H.J., Hopkins, A., Perovic, D.D., 1994. Effects of thermomechanical processing on the microstructure of Al–Li–Cu–Mg alloy (with < 1% Cr, Mn and Zr). In: Sanders Jr., T.H., Starke Jr., E.A. (Eds.), Proceedings of the Fourth International Conference on Aluminum Alloys: Their Physical and Mechanical Properties, vol. 2. Georgia Institute of Technology, Atlanta, GA, USA, pp. 258–265.

Babel, H., Parrish, C., 2004. Manufacturing considerations for aluminum–lithium alloys. Paper 03H0108, WESTEC 2004, March 22–25, 2004, Los Angeles, CA, USA.

Bach, R.O., Aluminium–lithium dispersion alloys. In: Sanders Jr., T.H., Starke Jr., E.A. (Eds.), Proceedings of the First International Conference on Aluminum–Lithium, Alloys. The Metallurgical Society of AIME, Warrendale, PA, USA, pp. 29–36.

Bagaryatsky, Y.A., 1948a. Rentgenograficheskoe issledovanie stareniya alyuminievykh splavov 1. Primenenie monokhromaticheskikh rentgenovykh luchei dlya izucheniya struktury ostarennykh splavov. Zhur. Tech. Fiziki. 18, 828–830.

Bagaryatsky, Y.A., 1948b. Zhur. Tech. Fiziki. 18, 827.

Bairwa, M.L., Date, P.P., 2004. Effect of heat treatment on the tensile properties of Al–Li alloys. J. Mater. Process. Technol. 153–154, 603–607.

Bairwa, M.L., Desai, S.G., Date, P.P., 2005. Identification of heat treatments for better formability in an aluminum–lithium alloy sheet. J. Mater. Eng. Perform. 14, 623–633.

Baker, C., Gregson, P.J., Harris, S.J., Peel, C.J. (Eds.), 1986. Proceedings of the Third International Conference on Aluminium–Lithium Alloys. The Institute of Metals, London, UK.

Balasubramaniam, R., Duquette, D.J., 1989. The stress corrosion cracking susceptibility of an aluminium–lithium copper alloy. In: Sanders Jr., T.H., Starke Jr., E.A. (Eds.), Proceedings of the Fifth International Conference on Aluminum–Lithium Alloys, vol. 3. Materials and Component Engineering Publications Ltd., Birmingham, UK, pp. 1271–1280.

Balasubramaniam, R., Duquette, D.J., Rajan, K., 1991. On stress corrosion cracking in aluminum–lithium alloys. Acta Metall. Mater. 39, 2597–2605.

Ball, M.D., Lagace, H., 1986. Characterisation of coarse precipitates in an overaged Al−Li−Cu−Mg alloy. In: Baker, C., Gregson, P.J., Harris, S.J., Peel, C.J. (Eds.), Proceedings of the Third International Conference on Aluminium−Lithium Alloys. The Institute of Metals, London, UK, pp. 555−564.

Ball, M.D., Lloyd, D.J., 1985. Particles apparently exhibiting five-fold symmetry in Al−Li−Cu−Mg alloys. Scr. Metall. 19, 1065−1068.

Balmuth, E.S., 1984. Particle size determination in an Al−3Li alloy using DSC. Scr. Metall. 18, 301−304.

Balmuth, E.S., 1994. The status of Al−Li alloys, in aluminium alloys: their physical and mechanical properties. In: Sanders Jr., T.H., Starke Jr., E.A. (Eds.), Proceedings of the Fourth International Conference on Aluminium−Lithium Alloys. Georgia Institute of Technology, Atlanta, GA, USA, pp. 82−89.

Balmuth, E.S. and Chellman, D.J., 1987, Alloy design for overcoming the limitations of Al−Li alloy plate. In: Champier, G., Dubost, B., Miannay, D., Sabatay, L. (Eds.), Proceedings of the Fourth International Conference on Aluminium−Lithium Alloys, J. Phys., 48, C3, 293−C3, 299.

Balmuth, E.S., Schmidt, R., 1981. A perspective on the development of aluminum−lithium alloys. In: Sanders Jr., T.H., Starke Jr., E.A. (Eds.), Proceedings of the First International Conference on Aluminium−Lithium Alloys. The Metallurgical Society of AIME, Warrendale, PA, USA, pp. 69−88.

Banerjee, S., Arya, A., Das, G.P., 1997. Formation of an ordered intermetallic phase from a disordered solid solution—a study using first-principle calculations in Al−Li alloys. Acta Mater. 45, 601−609.

Bang, H.J., Kim, S., Prakash, J., 2001. Electrochemical investigations of lithium−aluminum alloy anode in Li/polymer cells. J. Power Sources 92, 45−49.

Baram, O., 1991. Structure and properties of a rapidly solidified Al−Li−Mn−Zr alloy for high-temperature applications: Part II. Spray atomization and deposition processing. Metall. Trans. A, 22A, 2515−2522.

Barbaux, Y., 1989. Properties of commercially available Al−Li alloys possible use on civil aircraft. In: Sanders Jr., T.H., Starke Jr., E.A. (Eds.), Proceedings of the Fifth International Conference on Aluminium−Lithium Alloys, vol. 3. Materials and Component Engineering Publications Ltd., Birmingham, UK, pp. 1667−1676.

Bardi, U., Caporali, S., Craig, M., Giorgetti, A., Perissi, I., Nicholls, P.R., 2009. Electrodeposition of aluminium film on P90 Li−Al alloy as protective coating against corrosion. Surf. Coat. Technol. 203, 1373−1378.

Barlat, F., Brem, J.C., Liu, J., 1992. On crystallographic texture gradient and its mechanical consequence in rolled aluminum−lithium sheet. Scr. Metall. Mater. 27, 1121−1126.

Barlat, F., Miyasato, S.M., Liu, J., Brem, J.C., 1994. On crystallographic texture and anisotropy in Al−Li sheet. In: Sanders Jr., T.H., Starke Jr., E.A. (Eds.), Proceedings of the Fourth International Conference on Aluminum Alloys: Their Physical and Mechanical Properties, vol. 2. Georgia Institute of Technology, Atlanta, GA, USA, pp. 389−396.

Bartges, C., Tosten, M.H., Howell, P.R., Ryba, E.R., 1987. A combined single crystal X-ray diffraction and electron diffraction study of the T_2 phase in Al−Li−Cu alloys. J. Mater. Sci. 22, 1663−1669.

Bartges, C.W., Ryba, E., 1989. The crystal structure of Al_5CuLi_3. In: Sanders Jr., T.H., Starke Jr., E.A. (Eds.), Proceedings of the Fifth International Conference on Aluminum−Lithium Alloys, vol. 2. Materials and Component Engineering Publications Ltd., Birmingham, UK, pp. 711−720.

Bate, P.S., 1992. Plastic anisotropy in a superplastic Al–Li–Mg–Cu alloy. Metall. Trans. A, 23A, 1467–1478.

Baumann, S.F., Williams, D.B., 1984. The effect of ternary additions on the δ'/α misfit and the δ' solvus line Al–Li alloys. In: Sanders Jr., T.H., Starke Jr., E.A. (Eds.), Proceedings of the Second International on Conference Aluminum–Lithium Alloys. The Metallurgical Society of AIME, Warrendale, PA, USA, pp. 17–29.

Baumann, S.F., Williams, D.B., 1985a. Effects of capillarity and coherency on δ' (Al$_3$Li) precipitation in dilute Al–Li alloys at low undercoolings. Acta Metall. 33, 1069–1078.

Baumann, S.F., Williams, D.B., 1985b. Experimental observations of the nucleation and growth of δ' (Al$_3$Li) in dilute Al–Li alloys. Metall. Trans. A, 16A, 1203–1211.

Baumert, B.A., Ricker, R.E., 1986. Effect of heat treatment on corrosion resistance of Al–Mg–Li alloy. In: Baker, C., Gregson, P.J., Harris, S.J., Peel, C.J. (Eds.), Proceedings of the Third International Conference on Aluminium–Lithium Alloys. The Institute of Metals, London, UK, pp. 282–286.

Bauri, R., Surappa, M.K., 2007. Processing and properties of Al–Li–SiC$_p$ composites. Sci. Technol. Adv. Mater. 8, 494–502.

Bavarian, B., Becker, J., Parikh, S.N., Zamanzadeh, M., 1989. Localized corrosion of 2090 and 2091 Al–Li alloys. In: Sanders Jr., T.H., Starke Jr, E.A. (Eds.), Proceedings of the Fifth International Conference on Aluminium–Lithium Alloys, vol. 3. Materials and Component Engineering Publications Ltd., Birmingham, UK, pp. 1227–1236.

Behnood, N., Evans, J.T., 1989. Plastic deformation and the flow stress of aluminium–lithium alloys. Acta Metall., 687–695.

Belov, N.A., Eskin, D.G., Aksenov, A., 2005. Multicomponent Phase Diagrams: Applications for Aluminium Alloys. Elsevier Publications, New York, NY, USA, pp. 217–222.

Bencze, K., Pelikan, C.H., Bahemann-Hoffmeister, A., Kronseder, A., 1991. Lithium/aluminium alloys a problem material for biological monitoring. Sci. Total Environ. 101, 83–90.

Bennet, C.G., Webster, D., 1994. Vaclite^{TM} Al–Li alloy products—a new approach to Al–Li alloy production. In: Sanders Jr., T.H., Starke Jr., E.A. (Eds.), Proceedings of the Fourth International Conference on Aluminum Alloys: Their Physical and Mechanical Properties, vol. 2. Georgia Institute of Technology, Atlanta, GA, USA, pp. 98–105.

Bero, B., Ukolova, G., Lukina, E.A., Bykova, S.V., Samarina, M.V., 2002. Investigation of phase transformations in the process of long-term low-temperature heating of aluminum alloys with lithium. Metal Sci. Heat Treat. 44, 50.

Bhalla, P., Garg, M.L., Dhawan, D.K., 2010. Protective role of lithium during aluminium-induced neurotoxicity. Neurochem. Int. 56, 256–262.

Binsfeld, F., Habashi, M., Galland, J., Fidelle, J.P., Miannay, D. and Rofidal, P., Hydrogen embrittlement in Al–Li–Cu–Mg alloys (8090-T651). In: Champier, G., Dubost, B., Miannay, D., Sabetay, L. (Eds.), Proceedings of the Fourth International Conference on Aluminium–Lithium Alloys, J. Phys., 48, C3, 587–C3, 596.

Birch, M.E.J., 1986. Grain refining of aluminium–lithium based alloys with titanium boron aluminium. In: Baker, C., Gregson, P.J., Harris, S.J., Peel, C.J. (Eds.), Proceedings of the Third International Conference on Aluminium–Lithium Alloys. The Institute of Metals, London, UK, pp. 152–158.

Birch, M.E.J., Cowell, A.J.J., 1987. Titanium–carbon–aluminium: a novel grain refiner for aluminium–lithium alloys. In: Champier, G., Dubost, B., Miannay, D. and Sabatay, L. (Eds.), Proceedings of the Fourth International Conference on Aluminium–Lithium Alloys, J. Phys., 48, C3, 103–C3, 108.

Bird, R.K., Discus, D.L., Fridyander, I.N., Sandler, V.S., 2000. Al–Li alloy 1441 for fuselage applications. Mater. Sci. Forum 331–337, 907–912.

Bird, R.K., Discus, D.L., Fridlyander, N.I., Sandler, V.S., 2001. Aluminum–lithium alloy 1441 as a promising material for fuselage. Metal Sci. Heat Treat. 43, 298–301.

Birt, M.J., Beevers, C.J., 1989. The fatigue response of the aluminium–lithium alloy, 8090. In: Sanders Jr., T.H., Starke Jr., E.A. (Eds.), Proceedings of the Fifth International Conference on Aluminum–Lithium Alloys, vol. 2. Materials and Component Engineering Publications Ltd., Birmingham, UK, pp. 983–992.

Bischler, P.J.E., Martin, J.W., 1986. Effect of heat treatment upon tensile strength and fracture properties of an Al–Li–Cu–Mg alloy. In: Baker, C., Gregson, P.J., Harris, S.J., Peel, C.J. (Eds.), Proceedings of the Third International Conference on Aluminium–Lithium Alloys. The Institute of Metals, London, UK, pp. 539–546.

Bischler, P.J.E. and Martin, J.W., 1987. The temperature dependence of the high cycle fatigue properties of an Al–Li–Cu–Mg alloy. In: Champier, G., Dubost, B., Miannay, D., Sabatay, L. (Eds.), Proceedings of the Fourth International Conference on Aluminium–Lithium Alloys, J. Phys., 48, C3, 761–C3, 767.

Bishopp, J.A., Jobling, D., Thompson, G.E., 1990. The surface pretreatment of aluminium–lithium alloys for structural bonding. Int. J. Adhes. Adhes. 10, 153–160.

Blackburn, L.B., Starke Jr., E.A., 1989. Effect of In additions on microstructure mechanical property relationships for an Al–Cu–Li alloy. In: Sanders Jr., T.H., Starke Jr., E.A. (Eds.), Proceedings of the Fifth International Conference on Aluminium–Lithium Alloys, vol. 2. Materials and Component Engineering Publications Ltd., Birmingham, UK, pp. 751–766.

Blackwell, P.L., Bate, P.S., 1993. The absence of relative grain translation during superplastic deformation of an Al–Li–Mg–Cu–Zr alloy. Metall. Trans. A, 24A, 1085–1093.

Blandin, J.J., 1989. Lithium effects in high temperature deformation of an Al–Li alloy: application to superplasticity. Mater. Sci. Eng. A, A122, 215–225.

Blankenship, C.P., Starke, E.A., 1992. Fracture behavior of aluminum–lithium–X alloys. In: Peters, M., Winkler, P.J. (Eds.), Aluminum–Lithium Alloys VI, vol. 1. Deutsche Gesellschaft für Metallkunde, Frankfurt, Germany, pp. 187–201.

Blom, A.F., 1990, Short crack growth under realistic flight loading: Model predictions and experimental results for Al 2024 and Al–Li 2090. In: Short-Crack Growth Behaviour in Various Aircraft Materials. AGARD Report No. 767, Advisory Group for Aerospace Research and Development, Neuilly-sur-Seine, France, pp. 6-1–6-15.

Bohlen, J.W., Chanani, G.R., 1984. Investigation of Al–Li based alloys at Northrop. In: Sanders Jr., T.H., Starke Jr., E.A. (Eds.), Proceedings of the Second International Conference on Aluminum–Lithium Alloys. The Metallurgical Society of AIME, Warrendale, PA, USA, pp. 407–418.

Bolam, V.J., Gregson, P.J., Gray, A., 1989. The influence of surface treatments on the fatigue performance of 8090. In: Sanders Jr., T.H., Starke Jr., E.A. (Eds.), Proceedings of the Fifth International Conference on Aluminum–Lithium Alloys, vol. 2. Materials and Component Engineering Publications Ltd., Birmingham, UK, pp. 1097–1104.

Boselli, J., Bray, G., Rioja, R.J., Mooy, D., Venema, G., Feyen, G., et al., 2012. The metallurgy of high fracture toughness aluminum-based plate products for aircraft internal structure. In: Weiland, H., Rollett, A.D., Cassada, W.A. (Eds.), Proceedings of the 13th International Conference on Aluminum Alloys (ICAA13). The Minerals, Metals and Materials Society (TMS) and John Wiley & Sons, Hoboken, NJ, pp. 581–586.

Boukos, N., Rocofyllou, E., Papastaikoudis, C., 1998a. Microstructure of Al–Li–Cu–Mg–Zr alloys with In additions. Mater. Sci. Eng. A, A256, 280–288.

Boukos, N., Flouda, E., Papastaikoudis, C., 1998b. The effect of Ag additions on the microstructure of aluminium−lithium alloys. J. Mater. Sci. 33, 3213−3218.

Bourgasser, P., Wert, J.A., Starke Jr., E.A., 1989. Intergranular fracture of Al−Li−Cu−Mg alloy resulting from non-equilibrium eutectic melting during solution treatment. Mater. Sci. Technol. 5, 1102−1108.

Bowden, D.M., Meschter, P.J., 1984. Electron beam weld solidification structures and properties in Al−3Li−X alloys. Scr. Metall. 18, 963−968.

Bowen, A.W., 1990. Annex: texture analysis of 2090-T8E41aluminium−lithium alloy sheet. In: Short-Crack Growth Behaviour in Various Aircraft Materials, AGARD Report No. 767, Advisory Group for Aerospace Research and Development, Neuilly-sur-Seine, France, pp. 11-1−11−5.

Bradsky, V.G., Ricks, R.A., 1987. Solidification microstructures in rapidly solidified, gas atomized aluminium−lithium alloy powders. J. Mater. Sci. 22, 1469−1476.

Braun, R., 1994a. Exfoliation corrosion and stress corrosion cracking behaviour of 8090-T81 sheet. In: Sanders Jr., T.H., Starke Jr., E.A. (Eds.), Proceedings of the Fourth International Conference on Aluminum Alloys: Their Physical and Mechanical Properties, vol. 2. Georgia Institute of Technology, Atlanta, GA, USA, pp. 511−518.

Braun, R., 1994b. Stress corrosion cracking behaviour of Al−Li alloy 8090-T8171. In: Sanders Jr., T.H., Starke Jr., E.A. (Eds.), Proceedings of the Fourth International Conference on Aluminum Alloys: Their Physical and Mechanical Properties, vol. 2. Georgia Institute of Technology, Atlanta, GA, USA, pp. 519−528.

Braun, R., Buhl, H., 1987. Corrosion behaviour of Al−Li−Cu−Mg alloy 8090-T651. In: Champier, G., Dubost, B., Miannay, D., Sabetay, L. (Eds.), Proceedings of the Fourth International Conference on Aluminium−Lithium Alloys, J. Phys., 48, C3, 843−C3, 850.

Brechet, Y., Livet, F., 1987. Low cycle fatigue of binary Al−Li alloys: III-Coalescence of δ' precipitates in fatigue: X-ray low angle scattering investigation. In: Champier, G., Dubost, B., Miannay, D. and Sabetay, L. (Eds.), Proceedings of the Fourth International Conference on Aluminium−Lithium Alloys, J. Phys., 48, C3, 717−C3, 720.

Brechet, Y., Louchet, F., Marchionni, C., Gaugry, V.J.L., 1987a. Experimental (TEM and STEM) investigation and theoretical approach to the fatigue-induced dissolution of δ' precipitates in a 2.5 wt% Al−Li alloy. Philos. Mag. 56A, 353−366.

Brechet, Y., Louchet, F., Verger-Gaugry, J.L., 1987b. Low cycle fatigue of binary Al−Li alloys. II-Microstructural characterization and in situ fatigue experiments. In: Champier, G., Dubost, B., Miannay, D., Sabetay, L. (Eds.), Proceedings of the Fourth International Conference on Aluminium−Lithium Alloys, J. Phys., 48, C3, 709−C3, 716.

Brechet, Y., Louchet, F., Magnin, T., 1990. Fatigue of Ai−Li alloys: mechanical properties, microstructure (TEM and SEM) and strain localization. J. Mater. Sci. 25, 3053−3060.

Bregianos, A.C., Crosky, A.G., Munroe, P.R., Hellier, A.K., 2010. A study aimed at determining and understanding the fracture behaviour of an Al−Li−Cu−Mg−Zr alloy 8090. Int. J. Fract. 161, 141−159.

Bretz, P.E., 1987. Aluminum−lithium alloy development and production. In: Champier, G., Dubost, B., Miannay, D., Sabetay, L. (Eds.), Proceedings of the Fourth International Conference on Aluminium−Lithium Alloys, J. Phys., 48, C3, 25−C3, 31.

Bretz, P.E., Sawtell, R.R., 1986. Alithalite' alloys: progress, products and properties. In: Baker, C., Gregson, P.J., Harris, S.J., Peel, C.J. (Eds.), Proceedings of the Third International Conference on Aluminium−Lithium Alloys. The Institute of Metals, London, UK, pp. 47−56.

Bretz, P.E., Mueller, L.N., Vasudevan, A.K., 1984. Fatigue properties of 2020-T651 aluminium alloy. In: Sanders Jr., T.H., Starke Jr., E.A. (Eds.), Proceedings of the Second International Conference on Aluminium−Lithium Alloys. The Metallurgical Society of AIME, Warrendale, PA, USA, pp. 543−559.

Broussaud, F., Diot, C., 1987. Influence of grain morphology and texture on the deformation and fracture of extruded CP276 Al−Li alloy. In: Champier, G., Dubost, B., Miannay, D., Sabatay, L. (Eds.), Proceedings of the Fourth International Conference on Aluminium−Lithium Alloys, J. Phys., 48, C3, 597−C3, 603.

Broussaud, F., Thomas, M., 1986. Influence of δ' phase coalescence on Young's modulus in an Al−2.5 wt% Li alloy. In: Baker, C., Gregson, P.J., Harris, S.J., Peel, C.J. (Eds.), Proceedings of the Third International Conference on Aluminium−Lithium Alloys. The Institute of Metals, London, UK, pp. 442−447.

Bucci, R.J., Malcolm, R.C., Colvin, E.L., Murtha, S.J., James, R.S., 1989. Cooperative test programme for the evaluation of engineering properties of Al−Li alloy 2090-T8X sheet, plate and extrusion products. Naval Surface Warfare Center Technical Report 89-106, Silver Spring, MD, USA.

Buchheit, E.T., Wert, J.A., 1997. Microstructure−property relationships in low-density Al−Li−Mg alloys. Metall. Trans. A, 24A, 853−863.

Buchheit, R.G., Stoner, G.E., 1989. The corrosion behaviour of the T_1 (Al_2CuLi) intermetallic compound in aqueous environments. In: Sanders Jr., T.H., Starke Jr., E.A. (Eds.), Proceedings of the Fifth International Conference on Aluminum−Lithium Alloys, vol. 3. Materials and Component Engineering Publications Ltd., Birmingham, UK, pp. 1347−1359.

Bull, M.J., Lloyd, D.J., 1986. Textures developed in Al−Li−Cu−Mg alloy. In: Baker, C., Gregson, P.J., Harris, S.J., Peel, C.J. (Eds.), Proceedings of the Third International Conference Aluminium−Lithium Alloys. The Institute of Metals, London, UK, pp. 402−410.

Burke, M., Papazian, J.M., 1986. Elevated temperature oxidation of Al−Li alloys. In: Baker, C., Gregson, P.J., Harris, S.J., Peel, C.J. (Eds.), Proceedings of the Third International Conference on Aluminium−Lithium Alloys. The Institute of Metals, London, UK, pp. 287−293.

Butts, D.A., Gale, W.F., 2004. In: Gale, W.F., Totemeier, T.C. (Eds.), Equilibrium Diagrams, Smithells Metals Reference Book, VIII ed., pp. 1−534.

Canaby, J.L., Blazy, F., Fries, J.F., Traverse, J.P., 1991. Effects of high temperature surface reactions of aluminium−lithium alloy on the porosity of welded areas. Mater. Sci. Eng. A, A136, 131−139.

Cassada, W.A., Shiflet, G.J., Starke, E.A., 1986a. Grain boundary precipitates with five-fold diffraction symmetry in an Al−Li−Cu alloy. Scr. Metall. 20, 751−756.

Cassada, W.A., Shiflet, G.J., Starke Jr., E.A., 1986b. The effect of germanium on the precipitation and deformation behavior of Al−2Li alloys. Acta Metall. 34, 367−378.

Cassada, W.A., Shiflet, G.J., Starke Jr. E.A., 1987. The effect of plastic deformation on T_1 precipitation. In: Champier, G., Dubost, B., Miannay, D., Sabetay, L. (Eds.), Proceedings of the Fourth International Conference on Aluminium−Lithium Alloys, J. Phys., 48, C3, 397−C3, 406.

Cassada, W.A., Shiflet, G.J., Starke Jr., E.A., 1991a. Mechanism of Al_2CuLi (T_1) nucleation and growth. Metall. Trans. A, 22A, 287−296.

Cassada, W.A., Shiflet, G.J., Starke Jr., E.A., 1991b. The effect of plastic deformation on Al_2CuLi (T_1) precipitation. Metall. Trans. A, 22A, 299−306.

Castillo, D.L., Wu, Y., Hu, H.M., Lavernia, E.J., 1999. Microstructure and mechanical behavior of spray-deposited high-Li Al–Li alloys. Metall. Mater. Trans. A, 30A, 1381–1389.

Castro, M.R.S., Nogueira, J.C., Thim, G.P., Oliveira, M.A.S., 2004. Adhesion and corrosion studies of a lithium based conversion coating film on the 2024 aluminum alloy. Thin Solid Films 457, 307–312.

Cavaliere, P., Cabibbo, M., Panella, F., Squillace, A., 2009. 2198 Al–Li plates joined by friction stir welding: mechanical and microstructural behavior. Mater. Design 30, 3622–3631.

Cerasara, S., Girada, A., Sanchez, A., 1977. Annealing of vacancies and ageing in Al–Li alloys. Philos. Mag. 35, 97.

Chai, Z.G., Xu, Y., Meng, F.L., 1999. The interface between δ' and matrix in an 8090 Al–Li alloy studies by SAXS. Mater. Charact. 42, 27–30.

Chakravorthy, C.R., 1988. Studies on the Characteristics of Cast Binary Aluminium–Lithium Alloys (Doctoral Thesis). Indian Institute of Technology, Kharagpur, India.

Chakravorty, C.R., Chakraborty, M., 1991. Grain refining of aluminium–lithium alloy with Al–Ti–B. Cast Met. 4, 98–100.

Champier, G., Samuel, F.H., 1986. Microstructure of Al–Li–Si alloys rapidly solidified. In: Baker, C., Gregson, P.J., Harris, S.J., Peel, C.J. (Eds.), Proceedings of the Third International Conference on Aluminium–Lithium Alloys. The Institute of Metals, London, UK, pp. 131–136.

Chang, S.C., Huang, J.H., 1986. The fracture and shear band formation in an Al–Cu–Li–Mg–Zr alloy. Acta Metall. 34, 1657–1662.

Chang, Y.C., Wu, Y., Hao, Y.Y., Liu, Z.F., Chen, G.Y., 1989. Structure and properties of two rapidly solidified Al–Li–Cu–Mg–Zr alloys produced by liquid dynamic compaction. In: Sanders Jr., T.H., Starke Jr., E.A. (Eds.), Proceedings of the Fifth International Conference on Aluminium–Lithium Alloys, vol. 1. Materials and Component Engineering Publications Ltd., Birmingham, UK, pp. 95–104.

Chaudhuri, J., Gondhalekar, V., Inchekel, A., Talia, J.E., 2003. Study of precipitation and deformation characteristics of the aluminium–lithium alloy by X-ray double crystal diffractometry. J. Mater. Sci. 25, 3938–3940.

Chellman, D.J., Bayha, T.D., Qiong, L., Wawner, F.E., 1994. Microstructure/properties of high temperature spray deposited Al–Cu–Mg–X alloys. In: Sanders Jr., T.H., Starke Jr., E.A. (Eds.), Proceedings of the Fourth International Conference on Aluminum Alloys: Their Physical and Mechanical Properties, vol. 2. Georgia Institute of Technology, Atlanta, GA, USA, pp. 2–9.

Chen, C.Q., Li, H.X., 1989. Crack path profiles of Al–Li single crystals under monotonic and cyclic loading. In: Sanders Jr., T.H., Starke Jr., E.A. (Eds.), Proceedings of the Fifth International Conference on Aluminium–Lithium Alloys, vol. 2. Materials and Component Engineering Publications Ltd., Birmingham, UK, pp. 973–981.

Chen, C.Q., Dong, Y.J., Li, H.X., Zhang, Y.G., 1994a. Mechanical behaviour of aluminium–lithium single crystals at 77K. In: Sanders Jr., T.H., Starke Jr., E.A. (Eds.), Proceedings of the Fourth International Conference on Aluminum Alloys: Their Physical and Mechanical Properties, vol. 2. Georgia Institute of Technology, Atlanta, GA, USA, pp. 290–296.

Chen, D.L., Chaturvedi, M.C., 2000. Near-threshold fatigue crack growth behaviour of 2195 aluminium–lithium alloy—prediction of crack propagation direction and influence of stress ratio. Metall. Mater. Trans. A, 31A, 1531–1541.

Chen, D.L., Chaturvedi, M.C., 2001. Effects of welding and weld heat-affected zone simulation on the microstructure and mechanical behavior of AA 2195 aluminum–lithium alloy. Metall. Mater. Trans. A, 32a, 2729–2741.

Chen, D.L., Chaturvedi, M.C., Goel, N., Richards, N.L., 1999. Fatigue crack growth behavior of X2095 Al–Li alloy. Inter. J. Fatigue 21, 1079–1086.

Chen, G.S., Duquette, D.J., 1993. The effect of aging on the hydrogen-assisted fatigue cracking of a precipitation-hardened Al–Li–Zr alloy. Metall. Trans. A, 25A, 1551–1572.

Chen, G.S., Duquette, G.S., 1992. Corrosion fatigue of a precipitation-hardened Al–Li–Zr alloy in a 0.5 M sodium chloride solution. Metall. Trans. A, 23A, 1563–1572.

Chen, J., Mady, Y., Morgeneyer, Th.F., Besson, J., 2011. Plastic flow and ductile rupture of a 2198 Al–Cu–Li aluminium alloy. Comput. Mater. Sci. 50, 1365–1371.

Chen, L., Chen, W., Liu, Z., Shao, Y., Hu, Z., 1993. Effects of hydrogen on mechanical properties and fracture mechanism of 8090 Al–Li alloy. Metall. Trans. A, 24A, 1355–1361.

Chen, P.S., Bhat, B.N., 2002. Time–temperature–precipitation behavior in Al–Li alloy 2195. In: National Aeronautics and Space Administration Technical Memorandum NASA/TM 2002-211548, NASA Marshall Space Flight Center, Huntsville, AL, USA.

Chen, R.T., Starke Jr., E.A., 1984. Microstructure and mechanical properties of mechanically alloyed, ingot metallurgy and powder metallurgy Al–Li–Cu–Mg alloys. Mater. Sci. Eng. 67, 229–245.

Chen, S.W., Chang, Y.A., Chu, M.G., 1989a. Phase equilibria and solidification of Al–Li alloys. In: Sanders Jr., T.H., Starke Jr., E.A. (Eds.), Proceedings of the Fifth International Conference on Aluminium–Lithium Alloys, vol. 2. Materials and Component Engineering Publications Ltd., Birmingham, UK, pp. 585–594.

Chen, S.W., Jan, C.H., Lin, J.C., Chang, Y.A., 1989b. Phase equilibria of the Al–Li binary system. Metall. Trans. A, 20A, 2247–2258.

Chen, T.R., Peng, G.J., Huang, J.C., 1996. Low quench sensitivity of superplastic 8090 Al–Li thin sheets. Metall. Mater. Trans. A, 27A, 2923–2933.

Chen, Z., Zheng, Z., Yi, H., 2005. Effects of trace silver and magnesium on the microstructure and mechanical properties of Al–Cu–Li alloys. J. Wuhan Univ. Tech. Mater. Sci. Ed. 20, 9–12, Pp.

Chen, Z.G., Zheng, Z.Q., Tan, C.Y., Yun, D.F., 1994b. The effect of small additions of silver on ageing behaviour of Al–Mg–Li alloys. In: Sanders Jr., T.H., Starke Jr., E.A. (Eds.), Proceedings of the Fourth International Conference on Aluminum Alloys: Their Physical and Mechanical Properties, vol. 2. Georgia Institute of Technology, Atlanta, GA, USA, pp. 177–182.

Cheng-Yu, T., Zi-Qiao, Z., Chang-Qing, X., Ying, H., 2005. The aging feature of Al–Li–Cu–Mg–Zr alloy containing Sc. J. Cent. South Univ. Technol. 7, 65–67.

Chiem, C.Y., Zhou, X.W., Lee, W.S., 1987. High-strain-rate behaviour of CP-271 aluminium–lithium. In: Champier, G., Dubost, B., Miannay, D., Sabatay, L. (Eds.), Proceedings of the Fourth International Conference on Aluminium–Lithium Alloys, J. Phys., 48, C3, 577–C3, 586.

Cho, A., Ashton, R.F., Steele, W., Kirby, J.L., 1989. Status of weldable Al–Li alloy (Weldalite 049™) development at Reynolds metals company. In: Sanders Jr., T.H., Starke Jr., E.A. (Eds.), Proceedings of the Fifth International Conference on Aluminium–Lithium Alloys, vol. 3. Materials and Component Engineering Publications Ltd., Birmingham, UK, pp. 1359–1364.

Cho, C.W., Cheney, B.A., Lege, D.J., Petit, J.I., 1987. Superplasticity of 2090 SPF sheet at hot rolled gauge. In: Champier, G., Dubost, B., Miannay, D., Sabatay, L. (Eds.), Proceedings of the Fourth International Conference on Aluminium–Lithium Alloys, J. Phys., 48, C3, 277–C3, 284.

Cho, K.K., Chung, Y.H., Lee, C.W., Kwun, S.I., Shin, M.C., 1999. Effects of grain shape and texture on the yield strength anisotropy of Al–Li alloy sheet. Scr. Mater. 40, 651–657.

Chokshi, A.H., Mukherjee, A.K., 1988. A topological study of superplastic deformation in an Al−Li alloy with a bimodal grain size distribution. Metall. Trans. A, 19A, 1621−1623.

Chokshi, A.H., Mukherjee, A.K., 1989. The cavitation and fracture characteristics of a superplastic Al−Cu−Li−Zr alloy. Mater. Sci. Eng. A, A110, 49−60.

Christoulou, L., Struble, L., Pickens, J.R., 1984. Stress−corrosion cracking in Al−Li binary alloys. In: Sanders Jr., T.H., Starke Jr., E.A. (Eds.), Proceedings of the Second International Conference on Aluminium−Lithium Alloys. ASM International, Materials Park, OH, USA, pp. 561−581.

Chung, L.C., Cheng, J.H., 2002. Fracture criterion and forming pressure design for superplastic bulging. Mater. Sci. Eng. A, A333, 146−154.

Cingara, A.G., Perovic, D.D., McQueen, H.J., 1996. Hot deformation mechanisms of a solution-treated Al−Li−Cu−Mg−Zr alloy. Metall. Mater. Trans. A, 27A, 3478−3490.

Clarke, E.R., Gillespie, P., Page, F.M., 1986. Heat treatment of Li/Al alloys in salt baths. In: Baker, C., Gregson, P.J., Harris, S.J., Peel, C.J. (Eds.), Proceedings of the Third International Conference on Aluminium−Lithium Alloys. The Institute of Metals, London, UK, pp. 159−163.

Colvin, E.L., Murtha, S.J., 1989. Exfoliation corrosion testing of Al−Li alloys 2090 and 2091. In: Sanders Jr., T.H., Starke Jr., E.A. (Eds.), Proceedings of the Fifth International Conference on Aluminum−Lithium Alloys, vol. 3. Materials and Component Engineering Publications Ltd., Birmingham, UK, pp. 1251−1260.

Colvin, G.N., Starke Jr., E.A., 1988. Quench sensitivity of the Al−Li−Cu−Mg alloy 8090. SAMPE Q. 19, 10−21.

Constant, D., Doudeau, M., Mace, R., 1989. Development and industrialisation of Al−Li by Cegedur Pechiney Rhenalu. In: Sanders Jr., T.H., Starke Jr., E.A. (Eds.), Proceedings of the Fifth International Conference on Aluminum−Lithium Alloys, vol. 1. Materials and Component Engineering Publications Ltd., Birmingham, UK, pp. 65−74.

Cook, R., 1990. The growth of short fatigue cracks in 2024 and 2090 aluminium alloys under variable amplitude loading. In: Short-Crack Growth Behaviour in Various Aircraft Materials, AGARD Report No. 767, Advisory Group for Aerospace Research and Development, Neuilly-sur-Seine, France, pp. 5-1−5-11.

Couch, P.D., Bowen, P., 1994. Characteristics of quenched-in residual stresses generated in Al−Li matrix composites. In: Sanders Jr., T.H., Starke Jr., E.A. (Eds.), Proceedings of the Fourth International Conference on Aluminum Alloys: Their Physical and Mechanical Properties, vol. 2. Georgia Institute of Technology, Atlanta, GA, USA, pp. 690−697.

Couch, P.D., Burger, A., Suman, E.W., 1994. The effect of cooling rate from the homogenisation temperature on the microstructure of high strength 7000 series alloys. In: Sanders Jr., T.H., Starke Jr., E.A. (Eds.), Proceedings of the Fourth International Conference on Aluminum Alloys: Their Physical and Mechanical Properties, vol. 2. Georgia Institute of Technology, Atlanta, GA, USA, pp. 10−17.

Coyne, E.J., Sanders Jr., T.H., Starke Jr., E.A., 1981. The effect of microstructure and moisture on the low cycle fatigue and fatigue crack propagation of two Al−Li−X alloys. In: Sanders Jr., T.H., Starke Jr., E.A. (Eds.), Proceedings of the First International Conference on Aluminium−Lithium Alloys. The Metallurgical Society of AIME, Warrendale, PA, USA, pp. 293−305.

Craig, J.G., Newman, R.C., Jarrett, M.R., Holroyd, N.J.H., 1987. Local chemistry of stress−corrosion cracking in Al−Li−Cu−Mg alloys. In: Champier, G., Dubost, B., Miannay, D., Sabetay, L. (Eds.), Proceedings of the Fourth International Conference on Aluminium−Lithium Alloys. J. Phys., 48, C3, 825−C3, 834.

Criner, C.B., 1957. Aluminum Base Alloy. US Patent 2,784,126, issued March 5.

Criner, C.B.. 1959. Aluminum Base Alloy. US Patent 2,915,391, issued December 1.

Crookes, R.E., Kenik, E.A., Starke Jr., E.A., 1983. HVEM in situ deformation of Al–Li–X alloys. Scr. Metall. 17, 643–647.

Crookes, R.E., Kenik, E.A., Starke Jr., E.A., 1984. The microstructure and tensile properties of a splat-quenched Al–Cu–Li–Mg–Zr alloy. Metall. Trans. A, 15A, 1367–1377.

Crooks, R., Mitchell, M.R., 1989. The fracture of 2090 plate in bending fatigue. In: Sanders Jr., T.H., Starke Jr., E.A. (Eds.), Proceedings of the Fourth International Conference on Aluminum–Lithium Alloys, vol. 2. Materials and Component Engineering Publications Ltd., Birmingham, UK, pp. 1033–1042.

Crooks, R., Wang, Z., Levit, A.V.I., Shenoy, R.N., 1998. Microtexture, microstructure and plastic anisotropy of AA 2195. Mater. Sci. Eng. A, A257, 145–152.

Crosby, K.E., 2000. Development of texture and texture gradient in Al–Cu–Li (2195) thick plate. J. Mater. Sci. 35, 3189–3195.

Cross, C.E., Olso, D.L., Edwards, G.R., Capes, J.F., 1984. Weldability of aluminium–lithium alloys. In: Sanders Jr., T.H., Starke Jr., E.A. (Eds.), Proceedings of the Second International Conference on Aluminum–Lithium Alloys, The Metallurgical Society of AIME, Warrendale, PA, USA, pp. 675–682.

Csontos, A.A., Starke, E.A., 2000. The effect of processing and microstructure development on the slip and fracture behavior of the 2.1 wt.pct Li AF/C-489 and 1.8 wt.pct Li AF/C-458 Al–Li–Cu–X alloys. Metall. Mater. Trans. A, 31A, 1965–1977.

Csontos, A.A., Starke Jr., E.A., 2005. The role of inhomogeneous plastic deformation on the fracture behavior of age-hardenable Al alloys. Int. J. Plast. 21 (6), 1097–1118.

Csontos, A.A., Gable, B.M., Starke Jr., E.A., 2004. Effect of quench rate, pre-age stretch, and artificial aging on the Al–Li–Cu–X alloy AF/C-458. In: Tabereaux, A.T. (Ed.), Light Metals. The Minerals, Metals and Materials Society, Warrendale, PA, USA, pp. 891–896.

Cui, J., Fu, Y., Li, N., Sun, J., He, J., Dai, Y., 2000. Study on fatigue crack propagation and extrinsic toughening of an Al–Li alloy. Mater. Sci. Eng. A, A281, 126–131.

Cui, Z., Zhong, W., Wei, Q., 1994. Superplastic behavior at high strain rate of rapidly solidified powder metallurgy Al–Li alloy. Scr. Metall. 30, 123–128.

Cutler, L.R., Hajjaji, M., Lesperance, G., 1989. Influence of composition and processing history on the macrostructure and microstructure of melt-spun Al–Li alloys. In: Sanders Jr., T.H., Starke Jr., E.A. (Eds.), Proceedings of the Fifth International Conference on Aluminum–Lithium Alloys, vol. 1. Materials and Component Engineering Publications Ltd., Birmingham, UK, pp. 113–122.

Dahl, N., Wada, H., Kitamura, T., 1994. Aluminium borate reinforced Al–alloys. In: Sanders Jr., T.H., Starke Jr., E.A. (Eds.), Proceedings of the Fourth International Conference on Aluminum Alloys: Their Physical and Mechanical Properties. Georgia Institute of Technology, Atlanta, GA, USA, pp. 582–589.

Dai, S.L., Wu, Y., Castillo, L.D., Lavernia, E.J., 1997. Microstructure and properties of spray deposited high Li Al–Li–Mg–Ge–Zr alloys. Scr. Metall. 37, 265–270.

Dalins, I., Karimi, M., Ila, D., 1991. Violent oxidation of lithium-containing aluminium alloys in liquid oxygen. Appl. Surf. Sci. 48–49, 509–516.

Damerval, C., Raviart, J.L., Lapasset, G., 1987. Influence of processing conditions on the monotonic properties of 8090 alloy. In: Champier, G., Dubost, B., Miannay, D., Sabetay, L. (Eds.), Proceedings of the Fourth International on Conference on Aluminium–Lithium Alloys, J. Phys., 48, C3, 661–C3, 668.

Damerval, C., Lapasset, G., Kubin, L.P., 1989. Plastic instabilities in some aluminium–lithium alloys. In: Sanders Jr., T.H., Starke Jr., E.A. (Eds.), Proceedings of the Fifth International Conference on Aluminum–Lithium Alloys, vol. 2. Materials and Component Engineering Publications Ltd., Birmingham, UK, pp. 859–868.

Daniélou, A., Ronxin, J.P., Nardin, C., Ehrström, J.C., 2012. Fatigue resistance of Al–Cu–Li and comparison with 7XXX aerospace alloys. In: Weiland, H., Rollett, A.D., Cassada, W.A. (Eds.), Proceedings of the 13th International Conference on Aluminum Alloys (ICAA13). The Minerals, Metals and Materials Society (TMS) and John Wiley & Sons, Hoboken, NJ, USA, pp. 511–516.

Davenport, S.B., Gregson, P.J., Moreton, R., Peel, C.J., 1994. The mechanisms of fatigue crack and delamination zone growth in fibre reinforced aluminium–lithium laminates. In: Sanders Jr., T.H., Starke Jr., E.A. (Eds.), Proceedings of the Fourth International Conference on Aluminum Alloys: Their Physical and Mechanical Properties, vol. 2. Georgia Institute of Technology, Atlanta, GA, USA, pp. 590–598.

Davies, C.K.L., Poolay-Mootien, S., Stevens, R.N., Tetlow, P.L., 1992a. High-temperature creep of the particle hardened commercial Al–Li–Cu–Mg alloy 8090. J. Mater. Sci. 27, 3953–3962.

Davies, C.K.L., Poolay-Mootien, S., Stevens, R.N., 1992b. Effect of precipitate type and morphology on high-temperature creep and internal stress in Al–Li based alloys. J. Mater. Sci. 27, 6159–6170.

Davó, B., Conde, A., De Damborenea, J., 2006. Stress corrosion cracking of B13, a new high strength aluminium–lithium alloy. Corros. Sci. 48, 4113–4126.

De, P.S., Mishra, R.S., Baumann, J.A., 2011. Characterization of high cycle fatigue behavior of a new generation aluminum–lithium alloy. Acta Mater. 59, 5946–5960.

De Hosson, J., Th, M., Kanert, O., Tamler, H., 1984. Dislocation dynamics in aluminium–lithium alloys: a transmission electron microscopic and nuclear resonance study. In: Sanders Jr., T.H., Starke Jr., E.A. (Eds.), Proceedings of the Second International Conference on Aluminum–Lithium alloys II. The Metallurgical Society of AIME, Warrendale, PA, USA, pp. 169–180.

De Jonge, J.B., Schütz, D., Lowak, H., Schijve, J., 1973. A standardized load sequence for flight simulation tests on transport aircraft wing structures, National Aerospace Laboratory NLR Technical Report NLR-TR-73029, Amsterdam, the Netherlands.

Degreve, F., Dubost, B., Dubus, A., Thorne, N.A., Bodart, F., Demotier, G., 1987. Quantitative analysis of intermetallic phases in Al–Li alloys by electron, ion and nuclear microprobes. In: Champier, G., Dubost, B., Miannay, D., Sabetay, L. (Eds.), Proceedings of the Fourth International Conference on Aluminium–Lithium Alloys, J. Phys., 48, C3, 505–C3, 512.

Deschamps, A., Sigli, C., Mourey, T., de Geuser, F., Lefebvre, W., Davo, B., 2012. Experimental and modelling assessment of precipitation kinetics in an Al–Li–Mg alloy. Acta Mater. 60, 1917–1928.

Dhers, J., Driver, J., Fourdeux, A., 1986. Cyclic deformation of binary Al–Li alloys. In: Baker, C., Gregson, P.J., Harris, S.J., Peel, C.J. (Eds.), Proceedings of the Third International Conference on Aluminium–Lithium Alloys. The Institute of Metals, London, UK, pp. 233–238.

Di, Z., Sajji, S., Hori, S., 1987. Effect of microstructure on high cycle fatigue behaviour of Al–Li binary alloy. In: Champier, G., Dubost, B., Miannay, D., Sabatay, L. (Eds.), Proceedings of the Fourth International Conference on Aluminium–Lithium Alloys, J. Phys., 48, C3, 753–C3, 759.

Dickson, R.C., Lawless, K.R., Wefers, K., 1988. Internal LiH and hydrogen porosity in solutionised Al–Li alloys. Scr. Metall. 22, 917–922.

Dickson, R.C., Wefers, K., Lawless, K.R., 1989. Internal lithium hydride precipitation in Al–Li alloys. In: Sanders Jr., T.H., Starke Jr., E.A. (Eds.), Proceedings of the Fifth International Conference on Aluminum–Lithium Alloys, vol. 3. Materials and Component Engineering Publications Ltd., Birmingham, UK, pp. 1337–1346.

Diego, N.D., Río, J.D., Romero, R., Somoza, A., 1997. A positron study on the microstructural evolution of Al–Li based alloys in the early stages of plastic deformation. Scr. Metall. 37, 1367–1371.

Dinsdale, K., Harris, S.J., Noble, B., 1981. Relationship between microstructure and mechanical properties of aluminium–lithium–magnesium alloys. In: Sanders Jr., T.H., Starke Jr., E.A. (Eds.), Proceedings of the First International Conference on Aluminium–Lithium Alloys. The Metallurgical Society of AIME, Warrendale, PA, USA, pp. 101–118.

Dinsdale, K., Noble, B., Harris, S.J., 1988. Development of mechanical properties in Al–Li–Zn–Mg–Cu alloys. Mater. Sci. Eng. A, A104, 75–84.

Disson, G., Reboul, M., Fiaud, C., 1989. A stress corrosion cracking study of the 2091 Al–Li alloy. In: Sanders Jr., T.H., Starke Jr., E.A. (Eds.), Proceedings of the Fifth International Conference on Aluminum–Lithium Alloys, vol. 3. Materials and Component Engineering Publications Ltd., Birmingham, UK, pp. 1261–1270.

Divecha, A.P., Karmarkar, S.D., 1981. Casting problems specific to aluminium–lithium alloys. In: Sanders Jr., T.H., Starke Jr., E.A. (Eds.), Proceedings of the First International Conference on Aluminium–Lithium Alloys. The Metallurgical Society of AIME, Warrendale, PA, USA, pp. 49–62.

Dodd, B., Kobayashi, H., 1990. Dynamic fracture of binary Al–Li alloy in torsion materials science and technology. Mater. Sci. Technol. 6, 447–452.

Doglione, R., Ilia, E., Firrao, D., 1994. K-R curves of 2090 high strength tempers thick sheets. In: Sanders Jr., T.H., Starke Jr., E.A. (Eds.), Proceedings of the Fourth International Conference on Aluminum Alloys: Their Physical and Mechanical Properties. Georgia Institute of Technology, Atlanta, GA, USA, pp. 374–381.

Dolega, L., Mizera, J., Boguslawa, A.C., Kurzydlowski, K.J., 2011. Corrosion resistance of model ultrafine-grained Al–Li alloys produced by severe plastic deformation. J. Mater. Sci. 46, 1–8.

Donachie, S.J., Gilman, P.S., 1984. The microstructure and properties of Al–Mg–Li alloys prepared by mechanical alloying. In: Sanders Jr., T.H., Starke Jr., E.A. (Eds.), Proceedings of the Second International Conference on Aluminum–Lithium Alloys. The Metallurgical Society of AIME, Warrendale, PA, USA, pp. 507–516.

Donkor, F., Edwards, M.R., Wicks, D.C., 1989. Effect of prior cold rolling on the mechanical properties and ageing behaviour of the aluminium–lithium alloy 8090. In: Sanders Jr., T.H., Starke Jr., E.A. (Eds.), Proceedings of the Fifth International Conference on Aluminum–Lithium Alloys, vol. 1. Materials and Component Engineering Publications Ltd., Birmingham, UK, pp. 397–406.

Donnadieu, P., 1994. Al$_5$Li$_3$Cu (R) and Al6Li3Cu (T$_2$) phases: from planar defects to icosahedral order. In: Sanders Jr., T.H., Starke Jr., E.A. (Eds.), Proceedings of the Fourth International Conference on Aluminum Alloys: Their Physical and Mechanical Properties, vol. 2. Georgia Institute of Technology, Atlanta, GA, USA, pp. 222–229.

Donnadieu, P., Shao, F.e.Y., De Geuser, F., Botton, G.A., Lazar, S., Cheynet, M., et al., 2011. Atomic structure of T$_1$ precipitates in Al–Cu–Li alloys revisited with HAADF-STEM imaging and small angle X-ray scattering. Acta. Mater. 59, 462–472.

Donnellan, M.E., Frazier, W.E., 1989. An examination of the quench sensitivity of alloy 2090. In: Sanders Jr., T.H., Starke Jr., E.A. (Eds.), Proceedings of the Fifth International

Conference on Aluminum—Lithium Alloys, vol. 1. Materials and Component Engineering Publications Ltd., Birmingham, UK, pp. 355—364.

Doorbar, P.J., Borradaile, J.B., Driver, D., 1986. Evaluation of aluminium—lithium—copper—magnesium—zirconium alloy as forging material. In: Baker, C., Gregson, P.J., Harris, S.J., Peel, C.J. (Eds.), Proceedings of the Third International Conference on Aluminium—Lithium Alloys. The Institute of Metals, London, UK, pp. 496—508.

Dorward, R.C., 1988. Solidus and solvus isotherms for quaternary Al—Li—Cu—Mg alloys. Metall. Trans. A, 19A, 1631—1634.

Du, Z.W., Zhou, T.T., Chen, C.Q., Liu, P.Y., Dong, B.Z., 2005. Quantitative analysis of precipitation in an Al—Zn—Mg—Cu—Li alloy. Mater. Charact. vol. 55, 75—82.

Dubost, B., Lang, J.M., Degreve, F., 1986. Metallography and microanalysis of aluminium—lithium alloys by secondary ion mass spectrometry (SIMS). In: Baker, C., Gregson, P.J., Harris, S.J., Peel, C.J. (Eds.), Proceedings of the Third International Conference on Aluminium—Lithium Alloys. The Institute of Metals, London, UK, pp. 355—359.

Dubost, B., Bompard, P., Ansara. I., 1987a. Experimental study and thermodynamic calculation of the Al—Li—Mg equilibrium phase diagram. In: Champier, G., Dubost, B., Miannay, D., Sabetay, L. (Eds.), Proceedings of the Fourth International Conference on Aluminium—Lithium Alloys, J. Phys., 48, C3, 473—C3, 480.

Dubost, B., Audier, M., Jeanmart, P., Lang, J.M., Sainfort, P., 1987b. Structure of stable intermetallic compounds of the Al—Li—Cu—(Mg) and Al—Li—Zn—(Cu) systems. In: Champier, G., Dubost, B., Miannay, D., Sabetay, L. (Eds.), Proceedings of the Fourth International Conference on Aluminium—Lithium Alloys, J. Phys., 48, C3, 497—C3, 504.

Dubost, B., Colinet, C., Ansara, I., 1989. An experimental and thermodynamic study of the Al—Li—Cu equilibrium phase diagram. In: Sanders Jr., T.H., Starke Jr., E.A. (Eds.), Proceedings of the Fifth International Conference on Aluminium—Lithium Alloys, vol. 2. Materials and Component Engineering Publications Ltd., Birmingham, UK, pp. 623—632.

Dunford, D.V., Partridge, P.G., 1990. Strength and fracture behaviour of diffusion-bonded joints in Al—Li (8090) alloy. Part 1: shear strength. J. Mater. Sci. 25, 4957—4964.

Dunford, D.V., Partridge, P.G., 1991. Strength and fracture behaviour of diffusion-bonded joints in Al—Li (8090) alloy. Part 2: fracture behaviour. J. Mater. Sci. 26, 2625—2629.

Dunford, D.V., Partridge, P.G., 1992a. Critical strength criteria for DB/SPF processing of Ai—Li 8090 alloy. J. Mater. Sci. 27, 3389—3394.

Dunford, D.V., Partridge, P.G., 1992b. Diffusion bonding of Al—Li alloys. Mater. Sci. Technol. 8, 385—398.

Dunford, D.V., Partridge, P.G., 1992c. Effect of joint stiffness on peel strength of diffusion bonded joints between Al—Li 8090 alloy sheet. Mater. Sci. Technol. 8, 1131—1140.

Dunford, D.V., Partridge, P.G., 1992d. Strength and fracture behaviour of diffusion bonded joints in Ai—Li (8090) alloy. Part III: peel strength. J. Mater. Sci. 27, 5769—5776.

Dutkiewicz, J., Simmich, O., Scholz, R., Ciach, R., 1997. Evolution of precipitates in Al—Li—Cu and Al—Li—Cu—Sc alloys after age-hardening treatment. Mater. Sci. Eng. A, A234—236, 253—257.

Duva, J.M., Daubler, M.A., Starke Jr., E.A., Lütjering, G., 1988. Large shearable particles lead to coarse slip in particle reinforced alloys. Acta Metall. 36, 585—589.

Dvornak, M.J., Frost, R.H., Olson, D.L., 1989. The weldability and grain refinement of Al-2-2.2Li-2.7Cu. Weld. J. 68, 327—335.

Eddahbi, M., Thomson, C.B., Carreño, F., Ruano, O.A., 2000. Grain structure and microtexture after high temperature deformation of an Al—Li (8090) alloy. Mater. Sci. Eng. A, A284, 292—300.

Edwards, G.A., Couper, M.J., Dunlop, G.L., 1994. Age hardening and precipitation in an Al−Al$_2$O$_3$ based metal matrix composite. In: Sanders Jr., T.H., Starke Jr., E.A. (Eds.), Proceedings of the Fourth International Conference on Aluminum Alloys: Their Physical and Mechanical Properties, vol. 2. Georgia Institute of Technology, Atlanta, GA, USA, pp. 629−636.

Edwards, M.R., Stoneham, V.E., 1987. The fusion welding of Al−Li−Cu−Mg (8090) alloy. In: Champier, G., Dubost, B., Miannay, D., Sabatay, L. (Eds.), Proceedings of the Fourth International Conference on Aluminium−Lithium Alloys, J. Phys., 48, C3, 293−C3, 299.

Edwards, M.R., Whiley, M.J., 1994. The effect of prior stretching on the elevated temperature tensile behaviour of the aluminium−lithium alloy 8090. In: Sanders Jr., T.H., Starke Jr., E. A. (Eds.), Proceedings of the Fourth International Conference on Aluminum Alloys: Their Physical and Mechanical Properties, vol. 2. Georgia Institute of Technology, Atlanta, GA, USA, pp. 473−480.

Edwards, M.R., Klinkli, E., Stoneham, V.E., 1989. Studies of the diffusion bonding of the aluminium−lithium alloy 8090. In: Sanders Jr., T.H., Starke Jr., E.A. (Eds.), Proceedings of the Fifth International Conference on Aluminum−Lithium Alloys, vol. 1. Materials and Component Engineering Publications Ltd., Birmingham, UK, pp. 431−440.

Edwards, M.R., Moore, A., Mustey, A.J., 1994. The mechanical properties of unstretched aluminium−lithium alloy 2095 (weldalite ^{TM} 049) in the low temperature aged condition. In: Sanders Jr., T.H., Starke Jr., E.A. (Eds.), Proceedings of the Fourth International Conference on Aluminum Alloys: Their Physical and Mechanical Properties, vol. 2. Georgia Institute of Technology, Atlanta, GA, USA, pp. 334−341.

Ehab, A., El-Danaf, 2012. Mechanical properties, microstructure and micro-texture evolution for 1050AA deformed by equal channel angular pressing (ECAP) and post ECAP plane strain compression using two loading schemes. Mater. Des. 34, 793−807.

Ehrstrom, J.C., Shahani, R., Reeves, Sainfort, P., 1994. Texture and recrystallization in 7010 aluminium alloy thick plate. In: Sanders Jr., T.H., Starke Jr., E.A. (Eds.), Proceedings of the Fourth International Conference on Aluminum Alloys: Their Physical and Mechanical Properties, vol. 2. Georgia Institute of Technology, Atlanta, GA, USA, pp. 32−39.

Ekvall, J.C., Rhodes, J.E., Wald, G.G., 1982. Methodology for evaluating weight savings from basic material properties. Design of Fatigue and Fracture Resistant Structures, ASTM STP 761. American Society for Testing and Materials, Philadelphia, PA, USA, pp. 328−341.

Elisson, J., Sandstrom, R., 1994. The role of the value of weight savings in the use of aluminium matrix composites. In: Sanders Jr., T.H., Starke Jr., E.A. (Eds.), Proceedings of the Fourth International Conference on Aluminum Alloys: Their Physical and Mechanical Properties, vol. 2. Georgia Institute of Technology, Atlanta, GA, USA, pp. 535−542.

Engler, O., Lücke, K., 1991. Influence of the precipitation state on the cold rolling texture in 8090 Al−Li material. Mater. Sci. Eng. A, A148, 15−23.

Es-Said, O.S., Parrish, C.J., Bradberry, C.A., Hassoun, J.Y., Parish, R.A., Nash, A., et al., 2011. Effect of stretch orientation and rolling orientation on the mechanical properties of 2195 Al−Cu−Li alloy. J. Mater. Eng. Perform. 20, 1171−1179.

Eswara Prasad, N., 1993. In-Plane Anisotropy in the Fatigue and Fracture Properties of Quaternary Al−Li−Cu−Mg Alloys (Doctoral Thesis). Indian Institute of Technology (formerly Institute of Technology), Banaras Hindu University, Varanasi, India.

Eswara Prasad, N., 1997. The metallurgy of Al−Li alloys—an overview. Banaras Metall. 14−15, 69−87.

Eswara Prasad, N., Kamat, S.V., 1995. Fracture behaviour of quaternary Al−Li−Cu−Mg alloys under mixed-mode I/III loading. Metall. Mater. Trans. A, A26, 1823−1833.

Eswara Prasad, N., Malakondaiah, G., 1992. Anisotropy in the mechanical properties of quaternary Al–Li–Cu–Mg alloys. Bull. Mater. Sci. 15, 297–310.

Eswara Prasad, N., Rama Rao, P., 2000. Low cycle fatigue resistance in Al–Li alloys. Mater. Sci. Technol. 16, 408–426.

Eswara Prasad, N., Malakondaiah, G., Raju, K.N., Rama Rao, P., 1989a. Low cycle fatigue behaviour of an Al–Li alloy. In: Salma, K., Ravi-Chander, K., Taplin, D.M.R., Rama Rao, P. (Eds.), Advances in Fracture Research. Pergamon Press, New York, NY, USA, pp. 1103–1112.

Eswara Prasad, N., Prasad, K.S., Malakondaiah, G., Banerjee, D., Gokhale, A.A., Sundararajan, G., et al., 1989b. Microstructure and mechanical properties of 8090 Al–Li alloys, DMRL Technical Report No. 8995, Defence Metallurgical Research Laboratory, Hyderabad.

Eswara Prasad, N., Satya Prasad, K., Malakondaiah, G., Banerjee, D., Gokhale, A.A., Sundarajan, G., et al., 1990. Microstructure and mechanical properties of 2090 Al–Li alloys, DMRL Technical Report No. 90104, Defence Metallurgical Research Laboratory, Hyderabad.

Eswara Prasad, N., Kamat, S.V., Malakondaiah, G., 1991a. R-curve evaluation of quaternary Al–Li alloy 8090 sheet products. In: Dwarakadasa, E.S., Seshan, S., Abraham, K.K.P. (Eds.), Proceedings of the International Conference on Aluminium—Strategies for the Nineties and Beyond, vol. 2. Aluminium Association of India, Bangalore, India, pp. 613–620.

Eswara Prasad, N., Prasad, K.S., Malakondaiah, G., Banerjee, D., Gokhale, A.A., Sundararajan, G., et al., 1991b. Microstructure and mechanical properties of 1420 Al–Mg–Li alloy sheet product, DMRL Technical Report No. 91128, Defence Metallurgical Research Laboratory, Hyderabad.

Eswara Prasad, N., Malakondaiah, G., Rama Rao, P., 1992. Strength differential in Al–Li alloy 8090. Mater. Sci. Eng. A, A150, 221–229.

Eswara Prasad, N., Kammat, S.N., Malakondaiah, G., 1993a. Effect of crack deflection and branching on the R-curve behaviour of an Al–Li alloy 2090 sheet. Inter. J. Fract. 61, 55–69.

Eswara Prasad, N., Kamat, S.V., Prasad, K.S., Malakondaiah, G., Kutumbarao, V.V., 1993b. In-plane anisotropy in fracture toughness of an Al–Li 8090 plate. Eng. Fract. Mech. 46, 209–223.

Eswara Prasad, N., Kamat, S.V., Malakondaiah, G., Kutumbarao, V.V., 1994a. Static and dynamic fracture toughness of an Al–Li 8090 alloy plate. Fatigue Fract. Eng. Mater. Struct. 17, 441–450.

Eswara Prasad, N., Paradkar, A.G., Malakondaiah, G., Kutumbarao, V.V., 1994b. An analysis based on plastic strain energy for bilinearity in Coffin-Manson plots in an Al–Li alloy. Scr. Metall. Mater. 30, 1497–1502.

Eswara Prasad, N., Kamat, S.V., Prasad, K.S., Malakondaiah, G., Kutumbarao, V.V., 1994c. Fracture toughness of quaternary Al–Li–Cu–Mg alloy under mode I, mode II and mode III loading conditions. Metall. Mater. Trans. A, 25A, 2439–2452.

Eswara Prasad, N., Prasad, K.,S., Kamat, S.V., Malakondaiah, G., 1995. Influence of microstructural features on the fracture resistance of aluminium–lithium alloy sheets. Eng. Fract. Mech. 51, 87–96.

Eswara Prasad, N., Malakondaiah, G., Kutumbarao, V.V., Rama Rao, P., 1996a. In-plane anisotropy in low cycle fatigue properties of and bilinearity in Coffin-Manson plots for quaternary Al–Li–Cu–Mg 8090 alloy plate. Mater. Sci. Technol. 12, 563–577.

Eswara Prasad, N., Malakondaiah, G., Kutumbarao, V.V., 1996b. On the bilinearity in fatigue power–law relationships in Al–Li alloys. Trans. Indian Inst. Met. 49, 465–469.

Eswara Prasad, N., Malakondaiah, G., Kutumbarao, V.V., 1996c. Fracture mode transition corresponding to dual slope Coffin-Manson relationship in Al–Li alloys. In: Somashekar, B.R., Parida, B.K., Dattaguru, B., Rajaiah, K. (Eds.), Fatigue and Fracture of Materials and Structures: Proceedings of Sixth National Seminar on Aerospace Structures (6th NASAS), vol. 1. Allied Publishers Ltd., New Delhi, India, pp. 195–202.

Eswara Prasad, N., Malakondaiah, G., Kutumbarao, V.V., 1997a. On the micromechanisms responsible for bilinearity in fatigue power–law relationships in aluminium–lithium alloys. Scr. Metall. 37, 581–587.

Eswara Prasad, N., Ramulu, M., Malakondiah, G., 1997b. R-curve behaviour of Al–Li–Cu–Mg alloys. In: Karihaloo, B.L., Mai, Y.W., Ripley, M.I., Richie, R.O. (Eds.), Advances in Fracture Research, Proceedings of the Ninth International Conference on Fracture (ICF 9), vol. 5. Pergamon Press, New York, NY, USA, pp. 2487–2494.

Eswara Prasad, N., Gokhale, A.A., Rama Rao, P., 2003. Mechanical behaviour of Al–Li alloys. Sadhana 28, 209–246.

Eswara Prasad, N., Malakondaiah, G., Rama Rao, P., 2004. Low cycle fatigue behaviour of an underaged Al–Li–Cu–Mg alloy. Trans. Indian Inst. Met. 57, 181–194.

Eswara Prasad, N., Ram Mohan, R., Malakondaiah, G., 2006. Mode-I and mixed-mode I/III fracture behaviour of quaternary Al–Li–Cu–Mg alloy plate AA 8090-T6 in through-thickness orientations. Trans. Indian Inst. Met. 59, 321–336.

Evans, R.K., 1986. Western world lithium reserves and resources. In: Baker, C., Gregson, P.J., Harris, S.J., Peel, C.J. (Eds.), Proceedings of the Third International Conference on Aluminium–Lithium Alloys III. The Institute of Metals, London, UK, pp. 22–25.

Fager, D.N., Hyatt, M.V., Diep, H.T., 1986. A preliminary report on cleavage fracture in Al–Li alloys. Scr. Metall. 20, 1159–1164.

Fan, M.Q., Sun, L.X., Xu, F., Mei, D., Chen, D., Chai, W.X., et al., 2011. Microstructure of Al–Li alloy and its hydrolysis as portable hydrogen source for proton-exchange membrane fuel cells. Int. J. Hydrogen Energy 36, 9791–9798.

Fan, W., Kashyap, B.P., Chaturvedi, M.C., 2001. Effects of strain rate and test temperature on flow behaviour and microstructural evolution in AA 8090 Al–Li alloy. Mater. Sci. Technol. 17, 431–438.

Fang, W., Jinshan, L., Rui, H., Hongchao, K., 2008. Influence of 1.0 wt%Li on precipitates in Al–Zn–Mg–Cu alloy. Chin. J. Aeronaut. 21, 565–570.

Farcy, L., Clavel, M., 1990. Small fatigue crack growth behaviour in a 2091 aluminium–lithium alloy in comparison with 2024 alloy. In: Khan, T., Effenberg, G. (Eds.), Advanced Aluminium and Magnesium Alloys: Proceedings of the International Conference on Light Metals. ASM International European Council, Brussels, Belgium, pp. 173–180.

Farcy, L., Carre, C., Clavel, M., Barbaux, Y., Aliaga, D., 1987. Factors of crack initiation and microcrack propagation in aluminium lithium 2091 and in aluminium 2024. In: Champier, G., Dubost, B., Miannay, D., Sabatay, L. (Eds.), Proceedings of the Fourth International Conference on Aluminium–Lithium Alloys, J. Phys., 48, C3, 769–C3, 775.

Farcy, L., Hautefeuille, L., Carre, C., Clavel, M., Aliga, D., 1989. Microcrack initiation mechanisms in 2024 T3 and 2091 T8X sheets. Mechanical and microstructural approach. In: Sanders Jr., T.H., Starke Jr., E.A. (Eds.), Proceedings of the Fifth International Conference on Aluminum–Lithium Alloys, vol. 2. Materials and Component Engineering Publications Ltd., Birmingham, UK, pp. 921–930.

Fellner, P., Híveš, J., Korenko, M., Thonstad, J., 2001. Cathodic overvoltage and the contents of sodium and lithium in molten aluminium during electrolysis of cryolite-based melts. Electrochim. Acta 46, 2379–2384.

Feng, W.X., Lin, F.S., Starke Jr., E.A., 1984. The effect of minor alloying elements on the mechanical properties of Al–Cu–Li alloys. Metall. Trans. A, 15A, 1209–1220.

Fernandes, J.C.S., Ferreira, M.G.S., 1992. Effect of carbonate and lithium ions on the corrosion performance of pure aluminium. Electrochim. Acta 37, 2659–2661.

Field, D.J., Butler, E.P., 1984. Liquid metal oxidization of Al-3wt%Li. In: Sanders Jr., T.H., Starke Jr., E.A. (Eds.), Proceedings of the Second International Conference on Aluminum–Lithium Alloys. The Metallurgical Society of AIME, Warrendale, PA, USA, pp. 667–674.

Field, D.J., Butler, E.P., Scamans, G.M., 1981. High temperature oxidation studies of Al-3wt% Li, Al-4.2wt%Mg, and Al-3wt%Mg alloys. In: Sanders Jr., T.H., Starke Jr., E.A. (Eds.), Proceedings of the First International Conference on Aluminium–Lithium Alloys, vol. 1. The Metallurgical Society of AIME, Warrendale, PA, USA, pp. 325–346.

Fink, D., Hnatowicz, V., Kvitek, J., Havranek, V., Zhou, J.T., 1992. External oxidation of aluminium–lithium alloys. Surf. Coat. Tech. 51, 57–64.

Firrao, D., Scavio, G., Doglione, R., 1989. Precipitation phenomena during an Al–Li–Cu alloy ageing. In: Sanders Jr., T.H., Starke Jr., E.A. (Eds.), Proceedings of the Fifth International Conference on Aluminium–Lithium Alloys, vol. 2. Materials and Component Engineering Publications Ltd., Birmingham, UK, pp. 671–680.

Floriano, M.A., Triolo, A., Caponetti, E., Triolo, R., 1996. On the nature of phase separation in a commercial aluminium–lithium alloy. J. Mol. Struct. 383, 277–282.

Flower, H.M., Gregson, P.J., 1987. Solid state phase transformations in aluminium alloys containing lithium. Mater. Sci. Technol. 3, 81–90.

Fox, A.G., Fisher, R.M., 1988. The origin of the high elastic modulus in Al–Li alloys. J. Mater. Sci. Lett. 7, 301–303.

Fox, S., Flower, H.M., McDarmaid, D.S., 1986. Formation of solute-depleted surfaces in Al–Li–Cu–Mg–Zr alloys and their influence on mechanical properties. In: Baker, C., Gregson, P.J., Harris, S.J., Peel, C.J. (Eds.), Proceedings of the Third International Conference on Aluminium–Lithium Alloys, vol. 3. The Institute of Metals, London, UK, pp. 263–272.

Fragomeni, J., Wheeler, R., Jata, K.V., 2005. Effect of single and duplex aging on precipitation response, microstructure, and fatigue crack behavior in Al–Li–Cu alloy Af/C-458. J. Mater. Eng. Perform., 18–27.

Fragomeni, J.M., Hillberry, B.M., 1999. Determining the effect of microstructure and heat treatment on the mechanical strengthening behavior of an aluminum alloy containing lithium precipitation hardened with the D8 Al–Li intermetallic phase. J. Mater. Eng. Perform. 9, 428–440.

Fragomeni, J.M., Hillbery, B.M., Sanders Jr., T.H., 1989. An investigation of the δ′ particle strengthening mechanisms and microstructure for an Al–Li–Zr alloy. In: Sanders Jr., T.H., Starke Jr., E.A. (Eds.), Proceedings of the Fifth International Conference on Aluminium–Lithium Alloys, vol. 2. Materials and Component Engineering Publications Ltd., Birmingham, UK, pp. 837–848.

Frazer, E.J., 1981. Electrochemical formation of lithium–aluminium alloys in propylene carbonate electrolytes. J. Electroanal. Chem. Interf. Electrochem. 121, 329–339.

Fridlyander, I.N., 1965. Phase composition and mechanical properties of aluminium alloys containing magnesium and lithium. Russ. Metall. (Metally) (Engl. Transl.) 2, 83–90.

Fridlyander, I.N., 1989. Aluminium–lithium weldable alloy 1420. In: Sanders Jr., T.H., Starke Jr., E.A. (Eds.), Proceedings of the Fifth International Conference on Aluminium–Lithium Alloys, vol. 3. Materials and Component Engineering Publications Ltd., Birmingham, UK, pp. 1359–1364.

Fridlyander, I.N., 2006. Memories on the Establishment of an Aerospace and Nuclear Technology Aluminium Alloy, second ed. Russian Academy of Sciences, Department of Chemistry and Materials Sciences (in Russian).

Fridlyander, I.N., Kolobnev, N.I., Berezina, A.L., Chuistov, K.V., 1992. The effect of scandium on decomposition kinetics in aluminum–lithium alloys. In: Peters, M., Winkler, P.J. (Eds.), Aluminum–Lithium Alloys VI. Deutsche Gesellschaft für Metallkunde, Frankfurt, Germany, pp. 107–112.

Fridlyander, I.N., Tempus, G., Khokhlatova, B., Kolobnev, N.I., Rendiks, K., 2002. Thermally stable aluminum–lithium alloy 1424 for application in welded fuselage. Met. Sci. Heat Treat. 44, 4–8.

Fridlyander, I.N., Grushko, O.E., Shamrai, V.F., Klochkov, G.G., 2003. High-strength structural silver-alloyed underdensity Al–Cu–Li–Mg alloy. Met. Sci. Heat Treat. 46, 279–283.

Fridlyander, J., 1994. Advanced Russia aluminium alloys. In: Sanders Jr., T.H., Starke Jr., E.A. (Eds.), Proceedings of the Fourth International conference on Aluminum Alloys: Their Physical and Mechanical Properties, vol. 2. Georgia Institute of Technology, Atlanta, GA, USA, pp. 80–87.

Fridlyander, J.N., Kolobnev, N.I., Khokhlatova, L.B., Lovchelt, E., Winkler, P.J., Pfannenmuller, T., 1998. Properties of new Al–Li–Mg Alloy. In: Sata, T., Kumai, T., Murakami, Y. (Eds.), Aluminium Alloys: Their Physical and Mechanical Properties, vol. 2. Japan Institute of Metals, Tokyo, Japan, pp. 2055–2060.

Fridlyander, L.N., Rokhlin, L.L., Dobatkina, T.V., Nikitina, N.L., 1993. Metalloved. Term. Obrab. Met. 10, 16–19.

Friedrich, H.E., Mussack, R., Tensi, H.M., Wittmann, M., 1991. Al–Li alloys on the way to their application in aircraft manufacturing. In: Sanders Jr., T.H., Starke Jr., E.A. (Eds.), Proceedings of the Fifth International Conference on Aluminium–Lithium Alloys, vol. 3. Materials and Component Engineering Publications Ltd., Birmingham, UK, pp. 1615–1624.

Fujikawa, S., Izeki, Y., Hirano, K., 1986. Determination of equilibrium solute content in matrix, precipitate-matrix interfacial free energy and effective diffusivity in Al–Li alloys using coarsening data alone for δ' (Al_3Li) precipitates. Scr. Metall. 20, 1275–1280.

Fujikawa, S., Furusaka, M., Sakauchi, M., Hirano, K., 1987. Studies on early and late stages of formation of δ' phases in Al–Li alloys by neutron small angle scattering. In: Champier, G., Dubost, B., Miannay, D., Sabetay, L. (Eds.), Proceedings of the Fourth International Conference on Aluminium–Lithium Alloys, J. Phys., 48, C3, 365–C3, 372.

Furukawa, M., Miura, Y. , Nemoto, M., 1987. Acoustic emission during deformation of Al–Li alloys. In: Champier, G., Dubost, B., Miannay, D., Sabetay, L. (Eds.), Proceedings of the Fourth International Conference on Aluminium–Lithium Alloys, J. Phys., 48, C3, 557–C3, 564.

Gaber, A., Afify, N., 1997. Investigation of the structural transformations in the Al – Li – Cu – Mg (8090) alloy. Appl. Phys. A, 65, 57–62.

Gaber, A., Afify, N., 2002. Characterization of the precipitates in Al–Li (8090) alloy using thermal measurements and TEM examinations. Phys. B Condens. Mat. 315, 1–6.

Gable, B.M., Zhu, A.W., Csontos, A.A., Starke Jr., E.A., 2001. The role of plastic deformation on the competitive microstructural evolution and mechanical properties of a novel Al–Li–Cu–X alloy. J. Light Met. 1, 1–14.

Galbraith, J.M., Tosten, M.H., Howell, P.R., 1987. On the nucleation of δ′ and T₁ on Al₃Zr precipitates in Al−Li−Cu−Zr alloys. J. Mater. Sci. 22, 27−36.

Gandhi, C., Bampton, C.C., Ghosh, A.K., Anton, C.E., 1989. Superplastic behaviour of 8091 aluminium−lithium alloy. In: Sanders Jr., T.H., Starke Jr., E.A. (Eds.), Proceedings of the Fifth International Conference on Aluminium−Lithium Alloys, vol. 1. Materials and Component Engineering Publications Ltd., Birmingham, UK, pp. 141−150.

Gao, Y.J., Mo, Q.F., Chen, H.N., Huang, C.G., Zhang, L.N., 2009. Atomic bonding and mechanical properties of Al−Li−Zr alloy. Mater. Sci. Eng. A, A499, 299−303.

Garcia, J., Ponthiaux, P., Habashi, M., Galland, J., 1987. Correlation between long time and short time tests to detect pitting and exfoliation corrosion in Al−Li (8090) alloys. In: Champier, G., Dubost, B., Miannay, D., Sabetay, L. (Eds.), Proceedings of the Fourth International Conference on Aluminium−Lithium Alloys, J. Phys., 48, C3, 861−C3, 870.

Garcia, J., Habashi, M., Galland, J., 1989. Differential scanning calorimetric studies and stress-corrosion cracking in Al−Li alloys. In: Sanders Jr., T.H., Starke Jr., E.A. (Eds.), Proceedings of the Fifth International Conference on Aluminium−Lithium Alloys, vol. 3. Materials and Component Engineering Publications Ltd., Birmingham, UK, pp. 1293−1308.

Garg, A., Howe, J.M., 1991. Convergent beam electron diffraction analysis of the Ω phase in an Al-4.0Cu-0.5Mg-0.5Ag alloy. Acta Metall. Mater. 39, 1939−1946.

Garmestani, H., Kal, P.U., Dingle, D.Y., 1998. Characterization of Al-8090 superplastic materials using orientation imaging microscopy. Mater. Sci. Eng. A, A242, 284−291.

Gasior, W., Moser, Z., Pstrus, J., 1998. Densities of solid aluminum−lithium (Al−Li) alloys. J. Phase Equilib. 19, 234−238.

Gasparotto, L.H.S., Prowald, A., Borisenko, N., Abedin, Z.E.L., Garsuch, A., Endres, F., 2011. Electrochemical synthesis of macroporous aluminium films and their behavior towards lithium deposition/stripping. J. Power Sources 196, 2879−2883.

Gatenby, K.M., Reynolds, M.A., White, J., Palmer, I.G., 1989. The role of microstructure in controlling the mechanical properties of 8090 in damage tolerant tempers. In: Sanders Jr., T.H., Starke Jr., E.A. (Eds.), Proceedings of the Fifth International Conference on Aluminium−Lithium Alloys, vol. 2. Materials and Component Engineering Publications Ltd., Birmingham, UK, pp. 909−920.

Gault, B., Cui, X.Y., Moody, M.P., De Geuser, F., Sigli, C., Ringer, S.P., et al., 2012. Atom probe microscopy investigation of Mg site occupancy within δ′ precipitates in an Al−Mg−Li alloy. Scr. Metall. 66, 903−906.

Gayle, F.W., 1981. Alloying additions and property modification in aluminium−lithium−X systems. In: Sanders Jr., T.H., Starke Jr., E.A. (Eds.), Proceedings of the First International Conference on Aluminium−Lithium Alloys. The Metallurgical Society of AIME, Warrendale, PA, USA, pp. 119−140.

Gayle, F.W., 1987. The icosahedral Al−Li−Cu phase. In: Champier, G., Dubost, B., Miannay, D., Sabetay, L. (Eds.), Proceedings of the Fourth International Conference on Aluminium−Lithium Alloys, J. Phys., 48, C3, 481−C3, 488.

Gayle, F.W., Vander Sande, J.B., 1984. Composite precipitates in an Al−Li−Zr alloy. Scr. Metall. 18, 473−478.

Gayle, F.W., Vander Sande, J.B., 1986. Al₃(Li,Zr) or α′ phase in Al−Li−Zr system. In: Baker, C., Gregson, P.J., Harris, S.J., Peel, C.J. (Eds.), Proceedings of the Third International Conference on Aluminum−Lithium Alloys. The Institute of Metals, London, UK, pp. 376−385.

Gayle, F.W., Vander Sande, J.B., 1989. Phase transformations in the Al−Li−Zr system. Acta Metall. 37, 1033−1046.

Gayle, F.W., Heubaum, F.H., Pickens, J.R., 1989. Natural ageing and reversion behaviour of Al–Cu–Li–Ag–Mg alloy weldalite™049. In: Sanders Jr., T.H., Starke Jr., E.A. (Eds.), Proceedings of the Fifth International Conference on Aluminium–Lithium Alloys, vol. 2. Materials and Component Engineering Publications Ltd., Birmingham, UK, pp. 701–710.

Gayle, F.W., Vander Sande, J.B., McAlister, A.J., 1984. The Al–Li (aluminum–lithium) system. Bull. Alloy Phase Diagrams 5, 19–20.

Gayle, F.W., Tack, W.T., Swanson, G., Heubaum, F.H., Pickens, J.R., 1994. Composition and anisotropy in Al–Cu–Li–Ag–Mg–Zr alloys. Scr. Metall. Mater. 30, 761–766.

Gentzbittel, J.M., Esnouf, C., Fougeres, R., 1987a. Interactions between dislocations and crystalline defects during the cyclic deformation of 0.7 wt% Al–Li alloy. In: Champier, G., Dubost, B., Miannay, D., Sabetay, L. (Eds.), Proceedings of the Fourth International Conference on Aluminium–Lithium Alloys, J. Phys., 48, C3, 721–C3, 729.

Gentzbittel, J.M., Vigier, G., Fougeres, R., 1987b. The phenomenon of stress instabilities in Al–Li binary alloys and microscopic mechanisms connected. In: Champier, G., Dubost, B., Miannay, D., Sabetay, L. (Eds.), Proceedings of the Fourth International Conference on Aluminium–Lithium Alloys, J. Phys., 48, C3, 729–C3, 736.

Geronov, Y., Zlatilova, P., Staikov, G., 1984. The secondary lithium–aluminium electrode at room temperature: II. Kinetics of the electrochemical formation of the lithium–aluminium alloy. J. Power Sources 12, 155–165.

Ghosh, G., 1993. A comprehensive compendium of evaluated constitutional data and phase diagrams. In: Petzow, G., Effenberg, G. (Eds.), Ternary Alloys, vol. 6. Verlag Chemie (VCH), Weinheim, Germany, pp. 356–375.

Ghosh, K.S., Das, K., Chatterjee, U.K., 2004. Characterization of retrogression and reaging behavior of 8090 Al–Li–Cu–Mg–Zr alloy. Metall. Mater. Trans. A, 35A, 3681–3691.

Ghosh, K.S., Das, K., Chatterjee, U.K., 2005. Studies of retrogression and reaging behavior in a 1441 Al–Li–Cu–Mg–Zr alloy. Metall. Mater. Trans. A, 36A, 3477–3487.

Ghosh, K.S., Das, K., Chatterjee, U.K., 2006. Electrochemical behaviour of retrogressed and reaged (RRA) 8090 Al–Li–Cu–Mg–Zr alloys. J. Appl. Electrochem. 36, 1057–1068.

Ghosh, K.S., Das, K., Chatterjee, U.K., 2007. Kinetics of solid-state reactions in Al–Li–Cu–Mg–Zr alloys from calorimetric studies. Metall. Mater. Trans. A, 38A, 1965–1975.

Giles, T.L., Oh-Ishi, K., Zhilyaev, A.P., Swaminathan, S., Mahoney, M.W., 2009. The effect of friction stir processing on the microstructure and mechanical properties of an aluminum–lithium alloy. Metall. Mater. Trans. A, 40A, 104–115.

Gilman, P.S., 1984. The physical metallurgy of mechanically-alloyed, dispersion-strengthened Al–Li–Mg and Al–Li–Cu alloys. In: Sanders Jr., T.H., Starke Jr., E.A. (Eds.), Proceedings of the Second International Conference on Aluminum–Lithium Alloys. The Metallurgical Society of AIME, Warrendale, PA, USA, pp. 485–506.

Gilman, P.S., Brooks, J.W., Bridges, P.J., 1986. High temperature tensile properties of mechanically alloyed Al–Mg–Li alloys. In: Baker, C., Gregson, P.J., Harris, S.J., Peel, C.J. (Eds.), Proceedings of the Third International Conference on Aluminium–Lithium Alloy. The Institute of Metals, London, UK, pp. 112–120.

Gilmore, C.J., Dunford, D.V., Partridge, P.G., 1991. Microstructure of diffusion-bonded joints in Al–Li 8090 alloy. J. Mater. Sci. 26, 3119–3124.

Gilmore, D.L., Starke Jr., E.A., 1994. Trace element effects on precipitation processes and mechanical properties in Al–Cu–Li–X alloys. In: Sanders Jr., T.H., Starke Jr., E.A. (Eds.), Proceedings of the Fourth International Conference on Aluminum Alloys: Their Physical and Mechanical Properties, vol. 2. Georgia Institute of Technology, Atlanta, GA, USA, pp. 313–320.

Gilmore, D.L., Starke Jr., E.A., 1997. Trace element effects on precipitation processes and mechanical properties in an Al–Cu–Li alloy. Metall. Mater. Trans. A, 28A, 1399–1415.

Gittos, M.F., 1987. Gas shielded arc welding of the aluminum–lithium alloy 8090, Report 7944.01/87/556.2, The Welding Institute, Cambridge, UK.

Giummarra, C., Thomas, B., Rioja, R.J., New aluminum lithium alloys for aerospace applications. In: Sadayappan, K., Sahoo, M. (Eds.), Proceedings of the Third International Conference on Light Metals Technology, CANMET, Ottawa, Canada, pp. 41–46.

Glazer, J., Morris Jr., J.W., 1986. Thermomechanical processing of two-phase Al–Cu–Li–Zr alloy. In: Baker, C., Gregson, P.J., Harris, S.J., Peel, C.J. (Eds.), Proceedings of the Third International Conference on Aluminium–Lithium. The Institute of Metals, London, UK, pp. 191–198.

Glazer, J., Morris Jr., J.W., 1989. The strength–toughness relationship at cryogenic temperature in aluminium–lithium alloy plate. In: Sanders Jr., T.H., Starke Jr., E.A. (Eds.), Proceedings of the Fifth International Conference on Aluminium–Lithium Alloys, vol. 3. Materials and Component Engineering Publications Ltd., Birmingham, UK, pp. 1471–1480.

Glazer, J., Edgecumbe, T.S., Morris Jr., J.W., 1986. Theoretical analysis of aging response of Al–Li alloys strengthened by Al$_3$Li precipitates. In: Baker, C., Gregson, P.J., Harris, S.J., Peel, C.J. (Eds.), Proceedings of the Third International Conference on Aluminium–Lithium Alloys. The Institute of Metals, London, UK, pp. 369–375.

Gokhale, A.A., Ramachandran, T.R., 1990. Development of aluminum–lithium alloys. Indian J. Tech. 28, 235–246.

Gokhale, A.A., Singh, V., 2005. Effect of Zr content and mechanical working on the structure and tensile properties of AA8090 alloy plates. J. Mater. Process. Tech. 159, 369–376.

Gokhale, A.A., Singh, V., Chakravorthy, C.R., Satya Prasad, K., Banerjee, D., 1990. Effects of impurities on the microstructure and properties of an aluminium–lithium alloy. Met. Mater. Process. 2, 267–278.

Gokhale, A.A., Satya Prasad, K., Kumar, V., Chakravorthy, C.R., Leschiner, L.N., Mozharovsky, S.M., et al., 1994a. An aluminium–lithium alloy with improved in-plane anisotropy. In: Sanders Jr., T.H., Starke Jr., E.A. (Eds.), Proceedings of the Fourth International Conference on Aluminum Alloys: Their Physical and Mechanical Properties, vol. 2. Georgia Institute of Technology, Atlanta, GA, USA, pp. 428–435.

Gokhale, A.A., Singh, V., Esawara Prasad, N., Chakravorthy, C.R., Prasad, Y.V.R.K., 1994b. Processing maps for Al–Li alloys. In: Sanders Jr., T.H., Starke Jr., E.A. (Eds.), Proceedings of the Fourth International conference on Aluminum Alloys: Their Physical and Mechanical Properties. Georgia Institute of Technology, Atlanta, GA, USA, pp. 242–249.

Gomba, J.M., D'Angelo, C., Bertuccelli, D., Bertuccelli, G., 2001. Spectroscopic characterization of laser induced breakdown in aluminium–lithium alloy samples for quantitative determination of traces. Spectrochim. Acta B At. Spectrosc. 56, 695–705.

Gomiero, P., Livet, F., 1989. In situ small angle scattering study of the decomposition in a 2091 alloy. In: Sanders Jr., T.H., Starke Jr., E.A. (Eds.), Proceedings of the Fifth International Conference on Aluminium–Lithium Alloys. Materials and Component Engineering Publications Ltd., Birmingham, UK, pp. 641–650.

Gomiero, P., Livet, F., Lyon, O., Simon, J.P., 1991. Double structural hardening in an Al–Li–Cu–Mg alloy studied by anomalous small angle X-ray scattering. Acta Metall. 39, 3007–3014.

Gomiero, P., Livet, F., Brechet, Y., Louchet, F., 1992. Microstructure and mechanical properties of a 2091 Al−Li alloy−I. Microstructure investigated by SAXS and TEM. Acta Metall. 40, 847−855.

Gonclaves, M., Sellars, C.M., 1987. Static recrystallisation after hot working of Al−Li alloys. In: Champier, G., Dubost, B., Miannay, D., Sabatay, L. (Eds.), Proceedings of the Fourth International Conference on Aluminum−Lithium Alloys, J. Phys., 48, C3, 171−C3, 177.

Gonclaves, M., Sellars, C.M., 1989. Processing structure ad tensile properties of hot worked 8090 ad 8091 alloy. In: Sanders Jr., T.H., Starke Jr., E.A. (Eds.), Proceedings of the Fifth International Conference on Aluminium−Lithium Alloys, vol. 1. Materials and Component Engineering Publications Ltd., Birmingham, UK, pp. 305−314.

González, A., Martín, A., Llorca, J., 2004. Effect of temperature on the fracture mechanisms of 8090 Al−Li alloy and 8090 Al−Li/Si−C composite. Scr. Metall. 51, 1111−1115.

Grady, H.R., 1981. Distribution, consumption, pricing, and out look of the lithium industry. In: Sanders Jr., T.H., Starke Jr., E.A. (Eds.), Proceedings of the First International Conference on Aluminium−Lithium Alloys. The Metallurgical Society of AIME, Warrendale, PA, USA, pp. 1−8.

Grant, N.J., Kang, S., Wang, W., 1981. Structure and properties of rapidly solidified 2000 series Al−Li alloys. In: Sanders Jr., T.H., Starke Jr., E.A. (Eds.), Proceedings of the First International Conference on Aluminium−Lithium Alloys. The Metallurgical Society of AIME, Warrendale, PA, USA, pp. 141−170.

Gray, A., 1987. Factors influencing the environmental behaviour of aluminium−lithium alloys. In: Champier, G., Dubost, B., Miannay, D., Sabatay, L. (Eds.), Proceedings of the Fourth International Conference on Aluminum−Lithium Alloys, J. Phys., 48, C3, 891−C3, 904.

Gray, A., Newton, G.J., 1992. In: Peters, M., Winkler, P.J. (Eds.), Aluminum−Lithium Alloys VI. Deutsche Gesellschaft für Metallkunde, Frankfurt, Germany, p. 691.

Gray, A., Holroyd, N.J.H., White, J., 1989. The influence of microstructure of the environmental cracking behaviour of Al−Li−Cu−Mg−Zr alloys. In: Sanders Jr., T.H., Starke Jr., E.A. (Eds.), Proceedings of the Fifth International Conference on Aluminium−Lithium Alloys, vol. 3. Materials and Component Engineering Publications Ltd., Birmingham, UK, pp. 1175−1186.

Greene, G.A., Cho, D.H., Hyder, M.L.M., Allison, D.K., Ellison, P.G., 1994. Rapid quenching of molten lithium−aluminum jets in water. Nucl. Eng. Des. 148, 317−326.

Gregson, P.J., 1984. (Ph.D.Thesis) The University of London, UK.

Gregson, P.J., Flower, H.M., 1984. δ' Precipitation in Al−Li−Mg−Cu−Zr alloys. J. Mater. Sci. Lett. 3, 829−834.

Gregson, P.J., Flower, H.M., 1985. Microstructural control of toughness in aluminium−lithium alloys. Acta Metall. 33, 527−537.

Gregson, P.J., Flower, H.M., 1986. In: Shepherd, T. (Ed.), Aluminium Technology. The Institute of Metals, London, UK, p. 423.

Gregson, P.J., Harris, S.J., 1986. Physical metallurgy of Al−Li alloys. In: Baker, C., Gregson, P. J., Harris, S.J., Peel, C.J. (Eds.), Proceedings of the Third International Conference on Aluminium−Lithium Alloys. The Institute of Metals, London, UK, pp. 327−328.

Gregson, P.J., Flower, H.M., Tete, C.N.J., Mukhopadhyay, A.K., 1986a. Role of vacancies in precipitation of δ'- and S- Phases in Al−Li−Cu−Mg alloys. Mater. Sci. Technol. 2, 349−353.

Gregson, P.J., Peel, C.J., Evans, B., 1986b. Development of properties within high-strength aluminium−lithium alloys. In: Baker, C., Gregson, P.J., Harris, S.J., Peel, C.J. (Eds.),

Proceedings of the Third International Conference on Aluminum–Lithium Alloys, vol. 1. The Institute of Metals, London, UK, pp. 516–523.

Gregson, P.J., Dinsdale, K., Harris, S.J., Noble, B., 1987. Evolution of microstructure in Al–Li–Zn–Mg–Cu alloys. Mater. Sci. Technol. 3, 7–13.

Gregson, P.J., McDarmaid, D.S., Hunt, E., 1988. Post-yield deformation characteristics in Al–Li alloys. Mater. Sci. Technol. 4, 713–718.

Griffee, C.C., Jensen, G.A., Reinhart, T.L., 1989. Factors influencing the quality and properties of aluminium–lithium alloy welds. In: Sanders Jr., T.H., Starke Jr., E.A. (Eds.), Proceedings of the Fifth International Conference on Aluminium–Lithium Alloys, vol. 3. Materials and Component Engineering Publications Ltd., Birmingham, UK, pp. 1425–1434.

Griger, A., Turmezey, T., 1994. Microstructural evolution in mechanically alloyed Al based composites. In: Sanders Jr., T.H., Starke Jr., E.A. (Eds.), Proceedings of the Fourth International conference on Aluminum Alloys: Their Physical and Mechanical Properties, vol. 2. Georgia Institute of Technology, Atlanta, GA, USA, pp. 575–581.

Grimes, R., 1990. Aluminium–lithium based alloys. In: Cahn, R.W. (Ed.), Encyclopedia of Materials Science and Engineering, vol. 2. Pergamon Press, New York, NY, USA, pp. 667–678.

Grimes, R., Miller, W.S., 1984. Superplasticity in lithium-containing aluminium alloys. In: Sanders Jr., T.H., Starke Jr., E.A. (Eds.), Proceedings of the Second International Conference on Aluminum–Lithium Alloys. The Metallurgical Society of AIME, Warrendale, PA, USA, pp. 153–168.

Grimes, R., Cornish, A.J., Miller, W.S., Reynold, M.A., 1985. Aluminium–lithium based alloys for aerospace applications. Met. Mater. 1, 357–363.

Grimes, R., Miller, W.S., Butler, R.G., 1987a. Development of superplastic 8090 and 8091 sheet. In: Champier, G., Dubost, B., Miannay, D., Sabetay, L. (Eds.), Proceedings of the Fourth International Conference on Aluminium–Lithium Alloys, J. Phys., 48, C3, 239–C3, 250.

Grimes, R., Davis, T., Saxty, H.J., Fearon, J.E., 1987b. Progress to aluminium–lithium semi-fabricated products. In: Champier, G., Dubost, B., Miannay, D. and Sabatay, L. (Eds.), Proceedings of the Fourth International Conference on Aluminum–Lithium Alloys, J. Phys., 48, C3, 11–C3, 24.

Gu, B.P., Liedl, G.L., Sanders Jr., T.H., Welpmann, K., 1985a. The influence of zirconium on the coarsening of δ' (Al_3Li) in an Al-2.8wt%Li-0.14wt%Zr alloy. Mater. Sci. Eng. 76, 147–157.

Gu, B.P., Liedl, G.L., Kulwicki, J.H., Sanders Jr., T.H., 1985b. Coarsening of δ' (Al_3Li) precipitates in an Al-2.8Li-0.3Mn alloy. Mater. Sci. Eng. 70, 217–228.

Gu, B.P., Liedl, G.L., Mahalingam, K., Sanders Jr., T.H., 1986a. Application of the Weibull density function to describe the δ' (Al_3Li) particle size distribution in binary Al–Li alloys. Mater. Sci. Eng. 78, 71–85.

Gu, B.P., Mahalingam, K., Liedl, G.L., Sanders Jr., T.H., 1986b. The δ' (Al_3Li) particle size distributions in a variety of Al–Li alloys. In: Baker, C., Gregson, P.J., Harris, S.J., Peel, C.J. (Eds.), Proceedings of the Third International Conference on Aluminum–Lithium Alloys. The Institute of Metals, London. UK, pp. 360–368.

Gu, X., Hand, R.J., 1996. The use of lithium as a dopant in the directed melt oxidation of aluminium. J. Eur. Ceram. Soc. 16, 929–935.

Gudladt, H.-J., Lendvai, J., Schneider, J., Wunderlich, W.E., Gerold, V., 1989. Planar slip and deformation-induced particle dissolution in cyclically deformed Al–Li single crystals. In: Sanders Jr., T.H., Starke Jr., E.A. (Eds.), Proceedings of the Fifth International

Conference on Aluminium—Lithium Alloys, vol. 2. Materials and Component Engineering Publications Ltd., Birmingham, UK, pp. 1105—1114.

Gui, J., Devine, T.M., 1987. Influence of lithium on the corrosion of aluminum. Scr. Metall. 21, 853—857.

Gui, Q.H., Jiang, X.J., Ma, L.M., Li, Y.Y., 1993. Effect of 0.1%Y on microstructure and properties of 8090 Al—Li alloy. Scr. Metall. 28, 297—300.

Guo, C., Liang, Y., Li, C., Du, Z., 2011. Thermodynamic description of the Al—Li—Zn system. Calphad 35, 54—65.

Guo, J.Q., Ohtera, K., Kita, K., Nagahora, J., Kazama, N.S., Inoue, A., et al., 1994. Crystallization behaviour of rapidly solidified Al—Sm alloys. In: Sanders Jr., T.H., Starke Jr., E.A. (Eds.), Proceedings of the Fourth International conference on Aluminum Alloys: Their Physical and Mechanical Properties. Georgia Institute of Technology, Atlanta, GA, USA, pp. 753—759.

Gupta, M., Mohammed, F., Lavernia, E.J., 1989. Spray atomization and deposition of Al—Cu—Li—Zr alloy. In: Sanders Jr., T.H., Starke Jr., E.A. (Eds.), Proceedings of the Fifth International Conference on Aluminium—Lithium Alloys, vol. 1. Materials and Component Engineering Publications Ltd., Birmingham, UK, pp. 75—84.

Guterman, V., Yu, A., Grigorev, V., 1999. The effect of the composition of aluminum alloys on electrochemical incorporation of lithium. Electrochim. Acta. 45, 873—880.

Gutierrez, A., Lippold, J.C., Lin, W., 1996. Nondendritic equiaxed zone formation in aluminum—lithium welds. Mater. Sci. Forum 217—212, 1891—1893.

Gutierrez-Urrutia, I., 2011. Study of isothermal δ(Al₃Li) precipitation in an Al—Li alloy by thermoelectric power. J. Mater. Sci. 46, 3144—3150.

Guvenilir, A., Stock, S.,R., Barker, M.D., Betz, R.A., 1994. Physical processes of crack closure observed in the interior of an Al—Li 2090 compact tension sample. In: Sanders Jr., T.H., Starke Jr., E.A. (Eds.), Proceedings of the Fourth International Conference on Aluminum Alloys: Their Physical and Mechanical Properties. Georgia Institute of Technology, Atlanta, GA, USA, pp. 413—419.

Gysler, A., Crookes, R., Starke Jr., E.A., 1981. A comparison of microstructure and tensile properties of P/M and I/M Al—Li—X alloys. In: Sanders Jr., T.H., Starke Jr., E.A. (Eds.), Proceedings of the First International Conference on Aluminium—Lithium Alloys. The Metallurgical Society of AIME, Warrendale, PA, USA, pp. 263—291.

Ha, T.K., Chang, Y.W., 1995. Superplastic deformation behavior of 8090 aluminum—lithium alloy. Scr. Metall. Mater. 32, 809—814.

Haddenhorst, H., Hornbogen, E., 1994. Laser surfacing of aluminium cast alloys with ceramic particles. In: Sanders Jr., T.H., Starke Jr., E.A. (Eds.), Proceedings of the Fourth International Conference on Aluminum Alloys: Their Physical and Mechanical Properties. Georgia Institute of Technology, Atlanta, GA, USA, pp. 721—728.

Haddleton, F.L., Murphy, S., Griffin, T.J., 1987. Fatigue and corrosion fatigue of 8090 Al—Li—Cu—Mg alloy. In: Champier, G., Dubost, B., Miannay, D., Sabetay, L. (Eds.), Proceedings of the Fourth International Conference on Aluminium—Lithium Alloys, J. Phys., 48. C3, 809—C3, 816.

Hagiwara, T., Kazunori Kobayashi, K., Toshimasa Sakamoto, T., 1994. Effect of Li and Cu contents on the mechanical properties of Al—Cu—Li alloys. In: Sanders Jr., T.H., Starke Jr., E. A. (Eds.), Proceedings of the Fourth International Conference on Aluminum Alloys: Their Physical and Mechanical Properties, vol. 2. Georgia Institute of Technology, Atlanta, GA, USA, pp. 297—304.

Hajjaji, M., Cutler, L., L'Espérance, G., 1992. Effects of processing parameters on aluminium–lithium ribbon production. J. Alloys Compd. 188, 194–197.

Hales, S.J., Hafley, R.A., 1998. Texture and anisotropy in Al–Li alloy 2095 plate and near-net-shape extrusions. Mater. Sci. Eng. A, A257, 153–164.

Hales, S.J., Tayon, W.A., 2011. Heat treatment of a friction-stir-welded and spin-formed Al–Li alloy. Proc. Eng. 10, 2496–2501.

Hales, S.J., Oster, S.B., Sanchez, B.W., McNelley, T.R., 1987. Grain refinement and superplasticity in a lithium-containing Al–Mg alloy by thermomechanical processing. In: Champier, G., Dubost, B., Miannay, D., Sabetay, L. (Eds.), Proceedings of the Fourth International Conference on Aluminium–Lithium Alloys, J. Phys., 48, C3, 285–C3, 292.

Hales, S.J., McNelley, T.R., Groh, G.E., 1989. Intermediate temperature thermomechanical processing of Al 2090 for superplasticity. In: Sanders Jr., T.H., Starke Jr., E.A. (Eds.), Proceedings of the Fifth International Conference on Aluminium–Lithium Alloys, vol. 1. Materials and Component Engineering Publications Ltd., Birmingham, UK, pp. 211–222.

Hallstedt, B., Kim, O., 2007. Thermodynamic assessment of the Al–Li system. Int. J. Mater. Res. 98, 961–969.

Han, S., Tourabi, A., Wack, B., 1987. Low cycle fatigue of binary Al–Li alloy: 1-experimental results and mechanical modelization. In: Champier, G., Dubost, B., Miannay, D., Sabetay, L., (Eds.), Proceedings of the Fourth International Conference on Aluminium–Lithium Alloys, J. Phys., 48, C3, 701–C3, 708.

Hanada, K., Khor, K.A., Tan, M.J., Murakoshi, Y., Negishi, H., Sano, T., 1997. Aluminium–lithium/SiCp composites produced by mechanically milled powders. J. Mater. Process. Technol. 67, 8–12.

Hardy, H.K., Silcock, J.M., 1955. The phase sections at 500°C and 350°C of aluminium-rich aluminium–copper–lithium alloys. J. Inst. Met. 84, 423–428.

Hargarter, H., Lütjering, G., Becker, J., Fischer, G., 1994. Fatigue properties of Al 8090. In: Sanders Jr., T.H., Starke Jr., E.A. (Eds.), Proceedings of the Fourth International Conference on Aluminum Alloys: Their Physical and Mechanical Properties. Georgia Institute of Technology, Atlanta, GA, USA, pp. 420–427.

Harmelin, M., Legendre, B., 1991. In: Petzow, G., Effenberg, G. (Eds.), Ternary Alloys, vol. 4. Verlag Chemie (VCH), Weinheim, Germany, pp. 538–545.

Harris, S.J., Noble, B., Dinsdale, K., 1984. Effect of composition and heat treatment on strength and fracture characteristics of Al–Li–Mg alloys. In: Sanders Jr., T.H., Starke Jr., E.A. (Eds.), Proceedings of the Second International Conference on Aluminum–Lithium Alloys. The Metallurgical Society of AIME, Warrendale, PA, USA, pp. 219–233.

Harris, S.J., Noble, B., Dinsdale, K., Peel, C.J., Evans, B., 1986. Mechanical properties of Al–Li–Zn–Mg alloys. In: Baker, C., Gregson, P.J., Harris, S.J., Peel, C.J. (Eds.), Proceedings of the Third International Conference on Aluminum–Lithium Alloys, vol. 1. The Institute of Metals, London, UK, pp. 610–620.

Harris, S.J., Noble, B., Dinsdale, K., 1987. The role of magnesium in Al–Li–Cu–Mg–Zr alloys. In: Champier, G., Dubost, B., Miannay, D., Sabatay, L. (Eds.), Proceedings of the Fourth International Conference on Aluminum–Lithium Alloys, J. Phys., 48, C3, 415–C3, 423.

Harris, S.J., Noble, B., Dodd, A., 1989. The effect of texture on the tensile and fatigue properties of 8090 plate alloys. In: Sanders Jr., T.H., Starke Jr., E.A. (Eds.), Proceedings of the Fifth International Conference on Aluminium–Lithium Alloys, vol. 2. Materials and Component Engineering Publications Ltd., Birmingham, UK, pp. 1061–1076.

't Hart, W.G.J., Schra, L., McDarmaid, D.S., Peters, M., 1989. Mechanical properties and fracture toughness of 8090-T651 plate and 2091 and 8090 sheet. In: New Light Alloys, AGARD Conference Proceedings No. 444, Advisory Group for Aerospace Research and Development, Neuilly-sur-Seine, France, pp. 5-1–5-17.

Hatamleh, O., Rivero, I.V., Swain, S.E., 2009. An investigation of the residual stress characterization and relaxation in peened friction stir welded aluminum–lithium alloy joints. Mater. Des. 30, 3367–3373.

Hautefeuille, I., Rahouadj, R., Barbaux, Y., Clavel, M., 1987. Toughness and heat treatment relationship in a 2091 aluminium alloy. In: Champier, G., Dubost, B., Miannay, D. and Sabatay, L. (Eds.), Proceedings of the Fourth International Conference on Aluminum–Lithium Alloys, J. Phys., 48, C3, 669–C3, 675.

Hayakawa, H., Kaneko, J., Sugamata, M., 1994. Directional annealing of SiCw/6061Al extruded composites. In: Sanders Jr., T.H., Starke Jr., E.A. (Eds.), Proceedings of the Fourth International Conference on Aluminum Alloys: Their Physical and Mechanical Properties. Georgia Institute of Technology, Atlanta, GA, USA, pp. 661–667.

Hayashi, T., Okuno, K., Kudo, H., 1988. Tritium diffusion and trapping associated with lithium in Al–Li systems. J. Less Common Met. 141, 169–176.

Hayashi, T., Okuno, K., Naruse, Y., 1992. Permeation behavior of deuterium implanted into aluminium–lithium alloys. J. Nucl. Mater. 191–194, 1065–1069.

Haynes, T.G., Wyte, M., Webster, D., 1987. Mechanical properties and microstructure of Al–Li investment castings. In: Champier, G., Dubost, B., Miannay, D., Sabatay, L. (Eds.), Proceedings of the Fourth International Conference on Aluminum–Lithium Alloys, J. Phys., 48, C3, 123–C3, 128.

He, M., Wang, Z., Zhang, B., Wang, Z., 1989. Research on the role of minor Ce in Al–Li–Cu alloys. In: Sanders Jr., T.H., Starke Jr., E.A. (Eds.), Proceedings of the Fifth International Conference on Aluminium–Lithium Alloys, vol. 2. Materials and Component Engineering Publications Ltd., Birmingham, UK, pp. 741–750.

He, Y., Wang, Z., Zhu, Y., He, Y., Liang, M., 1994. Effect of alloying elements on solidification structure of 8090 Al–Li alloy welds. In: Sanders Jr., T.H., Starke Jr., E.A. (Eds.), Proceedings of the Fourth International Conference on Aluminum Alloys: Their Physical and Mechanical Properties, vol. 2. Georgia Institute of Technology, Atlanta, GA, USA, pp. 481–487.

Henshall, C.A., Wadsworth, J., Reynolds, M.J., Barnes, A.J., 1987. Design and manufacture of a superplastic-formed aluminum–lithium component. Mater. Des. 8, 324–330.

Hermann, R., 1995. Crack growth and residual stress in Al–Li metal matrix composites under far-field cyclic compression. J. Mater. Sci. 30, 3782–3790.

Hermann, R., Reid, C.D., 1989. Enhancing the fatigue life of fastener holes in aluminium–lithium alloy 8090. In: Sanders Jr., T.H., Starke Jr., E.A. (Eds.), Proceedings of the Fifth International Conference on Aluminium–Lithium Alloys, vol. 3. Materials and Component Engineering Publications Ltd., Birmingham, UK, pp. 1607–1614.

Higashi, K., Okada, T., Mukal, T., Tanimura, S., 1992. Superplastic behavior at high strain rates of mechanically alloyed Al–Mg–Li alloy. Scr. Metall. 26, 761–766.

Hill, D.P., Williams, D.N., Mobley, C.E., 1984. The effect of hydrogen on the ductility, toughness and yield strength of an Al–Mg–Li alloy. In: Sanders Jr., T.H., Starke Jr., E.A. (Eds.), Proceedings of the Second International Conference on Aluminum–Lithium Alloys. The Metallurgical Society of AIME, Warrendale, PA, USA, pp. 201–218.

Hirano, S., Yoshida, H., Uno, T., 1989. Quench sensitivity in Al–Li based alloys. In: Sanders Jr., T.H., Starke Jr., E.A. (Eds.), Proceedings of the Fifth International Conference on

Aluminium—Lithium Alloys, vol. 1. Materials and Component Engineering Publications Ltd., Birmingham, UK, pp. 335—344.

Hirosawa, S., Sato, T., Kamio, A., 1994. Effect of Mg addition on precipitation hardening of Al—Li—Cu—Zr alloys containing Ag. In: Sanders Jr., T.H., Starke Jr., E.A. (Eds.), Proceedings of the Fourth International Conference on Aluminum Alloys: Their Physical and Mechanical Properties, vol. 2. Georgia Institute of Technology, Atlanta, GA, USA, pp. 199—206.

Hirosawa, S., Sato, T., Kamio, A., 1998. Effects of Mg addition on the kinetics of low-temperature precipitation in Al—Li—Cu—Ag—Zr alloys. Mater. Sci. Eng. A, A242, 195—201.

Hirsch, J., Engler, O., Lücke, K., Peters, M., Welpmann, K., 1987. The rolling texture development in an 8090 Al—Li alloy. In: Champier, G., Dubost, B., Miannay, D., Sabatay, L. (Eds.), Proceedings of the Fourth International Conference on Aluminum—Lithium Alloys, J. Phys., 48, C3, 605—C3, 611.

Holroyd, N.J.H., Gray, A., Scamans, G.M., Hermann, R., 1986. Environment-sensitive fracture of Al—Li—Cu—Mg alloys. In: Baker, C., Gregson, P.J., Harris, S.J., Peel, C.J. (Eds.), Proceedings of the Third International Conference on Aluminum—Lithium Alloys. The Institute of Metals, London, UK, pp. 310—320.

Homeny, J., Buckley, M.M., 1990. Oxidation-sensitive low-energy Auger peaks in an aluminum oxide fiber/aluminum—lithium alloy matrix composite. Mater. Lett. 9, 443—446.

Hong, S.-K., Hioyasu Tezuka, H., Kamio, A., 1994. Effect of thermal stress on mechanical properties of SiCw reinforced Al matrix composites. In: Sanders Jr., T.H., Starke Jr., E.A. (Eds.), Proceedings of the Fourth International Conference on Aluminum Alloys: Their Physical and Mechanical Properties, vol. 2. Georgia Institute of Technology, Atlanta, GA, USA, pp. 698—705.

Howe, J.M., Laughlin, D.E., Vasudevan, A.K., 1988a. A high-resolution transmission electron microscopy investigation of the δ'- θ' precipitate structure in an Al-2 wt%Li-1 wt% Cu alloy. Philos. Mag. A, 57, 955—969.

Howe, J.M., Lee, J., Vasudevan, A.K., 1988b. Structure and deformation behavior of T_1 precipitate plates in an Al—2Li—1Cu alloy. Metall. Trans. A, 19A, 2911—2920.

Howell, P.R., Michel, D.J., Ryba, E., 1989. The nature of micro-crystalline regions produced by an in situ transformation of T2 particles in a ternary Al-2.5%Li-2.5%Cu alloy. Scr. Metall. 23, 825—828.

Hu, D., Zhang, Y., Liu, Y.L., Zhu, Z.Y., 1993. Corrosion behaviour of 8090 Al—Li alloy. Corrosion 49, 491—498.

Hua, D., Caibei, Z., Jianzhong, C., 1994. A study of recrystallization of an Al—Li alloys. In: Sanders Jr., T.H., Starke Jr., E.A. (Eds.), Proceedings of the Fourth International Conference on Aluminum Alloys: Their Physical and Mechanical Properties, vol. 2. Georgia Institute of Technology, Atlanta, GA, USA, pp. 238—241.

Huang, B.P., Zheng, Z.Q., 1998. Independent and combined roles of trace Mg and Ag additions in properties precipitation process and precipitation kinetics of Al—Cu—Li—(Mg)—(Ag)—Zr—Ti alloys. Acta Mater. 46, 381—393.

Huang, J.C., 1992. On the crystal structure of the T_1 phase in Al—Li—Cu alloys. Scr. Metall. 27, 755—760.

Huang, J.C., Ardell, A.J., 1986. Microstructure evolution in two Al—Li—Cu alloys. In: Baker, C., Gregson, P.J., Harris, S.J., Peel, C.J. (Eds.), Proceedings of the Third International Conference on Aluminum—Lithium Alloys. The Institute of Metals, London, UK, pp. 455—470.

Huang, J.C., Ardell, A.J., 1987. Strengthening mechanisms associated with T1 particles in two Al—Li—Cu alloys. In: Champier, G., Dubost, B., Miannay, D., Sabetay, L. (Eds.), Proceedings of the Fourth International Conference on Aluminium—Lithium Alloys, J. Phys., 48, C3, 373—C3, 384.

Huang, J.C., Ardell, A.J., 1988. Precipitation strengthening of binary Al—Li alloys by δ' precipitates. Mater. Sci. Eng. A, A104, 149—156.

Huang, R.D., Gray, G.T., 1989. Influence of precipitates on the mechanical response and substructure evolution of shock-loaded and quasi-statically deformed Al—Li—Cu alloys. Metall. Trans. A, 20A, 1061—1075.

Hunt, E.P., Gregson, P.J., Moreton, R., Peel, C.J., 1992. The fatigue behaviour of fibre reinforced aluminium—lithium alloy and aluminium—lithium MMC based laminates. Scr. Metall. Mater. 27, 1817—1822.

Hwang, Y.M., Lay, H.S., Huang, J.C., 2002. Study on superplastic blow-forming of 8090 Al—Li sheets in an ellip-cylindrical closed-die. Int. J. Mach. Tools Manuf. 42, 1363—1372.

Il'in, A.A., Zakharov, V.V., Betsofen, M.S., Osintsev, O.E., Rostova, T.D., 2008. Texture and anisotropy of the mechanical properties of an Al—Mg—Li—Zn—Sc—Zr alloy. Russ. Metall. 41, 406—412.

Il'in, A.A., Nikitin, S.L., Osintsev, O.E., Borisov, Y.V., 2009. Effect of alloying elements on the structure and properties of Al—Li—Cu cast alloys. Russ. Metall. 42, 338—344.

Islamgaliev, R.K., Yunusova, N.F., Nurislamova, G.V., Krasil'nikov, N.A., Valiev, R.Z., Ovid'ko, I.A., 2009. Structure and mechanical properties of strips and shapes from ultrafine-grained aluminium alloy 1421. Met. Sci. Heat Treat. 51, 82—86.

Istomin-Kastrovskii, V.V., Shamrai, V.F., Grushko, O.E., 2010. Effect of silver, magnesium, and zirconium additions on ageing of a V1469 alloy (Al—Cu—Li system). Russ. Metall. 43, 819—823.

Itoh, G., Cui, Q., Kanno, M., 1996. Effects of small additions of magnesium and silver on the precipitation of T1 phase in an Al-4%Cu-1.1Li-0.2Zr alloy. Mater. Sci. Eng. A, A211, 128—137.

Ivanov, R., Boselli, J., Denzer, D., Larouche, D., Gauvin, R., Brochu, M., 2012. Hardening potential of an Al—Cu—Li friction stir weld. In: Weiland, H., Rollett, A.D., Cassada, W.A. (Eds.), Proceedings of the 13th International Conference on Aluminum Alloys (ICAA13). The Minerals, Metals and Materials Society (TMS) and John Wiley & Sons, Hoboken, NJ, USA, pp. 659—664.

Jagan Reddy, G., 2010. Study on High Temperature Flow Properties of Al—Li Alloy UL40 and Development of Processing Maps (Doctoral Thesis), Indian Institute of Technology, Bombay, India.

Jagan Reddy, G., Srinivasan, N., Gokhale, A.A., Kashyap, B.P., 2008. Characterization of dynamic recovery during hot deformation of spray cast Al—Li alloy UL40 alloy. Mater. Sci. Technol. 24, 725—733.

Jagan Reddy, G., Srinivasan, N., Gokhale, A.A., Kashyap, B.P., 2009. Processing map for hot working of spray formed and hot isostatically pressed Al—Li alloy (UL40). J. Mater. Process. Technol. 209, 5964—5972.

Jata, K.V., 1989. Creep crack growth resistance of 2091 Al—Li alloy. In: Sanders Jr., T.H., Starke Jr., E.A. (Eds.), Proceedings of the Fifth International Conference on Aluminium—Lithium Alloys, vol. 2. Materials and Component Engineering Publications Ltd., Birmingham, UK, pp. 1163—1173.

Jata, K.V., Semiatin, S.L., 2000. Continuous dynamic recrystallization during friction stir welding of high strength aluminum alloys. Scr. Mater. 43, 743.

Jata, K.V., Starke Jr., E.A., 1986. Fatigue crack growth and fracture toughness behaviour of an Al−Li−Cu alloy. Metall. Trans. A, 17A, 1011−1026.

Jata, K.V., Starke Jr., E.A., 1988. Fracture toughness of Al−Li-X alloys at ambient and cryogenic temperatures. Scr. Metall. 22, 1553−1556.

Jata, K.V., Hopkins, A.K., Rioja, R.J., 1996. The anisotropy and texture of Al−Li alloys. Mater. Sci. Forum 217−222, 647−652.

Jata, K.V., Panchandeeswaran, S., Vasudevan, A.K., 1998. Evolution of texture, microstructure and mechanical property anisotropy in an Al−Li−Cu alloy. Mater. Sci. Eng. A, A257, 37−46.

Jensrud, O., 1986. Hardening mechanisms and ductility of an Al + 3.0 wt% Li alloy. In: Baker, C., Gregson, P.J., Harris, S.J., Peel, C.J. (Eds.), Proceedings of the Third International Conference on Aluminum−Lithium Alloys. The Institute of Metals, London, UK, pp. 411−419.

Jeon, S.M., Park, J.K., 1994. Transition behaviour of deformation mode from shearing to looping in Al−Li single crystals. Philos. Mag. A, 70, 493−504.

Jeon, S.M., Park, J.K., 1996. Precipitation strengthening behavior of Al−Li single crystals. Acta Mater. 44, 1449−1455.

De Jesus, R., Ardell, A.J., 1989. Characterization of the precipitate microstructures of solution treated and aged 2090 and 2091. In: Sanders Jr., T.H., Starke Jr., E.A. (Eds.), Proceedings of the Fifth International Conference on Aluminium−Lithium Alloys. Materials and Component Engineering Publications Ltd., Birmingham, UK, pp. 661−670.

Jha, S., Sanders Jr., T.H., Dayananda, M.A., 1986. Grain boundary precipitate free zone in Al−Li alloys. Acta Metall. 35, 473−482.

Ji, D.X., Tian, S.X., Chen, C.Q., 1992. In: Peters, M., Winkler, P.J. (Eds.), Aluminum−Lithium Alloys VI. Deutsche Gesellschaft für Metallkunde, Frankfurt, Germany, p. 863.

Jiang, B., Zhang, C., Wang, T., Qu, Z., Wu, R., Zhang, M., 2012. Creep behaviors of Mg−5Li−3Al−(0, 1) Ca alloys. Mater. Des. 34, 863−866.

Jiang, D., Wang, L., 2005. Deformation and fracture behavior of an Al−Li alloy 8090. J. Mater. Sci. 40, 2745−2747.

Jiang, J.Q., Bate, P.S., 1996. Use of microstructural gradients in hot gas-pressure forming of Zn−Al sheet. Metall. Mater. Trans. A, 27A, 3250−3258.

Jiang, N., Gao, X., Zheng, Z.Q., 2010. Microstructure evolution of aluminum−lithium alloy 2195 undergoing commercial production. Trans. Nonferrous Met. Soc. China 20, 740−745.

Jiang, X.D., Qi, R.C., 1989. The effect of cold working on the tensile properties of Al−Li-X alloy. In: Sanders Jr., T.H., Starke Jr., E.A. (Eds.), Proceedings of the Fifth International Conference on Aluminium−Lithium Alloys, vol. 1. Materials and Component Engineering Publications Ltd., Birmingham, UK, pp. 407−416.

Jiang, X.J., Gui, Q.H., Li, H.Y., Ma, L.M., Liang, G.J., Shi, C.X., 1993. Effects of minor additions on precipitation and properties of Al−Li−Cu−Mg−Zr alloy. Scr. Metall. Mater. 29, 211−216.

Jo, H.H., Hirano, K.I., 1987. Precipitation processes in Al−Cu−Li alloy studying by DSC. Mater. Sci. Forum 13/14, 377−382.

Johansen, A., Johnson, E., Sarholt-Kristensen, L., Steenstrup, S., Andersen, H.H., Buhanov, V. M., et al., 1991. Angular distributions of sputtered particles from lithium-implanted aluminium and copper crystals. Nucl. Instrum. Methods Phys. Res. B 61, 21−26.

Johnson, E., Johansen, A., Sarholt-Kristensen, L., 1985. Formation of ordered Al_3Li by ion implantation of lithium into aluminium. Mater. Lett. 4, 33−38.

Joshi, N.V., Gokhale, A.A., Prasad, K.S., Ramachandran, T.R., 1991. Effect of aluminium on precipitation in 8090 aluminum alloy, in aluminum: strategies for the nineties and beyond. In: Dwarakadasa, E.S., Sesan, S., Abraham, K.P. (Eds.), Proceedings of the Second International Conference on Aluminum (INCAL 2), vol. 2. Aluminum Association in India, Bangalore, India, pp. 569−574.

Jung, W.S., Park, J.K., 1989. Precipitation kinetics in Al−Li binary alloys. In: Sanders Jr., T.H., Starke Jr., E.A. (Eds.), Proceedings of the Fifth International Conference on Aluminium−Lithium Alloys. Materials and Component Engineering Publications Ltd., Birmingham, UK, pp. 595−604.

Kachaturyan, A.G., Lindsey, T.F., Morris, J.W., 1988. Theoretical investigation of the precipitation of δ' in Al−Li. Metall. Trans. A, 19A, 249−258.

Kaigorodova, L.I., Pushin, V.G., Rasposienko, Y.u.D., Pilyugin, V.P., 2011. Effect of severe plastic deformation on the formation of the nanocrystalline structure and the aging of the multicomponent aluminium−lithium alloy with small additions of Sc and Mg. Phys. Met. Metall. 111, 72−79.

Kalogeridis, A., Pesicka, J., Nembach, E., 1999a. On the increase of the precipitated volume fraction during Ostwald ripening, exemplified for aluminium−lithium alloys. Mater. Sci. Eng. A, A268, 197−201.

Kalogeridis, A., Pesicka, J., Nembach, E., 1999b. The critical resolved shear stress of peak-aged aluminium−lithium single crystals. Acta Mater. 47, 1953−1964.

Kalyanam, S., Beaudoin, A.J., Dodds Jr., R.H., Barlat, F., 2009. Delamination cracking in advanced aluminum−lithium alloys—experimental and computational studies. Eng. Fract. Mech. 76, 2174−2191.

Kamat, S.V., Eswara Prasad, N., 1990. Mixed-mode analysis of crack deflection and its influence on R-curve behaviour of aluminium−lithium 8090 alloy sheets. Scr. Metall. Mater. 24, 1907−1912.

Kamat, S.V., Eswara Prasad, N., 1991. Effect of notch root radius on fracture toughness of an 8090 Al−Li alloy. Scr. Metall. Mater. 25, 1519−1523.

Kamat, S.V., Eswara Prasad, N., 1993. Fatigue crack growth behaviour of an 8090 Al−Li alloy under mixed mode I−III loading conditions. Scr. Metall. Mater. 29, 1371−1376.

Kamat, S.V., Eswara Prasad, N., Malakondaiah, G., 1991. Comparison of mode I and mode II fracture toughness of an 8090 Al−Li alloy. Mater. Sci. Eng. A, A149, L1−L3.

Kang, S., Grant, N.J., 1984. Mechanical properties of rapidly solidified X2020 aluminium alloys. In: Sanders Jr., T.H., Starke Jr., E.A. (Eds.), Proceedings of the Second International Conference on Aluminum−Lithium Alloys. The Metallurgical Society of AIME, Warrendale, PA, USA, pp. 469−484.

Kar, R.J., Bohlen, J.W., Chanani, G.R., 1984. Correlation of microstructure, aging treatments, and properties of Al−Li−Cu−Mg−Zr I/M and P/M alloys. In: Sanders Jr., T.H., Starke Jr., E.A. (Eds.), Proceedings of the Second International Conference on Aluminum−Lithium Alloys. The Metallurgical Society of AIME, Warrendale, PA, USA, pp. 255−286.

Kar, R.J., Agarwal, S.P., Quist, W.E. (Eds.), In: Aluminum−Lithium Alloys: Design, Development and Application Update. ASM International, Materials Park, OH, USA.

Karabin, L.M., Bray, G.H., Rioja, R.J., Venema, G.B., 2012. Al−Li−Cu−Mg−(Ag) products for lower wing skin applications. In: Weiland, H., Rollett, A.D., Cassada, W.A. (Eds.), Proceedings of the 13th International Conference on Aluminum Alloys (ICAA13). The Minerals, Metals and Materials Society (TMS) and John Wiley & Sons, Hoboken, NJ, USA, pp. 529−534.

Katsikis, S., 2001. Effect of Copper and Magnesium on the Precipitation Characteristics of Al–Li–Mg, Al–Li–Cu and AI–Li–Cu–Mg Alloys (Ph.D. Thesis), University of Nottingham, Nottingham, UK.

Katsikis, S., Noble, B., Harris, S.J., 2008. Microstructural stability during low temperature exposure of alloys within the Al–Li–Cu–Mg system. Mater. Sci. Eng. A, A485, 613–620.

Kaufman, M.J., Morrone, A.A., Lewis, R.E., 1992. Complications concerning TEM analysis of the δ-AlLi phase in aluminum–lithium alloys. Scr. Metall. Mater. 27, 1265–1270.

Kerr, J.R., Merino, R.E., 1989. Cryogenic properties of VPPA welded Al–Li alloys. In: Sanders Jr., T.H., Starke Jr., E.A. (Eds.), Proceedings of the Fifth International Conference on Aluminium–Lithium Alloys, vol. 3. Materials and Component Engineering Publications Ltd., Birmingham, UK, pp. 1491–1500.

Khachaturyan, A.G., Lindsey, T.F., Morris, J.W., 1988. Theoretical investigation of the precipitation of δ' in Al–Li. Metall. Trans. A, 19A, 249–258.

Khatsinkaya, I.M., Grushko, O.E., Sheveleva, L.M., Kolobnev, N.I., Couch, P.D., Mille, W.S., 1994. The structure and phase transformations in Al–Li–Mg alloys containing Zr and Sc. In: Sanders Jr., T.H., Starke Jr., E.A. (Eds.), Proceedings of the Fourth International Conference on Aluminum Alloys: Their Physical and Mechanical Properties, vol. 2. Georgia Institute of Technology, Atlanta, GA, USA, pp. 183–190.

Khireddine, D., Rahouadj, R., Clavel, M., 1988. Evidence of S′ phase shearing in an aluminium–lithium alloy. Scr. Metall. 22, 167–172.

Khireddine, D., Rahouadj, R., Clavel, M., 1989. The influence of δ' and S′ precipitation on low cycle fatigue behaviour of an aluminium alloy. Acta Metall. 37, 191–201.

Kilmer, R.J., Stoner, G.E., 1991. Effect of Zn additions on precipitation during aging of alloy 8090. Scr. Metall. 25, 243.

Kilmer, R.J., Witters, J.J., Stoner, G.E., 1992. In: Peters, M., Winkler, P.J. (Eds.), Aluminum–Lithium Alloys VI. Deutsche Gesellschaft für Metallkunde, Frankfurt, Germany, pp. 755–755.

Kim, N.J., Skier, D.J., Okazaki, K., Adam, C.M., 1986. Development of low density aluminium–lithium alloys using rapid solidification technology. In: Baker, C., Gregson, P.J., Harris, S.J., Peel, C.J. (Eds.), Proceedings of the Third International Conference on Aluminium–Lithium Alloys. The Institute of Metals, London, UK, pp. 78–84.

Kim, N.J., Bye, R.L., Das, S.K., 1987a. Structure and properties of rapid solidification processed aluminium–lithium alloys. In: Champier, G., Dubost, B., Miannay, D., Sabetay, L. (Eds.), Proceedings of the Fourth International Conference on Aluminium–Lithium Alloys, J. Phys., 48, C3, 309–C3, 316.

Kim, N.J., Howe, J.M., Boden, E.G., 1987b. High resolution TEM study of precipitates in a rapidly solidified Al–Li–Zr alloy. In: Champier, G., Dubost, B., Miannay, D., Sabetay, L. (Eds.), Proceedings of the Fourth International Conference on Aluminium–Lithium Alloys, J. Phys., 48, C3, 457–C3, 464.

Kim, N.J., Raybould, D., Bye, R.L., Das, S.K., 1989a. Microstructure and mechanical properties of rapidly solidified Al–Li–Cu–Mg–Zr alloy die forgings. In: Sanders Jr., T.H., Starke Jr., E.A. (Eds.), Proceedings of the Fifth International Conference on Aluminium–Lithium Alloys, vol. 1. Materials and Component Engineering Publications Ltd., Birmingham, UK, pp. 123–130.

Kim, N.J., Bye, R.L., Das, S.K., 1989b. Recent developments in rapidly solidified aluminium–lithium alloys. In: Kar, R.J., Agarwal, S.P., Quist, W.E. (Eds.), Aluminum–Lithium Alloys: Design, Development and Application Update. ASM International, Materials Park, OH, USA, pp. 63–76.

Kim, S.S., Shin, K.S., 1999. Closure-affected fatigue crack propagation behaviors of powder metallurgy—processed Al—Li alloys in various environments. Metall. Mater. Trans. A, 30A, 2254—2258.

Kim, S.S., Shin, K.S., Kim, N.J., 1999. Tensile behaviour of rapidly solidified Al—Li—Zr alloys. Scr. Mater. 41, 333—340.

Kiselev, I., 2008. Dynamic and kinetic properties of Al—Li melts. Russ. Metall., 523—528.

Kiselev, I., 2010. Dynamic properties of aluminium—lithium and aluminium—magnesium melts. Russ. Metall., 726—732.

Kitabjian, P., Lee, E.W., 1994. Microstructure evolution during superplastic deformation of 2095 Al—Li alloy. In: Sanders Jr., T.H., Starke Jr., E.A. (Eds.), Proceedings of the Fourth International Conference on Aluminum Alloys: Their Physical and Mechanical Properties, vol. 2. Georgia Institute of Technology, Atlanta, GA, USA, pp. 268—273.

Kitahama, K., Hiratani, M., Kawai, S., 1983. Growth of lithium—aluminum single crystal by the floating zone technique. J. Cryst. Growth 62, 177—182.

Knowles, K.M., Stobbs, W.M., 1988. The structure of {111} age-hardening precipitates in Al—Cu—Mg—Ag alloys. Acta Crystallogr. B44, 207—227.

Kobayashi, K., Ohsaki, S., Kamio, A., Tsuji, Y., 1992. Effect of Zn addition on corrosion resistance of 2090 and 2091 alloys. In: Peters, M., Winkler, P.J. (Eds.), Aluminum—Lithium Alloys VI, vol. 2. Deutsche Gesellschaft für Metallkunde, Frankfurt, Germany, pp. 673—678.

Kobayashi, K., Hagiwara, T., Sakamoto, T., 1994. Resistance spot weldability and TIG weldability of Al—Li—Mg alloys. In: Sanders Jr., T.H., Starke Jr., E.A. (Eds.), Proceedings of the Fourth International Conference on Aluminum Alloys: Their Physical and Mechanical Properties, vol. 2. Georgia Institute of Technology, Atlanta, GA, USA, pp. 488—495.

Kobayashi, T., Niimoni, M., Degawa, K., 1989. Temperature dependence of impact toughness in Al—Li—Cu—Mg—Zr alloys. Mater. Sci. Technol. 5, 1013—1019.

Koenigsmann, H.J., Starke Jr., E.A., 1994. Microstructural stability and fracture behaviour in Al—Si—Ge alloys. In: Sanders Jr., T.H., Starke Jr., E.A. (Eds.), Proceedings of the Fourth International Conference on Aluminum Alloys: Their Physical and Mechanical Properties, vol. 2. Georgia Institute of Technology, Atlanta, GA, USA, pp. 24—31.

Koh, H.J., Kim, N.J., Lee, S., Lee, E.W., 1998. Superplastic deformation behavior of a rapidly solidified Al—Li alloy. Mater. Sci. Eng. A, A256, 208—213.

Kojima, K.A., Lewis, R.E., Kaufman, M.J., 1989. Microstructural characterization and mechanical properties of a spray-cast Al—Li—Cu—Mg—Zr alloy. In: Sanders Jr., T.H., Starke Jr., E.A. (Eds.), Proceedings of the Fifth International Conference on Aluminium—Lithium Alloys, vol. 1. Materials and Component Engineering Publications Ltd., Birmingham, UK, pp. 85—94.

Kolobney, N.I., Grushko, O.E., Cherkasov, V.V., Dolzhansky, Y.M., Miller, W.S., Couch, P.D., 1994. Aluminium—lithium alloys based on the Al—Mg—Li—Zr—Sc system. In: Sanders Jr., T.H., Starke Jr., E.A. (Eds.), Proceedings of the Fourth International Conference on Aluminum Alloys: Their Physical and Mechanical Properties, vol. 2. Georgia Institute of Technology, Atlanta, GA, USA, pp. 305—312.

Kooi, D.C.V., Park, W., Hilton, M.R., 1999. Characterization of cryogenic mechanical properties of aluminum—lithium alloy C-458. Scr. Mater. 30, 1185—1190.

Kortan, A.R., Chen, H.S., Parsey, J.M., Kimerling, L.C., 1989. Morphology and microstructure of Al—Li—Cu quasicrystals. J. Mater. Sci. 24, 199—205.

Kostrivas, A., Lippold, J.C., 1999. Weldability of Li-bearing aluminium alloys. Int. Mater. Rev. 44, 217—237.

Kramer, L.S., Pickens, J.R., 1992. Microstructure and properties of a welded Al–Cu–Li alloy. Weld. J. 71, 115s–121s.

Kramer, L.S., Heubaum, F.H., Pickens, J.R., 1989. The weldability of high strength Al–Cu–Li alloys. In: Sanders Jr., T.H., Starke Jr., E.A. (Eds.), Proceedings of the Fifth International Conference on Aluminium–Lithium Alloys, vol. 3. Materials and Component Engineering Publications Ltd., Birmingham, UK, pp. 1415–1424.

Kridli, G.T., El-Gizawy, A.S., Lederich, R., 1998. Development of process maps for superplastic forming of Weldalite™ 049. Mater. Sci. Eng. A, A244, 224–232.

Kulkarni, G.J., Banerjee, D., Ramachandran, T.R., 1989. Physical metallurgy of aluminium–lithium alloys. Bull. Mater. Sci. 12, 325–340.

Kulwicki, J.H., Sanders Jr., T.H., 1984. Coarsening of δ'(Al₃Li) precipitates in a Al-2.7Li-0.3 Mn alloy. In: Sanders Jr., T.H., Starke Jr., E.A. (Eds.), Proceedings of the Second International Conference on Aluminum–Lithium Alloys. The Metallurgical Society of AIME, Warrendale, PA, USA, pp. 31–51.

Kumai, C., Kusinski, J., Thomas, G., Devine, T.M., 1989. Influence of aging at 200°C on the corrosion resistance of Al–Li and Al–Li–Cu alloys. Corrosion 45, 294–303.

Kumar, S., McShane, H.B., Sheppard, T., 1994a. Effect of extrusion parameters on the microstructure and properties of an Al–Li–Mg–Zr alloy. J. Mater. Sci. 29, 1067–1074.

Kumar, S., McShane, H.B., Sheppard, T., 1994b. Effect of zirconium and magnesium additions on properties of Al–Li based alloy. Mater. Sci. Technol. 10, 162–170.

Kumar, S., Król, J., Pink, E., 1996. Mechanism of serrated flow in binary Al–Li alloys. Scr. Mater. 35, 775–780.

Kuokkala, V.T., Virta, J., 1994. Elastic constants of Al–SiCp MMC's at elevated temperatures. In: Sanders Jr., T.H., Starke Jr., E.A. (Eds.), Proceedings of the Fourth International Conference on Aluminum Alloys: Their Physical and Mechanical Properties, vol. 2. Georgia Institute of Technology, Atlanta, GA, USA, pp. 676–681.

Kwon, Y.N., Koh, H.J., Lee, S., Kim, N.J., Chang, Y.W., 2001. Effects of microstructural evolution on superplastic deformation characteristics of a rapidly solidified Al–Li alloy. Metall. Mater. Trans. A, 32A, 1649–1658.

Laberre, L.C., James, R.S., Witters, J.J., O'Malley, R.J., Emptage, M.R., 1987. Difficulties in grain refining aluminium–lithium alloys using commercial Al–Ti and Al–Ti–B master alloys. In: Champier, G., Dubost, B., Miannay, D., Sabatay, L. (Eds.), Proceedings of the Fourth International Conference on Aluminum–Lithium Alloys, J. Phys., 48, C3, 93–C3, 102.

Laffin, C., Raghunath, C.R., Lopez, H.F., 1993. Hydrogen induced surface cracking in an 8090 Al–Li alloy during high cycle fatigue. Scr. Metall. Mater. 29, 993–998.

Lagenbeck, S.L., Sakata, I.F., Ekvall, J.C., Rainen, R.A., 1987. Design considerations of new materials for aerospace vehicles. In: Kar, R.J., Agarwal, S.P., Quist, W.E. (Eds.), Aluminum–Lithium Alloys: Design, Development and Application Update. ASM International, Materials Park, OH, USA, pp. 293–314.

Lambri, O.A., Pérez-Landazábal, J.I., Nó, M.L., Juan, S.J., 1997. Study of the δ reversion process in 8090 alloys. Scr. Mater. 37, 851–859.

Lane, P.J., Gray, J.A., Smith, C.J.E., 1986. Comparison of corrosion behaviour of lithium-containing aluminium alloys and conventional aerospace alloys. In: Baker, C., Gregson, P.J., Harris, S.J., Peel, C.J. (Eds.), Proceedings of the Third International Conference on Aluminium–Lithium Alloys. The Institute of Metals, London, UK, pp. 273–281.

Lang, J.M., Degreve, F., Thorne, N.A., 1987. Grain boundary chemical analysis by SIMS. In: Champier, G., Dubost, B., Miannay, D., Sabatay, L. (Eds.), Proceedings of the Fourth International Conference on Aluminium–Lithium Alloys, J. Phys., 48, C3, 693–C3, 702.

Langan, P.J., Pickens, J.R., 1989. Identification of strengthening phases in Al−Cu−Li alloy Weldalite™ 049. In: Sanders Jr., T.H., Starke Jr., E.A. (Eds.), Proceedings of the Fifth International Conference on Aluminium−Lithium Alloys, vol. 2. Materials and Component Engineering Publications Ltd., Birmingham, UK, pp. 691−700.

Langan, T.J., Pickens, J.R., 1991. The effect of TiB_2 reinforcement on the mechanical properties of an Al−Cu−Li alloy-based metal−matrix composite. Scr. Metall. Mater. 25, 1587−1591.

Lapasset, G., Loiseau, A., 1987. A TEM Study of icosahedral and near icosahedral phases in 8090 alloy. In: Champier, G., Dubost, B., Miannay, D., Sabatay, L. (Eds.), Proceedings of the Fourth International Conference on Aluminum−Lithium Alloys, J. Phys., 48, C3, 489−C3, 495.

Lapasset, G., Damerval, C., Doudeau, M., 1989. Quench sensitivity of commercial Al−Li−Cu−Mg−Zr alloys. In: Sanders Jr., T.H., Starke Jr., E.A. (Eds.), Proceedings of the Fifth International Conference on Aluminium−Lithium Alloys, vol. 1. Materials and Component Engineering Publications Ltd., Birmingham, UK, pp. 365−374.

Lapasset, G., Octor, C., Sanchez, C., Barbaux, Y., Pons, G., 1994. Thermal stability and creep behaviour of four Al alloys. In: Sanders Jr., T.H., Starke Jr., E.A. (Eds.), Proceedings of the Fourth International Conference on Aluminum Alloys: Their Physical and Mechanical Properties, vol. 2. Georgia Institute of Technology, Atlanta, GA, USA, pp. 72−79.

Larson, L.A., Avalos-Borja, M., Pizzo, P.P., 1984. A surface analytical examination of stringer particles in aluminium−lithium−copper alloys. In: Sanders Jr., T.H., Starke Jr., E.A. (Eds.), Proceedings of the Second International Conference on Aluminum−Lithium Alloys. The Metallurgical Society of AIME, Warrendale, PA, USA, pp. 303−312.

Lavernia, E.J., Grant, N.J., 1987. Aluminum−lithium alloys: a review. J. Mater. Sci. 22, 1521−1529.

Lavernia, E.J., Srivatsan, T.S., Mohamed, F.A., 1990. Review: strength, deformation, fracture behaviour and ductility of aluminium−lithium alloys. J. Mater. Sci. 25, 1137−1158.

LeBaron, I.M. US Patent No. 2,381,219, Application date 1942, granted 1945.

Lederich, R.J., Sastry, S.M.L., 1984. Superplastic deformation of P/M and I/M Al−Li based alloys. In: Sanders Jr., T.H., Starke Jr., E.A. (Eds.), Proceedings of the Second International Conference on Aluminum−Lithium Alloys. The Metallurgical Society of AIME, Warrendale, PA, USA, pp. 137−152.

Lederich, R.J., Sastry, S.M.L., Meschter, P.J., 1985. Improved cavitation resistance in rapid solidification processed Al−Li alloys during superplastic forming. Scr. Metall. 19, 177−180.

Le Poac, P., Nomine, A.M., Miannay, D., 1987. Mechanical properties of beam electron weldings in 8090 alloy (CP 271). In: Champier, G., Dubost, B., Miannay, D., Sabetay, L. (Eds.), Proceedings of the Fourth International Conference on Aluminium−Lithium Alloys, J. Phys., 48, C3, 301−C3, 308.

Le Roy, G., Mace, R., Marchive, D., Meyer, P., Nossent, R., Schlecht, F., 1987. Status report on the development of aluminium−lithium at PECHINEY. In: Champier, G., Dubost, B., Miannay, D., Sabatay, L. (Eds.), Proceedings of the Fourth International Conference on Aluminum−Lithium Alloys, J. Phys., 48, C3, 33−C3, 39.

Lee, C.G., Lee, S., 1998. Correlation of dynamic torsional properties with adiabatic shear banding behavior in ballistically impacted aluminum−lithium alloys. Metall. Mater. Trans. A, 29A, 227−235.

Lee, C.G., Lee, Y.J., Lee, S., 1995. Observation of adiabatic shear bands formed by ballistic impact in aluminum−lithium alloys. Scr. Metall. Mater. 32, 821−826.

Lee, C.G., Kim, K.J., Lee, S., Cho, K., 1998. Effect of test temperature on the dynamic torsional deformation behavior of two aluminum−lithium alloys. Metall. Mater. Trans. A, 29A, 469−476.

Lee, C.S., Kim, S.S., Shin, K.S., 1997a. Effect of microstructure and load ratio on fatigue crack growth behavior of advanced Al−Cu−Li−Mg−Ag alloys. Met. Mater. 3, 51−59.

Lee, C.S., Park, K.J., Li, D., Kim, N.J., 1997b. Effects of temperature on the fatigue crack growth of an Al−Li 8090 alloy with δ' microstructure. Metall. Mater. Trans. A, 28A, 1089−1093.

Lee, E.W., Kim, N.J., 1989a. Microstructural effects on anisotropy of Al−Li alloy 2090. In: Sanders Jr., T.H., Starke Jr., E.A. (Eds.), Proceedings of the Fifth International Conference on Aluminium−Lithium Alloys, vol. 2. Materials and Component Engineering Publications Ltd., Birmingham, UK, pp. 809−816.

Lee, E.W., Kim, N.J., 1989b. Superplastic behaviour of a rapidly solidified Al−Cu−Mg−Zr alloy. In: Sanders Jr., T.H., Starke Jr., E.A. (Eds.), Proceedings of the Fifth International Conference on Aluminium−Lithium Alloys, vol. 1. Materials and Component Engineering Publications Ltd., Birmingham, UK, pp. 151−158.

Lee, E.W., Kalu, P.N., Brandao, L., Es-Said, O.S., Foyos, J., Garmestani, H., 1999. The effect of off-axis thermomechanical processing on the mechanical behavior of textured 2095 Al−Li alloy. Mater. Sci. Eng. A, 265, 100−109.

Lee, H.Y., 1992. Time-resolved scattering study of the early-stage decomposition and ordering processes in an Al-12 at.% Li alloy. Mater. Chem. Phys. 32, 336−341.

Lee, M.F., Huang, J.C., Ho, N.J., 1996. Microstructural and mechanical characterization of laser beam welding of 8090 Al−Li thin sheet. J. Mater. Sci. 31, 1455−1468.

Lee, S., Horita, Z., Hirosawa, S., Matsuda, K., 2012. Age-hardening of an Al−Li−Cu−Mg alloy (2091) processed by high-pressure torsion. Mater. Sci. Eng. A, A546, 82−89.

Leighly Jr., H.P., Coleman, P.G., West, R.N., 1989. The effect of Li content and the pre-annealing atmosphere on the ramp annealing behaviour of severely quenched Al−Li alloys. 1, 325−334.

Lendvai, J., Gudladt, H.J., Gerold, V., 1988. The deformation-induced dissolution of δ' precipitates in Al−Li alloys. Scr. Metall. 22, 1755−1760.

Lequeu, P., Smith, K.P., Danielou, A., 2010. Aluminum−copper−lithium alloy 2050 developed for medium to thick plate. J. Mater. Eng. Perform. 19, 841−848.

Lespinasse, C., Bathias, C., 1987. Fatigue crack growth of the 8090 alloy under overloading. In: Champier, G., Dubost, B., Miannay, D., Sabetay, L. (Eds.), Proceedings of the Fourth International Conference on Aluminium−Lithium Alloys, J. Phys., 48, C3, 793−C3, 800.

Levoy, N.F., Vander Sande, J.B., 1989. Phase transformations in the Al−Li−Hf and Al−Li−Ti systems. Metall. Trans. A, 20A, 999−1019.

Lewandowski, J.J., Holroyd, N.J.H., 1990. Intergranular fracture of Al−Li alloys: effects of aging and impurities. Mater. Sci. Eng. A, A123, 21−27.

Lewis, R.E., Yaney, D.L., Tanner, L.E., 1989. The effect of lithium in Al−Be alloys. In: Sanders Jr., T.H., Starke Jr., E.A. (Eds.), Proceedings of the Fifth International Conference on Aluminium−Lithium Alloys, vol. 2. Materials and Component Engineering Publications Ltd., Birmingham, UK, pp. 731−740.

Lewis, R.F., Starke, E.A., Coons, W.C., Shiflet, G.J., Willner, E., Bjeletich, J.G., et al., 1987. Microstructure and properties of Al−Li−Cu−Mg−Zr(8090) heavy section forgings. In: Champier, G., Dubost, B., Miannay, D., Sabetay, L. (Eds.), Proceedings of the Fourth International Conference on Aluminium−Lithium Alloys, J. Phys., 48, C3, 643−C3, 652.

Li, H., Tang, Y., Zeng, Z., Zheng, Z., Zheng, F., 2008a. Effect of ageing time on strength and microstructures of an Al−Cu−Li−Zn−Mg−Mn−Zr alloy. Mater. Sci. Eng. A, A498, 314−320.

Li, H.G., Tang, Y., Zeng, Z.D., Zheng, F., 2008b. Exfoliation corrosion of T6- and T8-aged Al$_x$Cu$_y$Li$_z$ alloy. Trans. Nonferrous Met. Soc. China 18, 778−783.

Li, H.X., Chen, C.Q., 1989. Deformation and fracture of Al−Li single crystals. In: Sanders Jr., T.H., Starke Jr., E.A. (Eds.), Proceedings of the Fifth International Conference on Aluminium−Lithium Alloys, vol. 2. Materials and Component Engineering Publications Ltd., Birmingham, UK, pp. 817−826.

Li, H.X., Chen, C.Q., 1990. Mechanism of anisotropy in fracture behaviour and fracture toughness of high strength aluminium alloy plate. Mater. Sci. Technol. 6, 850−856.

Li, H.X., Park, J.K., 2000. The serrated flow behavior of a binary Al−Li alloy tempered to conditions with and without δ' precipitates. Mater. Sci. Eng. A, A280, 156−160.

Li, J.F., Chen, W.J., Zhao, X.S., Ren, W.D., Zheng, Z.Q., 2006a. Corrosion behavior of 2195 and 1420 Al−Li alloys in neutral 3.5% NaCl solution under tensile stress. Trans. Nonferrous Met. Soc. China 16, 1171−1177.

Li, J.F., Zheng, Z.Q., Ren, W.D., Chen, W.J., Zhao, X.S., Li, S.C., 2006b. Simulation on function mechanism of Ti (Al$_2$CuLi) precipitate in localized corrosion of Al−Cu−Li alloys. Trans. Nonferrous Met. Soc. China 16, 1268−1273.

Li, S.P., Zhao, S.X., Pan, M.X., Zhao, D.Q., Chen, X.C., 2001. Eutectic reaction and microstructural characteristics of Al(Li)Mg$_2$Si alloys. J. Mater. Sci. 36, 1569−1576.

Liang, W., Pan, Q., He, Y., Li, Y., Zhou, Y., Lu, C., 2008. Effect of aging on the mechanical properties and corrosion susceptibility of an Al−Cu−Li−Zr alloy containing Sc. Rare Met. 27, 146−152.

Lieblich, M., Toralba, M., 1989. Microstructural evolution of rapidly solidified Al−Li−B alloys. In: Sanders Jr., T.H., Starke Jr., E.A. (Eds.), Proceedings of the Fifth International Conference on Aluminium−Lithium Alloys, vol. 2. Materials and Component Engineering Publications Ltd., Birmingham, UK, pp. 767−776.

Lieblich, M., Torralba, M., 1991. Cellular microstructure and heterogeneous coarsening of δ' in rapidly solidified Al−Li−Ti alloys. J. Mater. Sci. 26, 4361−4368.

Lieblich, M., Torralba, M., 1992. Ageing characteristics of rapidly solidified Al−Li−Ti alloys at 473 K. J. Mater. Sci. 27, 3474−3478.

Lieblich, M., Toralba, M., Champier, G., 1987. Microstructure of rapidly solidified Al−Li−Ti alloys. In: Champier, G., Dubost, B., Miannay, D., Sabetay, L. (Eds.), Proceedings of the Fourth International Conference on Aluminium−Lithium Alloys, J. Phys., 48, C3, 465−C3, 472.

Lin, D.C., Wang, G.X., Srivatsan, T.S., 2003. A mechanism for the formation of equiaxed grains in welds of aluminum−lithium alloy 2090. Mater. Sci. Eng. A, A351, 304−309.

Lin, F.S., Chakraborty, S.B., Starke Jr., E.A., 1982. Microstructure−property relationships of two Al−3Li−2Cu-0.2Zr-X Cd alloys. Metall. Trans. A, 13A, 401−410.

Lin, F.S., Moji, Y., Quist, W.E., Badner, D.V., 1987. Corrosion resistance of aluminium−lithium alloys. In: Champier, G., Dubost, B., Miannay, D., Sabatay, L. (Eds.), Proceedings of the Fourth International Conference on Aluminium−Lithium, J. Phys., 48, C3, 905−C3, 912.

Lipinski, P., Berveiller, M., Hihi, A., Sainfort, P., Mayer, P., 1987. Effect of crystallographic and morphologic textures on the anisotropy of mechanical properties of Al−Li alloys. In: Champier, G., Dubost, B., Miannay, D., Sabatay, L. (Eds.), Proceedings of the Fourth International Conference on Aluminum−Lithium Alloys, J. Phys., 48, C3, 613−C3, 619.

Lippold, J.C., 1989. Weldability of commercial aluminium–lithium alloys. In: Sanders Jr., T.H., Starke Jr., E.A. (Eds.), Proceedings of the Fifth International Conference on Aluminium–Lithium Alloys, vol. 3. Materials and Component Engineering Publications Ltd., Birmingham, UK, pp. 1365–1377.

Lippold, J.C., Lin, W., 1996. Weldability of commercial Al–Cu–Li alloys. Mater. Sci. Forum 217–222, 1685–1690.

Liu, D.R., Williams, D.B., 1988. Determination of the δ' solvus line in Al–Li alloys by measurement of the δ' volume fraction. Scr. Metall. 22, 1361–1365.

Liu, J., Chakrabarti, D.J., 1994. Microstructure evolution and micromechanisms of superplasticity in a high strength Al–Zn–Mg–Cu alloy. In: Sanders Jr., T.H., Starke Jr., E.A. (Eds.), Proceedings of the Fourth International Conference on Aluminum Alloys: Their Physical and Mechanical Properties. Georgia Institute of Technology, Atlanta, GA, USA, pp. 56–63.

Liu, Q., Yao, M., 1995. Comment on studies on activation energy of superplastic dynamically recrystallized aluminium–lithium alloys. Scr. Metall. Mater. 32, 1857–1859.

Liu, Q., Chen, C.Z., Cui, J.Z., 2000. Effect of copper content on mechanical properties and fracture behaviors of Al–Li–Cu alloy. Metall. Mater. Trans. A, 36A, 1389–1394.

Liu, S.M., Wang, Z.G., 2004. Low cycle fatigue properties of Al–Li–Cu–Mg–Zr alloy processed by equal-channel angular pressing. J. Mater. Sci. 40, 1753–1762.

Liu, W., 1997. A study on the ageing treatment of 2091 Al–Li alloy with an electric field. J. Mater. Sci. Lett. 16, 1410–1411.

Liu, W., Cui, J.Z., 1995. Effect of the homogenization treatment in an electric field on T_1 precipitation in 2091 Al–Li alloy. Scr. Metall. 33, 623–626.

Liu, W., Cui, J.Z., 1997. The effect of electric field on the recrystallization of 2091 Al–Li alloy. J. Mater. Sci. Lett. 16, 1400–1401.

Liu, W., Liang, K.M., Zheng, Y.K., 1998. Study of the diffusion of Al–Li alloys subjected to an electric field. J. Mater. Sci. 33, 1043–1047.

Liu, Y.L., Hu, Z.Q., Zhang, Y., Sh, C.X., 1993. The solidification behavior of 8090 Al–Li alloy. Metall. Trans. B 24B, 857–865.

Livenson, D.W., McPherson, D.J., 1956. Phase relations in magnesium–lithium–aluminium alloys. Trans. ASM 48, 689–706.

Livet, A., Bloch, D., 1985. A kinetic analysis of Al–Al$_3$Li unmixing. Scr. Metall. 19, 1147–1151.

Livet, F., Brechet, Y., 1987. Exploration of the phase diagram of Al–Li binary alloys by D.S.C. measurements and Monte Carlo methods. In: Champier, G., Dubost, B., Miannay, D., Sabetay, L. (Eds.), Proceedings of the Fourth International Conference on Aluminium–Lithium Alloys, J. Phys., 48, C3, 357–C3, 364.

Longzhou, M., 1997. A study on improving the cold-forming property of Al–Mg–Li alloy 01420. Adv. Perform. Mater. 4, 105–114.

Lorke, M., Stenzel, O.W., 1989. Melting of Al–Li alloys—the equipment manufacturer's view. In: Sanders Jr., T.H., Starke Jr., E.A. (Eds.), Proceedings of the Fifth International Conference on Aluminium–Lithium Alloys, vol. 1. Materials and Component Engineering Publications Ltd., Birmingham, UK, pp. 41–53.

Lu, L., Cui, J., Ma, J., 1989. Effect of heat treatment on the microstructures and mechanical properties of 2091 alloy. In: Sanders Jr., T.H., Starke Jr., E.A. (Eds.), Proceedings of the Fifth International Conference on Aluminium–Lithium Alloys, vol. 2. Materials and Component Engineering Publications Ltd., Birmingham, UK, pp. 889–898.

Lu, Z., Qiang, J., Li, Y.W., Wu, Y.L., 1994. Study on Al–Li 8090 extrudates—their mechanical properties and corrosion properties. In: Sanders Jr., T.H., Starke Jr., E.A. (Eds.),

Proceedings of the Fourth International Conference on Aluminum Alloys: Their Physical and Mechanical Properties, vol. 2. Georgia Institute of Technology, Atlanta, GA, USA, pp. 366–373.

Ludwiczak, E.A., Rioja, R.J., 1991. T_B precipitates in an Al–Cu–Li alloy. Scr. Metall. Mater. 25, 1415–1419.

Lukina, E.A., Alekseev, A.A., Antipov, V.V., Zaitsev, D.V., Yu, K., 2009. Application of the diagrams of phase transformations during ageing for optimizing the ageing conditions for V1469 and 1441 Al–Li alloys. Russ. Metall., 505–511.

Luo, A., Youdelis, W.V., 1993a. Microstructure and mechanical behavior of Al–Li–Cu–Mg alloy 8090 microalloyed with V and Be. Metall. Trans. A, 24A, 95–104.

Luo, A., Youdelis, W.V., 1993b. Effect of vanadium and beryllium on S′ (Al$_2$CuMg) precipitation in Al–Li–Cu–Mg alloy 8090. Mater. Sci. Technol. 9, 781–784.

Luo, A., Youdelis, W.V., 1993c. Clustering in a Be modified Al–Li–Cu–Mg alloy 8090. Scr. Metall. Mater. 28, 29–34.

Lynch, S.P., 1991a. Fracture of 8090 Al–Li plate, I. Short transverse fracture toughness. Mater. Sci. Eng. A, A136, 25–43.

Lynch, S.P., 1991b. Fracture of 8090 Al–Li plate II. Sustained-load crack growth in dry air at 50–200°C. Mater. Sci. Eng. A, A136, 45–57.

Lynch, S.P., Byrnes, R.T., 1994. Rate-controlling process for creep crack growth in Al–Li alloys. In: Sanders Jr., T.H., Starke Jr., E.A. (Eds.), Proceedings of the Fourth International Conference on Aluminum Alloys: Their Physical and Mechanical Properties, vol. 2. Georgia Institute of Technology, Atlanta, GA, USA, pp. 444–451.

Lynch, S.P., Wilson, A.R., Byrnes, R.T., 1993. Effects of ageing treatments on resistance to intergranular fracture of 8090 Al–Li alloy plate. Mater. Sci. Eng. A, A172, 79–93.

Lynch, S.P., Shekhter, A., Moutsos, S., Winkelman, G.B., 2003. Challenges in developing high performance Al–Li alloys. In: 'LiMAT 2003', Third International Conference in Light Materials for Transportation Systems. Published on a CD by the Center for Advanced Aerospace Materials, Pohang University of Science and Technology, Pohang, Korea.

Lynch, S.P., Knight, S.P., Birbilis, N., Muddle, B.C., 2008. Heat-treatment, grain-boundary characteristics and fracture resistance of some aluminium alloys. In: Hirsch, J., Skrotski, B., Gottstein, G. (Eds.), Proceedings of the 11th International Conference on Aluminium Alloys: Their Physical and Mechanical Properties. Wiley-VCH Verlag GmbH & Co. KGaA, Weinheim, Germany, pp. 1409–1415.

Lyttle, M.T., Wert, J.A., 1996. The plastic anisotropy of an Al–Li–Cu–Zr alloy extrusion in unidirectional deformation. Metall. Mater. Trans. A, 27A, 3503–3512.

Ma, Y., Staron, P., Fischer, T., Irving, P.E., 2011. Size effects on residual stress and fatigue crack growth in friction stir welded 2195-T8 aluminium—Part I: experiments. Int. J. Fatigue 33, 1417–1425.

Maddrell, E.R., Ricks, R.A., Wallach, E.R., 1989. Diffusion bonding of aluminium alloys containing lithium and magnesium. In: Sanders Jr., T.H., Starke Jr., E.A. (Eds.), Proceedings of the Fifth International Conference on Aluminium–Lithium Alloys, vol. 1. Materials and Component Engineering Publications Ltd., Birmingham, UK, pp. 451–460.

Madhusudhan Reddy, G., Gokhale, A.A., 1994. Pulsed current gas tungsten arc welding of Al–Li alloy 1441. In: Sanders Jr., T.H., Starke Jr., E.A. (Eds.), Proceedings of the Fourth International Conference on Aluminum Alloys: Their Physical and Mechanical Properties, vol. 2. Georgia Institute of Technology, Atlanta, GA, pp. 496–503.

Maehara, Y., Ohmori, Y., 1987. Microstructural change during superplastic deformation of δ-ferrite/austenite duplex stainless steel. Metall. Mater. Trans. A, 18A, 663.

Magnin, T., Rebiere, M., 1987. The effect of hydrogen during stress corrosion cracking and corrosion fatigue of Al−Li−Cu alloys in 3.5% NaCl solutions. In: Champier, G., Dubost, B., Miannay, D., Sabetay, L. (Eds.), Proceedings of the Fourth International Conference on Aluminium−Lithium Alloys, J. Phys., 48, C3, 835−C3, 842.

Magnin, T., Rieux, P., Lespinasse, C., Bathias, C., 1987. Fatigue crack initiation and propagation properties of Al−Li−Cu alloys in air and in aqueous corrosive solutions. In: Champier, G., Dubost, B., Miannay, D., Sabetay, L. (Eds.), Proceedings of the Fourth International Conference on Aluminium−Lithium Alloys, J. Phys., 48, C3, 817−C3, 824.

Magnusen, P.E., Mooy, D.C., Yocum, L.A., Rioja, R.J., 2012. Development of high toughness sheet and extruded products for airplane fuselage structures. In: Weiland, H., Rollett, A.D., Cassada, W.A. (Eds.), Proceedings of the 13th International Conference on Aluminum Alloys (ICAA13). The Minerals, Metals and Materials Society (TMS) and John Wiley & Sons, Hoboken, NY, USA, pp. 535−540.

Mahadev, V., Mahalingam, K., Liedl, G.L., Sanders, T.H., 1994. Very early stages of δ' precipitation in a binary Al−11.4 at.% Li alloy. Acta Metall. 42, 1039−1043.

Mahalingam, K., Gu, B.P., Liedl, G.L., Sanders, T.H., 1987. Coarsening of δ' precipitates in binary Al−Li alloys. Acta Metall. 35, 483−498.

Mahamoud, M.S., McShane, H.B., Sheppard, T., 1987. Extrusion of an Al−Li−Zr alloy prepared from atomized powder. In: Champier, G., Dubost, B., Miannay, D., Sabetay, L. (Eds.), Proceedings of the Fourth International Conference on Aluminium−Lithium Alloys, J. Phys., 48, C3, 327−C3, 334.

Maire, E., Verdu, C., Lormand, G., Fougeres, R., 1994. Influence of the aspect ratio distribution on the damage mechanisms in an Osprey Al alloy−SiCp composite. In: Sanders Jr., T.H., Starke Jr., E.A. (Eds.), Proceedings of the Fourth International Conference on Aluminum Alloys: Their Physical and Mechanical Properties, vol. 2. Georgia Institute of Technology, Atlanta, GA, USA, pp. 551−558.

Makin, P.L., Ralph, B., 1984. On the ageing of an aluminium−lithium−zirconium alloy. J. Mater. Sci. 19, 3835−3843.

Makin, P.L., Stobbs, W.M., 1986. Comparison of recrystallisation behaviour of an Al−Li−Zr alloy with related binary systems. In: Baker, C., Gregson, P.J., Harris, S.J., Peel, C.J. (Eds.), Proceedings of the Third International Conference on Aluminum−Lithium Alloys. The Institute of Metals, London, UK, pp. 392−401.

Malis, T., 1986. Characterisation of lithium distribution in aluminium alloys. In: Baker, C., Gregson, P.J., Harris, S.J., Peel, C.J. (Eds.), Proceedings of the Third International Conference on Aluminium−Lithium Alloys III. The Institute of Metals, London, UK, pp. 347−354.

Mallesham, P., Gokhale, A.A., Dutta, A., 2003. Interface microstructure and bond shear strength of aluminum alloy AA8090/AA7072 roll clad sheets. J. Mater. Sci. 22, 1793−1795.

Marchive, D., Charue, M.O., 1987. Processing and properties of 2091 and 8090 forgings. In: Champier, G., Dubost, B., Miannay, D., Sabetay, L. (Eds.), Proceedings of the Fourth International Conference on Aluminium−Lithium Alloys, J. Phys., 48, C3, 43−C3, 48.

Marisco, T.A., Kossowsky, R., 1989. Physical properties of laser-welded aluminium−lithium alloy 2090. In: Sanders Jr., T.H., Starke Jr., E.A. (Eds.), Proceedings of the Fifth International Conference on Aluminium−Lithium Alloys, vol. 3. Materials and Component Engineering Publications Ltd., Birmingham, UK, pp. 1447−1456.

Markey, D.T., Biederma, R.R., McCarthy, A.J., 1986. Effect of cold deformation on mechanical properties and microstructure of Alcan XXXA,. In: Baker, C., Gregson, P.J., Harris, S.J.,

Peel, C.J. (Eds.), Proceedings of the Third International Conference on Aluminium–Lithium Alloys. The Institute of Metals, London, UK, pp. 173–182.

Martukanitz, R.P., Natalie, C.A., Knoefel, J.O., 1987. The weldability of an Al–Li–Cu alloy. J. Met. 39, 38–42.

Masuda-Jindo, E., Terakura, K., 1989. First principle theoretical calculations on elastic moduli enhancement of Al–Li alloys. In: Sanders Jr., T.H., Starke Jr., E.A. (Eds.), Proceedings of the Fifth International Conference on Aluminium–Lithium Alloys, vol. 2. Materials and Component Engineering Publications Ltd., Birmingham, UK, pp. 809–816.

Matsuki, K., Tokizawa, M., Murakami, Y., 1994. Superplastic behaviour at high strain rates in SiC particulate reinforced 7075 aluminium composites. In: Sanders Jr., T.H., Starke Jr., E.A. (Eds.), Proceedings of the Fourth International Conference on Aluminum Alloys: Their Physical and Mechanical Properties, vol. 2. Georgia Institute of Technology, Atlanta, GA, USA, pp. 668–675.

Matsumuro, A., Sakai, K., 1993. High pressure phase diagram of an aluminium-rich Al–Li alloy at a pressure of 5.4 GPa. J. Mater. Sci. 28, 6567–6579.

McDarmaid, D.S., 1985. Fatigue and fatigue crack growth behaviour of medium and high strength Al–Li–Cu–Mg–Zr alloy plate. Royal Aircraft Establishment Technical Report 85016, Farnborough, Hampshire, UK.

McDarmaid, D.S., 1988. Effect of natural aging on the tensile properties of the Al–Li alloys 8090, 8091 and 2091. Mater. Sci. Eng. A, A101, 193–200.

McDarmaid, D.S., Peel, C.J., 1989. Aspects of damage tolerance in 8090 sheet. In: Sanders Jr., T.H., Starke Jr., E.A. (Eds.), Proceedings of the Fifth International Conference on Aluminium–Lithium Alloys, vol. 2. Materials and Component Engineering Publications Ltd., Birmingham, UK, pp. 993–1002.

McDarmaid, D.S., Shakesheff, A.J., 1987. The effect of superplastic deformation on the tensile and fatigue properties of Al–Li (8090) alloy. In: Champier, G., Dubost, B., Miannay, D., Sabetay, L. (Eds.), Proceedings of the Fourth International Conference on Aluminium–Lithium Alloys, J. Phys., 48, C3, 257–C3, 268.

McKeighan, P.C., Henkener, J.A., Kistler, G.P., Sanders Jr., T.H., Grandt Jr., A.F., Hillberry, B. M., 1989. Mechanical properties of a peek-aged Al–2.6Li–0.09Zr alloy. In: Sanders Jr., T. H., Starke Jr., E.A. (Eds.), Proceedings of the Fifth International Conference on Aluminium–Lithium Alloys, vol. 2. Materials and Component Engineering Publications Ltd., Birmingham, UK, pp. 849–858.

McMaster, F.J., Tabrett, C.P., Smith, D.J., 1998. Fatigue crack growth rates in Al–Li alloy 2090: influence of orientation, sheet thickness and specimen geometry. Fatigue Fract. Eng. Mater. Struct. 21, 139–150.

McNamara, D.K., Pickens, J.R., Heubaum, F.H., 1992. Forgings of weldalite (TM) 049 alloys X2094 and X2095. In: Peters, M., Winkler, P.J. (Eds.), Aluminium–Lithium Alloys VI, vol. 2. Deutsche Gesellschaft für Metallkunde, Frankfurt, Germany, pp. 921–926.

McQueen, H.J., Jain, V., Hopkins, A., Konopleva, E.V., Sakaris, P., 1994a. Comparison of compression and torsion results with sinh and power law constitutive analyses for an 8090 Cr/Mn alloy. In: Sanders Jr., T.H., Starke Jr., E.A. (Eds.), Proceedings of the Fourth International Conference on Aluminum Alloys: Their Physical and Mechanical Properties, vol. 2. Georgia Institute of Technology, Atlanta, GA, USA, pp. 250–257.

McQueen, H.J., Xia, X., Konopleva, E.V., Quin, Q., Sakaris, P., 1994b. Hot workability of particulate composites with various Al alloy matrices. In: Sanders Jr., T.H., Starke Jr., E.A. (Eds.), Proceedings of the Fourth International Conference on Aluminum Alloys: Their

Physical and Mechanical Properties, vol. 2. Georgia Institute of Technology, Atlanta, GA, USA, pp. 645−652.

McShane, H.B., Mahmoud, M.S., Sheppard, T., 1989. Extrusion processing of Al−Li−Zr RSP alloys with differing lithium and zirconium contents. In: Sanders Jr., T.H., Starke Jr., E.A. (Eds.), Proceedings of the Fifth International Conference on Aluminium−Lithium Alloys, vol. 1. Materials and Component Engineering Publications Ltd., Birmingham, UK, pp. 287−304.

Meletis, E.I., Huang, W., 1989. Pre-exposure embrittlement in Al−Li−Cu−Zr alloys. In: Sanders Jr., T.H., Starke Jr., E.A. (Eds.), Proceedings of the Fifth International Conference on Aluminium−Lithium Alloys, vol. 3. Materials and Component Engineering Publications Ltd., Birmingham, UK, pp. 1309−1318.

Meletis, E.I., Huang, W., 1991. The role of the T_1 phase in the pre-exposure and hydrogen embrittlement of Al−Li−Cu alloys. Mater. Sci. Eng. A, A148, 197−209.

Meletis, E.L., Sater, J.M., Sanders Jr., T.H., 1986. In: Sanders Jr., T.H., Starke Jr., E.A. (Eds.), Proceedings of the Fourth International Conference on Aluminum Alloys: Their Physical and Mechanical Properties, vol. 2. Georgia Institute of Technology, Atlanta, GA, USA, pp. 315−315.

Meng, F.L., Chai, Z.G., Li, B.J.X., Wang, H.J.Y., 2001. Small-angle X-ray scattering study of the d' phase growth kinetic in 1420 Al−Li alloy. Mater. Charact. 47, 43−46.

Meng, L., Tian, L., 2002. Stress concentration sensitivity of Al−Li based alloys with various contents of impurities and cerium addition. Mater. Sci. Eng. A, A323, 239−245.

Meng, L., Zheng, X.L., 1996. Tension characteristics of notched specimens for Al−Li−Cu−Zr alloy sheets with various cerium contents. Metall. Mater. Trans. A, 27A, 3089−3094.

Meng, L., Zheng, X.L., Tu, J.P., Liu, M.S., 1998. Effects of deleterious impurities and cerium modification on intrinsic and extrinsic toughening levels of Al−Li based alloys. Mater. Sci. Technol. 14, 585−591.

Menon, S.S., Rack, H.J., 1994. Flow characterisation of cast Al−2Li. In: Sanders Jr., T.H., Starke Jr., E.A. (Eds.), Proceedings of the Fourth International Conference on Aluminum Alloys: Their Physical and Mechanical Properties, vol. 2. Georgia Institute of Technology, Atlanta, GA, USA, pp. 230−237.

Meric, C., 2004. An investigation on the elastic modulus and density of vacuum casted aluminum alloy 2024 containing lithium additions. J. Mater. Eng. Perform. 9, 266−271.

Meschter, P.J., O'Neal, J.E., Lederich, R.J., 1984. Effect of rapid solidification method on the microstructure and properties of Al−3Li−1.5Cu−0.5Co−0.2Zr. In: Sanders Jr., T.H., Starke Jr., E.A. (Eds.), Proceedings of the Second International Conference on Aluminum−Lithium Alloys. The Metallurgical Society of AIME, Warrendale, PA, USA, pp. 419−432.

Meschter, P.J., Ledrich, R.J., O'Neal, J.E., 1986. Micostructure and properties of rapid solidification processed (RSP) Al−4Li and Al−5Li alloys. In: Baker, C., Gregson, P.J., Harris, S.J., Peel, C.J. (Eds.), Proceedings of the Third International Conference on Aluminium−Lithium Alloys. The Institute of Metals, London, UK, pp. 85−96.

Meschter, P.J., Gregory, J.K., Lederich, R.J., O'Neal, J.E., Laveria, E.J., Grant, N.J., 1987. Microstructures and properties of rapid solidification processed aluminium−high lithium alloys. In: Champier, G., Dubost, B., Miannay, D., Sabetay, L. (Eds.), Proceedings of the Fourth International Conference on Aluminium−Lithium Alloys, J. Phys., 48, C3, 317−C3, 326.

Meyer, P., Dubost, B., 1986. Production of aluminium−lithium alloy with high specific properties. In: Baker, C., Gregson, P.J., Harris, S.J., Peel, C.J. (Eds.), Proceedings of the Third

International Conference on Aluminum–Lithium Alloys. The Institute of Metals, London, UK, pp. 37–46.

Meyer, P., Cans, Y., Ferton, D., Reboul, M., 1987. The metallurgy of industrial Al–Li alloys. In: Champier, G., Dubost, B., Miannay, D., Sabetay, L. (Eds.), Proceedings of the Fourth International Conference on Aluminum–Lithium Alloys, J. Phys., 48, C3, 131–C3, 138.

Miller, W.S., Cornish, A.J., Titchener, A.P., Bennet, D.A., 1984. Development of lithium containing aluminium alloys for the ingot metallurgy production route. In: Sanders Jr., T.H., Starke Jr., E.A. (Eds.), Proceedings of the Second International Conference on Aluminum–Lithium Alloys. The Metallurgical Society of AIME, Warrendale, PA, USA, pp. 335–362.

Miller, W.S., Thomas, M.P., Lloyd, D.J., Creber, D., 1986. Deformation and fracture in Al–Li base alloys. In: Baker, C., Gregson, P.J., Harris, S.J., Peel, C.J. (Eds.), Proceedings of the Third International Conference on Aluminum–Lithium Alloys. The Institute of Metals, London, UK, pp. 584–594.

Miller, W.S., White, J., Reynolds, M.A., McDarmaid, D.S., Starr, G.M., 1987a. Aluminium–lithium–copper–magnesium–zirconium alloys with high strength and high toughness–solving the perceived dichotomy. In: Champier, G., Dubost, B., Miannay, D., Sabetay, L. (Eds.), Proceedings of the Fourth International Conference on Aluminum–Lithium Alloys, J. Phys., 48, C3, 151–C3, 162.

Miller, W.S., White, J., Lloyd, D.J., 1987b. The physical metallurgy of aluminium–lithium–copper–magnesium–zirconium alloys–8090 and 8091. In: Champier, G., Dubost, B., Miannay, D., Sabetay, L. (Eds.), Proceedings of the Fourth International Conference on Aluminum–Lithium Alloys, J. Phys., 48, C3, 139–C3, 149.

Miller, W.S., Lenssen, L.A., Humphreys, F.J., 1989. The strength, toughness and fracture behaviour in aluminium–lithium based metal matrix composites. In: Sanders Jr., T.H., Starke Jr., E.A. (Eds.), Proceedings of the Fifth International Conference on Aluminium–Lithium Alloys, vol. 2. Materials and Component Engineering Publications Ltd., Birmingham, UK, pp. 931–942.

Minay, J., Dashwood, R., McShane, H., 2001. Elevated temperature deformation behavior of dispersion-strengthened Al and Al–Li–Mg alloys. J. Mater. Eng. Perform. 10, 136–142.

Miura, Y., Matsui, A., Furukawa, M., Nemoto, M., 1986. Plastic deformation of Al–Li single crystals. In: Baker, C., Gregson, P.J., Harris, S.J., Peel, C.J. (Eds.), Proceedings of the Third International Conference on Aluminum–Lithium Alloys. The Institute of Metals, London, UK, pp. 427–434.

Miura, Y., Yusu, K., Furukawa, M., Nemoto, M., 1987. Temperature dependence of yield strength of Al–Li single crystals. In: Champier, G., Dubost, B., Miannay, D., Sabetay, L. (Eds.), Proceedings of the Fourth International Conference on Aluminium–Lithium Alloys, J. Phys., 48, C3, 549–C3, 556.

Miura, Y., Yusu, K., Aibe, S., Furukawa, M., Nemoto, M., 1989. Formation and stability of Orowan loops in Al–Li single crystals. In: Sanders Jr., T.H., Starke Jr., E.A. (Eds.), Proceedings of the Fifth International Conference on Aluminium–Lithium Alloys, vol. 2. Materials and Component Engineering Publications Ltd., Birmingham, UK, pp. 827–836.

Miura, Y., Horikawa, K., Yamada, K., Nakayama, M., 1994. Precipitation hardening in an Al–2.4Li–0.19Sc alloy. In: Sanders Jr., T.H., Starke Jr., E.A. (Eds.), Proceedings of the Fourth International Conference on Aluminum Alloys: Their Physical and Mechanical Properties, vol. 2. Georgia Institute of Technology, Atlanta, GA, USA, pp. 161–168.

Miyasato, S., Thomas, G., 1989. Microstructural evolution in Al–Li and Al–Li–Cu–Mg alloys. In: Sanders Jr., T.H., Starke Jr., E.A. (Eds.), Proceedings of the Fifth International

Conference on Aluminium—Lithium Alloys, vol. 2. Materials and Component Engineering Publications Ltd., Birmingham, UK, pp. 565—574.

Mogucheva, A.A., Kaibyshev, R.O., 2008. Ultrahigh superplastic elongations in an aluminum—lithium alloy. Physics 53, 431—433.

Molian, P.A., Srivatsan, T.S., 1989. Weldability of Al—Li—Cu alloy 2090 using laser welding. In: Sanders Jr., T.H., Starke Jr., E.A. (Eds.), Proceedings of the Fifth International Conference on Aluminium—Lithium Alloys, vol. 3. Materials and Component Engineering Publications Ltd., Birmingham, UK, pp. 1435—1446.

Molian, P.A., Srivatsan, T.S., 1990. Laser-beam weld microstructures and properties of aluminum—lithium alloy 2090. Mater. Lett. 9, 245—251.

Monachon, C., Krug, M.E., Seidman, D.N., Dunand, D.C., 2011. Chemistry and structure of core/double-shell nanoscale precipitates in Al—6.5Li—0.07Sc—0.02Yb (at.%). Acta. Mater. 59, 3396—3409.

Montoya, K.A., Heubaum, F.H., Kumar, K.S., Pickens, J.R., 1991. Compositional effects on the solidus temperature of an Al—Cu—Li—Ag—Mg alloy. Scr. Metall. Mater. 25, 1489—1494.

Moore, K., Langan, T.J., Heubaum, F.H., Pickens, J.R., 1989. Effect of Cu content on the corrosion and stress-corrosion behaviour of Al—Cu—Li weldalite alloys. In: Sanders Jr., T.H., Starke Jr., E.A. (Eds.), Proceedings of the Fifth International Conference on Aluminium—Lithium Alloys, vol. 3. Materials and Component Engineering Publications Ltd., Birmingham, UK, pp. 1281—1292.

Moran, J.P., Stoner, G.E., 1989. Solution chemistry effects on the stress corrosion cracking behaviour of alloy 2090 (Al—Li—Cu) and alloy 2024 (Al—Cu—Mg). In: Sanders Jr., T.H., Starke Jr., E.A. (Eds.), Proceedings of the Fifth International Conference on Aluminium—Lithium Alloys, vol. 3. Materials and Component Engineering Publications Ltd., Birmingham, UK, pp. 1187—1196.

Moreira, P.M.G.P., De Jesus, A.M.P., De Figueiredo, M.A.V., Windisch, M., Sinnema, G., De Castro, P.M.S.T., 2012. *Fatigue and fracture behaviour of friction stir welded aluminium—lithium 2195*. Theor. Appl. Fract. Mech. 60, 1—9.

Moreton, R., Hunt, E., Gregson, P.J., Peel, C.J., 1989. Fibre reinforced Al—Li laminates. In: Sanders Jr., T.H., Starke Jr., E.A. (Eds.), Proceedings of the Fifth International Conference on Aluminium—Lithium Alloys, vol. 3. Materials and Component Engineering Publications Ltd., Birmingham, UK, pp. 1685—1694.

Moser, Z., Gasior, W., Onderka, B., Sommer, F., Kim, Z., 2002. Al—Cu—Li system electromotive force and calorimetric studies—phase diagram calculations of the Al-rich part. J. Phase Equilib. 23, 127—133.

Mou, Y., Howe, J.M., Starke Jr., E.A., 1995. Grain boundary precipitation and fracture behavior of an Al—Cu—Li—Mg—Ag alloy. Metall. Mater. Trans. A, 26A, 1591—1595.

Mukhopadhyay, A.K., Flower, H.M., Sheppard, T., 1987a. Variation in structure and properties in an Al—Li—Cu—Mg—Zr alloy produced by extrusion processing. In: Champier, G., Dubost, B., Miannay, D., Sabetay, L. (Eds.), Proceedings of the Fourth International Conference on Aluminium—Lithium Alloys, J. Phys., 48, C3, 219—C3, 230.

Mukhopadhyay, A.K., Tite, C.N.J., Flower, H.M., Gregson, P.J., Sale, F., 1987b. Thermal analysis study of the precipitation reactions in Al—Li—Cu—Mg—Zr alloys. In: Champier, G., Dubost, B., Miannay, D., Sabetay, L. (Eds.), Proceedings of the Fourth International Conference on Aluminum—Lithium Alloys, J. Phys., 48, C3, 439—C3, 446.

Mukhopadhyay, A.K., Flower, H.M., Sheppard, T., 1990a. Development of microstructure in AA 8090 alloy produced by extrusion processing. Mater. Sci. Technol. 6, 461—468.

Mukhopadhyay, A.K., Flower, H.M., Sheppard, T., 1990b. Development of mechanical properties in AA 8090 alloy produced by extrusion processing. Mater. Sci. Technol. 6, 611–620.

Mukhopadhyay, A.K., Zhou, D.S., Yang, Q.B., 1992. Effect of variation in the Cu:Mg ratios on the formation of T_2 and C phases in AA 8090 alloys. Scr. Metall. Mater. 26, 237–242.

Muller, W., Bubeck, E., Gerold, V., 1986. Elastic constants of Al–Li solid solutions and δ' precipitates. In: Baker, C., Gregson, P.J., Harris, S.J., Peel, C.J. (Eds.), Proceedings of the Third International Conference on Aluminium–Lithium Alloys. The Institute of Metals, London, UK, pp. 435–441.

Murayama, M., Hono, K., 2001. Role of Ag and Mg on precipitation of T_1 phase in an Al–Cu–Li–Mg–Ag alloy. Scr. Mater. 44, 701–706.

Nagahama, H., Kawazoe, M., Ohtera, K., Inoue, A., Masumoto, T., 1994. Mechanical properties of rapidly solidified aluminium alloy extruded from nanocrystalline powders. In: Sanders Jr., T.H., Starke Jr., E.A. (Eds.), Proceedings of the Fourth International Conference on Aluminum Alloys: Their Physical and Mechanical Properties, vol. 2. Georgia Institute of Technology, Atlanta, GA, USA, pp. 760–765.

Namba, K., Sano, H., 1986. Fusion weldabilities of Al–4.7 Mg–0.3–1.3Li alloys for fusion reactor. J. Light Met. Weld. Constr. 24, 243–250.

Narayanan, G.H., Quist, W.E., Wilson, B.L., Wingert, A.L., 1982. Low density aluminum alloy development. United States Air Force Wright Aeronautical Laboratories First Interim Technical Report, AFWAL Contract No. F33615-81-C-5053, Dayton, OH, USA.

Narayanan, G.H., Wilson, B.L., Quist, W.E., 1984. P/M aluminium–lithium alloys by the mechanical alloying process. In: Sanders Jr., T.H., Starke Jr., E.A. (Eds.), Proceedings of the Second International Conference on Aluminum–Lithium Alloys. The Metallurgical Society of AIME, Warrendale, PA, USA, pp. 517–543.

Narukawa, S., Amazutsumi, T., Fukuda, H., Itou, K., Tamaki, H., Yamauchi, Y., 1998. Development of prismatic lithium-ion cells using aluminium alloy casing. J. Power Sources 76, 186–189.

Neugebauer, R., Bouzakis, K.D., Denkena, B., Klocke, F., Sterzing, A., Vaidya, R.U., et al., 1994. Ageing response and mechanical properties of a SiC$_p$/Al–Li (8090) composite. J. Mater. Sci. 29, 2944–2950.

Newbury, D.E., Levi-Setti, R., Chabala, J.M., Wang, Y.L., Williams, D.B., 1989. Microanalysis of precipitates in aluminum–lithium alloys with a scanning ion microprobe. Appl. Surf. Sci. 37, 78–94.

Newman, J.M., Gregson, P.J., Pitcher, P.D., Forsyth, P.J.E., 1989. The effect of grain structure on fatigue crack growth in Al–Li alloys. In: Sanders Jr., T.H., Starke Jr., E.A. (Eds.), Proceedings of the Fifth International Conference on Aluminium–Lithium Alloys, vol. 2. Materials and Component Engineering Publications Ltd., Birmingham, UK, pp. 1043–1052.

Nicholls, D.J., Martin, J.W., 1989. Microstructural effects on small cack growth in aluminium–lithium alloys. In: Sanders Jr., T.H., Starke Jr., E.A. (Eds.), Proceedings of the Fifth International Conference on 'Aluminium–Lithium Alloys, vol. 2. Materials and Component Engineering Publications Ltd., Birmingham, UK, pp. 1003–1012.

Nicholls, D.J., Martin, J.W., 1991. Slip homogenization in Al–Li alloys by thermomechanical treatment and its effect on mechanical behaviour. J. Mater. Sci. 26, 552–558.

Nieh, T.G., Wadsworth, J., 1994. High strain rate superplasticity of a powder metallurgy, Zr-modified 2124 aluminium. In: Sanders Jr., T.H., Starke Jr., E.A. (Eds.), Proceedings of the Fourth International Conference on Aluminum Alloys: Their Physical and Mechanical Properties. Georgia Institute of Technology, Atlanta, GA, USA, pp. 48–55.

Nieh, T.G., Pelton, A.R., Oliver, W.C., Wadsworth, J., 1998. Characterization of the aging response of a melt-spun Al−Be−Li alloy ribbon. Metall. Trans. A, 19A, 1173−1178.

Niikura, M., Takahashi, K., Ouchi, C., Hot deformation behaviour in Al−Li−Cu−Mg−Zr alloys. In: Baker, C., Gregson, P.J., Harris, S.J., Peel, C.J. (Eds.), Proceedings of the Third International Conference on Aluminium−Lithium Alloys. The Institute of Metals, London, UK, pp. 213−221.

Niinomi, M., Degawa, K., Kobayashi, T., 1987. Effect of thermomechanical treatment on toughening of Al−Li−Cu−Mg−Zr alloy and its toughness. In: Champier, G., Dubost, B., Miannay, D., Sabetay, L. (Eds.), Proceedings of the Fourth International Conference on Aluminum−Lithium Alloys, J. Phys., 48, C3, 653−C3, 659.

Niinomi, M., Kobayashi, T., Yamada, H., Irisa, T., Hagiwara, T., Sakamoto, T., 1994. Strengthening and toughening by microstructural control in 2091 Al−Li alloy. In: Sanders Jr., T.H., Starke Jr., E.A. (Eds.), Proceedings of the Fourth International Conference on Aluminum Alloys: Their Physical and Mechanical Properties. Georgia Institute of Technology, Atlanta, GA, USA, pp. 342−350.

Niskanen, P., Sanders, T.H., Marek, M., Rinker, J.G., 1981. The influence of microstructure on the corrosion of Al−Li, Al−Li−Mn, Al−Li−Mg and Al−Li−Cu alloys in 3.5% NaCl solution. In: Sanders Jr., T.H., Starke Jr., E.A. (Eds.), Proceedings of the First International Conference on Aluminum−Lithium, Alloys. The Metallurgical Society of AIME, Warrendale, PA, USA, pp. 347−376.

Niskanen, P., Sanders Jr., T.H., Rinker, J.G., Marek, M., 1982. Corrosion of aluminium alloys containing lithium. Corros. Sci. 22, 283−304.

Noble, B., Bray, S.E., 1998. On the α(Al)/δ'(Al₃Li) metastable solvus in aluminium−lithium alloys. Acta Mater. 46, 6163−6171.

Noble, B., Thompson, G.E., 1972. T_1 (Al₂CuLi) precipitation in aluminium−copper−lithium alloys. Met. Sci. J. 6, 167−174.

Noble, B., Thompson, G.E., 1973. Precipitation characteristics of aluminum−lithium alloys containing magnesium. J. Inst. Met. 101, 111−115.

Noble, B., Thompson, G.E., Precipitation characteristics of aluminium−lithium alloys. Met. Sci. J. 5, 114−120.

Noble, B., Trowsdale, A.J., 1995. Precipitation in an aluminium−14 at.% lithium alloy. Philos. Mag. A, 71, 1345−1362.

Noble, B., McClaughlin, I.R., Thompson, G.E., 1970. Solute atom clustering processes in aluminium−copper−lithium alloys. Acta Metall. 18, 339−345.

Noble, B., Harris, S.J., Dinsdale, K., 1982a. The elastic modulus of aluminium−lithium alloys. J. Mater. Sci. 17, 461−468.

Noble, B., Harris, S.J., Dinsdale, K., 1982b. Yield characteristics of aluminium−lithium alloys. Met. Sci. J. 16, 425−430.

Noble, B., Harris, S.J., Harlow, K., 1984. Mechanical properties of Al−Li−Mg alloys at elevated temperature. In: Sanders Jr., T.H., Starke Jr., E.A. (Eds.), Proceedings of the Second International Conference on Aluminum−Lithium Alloys. The Metallurgical Society of AIME, Warrendale, PA, USA, pp. 65−78.

Noble, B., Harris, S.J., Dinsdale, K., 1994. Low temperature embrittlement of 8090 in the damage tolerant condition. In: Sanders Jr., T.H., Starke Jr., E.A. (Eds.), Proceedings of the Fourth International Conference on Aluminum Alloys: Their Physical and Mechanical Properties. Georgia Institute of Technology, Atlanta, GA, USA, pp. 460−466.

Noble, B., Harris, S.J., Dinsdale, K., 1997. Microstructural stability of binary Al−Li alloys at low temperatures. Acta Mater. 45, 2069−2078.

Nozato, R., Nakai, G., 1977. Thermal analysis of precipitation in Al–Li alloys. Trans. Jpn. Inst. Met. 18, 679–689.

O'Dowd, M.E., Ruch, W., Starke, E.A., Jr., 1987. Dependence of elastic modulus on microstructure in 2090-type alloys. In: Champier, G., Dubost, B., Miannay, D., Sabetay, L. (Eds.), Proceedings of the Fourth International Conference on Aluminium–Lithium Alloys, J. Phys., 48, C3, 563–C3, 576.

Oh, Y.J., Lee, B.S., Kwon, S.C., Hong, J.H., Nam, S.W., 1999. Low-cycle fatigue crack initiation and break in strain-life curve of Al–Li 8090 alloy. Metall. Mater. Trans. A, 30A, 887–890.

Ohara, K., Kurino, S., Toyoshima, M., Wakasaki, O., 1994. Hot-tearing of Al–Li alloys in DC-casting. In: Sanders Jr., T.H., Starke Jr., E.A. (Eds.), Proceedings of the Fourth International Conference on Aluminum Alloys: Their Physical and Mechanical Properties. Georgia Institute of Technology, Atlanta, GA, USA, pp. 122–129.

Ohmori, Y., Ito, S., Nakai, K., 1999. Aging behavior of an Al–Li–Cu–Mg–Zr alloy. Metall. Mater. Trans. A, 30A, 741–749.

Ohrloff, N., Gysler, A., Lütjering, G., 1987. Fatigue crack propagation behaviour of 2091 T8 and 2024 T3 under constant and variable amplitude loading. In: Champier, G., Dubost, B., Miannay, D., Sabetay, L. (Eds.), Proceedings of the Fourth International Conference on Aluminium–Lithium Alloys, J. Phys., 48, C3, 801–C3, 808.

Ohsaki, S., Sato, T., Takahashi, T., 1990. Effect of aging on the pitting corrosion behaviour in Al–Li alloys. Aluminium 66, 565–572.

Ohsaki, S., Kobayashi, K., Iino, M., Sakamoto, T., 1996. Fracture toughness and stress corrosion cracking of aluminium–lithium alloys 2090 and 2091. Corros. Sci. 38, 793–802.

Okada, H., Itoh, G., Kanno, M., 1992. Hydrogen segregation in an Al–Li alloy. Scr. Metall. Mater. 26, 69–74.

Orio, L.L., Tillard, M., Belin, C., 2004. Exploration of the lithium–aluminum–silver system. Crystal and electronic structure analysis of new phases Li6.98Al4.15Ag0.87 and LiAlAg2. Solid State Sci. 6, 1429–1437.

Ortiz, D., Brown, J., Abdelshehid, M., DeLeon, P., Dalton, R., Mendez, L., et al., 2006. The effects of prolonged thermal exposure on the mechanical properties and fracture toughness of C458 aluminum–lithium alloy. Eng. Fail. Anal. 13, 170–180.

Osamura, K., Okuda, H., 1993, Phase decomposition and reversion in Al–Li alloys. In: Champier, G., Dubost, B., Miannay, D., Sabetay, L. (Eds.), Proceedings of the Fourth International Conference on Aluminium–Lithium Alloys, J. Phys., 48, C3, 11–C3, 16.

Osman, T.M., Lewandowski, J.J., Lesuer, D.R., Syn, C.K., Hunt Jr., W.H., 1994. Fracture behaviour of laminated discontinuously reinforced aluminium material. In: Sanders Jr., T.H., Starke Jr., E.A. (Eds.), Proceedings of the Fourth International Conference on Aluminum Alloys: Their Physical and Mechanical Properties, vol. 2. Georgia Institute of Technology, Atlanta, GA, USA, pp. 706–713.

Ou, B.L., Itoh, G., Kanno, M., 1989. Precipitation of zirconium in an Al–Li–Cu–Mg–Zr alloy. In: Sanders Jr., T.H., Starke Jr., E.A. (Eds.), Proceedings of the Fifth International Conference on Aluminium–Lithium Alloys, vol. 2. Materials and Component Engineering Publications Ltd., Birmingham, UK, pp. 721–730.

Owen, J.R., Maskell, W.C., Steele, B.C.H., Steen Nielsen, T., Sørensen, O.T., 1984. Thin film lithium aluminium negative plate material. Solid State Ionics 13, 329–334.

Owen, N.J., Field, D.J., Butler, E.P., 1986. Initiation of voiding at second-phase particles in a quaternary Al–Li alloy. In: Baker, C., Gregson, P.J., Harris, S.J., Peel, C.J. (Eds.), Proceedings of the Third International Conference on Aluminum–Lithium Alloys. The Institute of Metals, London, UK, pp. 576–583.

Ozbilen, S., 1997. Fractography and transmission electron microscopy analysis of an Al–Li–Cu–Mg–Zr alloy displaying an improved fracture toughness. J. Mater. Sci. 32, 4127–4131.

Ozbilen, S., Flower, H.M., 1989. The influence of Li on phase transformations in Al–Cu–Mg alloys. In: Sanders Jr., T.H., Starke Jr., E.A. (Eds.), Proceedings of the Fifth International Conference on Aluminium–Lithium Alloys, vol. 2. Materials and Component Engineering Publications Ltd., Birmingham, UK, pp. 651–660.

Page, F.M., Chamberlain, A.T., Grimes, R., 1987. The safety of molten aluminium–lithium alloys in the presence of coolants. In: Champier, G., Dubost, B., Miannay, D. and Sabetay, L. (Eds.), Proceedings of the Fourth International Conference on Aluminum–Lithium Alloys, J. Phys., 48, C3, 63–C3, 73.

Palmer, I.G., Lewis, R.E., Crooks, D.D., 1981. The design and mechanical properties of rapidly solidified Al–Li–X alloys. In: Sanders Jr., T.H., Starke Jr., E.A. (Eds.), Proceedings of the First International Conference on Aluminium–Lithium Alloys. The Metallurgical Society of AIME, Warrendale, PA, USA, pp. 241–262.

Palmer, I.G., Lewis, R.E., Crooks, D.D., Starke, E.A., Crooks, R.E., 1984. Effect of processing variables on microstructure and properties of two Al–Li–Cu–Mg–Zr alloys. In: Sanders Jr., T.H., Starke Jr., E.A. (Eds.), Proceedings of the Second International Conference on Aluminum–Lithium Alloys. The Metallurgical Society of AIME, Warrendale, PA, USA, pp. 91–110.

Palmer, I.G., Miller, W.S., Lloyd, D.J., Bull, M.J., 1986. Effect of grain structure and texture on mechanical properties of Al–Li base alloys. In: Baker, C., Gregson, P.J., Harris, S.J., Peel, C.J. (Eds.), Proceedings of the Third International Conference on Aluminium–Lithium Alloys. The Institute of Metals, London, UK, pp. 565–575.

Pan, Z., Zheng, Z., Liao, Z., Li, S., 2010. Effects of indium on precipitation in Al–3.3Cu–0.8Li alloy. Acta Metall. Sin. (English Letters) 23, 285–292.

Pancholi, V., Kashyap, B.P., 2003. Effect of layered microstructure on superplastic forming property of AA8090 Al–Li alloy. Mater. Sci. Eng. A, A351, 174–182.

Pandey, M.C., 1998. Particle characterization and cavity nucleation during superplastic deformation in Al–Li–Based alloys. J. Mater. Sci. 23, 3509–3514.

Pandey, M.C., Wadsworth, J., Mukherjee, A.K., 1985. Superplastic aluminum–lithium based alloys. Acta Metall. 19, 1229–1234.

Pandey, M.C., Wadsworth, J., Mukherjee, A.K., 1986a. Cavitation study in ingot and powder metallurgically processed superplastic Al–Li alloys. Mater. Sci. Eng. A, A78, 115–125.

Pandey, M.C., Wadsworth, J., Mukherjee, A.K., 1986b. Superplastic deformation behavior in ingot and powder metallurgically processed Al–Li-based alloys. Mater. Sci. Eng. A, A80, 169–179.

Paniwnyk, L., Perry, M.C., McWhinnie, M.R., Homer, J., Gelder, A., 1997. Do the measurements of 27Al and 7Li Knight Shift have any analytical role in studies of freshly prepared lithium aluminium alloys. Polyhedron 16, 2963–2969.

Pao, P.S., Sankaran, K.K., O'Neal, J.E., 1981. Microstructure, deformation and corrosion-fatigue behaviour of a rapidly solidified Al–Li–Cu–Mn alloy. In: Sanders Jr., T.H., Starke Jr., E.A. (Eds.), Proceedings of the First International Conference on Aluminum Lithium Alloys. The Metallurgical Society of AIME, Warrendale, PA, USA, pp. 307–324.

Pao, P.S., Cooley, L.A., Imam, M.A., Yoder, G.R., 1989a. Fatigue crack growth in 2090 Al–Li alloy. Scr. Metall. 23, 1455–1460.

Pao, P.S., Imam, M.A., Cooley, L.A., Yoder, G.R., 1989b. Corrosion fatigue crack growth in Al–Li alloy 2090. In: Sanders Jr., T.H., Starke Jr., E.A. (Eds.), Proceedings of the Fifth

International Conference on Aluminium–Lithium Alloys, vol. 2. Materials and Component Engineering Publications Ltd., Birmingham, UK, pp. 1125–1134.

Papazian, J.M., Sigli, C., Sanchez, J.M., 1986. New evidence of GP zones in binary Al–Li alloys. Scr. Metall. 20, 201–206.

Papazian, J.M., Bott, G.G., Shaw, P., 1987a. Effects of lithium loss on strength and formability of aluminum–lithium alloys 8090 and 8091. Mater. Sci. Eng. A, A94, 219–224.

Papazian, J.M., Bott, G.G., Shaw, P., 1987b. Influence of forming in the T3 condition on properties of 2090-T8X, 2091-T8X and 8090-T8X. In: Champier, G., Dubost, B., Miannay, D., Sabetay, L. (Eds.), Proceedings of the Fourth International Conference on Aluminium–Lithium Alloys, J. Phys., 48, C3, 231–C3, 238.

Papazian, J.M., Wagner, J.P.W., Rooney, W.D., 1987c. Porosity development during heat treatment of aluminium–lithium alloys. In: Champier, G., Dubost, B., Miannay, D., Sabetay, L. (Eds.), Proceedings of the Fourth International Conference on Aluminium–Lithium Alloys, J. Phys., 48, C3, 513–C3, 520.

Park, K.T., Lavernia, E.J., Mohamed, F.A., 1989. Creep behaviour of an aluminium–lithium alloy. In: Sanders Jr., T.H., Starke Jr., E.A. (Eds.), Proceedings of the Fifth International Conference on Aluminium–Lithium Alloys, vol. 2. Materials and Component Engineering Publications Ltd., Birmingham, UK, pp. 1155–1162.

Park, S.D., Yoo, B.H., Chung, D.S., Cho, H.K., 1994. The effect of be addition on precipitation response in Al–Li alloys. In: Sanders Jr., T.H., Starke Jr., E.A. (Eds.), Proceedings of the Fourth International Conference on Aluminum Alloys: Their Physical and Mechanical Properties, vol. 2. Georgia Institute of Technology, Atlanta, GA, USA, pp. 207–214.

Parson, N.C., Sheppard, T., 1986. Extrusion processing of Al–Mg–Li alloys. In: Baker, C., Gregson, P.J., Harris, S.J., Peel, C.J. (Eds.), Proceedings of the Third International Conference on Aluminium–Lithium Alloys. The Institute of Metals, London, UK, pp. 222–232.

Partridge, P.G., 1990. Oxidation of aluminium–lithium alloys in the solid and liquid states. Int. Mater. Rev. 35, 37–58.

Partridge, P.G., Chadbourne, N.C., 1989. Effect of short-time thermal oxidation on the surface of Ai–Li–Cu–Mg Alloy (8090) sheet. J. Mater. Sci. 24, 2765–2774.

Pasang, T., Symonds, N., Moutsos, S., Wanhill, R.J.H., Lynch, S.P., 2012. Low-energy intergranular fracture in Al–Li alloys. Eng. Fail. Anal. 22, 166–178.

Paul, S.F., 1996. Aluminium lithium for aerospace. Adv. Mater. Process. 10, 21–24.

Peacock, H.D., Martin, J.W., 1989. The effect of S-phase distribution on fatigue crack growth in peak-aged 8090 and 8091 Al–Li alloys. In: Sanders Jr., T.H., Starke Jr., E.A. (Eds.), Proceedings of the Fifth International Conference on Aluminium–Lithium Alloys, vol. 2. Materials and Component Engineering Publications Ltd., Birmingham, UK, pp. 1013–1032.

Peel, C.J., 1990. The development of aluminium–lithium alloys: an overview. In: New Light Alloys, AGARD Lecture Series No. 174, Advisory Group for Aerospace Research and Development, Neuilly-sur-Seine, France, pp. 1-1–1-55.

Peel, C.J., Evans, B., Baker, C.A., Bannet, D.A., Gregson, P.J., Flower, H.M., 1983. The development and application of improved aluminum–lithium alloys. In: Sanders Jr., T.H., Starke Jr., E.A. (Eds.), Proceedings of the Second International Conference on Aluminum–Lithium Alloys. The Metallurgical Society of AIME, Warrendale, PA, USA, pp. 363–392.

Peel, C.J., Evans, B., McDarmaid, D., 1986. Current status of UK lightweight lithium-containing aluminium alloys. In: Baker, C., Gregson, P.J., Harris, S.J., Peel, C.J. (Eds.), Proceedings of the Third International Conference on Aluminium–Lithium Alloys. The Institute of Metals, London, UK, pp. 26–36.

Peel, C.J., McDarmaid, D., Evans, B., 1988. Considerations of critical factors for the design of aerospace structures using current and future Al–Li alloys. In: Kar, R.J., Agarwal, S.P., Quist, W.E. (Eds.), Aluminum–Lithium Alloys: Design, Development and Application Update. ASM International, Materials Park, OH, USA, pp. 315–337.

Pegram, J.E., Sanders Jr., T.H., Hatshorn, C.J., McKeighan, P.C., Valentine, M.G., Hillberry, B. M., 1989. Tensile properties of an Al–2.6Li–0.09 Zr alloy. In: Sanders Jr., T.H., Starke Jr., E.A. (Eds.), Proceedings of the Fifth International Conference on Aluminium–Lithium Alloys, vol. 1. Materials and Component Engineering Publications Ltd., Birmingham, UK, pp. 261–272.

Peters, M., Welpmann, K., Zink, W., Sanders, T.H., 1986a. Fatigue behaviour of Al–Li–Cu–Mg alloy. In: Baker, C., Gregson, P.J., Harris, S.J., Peel, C.J. (Eds.), Proceedings of the Third International Conference on Aluminum–Lithium Alloys. The Institute of Metals, London, UK, pp. 239–246.

Peters, M., Eschweiler, J., Welpmann, K., 1986b. Strength profile in Al–Li plate material. Scr. Metall. 20, 259–264.

Peters, M., Bachmann, V., Welpmann, K., 1987. Fatigue crack propagation behaviour of the Al–Li alloy 8090 compared to 2024. In: Champier, G., Dubost, B., Miannay, D., Sabetay, L. (Eds.), Proceedings of the Fourth International Conference on Aluminium–Lithium Alloys, J. Phys., 48, C3, 785–C3, 792.

Peters, M., Welpmann, K., McDarmaid, D.S., 't Hart, W.G.J., 1989. Fatigue properties of Al–Li alloys. In: New Light Alloys, AGARD Conference Proceedings No. 444, Advisory Group for Aerospace Research and Development, Neuilly-sur-Seine, France, pp. 6-1–6-18.

Petit, J., Suresh, S., Vasudevan, A.K., Malcolm, R.C., 1986. Constant amplitude and post-overload fatigue crack growth in Al–Li alloys. In: Baker, C., Gregson, P.J., Harris, S.J., Peel, C.J. (Eds.), Proceedings of the Third International Conference on Aluminum–Lithium Alloys. The Institute of Metals, London, UK, pp. 257–262.

Pfost, D.P., Crawford, P., Mendez, J., Barrosa, R., Quintanilla, F., Flores, F., et al., 1996. The effect of solution treatment and rolling mode on the mechanical properties of 2090 Al–Li alloy. J. Mater. Process. Technol. 56, 542–551.

Piascik, R.S., Gangloff, R.P., 1991. Environmental fatigue of an Al–Li–Cu alloy. Part I— Intrinsic crack propagation kinetics in hydrogenous environments. Metall. Trans. A, 22A, 2415–2428.

Piascik, R.S., Gangloff, R.P., 1993. Environmental fatigue of an Ai–Li–Cu alloy—Part II. Microscopic hydrogen cracking processes. Metall. Trans. A, 24A, 2751–2762.

Pickens, J.R., 1985. The weldability of lithium-containing aluminium alloys. J. Mater. Sci. 20, 4247–4258.

Pickens, J.R., 1990. Review: recent developments in the weldability of lithium-containing aluminium alloys. J. Mater. Sci. 25, 3035–3047.

Pickens, J.R., Langa, T.J., Barta, E., 1986. Weldability of Al–5Mg–2Li–0.1Zr alloy 01420. In: Baker, C., Gregson, P.J., Harris, S.J., Peel, C.J. (Eds.), Proceedings of the Third International Conference on Aluminium–Lithium Alloys. The Institute of Metals, London, UK, pp. 137–147.

Pickens, J.R., Heubaum, F.H., Langan, T.J., Kramer, L.S., 1989. Al–(4.5–6.3) Cu–1.3 Li–0.4 Ag–0.4 Mg–0.14 Zr alloy weldalite 049. In: Sanders Jr., T.H., Starke Jr., E.A. (Eds.), Proceedings of the Fifth International Conference on Aluminium–Lithium Alloys, vol. 3. Materials and Component Engineering Publications Ltd., Birmingham, UK, pp. 1397–1411.

Pilling, J., Ridley, N., 1986. Role of hydrostatic pressure on cavitation during superplastic flow of Al–Li alloy. In: Baker, C., Gregson, P.J., Harris, S.J., Peel, C.J. (Eds.), Proceedings of the Third International Conference on Aluminium–Lithium Alloys. The Institute of Metals, London, UK, pp. 183–190.

Pitcher, P.D., 1988a. Ageing of forged aluminium–lithium 8091 alloy. Scr. Metall. 22, 1301–1306.

Pitcher, P.D., 1988b. Determination of the δ' solvus line in Al–Li alloys by measurement of the δ' volume fraction. Scr. Metall. 22, 1361–1365.

Pitcher, P.D., 1988c. Ageing of forged aluminium–lithium 8091 alloy. Royal Aircraft Establishment Technical Memorandum MAT/STR 1120, Farnborough, UK.

Pizzo, P.P., Galvin, R.P., Howard Nelson, G., 1984. Stress-corrosion behaviour of aluminium–lithium alloys in aqueous salt environments. In: Sanders Jr., T.H., Starke Jr., E.A. (Eds.), Proceedings of the Second International Conference on Aluminum–Lithium Alloys. The Metallurgical Society of AIME, Warrendale, PA, USA, pp. 627–656.

Poduri, R., Chen, L.Q., 1997. Computer simulation of the kinetics of order-disorder and phase separation during precipitation of δ' in Al–Li alloys. Acta Metall. 45, 245–255.

Pollak, E., Ivan, T., Lucas, Kostecki, R., 2010. A study of lithium transport in aluminium membranes. Electrochem. Commun. 12, 198–201.

Pollock, W.D., Hales, S.J., 1994. Tensile fracture of a 2090 extrusion. In: Sanders Jr., T.H., Starke Jr., E.A. (Eds.), Proceedings of the Fourth International Conference on Aluminum Alloys: Their Physical and Mechanical Properties, vol. 2. Georgia Institute of Technology, Atlanta, GA, USA, pp. 358–365.

Polmear, I.G., Miller, W.S., Lloyd, D.J., Bull, M.J., 1986. Effect of grain structure and texture on mechanical properties of Al–Li base alloys. In: Baker, C., Gregson, P.J., Harris, S.J., Peel, C.J. (Eds.), Proceedings of the Third International Conference on Aluminium–Lithium Alloys. The Institute of Metals, London, UK, pp. 565–575.

Polmear, I.J., 1986. Development of an experimental wrought aluminum alloy for use at elevated temperatures. In: Sanders Jr., T.H., Starke Jr., E.A. (Eds.), Proceedings of the Fourth International Conference on Aluminum Alloys: Their Physical and Mechanical Properties. Georgia Institute of Technology, Atlanta, GA, USA, pp. 661–674.

Polmear, I.J., 2005. Light Alloys, fourth ed. Butterworth-Heinemann.

Potti, P.K.G., Rao, B.N., Srivastava, V.K., 2000. Residual strength of aluminum–lithium alloy center surface crack tension specimens at cryogenic temperatures. Cryogenics 40, 789–795.

Powers, J.H., Johnson, W.F., Meher, P.L., 1989. Finishing characteristics of aluminium–lithium alloys. In: Sanders Jr., T.H., Starke Jr., E.A. (Eds.), Proceedings of the Fifth International Conference on Aluminium–Lithium Alloys, vol. 1. Materials and Component Engineering Publications Ltd., Birmingham, UK, pp. 461–472.

Poza, P., Llorca, J., 1999. Mechanical behavior of Al–Li/SiC composites: part II. cyclic deformation. Metall. Mater. Trans. A, 30A, 857–867.

Prangnelli, P.B., Ozkaya, D., Stobbs, W., 1994a. Discontinuous precipitation in high lithium content Al–Li–Zr alloys. Acta Metall. 42, 419–433.

Prangnell, P.B., Hulley, S.I., Palmer, I.G., 1994b. Microstructure and mechanical properties of hot oiled high Li content Aluminium alloys produced by spray-casting. In: Sanders Jr., T.H., Starke Jr., E.A. (Eds.), Proceedings of the Fourth International Conference on Aluminum Alloys: Their Physical and Mechanical Properties. Georgia Institute of Technology, Atlanta, GA, USA, pp. 106–113.

Prasad, K.S., 1999. Solid State Phase Transformations in AA 8090 Al–Li Alloys (Doctoral Thesis), University of Roorkee, Roorkee, India.

Prasad, K.S., Gokhale, A.A., Banerjee, D., Goel, D.B., 1992. New nanocrystalline regions in 1420 Al–Li alloy. Scr. Metall. Mater. 26, 1803–1808.

Prasad, K.S., Mukhopadhyay, A.K., Gokhale, A.A., Banerjee, D., Goel, D.B., 1994. δ Precipitation in an Al−Li−Cu−Mg−Zr alloy. Scr. Metall. 30, 1299−1304.

Prasad, K.S., Mukhopadhyay, A.K., Gokhale, A.A., Banerjee, D., Goel, D.B., 1999. On the formation of facetted $Al_3Zr(β')$ precipitates in Al−Li−Cu−Mg−Zr alloys. Acta Metall. 47, 2581−2592.

Prasad, Y.V.R.K., Ravichandran, N., 1991. Effect of stacking fault energy on the dynamic recrystallization during hot working of FCC metals: a study using processing maps. Bull. Mater. Sci. 14, 1241−1248.

Prasad, Y.V.R.K., Sashidhara, S., 1997. Aluminium alloys in hot working guide: a compendium of processing maps. Prasad, Y.V.R.K., Sashidhara, S. (Eds.), ASM International, Materials Park, OH, USA, pp. 160−177.

Pratt, R.C.A., Tsakiropoulos, P., Jones, H., Gardiner, R.W., Restall, J.E., 1987. Effect of dispersoid-forming additions on the response to heat treatment of splat−quenched Al−Li−X alloys. In: Champier, G., Dubost, B., Miannay, D., Sabetay, L. (Eds.), Proceedings of the Fourth International Conference on Aluminium−Lithium Alloys, J. Phys., 48, C3, 341−C3, 346.

Pridham, M., Noble, B., Harris, S.J., 1986. Elevated temperature strength of Al−Li−Cu−Mg alloys. In: Baker, C., Gregson, P.J., Harris, S.J., Peel, C.J. (Eds.), Proceedings of the Third International Conference on Aluminium−Lithium Alloys. The Institute of Metals, London, UK, pp. 547−554.

Pu, H.P., Huang, J.C., 1993. Low temperature superplasticity in 8090 Al−Li alloys. Scr. Metall. 28, 1125−1130.

Pu, H.P., Huang, J.C., 1995. Processing routes for intertransformation between low-and high-temperature 8090 Al−Li superplastic sheets. Scr. Metall. Mater. 33, 383−389.

Pu, H.P., Liu, F.C., Huang, J.C., 1995. Characterization and analysis of low-temperature superplasticity in 8090 Al−Li Alloys. Metall. Mater. Trans. A, 34A, 1153−1161.

Puhakainen, K., Boström, M., Groy, T.L., Haussermann, U., 2010. A new phase in the system lithium−aluminum: characterization of orthorhombic Li_2Al. J. Solid State Chem. 183, 2528−2533.

Pyun, S.I., Chun, Y.G., 1993. Environmental effects on crack closure in the corrosion fatigue of an aluminium−lithium alloy. Corros. Sci. 35 (619), 611−617.

Pyun, S.I., Chun, Y.G., 1996. Effects of aging and passivation on corrosion fatigue crack propagation of an Al−Li−Zr alloy. Mater. Lett. 26, 137−143.

Qiang, J., Lu, Z., Li, Y.W., Wu, Y.L., Liu, B.C., 1994. Effect of thermo-exposure on properties of 8090 Al−Li extrusion. In: Sanders Jr., T.H., Starke Jr., E.A. (Eds.), Proceedings of the Fourth International Conference on Aluminum Alloys: Their Physical and Mechanical Properties. Georgia Institute of Technology, Atlanta, GA, USA, pp. 467−472.

Qing, L., Xiaoxu, H., Mei, Y.Z., Jinfeng, Y., 1990. On deformation-induced continuous recrystallization in a superplastic Al−Li−Cu−Mg−Zr alloy. Acta Metall. Mater. 40, 1753−1762.

Qiong Li, Wawner, F.E., 1994. Investigation of the microstructure and mechanical properties in modified 2009 Al/SiCp composites. In: Sanders Jr., T.H., Starke Jr., E.A. (Eds.), Proceedings of the Fourth International Conference on Aluminum Alloys: Their Physical and Mechanical Properties. Georgia Institute of Technology, Atlanta, GA, USA, pp. 682−989.

Quadrini, E., Mengucci, P., 1996. Influence of microstructure on the hydrogen embrittlement of Al−Li−Cu−Mg−Zr alloys. J. Mater. Sci. 27, 1391−1396.

Quist, W.E., Narayanan, G.H., 1989. Aluminium−lithium alloys. Treatise Mater. Sci. Technol. 31, 219−254.

Quist, W.E., Narayanan, G.H., Wingert, A.L., 1984. Aluminum–lithium alloys for aircraft struc-
tures—an overview. In: Sanders Jr., T.H., Starke Jr., E.A. (Eds.), Proceedings of the Second
International Conference on Aluminum–Lithium Alloys. The Metallurgical Society of
AIME, Warrendale, PA, USA, pp. 313–334.

Quist, W.E., Narayanan, G.H., Wingert, A.L., Fronald, T.M., 1986. Analysis of powder metallurgy and related techniques for the production of aluminium–lithium alloys. In: Baker, C.,
Gregson, P.J., Harris, S.J., Peel, C.J. (Eds.), Proceedings of the Third International
Conference on Aluminium–Lithium Alloys. The Institute of Metals, London, UK,
pp. 625–638.

Quist, W.E., Bevers, C.Y., Narayanan, G.H., 1989. Microstructure and engineering properties of
alloy 644B. In: Sanders Jr., T.H., Starke Jr., E.A. (Eds.), Proceedings of the Fifth
International Conference on Aluminium–Lithium Alloys, vol. 3. Materials and Component
Engineering Publications Ltd., Birmingham, UK, pp. 1695–1706.

Rack, H.J., Piper, G.R., 1994. Age hardening response of SiC whisker reinforced dilute
Al–Cu–Mg–Li composites. In: Sanders Jr., T.H., Starke Jr., E.A. (Eds.), Proceedings of
the Fourth International Conference on Aluminum Alloys: Their Physical and Mechanical
Properties. Georgia Institute of Technology, Atlanta, GA, USA, pp. 613–620.

Radhakrishnan, V.M., 1992. On the bilinearity of the Coffin-Manson low-cycle fatigue relationship. Int. J. Fatigue 14, 305–311.

Rading, G.O., Berry, J.T., 1996. On deviated and branched crack paths in Al–Li–X alloys.
Mater. Sci. Eng. A, A219, 192–201.

Rading, G.O., Berry, J.T., Carden, A.E., 1993. An investigation of fatigue crack growth in
Al–Li–Cu alloy 2095 and associated weldment. In: Zacharia, T. (Ed.), Proceedings of the
International Conference on Modelling and Control of Joining Processes. American Welding
Society, Doral, FL, USA, pp. 616–623.

Radmilovic, V., Thomas, G., 1987. Atomic resolution imaging in Al–Li–Cu alloy. In:
Champier, G., Dubost, B., Miannay, D., Sabetay, L. (Eds.), Proceedings of the Fourth
International Conference on Aluminium–Lithium Alloys, J. Phys., 48, C3, 385–C3, 396.

Radmilovic, V., Thomas, G., 1994. Nonstoichiometry of Al–Zr intermetallic phases. In: Sanders
Jr., T.H., Starke Jr., E.A. (Eds.), Proceedings of the Fourth International Conference on
Aluminum Alloys: Their Physical and Mechanical Properties, vol. 2. Georgia Institute of
Technology, Atlanta, GA, USA, pp. 153–160.

Radmilovic, V., Fox, A.G., Fisher, R.M., Thomas, G., 1989a. Lithium depletion in precipitate
free zones (PFZ's) in Al–Li base alloys. Scr. Metall. 23, 75–79.

Radmilovic, V., Fox, A.G., Thomas, G., 1989b. Spinodal decomposition of Al-rich Al–Li alloys.
Acta Metall. 37, 2385–2394.

Radmilovic, V., Thomas, G., Shiflet, G.J., Starke Jr., E.A., 1989c. On the nucleation and growth
of $Al_2CuMg(S')$ in $Al–Li–Cu–Mg$ and $Al–Cu–Mg$ alloys. Scr. Metall. 23 (1989),
1141–1146.

Radmilovic, V., Rossell, M.D., Tolley, A., Marqui, E.A., Dahmen, U., 2008. Core/shell precipitates in Al–Li–Sc–Zr alloys. Mater. Sci. 2, 473–474.

Raghavan, V., 2007. Al–Li–Si (aluminum–lithium–silicon). J. Phase Equilib. Diffus. 28,
549–551.

Raghavan, V., 2009. Al–Li–Zr (aluminum–lithium–zirconium). J. Phase Equilib. Diffus. 30,
624–625.

Raghavan, V., 2010. Phase diagram evaluations: Al–Li–Cu. J. Phase Equilib. Diffus. 31,
288–290.

Raji, N., Sheshan, S., 1989. Casting Al—Li alloys in open atmosphere. In: Sanders Jr., T.H., Starke Jr., E.A. (Eds.), Proceedings of the Fifth International Conference on Aluminium—Lithium Alloys, vol. 1. Materials and Component Engineering Publications Ltd., Birmingham, UK, pp. 55—64.

Ramamurty, U., Bandyopadhyay, A., Dwarakadasa, E.S., Effect of heat treatment environment on Li depletion and on mechanical properties in Al—Li alloy sheets. J. Mater. Sci. 28, 6330—6336.

Raman, K.S., Das, E.S.D., Vasu, K.I., 1970. Values of solute-vacancy binding energy in aluminium matrix for Ag, Be, Ce, Dy, Fe, Li, Mn, Nb, Pt, Sb, Si, Y and Yb. Scr. Metall. 4, 291—294.

Ramani, A., Tosten, M.H., Michel, D.,J., Bartges, C.W., Ryba, E., Howell, P.R., 1989. *Phase transformations in "Single Crystal" fragments of the $T_2(Al_6CuLi_3)$ phase).* In: Sanders Jr., T. H., Starke Jr., E.A. (Eds.), Proceedings of the Fifth International Conference on Aluminium—Lithium Alloys, vol. 2. Materials and Component Engineering Publications Ltd., Birmingham, UK, pp. 777—788.

Ranganathan, N., Ait Abdedaim, M., Petit, J., 1990. Microscopic load interaction effects observed in an Al—Li alloy as compared to classical damage tolerant alloys. In: Khan, T., Effenberg, G. (Eds.), Proceedings of the International Conference on Light Metals Advanced Aluminium and Magnesium Alloys. ASM International European Council, Brussels, Belgium, pp. 165—172.

Ranganathan, N., Li, S.Q., Bailon, J.P., Petit, J., 1994. On micromechanisms of fatigue crack growth in the 8090 T651 aluminium—lithium alloy. Mater. Sci. Eng. A, A187, 37—42.

Ranganathan, N., Adiwijayanto, V., Petit, J., Bailon, J.P., 1995. Fatigue crack propagation mechanisms in an aluminium—lithium alloy. Acta Metall. Mater. 43, 1029—1035.

Rangel, C.M., Travassos, M.A., 2006. Li-based conversion coatings on aluminium: an electrochemical study of coating formation and growth. Surf. Coat. Technol. 200, 5823—5828.

Rangel-Ortiz, T., Alcala, F.C., Hirata, V.M.L., Flores, J.F., Araujo-Osorio, J.E., Rosales, H.J.D., et al., 2005. Microstructure and tensile properties of a continuous-cast Al—Li—Hf alloy. J. Mater. Process. Technol. 159, 164—168.

Ravindra, A., Dwarakadasa, E.S., 1992. Effect of post-weld heat treatment on mechanical properties of gas tungsten arc welds of Al—Li alloy 8090. J. Mater. Sci. Lett. 11, 1543—1546.

Ravindra, A., Dwarakadasa, E.S., Srivatsan, T.S., Ramanath, C., Iyengar, K.V.V., Electron-beam weld microstructures and properties of aluminium—lithium alloy 8090. J. Mater. Sci. 28, 3173—3182.

Reboul, M., Meyer, P., 1987. Intergranular and exfoliation corrosion study of Al—Li—Cu—Mg—Zr alloys. In: Champier, G., Dubost, B., Miannay, D., Sabetay, L., (Eds.), Proceedings of the Fourth International Conference on Aluminium—Lithium Alloys, J. Phys., 48, C3, 881—C3, 889.

Reddy Madhusudhan, G., 1998. Studies on the Application of Pulsed Current and Arc Oscillation Techniques on Aluminium—Lithium alloy Welds (Ph.D. Thesis). Department of Metallurgical Engineering, Indian Institute of Technology, Chennai.

Reddy Madhusudhan, G., Gokhale, A.A., 1993. Gas tungsten arc welding of 8090 Al—Li alloy. Trans. Indian Inst. Met. 46, 21—30.

Reddy Madhusudhan, G., Gokhale, A.A., 1994. Pulsed current gas tungsten arc welding of Al—Li alloy 1441. In: Sanders, T.H., Starke, E.A. (Eds.), Proceedings of the Fourth International Conference on Aluminium Alloys: Their Physical and Mechanical Properties, vol. 2. Georgia Institute of Technology, Atlanta, GA, USA, pp. 496—503.

Reddy Madhusudhan, G., Gokhale, A.A., 2007. On weld solidification cracking in aluminum —lithium alloys. Met. Mater. Process. 19, 297—306.

Reddy Madhusudhan, G., Gokhale, A.A., Prasad Rao, K., 1997. Weld microstructure refinement in a 1441 grade Al—Li alloy. J. Mater. Sci. 32, 4117—4126.

Reddy Madhusudhan, G., Gokhale, A.A., Prasad Rao, K., 1998a. Porosity and hot cracking in aeronautical grade aluminum—lithium alloy 1441 welds. In: Sastry, D.H., Subramanian, S., Murthy, K.S.S., Abraham, K.P. (Eds.), Proceedings of the International Conference on Aluminum, INCAL 98, 11—13 February, New Delhi, vol. 2. The Aluminium Association of India.

Reddy Madhusudhan, G., Gokhale, A.A., Prasad Rao, K., 1998b. Effect of filler metal composition on weldability of Al—Li alloy 1441. Sci. Technol. Weld. Joining 3, 151.

Reddy Madhusudhan, G., Gokhale, A.A., Prasad, K.S., Prasad Rao, K., 1998c. Chill zone formation in Al—Li alloy welds. Sci. Technol. Weld. Joining 3, 208—212.

Reddy Madhusudhan, G., Gokhale, A.A., Prasad Rao, K., 1998d. Optimisation of pulse frequency in pulsed current gas tungsten arc welding of aluminium—lithium alloy weld. Mater. Sci. Technol. 14, 61—68.

Reddy Madhusudhan, G., Gokhale, A.A., Prasad Rao, K., 1999. Weldability aspects of an aluminium—lithium alloy. In: Proceedings of the International Welding Conference (IWC'99) on Welding and Allied Technology Challenges in 21st Century. pp. 659—671.

Reddy Madhusudhan, G., Gokhale, A.A., Narendra Janaki Ram, N., Prasad Rao, K., 2001. Influence of welding techniques on microstructure and pitting corrosion behaviour of 1441 grade Al—Li alloy gas tungsten arc welds. Br. Corros. J. 36, 304—309.

Reddy Madhusudhan, G., Gokhale, A.A., Prasad Rao, K., 2002. Effect of the ratio of peak and background current durations on the fusion zone microstructure of pulsed current gas tungsten arc welded Al—Li alloy. J. Mater. Sci. 21, 1623—1625.

Reddy Madhusudhan, G., Mohandas, T., Sobhana Chalam, P., 2003. Metallurgical and mechanical properties of AA 8090 Al—Li alloy friction welds. In: International Welding Symposium (IWS 2K3) on Emerging Trends in Welding organized by Indian Welding Society, 22—23 February 2003, Hyderabad, India, pp.147—158.

Reed, P.A.S., Sinclair, I., Gregson, P.J., 1994. The effect of orientation on short crack path and growth rate behaviour in Al—Li alloy AA8090. In: Sanders Jr., T.H., Starke Jr., E.A. (Eds.), Proceedings of the Fourth International Conference on Aluminum Alloys: Their Physical and Mechanical Properties, vol. 2. Georgia Institute of Technology, Atlanta, GA, USA, pp. 397—404.

Ren, B., Hamilton, C.H., Ash, B., 1989. An approach to rapid SPF of an Al—Li—Cu—Zr alloy. In: Sanders Jr., T.H., Starke Jr., E.A. (Eds.), Proceedings of the Fifth International Conference on Aluminium—Lithium Alloys, vol. 1. Materials and Component Engineering Publications Ltd., Birmingham, UK, pp. 131—140.

Reuleaux, O., 1924. Scleron alloys. J. Inst. Met. 33, 346.

Reynolds, M.A., Creed, E., 1987. The development of 8090 and 8091 alloy extrusions, In: Champier, G., Dubost, B., Miannay, D., Sabetay, L., (Eds.), Proceedings of the Fourth International Conference on Aluminium—Lithium Alloys, J. Phys., 48, C3, 195—C3, 207.

Reynolds, M.A., Gray, A., Creed, E., Jordan, R.M., Titchener, A.P., 1986. Processing and properties of Alcan medium and high strength Al—Li—Cu—Mg alloys in various product forms. In: Baker, C., Gregson, P.J., Harris, S.J., Peel, C.J. (Eds.), Proceedings of the Third International Conference on Aluminium—Lithium Alloys. The Institute of Metals, London, UK, pp. 57—65.

Richard, S., Sarazzin-Baudoux, C., Petit, J., 2012. Fatigue crack propagation in new generation aluminium alloys. Key Eng. Mater. 488–489, 476–479.

Ricker, R.E., Duquette, D.J., 1984. Potentiodynamic polarisation studies of an Al–Mg–Li alloy. In: Sanders Jr., T.H., Starke Jr., E.A. (Eds.), Proceedings of the Second International Conference on Aluminum–Lithium Alloys. The Metallurgical Society of AIME, Warrendale, PA, USA, pp. 581–596.

Ricks, R.A., Parson, N.C., 1989. Microstructural optimisation for superplasticity in an Al–Li–Cu–Mg–Zr alloy 8090. In: Sanders Jr., T.H., Starke Jr., E.A. (Eds.), Proceedings of the Fifth International Conference on Aluminium–Lithium Alloys, vol. 1. Materials and Component Engineering Publications Ltd., Birmingham, UK, pp. 169–178.

Ricks, R.A., Budd, P.M., Goodhew, P.J., Kohler, V.L., Clyne, T.W., 1986. Production of fine, rapidly solidified aluminium–lithium powder by gas atomisation. In: Baker, C., Gregson, P. J., Harris, S.J., Peel, C.J. (Eds.), Proceedings of the Third International Conference on Aluminium–Lithium Alloys. The Institute of Metals, London, UK, pp. 97–104.

Ricks, R.A., Winkler, P.J., Stoklossa, H., Grimes, R., 1989. Transient liquid phase bonding of aluminium–lithium base alloy AA8090 using roll-clad Zn based interlayers. In: Sanders Jr., T.H., Starke Jr., E.A. (Eds.), Proceedings of the Fifth International Conference on Aluminium–Lithium Alloys, vol. 1. Materials and Component Engineering Publications Ltd., Birmingham, UK, pp. 441–450.

Ridley, N., Amichi, R., 1994. Superplastic behaviour in dynamically recrystallizing Al–Li alloy 8090. In: Sanders Jr., T.H., Starke Jr., E.A. (Eds.), Proceedings of the Fourth International Conference on Aluminum Alloys: Their Physical and Mechanical Properties, vol. 2. Georgia Institute of Technology, Atlanta, GA, USA, pp. 274–281.

Ridley, N., Livesey, D.W., Pilling, J., 1987. Optimisation of strain rate sensitivity during super-plastic deformation of Al–Li alloy Lital A, In: Champier, G., Dubost, B., Miannay, D,. Sabetay, L., (Eds.), Proceedings of the Fourth International Conference on Aluminium–Lithium Alloys, J. Phys., 48, C3, 251–C3, 256.

Rinker, J.G., Marek, M., Sanders Jr., T.H., Microstructure, toughness and SCC behaviour of 2020. In: Sanders Jr., T.H., Starke Jr., E.A., (Eds.), Proceedings of the Second International Conference on Aluminum–Lithium Alloys, The Metallurgical Society of AIME, Warrendale, PA, USA, pp. 597–626.

Río, J.D., Plazaola, F., Diego, N.D., 1994. The influence of Li on the nucleation of defects of quenched Al–Li alloys. Acta Metall. Mater. 42, 2267–2273.

Rioja, R.J., 1998. Fabrication methods to manufacture isotropic Al–Li alloys and products for space and aerospace applications. Mater. Sci. Eng. A, A257, 100–107.

Rioja, R.J., Liu, J., 2012. The evolution of Al–Li base products for aerospace and space applications. Metall. Mater. Trans. A, 43A, 3325–3337.

Rioja, R.J., Cho, A., Colvin, E.L., Vasudevan, A.K., 1992, Aluminum–Lithium Alloy, US Patent No. 5137686, Issued 11th August 1992.

Rioja, R.J., Giummarra, C., Cheong, S., 2008. The roll of crystallographic texture on the performance of flat rolled aluminum products for aerospace applications. In: DeYoung, D.H. (Ed.), 'Light Metals 2008', TMS, vol. 3. The Metallurgical Society of AIME, Warrendale, PA, USA, pp. 1065–1069.

Riontino, G., Zanada, A., 1994. A DSC and TEM investigation on a 2014 + 20%Al_2O_3 composite. In: Sanders Jr., T.H., Starke Jr., E.A. (Eds.), Proceedings of the Fourth International Conference on Aluminum Alloys: Their Physical and Mechanical Properties, vol. 2. Georgia Institute of Technology, Atlanta, GA, USA, pp. 637–644.

Riping, L., Jianhua, Z., Xiangyi, Z., 1999. Differences in microstructure of Pd77.5Au6Si16.5 alloy solidified under microgravity and gravity conditions. Sci. Chin. 42, 74−79.

Rivet, F.C., Swanson, R.E., 1989. Influence of dissolved hydrogen on aluminium−lithium alloy fracture behaviour. In: Sanders Jr., T.H., Starke Jr., E.A. (Eds.), Proceedings of the Fifth International Conference on Aluminium−Lithium Alloys, vol. 3. Materials and Component Engineering Publications Ltd., Birmingham, UK, pp. 1329−1336.

Roberge, P.R., Halliop, E., Lenard, D.R., Moores, J.G., 1993. Electrochemical characterization of the corrosion resistance of aluminum−lithium alloys. Corros. Sci. 35, 213−221.

Roder, O., Albrecht, J., Lütjering, G., 1994. Microstructure and deformation behaviour of high-temperature P/M Al-alloys. In: Sanders Jr., T.H., Starke Jr., E.A. (Eds.), Proceedings of the Fourth International Conference on Aluminum Alloys: Their Physical and Mechanical Properties, vol. 2. Georgia Institute of Technology, Atlanta, GA, USA, pp. 774−781.

Roeder, J.P., 1987. Keynote lecture to the fourth Al−Li conference. In: Champier, G., Dubost, B., Miannay, D., Sabetay, L., (Eds.), Proceedings of the Fourth International Conference on Aluminium−Lithium, J. Phys., 48, C3, 1−C3, 7.

Rohatgi, P.K., Guo, R.Q., Keshavaram, B.N., Golden, D.M., 1994a. Aluminium alloy-fly ash composite (ashalloy)—future foundry products for automotive applications. In: Sanders Jr., T.H., Starke Jr., E.A. (Eds.), Proceedings of the Fourth International Conference on Aluminum Alloys: Their Physical and Mechanical Properties, vol. 2. Georgia Institute of Technology, Atlanta, GA, USA, pp. 567−574.

Rohatgi, P.K., Narendranath, C.S., Cole, G.S., Bin, F., 1994b. Solidification microstructure and tribological properties of centrifugally cast aluminium alloy−graphite composites. In: Sanders Jr., T.H., Starke Jr., E.A. (Eds.), Proceedings of the Fourth International Conference on Aluminum Alloys: Their Physical and Mechanical Properties, vol. 2. Georgia Institute of Technology, Atlanta, GA, USA, pp. 606−612.

Rokhlin, L.L., Dobatkina, T.V., Muratova, E.V., Korol'kova, I.G., 1994. Izvestia RAN Metally 1, 113−118.

Romios, M., Tiraschi, R., Parrish, C., Babel, H.W., Ogren, J.R., Es-Said, O.S., 2005. Design of multistep aging treatments of 2099 (C458) Al−Li alloy. J. Mater. Eng. Perform. 14, 641−646.

Rooney, W.D., Papazian, J.M., Balmuth, E.S., Davis, R.C., Adler, P.N., 1989. Elastic anisotropy in Al−Li alloys. In: Sanders Jr., T.H., Starke Jr., E.A. (Eds.), Proceedings of the Fifth International Conference on Aluminium−Lithium Alloys, vol. 2. Materials and Component Engineering Publications Ltd., Birmingham, UK, pp. 799−808.

Rostova, T.D.T., Zakharov, V.V., 1997. Shear bands in aluminum−lithium alloys. Met. Sci. Heat Treat. 39, 236−239.

Roth, A., Kaesche, H., 1989a. Electrochemical investigation of technical aluminium−lithium alloys—part I: general aspects. In: Sanders Jr., T.H., Starke Jr., E.A. (Eds.), Proceedings of the Fifth International Conference on Aluminium−Lithium Alloys, vol. 3. Materials and Component Engineering Publications Ltd., Birmingham, UK, pp. 1197−1206.

Roth, A., Kaesche, H., 1989b. Electrochemical investigation of technical aluminium−lithium alloys—part II: the lithium depleted zone. In: Sanders Jr., T.H., Starke Jr., E.A. (Eds.), Proceedings of the Fifth International Conference on Aluminium−Lithium Alloys, vol. 3. Materials and Component Engineering Publications Ltd., Birmingham, UK, pp. 1207−1216.

Roven, H.J., Starke Jr., E.A., Sodahl, O., Hjelen, J., 1990. Effects of texture on delamination behaviour of AA 8090 type Al−Li alloy at cryogenic and room temperatures. Scr. Metall. Mater. 24, 421−426.

Röyset, J., 2007. Scandium in aluminium alloys—overview: physical metallurgy, properties and applications. Metall. Sci. Technol. 25, 11−21, www.TeksidAluminum.com.

Ruch, W., Starke Jr., E.A., 1986. Fatigue crack propagation in mechanically alloyed Al−Li−Mg alloys. In: Baker, C., Gregson, P.J., Harris, S.J., Peel, C.J. (Eds.), Proceedings of the Third International Conference on Aluminium−Lithium Alloys. The Institute of Metals, London, UK, pp. 121−130.

Ruhr, M., Baram, J., 1991. Structure and properties of a rapidly solidified Al−Li−Mn−Zr alloy for high-temperature applications: part I. Inert gas atomization processing. Metall. Trans. A, 22A, 2503−2514.

Ruhr, M., Lavernia, E.J., Baram, J.C., 1990. Extended Al (Mn) solution in a rapidly solidified Al−Li−Mn−Zr alloy. Metall. Trans. A, 21A, 1785−1789.

Sadananda, K., Jata, K.V., 1988. Creep crack growth behaviour of two Al−Li alloys. Metall. Trans. A, 19A, 847−854.

Sainfort, P., Dubost, B., 1987. Coprecipitation hardening in Al−Li−Cu−Mg alloys. In: Champier, G., Dubost, B., Miannay, D, Sabetay, L., (Eds.), Proceedings of the Fourth International Conference on Aluminium−Lithium Alloys, J. Phys., 48, C3, 407−C3, 413.

Sainfort, P. and Guyot, P., Fundamental aspects of hardening in Al−Li and Al−Li−Cu alloys. In: Baker, C., Gregson, P.J., Harris, S.J., Peel, C.J. (Eds.), Proceedings of the Third International Conference on Aluminium−Lithium Alloys, The Institute of Metals, London, UK, pp. 420−426.

Sainfort, P., Guyot, P., 1985. High-spatial-resolution STEM analysis of transition micro-phases in Al−Li and Al−Li−Cu alloys. Philos. Mag. A51, 575−588.

Saito, H., Ishida, Y., Yoshida, H., 1987a. Analysis of hydrogen behaviour in an Al−Li alloy by tritium analysis and transmission electron microscopic autoradiography. In: Champier, G., Dubost, B., Miannay, D., Sabetay, L., (Eds.), Proceedings of the Fourth International Conference on Aluminium−Lithium Alloys, J. Phys., 48, C3, 535−C3, 540.

Saito, H., Ishida, Y., Yoshida, H., 1987b. Tritium release characteristics of Al−Li alloys examined by liquid scintillation technique. In: Champier, G., Dubost, B., Miannay, D., Sabetay, L., (Eds.), Proceedings of the Fourth International Conference on Aluminium−Lithium Alloys, J. Phys., 48, C3, 541−C3, 548.

Saka, K.I., Matsumuro, A., 1996. Elastic moduli of Al−Li alloys treated at a high pressure of 5.4 GPa. J. Mater. Sci. 31, 3309−3313.

Sakurai, T., Kobayashi, A., Hasegawa, Y., Sakai, A., Pickering, H.W., 1986. Atomistic study of metastable phases in an Al-3 wt.%-Li-0.12 wt.%-Zr alloy. Scr. Metall. 20, 1131−1136.

Salem, H.G., Lyon, J.S., 2002. Effect of equal channel angular extrusion on the microstructure and superplasticity of an Al−Li alloy. J. Mater. Eng. Perform. 11, 384−391.

Salem, H.G., Goforth, R.E., Hartwig, K.T., 2003. Influence of intense plastic straining on grain refinement, precipitation and mechanical properties of Al−Cu−Li-based alloys. Metall. Mater. Trans. A, 34A, 1153−1161.

Samuel, F.H., 1986. Microstructural characterization of rapidly solidified Al−Li−Cu powders. Metall. Trans. A, 17A, 73−91.

Samuel, F.H., Champier, G., 1988. Effect of silicon, magnesium, cobalt and iron additions on the microstructure of Al−Li centrifugally atomized powders. J. Mater. Sci. 23, 541−546.

Samuel, F.H., Champier, G., Todeschini, P., Torres, J.H., 1992. Comparative study on the effect of iron and silicon addition on the microstructure and mechanical properties of aluminium−lithium powder atomized alloys. J. Mater. Sci. 27, 4917−4929.

Sanders Jr., T.H., 1981. Al—Li—X alloys: an overview. In: Sanders Jr., T.H., Starke Jr., E.A. (Eds.), Proceedings of the First International Conference on Aluminum—Lithium Alloys. The Metallurgical Society of AIME, Warrendale, PA, USA, pp. 63—68.

Sanders Jr., J.H., 1996. Investigation of grain boundary chemistry in Al—Li 2195 welds using Auger electron spectroscopy. Thin Solid Films 277, 121—127.

Sanders Jr., T.H., Balmuth, E.S., 1978. Aluminum—lithium alloys: low density. Met. Prog., 32—37.

Sanders Jr., T.H., Niskanen, P.W., 1981. *Microstructure, mechanical properties and corrosion resistance of Al—Li—X alloys—an overview*. Res. Mech. Lett. 1, 363—370.

Sanders Jr., T.H., Starke Jr., E.A., 1982. The effect of slip distribution on the monotonic and cyclic ductility of Al—Li binary alloys. Acta Metall. 30, 927—939.

Sanders Jr., T.H., Starke Jr., E.A., 1984. Overview of the physical metallurgy in the Al—Li—X systems. In: Sanders Jr., T.H., Starke Jr., E.A. (Eds.), Proceedings of the Second International Conference on Aluminum —Lithium Alloys. The Metallurgical Society of AIME, Warrendale, PA, USA, pp. 1—16.

Sanders Jr., T.H., Starke Jr., E.A., 1989. The physical metallurgy of aluminium—lithium alloys—a review. In: Sanders Jr., T.H., Starke Jr., E.A. (Eds.), Proceedings of the Fifth International Conference on Aluminium—Lithium Alloys, vol. 1. Materials and Component Engineering Publications Ltd., Birmingham, UK, pp. 1—37.

Sanders Jr., T.H., Ludwiczak, E.A., Sawtell, R.R., 1980. The fracture behavior of recrystallized Al-2.8% Li-0.3% Mn sheet. Mater. Sci. Eng. A, A43, 247—260.

Sanders Jr., T.H., Grandt Jr., A.F., Hillberry, B.M., Fragomeni, J.M., Henkener, J.A., Kistler, G. P., et al., 1989. An extrusion program designed to relate processing parameters to microstructure and properties of an Al—Li—Zr alloy. In: Sanders Jr., T.H., Starke Jr., E.A. (Eds.), Proceedings of the Fifth International Conference on Aluminium—Lithium Alloys, vol. 1. Materials and Component Engineering Publications Ltd., Birmingham, UK, pp. 273—286.

Sankaran, K.K., Grant, N.J., 1980a. The structure and properties of splat-quenched aluminium alloy 2024 containing lithium additions. Mater. Sci. Eng. A, A44, 213—227.

Sankaran, K.K., Grant, N.J., 1980b. Structure and properties of splat quenched 2024-aluminium alloy containing lithium additions. In: Sanders Jr., T.H., Starke Jr., E.A. (Eds.), Proceedings of the First International Conference on Aluminum Lithium Alloys. The Metallurgical Society of AIME, Warrendale, PA, USA, pp. 189—204.

Sankaran, K.K., O'Neal, E.A., 1984. Structure—property relationships in Al—Cu—Li alloys. In: Sanders Jr., T.H., Starke Jr., E.A. (Eds.), Proceedings of the Second International Conference on Aluminum—Lithium Alloys. The Metallurgical Society of AIME, Warrendale, PA, USA, pp. 393—405.

Sankaran, K.K., Sastry, S.M.L., O'Neal, J.E., 1980. Microstructure and deformation of rapidly solidified Al—3Li alloys containing incoherent dispersoids. In: Sanders Jr., T.H., Starke Jr., E.A. (Eds.), Proceedings of the First International Conference on Aluminum Lithium Alloys. The Metallurgical Society of AIME, Warrendale, PA, USA, pp. 171—188.

Sankaran, K.K., Vasey-Glandon, V.M., Tuegel, E.J., 1989. Application of Al—Li alloys to fighter aircraft. In: Sanders Jr., T.H., Starke Jr., E.A. (Eds.), Proceedings of the Fifth International Conference on Aluminium—Lithium Alloys, vol. 3. Materials and Component Engineering Publications Ltd., Birmingham, UK, pp. 1625—1634.

Sannino, A.P., Rack, H.J., 1994. Precipitation phenomenon in PM SiC whisker and particulate reinforced 2XXX Al composites. In: Sanders Jr., T.H., Starke Jr., E.A. (Eds.), Proceedings of the Fourth International Conference on Aluminum Alloys: Their Physical and Mechanical Properties. Georgia Institute of Technology, Atlanta, GA, USA, pp. 621—628.

Sastry, S.M.L., O'Neal, J.E., 1984. High temperature deformation behaviour and mechanical properties of rapidly solidified Al–Li–Co and Al–Li–Zr alloys. In: Sanders Jr., T.H., Starke Jr., E.A. (Eds.), Proceedings of the Second International Conference on Aluminum–Lithium Alloys. The Metallurgical Society of AIME, Warrendale, PA, USA, pp. 79–90.

Sater, J.M., Sanders Jr., T.H., 1989. Corrosion behaviour of binary Al–Li and ternary Al–Li–Zr alloys. In: Sanders Jr., T.H., Starke Jr., E.A. (Eds.), Proceedings of the Fifth International Conference on Aluminium–Lithium Alloys, vol. 3. Materials and Component Engineering Publications Ltd., Birmingham, UK, pp. 1217–1225.

Saunders, N., 1989. Calculated stable and metastable phase equilibria in Al–Li–Zr Alloys. Z. Metallkund. 80, 894–903.

Schelling, R.D., Kemppinen, A.I., Weber, J.H., 1989. IncoMAP alloy Al-905XL for aerospace forgings—production status. In: Sanders Jr., T.H., Starke Jr., E.A. (Eds.), Proceedings of the Fifth International Conference on Aluminium–Lithium Alloys, vol. 3. Materials and Component Engineering Publications Ltd., Birmingham, UK, pp. 1577–1586.

Schlecht, F., Doisne, R., Nossent, R., 1987. Aluminium–lithium extrusion. In: Champier, G., Dubost, B., Miannay, D., Sabetay, L., (Eds.), Proceedings of the Fourth International Conference on Aluminium–Lithium Alloys, J. Phys., 48, C3, 41–C3, 42.

Schlesier, C., Nembach, E., 1995. Strengthening of aluminium–lithium alloys by long-range ordered δ' precipitates. Acta Metall. Mater. 43, 3983–3990.

Schmitz, G., Haasen, P., 1992. Decomposition of an Al–Li alloy—the early stages observed by HREM. Acta Metall. 40, 2209–2217.

Schmitz, G., Hono, K., Haasen, P., 1994. High resolution electron microscopy of the early decomposition stage of Al–Li alloys. Acta Metall. Mater. 42, 201–211.

Schneider, J., Gudladt, H.J., Gerold, V., 1987. The effect of ageing on the fatigue behaviour of Al-2.3 wt% Li single crystals. In: Champier, G., Dubost, B., Miannay, D., Sabetay, L., (Eds.), Proceedings of the Fourth International Conference on Aluminium–Lithium Alloys, J. Phys., 48, C3, 745–C3, 752.

Schnuriger, S., Mankowski, G., Roques, Y., Chatainier, G., Dabosi, F., 1987. Statistical study of the pitting corrosion of the 8090 aluminium lithium alloy. In: Champier, G., Dubost, B., Miannay, D., Sabetay, L., (Eds.), Proceedings of the Fourth International Conference on Aluminium–Lithium Alloys, J. Phys., 48, C3, 851–C3, 860.

Schöberl, T., Kumar, S., 1997. Depletion of lithium due to surface oxidation: an investigation of an Al–Li-sheet by Auger-spectroscopy. J. Alloys Compd. 255, 135–141.

Schorn, R.P., Bay, H.L., Hintz, E., Schweer, B., 1987. Investigation of partial sputtering of lithium from a binary Al/Li alloy with laser induced fluorescence. Appl. Phys. A43, 147–151.

Semenchenkov, A.A., Fedosov, A.S., Davydov, V.G., 1992. In: Peters, M., Winkler, P.J. (Eds.), Aluminum–Lithium Alloys VI. Deutsche Gesellschaft für Metallkunde, Frankfurt, Germany, pp. 873–873.

Semenov, A.M., 2001. Effect of Mg additions and thermal treatment on corrosion properties of Al–Li–Cu-base alloys. Prot. Met. 37, 126–131.

Semenov, M., Sinyavskii, V.S., 2001. The effect of the copper to magnesium ratio and their summary content on the corrosion properties of Al–Li alloys. Prot. Met. 32, 132–137.

Setargew, N., Parker, B.A., Couper, M.J., 1994. Characteristics of the solidification in Al–7Si–0.5 Mg based metal matrix composites by thermal analysis. In: Sanders Jr., T.H., Starke Jr., E.A. (Eds.), Proceedings of the Fourth International Conference on Aluminum Alloys: Their Physical and Mechanical Properties, vol. 2. Georgia Institute of Technology, Atlanta, GA, USA, pp. 543–550.

Shah, S.R., Wittig, J.E., Hahn, G.T., 1992. *Microstructural analysis of a high strength Al−Cu−Li (Weldalite™ 049) alloy weld.* In: David, S.A., Vitek, J.M. (Eds.), International Trends in Welding Science and Technology: Proceedings of the Third International Conference on Trends in Welding Research. ASM International, Materials Park, OH, USA, pp. 281−285.

Shaiu, B.J., Li, H.T., Chen, H., 1989. Decomposition and reversion kinetics in Al−Li binary alloys. In: Sanders Jr., T.H., Starke Jr., E.A. (Eds.), Proceedings of the Fifth International Conference on Aluminium−Lithium Alloys, vol. 2. Materials and Component Engineering Publications Ltd., Birmingham, UK, pp. 613−622.

Shaiu, B.J., Li, H.T., Lee, H.Y., Chen, H., 1990. Decomposition and dissolution kinetics of precipitation in Al−Li binary alloys. Metall. Mater. Trans. A, 21A, 1133−1141.

Shakesheff, A.J., Partridge, P.G., 1986. Superplastic deformation of Al−Li−Cu−Mg alloy sheet. J. Mater. Sci. 21, 1368−1376.

Shakesheff, A.J., McDarmaid, D.S., Gregson, P.J., 1989a. Microstructure−property relationships in superplastic alloy formed 8090 sheet. In: Sanders Jr., T.H., Starke Jr., E.A. (Eds.), Proceedings of the Fifth International Conference on Aluminium−Lithium Alloys, vol. 1. Materials and Component Engineering Publications Ltd., Birmingham, UK, pp. 201−210.

Shakesheff, A.J., McDarmaid, D.S., Gregson, P.J., 1989b. Effects of cooling rate and copper content on the properties of Al−Li−Cu−Mg−Zr alloy 8090. Mater. Lett. 7, 353−358.

Shamraia, V.F., Timofeeva, V.N., Grushko, O.E., 2010. *Investigation of the structure of compacts and sheets of an Al−Cu−Li alloy strengthened by Al₂CuLi (T₁) Particles.* Phys. Met. Metall. 109, 383−393.

Shan, Z.H., Wang, Z.G., Zhang, Y., 1997. Effect of notch geometry on short fatigue crack growth in 8090 Al−Li alloy. J. Mater. Sci. 32, 4673−4677.

Shashidhar, S.R., Kumar, A.M., Hirth, J.P., 1995. Fracture toughness of an Al−Li alloy at ambient and cryogenic temperatures. Metall. Mater. Trans. A, 26A, 2269−2274.

Shaw, P., 1985. WESTEC '85−aluminium−lithium alloy sessions, GAC Memorandum M&ME-TS441A-85K-01, Grumman Aerospace Corporation, Bethpage, NY, USA.

Shchegoleva, T.V., Rybalko, O.F., 1980. Structure of the metastable S′-phase in alloy Al−Mg−Li. Fiz. Met. Metalloved. 50, 86−90.

Sheppard, T., Parson, N.C., 1987. Corrosion resistance of Al−Li alloys. Mater. Sci. Technol. 3, 345−352.

Sheppard, T., Tan, M.J., 1989. Thermomechanical treatments of an extruded Al−Li alloy. In: Sanders Jr., T.H., Starke Jr., E.A. (Eds.), Proceedings of the Fifth International Conference on Aluminium−Lithium Alloys, 1. Materials and Component Engineering Publications Ltd., Birmingham, UK, pp. 233−260.

Sheshadri, M.R., Ramachandran, A., 1966. *Casting fluidity and fluidity of aluminium and its alloys.* Trans. Am. Foundrymen's Soc. 73, 292−304.

Shin, K.S., Kim, S.S., Lee, E.W., 1989a. Hydrogen embrittlement of 2090 Al−Li alloy. In: Sanders Jr., T.H., Starke Jr., E.A. (Eds.), Proceedings of the Fifth International Conference on Aluminium−Lithium Alloys, vol. 3. Materials and Component Engineering Publications Ltd., Birmingham, UK, pp. 1319−1328.

Shin, K.S., Kim, S.S., Kim, N.J., 1989b. Mechanical behaviour of RSP Al−Li−Zr alloys at 293 and 77 K. In: Sanders Jr., T.H., Starke Jr., E.A. (Eds.), Proceedings of the Fifth International Conference on Aluminium−Lithium Alloys, vol. 3. Materials and Component Engineering Publications Ltd., Birmingham, UK, pp. 1543−1552.

Shneider, G.L., 1998. Stability of supersaturated solid solution of aluminum-lithium alloy 1470. Met. Sci. Heat Treat. 40, 294−298.

Shpeizman, V.V., Myshlyaev, M.M., Kamalov, M.M., Myshlyaeva, M.M., 2003. Superplasticity of a microcrystalline aluminum–lithium alloy under torsion. Phys. Solid State 43, 865–870.

Shrimpton, G.R.D., Angus, H.C., 1989. Forged components in aluminium–lithium alloys for aerospace applications. In: Sanders Jr., T.H., Starke Jr., E.A. (Eds.), Proceedings of the Fifth International Conference on Aluminium–Lithium Alloys, vol. 3. Materials and Component Engineering Publications Ltd., Birmingham, UK, pp. 1565–1576.

Shukla, A.K., Baeslack, W.A., 2007. Study of microstructural evolution in friction-stir welded thin-sheet Al–Cu–Li alloy using transmission electron microscopy. Scr. Mater. 56, 513–516.

Shuncai, W.I., Chunzhi, L., Minggao, Y., 1994. A transmission-electron-microscopy study of a face-centred-cubic phase in an Al–Li–Cu–Mg–Zr alloy. J. Mater. Sci. 29, 384–388.

Sigli, C., Sanchez, J.M., 1986. Calculation of phase equilibrium in aluminium lithium alloys. Acta Metall. 34, 1021–1028.

Silcock, J.M., 1959. The structural ageing characteristics of Al–Cu–Li alloys. J. Inst. Met. 88, 357–364.

Sinclair, I., Gregson, P.J., 1994. Microstructural and micromechanical influences on fatigue crack path behaviour in Al–Li alloy AA8090. In: Sanders Jr., T.H., Starke Jr., E.A. (Eds.), Proceedings of the Fourth International Conference on Aluminum Alloys: Their Physical and Mechanical Properties, vol. 2. Georgia Institute of Technology, Atlanta, GA, USA, pp. 436–443.

Singh, A., Fiset, M., Lapointe, R., Knystautas, E.J., Surface microhardening in a lithium implanted aluminium alloy. Scr. Metall., 18, 995–997.

Singh, A.K., Saha, G.G., Gokhale, A.A., Ray, R.K., 1998. Evolution of texture and microstructure in a thermomechanically processed Al–Li–Cu–Mg alloy. Metall. Mater. Trans. A, 29A, 665–675.

Singh, A.K., Gokhale, A.A., Saha, G.G., Ray, R.K., 1999. Texture evolution and anisotropy in Al–Li–Cu–Mg alloys. In: Ray, R.K., Singh, R.K. (Eds.), Textures in Materials Research. Oxford & IBH Pub. Co. Private Ltd., New Delhi, India, pp. 219–234.

Singh, V., 1997. Preparation and Characteristics of Al–Li–Cu–Mg–Zr Based Alloys (Doctoral Thesis). Banaras Hindu University, Varanasi, India.

Singh, V., Chakravorthy, C.R., 1989. Melting and casting of Al–Li alloys—a review. Proceedings of the Science and Technology of Al–Li Alloys, Seminar. Hindustan Aeronautical Ltd., Bangalore, India, pp. 83–91.

Sinko, R.J., Ahrens, T., Shifler, G.J., Starke, E.A., 1989. Effect of stretch on nucleation and growth of S′ in an 8090 Al–Li alloy. In: Sanders Jr., T.H., Starke Jr., E.A. (Eds.), Proceedings of the Fifth International Conference on Aluminium–Lithium Alloys, vol. 1. Materials and Component Engineering Publications Ltd., Birmingham, UK, pp. 375–384.

Skillingberg, M.H., 1986. Fusion welding of Al–Li–Cu–(Mg)–Zr plate. In: Sheppard, T. (Ed.), Proceedings of the International Conference on Aluminium Technology 86'. Institute of Metals, London, UK, pp. 509–515.

Skillingberg, M.H., Ashton, R.F., 1987. Processing and performance of Al–Li–Cu–X extrusions. In: Champier, G., Dubost, B., Miannay, D., Sabetay, L., (Eds.), Proceedings of the Fourth International Conference on Aluminium–Lithium Alloys, J. Phys., 48, C3, 179–C3, 186.

Skrotzki, B., Starke Jr., E.A., Shiflet, G.J., 1994. Effect of texture and precipitates on mechanical property anisotropy of Al–Cu–Mg–X alloys. In: Sanders Jr., T.H., Starke Jr., E.A. (Eds.), Proceedings of the Fourth International Conference on Aluminum Alloys: Their Physical and Mechanical Properties, vol. 2. Georgia Institute of Technology, Atlanta, GA, USA, pp. 40–47.

Slak, A.C.B., Mizera, J., Kurzydlowski, K.J., 2010. Thermal stability of model Al–Li alloys after severe plastic deformation—effect of the solute Li atoms. Mater. Sci. Eng. A, A527, 4716–4722.

Slavik, D.C., Blankenship Jr., C.P., Starke Jr., E.A., Gangloff, R.P., 1993. Intrinsic fatigue crack growth rates for Al–Li–Cu–Mg alloys in vacuum. Metall. Mater. Trans. A, 24A, Pp. 1807–1517.

Smith, A.F., 1986. Microstructure of cast aluminium–lithium–copper alloy. In: Baker, C., Gregson, P.J., Harris, S.J., Peel, C.J. (Eds.), Proceedings of the Third International Conference on Aluminium–Lithium Alloys. The Institute of Metals, London, UK, pp. 164–172.

Smith, A.F., 1987a. The metallurgical aspects of aluminium–lithium alloys in various product forms for helicopter structural applications. In: Champier, G., Dubost, B., Miannay, D., Sabetay, L., (Eds.), Proceedings of the Fourth International Conference on Aluminium–Lithium Alloys, J. Phys., 48, C3, 49–C3, 62.

Smith, A.F., 1987b. A study of the microstructure and properties of die forgings in aluminium–lithium alloys 2091 abd 8090. In: Champier, G., Dubost, B., Miannay, D., Sabetay, L., (Eds.), Proceedings of the Fourth International Conference on Aluminium–Lithium Alloys, J. Phys., 48, C3, 629–C3, 642.

Smith, A.F., 1987c. Aluminium–lithium alloys for helicopter structures. Met. Mater. 3, 438–444.

Smith, A.F., 1989a. Structure and properties of Al–Li–Cu–Mg–Zr alloy AA 2091 in sheet form. Mater. Sci. Technol. 5, 533–541.

Smith, A.F., 1989b. A comparison of large AA8090, AA8091 and AA7010 die forgings for helicopter structural applications. In: Sanders Jr., T.H., Starke Jr., E.A. (Eds.), Proceedings of the Fifth International Conference on Aluminium–Lithium Alloys, vol. 3. Materials and Component Engineering Publications Ltd., Birmingham, UK, pp. 1587–1596.

Smith, H.H., Reed, J.R., Michel, D.J., McNelley, T.R., 1989a. Tensile behaviour and microstructure of an aluminium–lithium–copper alloy. In: Sanders Jr., T.H., Starke Jr., E.A. (Eds.), Proceedings of the Fifth International Conference on Aluminium–Lithium Alloys, vol. 2. Materials and Component Engineering Publications Ltd., Birmingham, UK, pp. 943–954.

Smith, I.C., Avramovic-Cingara, G., McQueen, H.J., 1989b. Hot torsion strength of commercial Al–Li–Cu–Mg alloys. In: Sanders Jr., T.H., Starke Jr., E.A. (Eds.), Proceedings of the Fifth International Conference on Aluminium–Lithium Alloys, vol. 1. Materials and Component Engineering Publications Ltd., Birmingham, UK, pp. 223–232.

Smith, I.O., Russell, B., 1970. The dependence of the tensile properties of irradiated aluminium–lithium alloys upon gas bubble size and distribution. J. Nucl. Mater. 37, 96–108.

Smith, I.O., Russell, B., 1971. The effect of radiation induced gas on the stress rupture properties of an aluminium–lithium alloy. J. Nucl. Mater. 38, 1–16.

Smith, R.K., 1991. The quest for excellence. In: Greenwood, J.T. (Ed.), Milestones of Aviation. Crescent Books, New York, NY, USA, pp. 222–296.

Smith, S.W., Scully, J.R., 2000. The identification of hydrogen trapping states in an Al–Li–Cu–Zr alloy using thermal desorption spectroscopy. Metall. Mater. Trans. A, 31A, 179–193.

Smith, W.D., Sharma, A., Gregson, P.J., Court, S.A., 1994. Grain boundary precipitation of the I-phase in an Al–Li–Cu–Mg–(Zr) alloy (8090). In: Sanders Jr., T.H., Starke Jr., E.A. (Eds.), Proceedings of the Fourth International Conference on Aluminum Alloys: Their

Physical and Mechanical Properties, vol. 2. Georgia Institute of Technology, Atlanta, GA, USA, pp. 215–221.

Song, G.S., Staiger, M., Kral, M., 2004. Some new characteristics of the strengthening phase in β-phase magnesium–lithium alloys containing aluminum and beryllium. Mater. Sci. Eng. A, A371, 371–376.

Soni, K.K., Williams, D.B., Chabala, J.M., Levi-Setti, R., Newbury, D.E., 1992. Electron and ion microscopy studies of Fe-rich second-phase particles in Al–Li alloys. Acta Metall. Mater. 40, 663–671.

Soni, K.K., Setti, R.L., Shah, S., Gentz, S., 1996. SIMS imaging of Al–Li alloy welds. Adv. Mater. Process. 4, 35–36.

Sozaev, V.A., Chernyshova, R.A., 2005. The effect of a dielectric coating on the surface energy of thin metal films of the Al–Li alloy system. Tech. Phys. Lett. 31, 403–404.

Spooner, S., Williams, D.B., Sung, C.M., 1986. Combined small angle X-ray scattering and transmission electron microscopy studies of Al–Li alloys. In: Baker, C., Gregson, P.J., Harris, S.J., Peel, C.J. (Eds.), Proceedings of the Third International Conference on Aluminium–Lithium Alloys. The Institute of Metals, London, UK, pp. 329–336.

Spowage, A.C., Bray, S., 2011. Characterization of nanoprecipitation mechanisms during isochronal aging of a pseudo-binary Al-8.7at.pct Li alloy. Metall. Mater. Trans. A, 42A, 227–230.

Srinivasan, M.N., Goforth, R.E., 1995. Response to comment on studies on activation energy of superplastic dynamically recrystallized aluminium–lithium alloys. Scr. Metall. 32, 1861–1864.

Srinivasan, M.N., Goforth, R.E., Balasubramanian, R., 1992. Microstructural evaluation of a dynamically recrystallized superplastic aluminum–lithium alloy. Mater. Charact. 29, 397–406.

Srivatsan, T.S., 1986a. The influence of processing on the microstructure of an aluminum–lithium alloy. Mater. Lett. 4, 201–206.

Srivatsan, T.S., 1986b. Microstructural characterization of two lithium-containing aluminium alloys. J. Mater. Sci. 21, 1553–1561.

Srivatsan, T.S., 1987. Relationship between annealing and recrystallization in an aluminium–lithium alloy. J. Mater. Sci. Lett. 6, 948–950.

Srivatsan, T.S., 1988a. The effect of grain-refining additions to lithium-containing aluminium alloys. J. Mater. Sci. Lett. 1, 940–943.

Srivatsan, T.S., 1988b. Mechanisms of damage in high-temperature, low cycle fatigue of an aluminium alloy. Int. J. Fatigue 2, 91–99.

Srivatsan, T.S., Coyne, E.J., 1986a. Cyclic stress response and deformation behaviour of precipitation-hardened aluminium–lithium alloys. Int. J. Fatigue 8, 201–208.

Srivatsan, T.S., Coyne, E.J., 1986b. Mechanisms governing the high strain fatigue behavior of Al–Li–X alloys, In: Proceedings ISTFA 1986: International Symposium for Testing and Failure Analysis 1986, ASM International, Materials Park, OH, USA, pp. 281–293.

Srivatsan, T.S., Coyne, E.J., 1987. Mechanisms governing cyclic fracture in an Al–Cu–Li alloy. Mater. Sci. Technol. 3, 130–138.

Srivatsan, T.S., Lavernia, J., 1991. The presence and consequences of precipitate free zones in an aluminium–copper–lithium alloy. J. Mater. Sci. 26, 940–950.

Srivatsan, T.S., Place, T.A., 1989. Microstructure, tensile properties and fracture behaviour of an Al–Cu–Li–Mg–Zr alloy 8090. J. Mater. Sci. 24, 1543–1551.

Srivatsan, T.S., Yamaguchi, K., Starke, E.A., 1986a. The effect of environment and temperature on the low cycle fatigue behavior of aluminum alloy 2020. Mater. Sci. Eng. 83, 87–107.

Srivatsan, T.S., Coyne, E.J., Strake, E.A., 1986b. Microstructural characterization of two lithium-containing aluminium alloys. J. Mater. Sci. 21, 1553–1560.

Srivatsan, T.S., BoBeck, G.E., Sudarshan, T.S., Mollan, P.A., 1989. Environmental factors affecting localized corrosion of Al–Li–Cu–Mg alloys. In: Sanders Jr., T.H., Starke Jr., E. A. (Eds.), Proceedings of the Fifth International Conference on Aluminium–Lithium Alloys, vol. 3. Materials and Component Engineering Publications Ltd., Birmingham, UK, pp. 1237–1250.

Srivatsan, T.S., Hoff, T., Prakash, A., 1991a. Cyclic stress response characteristics and fracture behavior of aluminum alloy 2090. Mater. Sci. Technol. 7, 991–997.

Srivatsan, T.S., Hoff, T., Prakash, A., 1991b. The high strain cyclic fatigue behaviour of 2090 aluminum alloy. Eng. Fract. Mech. 40, 297–309.

Srivatsan, T.S., Soni, K., Sudarshan, T.S., 1992. General corrosion behaviour of an Al-2Li-1.2Cu-3Mg-0.12Zr alloy in aggressive aqueous environments. Met. Mater. Process. 4, 77–92.

Staikov, G., Yankulov, P.D., Mindjov, K., Aladjov, B., Budevski, E., 1984. Phase formation and acoustic emission during electrochemical incorporation of lithium into aluminium from molten LiCl–KCl. Electrochim. Acta 29, 661–665.

Staley, J.T., Doherty, R.D., 1989. Quench sensitivity in Al–Cu–Li alloy plate. In: Sanders Jr., T.H., Starke Jr., E.A. (Eds.), Proceedings of the Fifth International Conference on Aluminium–Lithium Alloys, vol. 1. Materials and Component Engineering Publications Ltd., Birmingham, UK, pp. 345–354.

Staley, J.T., Lege, D.J., 1993. Advances in aluminum alloy products for structural applications in transportation. J. Phys. IV, Colloque C7 3, 179–190.

Starink, M.J., Gregson, P.J., 1995. A quantitative interpretation of DSC experiments on quenched and aged SiC reinforced 8090 alloys. Scr. Met. Mater. 33, 893–900.

Starink, M.J., Hobson, A.J., Sinclair, I., Gregson, P.J., 2000. Embrittlement of Al–Li–Cu–Mg alloys at slightly elevated temperatures: microstructural mechanisms of hardening. Mater. Sci. Eng. A, A289, 130–142.

Starke Jr., E.A., 1977. Aluminium alloys of the 70's: scientific solutions to engineering problems. Mater. Sci. Eng. 29, 99–114.

Starke Jr., E.A., Lin, F.S., 1982. The influence of grain structure on the ductility of the Al–Cu–Li–Mn–Cd alloy 2020. Metall. Trans. A, 13A, 2259–2269.

Starke Jr., E.A., Quist, W.E., 1989. The microstructure and properties of aluminium–lithium alloys. 'New Light Alloys', AGARD Conference Proceedings No. 444. Advisory Group for Aerospace Research and Development, Neuilly-sur-Seine, France, pp. 4-1–4-23.

Starke Jr., E.A., Staley, J.T., 1996. Application of modern aluminum alloys to aircraft. Prog. Aerosp. Sci. 32, 131–172.

Starke Jr., E.A., Sanders Jr., T.H., Palmer, I.G., 1981. New approaches to alloy development in the Al–Li system. J. Met. 33, 24–33.

Steuwer, A., Dumont, M., Altenkirch, J., Birosca, S., Deschamps, A., Prangnell, P.B., 2011. A combined approach to microstructure mapping of an Al–Li AA2199 friction stir weld. Acta Mater. 59, 3002–3011.

Stiltz, S., 1993. In: Petzow, G, Effenberg, G. (Eds.), Ternary Alloys, vol. 6. Verlag Chemie (VCH), Weinheim, Germany, pp. 411–417.

Stimson, W., Tosten, M.H., Howell, P.R., Williams, D.B., 1986. Precipitation and lithium segregation studies in Al-2wt%Li-0.1wt%Zr. In: Baker, C., Gregson, P.J., Harris, S.J., Peel, C.J. (Eds.), Proceedings of the Third International Conference on Aluminium–Lithium Alloys. The Institute of Metals, London, UK, pp. 386–391.

Stokes, K.R., Moth, D.A., Sherwood, P.J., 1986. Hard anodising and marine corrosion character-istics of 8090 Al–Li–Cu–Mg–Zr alloy. In: Baker, C., Gregson, P.J., Harris, S.J., Peel, C.J. (Eds.), Proceedings of the Third International Conference on Aluminium–Lithium Alloys. The Institute of Metals, London, UK, pp. 294–302.

Stowell, M.J., Livesey, D.W., Ridley, N., 1984. Cavity coalescence in superplastic deformation. Acta Metall. 32, 35–42.

Stoyanova, R., Zhecheva, E., Kuzmanova, E., Alcántara, R., Lavela, P., Tirado, J.L., 2000. Aluminium coordination in $LiNi_{1-y}Al_yO_2$ solid solutions. Solid State Ionics 128, 1–10.

Sugamata, M., Blankenship Jr., C.P., Starke Jr., E.A., 1993. Predicting plane strain fracture toughness of Al–Li–Cu–Mg alloys. Mater. Sci. Eng. A, A163, 1–10.

Sugisaki, M., Furuya, H., Ichigi, T., Koori, N., Kumabe, I., Kawamura, K., 1988. Tritium release from an Al–Mg–Li alloy irradiated by 14 MeV neutrons. J. Nucl. Mater. 158, 202–209.

Sugisaki, M., Furuya, H., Ichigi, T., Koori, N., Kumabe, I., 1991. Tritium release behavior of an Al–Mg–Li alloy irradiated by thermal neutrons. J. Nucl. Mater. 179–181, 312–315.

Sun, D.L., Yang, D.Z., Lei, T.Q., 1990. A study of serrated plastic flow behavior in an alumi-num–lithium binary alloy. Mater. Chem. Phys. 25, 307–313.

Sun, D.L., Chen, S.Q., Mao, J.F., Yang, D.Z., 1994. Microstructure and deformation behavior of a SiC whisker reinforced aluminum–lithium alloy. Mater. Chem. Phys. 36, 217–221.

Sun, Z.Q., Huang, M.H., Hu, G.H., 2012. Surface treatment of new type aluminum lithium alloy and fatigue crack behaviors of this alloy plate bonded with Ti–6Al–4V alloy strap. Mater. Des. 35, 725–730.

Sung, C.M., Chan, H.M., Williams, D.B., 1986. Quantitative microanalysis of Li in binary Al–Li alloys. In: Baker, C., Gregson, P.J., Harris, S.J., Peel, C.J. (Eds.), Proceedings of the Third International Conference on Aluminium–Lithium Alloys. The Institute of Metals, London, UK, pp. 337–346.

Sunwoo, A.J., Morris, J.W., 1989. Cryogenic tensile properties of AA 2090 weldments. In: Sanders Jr., T.H., Starke Jr., E.A. (Eds.), Proceedings of the Fifth International Conference on Aluminium–Lithium Alloys, vol. 3. Materials and Component Engineering Publications Ltd., Birmingham, UK, pp. 1481–1490.

Suresh, S., Vasudevan, A.K., 1986. Influence of composition and aging treatment on frature toughness of lithium-containing aluminium alloys. In: Baker, C., Gregson, P.J., Harris, S.J., Peel, C.J. (Eds.), Proceedings of the Third International Conference on Aluminium–Lithium Alloys. The Institute of Metals, London, UK, pp. 595–601.

Suresh, S., Vasudevan, A.K., Tosten, M., Howell, P.R., 1987. Microscopic and macroscopic aspects of fracture in lithium-containing aluminium alloys. Acta Metall. 35, 25–46.

Susman, S., Brun, T.O., 1980. The growth of single crystal lithium–aluminum, 7LiAl. Solid State Ionics 1, 133–143.

Suzuki, H., Araki, I., Kanno, M., 1977. Effects of indium on precipitation in Al-3.3Cu-0.8Li alloy. J. Jpn. Inst. Light Met. 27, 239–245 (in Japanese).

Sverdlin, A., Drits, A.M., Krimova, T.V., Sergeev, K.N., Ginko, I.B., 1998. Aluminium–lithium alloys for aerospace. Adv. Mater. Process. 153 (6), 49–51.

Sweet, E.D., Lynch, S.P., Bennett, C.G., Nethercott, R.B., Musulin, I., 1996. Effects of alka-li–metal impurities on S-L fracture toughness of 2090 Al–Li–Cu extrusions. Metall. Mater. Trans. A, 27A, 3530–3541.

Tack, W.T., Loechel, L.W., 1989. Weldalite ™ 049: applicabilty of a new high strength, weld-able Al–Li–Cu alloy. In: Sanders Jr., T.H., Starke Jr., E.A. (Eds.), Proceedings of the Fifth International Conference on Aluminium–Lithium Alloys, vol. 3. Materials and Component Engineering Publications Ltd., Birmingham, UK, pp. 1457–147.

Takahashi, K., Minakawa, K. Ouchi, C., 1987. The effect of thermomechanical processing variables on anisotropy in mechanical properties of Al—Li alloys. In: Champier, G., Dubost, B., Miannay, D., Sabetay, L. (Eds.), Proceedings of the Fourth International Conference on Aluminium—Lithium Alloys, J. Phys., 48, C3, 163—C3, 169.

Talbot, J.B., Wiffen, F.W., 1979. Recovery of tritium from lithium-sintered aluminum product (SAP) and lithium—aluminum alloys. J. Inorg. Nucl. Chem. 41, 439—444.

Talbot, J.B., Smith, F.J., Land, J.F., Barton, P., 1976. Tritium sorption in lithium—bismuth and lithium—aluminum alloys. J. Less Common Met. 50, 23—28.

Talia, J.E., Mazumdar, P.K., 1989. Detrimental influence of oxide films on the fatigue behaviour of an aluminium—lithium alloy. In: Sanders Jr., T.H., Starke Jr., E.A. (Eds.), Proceedings of the Fifth International Conference on Aluminium—Lithium Alloys, vol. 2. Materials and Component Engineering Publications Ltd., Birmingham, UK, pp. 1053—1060.

Talia, J.E., Eftekhari, A., Mazumdar, P.K., 1995. Influence of surface condition on the fatigue of an aluminum—lithium alloy (2090-T3). Mater. Sci. Eng. A, A199, L3—L6.

Tan, C.Y., Zheng, Z.Q., Liang, S.Q., 1994. The ageing behaviour and tensile properties of Al—Li alloy containing Sc. In: Sanders Jr., T.H., Starke Jr., E.A. (Eds.), Proceedings of the Fourth International Conference on Aluminum Alloys: Their Physical and Mechanical Properties, vol. 2. Georgia Institute of Technology, Atlanta, GA, USA, pp. 329—333.

Tan, M.J., Koh, L.H., Khor, F.Y., Boey, C., Murakoshi, Y., Sano, T., 1993. Discontinuous reinforcements in extruded aluminium—lithium matrix composites. J. Mater. Process. Tech. 37, 391—403.

Tan, M.J., Koh, L.H., Khor, K.A., Murakoshi, Y., Sano, T., 1995. Heat treatments in aluminium—lithium composites extrusion. J. Mater. Process. Tech. 48, 747—755.

Tarasenko, V., Kolobnev, N.I., Khokhlatova, L.B., 2008. Phase composition and mechanical properties of alloys of the Al—Mg—Li—Me system. Met. Sci. Heat Treat. 50, 80—82.

Tatsuo Sato, T., Akihiko Kamio, A., 1994. Effects of additional elements on δ' phase stability in Al—Li alloys. In: Sanders Jr., T.H., Starke Jr., E.A. (Eds.), Proceedings of the Fourth International Conference on Aluminum Alloys: Their Physical and Mechanical Properties, vol. 2. Georgia Institute of Technology, Atlanta, GA, USA, pp. 169—176.

Tchitembo Goma, F.A., Larouche, D., Bois-Brochu, A., Blais, C., Boselli, J., Brochu, M., 2012. Fatigue crack growth behavior of 2099-T83 extrusions in two different environments. In: Weiland, H., Rollett, A.D., Cassada, W.A. (Eds.), Proceedings of the 13th International Conference on Aluminum Alloys (ICAA13). The Minerals, Metals and Materials Society (TMS) and John Wiley & Sons, Hoboken, NJ, USA, pp. 517—522.

Tempus, G., Scharf, G., Calles, W., 1987. Influence of extrusion process parameters on the mechanical properties of Al—Li extrusions. In: Champier, G., Dubost, B., Miannay, D., Sabetay, L. (Eds.), Proceedings of the Fourth International Conference on Aluminium—Lithium Alloys, J. Phys., 48, C3, 187—C3, 194.

Tempus, G., Calles, W., Scharf, G., 1991. Influence of extrusion process parameters and texture on mechanical properties of Al—Li extrusions. Mater. Sci. Technol. 7, 937—945.

Terlinde, G., Sauer, D., Fischer, G., Smith, A., 1992. Development of Al—Li forgings and extrusions for aerospace applications. In: Peters, M., Winkler, P.J. (Eds.), Aluminium—Lithium Alloys VI, vol. 2. Deutsche Gesellschaft für Metallkunde, Frankfurt, Germany, pp. 927—932.

Terrones, L.A.H., Monteiro, S.N., 2007. Composite precipitates in a commercial Al—Li—Cu—Mg—Zr alloy. Mater. Charact. 58, 156—161.

Thakur, C., Balasubramaniam, C., 1998. Surface film characteristics of Al—Li—Cu—Mg alloys in 0.1N NaOH. Bull. Mater. Sci. 21, 485—492.

Thakur, C., Balasubramaniam, R., 1997. Effect of electrolyte temperature on the polarization characteristics of an Al–Li–Cu–Mg alloy in NaOH. Bull. Mater. Sci. 20, 125–133.

Thomas, J.P., Fallavier, M., Beurton, G., Berlioux, G., 1987. Nuclear reaction analysis of hydrogen surface contamination of Al(Li) alloy. In: Champier, G., Dubost, B., Miannay, D., Sabetay, L. (Eds.), Proceedings of the Fourth International Conference on Aluminium–Lithium Alloys, J. Phys., 48, C3, 527–C3, 534.

Thompson, G.E., Noble, B., 1973. Precipitation characteristics of Al–Li alloys containing Mg. J. Inst. Met. 101, 111–115.

Thorne, N., Dubus, A., Lang, J.M., Degreve, F., Meyer, P., 1987. SIMS determination of the surface lithium depletion zone in Al–Li alloys by quantitative image analysis. In: Champier, G., Dubost, B., Miannay, D., Sabetay, L., (Eds.), Proceedings of the Fourth International Conference on Aluminium–Lithium Alloys, J. Phys., 48, C3, 521–C3, 526.

Tian, B., Schöberl, T., Pink, E., Fratzla, P., 2000. Local mechanical properties of tensile-deformed Al-8.4 at.%Li alloys examined by nanoindentation under an atomic force microscope. Scr. Mater. 43, 15–20.

Tian, B.H., Prem, P., 1997. Serrated flow and related microstructures in an Al-8.4 at.% Li alloy. J. Mater. Sci. Lett. 37, 1355–1361.

Tian, B.H., Zhang, B.H., Chen, C.Q., 1997a. The cryogenic properties of Al–Li single crystals. J. Mater. Sci. Lett. 16, 332–334.

Tian, B.H., Li, H.X., Zhang, Y.G., Chen, C.Q., 1997b. Types of serrations in Al–Li single crystals. J. Mater. Sci. Lett. 16, 611–612.

Ting, E.Y., Ward, B., Williams, T., 1989. Optimization of SPF using TMP and deformation control in 2090 Al–Li alloy. In: Sanders Jr., T.H., Starke Jr., E.A. (Eds.), Proceedings of the Fifth International Conference on Aluminium–Lithium Alloys, vol. 1. Materials and Component Engineering Publications Ltd., Birmingham, UK, pp. 189–200.

Tintillier, R., Yang, H.S., Raganathan, N., Petit, J., 1987. Near threshold fatigue crack growth in a 8090 lithium containing Al alloy. In: Champier, G., Dubost, B., Miannay, D., Sabetay, L. (Eds.), Proceedings of the Fourth International Conference on Aluminium–Lithium Alloys, J. Phys., 48, C3, 777–C3, 785.

Tintillier, R., Gudladt, H.J., Gerold, V., Petit, J., 1989. Near threshold fatigue growth in high purity binary Al–Li single crystals. In: Sanders Jr., T.H., Starke Jr., E.A. (Eds.), Proceedings of the Fifth International Conference on Aluminium–Lithium Alloys, vol. 2. Materials and Component Engineering Publications Ltd., Birmingham, UK, pp. 1135–1146.

Tiryakioglu, M., Staley, J.T., 2003. Physical metallurgy and the effect of alloying additions in aluminum alloys. In: Totten, G.E., Mackenzie, D.S. (Eds.), Handbook of Aluminum, Vol. 1, Physical Metallurgy and Processes. Marcel Dekker Inc., New York, NY, USA, pp. 82–210.

Tobler, R.L., Han, J.K., Ma, L., Walsh, R.P., Reed, R.P., 1989. Tensile fracture and fatigue properties of notched aluminium alloy sheets at liquid nitrogen temperature. In: Sanders Jr., T.H., Starke Jr., E.A. (Eds.), Proceedings of the Fifth International Conference on Aluminium–Lithium Alloys, vol. 2. Materials and Component Engineering Publications Ltd., Birmingham, UK, pp. 1115–1124.

Tong, C.H., Yao, I.G., Nieh, C.Y., Chang, C.P., Hsu, S.E., 1987. Castability of Al–Li–Mg and Al–Li–Cu–Mg alloys. In: Champier, G., Dubost, B., Miannay, D., Sabetay, L. (Eds.), Proceedings of the Fourth International Conference on Aluminium–Lithium Alloys, J. Phys., 48, C3, 117–C3, 122.

Toriyama, T., Mazumdar, P.K., Talia, J.E., 1989. A study of the fatigue behaviour in scratched samples of Al–Li (2090-T3) alloys. In: Sanders Jr., T.H., Starke Jr., E.A. (Eds.),

Proceedings of the Fifth International Conference on Aluminium–Lithium Alloys, vol. 2. Materials and Component Engineering Publications Ltd., Birmingham, UK, pp. 1077–1086.

Toshiro, K., Heonjoo, K., 1999. Toughness and microstructural parameters in thermomechanically processed 2091 Ai–Li alloy. Met. Mater. 5, 303–308.

Tosten, M.H., Vasudevan, A.K., Howell, P.R., 1986a. Microstructural development in Al-2%Li-3%Cu alloy. In: Baker, C., Gregson, P.J., Harris, S.J., Peel, C.J. (Eds.), Proceedings of the Third International Conference on Aluminium–Lithium Alloys. The Institute of Metals, London, UK, pp. 483–489.

Tosten, M.H., Vasudevan, A.K., Howell, P.R., 1986b. Grain boundary precipitation in Al–Li–Cu alloys. In: Baker, C., Gregson, P.J., Harris, S.J., Peel, C.J. (Eds.), Proceedings of the Third International Conference on Aluminium–Lithium Alloys. The Institute of Metals, London, UK, pp. 490–495.

Tosten, M.H., Ramani, A., Bartges, C.W., Michel, D.J., Ryba, E., Howell, P.R., 1989. On the origin and nature of microcrystalline regions in an Al–Li–Cu–Zr alloy. Scr. Metall. 23, 829–834.

Toyoshima, M., Kurino, S., Ohara, K., Wakasaki, O., 1994. Behaviour of refractory materials for Al–Li alloy. In: Sanders Jr., T.H., Starke Jr., E.A. (Eds.), Proceedings of the Fourth International Conference on Aluminum Alloys: Their Physical and Mechanical Properties, vol. 2. Georgia Institute of Technology, Atlanta, GA, USA, pp. 114–121.

Triolo, A., 2005. Early and late stages of demixing of a commercial Al–Li Alloy. J. Mater. Sci. 37, 1207–1213.

Tsao, C.S., Lin, T.L., Yu, M.S., 1999. An improved small-angle X-ray scattering analysis of δ' precipitation in Al–Li alloy with hard-sphere interaction. Scr. Mater. 41, 81–87.

Tuegel, E.J., Vasey-Glandon, V.M., Pruitt, M.O., Sankaran, K.K., 1989. Forming of aluminium–lithium sheet for fighter aircraft applications. In: Sanders Jr., T.H., Starke Jr., E.A. (Eds.), Proceedings of the Fifth International Conference on Aluminium–Lithium Alloys, vol. 3. Materials and Component Engineering Publications Ltd., Birmingham, UK, pp. 1597–1606.

Uenishi, K., Kobayashi, K.F., 1994. Improvement in wear property of aluminium by laser cladding of Al_3Ti /ceramics composite layers. In: Sanders Jr., T.H., Starke Jr., E.A. (Eds.), Proceedings of the Fourth International Conference on Aluminum Alloys: Their Physical and Mechanical Properties, vol. 2. Georgia Institute of Technology, Atlanta, GA, USA, pp. 714–720.

Urena, A., Gomez de Salazar, J.M., Quinones, J., Martin, J.J., 1996. TEM characterization of diffusion bonding of superplastic 8090 Al–Li alloy. Scr. Mater. 34, 617–623.

Vadon, A., Larulle, C., Heizmann, J.J., Jeamart, P., 1987. Quantitative determination of the texture of Al–Li alloys. In: Champier, G., Dubost, B., Miannay, D., Sabetay, L. (Eds.), Proceedings of the Fourth International Conference on Aluminium–Lithium Alloys, J. Phys., 48, C3, 621–C3, 628.

Valentine, M.G., Sanders Jr., T.H., 1989. The influence of temperature and composition on the distribution of δ' (Al_3Li). In: Sanders Jr., T.H., Starke Jr., E.A. (Eds.), Proceedings of the Fifth International Conference on Aluminium–Lithium Alloys, vol. 2. Materials and Component Engineering Publications Ltd., Birmingham, UK, pp. 575–584.

Van Smaalen, S., Meetsma, A., De Boer, J.L., Bronsveld, P.M., 1990. Refinement of the crystal structure of hexagonal Al_2CuLi. J. Solid State Chem. 85, 293–298.

Vasudevan, A.K., Doherty, R.D., 1987. Grain boundary ductile fracture in precipitation hardened aluminum alloys. Acta Metall. 35, 1193–1218.

Vasudevan, A.K., Suresh, S., 1985a. Lithium-containing aluminium alloys cyclic fracture. Metall. Mater. Trans. A, 16A, 475−477.

Vasudevan, A.K., Suresh, S., 1985b. Microstructural effects on quasi-static fracture mechanisms in Al−Li alloys: the role of crack geometry. Mater. Sci. Eng. 72, 37−49.

Vasudevan, A.K., Miller, A.C., Kersker, M.M., 1984a. Contribution of Na segregation to fracture behaviour of Al-11.4 at. % Li alloys. In: Sanders Jr., T.H., Starke Jr., E.A. (Eds.), Proceedings of the Second International Conference on Aluminum−Lithium Alloys. The Metallurgical Society of AIME, Warrendale, PA, USA, pp. 181−199.

Vasudevan, A.K., Bretz, P.E., Miller, A.C., Suresh, S., 1984b. Fatigue crack growth behaviour of aluminium alloy 2020 (Al−Cu−Li−Mn−Cd). Mater. Sci. Eng. 64, 113−122.

Vasudevan, A.K., Ziman, P.R., Jha, S.C., Sanders Jr., T.H., 1986a. Stress corrosion resistance of Al−Cu−Li−Zr alloys. In: Baker, C., Gregson, P.J., Harris, S.J., Peel, C.J. (Eds.), Proceedings of the Third International Conference on Aluminium−Lithium Alloys. The Institute of Metals, London, UK, pp. 303−309.

Vasudevan, A.K., Ludwiczak, E.A., Baumann, S.F., Howell, P.R., Doherty, R.D., Kersker, M. M., 1986b. Grain boundary fracture in Al−Li alloys. Mater. Sci. Technol. 2, 1205−1209.

Vasudevan, A.K., Fricke Jr., W.G., Malcolm, R.C., Bucci, R.J., Przystupa, M.A., Barlat, F., 1988. On through thickness crystallographic texture gradient in Al−Li−Cu−Zr alloy. Metall. Trans. A, 19A, 731−732.

Vecchio, K.S., Williams, D.B., 1987. Convergent electron beam diffraction study of Al_3Zr in Al−Zr and Al−Li−Zr alloys. Acta Metall. 35, 2959−2970.

Vecchio, K.S., Williams, D.B., 1988a. Convergent beam electron diffraction analysis of the (Al_2CuLi) phase in Al−Li−Cu alloys. Metall. Trans. A, 19A, 2885−2891.

Vecchio, K.S., Williams, D.B., 1988b. A non-icosahedral T_2 (Al_6Li_3Cu) phase. Philos. Mag. B. 57, 535−546.

Vecchio, K.S., Williams, D.B., 1988c. The apparent 'five-fold' nature of large T2 (Al6Li3Cu) crystals. Metall. Trans. A, 19A, 2875−2884.

Veldman, N.L.M., Huynh, C.N., Yao, J.Y., 1994. Sedimentation and directional solidification of a particulate $Al−Al_2O_3$ metal matrix composite. In: Sanders Jr., T.H., Starke Jr., E.A. (Eds.), Proceedings of the Fourth International Conference on Aluminum Alloys: Their Physical and Mechanical Properties, vol. 2. Georgia Institute of Technology, Atlanta, GA, USA, pp. 598−605.

Venables, D., Christodoulou, L., Pickens, J.R., 1983. On the $\delta' \rightarrow \delta$ transformation in Al−Li alloys. Scr. Metall. 17, 1263−1268.

Venkateswara Rao, K.T., Ritchie, R.O., 1988a. On the behavior of small fatigue cracks in commercial aluminum−lithium alloys. Eng. Fract. Mech. 31, 623−635.

Venkateswara Rao, K.T., Ritchie, R.O., 1988b. Effect of prolonged high-temperature exposure on the fatigue and fracture behavior of aluminum−lithium alloy 2090. Mater. Sci. Eng. A, A100, 23−30.

Venkateswara Rao, K.T., Ritchie, R.O., 1988c. Mechanisms for the retardation of fatigue cracks following single tensile overloads: behavior in aluminum−lithium alloys. Acta Metall. 36, 2849−2862.

Venkateswara Rao, K.T., Ritchie, R.O., 1989a. Mechanical properties of Al−Li alloys. Part I: fracture toughness and microstructure. Mater. Sci. Technol. 5, 882−895.

Venkateswara Rao, K.T., Ritchie, R.O., 1989b. Mechanical properties of Al−Li alloys. Part II: fatigue crack propagation. Mater. Sci. Technol. 5, 896−907.

Venkateswara Rao, K.T., Ritchie, R.O., 1990. Mechanisms influencing the cryogenic fracture-toughness behaviour of aluminium−lithium alloys. Acta Metall. 38, 2309−2326.

Venkateswara Rao, K.T., Ritchie, R.O., 1991. Influence of extrinsic crack deflection and delamination mechanisms on the cryogenic toughness of aluminium–lithium alloy 2090: behaviour in plate (T81) vs. sheet (T83) material. In: Sanders Jr., T.H., Starke Jr., E.A. (Eds.), Proceedings of the Fifth International Conference on Aluminium–Lithium Alloys, vol. 3. Materials and Component Engineering Publications Ltd., Birmingham, UK, pp. 1501–1512.

Venkateswara Rao, K.T., Ritchie, R.O., 1992. Fatigue in aluminium–lithium alloys. Int. Mater. Rev. 37, 153–185.

Venkateswara Rao, K.T., Yu, W., Ritchie, R.O., 1988a. On the behavior of small fatigue cracks in commercial aluminum–lithium alloys. Eng. Fract. Mech. 31, 623–635.

Venkateswara Rao, K.T., Yu, W., Ritchie, R.O., 1988b. Fatigue crack propagation in aluminium–lithium alloy 2090: part I. Long crack behaviour. Metall. Trans. A, 19A, 549–561.

Venkateswara Rao, K.T., Yu, W., Ritchie, R.O., 1988c. Fatigue crack propagation in aluminium–lithium alloy 2090: part II. Small crack behaviour. Metall. Trans. A, 19A, 563–569.

Venkateswara Rao, K.T., Hayashigatani, H.F., Yu, W., Ritchie, R.O., 1988d. On the fracture toughness of aluminum–lithium alloy 2090-T8E41 at ambient and cryogenic temperatures. Scr. Metall. 22, 93–98.

Venkateswara Rao, K.T., Piascik, R.S., Gangloff, R.P., Ritchie, R.O., 1989. Fatigue crack propagation in aluminum–lithium alloys. In: Sanders Jr., T.H., Starke Jr., E.A. (Eds.), Proceedings of the Fifth International Conference on Aluminium–Lithium Alloys, vol. 2. Materials and Component Engineering Publications Ltd., Birmingham, UK, pp. 955–971.

Venkateswara Rao, K.T., Bucci, R.J., Ritchie, R.O., 1990. On the micromechanisms of fatigue-crack propagation in aluminium–lithium alloys: sheet versus plate material. In: Kitagawa, H., Tanaka, T. (Eds.), 'Fatigue 90', Proceedings of the Fourth International Conference on Fatigue and Fatigue Thresholds, vol. 2. Engineering Materials Advisory Services, Warley, UK, pp. 963–970.

Venkateswara Rao, K.T, Bucci, R.J., Jata, K.V., Ritchie, R.O., 1991. A comparison of fatigue-crack propagation behaviour in sheet and plate aluminum–lithium alloys. Mater. Sci. Eng. A, A141, 39–48.

Verdu, C., Gentzbittel, J.M., Fougeres, R., 1989. A microstructural study of intergranular fracture in a 8090 Al–Li alloy. In: Sanders Jr., T.H., Starke Jr., E.A. (Eds.), Proceedings of the Fifth International Conference on Aluminium–Lithium Alloys, vol. 2. Materials and Component Engineering Publications Ltd., Birmingham, UK, pp. 899–908.

Verzasconi, S.L., Morris Jr., J.W., 1989. Cryogenic mechanical properties of low density superplastically formable Al–Li alloys. In: Sanders Jr., T.H., Starke Jr., E.A. (Eds.), Proceedings of the Fifth International Conference on Aluminium–Lithium Alloys, vol. 3. Materials and Component Engineering Publications Ltd., Birmingham, UK, pp. 1523–1532.

Vietz, J.T., Polmear, I.J., 1966. The influence of small additions of silver on the ageing of aluminium alloys: observations on Al–Cu–Mg alloys. J. Inst. Met. 94, 410.

Vijaya Singh, Chakravorty, C.R., 1989. Melting and casting of Al–Li alloys—a review. Science and Technology of Aluminium–Lithium Alloys, Conference Proceedings. Hindustan Aeronautics Limited, Bangalore, India, pp. 83–91.

Vijaya Singh, Prasad, K.S., Eswara Prasad, N., Singh, R.K., Gokhale, A., Banerjee, D., et al., 1991. Microstructure and mechanical properties of an Al–Li alloy developed at DMRL. DMRL Technical Report No. 91127, Defence Metallurgical Research Laboratory, Hyderabad.

Villars, P., Calvert, L.D., 1985. Pearson's Handbook of Crystallographic Data for Intermetallic Phases, vol. 2. ASM International, Materials Park, OH, USA.

Vives, C., Bas, J., Cans, Y., 1987. Grain refining in Al–Li alloys by electromagnetic stirring. In: Champier, G., Dubost, B., Miannay, D., Sabetay, L. (Eds.), Proceedings of the Fourth International Conference on Aluminium–Lithium Alloys, J. Phys., 48, C3, 109–C3, 115.

Wack, B., Tourabi, A., 1995. A new method to quantify the Portevin-Le Chatelier instabilities: application to aluminium–lithium alloys. Mater. Sci. Eng. A, A196, 79–87.

Wadsworth, D., Crooks, D., Lewis, R.E., Vidoz, A.E., 1986b. Microstructural evaluation of arc-melted Al–Li–Be alloys. J. Mater. Sci. 21, 3843–3849.

Wadsworth, J., Pelton, A.R., 1984. Superplastic behavior of a powder-source aluminium–lithium based alloy. Scr. Metall. 18, 387–392.

Wadsworth, J., Palmer, I.G., Crooks, D.D., 1983. Superplasticity in Al–Li based alloys. Scr. Metall. 17, 347–352.

Wadsworth, J., Palmer, I.G., Crookes, D.D., Lewis, R.E., 1984. Superplastic behaviour of aluminium–lithium alloys. In: Sanders Jr., T.H., Starke Jr., E.A. (Eds.), Proceedings of the Second International Conference on Aluminum–Lithium Alloys II. The Metallurgical Society of AIME, Warrendale, PA, USA, pp. 111–135.

Wadsworth, J., Pelton, A.R., Lewis, R.E., 1985. Superplastic Al–Cu–Li–Mg–Zr alloys. Metall. Mat. Trans. A, 16A, 2319–2332.

Wadsworth, J., Heusahall, C.A., Nieh, T.E., 1986. Superplastic aluminium–lithium alloys. In: Baker, C., Gregson, P.J., Harris, S.J., Peel, C.J. (Eds.), Proceedings of the Third International Conference on Aluminium–Lithium Alloys. The Institute of Metals, London, UK, pp. 199–212.

Waheed, A., 1996. Pinning of subgrain boundaries by Al_3Zr dispersoids during annealing in Al–Li commercial alloys. J. Mater. Sci. Lett. 16, 1643–1646.

Waheed, A., Lorimer, G.W., 1997. Dispersoids in Al–Li AA8090 series alloys. J. Mater. Sci. 32, 3341–3347.

Wakasaki, O., Kurino, S., Toyoshima, M., Ohara, K., 1994. Hydrogen solubility in Li content molten Al alloys. In: Sanders Jr, T.H., Starke Jr, E.A. (Eds.), Proceedings of the Fourth International Conference on Aluminum Alloys: Their Physical and Mechanical Properties, vol. 2. Georgia Institute of Technology, Atlanta, GA, USA, pp. 130–136.

Walsh, J.A., Jata, K.V., Starke Jr., E.A., 1989. The influence of Mn dispersoid content and stress state on ductile fracture of 2134 type Al alloys. Acta Metall. 37, 2861–2871.

Wan, C.C., Smalle, H., Carter, R.V., 1989. Tensile properties of 8090 Al–Li alloy at cryogenic and elevated temperatures. In: Sanders Jr., T.H., Starke Jr., E.A. (Eds.), Proceedings of the Fifth International Conference on Aluminium–Lithium Alloys, vol. 3. Materials and Component Engineering Publications Ltd., Birmingham, UK, pp. 1553–1564.

Wang, S.C., Starink, M.J., 2005. Review of precipitation in Al–Cu–Mg(-Li) alloys. Int. Mater. Rev. 50, 193–215.

Wang, Y., Ridley, N., Lorimer, G.W., 1994. The microstructure of a rapidly solidified Al–Fe–V–Si alloy produced by spray forming. In: Sanders Jr, T.H., Starke Jr, E.A. (Eds.), Proceedings of the Fourth International Conference on Aluminum Alloys: Their Physical and Mechanical Properties, vol. 2. Georgia Institute of Technology, Atlanta, GA, USA, pp. 766–773.

Wang, Z.F., Zhu, Z.Y., Zhang, Y., Ke, W., 1992. Stress corrosion cracking of an Al–Li alloy. Metall. Trans. A, 23A, 3337–3341.

Wang, Z.M., Shiflet, G.J., 1994. δ′ Growth on dislocations in Al-2.27 wt% Li. In: Sanders Jr, T.H., Starke Jr, E.A. (Eds.), Proceedings of the Fourth International Conference on Aluminum Alloys: Their Physical and Mechanical Properties, vol. 2. Georgia Institute of Technology, Atlanta, GA, USA, pp. 145–152.

Wang, Z.M., Shiflet, G.J., 1998. Growth of δ' on dislocations in a dilute Al—Li alloy. Metall. Mater. Trans. A, 29A, 2073—2085.

Wanhill, R.J.H., 1979. Gust spectrum fatigue crack propagation in candidate skin materials. Fatigue Eng. Mater. Struct. 1, 5—19.

Wanhill, R.J.H., 1994. Flight simulation fatigue crack growth testing of aluminium alloys. Specific issues and guidelines. Int. J. Fatigue 16, 99—110.

Wanhill, R.J.H., Hattenberg, T., 2006. Fractography-based estimation of fatigue crack "initiation" and growth lives in aircraft components. National Aerospace Laboratory NLR Technical Publication NLR-TP-2006-184, Amsterdam, the Netherlands.

Wanhill, R.J.H., Jacobs, F.A., Schijve, J., 1976. Environmental fatigue under gust spectrum loading for sheet and forging aircraft materials. In: Bathgate, R.G. (Ed.), Fatigue Testing and Design, vol. 1. The Society of Environmental Engineers, Buntingford, UK, pp. 8.1—8.33.

Wanhill, R.J.H., Schra, L., 't Hart, W.G.J., 1990. Fracture and fatigue evaluation of damage tolerant Al—Li alloys for aerospace applications. In: Firrao, D. (Ed.), Fracture Behaviour and Design of Materials and Structures: Proceedings of the Eighth European Conference on Fracture, ECF 8, vol. 1. Engineering Materials Advisory Services, Warley, UK, pp. 257—271.

Wanhill, R.J.H., 't Hart, W.G.J., Schra, L., 1991. Flight simulation and constant amplitude fatigue crack growth in aluminium—lithium sheet and plate. Aeronautical Fatigue: Key to Safety and Structural Integrity. Engineering Material Advisory Services, pp.393—430.

Wanhill, R.J.H., Molent, L., Barter, S.A., 2013. Fracture mechanics in aircraft failure analysis: uses and limitations. Eng. Fail. Anal. [In Press].

Warner, T.J., Newcomb, S.B., Stobbs, W.M., 1994. Stress corrosion crack propagation in aluminium alloy 7150. In: Sanders Jr, T.H., Starke Jr, E.A. (Eds.), Proceedings of the Fourth International Conference on Aluminum Alloys: Their Physical and Mechanical Properties, vol. 2. Georgia Institute of Technology, Atlanta, GA, pp. 88—96.

Warren, C.J., Rioja, R.J., 1989. Forming characteristics and post-formed properties of Al—Li alloys. In: Sanders Jr., T.H., Starke Jr., E.A. (Eds.), Proceedings of the Fifth International Conference on Aluminium—Lithium Alloys, 1. Materials and Component Engineering Publications Ltd., Birmingham, UK, pp. 417—430.

Watanabe, Y., Toyoshima, M., Itoh, K., 1987. High purity Al—Li master alloy by molten salt electrolysis. In: Champier, G., Dubost, B., Miannay, D., Sabetay, L. (Eds.), Proceedings of the Fourth International Conference on Aluminium—Lithium Alloys, J. Phys., 48, C3, 85—C3, 91.

Waterloo, G., Arnberg, L., Høier, R., 1994. Crystallization of an amorphous $Al_{85}Y_8Ni_5Co_2$ alloy. In: Sanders, T.H., Starke, E.A. (Eds.), Proceedings of the Fourth International Conference on Aluminum Alloys: Their Physical and Mechanical Properties, vol. 2. Georgia Institute of Technology, Atlanta, GA, pp. 746—752.

Watts, B.M., Stowell, M.J., Baikie, B.L., Owen, D.G.E, 1976. Superplasticity in Al—Cu—Zr alloys. Part I: material preparation and properties. Met. Sci. 6, 189—197.

Webster, D., 1980. Toughness and ductility of aluminium—lithium alloys prepared by powder metallurgy and ingot metallurgy. In: Sanders Jr., T.H., Starke Jr., E.A. (Eds.), Proceedings of the First International Conference on Aluminium—Lithium Alloys. The Metallurgical Society of AIME, Warrendale, PA, USA, pp. 205—240.

Webster, D., 1981. Toughness and ductility of aluminium—lithium alloys prepared by powder metallurgy and ingot metallurgy. In: Sanders Jr., T.H., Starke Jr., E.A. (Eds.), Proceedings of the First International Conference on Aluminum—Lithium Alloys. The Metallurgical Society of AIME, Warrendale, PA, USA, pp. 228—240.

Webster, D., 1986. Temperature dependence of toughness in various aluminium–lithium alloys. In: Baker, C., Gregson, P.J., Harris, S.J., Peel, C.J. (Eds.), Proceedings of the Third International Conference on Aluminium–Lithium Alloys. The Institute of Metals, London, UK, pp. 602–609.

Webster, D., 1987a. A technique for improving the toughness of Al–Li powder metallurgy alloys. In: Champier, G., Dubost, B., Miannay, D., Sabetay, L. (Eds.), Proceedings of the Fourth International Conference on Aluminium–Lithium Alloys, J. Phys., 48, C3, 335–C3, 340.

Webster, D., 1987b. The effect of low melting point impurities on aluminium–lithium alloys. In: Champier, G., Dubost, B., Miannay, D., Sabetay, L. (Eds.), Proceedings of the Fourth International Conference Aluminium–Lithium Alloys, J. Phys., 48, C3, 685–C3, 692.

Webster, D., 1987c. The effect of low melting point impurities on the properties of aluminum–lithium alloys. Metall. Mater. Trans. A, 18A, 2187–2193.

Webster, D., 1988. Aluminum–lithium powder metallurgy alloys with improved toughness. Metall. Trans. A, 19A, 603–615.

Webster, D., 1989a. Mechanical properties of an Al-2.3%Li-15%SiC composite with a duplex microstructure. In: Sanders Jr., T.H., Starke Jr., E.A. (Eds.), Proceedings of the Fifth International Conference on Aluminium–Lithium Alloys, vol. 3. Materials and Component Engineering Publications Ltd., Birmingham, UK, pp. 1657–1666.

Webster, D., 1989b. Aluminium–lithium alloys with very high toughness. In: Sanders Jr., T.H., Starke Jr., E.A. (Eds.), Proceedings of the Fifth International Conference on Aluminium–Lithium Alloys, vol. 3. Materials and Component Engineering Publications Ltd., Birmingham, UK, pp. 1677–1684.

Webster, D., 1989c. Effect of alkali metal impurities on the toughness of aluminium–lithium alloys. In: Sanders Jr., T.H., Starke Jr., E.A. (Eds.), Proceedings of the Fifth International Conference on Aluminium–Lithium Alloys, 1. Materials and Component Engineering Publications Ltd., Birmingham, UK, pp. 497–518.

Webster, D., Kirkbride, R., 1986. Mechanical properties and microstructure of Al–Li–Cu–Mg–Zr die forgings. Metall. Trans. A, 17A, 2007–2016.

Webster, D., Wald, G., Cremens, W.S., 1981. Mechanical properties and microstructure of argon atomized aluminium–lithium powder metallurgy alloys. Metall. Trans. A, 12A, 1495–1502.

Wei, B.C., Chen, C.Q., Huang, Z., Zhang, Y.G., Aging behavior of Li containing Al–Zn–Mg–Cu alloys. Mater. Sci. Eng. A, A280, 161–167.

Wei, Y.H., Wang, S.T., 1996. Experimental evidence for spinodal decomposition along with simultaneous ordering in Al-12.7 at% Li alloy. Mater. Lett., 123–127.

Wellstead, P.E., 1979. Introduction to Physical System Modelling. Academic Press Ltd., London, UK.

Welpmann, K., Peters, M., Sanders Jr., T.H., 1986. Age hardening behaviour of DTD XXXA,. In: Baker, C., Gregson, P.J., Harris, S.J., Peel, C.J. (Eds.), Proceedings of the Third International Conference on Aluminium–Lithium Alloys. The Institute of Metals, London, UK, pp. 524–529.

Welpmann, K., Buhl, H., Braun, R., Peters, M., 1987. Mechanical properties and corrosion behaviour of 2091 sheet material. In: Champier, G., Dubost, B., Miannay, D., Sabetay, L. (Eds.), Proceedings of the Fourth International Conference on Aluminium–Lithium Alloys, J. Phys., 48, C3, 677–C3, 684.

Welpmann, K., Lee, Y.T., Peters, M., 1991. Low temperature deformation behaviour of 8090. In: Sanders Jr., T.H., Starke Jr., E.A. (Eds.), Proceedings of the Fifth International

Conference on Aluminium–Lithium Alloys, vol. 3. Materials and Component Engineering Publications Ltd., Birmingham, UK, pp. 1513–1522.

Wen-Jie, L., Qing-Lin, P., Yun-Bin, H., Yun-Chun, L., Xiao-Gang, Z., 2008a. Flow stress behavior of Al–Cu–Li–Zr alloy containing Sc during hot compression deformation. J. Cent. South Univ. Tech. 15, 289–294.

Wen-Jie, L., Qing-Lin, P., Yun-Bin, H., Yun-Chun, L., Xiao-Gang, Z., 2008b. Formation mechanism of gradient-distributed particles and their effects on grain structure in 01420 Al–Li alloy. J. Cent. South Univ. Tech. 17, 659–665.

Wen-Jie, L., Qing-Lin, P., Yun-Bin, H., Yun-Chun, L., Xiao-Gang, Z., 2008c. Superplastic deformation behavior and mechanism of 1420 Al–Li alloy sheets with elongated grains. J. Cent. South Univ. Tech. 15, 147–152.

Wert, J.A., Lumsden, J.B., 1985. Intergranular fracture in an Al–Li–Cu–Mg–Zr alloy. Scr. Metall. 19, 205–209.

Wert, J.A., Ward, A.B., 1985. Use of eddy current conductivity measurements to detect Li depletion from Al–Li alloys. Scr. Metall. 19, 367–370.

Wert, J.A., Wycliffe, P.A., 1985. Correlation between S' precipitation and the Portevin–le Chatelier effect in an Al–Li–Cu–Mg–Zr alloy. Scr. Metall. 19, 463–466.

Whitaker, I.R., McCartney, D.G., 1993. Microstructural characterization of CO_2 laser welds in the Al–Li based alloy 8090. J. Mater. Sci. 28, 5469–5478.

Whitaker, I.R., McCartney, D.G., 1994. Fracture of bead-on-plate CO_2 laser welds in the Al–Li alloy 8090. Scr. Met. Mater. 31, 1717–1722.

White, C.L., 1981. Grain boundary segregation and intergranular failure. In: Sanders Jr., T.H., Starke Jr., E.A. (Eds.), Proceedings of the First International Conference on Aluminum–Lithium Alloys. The Metallurgical Society of AIME, Warrendale, PA, USA, pp. 119–140.

White, J., Miller, W.S., 1987. The effect of FTMT on the grain boundary microstructure of aluminium–lithium–copper–magnesium–zirconium alloys. In: Champier, G., Dubost, B., Miannay, D., Sabetay, L. (Eds.), Proceedings of the Fourth International Conference on Aluminum–Lithium Alloys, J. Phys., 48, C3, 425–C3, 432.

White, J., Miller, W.S., Palmer, I.G., Davis, R., Saini, T.S., 1986. Effect of precipitation on mechanical properties of Al–Li–Cu–Mg–Zr alloy. In: Baker, C., Gregson, P.J., Harris, S. J., Peel, C.J. (Eds.), Proceedings of the Third International Conference on Aluminium–Lithium Alloys III. The Institute of Metals, London, UK, pp. 530–538.

White, J., Hughes, I.R., Willis, T.C., Jordan, R.M., 1987. Metal matrix composites based on aluminium–lithium and silicon carbides. In: Champier, G., Dubost, B., Miannay, D., Sabetay, L. (Eds.), Proceedings of the Fourth International Conference on Aluminium–Lithium Alloys, J. Phys., 48, C3, 347–C3, 356.

White, J., Palmer, I.G., Hughes, I.R., Court, S.A., 1991. Development of metal matrix composite systems based on aluminium–lithium. In: Sanders Jr., T.H., Starke Jr., E.A. (Eds.), Proceedings of the Fifth International Conference on Aluminium–Lithium Alloys, vol. 3. Materials and Component Engineering Publications Ltd., Birmingham, UK, pp. 1635–1646.

Wiiliams, A., 2010. Microstructural Analysis of Aluminium Alloy 2096 as a Function of Heat Treatment (Master of Research Thesis). University of Birmingham, UK.

Williams, D.B., 1981. Microstructural characteristics of Al–Li alloys. In: Sanders Jr., T.H., Starke Jr., E.A. (Eds.), Proceedings of the First International Conference on Aluminium–Lithium Alloys. The Metallurgical Society of AIME, Warrendale, PA, USA, pp. 89–100.

Williams, D.B., 1989. The binary aluminium–lithium phase diagram. In: Sanders Jr., T.H., Starke Jr., E.A. (Eds.), Proceedings of the Fifth International Conference on

Aluminium–Lithium Alloys, vol. 2. Materials and Component Engineering Publications Ltd., Birmingham, UK, pp. 551–564.

Williams, D.B., Edington, J.W., 1974. Microanalysis of Al–Li alloys containing fine δ' (Al$_3$Li) precipitates. Philos. Mag. 30, 1147–1153.

Williams, D.B., Edington, J.W., 1975. The precipitation of δ' (Al$_3$Li) in dilute aluminium–lithium alloys. Met. Sci. J., 529–532.

Williams, D.B., Edington, J.W., 1976. The discontinuous precipitation reaction in dilute Al–Li alloys. Acta Metall. 24, 323–332.

Williams, D.B., Levi-Setti, R., Chabala, J.M., Wang, Y.L., Newbury, D.E., Soni, K.K., 1989. SIMS studies of binary Al–Li alloys. In: Sanders Jr., T.H., Starke Jr., E.A. (Eds.), Proceedings of the Fifth International Conference on Aluminium–Lithium Alloys, vol. 2. Materials and Component Engineering Publications Ltd., Birmingham, UK, pp. 605–612.

Williams, D.B., Hunt, J.A., Soni, K.K., 1994. Quantitative imaging of lithium in aluminium–lithium alloys. In: Sanders Jr., T.H., Starke Jr., E.A. (Eds.), Proceedings of the Fourth International Conference on Aluminum Alloys: Their Physical and Mechanical Properties, vol. 2. Georgia Institute of Technology, Atlanta, GA, USA, pp. 137–144.

Wilson, W.R., Worth, J., Short, E.P., Pygall, C.F., 1987. Recycling of aluminium–lithium process scrap. In: Champier, G., Dubost, B., Miannay, D., Sabetay, L. (Eds.), Proceedings of the Fourth International Conference on Aluminium–Lithium Alloys, J. Phys., 48, C3, 75–C3, 84.

Wilson, W.R., Allan, D.J., Stenzel, O., Lorke, M., Krone, K.-W., Seebauer, C., 1989a. Al–Li scrap recycling by vacuum-distillation. In: Sanders Jr., T.H., Starke Jr., E.A. (Eds.), Proceedings of the Fifth International Conference on Aluminium–Lithium Alloys, vol. 1. Materials and Component Engineering Publications Ltd., Birmingham, UK, pp. 473–498.

Wilson, W.R., Allan, D.J., Dalmijn, W.L., Brassinga, R.D., 1989b. Segregation of Al–Li from mixed alloy turnings. In: Sanders Jr., T.H., Starke Jr., E.A. (Eds.), Proceedings of the Fifth International Conference on Aluminium–Lithium Alloys, vol. 1. Materials and Component Engineering Publications Ltd., Birmingham, UK, pp. 497–518.

Woo, K.D., Kim, S.W., 2002. The mechanical properties and precipitation behavior of an Al–Cu–Li-(In, Be) alloy. J. Mater. Sci. 37, 411–416.

Wright, M.D., Beevers, C.J., 1989. A study of the fatigue and fracture behaviour of a mechanically alloyed Al–Mg–Li alloy In 905 XL. In: Sanders Jr., T.H., Starke Jr., E.A. (Eds.), Proceedings of the Fifth International Conference on Aluminium–Lithium Alloys, vol. 2. Materials and Component Engineering Publications Ltd., Birmingham, UK, pp. 1087–1096.

Wu, H.Y., 2000. Cavitation characteristics of a superplastic 8090 Al alloy during equi-biaxial tensile deformation. Mater. Sci. Eng. A, A291, 1–8.

Wu, X.J., Wallace, W., Raizenne, M.D., Koul, A.K., 1994a. The orientation dependence of fatigue-crack growth in 8090 Al–Li plate. Metall. Mater. Trans. A, 25A, 575–582.

Wu, X.J., Koul, A.K., Wallace, W., 1994c. The influence of microstructure on moist-air fatigue crack growth in 8090 aluminium–lithium alloy. In: Sanders Jr., T.H., Starke Jr., E.A. (Eds.), Proceedings of the Fourth International Conference on Aluminum Alloys: Their Physical and Mechanical Properties, vol. 2. Georgia Institute of Technology, Atlanta, GA, USA, pp. 529–533.

Wu, X.J., Wallace, W., Koul, A.K., Raizenne, M.D., 1995. Near-threshold fatigue crack growth in 8090 Al–Li alloy. Metall. Mater. Trans. A, 26A, 2973–2982.

Wu, Y.E., Lo, Y.L., 2002. Surface protection for AA8090 aluminum alloy by diffusion bonding. Theor. Appl. Fract. Mech. 38, 71–79.

Wu, Y.L., Quing, J., Li, Y.W., Lu, Z., Ouyang, R.G., Liu, B.C., et al., 1994a. Effect of Ce addition on corrosion behaviour of Al−Li alloy 1430. In: Sanders Jr., T.H., Starke Jr., E.A. (Eds.), Proceedings of the Fourth International Conference on Aluminum Alloys: Their Physical and Mechanical Properties, vol. 2. Georgia Institute of Technology, Atlanta, GA, USA, pp. 504−510.

Wu, Y.L., Qiang, J., Li, Y.W., Li, C.Y., 1994b. Investigation of microstructure and properties of advanced Al−Zn−Mg−Cu alloy. In: Sanders Jr., T.H., Starke Jr., E.A. (Eds.), Proceedings of the Fourth International Conference on Aluminum Alloys: Their Physical and Mechanical Properties, vol. 2. Georgia Institute of Technology, Atlanta, GA, USA, pp. 18−23.

Xiao, Y., Bompard, P., 1987. Low cycle fatigue and fatigue crack growth in Al−Li, Al−Li−Zr and 8090 alloys. In: Champier, G., Dubost, B., Miannay, D., Sabetay, L. (Eds.), Proceedings of the Fourth International Conference on Aluminium−Lithium Alloys, J. Phys., 48, C3, 737−C3, 745.

Xiaoxin, X., Martin, J.B., 1987. The effect of stretch upon S-phase distribution and dynamic recovery behaviour of an Al−Li−Cu−Mg alloy. In: Champier, G., Dubost, B., Miannay, D., Sabetay, L. (Eds.), Proceedings of the Fourth International Conference on Aluminium−Lithium Alloys, J. Phys., 48, C3, 433−C3, 438.

Xiaoxin, X., Martin, J.B., 1989. The micromechanisms of dynamic recovery in Al−Li−Cu−Mg−Zr alloys 8090-8091. In: Sanders Jr., T.H., Starke Jr., E.A. (Eds.), Proceedings of the Fifth International Conference on Aluminium−Lithium Alloys, vol. 1. Materials and Component Engineering Publications Ltd., Birmingham, UK, pp. 315−324.

Xiaoxin, X., Martin, J.W., 1992. High-temperature deformation of Al−Li−Cu−Mg−Zr alloys 8090 and 8091. J. Mater. Sci. 27, 592−598.

Xu, L., Wang, Y.B., Zhang, Y., Wang, Z.G., Hu, Q.Z., 1991a. Fatigue and fracture behavior of an aluminum−lithium alloy 8090-T6 at ambient and cryogenic temperature. Metall. Trans. A, 22A, 1723−1729.

Xu, Y., Wang, X., Yan, Z., Li, J., 2011. Corrosion properties of light-weight and high-strength 2195 Al−Li alloy. Chin. J. Aeronaut. 24, 681−686.

Xu, Y.B., Wang, Z.G., Zhang, Y., Zhao, H.H., Hu, Z.Q., 1989. Fatigue behaviour of aluminium−lithium alloys at ambient and cryogenic temperatures. In: Sanders Jr., T.H., Starke Jr., E.A. (Eds.), Proceedings of the Fifth International Conference on Aluminium−Lithium Alloys, vol. 2. Materials and Component Engineering Publications Ltd., Birmingham, UK, pp. 1147−1154.

Xu, Y.B., Wang, X., Wang, Z.G., Luo, L.M., Bai, Y.L., 1990. Localised shear deformation of an aluminium−lithium 8090 alloy during low cycle fatigue. Scr. Metall. 25, 1149−1154.

Xu, Y.B., Wang, L., Wang, Z.G., Zhang, Y., Hu, Z.Q., 1991b. Localized shear deformation of an aluminum−lithium 8090 alloy during low cycle fatigue. Scr. Met. Mater. 25, 1149−1154.

Xun, Y., Tan, M.J., Liew, K.M., 2005. EBSD characterization of cavitation during superplastic deformation of Al−Li alloy. J. Mater. Process. Tech. 162 & 163, 429−434.

Yamamoto, A., Tsubakino, H., Nozato, R., 1994. Precipitation in Al−Li−Zn ternary alloys. In: Sanders Jr., T.H., Starke Jr., E.A. (Eds.), Proceedings of the Fourth International Conference on Aluminum Alloys: Their Physical and Mechanical Properties, vol. 2. Georgia Institute of Technology, Atlanta, GA, USA, pp. 191−198.

Yan, Y.D., Tang, H., Zhang, M.L., Xue, Y., Han, W., Cao, D.X., et al., 2012. Extraction of europium and electrodeposition of Al−Li−Eu alloy from Eu_2O_3 assisted by $AlCl_3$ in LiCl−KCl melt. Electrochim. Acta 59, 531−537.

Yang, Y., Tan, G.Y., Chen, P.X., Zhang, Q.M., 2012. Effects of different aging statuses and strain rate on the adiabatic shear susceptibility of 2195 aluminum—lithium alloy. Mater. Sci. Eng. A, A546, 279—283.

Yao, D.P., Hu, Z.Q., Li, Y.Y., Zhang, Y., Shi, C.X., 1989a. The microstructure of rapidly solidified Al—Li—Cu—Zr alloy by laser glazing. In: Sanders Jr., T.H., Starke Jr., E.A. (Eds.), Proceedings of the Fifth International Conference on Aluminium—Lithium Alloys, vol. 1. Materials and Component Engineering Publications Ltd., Birmingham, UK, pp. 105—112.

Yao, D.P., Li, Y.Y., Hu, Z.Q., Zhang, Y., Shi, C.X., 1989b. Microstructures and properties of Al—Li—Cu—Zr alloy at cryogenic temperature. In: Sanders Jr., T.H., Starke Jr., E.A. (Eds.), Proceedings of the Fifth International Conference on Aluminium—Lithium Alloys, vol. 3. Materials and Component Engineering Publications Ltd., Birmingham, UK, pp. 1533—1542.

Yao, J.Y., Schmidt, R., Thorpe, W.R., 1994. Structural quality of squeeze castings of Al alloy AA603 and an Al_2O_3-603 Al composite. In: Sanders Jr., T.H., Starke Jr., E.A. (Eds.), Proceedings of the Fourth International Conference on Aluminum Alloys: Their Physical and Mechanical Properties, vol. 2. Georgia Institute of Technology, Atlanta, GA, USA, pp. 559—566.

Ye, L., Zhang, X., Zheng, D., Liu, S., Tang, J., 2009. Superplastic behavior of an Al—Mg—Li alloy. J. Alloys Compd. 487, 109—115.

Yeon, C.Y., Jeong, S.J., Ho, I.L., 1994. Effects of deformation condition on hot workability and microstructure evolution of SiC whisker reinforced AA2124 matrix composite. In: Sanders Jr., T.H., Starke Jr., E.A. (Eds.), Proceedings of the Fourth International Conference on Aluminum Alloys: Their Physical and Mechanical Properties, vol. 2. Georgia Institute of Technology, Atlanta, GA, USA, pp. 653—660.

Ying, J.K., Ohashi, T., 1999. Precipitation structures and mechanical properties of Al—Li—Zr alloy containing V. Scr. Mater. 41, 137—141.

Yoder, G.R., Pao, P.S., Imam, M.A., Cooley, L.A., 1989. Micromechanisms of fatigue fracture in Al—Li alloy 2090. In: Sanders Jr., T.H., Starke Jr., E.A. (Eds.), Proceedings of the Fifth International Conference on Aluminium—Lithium Alloys, vol. 2. Materials and Component Engineering Publications Ltd., Birmingham, UK, pp. 1033—1041.

Yoshida, H., Hirano, S., Baba, Y., Tsuzuku, T., Takahashi, A., 1987. The effect of grain boundary precipitation on the superplasticity of Al—Li alloys. In: Champier, G., Dubost, B., Miannay, D., Sabetay, L. (Eds.), Proceedings of the Fourth International Conference on Aluminium—Lithium Alloys, J. Phys., 48, C3, 269—C3, 276.

Yoshida, H., Tanakah, H., Suchida, T., Tsuzuku, T., Takahashi, A., 1989. Superplasticity of Al—Li alloys, 1, 179—188.

Yoshimura, R., Konno, T.J., Abe, E., Hiraga, K., 2003a. Transmission electron microscopy study of the early stage of precipitates in aged Al—Li—Cu alloys. Acta Mater. 51, 2891—2903.

Yoshimura, R., Konno, T.J., Abe, E., Hiraga, K., 2003b. Transmission electron microscopy study of the evolution of precipitates in aged Al—Li—Cu alloys: the θ' and T_1 phases. Acta Mater. 51, 4251—4266.

Yu, G.F., Zhang, S.Q., Feng, Y.S., Chai, S.C., 1991. The effects of aging on the properties of a rapidly solidified Al—Li—Mg—Zr alloy. Mater. Sci. Eng. A, A133, 274—278.

Yuan, G.S., Zhao, Z.Y., Zhu, X.D., Kuang, J.P., Liu, Y., Xing, Z.J., 1991. Structure and mechanical properties of rapidly solidified Al—Li—Cu—Mg alloys containing minor zirconium and rare earths. Mater. Sci. Eng. A, A134, 1179—1181.

Yuryev, V.A., Baranov, Yu.V., Stolyarov, V.V., Shulga, V.A., Kostina, I.V., 2008. Effect of electroplastic processing on the structure of aluminum—lithium alloy 1463. Bull. Russ. Acad. Sci. 28, 1248—1250.

Yusheng, C., Ziyong, Z., Sue, L., Wei, K., 1996. The corrosion behaviour and mechanisms of 1420 Al−Li alloy. Scr. Mater. 34, 781−786.

Yuwei, X., Yiyuan, Z., Wenfeng, M., Jainzhong, C., 1997. Superplastic forming technology of aircraft structures for Al−Li alloy and high-strength Al alloy. J. Mater. Process. Tech. 72, 183−187.

Zacharia, T., David, S.A., Vitek, J.M., Martukanitz, R.P., 1989. Weldability and microstructural characterization of Al−Li alloys. In: Sanders Jr., T.H., Starke Jr., E.A. (Eds.), Proceedings of the Fifth International Conference on Aluminium−Lithium Alloys, vol. 3. Materials and Component Engineering Publications Ltd., Birmingham, UK, pp. 1387−1396.

Zakharov, V.V., 1999. Thermal stability of Al−Li alloys. Met. Sci. Heat Treat. 41, 39−43.

Zakharov, V.V., 2003. Some problems of the use of aluminum−lithium alloys. Met. Sci. Heat Treat. 45, 49−54.

Zamiski, G.F., Ono, K., 1989. Micro-mechanism fracture toughness model for intergranular ductile fracture. In: Sanders Jr., T.H., Starke Jr., E.A. (Eds.), Proceedings of the Fifth International Conference on Aluminium−Lithium Alloys, vol. 2. Materials and Component Engineering Publications Ltd., Birmingham, UK, pp. 869−878.

Zedalis, M.S., Gilman, P.S., 1991. Ageing response of SiCp reinforced Al-3.1Li-1.0Cu-0.5Mg-0.5Zr. In: Sanders Jr., T.H., Starke Jr., E.A. (Eds.), Proceedings of the Fifth International Conference on Aluminium−Lithium Alloys, vol. 3. Materials and Component Engineering Publications Ltd., Birmingham, UK, pp. 1647−1656.

Zeng, X.H., Barlat, F.D., 1994. Effects of texture gradients on yield loci and forming limit diagrams in various aluminum−lithium sheet alloys. Metall. Mater. Trans. A, 25A, 2783−2795.

Zeng, X.H., Ericsson, T., 1993. Anisotropy of elastic properties in various aluminium−lithium sheet alloys. Acta Mater. 43, 1801−1812.

Zhai, T., 2006. Strength distribution of fatigue crack initiation sites in an Al−Li alloy. Metall. Mater. Trans. A, 37A, 3139−3147.

Zhang, B., Wang, Z., Zhu, W., Li, Y., Liang, Y., 1989a. Effects of minor Ce and La on properties of Al−Li−Cu alloys containing deleterious impurities. In: Sanders Jr., T.H., Starke Jr., E.A. (Eds.), Proceedings of the Fifth International Conference on Aluminium−Lithium Alloys, vol. 1. Materials and Component Engineering Publications Ltd., Birmingham, UK, pp. 529−538.

Zhang, J.J., Wang, Y.X., Chen, Z., Liu, B., 2007. Calculation of free energy of Al−Cu−Li alloy under electric field. Trans. Nonferr. Metals Soc. China 17, 575−580.

Zhang, J.Y., Ai, B., Liu, C., Zhu, R., Zhang, D., Ma, C., 1992. Microstructures of explosively consolidated rapidly solidified aluminium and Al−Li alloy powders. J. Mater. Sci. 27, 2298−2308.

Zhang, L.C., He, A.Q., Ye, H.Q., Zhang, Y.C., 1996. A high resolution electron microscopic study of the microstructure in rapidly solidified Al−Li−Cu−Mg−Zr alloys. J. Mater. Sci. 31, 224−231.

Zhang, M., Chang, Z., Yan, J., Zhihao, J., 2001. Investigation of the behaviour of rare earth element cerium in aluminium−lithium alloys by the method of internal friction. J. Mater. Process. Tech. 115, 294−297.

Zhang, Y., Zhao, H., Gao, G., Chen, J., Yao, D., Hu, Z., 1989b. Effect of rare earth elements on mechanical and chemical properties of aluminium−lithium alloy. In: Sanders Jr., T.H., Starke Jr., E.A. (Eds.), Proceedings of the Fifth International Conference on Aluminium−Lithium Alloys, vol. 1. Materials and Component Engineering Publications Ltd., Birmingham, UK, p. 539.

Zhao, Z., Liu, L., Chen, Z., 2006. Co-strengthening contribution of δ' and T1 precipitates in Al–Li alloys 2090 and 2090 + Ce. Rare Met. 25, 197–201.

Zhen, L., Yang, D.Z., Sun, S.W., Yu, G.F., Jerky flow behavior in a rapid solidification processed Al–Li alloy. Mater. Sci. Eng. A, A248, 221–229.

Zhen, L., Sun, S.W., Cui, Y.X., Yang, D.Z., Yu, G.F., Jiao, C.G., 1994. Influence of aging on the impact fracture behavior of a RSP Al–Li alloy. Scr. Met. Mater. 30, 529–533.

Zhen, L., Yang, D.Z., Yu, G.F., Cui, Y.X., 1996. Impact fracture of rapid solidification processed Al–Li alloys. Mater. Sci. Eng. A, A207, 87–96.

Zheng, Z., Zhao, Y., Liu, Y., Yin, D., 1993. Microstructure and mechanical properties of an Al–Li–Cu–Mg–Zr alloy containing minor lanthanum additions. Trans. Nonferr. Metals Soc. China 3, 37–42.

Zheng, Z.Q., Liu, M.G., Yin, D., Tan, C.Y., 1994. Effect of precipitation hardening on the serrated flow characteristics in an Al–Li alloy. In: Sanders Jr., T.H., Starke Jr., E.A. (Eds.), Proceedings of the Fourth International Conference on Aluminum Alloys: Their Physical and Mechanical Properties, vol. 2. Georgia Institute of Technology, Atlanta, GA, USA, pp. 351–357.

Zhou, Z.Q., Zhu, X.D., 1994. Damping behaviour study in RS/PM aluminium–lithium alloy. In: Sanders Jr., T.H., Starke Jr., E.A. (Eds.), Proceedings of the Fourth International Conference on Aluminum Alloys: Their Physical and Mechanical Properties, vol. 2. Georgia Institute of Technology, Atlanta, GA, USA, pp. 382–388.

Zhou, Z.Q., He, X., Ji, D.X., Chen, C.Q., 1989. Fracture mechanism and toughness model of Al–Li–Cu–Mg–Zr alloys. In: Sanders Jr., T.H., Starke Jr., E.A. (Eds.), Proceedings of the Fifth International Conference on Aluminium–Lithium Alloys, vol. 2. Materials and Component Engineering Publications Ltd., Birmingham, UK, pp. 879–888.

Zhu, T., Li, J, 2010. Ultra-strength materials. Prog. Mater. Sci. 55, 710–757.

Printed and bound by CPI Group (UK) Ltd, Croydon, CR0 4YY

08/05/2025

01864900-0003